Handbook of Experimental Pharmacology

Volume 262

The *Handbook of Experimental Pharmacology* is one of the most authoritative and influential book series in pharmacology. It provides critical and comprehensive discussions of the most significant areas of pharmacological research, written by leading international authorities. Each volume in the series represents the most informative and contemporary account of its subject available, making it an unrivalled reference source.

HEP is indexed in PubMed and Scopus.

More information about this series at http://www.springer.com/series/164

Paula H. Stern
Editor

Bone Regulators and Osteoporosis Therapy

Springer

Editor
Paula H. Stern
Department of Pharmacology
Northwestern University Feinberg
School of Medicine
Chicago, IL, USA

ISSN 0171-2004 ISSN 1865-0325 (electronic)
Handbook of Experimental Pharmacology
ISBN 978-3-030-57377-5 ISBN 978-3-030-57378-2 (eBook)
https://doi.org/10.1007/978-3-030-57378-2

This Springer imprint is published by the registered company Springer Nature Switzerland AG.
The registered company address is: Gewerbestrasse 11, 6330 Cham, Switzerland

Preface

Osteoporosis is a major health problem, affecting more than 200 million people worldwide (Kanis, JA, WHO Scientific Group: Assessment of Osteoporosis at the Primary Health Care Level, 2007). Osteoporotic fractures can result in disability and death. However, in recent years, there have been tremendous advances in knowledge of the cellular physiology of bone and the systemic and local factors that affect bone cells and mineral metabolism, leading to novel concepts and new therapies to slow or reverse the pathology. Tools including diagnostic techniques, genetic approaches, and animal models have been important in facilitating these advances. The goal of this volume is to bring this knowledge together in a form that is useful to students, researchers, and practitioners. Each of the 20 chapters is an entity in itself; however, there are also threads and themes that connect them.

The volume begins with *The Cells of Bone and Their Interactions*, by Ansari and Sims, a comprehensive overview of the development, functions, signaling, and coordination of the bone cell types, establishing their roles in modeling and remodeling. The next several chapters review the effects of critical systemic hormones long recognized for their importance to bone and mineral. In *PTH and PTHrP Actions on Bone,* Suva and Friedman review signaling through the PTH/PTHrP receptor and insights gained through structure–activity studies of hormone fragments. They present findings that establish distinct physiological roles of the two hormones. In *Vitamin D and Bone*, Christakos, Li, DeLa Cruz, Verlinden, and Carmeliet describe the biosynthesis of 1,25-$(OH)_2D_3$/calcitriol, the multiple actions of the hormone mediated through the vitamin D receptor (Vdr), and the insight gained through transgenic and knockout models. In *Gonadal Hormones and Bone*, Yoshida and Wang and I review gonadal hormones and their actions on bone, including receptor interactions, gene responses, and clinical observations. In the chapter *Thyroid Hormones, Glucocorticoids, Insulin, and Bone*, Lakatos, Szili, Bakos, Takacs, Putz, and Istenes review the actions of these endocrine agents on the skeleton, from basic mechanisms to the bone disorders resulting from excess thyroid and glucocorticoid hormones and types 1 and 2 diabetes.

The volume then moves to a series of chapters on local factors, each chapter taking a unique approach to the topic. In *Growth Factors, Carrier Materials, and Bone Repair*, Hsu and Stock focus on the growth factors important in bone regeneration and fracture healing. In addition to reviewing the basic bone biology of these growth factors, they describe how the factors are being applied to promote bone grafts and are combined with carrier systems and scaffolds. The carrier materials themselves are presented as an additional class of pharmacological agents. In the chapter *Prostaglandins and Bone*, Pilbeam reviews the receptors, signaling, and actions of prostaglandins affecting bone. The chapter describes the role and process by which cyclooxygenase-2 induction controls the skeletal response to PTH. In *Cytokines and Bone: Osteoimmunology*, Lorenzo presents a broad and comprehensive review of the large number of cytokines affecting bone, summarizing their effects on bone cells and animal models and contribution to disease states. In *Chemokines and Bone*, Gilchrist describes evidence for chemokine effects on bone cells and roles in diseases affecting bone.

In the chapter *Calcium and Bone*, Reid and Bristow review extracellular calcium homeostasis and provide critical analysis of the evidence regarding calcium supplementation and osteoporosis. In *FGF-23 and Bone and Mineral Metabolism*, Fukumoto reviews the phosphotropic hormone FGF23, its role to regulate the production of 1,25-$(OH)_2D_3$, and the disorders resulting from FGF-23 dysregulation. In the chapter *The Central Regulation of Bone Mass: Genetic Evidence and Molecular Bases*, Karsenty traces the research path leading to the recognition of the effects of leptin and the sympathetic nervous system on bone.

Genetic models are mentioned in many of the chapters, and in *Genetics of Skeletal Disorders*, Hannan, Newey, Whyte, and Thakker provide a comprehensive presentation of bone disorders with a genetic etiology and the use of molecular genetic tests together with other approaches in the diagnosis of bone diseases.

The next group of chapters relates most directly to osteoporosis. In *Pathogenesis of Osteoporosis*, Al Saedi, Stupka, and Duque bring us back to the cell overview presented in the first chapter, this time with an emphasis on the pathophysiologic consequences of disordered bone cell function. They also speculate on the potential role of marrow adiposity in osteoporotic bone loss. In *Structural and Metabolic Assessment of Bone*, Narla and Ott present a comprehensive and critical review of well-established and more novel structural and metabolic approaches for the assessment of bone loss as well as predictive tools. In the chapter *Osteoporosis Therapeutics* 2020, Kocijan, Klaushofer, and Misof review the individual therapies currently used for the treatment of osteoporosis. They also address bone material properties predicting bone fragility. In *Reduced Bone Modeling and Unbalanced Bone Remodeling: Rational Targets for Antiresorptive and Anabolic Therapy*, Ramchand and Seeman address the central roles of modeling and remodeling in the mechanisms and efficacy of osteoporosis treatment and discuss the status of combination and sequential therapy.

In *New Targets and Emergent Therapies for Osteoporosis*, Brommage provides a history of potential therapies that failed at some stage, describes concepts for improving existing therapies, and discusses novel approaches suggested by recent research advances. In *Drugs Causing Bone Loss*, Vestergaard reviews pharmacological agents having the unfortunate property of being harmful to bone. Finally, in *Natural Products as Potential Bone Therapies*, Wong, Poon, Zhou, and Xiao present an overview of a number of natural products identified in preclinical studies as having protective effects on bone through actions on bone cells, on the microbiome or through estrogenic, vitamin D, or metabolomic mechanisms.

In closing, I would like to thank the international group of authors for their efforts in producing a fascinating and valuable journey through cell biology, endocrinology, "paracrinology," pathophysiology, diagnostics, and pharmacology of bone as related to osteoporosis. I thank the editorial board of the *Handbook of Experimental Pharmacology* for inviting me to edit a volume on bone and osteoporosis and the editorial and production team at Springer for their always prompt and perceptive support.

Chicago, IL, USA Paula H. Stern

Contents

The Cells of Bone and Their Interactions

Niloufar Ansari and Natalie A. Sims

Contents

Abstract

Bone tissue is comprised of a collagen-rich matrix containing non-collagenous organic compounds, strengthened by mineral crystals. Bone strength reflects the amount and structure of bone, as well as its quality. These qualities are determined and maintained by osteoblasts (bone-forming cells) and osteoclasts (bone-resorbing cells) on the surface of the bone and osteocytes embedded within the bone matrix. Bone development and growth also involves cartilage cells

N. Ansari
Drug Delivery, Disposition and Dynamics, Monash Institute of Pharmaceutical Sciences, Monash University, Parkville, VIC, Australia

N. A. Sims (✉)
Bone Biology and Disease Unit, St. Vincent's Institute of Medical Research, Melbourne, VIC, Australia

Department of Medicine, The University of Melbourne, St. Vincent's Hospital, Melbourne, VIC, Australia
e-mail: nsims@svi.edu.au

P. H. Stern (ed.), *Bone Regulators and Osteoporosis Therapy*,
Handbook of Experimental Pharmacology 262, https://doi.org/10.1007/164_2019_343

(chondrocytes). These cells do not act in isolation, but function in a coordinated manner, including co-ordination within each lineage, between the cells of bone, and between these cells and other cell types within the bone microenvironment. This chapter will briefly outline the cells of bone, their major functions, and some communication pathways responsible for controlling bone development and remodeling.

Keywords
Bone modeling · Bone remodeling · Coupling factors · Osteoblast · Osteoclast · Osteocyte

1 Osteoclasts

Osteoclasts are large, multinucleated cells responsible for bone resorption. Osteoclasts form when circulating mononuclear hematopoietic progenitor cells migrate to the bone surface, fuse with each other, and attach to the bone surface to form active osteoclasts. This process is mediated by the interaction of progenitor cells expressing receptor activator of NF-kappa B (RANK) with supporting cells expressing RANK ligand (RANKL) (Martin and Sims 2015). Fusion is triggered by dendritic cell-specific transmembrane protein (DC-STAMP) (Kukita et al. 2004) and osteoclast-specific transmembrane protein (OC-STAMP) (Miyamoto et al. 2012). Osteoclasts are also capable of generating new osteoclasts by fission of osteoclasts, where the released mononuclear or multinucleated osteoclast detaches before joining another osteoclast or osteoclast precursor to continue the resorptive process (Jansen et al. 2012).

Once formed and attached to the bone surface, osteoclasts are polarized cells with specific functional domains responsible for attachment, resorption, and release of resorption products (Fig. 1). Attachment to bone is mediated by the vitronectin receptor $\alpha_v\beta_3$ integrin which binds to arginine-glycine-aspartic acid (RGD) sequences in bone matrix proteins including osteopontin and bone sialoprotein (Ross and Teitelbaum 2005). These attachment sites, known as podosomes, form an actin-rich ring-like sealing zone surrounding and isolating the resorption lacunae from the rest of the local environment. Podosomes are rapidly assembled and disassembled, allowing the highly motile osteoclasts to move across the bone surface by partially detaching and reattaching (Horne et al. 2005).

Resorption makes use of the osteoclast's characteristic ruffled border which increases the surface area on which ion exchange can occur. To initiate resorption, osteoclasts acidify the lacunae to dissolve bone mineral (bioapatite) and matrix proteins. This is achieved by carbonic anhydrase II (Sly et al. 1985) which generates protons (H^+) and the vacuolar-type H+ ATPase (V-ATPase) which pumps protons across the plasma membrane (Blair et al. 1989). Cellular equilibrium is maintained by passive chloride-bicarbonate exchangers (Teti et al. 1989) and chloride channel 7 (CLC-7), which exports chloride ions. Osteoclasts also express and release lysosomes and proteolytic enzymes, such as cathepsin K and matrix metalloproteinases (MMPs), which degrade the collagenous component of the bone

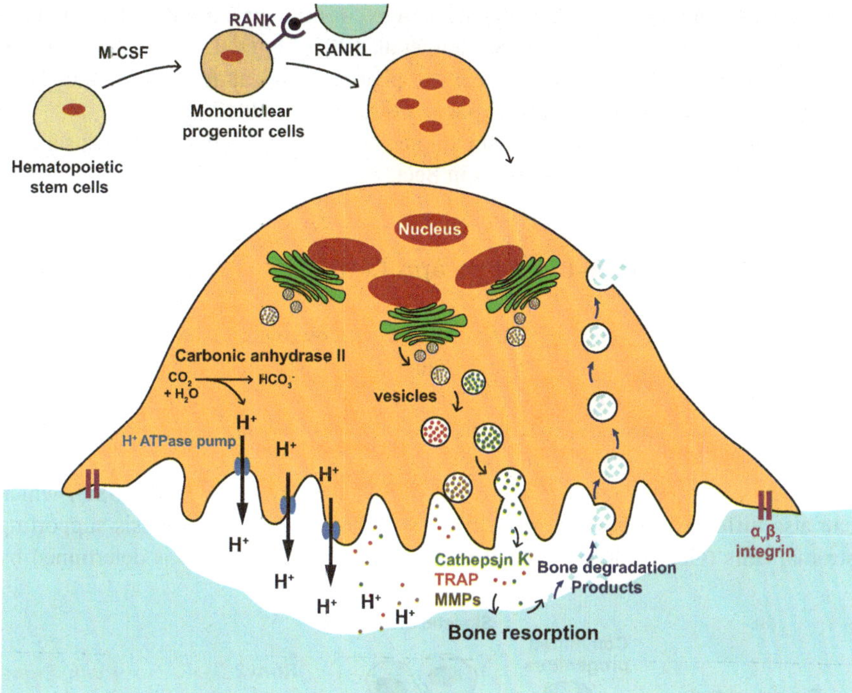

Fig. 1 Osteoclast formation and function. Multinucleated osteoclasts are formed by fusion of mononuclear precursors derived from hematopoietic stem cells; a process stimulated by macrophage colony-stimulating factor (M-CSF) and the interaction of receptor activator of NF-kappa B ligand (RANKL) provided by supporting cells (e.g., osteoblast lineage cells, T cells, etc.) with RANK on the surface of the osteoclast precursor. Multinucleated cells attach to the bone surface via $\alpha_v\beta_3$ integrin to form polarized osteoclasts with a sealing zone at the resorption area. The vacuolar-type H^+ ATPase in the ruffled border acidifies the resorption lacuna. Lysosomal proteases such as cathepsin K, tartrate-resistant acid phosphatase (TRAP), and matrix metalloproteinases (MMPs) are released by vesicle-mediated exocytosis to digest the organic matrix. Products of bone degradation are endocytosed in transcytotic vesicles, further digested inside the osteoclast, and released from the other side of the cell

matrix (Delaisse et al. 2003). Since bone resorption is a rapid energy-demanding process, osteoclasts have abundant mitochondria and a well-developed Golgi complex and endoplasmic reticulum.

During resorption, bone matrix degradation products are taken up from resorption lacunae by the osteoclast in transcytotic vesicles; they are then digested further within the cell and released by vesicular exocytosis at the cell surface opposite to the bone surface (Nesbitt and Horton 1997; Salo et al. 1997). One degradation product released is C-terminal telopeptide of type I collagen (CTX-1), which is measured in the serum as a bone resorption marker.

Bone resorption levels are determined both by the number and activity of osteoclasts. Reduced bone resorption, regardless of whether it is due to a reduction in osteoclast number, osteoclast enzyme activity, or osteoclast motility, leads to high bone mass, termed osteopetrosis (Soriano et al. 1991; Sly et al. 1985). The study of these conditions led to the development of pharmacological agents to reduce bone resorption (anti-resorptives) in osteoporosis (Lacey et al. 2012).

The function of osteoclasts is not limited to bone resorption. Osteoclasts also regulate bone formation as discussed in Sect. 6.

2 The Osteoblast Lineage and Matrix-Producing Osteoblasts

Osteoblasts produce bone's collagen matrix and regulate its mineralization. The osteoblast lineage includes matrix-producing osteoblasts, their pluripotent and lineage-committed precursors, bone lining cells, and matrix-embedded osteocytes; each stage has distinct functions, morphologies, locations relative to the bone surface, and increasingly well-defined markers (Fig. 2).

The osteoblast lineage arises from pluripotent skeletal stem cells (SSC) which can also differentiate into chondrocytes, adipocytes, and hematopoiesis-supporting stromal cells (Gehron Robey and Riminucci 2020); their lineage is determined by

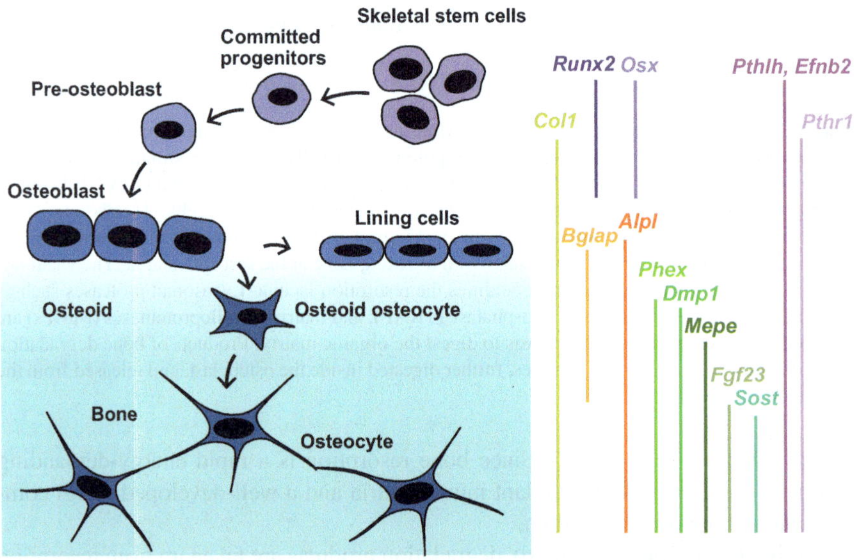

Fig. 2 Osteoblast/osteocyte lineage differentiation and gene expression. At early stages of commitment to the osteoblast lineage, expression of *Runx2* and *Osx* is high. Matrix-producing osteoblasts express PTH receptor (*Pth1r*), alkaline phosphatase (*Alpl*), and collagen type I (*Col1*). Osteocalcin (*Bglap*) is expressed by mature osteoblasts and newly embedded osteocytes close to bone-forming surfaces. Osteoid osteocytes and mature osteocytes express higher levels of *Phex*, *Mepe*, and *Dmp1*. Levels of *Fgf23* and *Sost* (sclerostin) are expressed at highest levels by the most mature osteocytes. Some genes, such as parathyroid hormone-related protein (*Pthlh*) and EphrinB2 (*Efnb2*), are expressed at similar levels through all stages of osteoblast differentiation. There are no known specific markers of bone lining cells

cytokines, hormones, growth factors, epigenetic factors, cell-matrix interactions, and direct cell-cell interactions within the lineage, each regulates expression of cell fate-specific transcription factors. Many of these promote commitment of SSC to the osteoblast lineage, but two are essential: runt-related transcription factor 2 (RUNX2) and osterix (OSX) (Ducy et al. 1999; Nakashima et al. 2002). RUNX2 stimulates osteoblast progenitor cell differentiation into preosteoblasts expressing low levels of collagen type I and alkaline phosphatase. OSX stimulates subsequent differentiation to mature osteoblasts producing high levels of collagen type I, alkaline phosphatase, and osteocalcin. Other transcription factors including activating transcription factor 4 (ATF4) (Yang et al. 2004), activator protein 1 (AP-1) (Sabatakos et al. 2000), and CCAAT/enhancer-binding proteins β and δ (C/EBPβ and C/EBPδ) (Gutierrez et al. 2002) also promote the transition to matrix-producing osteoblasts.

Mature matrix-producing osteoblasts normally appear as a single layer of aligned cuboidal cells on the bone surface, forming tight junctions with adjacent osteoblasts, but they also form clusters within mesenchymal condensations and can deposit bone on other surfaces, such as mineralized cartilage. Bone is a heterogeneous compound material; it includes a mineral phase containing bioapatite and an organic matrix comprising largely type I collagen (~90%), non-collagenous proteins (~5%), small amounts of lipid (~2%), proteoglycans, and water (Fratzl et al. 2004). Osteoblasts deposit the organic component of bone matrix and non-collagenous proteins including growth factors such as transforming growth factor-β (TGF-β), insulin-like growth factor-1 (IGF-1), and mineralization regulators, such as osteocalcin. For this high level of protein production, active osteoblasts have a prominent Golgi complex and abundant rough endoplasmic reticulum.

Prior to mineralization, the collagen-containing matrix substance is called osteoid; thus, osteoblasts do not produce "bone," but osteoid matrix, which is later mineralized to form bone. Osteoid is deposited in one of the two forms: woven or lamellar. Woven bone is deposited during bone development and fracture healing; it contains disordered, seemingly randomly oriented collagen fibers and osteocytes. In contrast, lamellar bone is highly organized, with parallel collagen fibers oriented in perpendicular planes in adjacent lamellae (Giraud-Guille 1988), adding strength. The collagen fibers are also stabilized and strengthened by inter- and intramolecular cross-links (Oxlund et al. 1995). Procollagen type 1 N-terminal propeptide (P1NP) is a specific marker of collagen deposition and is used as a serum bone formation marker. Defects in collagen composition, collagen fibril assembly, cross-linking, or posttranslational modifications lead to skeletal fragilities, including those observed in osteogenesis imperfecta (Forlino and Marini 2016).

After osteoid is deposited, osteoblasts have three possible fates: they may undergo apoptosis, be trapped within the bone matrix and differentiate into osteocytes (Sect. 3), or remain on the bone surface as bone lining cells. The latter are characterized by flat nuclei and reduced capacity to produce protein but, like other cells of the osteoblast lineage, lining cells retain connections with each other via gap junctions (Miller et al. 1989). Bone lining cells can act as osteoblast precursors (Matic et al. 2016), and bone anabolic agents, such as PTH, convert murine quiescent bone lining cells into active osteoblasts (Dobnig and Turner 1995;

Kim et al. 2017). In addition, bone lining cells retract from the bone surface to allow access by osteoclasts for remodeling (Sect. 7).

The osteoblast lineage also regulates osteoclast differentiation by producing RANKL and OPG (see Sect. 9). This communication between bone-forming and bone-resorbing cells ensures an integrated response of bone tissue to systemic, local, or mechanical stimuli. Osteoblast lineage cells also act as "reversal" cells during bone remodeling (Sect. 7).

3 Osteocytes

Osteocytes form when osteoblasts are embedded within the osteoid matrix during bone formation. Osteocytes do not divide and may live for decades in the bone matrix, with their life span determined by the rate of bone remodeling. Osteocytes are the most abundant cells in the skeleton comprising 90–95% of all bone cells, a total of ~42 billion in the adult human skeleton (Buenzli and Sims 2015). They are now understood to be central regulators of bone strength by signaling to cells on the bone surface (Dallas et al. 2013) and by modifying their local environment (Blank and Sims 2019; Tsourdi et al. 2018).

The mechanism by which some osteoblasts become entrapped in the matrix is not yet known, but as they mature into osteocytes, their morphology changes including an increase in dendritic length and a reduction in cell body size (Dallas et al. 2013). The osteocyte cell body and processes reside in fluid-filled spaces within bone matrix called lacunae and canaliculi, respectively, to form the lacunocanalicular network. Osteocyte lacunar shape depends on the bone type and region. In woven bone, lacunae are irregular, spherical, and randomly oriented, while in lamellar bone, lacunae are ellipsoid and aligned with the direction of the collagen fibers (Marotti et al. 1985). Osteocyte dendritic processes make intercellular contacts with each other, with osteoblasts and lining cells on the bone surface, and they have been reported to reach into the nearby vascular canals (Dallas et al. 2013). The network is highly complex – within osteocytes alone, there are 3.7 trillion connections, and the network is very densely arranged (with cell numbers in the order of 19,000–28,500 per mm^3) making up a large volume of the bone matrix itself (Buenzli and Sims 2015). This suggests the network could be an appealing target for therapeutic intervention. The large bone surface area of the osteocyte lacunocanalicular network (215 m^2) provides ample space for bone mineral exchange and regulation (Buenzli and Sims 2015). However, there is only ~85 nm canalicular space surrounding each process (Varga et al. 2015; Buenzli and Sims 2015). This limits the size of therapeutics able to pass through the lacunocanalicular network to act on osteocytes.

Gene expression also changes during the osteoblast-to-osteocyte differentiation. Osteocalcin expression is higher in less mature osteocytes close to bone-forming surfaces (Sims et al. 1997), while expression of phosphate-regulating endopeptidase homolog X-linked (PHEX), matrix extracellular phosphoglycoprotein (MEPE), dentin matrix acidic phosphoprotein 1 (DMP1), and sclerostin increases as osteocytes differentiate. Such osteocyte markers regulate bone formation (sclerostin

(Li et al. 2005; Van Bezooijen et al. 2005)), mineralization (DMP1), phosphate homeostasis (PHEX, MEPE, FGF23), and dendrite formation (E11) (Dallas et al. 2013). Osteocytes also control bone mass and quality through the controlled release of cytokines and cytokine receptors including tumor necrosis factor-α (TNF-α (Bakker et al. 2009)), oncostatin M (Walker et al. 2010), soluble IL-6 receptor (McGregor et al. 2019), and IL-1 (Bakker et al. 2009).

4 The Process of Bone Mineralization: Control by Osteoblasts and Osteocytes

After osteoid is deposited, it becomes hardened by deposition of calcium-phosphate mineral under the control of osteoblasts and osteocytes. This has two phases: a primary, rapid initiation (which lasts for 1–3 days), followed by slower, gradual secondary mineralization (Fuchs et al. 2008). The latter phase may continue for several years, until a maximal level of mineralization is reached or until that portion of bone is remodeled (Sect. 7).

The initiation of primary mineralization is controlled, at least in part, by non-collagenous proteins produced by osteoblasts on the bone surface, since defects in late-stage osteoblast differentiation delay its initiation (Takyar et al. 2013). These proteins include osteocalcin, MEPE, PHOSPHO-1, and alkaline phosphatase (Houston et al. 2004; Ling et al. 2005; Gowen et al. 2003; Poole et al. 2005). In the absence of these factors, initiation is slow, and the mineralization front is diffuse, as observed in hypophosphatemic rickets or osteomalacia.

Secondary mineralization is controlled by later osteoblasts and osteocytes (Vrahnas et al. 2019; Blank and Sims 2019) and their expression of proteins such as DMP1, MEPE, ectonucleotide pyrophosphatase/phosphodiesterase (ENPP1), and PHEX (Kalajzic et al. 2004; Dallas and Bonewald 2010; Paic et al. 2009). There is naturally some overlap between these functions.

During primary and secondary mineralization, many changes occur within the matrix as it matures. Mineral accumulates (Fuchs et al. 2008, 2011), mineral crystal size increases, its structure becomes more ordered, and carbonate is incorporated into the bioapatite lattice (Vrahnas et al. 2016). The collagen fibers become more cross-linked (Paschalis et al. 2004) and more compact (Vrahnas et al. 2016, 2018), and water content reduces (Granke et al. 2015). There is much variation in the degree of these changes at the tissue level because basic multicellular unit (BMU)-based remodeling (Sect. 7) leads to bone heterogeneity (Boivin and Meunier 2002; Roschger et al. 2003). A high bone remodeling rate results in lower total mineralization since secondary mineralization of the BMU is truncated by bone resorption (Boivin and Meunier 2002; Paschalis et al. 1997). In contrast, when remodeling is slow, for example, with anti-resorptive therapies, more bone completes secondary mineralization uninterrupted and reaches the maximal mineralization level, leading to a higher average level of mineralization throughout the bone (Fuchs et al. 2011).

The processes by which mineral is deposited and accumulates within osteoid remain poorly defined. It commences with the budding of matrix vesicles from cells facing the matrix (osteoblasts and possibly osteocytes) which attach to the surrounding collagen matrix. The vesicles accumulate ions outside the cell and rupture, releasing mineral into the surrounding matrix (Anderson 1967). This amorphous mineral forms ordered crystals through a nucleation process driven by contact with collagen (Rohde and Mayer 2007; Stanford et al. 1995). Matrix vesicles were originally reported to lack mineral and to accumulate poorly crystalline mineral only after budding from the cell and becoming immobilized in the collagen matrix (Anderson 1995). This remains the predominant view (Dillon et al. 2019), but others have reported the presence of mineral in matrix vesicles prior to their release (Rohde and Mayer 2007; Azari et al. 2008; Stanford et al. 1995). Approaches to treat hypophosphatasia and X-linked hypophosphatemia have focused on this process.

5 Chondrocytes

While not "bone cells" per se, chondrocytes (cells embedded in cartilage matrix) contribute to longitudinal bone growth and joint health in adults, so it is important to have a basic understanding of them when considering pharmacological effects on the skeleton.

In embryonic development, chondrocytes drive the initial stages of endochondral ossification (see Sect. 6). Their programmed proliferation, differentiation, cartilage matrix production, and eventual hypertrophy define the cartilaginous model of the skeleton to be formed. After ossification commences, continued proliferation, differentiation, and hypertrophy of chondrocytes at the epiphyseal growth plates facilitate longitudinal growth of the pediatric and adolescent skeleton, which ends with the removal of the growth plates in the adult skeleton. Growth plate chondrocytes also influence the early stages of vascularization and bone deposition during bone growth (Poulton et al. 2012), and this may determine adult bone mass, as well as bone size. Many factors influencing osteoblast differentiation and bone formation also regulate chondrocyte activity (e.g., parathyroid hormone-related protein (PTHrP), TGF-β, BMP2, Wnt/β-catenin signaling, fibroblast growth factor, growth hormone, and the insulin-like growth factors (IGFs), reviewed in detail elsewhere (Hartmann and Yang 2020)); this may relate to their derivation from a common precursor cell. These effects should be kept in mind, particularly in pediatric pharmacology.

Chondrocytes also form the cartilage matrix covering the ends of long bones at the joints. Here, cartilage protects subchondral bone (bone underlying the cartilage) from mechanical stress, and the close proximity of cartilage and bone provides biochemical and molecular cross talk between these two tissues. Since cartilage is avascular, bone tissue provides nutrition to support chondrocyte metabolism, and chondrocytes in turn may provide factors able to promote osteoblast activity. This is reviewed in detail elsewhere (Findlay and Atkins 2014).

6 Osteogenesis and Bone Modeling

The cells of bone (osteoclasts, osteoblasts, and osteocytes) and cartilage (chondrocytes) form the skeleton during embryogenesis, control its growth in childhood, and manage its renewal in adult life. The shape and size of bone, and its organization into compact cortical bone at the periphery, with the inner trabecular network require tight coordination within and between each cell lineage. During embryogenesis, the skeleton forms initially when mesenchymal progenitor cells aggregate at sites where bone is to be formed. These bones form by two ossification processes: intramembranous and endochondral.

In intramembranous ossification, the progenitors gather and differentiate directly into osteoblasts; these deposit osteoid, which forms membranous bone, including the flat bones of the skull, mandible, maxilla, and clavicles. These bones grow by further osteoblastic differentiation of mesenchymal cells at the periphery (periosteum) of the newly forming bone. This process requires tight regulation of osteoblast differentiation for the skull to form normally; for example, in mice and humans with defective osteoblast differentiation due to lack of the transcription factors *Runx2* and *Osterix*, early cessation of skull growth results in cleidocranial dysplasia (Ducy et al. 1997; Mundlos et al. 1997; Otto et al. 1997; Lee et al. 1997).

The majority of the skeleton forms by endochondral ossification; this also requires tight control of cellular differentiation and communication between multiple cell types. In this process, mesenchymal progenitors first differentiate to chondrocytes and form a cartilage template of the future bone, which is gradually replaced with mineralized bone (Fig. 3). As the chondrocytes proliferate and deposit cartilage, the template grows, and chondrocytes at the center become hypoxic and enter hypertrophy; this contributes to longitudinal expansion (Cooper et al. 2013). As the cells enter hypertrophy, their gene expression pattern changes, leading to three major changes: mineralization of the surrounding cartilage matrix, vascular invasion, and entry of osteoclast precursors. This is followed by resorption of the mineralized cartilage, which expands the marrow cavity, leading to a new morphology where the long bone houses two growth plates, in mirror image, separated by the primary ossification center. As the bone continues to grow, mineralization and vascular invasion occur in the two remaining cartilaginous ends of the bones, forming the secondary ossification centers.

Chondrocyte proliferation and subsequent hypertrophic expansion at the two growth plates continues longitudinal bone growth. As this occurs, mineralized cartilage remnants from the resorptive process form a template on which osteoblasts deposit new bone to form trabecular bone (Mackie et al. 2011). At the diaphyseal perichondrium, cortical bonc forms by the coalescence of trabecular bone with a thin periosteum formed by intramembranous ossification (Rauch 2012). In humans, the growth plates close and are fully resorbed in adulthood. In rodents, while the growth plates remain, they are essentially inactive as bones do not continue to lengthen.

Changes in bone shape continue as the skeleton adapts during growth and in adulthood. Growth continues to be mediated by bone formation on expanding external surfaces (such as the periosteum) and resorption on internal expanding

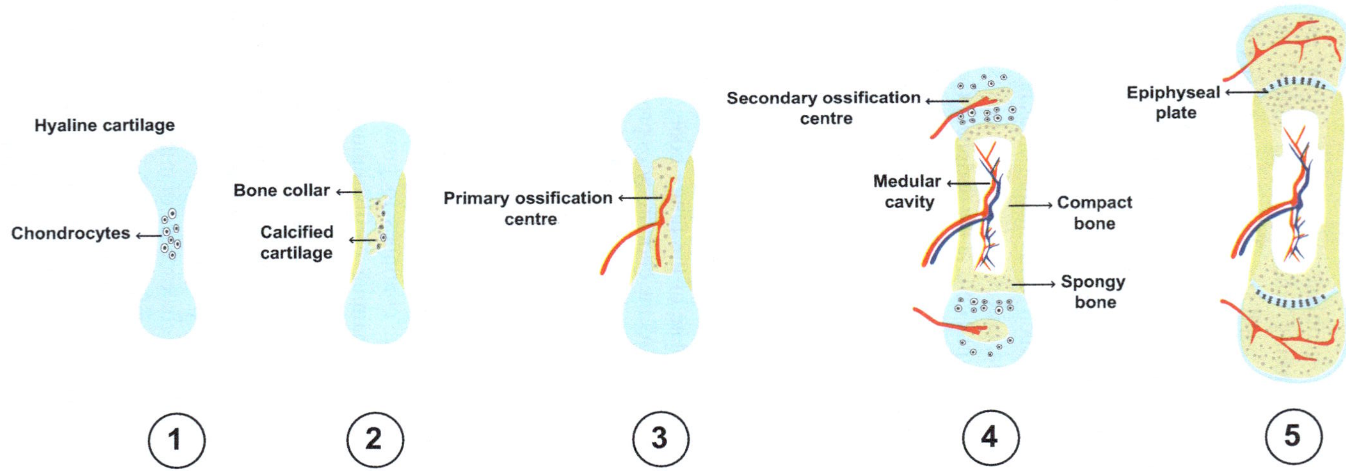

Fig. 3 Endochondral ossification. During skeletal development, long bones are formed by endochondral ossification. Chondrocytes at the center of the growing cartilage model of bone become hypertrophic (1) and produce calcified matrix (2). Chondrocytes embedded within the matrix undergo apoptosis due to hypoxia and lack of nutrients (2). Blood vessels invade the spaces remaining from the death of chondrocytes and import mesenchymal and hematopoietic progenitors into the tissue. These differentiate into osteoblasts and osteoclasts, forming the primary ossification center (3), which enlarges with the activity of osteoclasts and eventually becomes the medullary cavity. Cartilage continues to grow at the ends of the bone (epiphyses) and forms a secondary ossification center through the same process (4). When fetal skeletal development is completed, cartilage remains only at the joint surface and between the diaphysis and epiphysis (referred as the epiphyseal growth plate) (5). Epiphyseal plate chondrocytes drive the longitudinal growth of bones until puberty, when their proliferation ceases.

surfaces (such as the endocortical surface). In such situations, there is not direct communication between osteoblasts and osteoclasts, but these processes must be coordinated, perhaps by signals through the osteocyte network, termed osteotransmitters (Johnson et al. 2015). When bone formation and resorption occur on different surfaces, and lead to changes in bone shape, this is termed "modeling."

7 Bone Remodeling

Another process involving coordination of osteoclasts and osteoblasts is the renewal of cortical and trabecular bone throughout life by remodeling. This process occurs in small regions of bone asynchronously throughout the skeleton, allowing gradual renewal of the entire skeleton over time. Bone resorption and formation occur in sequence on the same surface in the remodeling cycle; this contrasts with modeling, in which the activities occur on different surfaces. Remodeling has five well-defined phases: (1) activation of remodeling, (2) bone resorption, (3) a reversal phase, (4) bone formation, and (5) quiescence (Fig. 4). The process by which growth plate cartilage is converted to bone during growth (Sect. 6) is analogous to bone remodeling, since it includes sequential cycles of bone resorption and formation. For bone remodeling to maintain bone mass, the amount of bone formed by osteoblasts must be matched to the amount resorbed by osteoclasts. If unbalanced, bone mass changes (e.g., greater resorption than formation leads to bone loss). A better understanding of coupling, both on the surface of bone and in the generation of new trabecular bone at the growth plate, will aid in developing agents to build bone mass by shifting the balance toward bone formation for osteoporosis therapy.

Activation of bone remodeling likely occurs when osteocytes respond to bone damage by apoptosis by releasing signals to nearby cells (Schaffler et al. 2014). This attracts osteoclast precursors which fuse to form multinucleated cells, attach, and resorb bone; the signals terminating resorption are not known. During the subsequent reversal phase (Baron 1977), the newly exposed surface is cleaned by reversal cells, now understood to be mononuclear osteoblast lineage cells (Everts et al. 2002). Recently it has been proposed that small osteoclasts reside among the reversal cells and continue bone resorption but at a slower rate (Lassen et al. 2017). Toward the end of the reversal phase, osteoblast precursors accumulate on and above the bone surface and differentiate into bone-forming osteoblasts (Lassen et al. 2017). These cells form new bone in the pit left by the resorbing osteoclasts. Following bone formation, lining cells cover the new bone surface which becomes quiescent, but mineralization in the underlying bone continues.

Remodeling occurs both on trabecular bone surfaces and in the cutting cones of osteonal cortical bone. In the latter case, osteoclasts resorb a tunnel into the bone, which is refilled by osteoblasts; the sequence is the same in both processes. Here we will review basic principles and examples and refer the reader to recent reviews for the details of the many communication pathways involved (Sims and Martin 2015, 2020).

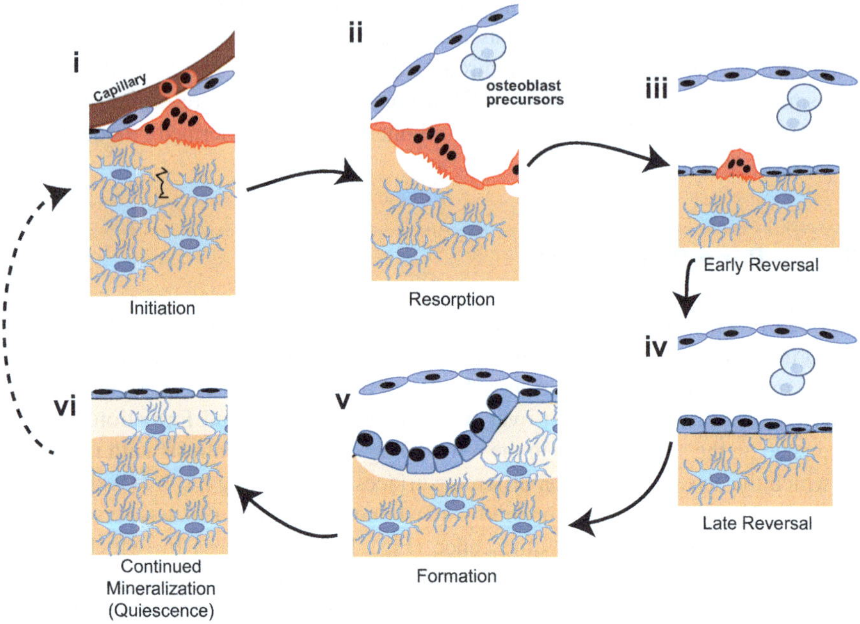

Fig. 4 The bone remodeling cycle. Bone remodeling is initiated (i) when osteocytes sense microdamage within the bone and signal to the bone surface, resulting in lifting of bone lining cells, egress of osteoclast precursors from the vasculature, and the attachment of multinucleated osteoclasts to the bone surface. Bone resorption (ii) follows, occurring within the space outlined by the canopy, and during bone resorption, osteoclasts also provide signals to osteoclast precursors within the remodeling space. The reversal phase follows. During the early reversal phase (iii), mononuclear osteoblast lineage cells cover the bone surface and, with small osteoclasts, complete the process of resorption. In the late reversal phase (iv), osteoblast progenitors accumulate on the bone surface, until there are sufficient for the bone formation phase to commence (v). After the osteoblasts have refilled the pit, bone mineralization continues during the "quiescence" phase (vi) until a maximum level is reached or until the cycle recommences

8 Coupling Factors: Signals Between Osteoclasts and the Osteoblast Lineage

Since bone remodeling occurs on a single surface, there must be local signals to control transitions between each phase. The coupling of bone resorption to subsequent bone formation has been studied extensively, leading to the identification of multiple classes of coupling mechanisms. These include (1) factors released from the bone matrix during osteoclast-mediated resorption (matrix-derived factors); (2) factors produced and expressed by osteoclasts themselves (secreted, released by exocytosis, or expressed on the cell surface); and (3) non-osteoclast-mediated mechanisms. Although most coupling factors released or expressed by osteoclasts described to date stimulate bone formation, they may also have inhibitory effects.

Release of non-collagenous factors from the bone matrix by active osteoclasts was the first mechanism of coupling described. During bone resorption, the acid pH generated by osteoclasts releases and activates proteins such as TGF-β and IGF-1 from the bone matrix (Howard et al. 1981). Since there is a time delay between bone resorption and subsequent formation, these factors must act on early osteoblast precursors to promote their recruitment and migration to the bone surface. Differentiation of these cells into osteoblasts and the level of bone formation they exert are likely controlled by later processes during the remodeling cycle (Sims and Martin 2014). The importance of released matrix-bound factors is highlighted by low bone formation levels in patients and mouse models with "osteoclast-poor" osteopetrosis (Sobacchi et al. 2007; Frattini et al. 2007; Grigoriadis et al. 1994).

However, in "osteoclast-rich" osteopetrosis, where bone resorption is impaired but osteoclasts are still present, bone formation is normal or even increased, rather than being reduced (Sobacchi et al. 1993; Marzia et al. 2000; Gil-Henn et al. 2007; Pennypacker et al. 2009; Del Fattore et al. 2006). Osteoclasts therefore also express coupling factors independent of their resorptive activity. This is one reason why anti-resorptive approaches to reduce osteoclast formation (such as RANKL inhibition) may have different effects on bone formation to anti-resorptives that block osteoclast activity. This was noted when experimental cathepsin K inhibitors, which blocked bone resorption without blocking osteoclast formation, retained osteoblast activity when used in animal models (reviewed in Sims and Ng 2014).

Many coupling factors synthesized by osteoclasts have now been identified, such as cardiotrophin-1 (Walker et al. 2008), sphingosine-1-phosphate (Pederson et al. 2008; Ryu et al. 2006), and contact-dependent molecules, such as EphrinB2 (Zhao et al. 2006) and semaphorin D (Negishi-Koga et al. 2011). None are uniquely expressed by osteoclasts (Sims and Martin 2015, 2020). It should also be noted that while it is plausible in vitro for membrane-bound factors to access mature osteoblasts, this is unlikely in vivo due to the time delay between bone resorption and formation during remodeling (Sims and Martin 2014). Osteoclasts are more likely to come into contact with osteoblast precursors, bone lining cells on the bone surface or in the remodeling canopy, or osteoblast lineage cells in the reversal phase.

Osteoclasts may also stimulate bone formation during remodeling by releasing microvesicles containing coupling factors (Ekstrom et al. 2013). These have been reported to stimulate osteoblast differentiation by their expression of RANK (Ikebuchi et al. 2018) and to inhibit osteoblast activity by providing micro-RNAs (Li et al. 2016). Although it is hard to know how these microvesicles could be provided to osteoblast precursors in a controlled manner, this is an intriguing possibility.

Coupling can also be regulated by osteoclast-independent mechanisms. One example is the size and shape of the resorption pit: osteoblast lineage cells, once attracted to the resorbed bone surface, can "sense" changes in topography and fill the space left by the resorbing osteoclast (Gray et al. 1996). While osteoclast-derived coupling factors may attract precursor cells to the surface, the topography of the bone itself regulates the matrix-depositing activity of the osteoblasts. Osteoblasts must sense the spatial limits and inform each other (via gap junctions or cell-contact-dependent communication processes such as EphrinB2:EphB4 signaling) of when the space has been filled (Tonna and Sims 2014).

The formation of a cellular canopy over the active BMU during the initiation of remodeling was proposed many years ago (Rasmussen and Bordier 1974) and identified much later in human biopsies (Hauge et al. 2001). Bone lining cells, confirmed as osteoblast lineage cells by immunohistochemistry (Kristensen et al. 2013), are suggested to lift from the bone surface at the start of the remodeling cycle to form a compartment that moves with the osteoclast during resorption. This zone provides an isolated space for coupling to occur and connects with the vasculature, providing a route for osteoclast precursors, including partially differentiated "quiescent osteoclast precursors" (Mizoguchi et al. 2009) to enter the remodeling space. Capillaries associated with the canopy also provide a mechanism for ingress of other cells, including mesenchymal precursors (Eghbali-Fatourechi et al. 2007) and immune and endothelial cells (see below).

The reversal phase may also mediate coupling independent of osteoclasts. Mononuclear cells lining the bone surface during the reversal phase activate matrix metalloproteinases to clean collagen remnants from the resorption pits and signal to osteoblasts to act in that space (Everts et al. 2002); this is controversial since osteoblasts even form bone in areas in which pits have been made mechanically (Gray et al. 1996). The lining cells express osteoblast lineage markers (Delaisse 2014) and become progressively more active as they accumulate (Lassen et al. 2017), suggesting a reversal phase during which osteoblast differentiation continues until a critical mass of mature osteoblasts is reached for matrix formation to commence.

9 RANKL/OPG as an Example of Signals from the Osteoblast Lineage to Osteoclasts

Communication between bone cells is not restricted to actions from osteoclasts to the osteoblast lineage. The osteoblast lineage also supports osteoclast formation and the process of bone resorption. The most intensely studied pathway is the production of RANKL and OPG by the osteoblast lineage, but the stage of osteoblast differentiation responsible remains controversial.

RANKL (gene name: *TNFSF11*) is a member of the TNF superfamily, expressed by a wide range of cells, including osteoblasts, osteocytes, hypertrophic chondrocytes, T cells, mammary gland epithelial cells, and stromal cells (Martin and Sims 2015); many of these reside within the local bone environment. RANKL stimulates osteoclast differentiation when it binds to RANK on hematopoietic progenitors and induces fusion of osteoclast precursors (Sect. 1, Fig. 1). RANKL binding to RANK is restricted by a decoy soluble receptor osteoprotegerin (OPG; gene name: *TNFRSF11B*), which is thereby a strong physiological inhibitor of osteoclast formation. Early studies in gene knockout mice showed RANKL is essential for osteoclastogenesis and OPG could inhibit it; mice overexpressing OPG (Simonet et al. 1997) and mice lacking RANKL (Kong et al. 1999; Dougall et al. 1999) both showed severe osteopetrosis. These discoveries led to the development of denosumab, an anti-RANKL monoclonal antibody now used for the treatment of osteoporosis (Lacey et al. 2012).

Elevated expression of RANKL stimulates osteoclast formation both in pathology and in normal remodeling. One example of its role in pathology is inflammatory arthritis, where T cells express elevated RANKL levels under the influence of inflammatory cytokines (Horwood et al. 1999). Pharmacologic inhibition of RANKL or of osteoclastogenesis therefore blocks both pathological joint erosion and the systemic bone loss associated with the inflammation (Romas et al. 2002; Sims et al. 2004).

RANKL-inducing cytokines, such as oncostatin M, are also required for physiological regulation of remodeling (Sims 2016). These locally produced cytokines stimulate osteoclast formation, not by acting directly on osteoclast precursors, but acting on by osteoblast lineage cells (Tamura et al. 1993; Jimi et al. 1996). The same is true for systemic factors like PTH and 1,25-dihydroxyvitamin-D_3. These early in vitro studies used a mixed stromal cell population including immature and mature osteoblast lineage cells. Importantly, the interaction of RANK and RANKL was contact-dependent (Tamura et al. 1993; Jimi et al. 1996), indicating it is membrane-bound RANKL, and not any soluble form, that induces osteoclastogenesis. Very recent mouse genetic studies have confirmed membrane-bound, and not soluble, RANKL is required for physiological osteoclastogenesis in vivo (Asano et al. 2019).

More recently, osteopetrosis due to impaired osteoclastogenesis was detected in mice with deletion of RANKL in the entire osteoblast lineage (Xiong et al. 2011), confirming the importance of RANKL in the osteoblast lineage indicated by those early stromal cell culture studies. Surprisingly, RANKL deletion in late osteoblasts and osteocytes also led to a reduction in osteoclastogenesis and high bone mass (Xiong et al. 2015; Nakashima et al. 2011). This suggested both early osteoblasts and osteocytes might support osteoclast formation by producing RANKL. The ability of osteocytes to support osteoclastogenesis has been controversial, since there is little evidence of direct contact between osteocytes and osteoclast precursors. However, even when cultured in contact with osteoclast precursors in vitro osteocytes do not fully support osteoclast formation (Chia et al. 2015; McGregor et al. 2019), and it must be remembered that the phenotype of the osteocyte-specific knockout is much less severe than that of the osteoblast-lineage knockout. This suggests that the physiological role of osteocyte-derived RANKL is less critical than RANKL derived from less differentiated osteoblasts and their precursors.

Although this is the most studied pathway, the osteoblast lineage also uses RANKL-independent mechanisms to regulate bone resorption, including signals to initiate remodeling (Schaffler et al. 2014) and to stimulate osteoblast precursor proliferation (Van Wesenbeeck et al. 2002; Wiktor-Jedrzejczak et al. 1990), chemo-attractants to enlist osteoclast precursors to the bone surface (Onan et al. 2009), and factors modulating mature osteoclast activity (Wong et al. 1999).

10 Signals Within the Osteoblast Lineage Regulate Bone Formation

The regulation of remodeling is not restricted to coupling factors and the RANKL/OPG system; many factors expressed in the osteoblast lineage also work within the lineage to regulate bone formation and resorption.

A key example of communication within the osteoblast lineage is sclerostin. It's importance is clear since genetic suppression of sclerostin, either in mouse models or in humans, leads to profoundly elevated bone mass (Balemans et al. 2001, 2002; Li et al. 2009). Sclerostin is synthesized and secreted by osteocytes yet acts at earlier stages of osteoblast differentiation to inhibit bone formation. This is achieved by binding of sclerostin to lipoprotein receptor-related protein (LRP) 4/5/6 and thereby antagonizing the Wnt/β-catenin signaling pathway earlier in the osteoblast lineage (Baron and Kneissel 2013). The profound high bone mass phenotypes of individuals with suppressed sclerostin expression led to the development of pharmacological agents to inhibit sclerostin action as treatments for osteoporosis and other conditions of skeletal destruction (Cosman et al. 2016). Sclerostin also appears to stimulate osteoclastogenesis (at least indirectly), since patients treated with anti-sclerostin exhibited decreased CTX1 levels (Cosman et al. 2016). The mechanism by which this occurs remains somewhat unclear: while mice lacking β-catenin in osteoblasts have reduced osteoclast formation and elevated OPG levels (Glass et al. 2005), sclerostin treatment of osteocytes in vitro suppresses OPG expression (Wijenayaka et al. 2011). Regardless of mechanism, if anti-sclerostin acts as a mixed anabolic and anti-catabolic agent, this is likely to improve its efficacy for low bone mass.

The osteoblast lineage also secretes locally acting proteins that stimulate bone formation, such as PTHrP (Ansari et al. 2018; Miao et al. 2005) and the IL-6 family cytokine oncostatin M (Walker et al. 2010). Both are secreted by osteoblasts and osteocytes, both promote bone formation by suppressing sclerostin and by promoting osteoblast progenitor proliferation and commitment to the osteoblast lineage. Furthermore, when acting early in the osteoblast lineage, both promote osteoclastogenesis by stimulating expression of RANKL.

Osteoblasts also express contact-dependent molecules regulating osteoblast differentiation. Examples of this include gap junction molecules such as connexin 43, which promotes osteoblast survival and function (Stains et al. 2014; Plotkin and Bellido 2013), and membrane-bound EphrinB2. Although the latter is expressed at all stages of the lineage, its function differs depending on the stage of differentiation. Early in the committed lineage, EphrinB2 promotes the late stages of osteoblast differentiation (Takyar et al. 2013) and initiation of osteoid mineralization (Tonna et al. 2014). In contrast, EphrinB2 expression in osteocytes limits secondary mineralization, without affecting initiation of osteoid mineralization (Vrahnas et al. 2019).

11 Interactions Between Bone Cells and Other Cells and Tissues in and Beyond Bone

Bone cells act as local regulators to maintain bone tissue homeostasis and to influence nearby tissues such as the marrow, vasculature, and skeletal muscle. Within bone itself, endochondral ossification and remodeling depend on contributions from the vasculature (see above) and the neuronal system (Idelevich and Baron 2018). Not only this, but a wide range of cells present in the BMU, including macrophages and T cells, regulate osteoblast differentiation (Sims and Martin 2014). The osteoblast lineage in turn regulates the hematopoietic stem cell niche (Calvi et al. 2003; Askmyr et al. 2009) and contributes to hematopoietic malignancies (Raaijmakers et al. 2010) and B cell homeostasis (Wu et al. 2008).

Cross talk between bone and muscle is also vital for skeletal development and maintenance. A well-studied example of this is the response of osteocytes to mechanical load and the anabolic effect of exercise on the skeleton, particularly during bone growth and development. However, the interaction is not limited to mechanical effects. Osteoblasts release muscle-active cytokines (myokines), and muscle also produces bone-active cytokines (Johnson et al. 2014; Brotto and Bonewald 2015).

The osteoblast lineage also acts as an endocrine system to regulate the function of other organs (Fig. 5). Through the release of FGF23, osteocytes act on the kidney

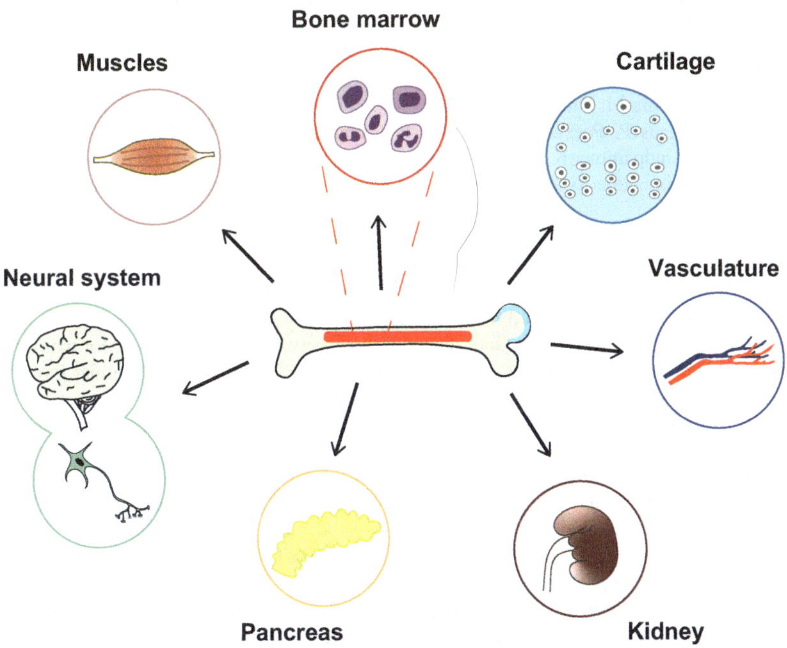

Fig. 5 An overview of the diversity of interactions regulating bone structure. Bone tissue interacts with other cells and tissues in their close proximity, such as cartilage, blood vessels, marrow cells, nerves, and muscles. It also affects distant organs, such as the pancreas, brain, and kidney, by producing and secreting different factors into circulation

to regulate phosphate homeostasis (Fukumoto and Martin 2009), and release of osteocalcin by the osteoblast lineage has been implicated in functions as diverse as glucose metabolism, fertility, memory, and the fight-or-flight response (Berger et al. 2019; Karsenty 2017). These effects and the molecules responsible for transmitting them must be considered when using pharmacological agents to modify bone cell function.

References

Anderson HC (1967) Electron microscopic studies of induced cartilage development and calcification. J Cell Biol 35:81–101

Anderson HC (1995) Molecular biology of matrix vesicles. Clin Orthop Relat Res:266–280

Ansari N, Ho PW, Crimeen-Irwin B, Poulton IJ, Brunt AR, Forwood MR, Divieti Pajevic P, Gooi JH, Martin TJ, Sims NA (2018) Autocrine and paracrine regulation of the murine skeleton by osteocyte-derived parathyroid hormone-related protein. J Bone Miner Res 33:137–153

Asano T, Okamoto K, Nakai Y, Tsutsumi M, Muro R, Suematsu A, Hashimoto K, Okamura T, Ehata S, Nitta T, Takayanagi H (2019) Soluble RANKL is physiologically dispensable but accelerates tumour metastasis to bone. Nat Metab 1:868–875

Askmyr M, Sims NA, Martin TJ, Purton LE (2009) What is the true nature of the osteoblastic hematopoietic stem cell niche? Trends Endocrinol Metab 20:303–309

Azari F, Vali H, Guerquin-Kern JL, Wu TD, Croisy A, Sears SK, Tabrizian M, McKee MD (2008) Intracellular precipitation of hydroxyapatite mineral and implications for pathologic calcification. J Struct Biol 162:468–479

Bakker AD, Silva VC, Krishnan R, Bacabac RG, Blaauboer ME, Lin YC, Marcantonio RA, Cirelli JA, Klein-Nulend J (2009) Tumor necrosis factor alpha and interleukin-1beta modulate calcium and nitric oxide signaling in mechanically stimulated osteocytes. Arthritis Rheum 60:3336–3345

Balemans W, Ebeling M, Patel N, Van Hul E, Olson P, Dioszegi M, Lacza C, Wuyts W, Van Den Ende J, Willems P, Paes-Alves AF, Hill S, Bueno M, Ramos FJ, Tacconi P, Dikkers FG, Stratakis C, Lindpaintner K, Vickery B, Foernzler D, Van Hul W (2001) Increased bone density in sclerosteosis is due to the deficiency of a novel secreted protein (SOST). Hum Mol Genet 10:537–543

Balemans W, Patel N, Ebeling M, Van Hul E, Wuyts W, Lacza C, Dioszegi M, Dikkers FG, Hildering P, Willems PJ, Verheij JB, Lindpaintner K, Vickery B, Foernzler D, Van Hul W (2002) Identification of a 52 kb deletion downstream of the SOST gene in patients with van Buchem disease. J Med Genet 39:91–97

Baron R (1977) Importance of the intermediate phase between resorption and formation in the measurement and understanding of the bone remodelling sequence. I. In: Meunier P (ed) Bone histomorphometry, 2nd int workshop. Lab Armour Montague, Paris

Baron R, Kneissel M (2013) WNT signaling in bone homeostasis and disease: from human mutations to treatments. Nat Med 19:179–192

Berger JM, Singh P, Khrimian L, Morgan DA, Chowdhury S, Arteaga-Solis E, Horvath TL, Domingos AI, Marsland AL, Kumal Yadav V, Rahmouni K, Gao X-B, Karsenty G (2019) Mediation of the acute stress response by the skeleton. Cell Metab 30:890–902

Blair HC, Teitelbaum SL, Ghiselli R, Gluck S (1989) Osteoclastic bone resorption by a polarized vacuolar proton pump. Science 245:855–857

Blank M, Sims NA (2019) Cellular processes by which osteoblasts and osteocytes control bone mineral deposition and maturation revealed by stage-specific EphrinB2 knockdown. Curr Osteoporos Rep 17:270–280

Boivin G, Meunier PJ (2002) Changes in bone remodeling rate influence the degree of mineralization of bone. Connect Tissue Res 43:535–537

Brotto M, Bonewald L (2015) Bone and muscle: interactions beyond mechanical. Bone 80:109–114
Buenzli PR, Sims NA (2015) Quantifying the osteocyte network in the human skeleton. Bone 75:144–150
Calvi LM, Adams GB, Weibrecht KW, Weber JM, Olson DP, Knight MC, Martin RP, Schipani E, Divieti P, Bringhurst FR, Milner LA, Kronenberg HM, Scadden DT (2003) Osteoblastic cells regulate the haematopoietic stem cell niche. Nature 425:841–846
Chia LY, Walsh NC, Martin TJ, Sims NA (2015) Isolation and gene expression of haematopoietic-cell-free preparations of highly purified murine osteocytes. Bone 72:34–42
Cooper KL, Oh S, Sung Y, Dasari RR, Kirschner MW, Tabin CJ (2013) Multiple phases of chondrocyte enlargement underlie differences in skeletal proportions. Nature 495:375
Cosman F, Crittenden DB, Adachi JD, Binkley N, Czerwinski E, Ferrari S, Hofbauer LC, Lau E, Lewiecki EM, Miyauchi A, Zerbini CA, Milmont CE, Chen L, Maddox J, Meisner PD, Libanati C, Grauer A (2016) Romosozumab treatment in postmenopausal women with osteoporosis. N Engl J Med 375:1532–1543
Dallas SL, Bonewald LF (2010) Dynamics of the transition from osteoblast to osteocyte. Ann N Y Acad Sci 1192:437–443
Dallas SL, Prideaux M, Bonewald LF (2013) The osteocyte: an endocrine cell . . . and more. Endocr Rev 34:658–690
Del Fattore A, Peruzzi B, Rucci N, Recchia I, Cappariello A, Longo M, Fortunati D, Ballanti P, Iacobini M, Luciani M, Devito R, Pinto R, Caniglia M, Lanino E, Messina C, Cesaro S, Letizia C, Bianchini G, Fryssira H, Grabowski P, Shaw N, Bishop N, Hughes D, Kapur RP, Datta HK, Taranta A, Fornari R, Migliaccio S, Teti A (2006) Clinical, genetic, and cellular analysis of 49 osteopetrotic patients: implications for diagnosis and treatment. J Med Genet 43:315–325
Delaisse JM (2014) The reversal phase of the bone-remodeling cycle: cellular prerequisites for coupling resorption and formation. Bonekey Rep 3:561
Delaisse JM, Andersen TL, Engsig MT, Henriksen K, Troen T, Blavier L (2003) Matrix metalloproteinases (MMP) and cathepsin K contribute differently to osteoclastic activities. Microsc Res Tech 61:504–513
Dillon S, Staines KA, Millan JL, Farquharson C (2019) How to build a bone: PHOSPHO1, biomineralization, and beyond. JBMR Plus 3:e10202
Dobnig H, Turner RT (1995) Evidence that intermittent treatment with parathyroid hormone increases bone formation in adult rats by activation of bone lining cells. Endocrinology 136:3632–3638
Dougall WC, Glaccum M, Charrier K, Rohrbach K, Brasel K, De Smedt T, Daro E, Smith J, Tometsko ME, Maliszewski CR, Armstrong A, Shen V, Bain S, Cosman D, Anderson D, Morrissey PJ, Peschon JJ, Schuh J (1999) RANK is essential for osteoclast and lymph node development. Genes Dev 13:2412–2424
Ducy P, Zhang R, Geoffroy V, Ridall AL, Karsenty G (1997) Osf2/Cbfa1: a transcriptional activator of osteoblast differentiation. Cell 89:747–754
Ducy P, Starbuck M, Priemel M, Shen J, Pinero G, Geoffroy V, Amling M, Karsenty G (1999) A Cbfa1-dependent genetic pathway controls bone formation beyond embryonic development. Genes Dev 13:1025–1036
Eghbali-Fatourechi GZ, Modder UI, Charatcharoenwitthaya N, Sanyal A, Undale AH, Clowes JA, Tarara JE, Khosla S (2007) Characterization of circulating osteoblast lineage cells in humans. Bone 40:1370–1377
Ekstrom K, Omar O, Graneli C, Wang X, Vazirisani F, Thomsen P (2013) Monocyte exosomes stimulate the osteogenic gene expression of mesenchymal stem cells. PLoS One 8:e75227
Everts V, Delaisse JM, Korper W, Jansen DC, Tigchelaar-Gutter W, Saftig P, Beertsen W (2002) The bone lining cell: its role in cleaning Howship's lacunae and initiating bone formation. J Bone Miner Res 17:77–90
Findlay DM, Atkins GJ (2014) Osteoblast-chondrocyte interactions in osteoarthritis. Curr Osteoporos Rep 12:127–134

Forlino A, Marini JC (2016) Osteogenesis imperfecta. Lancet 387:1657–1671
Frattini A, Vezzoni P, Villa A, Sobacchi C (2007) The dissection of human autosomal recessive osteopetrosis identifies an osteoclast-poor form due to RANKL deficiency. Cell Cycle 6:3027–3033
Fratzl P, Gupta H, Paschalis E, Roschger P (2004) Structure and mechanical quality of the collagen–mineral nano-composite in bone. J Mater Chem 14:2115–2123
Fuchs RK, Allen MR, Ruppel ME, Diab T, Phipps RJ, Miller LM, Burr DB (2008) In situ examination of the time-course for secondary mineralization of Haversian bone using synchrotron Fourier transform infrared microspectroscopy. Matrix Biol 27:34–41
Fuchs RK, Faillace ME, Allen MR, Phipps RJ, Miller LM, Burr DB (2011) Bisphosphonates do not alter the rate of secondary mineralization. Bone 49:701–705
Fukumoto S, Martin TJ (2009) Bone as an endocrine organ. Trends Endocrinol Metab 20:230–236
Gehron Robey P, Riminucci M (2020) Skeletal stem cells: tissue-specific stem/progenitor cells of cartilage, bone, stroma, and marrow adipocytes. In: Bilezikian JP, Martin TJ, Clemens TL, Rosen CJ (eds) Principles of bone biology, 4th edn. Elsevier, Cambridge
Gil-Henn H, Destaing O, Sims NA, Aoki K, Alles N, Neff L, Sanjay A, Bruzzaniti A, De Camilli P, Baron R, Schlessinger J (2007) Defective microtubule-dependent podosome organization in osteoclasts leads to increased bone density in Pyk2(−/−) mice. J Cell Biol 178:1053–1064
Giraud-Guille MM (1988) Twisted plywood architecture of collagen fibrils in human compact bone osteons. Calcif Tissue Int 42:167–180
Glass DA 2nd, Bialek P, Ahn JD, Starbuck M, Patel MS, Clevers H, Taketo MM, Long F, McMahon AP, Lang RA, Karsenty G (2005) Canonical Wnt signaling in differentiated osteoblasts controls osteoclast differentiation. Dev Cell 8:751–764
Gowen LC, Petersen DN, Mansolf AL, Qi H, Stock JL, Tkalcevic GT, Simmons HA, Crawford DT, Chidsey-Frink KL, Ke HZ, Mcneish JD, Brown TA (2003) Targeted disruption of the osteoblast/osteocyte factor 45 gene (OF45) results in increased bone formation and bone mass. J Biol Chem 278:1998–2007
Granke M, Does MD, Nyman JS (2015) The role of water compartments in the material properties of cortical bone. Calcif Tissue Int 97:292–307
Gray C, Boyde A, Jones SJ (1996) Topographically induced bone formation in vitro: implications for bone implants and bone grafts. Bone 18:115–123
Grigoriadis AE, Wang ZQ, Cecchini MG, Hofstetter W, Felix R, Fleisch HA, Wagner EF (1994) c-Fos: a key regulator of osteoclast-macrophage lineage determination and bone remodeling. Science 266:443–448
Gutierrez S, Javed A, Tennant DK, Van Rees M, Montecino M, Stein GS, Stein JL, Lian JB (2002) CCAAT/enhancer-binding proteins (C/EBP) beta and delta activate osteocalcin gene transcription and synergize with Runx2 at the C/EBP element to regulate bone-specific expression. J Biol Chem 277:1316–1323
Hartmann C, Yang Y (2020) Molecular and cellular regulation of intramembranous and endochondral bone formation during embryogenesis. In: Bilezikian JP, Martin TJ, Clemens TL, Rosen CJ (eds) Principles of bone biology, 4th edn. Elsevier, Cambridge
Hauge EM, Qvesel D, Eriksen EF, Mosekilde L, Melsen F (2001) Cancellous bone remodeling occurs in specialized compartments lined by cells expressing osteoblastic markers. J Bone Miner Res 16:1575–1582
Horne WC, Sanjay A, Bruzzaniti A, Baron R (2005) The role(s) of Src kinase and Cbl proteins in the regulation of osteoclast differentiation and function. Immunol Rev 208:106–125
Horwood NJ, Kartsogiannis V, Quinn JM, Romas E, Martin TJ, Gillespie MT (1999) Activated T lymphocytes support osteoclast formation in vitro. Biochem Biophys Res Commun 265:144–150
Houston B, Stewart AJ, Farquharson C (2004) PHOSPHO1-A novel phosphatase specifically expressed at sites of mineralisation in bone and cartilage. Bone 34:629–637
Howard GA, Bottemiller BL, Turner RT, Rader JI, Baylink DJ (1981) Parathyroid hormone stimulates bone formation and resorption in organ culture: evidence for a coupling mechanism. Proc Natl Acad Sci USA 78:3204–3208

Idelevich A, Baron R (2018) Brain to bone: what is the contribution of the brain to skeletal homeostasis? Bone 115:31–42

Ikebuchi Y, Aoki S, Honma M, Hayashi M, Sugamori Y, Khan M, Kariya Y, Kato G, Tabata Y, Penninger JM, Udagawa N, Aoki K, Suzuki H (2018) Coupling of bone resorption and formation by RANKL reverse signalling. Nature 561:195–200

Jansen ID, Vermeer JA, Bloemen V, Stap J, Everts V (2012) Osteoclast fusion and fission. Calcif Tissue Int 90:515–522

Jimi E, Nakamura I, Amano H, Taguchi Y, Tsurukai T, Tamura M, Takahashi N, Suda T (1996) Osteoclast function is activated by osteoblastic cells through a mechanism involving cell-to-cell contact. Endocrinology 137:2187–2190

Johnson RW, White JD, Walker EC, Martin TJ, Sims NA (2014) Myokines (muscle-derived cytokines and chemokines) including ciliary neurotrophic factor (CNTF) inhibit osteoblast differentiation. Bone 64C:47–56

Johnson RW, McGregor NE, Brennan HJ, Crimeen-Irwin B, Poulton IJ, Martin TJ, Sims NA (2015) Glycoprotein130 (Gp130)/interleukin-6 (IL-6) signalling in osteoclasts promotes bone formation in periosteal and trabecular bone. Bone 81:343–351

Kalajzic I, Braut A, Guo D, Jiang X, Kronenberg MS, Mina M, Harris MA, Harris SE, Rowe DW (2004) Dentin matrix protein 1 expression during osteoblastic differentiation, generation of an osteocyte GFP-transgene. Bone 35:74–82

Karsenty G (2017) Update on the biology of osteocalcin. Endocr Pract 23:1270–1274

Kim SW, Lu Y, Williams EA, Lai F, Lee JY, Enishi T, Balani DH, Ominsky MS, Ke HZ, Kronenberg HM, Wein MN (2017) Sclerostin antibody administration converts bone lining cells into active osteoblasts. J Bone Miner Res 32:892–901

Kong YY, Yoshida H, Sarosi I, Tan HL, Timms E, Capparelli C, Morony S, Oliveira-Dos-Santos AJ, Van G, Itie A, Khoo W, Wakeham A, Dunstan CR, Lacey DL, Mak TW, Boyle WJ, Penninger JM (1999) OPGL is a key regulator of osteoclastogenesis, lymphocyte development and lymph-node organogenesis. Nature 397:315–323

Kristensen HB, Andersen TL, Marcussen N, Rolighed L, Delaisse JM (2013) Increased presence of capillaries next to remodeling sites in adult human cancellous bone. J Bone Miner Res 28:574–585

Kukita T, Wada N, Kukita A, Kakimoto T, Sandra F, Toh K, Nagata K, Iijima T, Horiuchi M, Matsusaki H, Hieshima K, Yoshie O, Nomiyama H (2004) RANKL-induced DC-STAMP is essential for osteoclastogenesis. J Exp Med 200:941–946

Lacey DL, Boyle WJ, Simonet WS, Kostenuik PJ, Dougall WC, Sullivan JK, Martin JS, Dansey R (2012) Bench to bedside: elucidation of the OPG–RANK–RANKL pathway and the development of denosumab. Nat Rev Drug Discov 11:401

Lassen NE, Andersen TL, Ploen GG, Soe K, Hauge EM, Harving S, Eschen GET, Delaisse JM (2017) Coupling of bone resorption and formation in real time: new knowledge gained from human Haversian BMUs. J Bone Miner Res 32:1395–1405

Lee B, Thirunavukkarasu K, Zhou L, Pastore L, Baldini A, Hecht J, Geoffroy V, Ducy P, Karsenty G (1997) Missense mutations abolishing DNA binding of the osteoblast-specific transcription factor OSF2/CBFA1 in cleidocranial dysplasia. Nat Genet 16:307–310

Li X, Zhang Y, Kang H, Liu W, Liu P, Zhang J, Harris SE, Wu D (2005) Sclerostin binds to LRP5/6 and antagonizes canonical Wnt signaling. J Biol Chem 280:19883–19887

Li X, Ominsky MS, Warmington KS, Morony S, Gong J, Cao J, Gao Y, Shalhoub V, Tipton B, Haldankar R, Chen Q, Winters A, Boone T, Geng Z, Niu QT, Ke HZ, Kostenuik PJ, Simonet WS, Lacey DL, Paszty C (2009) Sclerostin antibody treatment increases bone formation, bone mass, and bone strength in a rat model of postmenopausal osteoporosis. J Bone Miner Res 24:578–588

Li D, Liu J, Guo B, Liang C, Dang L, Lu C, He X, Cheung HY, Xu L, Lu C, He B, Liu B, Shaikh AB, Li F, Wang L, Yang Z, Au DW, Peng S, Zhang Z, Zhang BT, Pan X, Qian A, Shang P, Xiao L, Jiang B, Wong CK, Xu J, Bian Z, Liang Z, Guo DA, Zhu H, Tan W, Lu A, Zhang G (2016) Osteoclast-derived exosomal miR-214-3p inhibits osteoblastic bone formation. Nat Commun 7:10872

Ling Y, Rios HF, Myers ER, Lu Y, Feng JQ, Boskey AL (2005) DMP1 depletion decreases bone mineralization in vivo: an FTIR imaging analysis. J Bone Miner Res 20:2169–2177
Mackie EJ, Tatarczuch L, Mirams M (2011) The skeleton: a multi-functional complex organ. The growth plate chondrocyte and endochondral ossification. J Endocrinol 211:109–121
Marotti G, Muglia MA, Zaffe D (1985) A SEM study of osteocyte orientation in alternately structured osteons. Bone 6:331–334
Martin TJ, Sims NA (2015) RANKL/OPG; critical role in bone physiology. Rev Endocr Metab Disord 16:131–139
Marzia M, Sims NA, Voit S, Migliaccio S, Taranta A, Bernardini S, Faraggiana T, Yoneda T, Mundy GR, Boyce BF, Baron R, Teti A (2000) Decreased c-Src expression enhances osteoblast differentiation and bone formation. J Cell Biol 151:311–320
Matic I, Matthews BG, Wang X, Dyment NA, Worthley DL, Rowe DW, Grcevic D, Kalajzic I (2016) Quiescent bone lining cells are a major source of osteoblasts during adulthood. Stem Cells 34:2930–2942
McGregor NE, Murat M, Elango J, Poulton IJ, Walker EC, Crimeen-Irwin B, Ho PWM, Gooi JH, Martin TJ, Sims NA (2019) IL-6 exhibits both cis- and trans-signaling in osteocytes and osteoblasts, but only trans-signaling promotes bone formation and osteoclastogenesis. J Biol Chem 294:7850–7863
Miao D, He B, Jiang Y, Kobayashi T, Soroceanu MA, Zhao J, Su H, Tong X, Amizuka N, Gupta A, Genant HK, Kronenberg HM, Goltzman D, Karaplis AC (2005) Osteoblast-derived PTHrP is a potent endogenous bone anabolic agent that modifies the therapeutic efficacy of administered PTH 1-34. J Clin Invest 115:2402–2411
Miller SC, De Saint-Georges L, Bowman BM, Jee WS (1989) Bone lining cells: structure and function. Scanning Microsc 3:953–960; discussion 960–961
Miyamoto H, Suzuki T, Miyauchi Y, Iwasaki R, Kobayashi T, Sato Y, Miyamoto K, Hoshi H, Hashimoto K, Yoshida S, Hao W, Mori T, Kanagawa H, Katsuyama E, Fujie A, Morioka H, Matsumoto M, Chiba K, Takeya M, Toyama Y, Miyamoto T (2012) Osteoclast stimulatory transmembrane protein and dendritic cell-specific transmembrane protein cooperatively modulate cell-cell fusion to form osteoclasts and foreign body giant cells. J Bone Miner Res 27:1289–1297
Mizoguchi T, Muto A, Udagawa N, Arai A, Yamashita T, Hosoya A, Ninomiya T, Nakamura H, Yamamoto Y, Kinugawa S, Nakamura M, Nakamichi Y, Kobayashi Y, Nagasawa S, Oda K, Tanaka H, Tagaya M, Penninger JM, Ito M, Takahashi N (2009) Identification of cell cycle-arrested quiescent osteoclast precursors in vivo. J Cell Biol 184:541–554
Mundlos S, Otto F, Mundlos C, Mulliken JB, Aylsworth AS, Albright S, Lindhout D, Cole WG, Henn W, Knoll JH, Owen MJ, Mertelsmann R, Zabel BU, Olsen BR (1997) Mutations involving the transcription factor CBFA1 cause cleidocranial dysplasia. Cell 89:773–779
Nakashima K, Zhou X, Kunkel G, Zhang Z, Deng JM, Behringer RR, De Crombrugghe B (2002) The novel zinc finger-containing transcription factor osterix is required for osteoblast differentiation and bone formation. Cell 108:17–29
Nakashima T, Hayashi M, Fukunaga T, Kurata K, Oh-Hora M, Feng JQ, Bonewald LF, Kodama T, Wutz A, Wagner EF, Penninger JM, Takayanagi H (2011) Evidence for osteocyte regulation of bone homeostasis through RANKL expression. Nat Med 17:1231–1234
Negishi-Koga T, Shinohara M, Komatsu N, Bito H, Kodama T, Friedel RH, Takayanagi H (2011) Suppression of bone formation by osteoclastic expression of semaphorin 4D. Nat Med 17:1473–1480
Nesbitt SA, Horton MA (1997) Trafficking of matrix collagens through bone-resorbing osteoclasts. Science 276:266–269
Onan D, Allan EH, Quinn JM, Gooi JH, Pompolo S, Sims NA, Gillespie MT, Martin TJ (2009) The chemokine Cxcl1 is a novel target gene of parathyroid hormone (PTH)/PTH-related protein in committed osteoblasts. Endocrinology 150:2244–2253
Otto F, Thornell AP, Crompton T, Denzel A, Gilmour KC, Rosewell IR, Stamp GW, Beddington RS, Mundlos S, Olsen BR, Selby PB, Owen MJ (1997) Cbfa1, a candidate gene for cleidocranial dysplasia syndrome, is essential for osteoblast differentiation and bone development. Cell 89:765–771

Oxlund H, Barckman M, Ørtoft G, Andreassen TT (1995) Reduced concentrations of collagen cross-links are associated with reduced strength of bone. Bone 17:S365–S371
Paic F, Igwe JC, Nori R, Kronenberg MS, Franceschetti T, Harrington P, Kuo L, Shin DG, Rowe DW, Harris SE, Kalajzic I (2009) Identification of differentially expressed genes between osteoblasts and osteocytes. Bone 45:682–692
Paschalis EP, Betts F, Dicarlo E, Mendelsohn R, Boskey AL (1997) FTIR microspectroscopic analysis of normal human cortical and trabecular bone. Calcif Tissue Int 61:480–486
Paschalis EP, Shane E, Lyritis G, Skarantavos G, Mendelsohn R, Boskey AL (2004) Bone fragility and collagen cross-links. J Bone Miner Res 19:2000–2004
Pederson L, Ruan M, Westendorf JJ, Khosla S, Oursler MJ (2008) Regulation of bone formation by osteoclasts involves Wnt/BMP signaling and the chemokine sphingosine-1-phosphate. Proc Natl Acad Sci USA 105:20764–20769
Pennypacker B, Shea M, Liu Q, Masarachia P, Saftig P, Rodan S, Rodan G, Kimmel D (2009) Bone density, strength, and formation in adult cathepsin K(−/−) mice. Bone 44:199–207
Plotkin LI, Bellido T (2013) Beyond gap junctions: connexin43 and bone cell signaling. Bone 52:157–166
Poole KE, Van Bezooijen RL, Loveridge N, Hamersma H, Papapoulos SE, Lowik CW, Reeve J (2005) Sclerostin is a delayed secreted product of osteocytes that inhibits bone formation. FASEB J 19:1842–1844
Poulton IJ, McGregor NE, Pompolo S, Walker EC, Sims NA (2012) Contrasting roles of leukemia inhibitory factor in murine bone development and remodeling involve region-specific changes in vascularization. J Bone Miner Res 27:586–595
Raaijmakers MH, Mukherjee S, Guo S, Zhang S, Kobayashi T, Schoonmaker JA, Ebert BL, Al-Shahrour F, Hasserjian RP, Scadden EO, Aung Z, Matza M, Merkenschlager M, Lin C, Rommens JM, Scadden DT (2010) Bone progenitor dysfunction induces myelodysplasia and secondary leukaemia. Nature 464:852–857
Rasmussen H, Bordier P (1974) The physiological basis of metabolic bone disease. Williams and Wilkins, Waverley, Baltimore
Rauch F (2012) The dynamics of bone structure development during pubertal growth. J Musculoskelet Neuronal Interact 12:1–6
Rohde M, Mayer H (2007) Exocytotic process as a novel model for mineralization by osteoblasts in vitro and in vivo determined by electron microscopic analysis. Calcif Tissue Int 80:323–336
Romas E, Sims NA, Hards DK, Lindsay M, Quinn JW, Ryan PF, Dunstan CR, Martin TJ, Gillespie MT (2002) Osteoprotegerin reduces osteoclast numbers and prevents bone erosion in collagen-induced arthritis. Am J Pathol 161:1419–1427
Roschger P, Gupta HS, Berzlanovich A, Ittner G, Dempster DW, Fratzl P, Cosman F, Parisien M, Lindsay R, Nieves JW, Klaushofer K (2003) Constant mineralization density distribution in cancellous human bone. Bone 32:316–323
Ross FP, Teitelbaum SL (2005) Alphavbeta3 and macrophage colony-stimulating factor: partners in osteoclast biology. Immunol Rev 208:88–105
Ryu J, Kim HJ, Chang EJ, Huang H, Banno Y, Kim HH (2006) Sphingosine 1-phosphate as a regulator of osteoclast differentiation and osteoclast-osteoblast coupling. EMBO J 25:5840–5851
Sabatakos G, Sims NA, Chen J, Aoki K, Kelz MB, Amling M, Bouali Y, Mukhopadhyay K, Ford K, Nestler EJ, Baron R (2000) Overexpression of DeltaFosB transcription factor(s) increases bone formation and inhibits adipogenesis. Nat Med 6:985–990
Salo J, Lehenkari P, Mulari M, Metsikko K, Vaananen HK (1997) Removal of osteoclast bone resorption products by transcytosis. Science 276:270–273
Schaffler MB, Cheung WY, Majeska R, Kennedy O (2014) Osteocytes: master orchestrators of bone. Calcif Tissue Int 94:5–24
Simonet WS, Lacey DL, Dunstan CR, Kelley M, Chang MS, Luthy R, Nguyen HQ, Wooden S, Bennett L, Boone T, Shimamoto G, Derose M, Elliott R, Colombero A, Tan HL, Trail G, Sullivan J, Davy E, Bucay N, Renshaw-Gegg L, Hughes TM, Hill D, Pattison W, Campbell P, Boyle WJ et al (1997) Osteoprotegerin: a novel secreted protein involved in the regulation of bone density. Cell 89:309–319

Sims NA (2016) Cell-specific paracrine actions of IL-6 family cytokines from bone, marrow and muscle that control bone formation and resorption. Int J Biochem Cell Biol 79:14–23

Sims NA, Martin TJ (2014) Coupling the activities of bone formation and resorption: a multitude of signals within the basic multicellular unit. Bonekey Rep 3:481

Sims NA, Martin TJ (2015) Coupling signals between the osteoclast and osteoblast: how are messages transmitted between these temporary visitors to the bone surface? Front Endocrinol (Lausanne) 6:41

Sims NA, Martin TJ (2020) Osteoclasts provide coupling signals to osteoblast lineage cells through multiple mechanisms. Annu Rev Physiol 82. (in press)

Sims NA, Ng KW (2014) Implications of osteoblast-osteoclast interactions in the management of osteoporosis by antiresorptive agents denosumab and odanacatib. Curr Osteoporos Rep 12:98–106

Sims NA, White CP, Sunn KL, Thomas GP, Drummond ML, Morrison NA, Eisman JA, Gardiner EM (1997) Human and murine osteocalcin gene expression: conserved tissue restricted expression and divergent responses to 1,25-dihydroxyvitamin D3 in vivo. Mol Endocrinol 11:1695–1708

Sims NA, Green JR, Glatt M, Schlict S, Martin TJ, Gillespie MT, Romas E (2004) Targeting osteoclasts with zoledronic acid prevents bone destruction in collagen-induced arthritis. Arthritis Rheum 50:2338–2346

Sly WS, Whyte MP, Sundaram V, Tashian RE, Hewett-Emmett D, Guibaud P, Vainsel M, Baluarte HJ, Gruskin A, Al-Mosawi M et al (1985) Carbonic anhydrase II deficiency in 12 families with the autosomal recessive syndrome of osteopetrosis with renal tubular acidosis and cerebral calcification. N Engl J Med 313:139–145

Sobacchi C, Villa A, Schulz A, Kornak U (1993) CLCN7-related osteopetrosis. In: Adam MP, Ardinger HH, Pagon RA, Wallace SE, Bean LJH, Stephens K, Amemiya A (eds) GeneReviews((R)). University of Washington, Seattle

Sobacchi C, Frattini A, Guerrini MM, Abinun M, Pangrazio A, Susani L, Bredius R, Mancini G, Cant A, Bishop N, Grabowski P, Del Fattore A, Messina C, Errigo G, Coxon FP, Scott DI, Teti A, Rogers MJ, Vezzoni P, Villa A, Helfrich MH (2007) Osteoclast-poor human osteopetrosis due to mutations in the gene encoding RANKL. Nat Genet 39:960–962

Soriano P, Montgomery C, Geske R, Bradley A (1991) Targeted disruption of the c-src proto-oncogene leads to osteopetrosis in mice. Cell 64:693–702

Stains JP, Watkins MP, Grimston SK, Hebert C, Civitelli R (2014) Molecular mechanisms of osteoblast/osteocyte regulation by connexin43. Calcif Tissue Int 94:55–67

Stanford CM, Jacobson PA, Eanes ED, Lembke LA, Midura RJ (1995) Rapidly forming apatitic mineral in an osteoblastic cell line (UMR 106-01 BSP). J Biol Chem 270:9420–9428

Takyar FM, Tonna S, Ho PW, Crimeen-Irwin B, Baker EK, Martin TJ, Sims NA (2013) EphrinB2/EphB4 inhibition in the osteoblast lineage modifies the anabolic response to parathyroid hormone. J Bone Miner Res 28:912–925

Tamura T, Udagawa N, Takahashi N, Miyaura C, Tanaka S, Yamada Y, Koishihara Y, Ohsugi Y, Kumaki K, Taga T, Kishimoto T, Suda T (1993) Soluble interleukin-6 receptor triggers osteoclast formation by interleukin 6. Proc Natl Acad Sci USA 90:11924–11928

Teti A, Blair HC, Teitelbaum SL, Kahn AJ, Koziol C, Konsek J, Zambonin-Zallone A, Schlesinger PH (1989) Cytoplasmic pH regulation and chloride/bicarbonate exchange in avian osteoclasts. J Clin Invest 83:227–233

Tonna S, Sims NA (2014) Talking among ourselves: paracrine control of bone formation within the osteoblast lineage. Calcif Tissue Int 94:35–45

Tonna S, Takyar FM, Vrahnas C, Crimeen-Irwin B, Ho PW, Poulton IJ, Brennan HJ, McGregor NE, Allan EH, Nguyen H, Forwood MR, Tatarczuch L, Mackie EJ, Martin TJ, Sims NA (2014) EphrinB2 signaling in osteoblasts promotes bone mineralization by preventing apoptosis. FASEB J 28:4482–4496

Tsourdi E, JÄhn K, Rauner M, Busse B, Bonewald LF (2018) Physiological and pathological osteocytic osteolysis. J Musculoskelet Neuronal Interact 18:292–303

Van Bezooijen RL, Ten Dijke P, Papapoulos SE, Lowik CW (2005) SOST/sclerostin, an osteocyte-derived negative regulator of bone formation. Cytokine Growth Factor Rev 16:319–327
Van Wesenbeeck L, Odgren PR, Mackay CA, D'Angelo M, Safadi FF, Popoff SN, Van Hul W, Marks SC Jr (2002) The osteopetrotic mutation toothless (tl) is a loss-of-function frameshift mutation in the rat Csf1 gene: evidence of a crucial role for CSF-1 in osteoclastogenesis and endochondral ossification. Proc Natl Acad Sci USA 99:14303–14308
Varga P, Hesse B, Langer M, Schrof S, Mannicke N, Suhonen H, Pacureanu A, Pahr D, Peyrin F, Raum K (2015) Synchrotron X-ray phase nano-tomography-based analysis of the lacunar-canalicular network morphology and its relation to the strains experienced by osteocytes in situ as predicted by case-specific finite element analysis. Biomech Model Mechanobiol 14:267–282
Vrahnas C, Pearson TA, Brunt AR, Forwood MR, Bambery KR, Tobin MJ, Martin TJ, Sims NA (2016) Anabolic action of parathyroid hormone (PTH) does not compromise bone matrix mineral composition or maturation. Bone 93:146–154
Vrahnas C, Buenzli PR, Pearson TA, Pennypacker BL, Tobin MJ, Bambery KR, Duong LT, Sims NA (2018) Differing effects of parathyroid hormone, alendronate, and odanacatib on bone formation and on the mineralization process in intracortical and endocortical bone of ovariectomized rabbits. Calcif Tissue Int 103:625–637
Vrahnas C, Blank M, Dite TA, Tatarczuch L, Ansari N, Crimeen-Irwin B, Nguyen H, Forwood MR, Hu Y, Ikegame M, Bambery KR, Petibois C, Mackie EJ, Tobin MJ, Smyth GK, Oakhill JS, Martin TJ, Sims NA (2019) Increased autophagy in EphrinB2-deficient osteocytes is associated with elevated secondary mineralization and brittle bone. Nat Commun 10:3436
Walker EC, McGregor NE, Poulton IJ, Pompolo S, Allan EH, Quinn JM, Gillespie MT, Martin TJ, Sims NA (2008) Cardiotrophin-1 is an osteoclast-derived stimulus of bone formation required for normal bone remodeling. J Bone Miner Res 23:2025–2032
Walker EC, McGregor NE, Poulton IJ, Solano M, Pompolo S, Fernandes TJ, Constable MJ, Nicholson GC, Zhang JG, Nicola NA, Gillespie MT, Martin TJ, Sims NA (2010) Oncostatin M promotes bone formation independently of resorption when signaling through leukemia inhibitory factor receptor in mice. J Clin Invest 120:582–592
Wijenayaka AR, Kogawa M, Lim HP, Bonewald LF, Findlay DM, Atkins GJ (2011) Sclerostin stimulates osteocyte support of osteoclast activity by a RANKL-dependent pathway. PLoS One 6:e25900
Wiktor-Jedrzejczak W, Bartocci A, Ferrante AW Jr, Ahmed-Ansari A, Sell KW, Pollard JW, Stanley ER (1990) Total absence of colony-stimulating factor 1 in the macrophage-deficient osteopetrotic (op/op) mouse. Proc Natl Acad Sci USA 87:4828–4832
Wong BR, Besser D, Kim N, Arron JR, Vologodskaia M, Hanafusa H, Choi Y (1999) TRANCE, a TNF family member, activates Akt/PKB through a signaling complex involving TRAF6 and c-Src. Mol Cell 4:1041–1049
Wu JY, Purton LE, Rodda SJ, Chen M, Weinstein LS, McMahon AP, Scadden DT, Kronenberg HM (2008) Osteoblastic regulation of B lymphopoiesis is mediated by Gs{alpha}-dependent signaling pathways. Proc Natl Acad Sci USA 105:16976–16981
Xiong J, Onal M, Jilka RL, Weinstein RS, Manolagas SC, O'Brien CA (2011) Matrix-embedded cells control osteoclast formation. Nat Med 17:1235–1241
Xiong J, Piemontese M, Onal M, Campbell J, Goellner JJ, Dusevich V, Bonewald L, Manolagas SC, O'Brien CA (2015) Osteocytes, not osteoblasts or lining cells, are the main source of the RANKL required for osteoclast formation in remodeling bone. PLoS One 10:e0138189
Yang X, Matsuda K, Bialek P, Jacquot S, Masuoka HC, Schinke T, Li L, Brancorsini S, Sassone-Corsi P, Townes TM, Hanauer A, Karsenty G (2004) ATF4 is a substrate of RSK2 and an essential regulator of osteoblast biology: implication for Coffin-Lowry syndrome. Cell 117:387–398
Zhao C, Irie N, Takada Y, Shimoda K, Miyamoto T, Nishiwaki T, Suda T, Matsuo K (2006) Bidirectional ephrinB2-EphB4 signaling controls bone homeostasis. Cell Metab 4:111–121

PTH and PTHrP Actions on Bone

Larry J. Suva and Peter A. Friedman

Contents

Abstract

Parathyroid hormone (PTH), PTH-related peptide (PTHrP), PTHR, and their cognate G protein-coupled receptor play defining roles in the regulation of extracellular calcium and phosphate metabolism and in controlling skeletal growth and repair. Acting through complex signaling mechanisms that in many instances proceed in a tissue-specific manner, precise control of these processes is achieved. A variety of direct and indirect disease processes, along with genetic anomalies, can cause these schemes to become dysfunctional. Here, we review the basic components of this regulatory network and present both the well-established elements and emerging findings and concepts with the overall objective to provide a framework for understanding the elementary aspects of how PTH and PTHrP behave and as a call to encourage further investigation that will

L. J. Suva
Department of Veterinary Physiology and Pharmacology, College of Veterinary Medicine and Biomedical Sciences, Texas Veterinary Medical Center, Texas A&M University, College Station, TX, USA

P. A. Friedman (✉)
Laboratory for GPCR Biology, Department of Pharmacology and Chemical Biology, University of Pittsburgh School of Medicine, Pittsburgh, PA, USA
e-mail: paf10@pitt.edu

P. H. Stern (ed.), *Bone Regulators and Osteoporosis Therapy*,
Handbook of Experimental Pharmacology 262, https://doi.org/10.1007/164_2020_362

yield more comprehensive understanding of the physiological and pathological steps at play, with a goal toward novel therapeutic interventions.

Keywords

Bone biology · Mineral metabolism · Parathyroid hormone · Parathyroid hormone-related peptide · Receptor signaling

1 Introduction

The bone and kidney form a mineral ion storage depot and regulatory axis that assures normal skeletal growth and development. PTHrP is a major hormonal regulator of bone formation, while PTH contributes to postnatal skeletal integrity and extracellular mineral ion homeostasis through actions on the kidneys. Despite these distinct actions and tissue effects, PTH and PTHR operate through a single canonical PTH/PTHrP receptor (PTHR). This chapter reviews both well-accepted and recent advances in our understanding of PTH and PTHrP actions and considers controversial or unsettled elements of their effects.

2 Receptors and Second Messenger Systems for PTH and PTHrP

PTH and PTHrP actions in bone are mediated by the type 1 PTH/PTHrP receptor (PTHR) (Abou-Samra et al. 1992; Jüppner et al. 1991). PTHR, a family B G protein-coupled receptor (GPCR), recognizes both PTH and PTHrP, as well as their biologically active amino-terminal peptides PTH(1–34) and PTHrP(1–36) and diverse engineered amino-terminally modified variants such as long-acting PTH (Shimizu et al. 2016) and abaloparatide, the 1–34 analog of PTHrP (Fig. 1). Mature parathyroid hormone is an 84-amino acid polypeptide that is synthesized and cleaved into its active form in the parathyroid gland. The initial precursor is pre-pro-PTH that contains 31 additional amino-terminal residues that are required for stringent processing. Five described mutations located in the pre-pro PTH leader sequence cause idiopathic hypoparathyroidism due to faulty hormone secretion. This 115 amino acid polypeptide is rapidly cleaved in the endoplasmic reticulum to form pro-PTH comprised of 90 amino acids. Following translocation to the trans-Golgi network, pro-PTH is cleaved at its amino-terminus to release mature PTH comprised of 84 amino acids (Sakwe et al. 2002). This is the primary hormone form that is stored and secreted and possesses full biological activity.

PTH and PTHrP bind the PTHR with equal affinity (although some differences in the affinity for G protein-coupled and uncoupled states have been reported (Dean et al. 2008)) and, in response, ligand binding signals through various cellular effector systems. The structure of the PTHR was recently solved (Zhao et al. 2019). This seminal contribution revealed that the extracellular PTHR domain assumes multiple conformations, which may help explain receptor activation and downstream

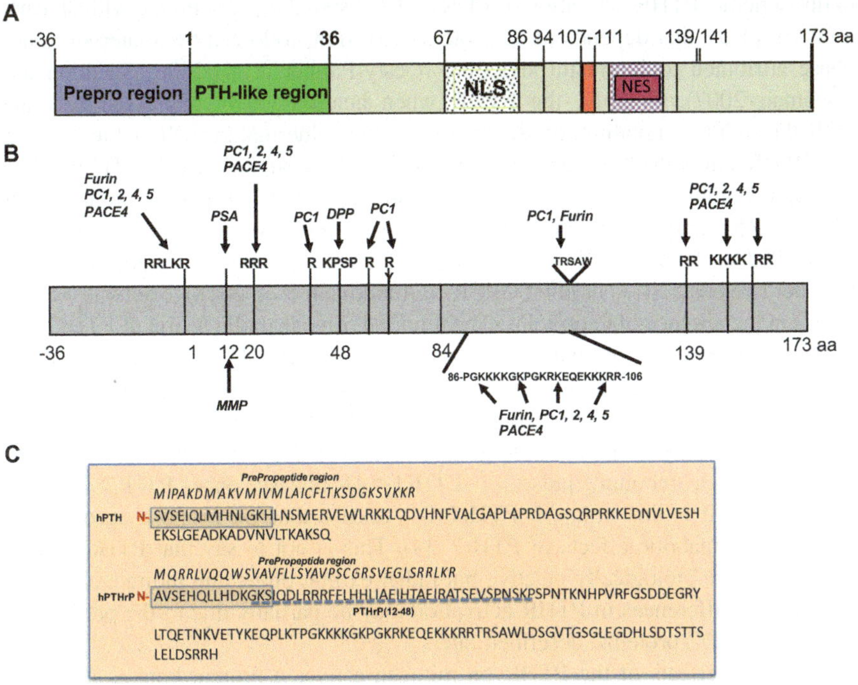

Fig. 1 Structural domains and amino acid sequence of human PTHrP and PTH. (**a**) Human PTHrP. Pre-pro leader sequence region (blue); PTH-like region (N-terminal 36 amino acids); putative nuclear localization signal (NLS; yellow); osteostatin region (red; amino acids 107–111); nuclear export signal (NES; purple). Amino acid numbering of human PTHrP (−36 to 173). (**b**) Schematic of the extensive putative (as yet unconfirmed) proteolytic sites within human PTHrP. Specific residue numbers and amino acids are indicated associated with potential cleavage sites; note there is no evidence for potential cleavage sites of PTH or PTHrP at position 34 or position 36. Potential cleavage sites exist for the production of PTHrP(12–48). Potential proteolytic enzymes are listed in italics (PC, prohormone convertases; PACE4, paired basic amino acid cleaving enzyme-4; PSA, prostate-specific antigen; DPP, dipeptidyl peptidase; MMP, matrix metalloproteinase). (**c**) Primary amino acid sequence of human PTH (84 amino acids) aligned with human PTHrP (141 amino acids). N identifies the amino-terminus of mature peptide; boxed region shows homology of the PTHR binding region of both peptides. PTHrP(12–48) sequence is underlined

signaling. Signal transduction by the PTHR is primarily mediated by G_s, $G_{q/11}$, G_i, and G_{12}/G_{13} heterotrimeric G proteins (Gardella and Vilardaga 2015). The particular coupling mechanism for distinct G proteins depends on the cell type in a manner that remains incompletely understood (Abou-Samra et al. 1992; Civitelli et al. 1989, 1990; Jüppner et al. 1991; Pines et al. 1996). Studies of the PDZ adapter protein NHERF1 revealed that it serves as a switch to control signaling by G_s, $G_{q/11}$, and G_i (Mahon et al. 2002; Wang et al. 2010).

Mitogenic signal transduction by extracellular signal-regulated kinases 1 and 2 (ERK1/ERK2) was the first example of G protein-independent, arrestin-dependent signaling (Shenoy et al. 2006). This paradigm was extended to the PTHR. According to this scheme, PTHR activation of ERK1/ERK2 signaling is bimodal, with an early transient phase that depends on G protein activation, followed by a later sustained phase attributed to ß-arrestin signaling (Gesty-Palmer et al. 2006; Sneddon and Friedman 2007). Notably, the PTHR when heterologously expressed activates ERK through Gq signaling, transactivation of the epidermal growth factor receptor, and PTHR internalization (Syme et al. 2005). It was supposed that PTH peptides lacking the amino-terminus, which do not stimulate cAMP formation, preferentially induce ERK signaling (Cupp et al. 2013). An impressive array of evidence was marshalled in defense of this notion. Other findings question the ability of PTHR (van der Lee et al. 2013) or other GPCRs (Grundmann et al. 2018) to activate ERK1/ERK2 in the absence of G proteins. The reported osteoanabolic action of PTH(7–34) is cited as an example of putative arrestin signaling (Gesty-Palmer et al. 2009). The peptide, however, was administered to mice with intact parathyroid glands, and we propose that its effect is a secondary consequence of transient suppression of PTH (1–84) secretion by the parathyroid gland (Huan et al. 2006; Slatopolsky et al. 2000) followed by a rebound pulse of PTH(1–84) release once PTH(7–34) was metabolized. It is this bolus of PTH(1–84) release that we propose accounts for the reported anabolic effects of PTH(7–34). This is not to say that PTH(7–34) or PTH(7–84) are biologically inactive but rather exhibit direct and indirect effects on the PTHR. Differences in PTHR activation may be partially due to the presence of modulating PDZ proteins, described later.

Important effects of the PTHR on mitogen-activated protein kinases (MAPK), especially ERK1 and ERK2, have been described (Cole 1999; Miao et al. 2001; Sneddon and Friedman 2007; Sneddon et al. 2000; Swarthout et al. 2001; Syme et al. 2005). The mechanisms of activation of MAPK signaling by the PTHR are complex and involve Src. This non-receptor tyrosine kinase interacts with the PTHR and contributes to PTH-mediated ERK activation (Rey et al. 2006). In addition, ERK activation may arise from EGFR transactivation (Syme et al. 2005). Polyubiquitination of specific lysine residues in the carboxy-terminal tail of the PTHR also regulates MAPK signaling, specifically activation of ERK1/ERK2 and p38 (Zhang et al. 2018).

More recently, a newer paradigm of PTHR signaling via Gs/cAMP has emerged that differentiates the signaling activities of PTH and PTHrP (and related analogs) (Cheloha et al. 2015) (Fig. 2). In response to PTHrP, the PTHR at the cell membrane generates short-lived cAMP signals. PTH also elicits membrane signals via Gs. However, PTH also stimulates the recruitment of β-arrestins by the PTHR followed by its internalization in endosomes. Here, the PTHR assembles a functional complex with β-arrestin and Gs to produce sustained cAMP signaling (Ferrandon et al. 2009). In endosomes, β-arrestin promotes cAMP signaling, contrary to its typical function at the cell membrane, which terminates cAMP signaling. This action involves stimulation of ERK1/ERK2, leading to inhibition of phosphodiesterase PDE4 and diminished cAMP degradation (Wehbi et al. 2013). Subsequent studies in animals show that this temporally discrete pattern of signaling by PTH and PTHrP results in markedly distinct calcemic responses (Shimizu et al. 2016). Thus, the induction of hypercalcemia by PTHrP is significantly lower than that elicited by

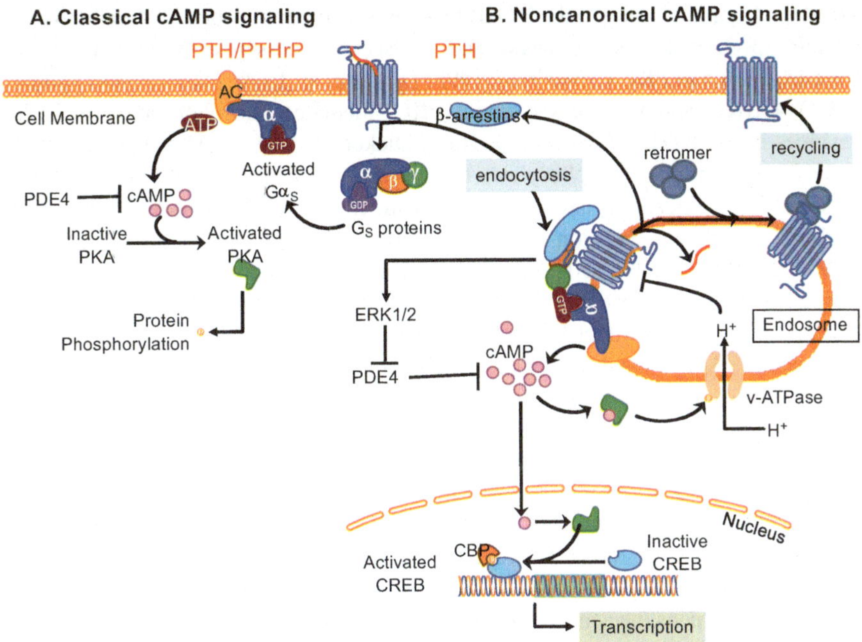

Fig. 2 Classic and non-canonical cAMP signaling by PTHR. PTH and PTHrP bind and stabilize distinct PTHR conformations resulting in different patterns and durations of cAMP formation and action. (**a**) PTH and PTHrP induce transient cAMP responses that are produced from plasma membrane-delimited PTHR. (**b**) PTH, in contrast, also provokes prolonged cAMP responses that arise from receptor internalization. Here, internalized PTHR complexes containing PTH and ß-arrestin promote rather than terminate cAMP signaling. The extended generation of cAMP depends on formation of PTH–PTHR–ß-arrestin complexes that stabilize Gs and also activate ERK1/ERK2. ERK1/ERK2 activation, in turn, inhibits PDE4, which normally metabolizes cAMP results in a protracted elevation of cAMP levels and stimulation of PKA. cAMP formation is terminated by a negative feedback mechanism, where PKA targets the endosomal V-ATPase and the attendant acidification disassembles the PTH–PTHR–ß-arrestin signaling complex. (Kindly provided by J-P Vilardaga, with modifications)

PTH. These observations are consistent with studies in humans showing that both PTHrP and its modified analog abaloparatide elicit significantly less hypercalcemia compared to PTH (Leder et al. 2015).

The magnitude of physiological responses mediated by the PTHR is tightly linked to the balance between signal generation and termination. The receptor normally behaves in a cyclical pattern of activation and inactivation, where PTHR desensitization guards cells against excessive stimulation and resensitization protects cells against prolonged hormone resistance. Rapid attenuation of PTHR signaling is mediated by receptor desensitization and internalization, while protracted reductions in responsiveness are due to downregulation and diminished receptor biosynthesis. Desensitization and internalization of the PTHR, as with other G protein-coupled receptors, are generally thought to be regulated primarily by phosphorylation and arrestins. Ligand binding to the PTHR promotes receptor phosphorylation both by G protein receptor kinases (GRKs) and by second messenger-dependent protein kinases, PKA and PKC (Blind et al. 1996; Dicker et al. 1999; Flannery and Spurney 2001). GRK2, and to a lesser extent PKC, mediates phosphorylation of serine residues in the PTHR (Castro et al. 2002; Dicker et al. 1999; Malecz et al. 1998). GRK2 preferentially phosphorylates the distal sites of the intracellular PTHR tail, whereas PKC phosphorylates more upstream residues (Blind et al. 1996). Mice harboring a phosphorylation-resistant PTHR, where serine residues at positions 489, 491, 492, 493, 495, 501, and 504 were mutated to alanine, exhibit essentially normal anabolic responses (Datta et al. 2012) suggesting that PTHR phosphorylation does not importantly affect PTHR internalization or bone anabolism. PTHR phosphorylation by the insulin-like growth factor type I receptor (IGF1R), however, promotes transition of osteoblasts to osteocytes (Qiu et al. 2018).

Following activation, the PTHR rapidly recruits ß-arrestins at the plasma membrane, an event that initiates dynamin-dependent endocytosis. Interestingly, PTHR phosphorylation is not required for the interaction of the receptor with ß-arrestins or for receptor internalization (Dicker et al. 1999; Ferrari et al. 1999; Malecz et al. 1998; Sneddon et al. 2003). However, receptor phosphorylation may stabilize the receptor-arrestin complex. As described previously, although the interaction with ß-arrestins dampens the acute cAMP generation at the plasma membrane, the PTHR, in complex with β-arrestins, remains active in endosomes and induces prolonged cAMP signaling. Although the PTHR lacks a canonical NPXXY internalization motif, other endocytotic signals have been identified within the carboxy terminus. Detailed analysis of the intracellular tail of the PTHR revealed bipartite sequences that negatively or positively regulate receptor endocytosis (Huang et al. 1995). An endocytic signal was detected within residues 475–494 of the opossum kidney PTHR (corresponding to D482-S501 of the human PTHR). Mutations or deletions within this region result in diminished PTH-induced receptor endocytosis (Huang et al. 1995).

In addition to governing PTHR signaling, NHERF1 is importantly involved in determining receptor endocytosis. Disrupting the interaction of NHERF1 with the PTHR by modifying its carboxy-terminal PDZ-binding motif, mutating NHERF1, or depolymerizing the actin cytoskeleton all cause important alterations in PTHR endocytosis (Sneddon et al. 2003). Notably, in the absence of NHERF1, both PTH (1–34) and PTH(7–34) (along with their corresponding full-length peptides) promoted efficient PTHR sequestration. In the presence of NHERF1, however, PTH (1–34)-induced receptor internalization was unaffected, whereas PTH(7–34)-initiated endocytosis was largely inhibited (Sneddon et al. 2003). NHERF1 contains two tandem PDZ domains and an ezrin-binding domain (EBD). PDZ core-binding domains and the NHERF1 EBD domain are required for inhibition of endocytosis (Wang et al. 2007).

3 Expression and Actions of PTHR in Bone

3.1 Effects of PTH and PTHrP on Bone Cells

In contrast to PTH, which has an entirely endocrine mechanism of action in virtually all target tissues, PTHrP exerts a plethora of physiological functions by acting as a paracrine, autocrine, or even intracrine regulator in a wide variety of target tissues. As far as we know, all the currently known PTH and PTHrP actions are mediated via specific interactions with PTHR.

Because PTH is the classical circulating endocrine hormone and parathyroid glands develop only after mesenchymal condensations are formed at sites of skeletal development, it is now well established and accepted that paracrine/autocrine PTHrP, secreted locally by chondrocytes, plays the primary role in endochondral bone formation during development (and likely during fracture repair postnatally). Indeed, multiple defects in skeletal development following PTHrP gene (*PTHLH*[1]) deletion in mice confirmed the importance of PTHrP in fetal bone development. In contrast, PTH gene (*PTH*) deletion results in a comparatively mild phenotype (Karaplis et al. 1994).

The deletion of *Pthlh* in mice resulted in perinatal death from respiratory failure due to defective rib development (Amizuka et al. 1994; Karaplis et al. 1994). The phenotype was rescued by targeting PTHrP production to cartilage via the collagen II promoter allowed and allowing examination of the PTHrP-null phenotype in other organs. With these studies, the central role of PTHrP (and not PTH) in endochondral bone formation was uncovered, as was the mechanism, namely, PTHrP and PTHR as downstream effectors of the Indian hedgehog pathway, that determines the rate and extent of long bone formation (Lanske et al. 1996; Vortkamp et al. 1996).

The predominant role of PTHrP in bone metabolism was demonstrated in mice deficient in PTHrP. Although mice lacking *Pthlh* die immediately after birth (Amizuka et al. 1994), haploinsufficient *Pthlh* $^{+/-}$ mice, that are phenotypically normal at birth, by 3 months of age have low bone mass, with a marked decrease in trabecular thickness and connectivity, and an abnormally high number of adipocytes in the bone marrow compared to wild-type littermates (Amizuka et al. 1996). The low bone mass is due to the decreased recruitment of bone marrow precursors coupled with increased osteoblast apoptosis compared to wild type (Amizuka et al. 1996). The critical role of osteoblast-derived PTHrP in the process of bone formation was confirmed when this phenotype was recapitulated in transgenic mice with an osteoblast-specific deletion of *Pthlh* (Miao et al. 2005). Notably, these mice also demonstrated reduced osteoclast formation that is likely due to the impaired ability of *Pthlh*-deficient osteoblasts to support osteoclast formation. Importantly, mice with *Pthlh*-deficient osteoblasts were normocalcemic, emphasizing the fact that PTHrP action in bone is not required for the maintenance of normal calcium homeostasis.

[1]By convention, gene names are italicized. Human genes and their protein products are in upper-case, for other species only the first letter is capitalized.

These and other genetic experiments too numerous to include revealed PTHrP actions that reproduce the known and very well-described anabolic effects of intermittent PTH, namely, the stimulation of bone formation by promoting the differentiation of committed osteoblast precursors and inhibiting the apoptosis of mature osteoblasts and osteocytes (Jilka 2007; Jilka et al. 1999). The potent effects of PTH as a skeletal anabolic therapy are well-established in osteoporosis (Black et al. 2008; Neer et al. 2001), an effect entirely dependent on the pharmacology of intermittent injections, inducing a sharp rise of serum PTH (Frolik et al. 2003). Indeed, sustained elevations of serum PTH levels (as seen in primary hyperparathyroidism) induce significant osteoclast formation and subsequent bone resorption through the production of RANKL by PTH target cells (Bilezikian et al. 2018).

With the recognition that PTHrP was the endogenous ligand for osteoblastic PTHR, several truncated forms of the molecule were investigated as anabolic agents. The anabolic effect of intermittent injection of PTHrP(1–36) in human subjects, as assessed by measurement of bone formation markers, has been suggested to be relatively free of the bone resorption effect of PTH (Horwitz et al. 2006, 2010). In these studies utilizing daily PTHrP(1–36) injections, the doses required to increase levels of anabolic markers were many-fold higher than those of PTH(1–34), even though the two are largely equipotent in acute infusion studies (Everhart-Caye et al. 1996). The possibility that PTHrP(1–36) is purely anabolic (Horwitz et al. 2010) is an interesting one, though somewhat difficult to explain. It might be related to different pharmacokinetics of the two, with PTHrP(1–36) degraded more rapidly following injection and thus less widely distributed to activate bone resorption, although this idea remains unproven.

Another potential plausible explanation is offered from studies using cells overexpressing PTHR, where PTHrP(1–36) and PTH(1–34) differed in their initial PTHR interaction(s). It appeared that PTHrP(1–36) action was restricted to the cell surface, whereas PTH(1–34) was more readily internalized and therefore more likely to activate G protein signaling interactions, so-called non-canonical signaling as discussed earlier (Fig. 2) (Ferrandon et al. 2009). It is this differential response that is suggested to explain why PTH(1–34) exhibits a strong bone resorption response, whereas PTHrP(1–36) did not, when the peptides are given in vivo via daily injection (Horwitz et al. 2010). Although such experiments may provide insight into the pharmacology of PTHR, the relative in vivo consequences of PTH and PTHrP action on bone remodeling are unclear. In this light, our lack of understanding of PTH and PTHrP action in vivo and the details of the exact form of the PTHrP protein (and PTH for that matter) presented in vivo makes resolution of the situation complex. There is no reason to assume that PTHrP(1–36) is presented in such a form to the PTHR. Clearly something akin to a full-length (amino-terminal containing) peptide or something very close to it could act as a paracrine agonist. In this case, circulating endogenous PTH might have only limited access to active remodeling sites, unless administered by injection at sufficient dose to compete with paracrine PTHrP, for example, during therapeutic dosing of PTH.

To activate PTHR, some amino-terminal-containing PTHrP peptide(s) must be produced in vivo. Presumably individual PTHrP isoforms are modified by posttranslational and enzymatic processing. Examination of the primary amino acid sequence of PTHrP in comparison with PTH (Fig. 1) identifies distinct peptide regions and numerous putative modification sites. The intracellular "pre-pro" and "pro" precursors are encoded in the first 36 amino acids of the primary sequence. The amino-terminal amino acid (Ser in PTH and Ala in PTHrP) encodes the start of mature peptide (Suva et al. 1987) (Fig. 1). The region showing the highest degree of primary sequence homology with PTH is indeed the first 13 amino acids, of which 8 of 13 residues are identical. This region is critical for almost all the agonist effects of both PTH and PTHrP on PTHR (Fig. 2). In fact, regions within PTH(1–14) and PTHrP(1–14), in particular His at position 5 of PTHrP, encode the ability for the ligand to discriminate between the PTHR and the related PTH-2 receptor (PTH2R) (Behar et al. 1996).

The PTHrP 14–36 region, with little or no primary amino acid sequence homology with PTH, appears to be required for the binding of PTHrP to PTHR and subsequent activation (Jüppner et al. 1991), illustrating the strong similarity in the secondary/tertiary structure of PTH and PTHrP, despite stark differences in primary amino acid sequence.

The importance and role of the sequence from amino acids 36 to 139 remain largely unknown. This region exists in all three isoforms of human PTHrP mRNA, with the primary sequence highly conserved across species through to amino acid 111, suggesting some important but as yet unknown function.

There is evidence supporting the idea that PTHrP residues 35–84 promote calcium transport across the placenta during fetal skeletal development (Abbas et al. 1989; Behar et al. 1996; Rodda et al. 1988). PTHrP(107–139) inhibits osteoclast activity and bone resorption both in vitro and in vivo (Cornish et al. 1997; Fenton et al. 1993). The same region was also shown to stimulate osteoblast proliferation and function (Alonso et al. 2008; Cornish et al. 1999). More recently, this PTHrP activity was proposed to be correlated with decreased sclerostin (Portal-Nunez et al. 2010). Similar experiments using mid- and distal PTHrP peptides have strong similarity with a series of studies using carboxy-terminus PTH fragments (Divieti et al. 2001; Duvos et al. 2006), where biological actions were reported yet no putative mechanism(s) identified or proposed.

The PTHrP amino acid sequence is littered with potential enzymatic processing sites and multiple mono- and multi-basic cleavage sites abound in the primary sequence (Fig. 1), indicative a potentially highly enzymatically remodeled peptide. Amino-terminal PTHrP has not been detected in normal patient serum; however, there is evidence in renal failure patients that PTHrP may circulate with a carboxy-terminal fragment removed by the kidney (Burtis et al. 1990; Orloff et al. 1992). Peptides such as PTHrP(1–36) and PTHrP(38–94), (38–95), and (38–101) as well as PTHrP(107–139) have been measured as post-translational cleavage products (Soifer et al. 1992; Wu et al. 1996) and PTHrP(12–48) identified in breast cancer patient serum (Washam et al. 2013). Defining the circulating forms of PTHrP has proven to be extremely difficult largely due to the reliability of notoriously poor antibodies for peptide detection, coupled with relatively low circulating PTHrP levels and the absence of assays with sufficient sensitivity and specificity. Indeed, the over reliance on antibodies alone remains a major confounder in much of current biology (Frohner et al. 2020).

With regard to PTHrP(12–48), this latter concern has been alleviated by the application of mass spectrometry methodologies for the identification of PTHrP peptides in breast cancer patient serum (Washam et al. 2013). The previously unknown PTHrP isoform (12–48) was undetectable using currently available PTHrP antisera. The peptide was measured by mass spectrometry at concentrations in the 50 ng/ml range, confirming the lack of cross-reactivity with existing PTHrP immunoassays, which may be expected to detect the fragment at this concentration (Washam et al. 2013). Subsequent studies using synthetic PTHrP(12–48), as well as specific PTHrP(12–48) antisera, identified expression in primary breast cancer and bone metastasis, with little expression in normal breast. PTHrP(12–48) also had a potent ability dose-dependently to inhibit osteoclastogenesis (via stimulation of apoptosis) in vitro (Kamalakar et al. 2017).

The implications of the identification of circulating PTHrP fragments, as well as the extensive post-translational processing of PTHrP (as outlined above), remain poorly understood (Orloff et al. 1994). However, it is clear that PTHrP mid- and carboxy-terminal peptides are biologically active, with actions beyond simply activating PTHR. As such, PTHrP must be considered another member of the family of polyhormones, similar to other endocrine proteins that give rise to mature secretory peptides with activities that can activate at the cell membrane, as well as access the nucleus (Fiaschi-Taesch and Stewart 2003).

Potential enzymatic processing sites abound in the PTHrP primary sequence (Fig. 1). Endoproteolytic processing at Arg-37 gives rise to the amino-terminal region and the various mid-region forms (Wu et al. 1996). Furin is an endoproteolytic enzyme resident in the Golgi cisterna that has been shown to be the enzyme responsible for the physiological processing of pro-PTH to PTH (Hendy et al. 1995). Other candidate furin sites in PTHrP have been identified at residues 97, 105, 106, and 108 (Diefenbach-Jagger et al. 1995). However, no enzymatic sites exist or have been identified that release PTH(1–34) as a physiologically relevant molecule. Indeed, the designation and utility of PTH(1–34) (and PTHrP(1–36) are solely based on the original peptide synthesis of PTH (Potts et al. 1971). At other sites, pro-hormone convertases such as PACE 4, PC1/3, PC2, and PSA have been suggested to process PTHrP in secretory vesicles (Cramer et al. 1996) (Fig. 1). In addition, cleavage sites for dipeptidyl peptidase at position 48/49 of PTHrP implicate this enzyme family in the processing of PTHrP(12–48) (Fig. 1), with enzymatic cleavage of the amino or carboxyl terminus of glycine 12 unclear.

Functional genetic studies have uncovered the relative physiologic roles of both the endogenous PTHR ligands, PTH and PTHrP, in calcium homeostasis and bone metabolism, although the specific peptide sequences mediating these processes are still largely unknown. *PTH*- and *PTHLH*-null mice of the same age have the opposite phenotype in newly formed trabecular bone, whereas haploinsufficiency of *PTHLH* in the *PTH*-null mice background results in some worsening of the phenotype (Miao et al. 2004). Such observations illustrate that PTHrP only partially compensates for the profound hypocalcemia of PTH loss. When adult *PTH*-null mice are crossed into the haploinsufficient *PTHLH* background (Miao et al. 2004), the offspring remain hypoparathyroid. However, the *PTHLH* haploinsufficiency reversed the high trabecular bone mass phenotype observed in *PTH* deficient mice (Miao et al. 2004). These data are consistent with the idea that during development, PTHrP directs growth plate formation via an exquisite control of chondrocyte proliferation and differentiation (Lanske et al. 1996; Vortkamp et al. 1996), while postnatal PTHrP serves largely as the major mediator of new bone formation (Miao et al. 2004). Despite the profound effects of PTHrP in malignant hypercalcemia (Johnson and Suva 2018), PTHrP does not significantly contribute to the regulation and maintenance of serum calcium (McCauley and Martin 2012). This fundamentally important process results entirely from the action of endocrine PTH, thus defining the distinct physiological role of PTHR ligands in fetal and postnatal development.

3.1.1 Molecular Mechanisms of Action in Osteoblasts

So how are these important physiologic effects of PTH and PTHrP mediated in adult tissues? In mammals, PTH is secreted by the parathyroid gland (Brown et al. 1981). As described earlier, PTH functions as the major endocrine regulator of calcium and phosphate metabolism (McCauley and Martin 2012). In response to low blood calcium, PTH enters the circulation and then acts in the bone and kidney and indirectly in the intestine interacting with local and systemic factors to restore normal serum levels in a classical endocrine feedback loop (Brown et al. 1993). The ensuing direct and indirect responses of these target cells help to maintain blood calcium concentrations to within narrow normal limits (Mannstadt et al. 1999). In contrast, PTHrP is not secreted by the parathyroid gland but from a variety of cells and with a wide spatial distribution (McCauley and Martin 2012). In particular, PTHrP in a autocrine/paracrine fashion primarily participates in the embryonic development of the skeleton (Kronenberg 2006) as well as promoting placental calcium mobilization as an endocrine factor during gestation and lactation (Abbas et al. 1989; Kovacs 2017; Neville et al. 2002; VanHouten et al. 2004) and regulating postnatal bone formation (McCauley and Martin 2012). PTHR is widely expressed in osteoblast linage cells, including osteocytes (Jilka 2007) in the skeleton which facilitates the broad effects of PTHR ligands on these cells in vitro and in vivo.

3.1.2 Adaptor Proteins

G Protein-Coupled Receptor Kinase 2 and β-Arrestins These two ubiquitously expressed proteins play important roles in the regulation of several G protein-coupled receptors, including the PTHR. It is not surprising that alterations in their expression or activity impact the function of osteoblasts and their responsiveness to PTH. G protein-coupled receptor kinase 2 (GRK2) phosphorylates agonist-occupied PTHR and promotes the binding of β-arrestins (Dicker et al. 1999; Ferrari et al. 1999; Flannery and Spurney 2001; Vilardaga et al. 2001). These combined actions result in decreased signaling at the cell membrane (Bisello et al. 2002; Dicker et al. 1999; Ferrari and Bisello 2001). As mentioned earlier, prolonged PTHR signaling in endosomes is promoted by β-arrestins through inhibition of PDE4 and reduction in cAMP degradation (Cheloha et al. 2015; Gardella and Vilardaga 2015).

Targeted overexpression of Grk2 in osteoblasts promotes bone loss (Wang et al. 2005). In contrast, inhibition of Grk2 activity by expression of a dominant negative mutant increases PTH-stimulated cAMP, and mice overexpressing Grk2 have increased bone remodeling with a net gain in bone content (Spurney et al. 2002). Similarly, intermittent PTH increased the number of osteoblasts in mice null for β-arrestin2 (Ferrari et al. 2005). However, no net increase in bone mass was observed, likely due to the intense stimulation of osteoclastogenesis. Another regulator of G protein-coupled receptor signaling, the regulator of G protein signaling-2 (RGS2), which increases the rate of hydrolysis of GTP bound to G proteins, thereby terminating signaling, has also been implicated in PTHR actions on bone cells. Rgs2 mRNA is rapidly and transiently increased by PTH in rat bones, as well as in osteoblast cultures (Miles et al. 2000), and its expression in bone cells decreased PTH-stimulated cAMP production (Thirunavukkarasu et al. 2002). Interestingly, Rgs2 upregulation was also observed in cells overexpressing Runx2 (Thirunavukkarasu et al. 2002), suggesting the possibility that mechanisms limiting PTHR signaling by G proteins may be activated upon differentiation of cells along the osteoblastic lineage.

Na^+/H^+ Exchange Regulatory Factor-1 NHERF1 is a PDZ domain scaffolding protein that interacts with the PTHR and regulates various functions, including preferential G protein coupling, membrane retention, and trafficking. Limited information exists on the role of NHERF1 in regulating PTH effects on bone cells. NHERF1 is expressed by actively mineralizing osteoblasts, where it modulates PTHR expression during differentiation (Liu et al. 2012). Mesenchymal stem cells from NHERF1$^{-/-}$ mice, when cultured in differentiating medium, show a dramatic reduction in PTHR mRNA accompanied by preferential differentiation to adipocytes. NHERF1 is required for osteoblast differentiation and matrix synthesis, but whether this is directly due to alterations in PTHR signaling is yet unknown. NHERF1-null mice and humans with NHERF1 polymorphisms display osteopenia and increased skeletal fractures (Karim et al. 2008; Morales et al. 2004; Shenolikar et al. 2002). Proliferating osteoblasts express Npt2a and Npt2b, PTHR, and NHERF1 (Wang et al. 2013). In cells from wild-type mice, PTH inhibits phosphate uptake, but this effect was not observed in cells from NHERF1$^{-/-}$ mice. In contrast, PTH increases phosphate uptake in differentiated osteoblasts, and this effect depends on NHERF1 expression.

4 Conclusion

Endocrine PTH and paracrine PTHrP regulate bone and mineral ion metabolism. These important functions are most complex during early development. Ongoing efforts are resulting in emerging information regarding the mechanisms by which

PTH and PTHrP share the use of their common PTHR in many tissues. Currently, PTH regulates calcium and phosphate homeostasis postnatally, whereas PTHrP directs growth plate development via the intimate control of chondrocyte proliferation and differentiation. Postnatally, PTH is solely responsible for endocrine mineral ion homeostasis, and PTHrP is the primary factor, locally derived in bone, to control bone remodeling via PTHR. These insights arise largely from mouse genetics as well as pharmacological studies in animals and the efficacy of PTH in the treatment of osteoporosis.

In terms of physiologic activity, the potential for post-translational processing of PTHrP is a striking feature of the molecule that may yield the generation of a number of biologically active peptides. Indeed, some (but not all) have been demonstrated to have biological activities that both require (or not) PTHR.

The shared function between a hormone (PTH) and a related molecule (PTHrP) is similar to the shared biology of growth hormone and IGF-1. In this case, the relationship is functional in many tissues (Anderson et al. 2018). Similarly, the actions of PTHrP and PTH on vasculature smooth muscle are also informative. In this tissue, local PTHrP enhances vascular smooth muscle activity throughout the vasculature, whereas pharmacologic PTH elicits widespread activation of smooth muscle, accompanied by vascular dilatation and decreased blood pressure (Wang et al. 1984).

Mouse genetics demonstrate the important role of local PTHrP production and paracrine action in bone remodeling (Karaplis et al. 1994; Miao et al. 2004, 2005). This PTHrP-specific activity occurs at multiple sites throughout the skeleton and is critical for the removal of damaged and/or old bone. If the vasculature scenario is recapitulated in the skeleton (as is presumably the case), PTHrP will be produced and act as needed at active sites of bone remodeling. This is indeed the situation as pharmacological administration of PTHrP elicits widespread activation of bone turnover. Biochemical markers of bone resorption rise slowly, whereas markers of bone formation increase early, exactly as reported when PTH is used clinically as an anabolic agent (Black et al. 2008, 2013; Frolik et al. 2003; Neer et al. 2001).

Molecular tools using mass spectrometry to characterize circulating PTH and PTHrP fragments will expand the accuracy of diagnostic testing and help reconcile anomalies where apparent PTH levels, for instance, are normal but the patient is clinically hypocalcemic and hypoparathyroid due to a mutation in PTH (Lee et al. 2015). In this example, commercial ELISA kits failed to distinguish a homozygous Arg-to-Cys mutation at position 25 of PTH(1–84). The current application of mass spectrometry approaches targeting sequencing of circulating PTH is limited by the use of the same or similar precipitating antibodies for purification that are applied for ELISA kits and hence bias outcomes. These limitations are likely to be overcome by using physical preparation techniques that do not suffer from these limitations. Further diagnostic prognostication and therapeutic intervention will likely emerge from application of CRISPR with improved associated guide RNAs (Lee et al. 2020).

Taken together, new information on PTH, PTHrP, PTHR, and its critical canonical and non-canonical signaling pathways will yield both new and more refined understanding of how these proteins are orchestrated in a pattern that allows control of skeletal growth and repair while maintaining the intra- and extracellular mineral ion environment. Continued interrogation of the actions of PTH and PTHrP is clearly warranted, since many of the intricacies and details of the specific players remain unknown. Such continued investigation is vital to reveal details of the selective activation and antagonism of PTHR and provide critical insights needed for development of small molecules with high therapeutic value.

Acknowledgments Original work cited herein was supported by grants DK105811 and DK111427 (PAF) and CA166060 and AA018282 (LJS) from the National Institutes of Health.

References

Abbas SK et al (1989) Stimulation of ovine placental calcium transport by purified natural and recombinant parathyroid hormone-related protein (PTHrP) preparations. Q J Exp Physiol 74:549–552

Abou-Samra AB et al (1992) Expression cloning of a common receptor for parathyroid hormone and parathyroid hormone-related peptide from rat osteoblast-like cells: a single receptor stimulates intracellular accumulation of both cAMP and inositol trisphosphates and increases intracellular free calcium. Proc Natl Acad Sci U S A 89:2732–2736

Alonso V, de Gortazar AR, Ardura JA, Andrade-Zapata I, Alvarez-Arroyo MV, Esbrit P (2008) Parathyroid hormone-related protein (107-139) increases human osteoblastic cell survival by activation of vascular endothelial growth factor receptor-2. J Cell Physiol 217:717–727

Amizuka N, Warshawsky H, Henderson JE, Goltzman D, Karaplis AC (1994) Parathyroid hormone-related peptide-depleted mice show abnormal epiphyseal cartilage development and altered endochondral bone formation. J Cell Biol 126:1611–1623

Amizuka N et al (1996) Haploinsufficiency of parathyroid hormone-related peptide (PTHrP) results in abnormal postnatal bone development. Dev Biol 175:166–176

Anderson LJ, Tamayose JM, Garcia JM (2018) Use of growth hormone, IGF-I, and insulin for anabolic purpose: pharmacological basis, methods of detection, and adverse effects. Mol Cell Endocrinol 464:65–74

Behar V, Nakamoto C, Greenberg Z, Bisello A, Suva LJ, Rosenblatt M, Chorev M (1996) Histidine at position 5 is the specificity "switch" between two parathyroid hormone receptor subtypes. Endocrinology 137:4217–4224

Bilezikian JP, Bandeira L, Khan A, Cusano NE (2018) Hyperparathyroidism. Lancet 391:168–178

Bisello A, Chorev M, Rosenblatt M, Monticelli L, Mierke DF, Ferrari SL (2002) Selective ligand-induced stabilization of active and desensitized parathyroid hormone type 1 receptor conformations. J Biol Chem 277:38524–38530

Black DM et al (2008) Randomized trial of once-weekly parathyroid hormone (1-84) on bone mineral density and remodeling. J Clin Endocrinol Metab 93:2166–2172

Black DM et al (2013) Improved adherence with PTH(1-84) in an extension trial for 24 months results in enhanced BMD gains in the treatment of postmenopausal women with osteoporosis. Osteoporos Int 24:1503–1511

Blind E, Bambino T, Huang ZM, Bliziotes M, Nissenson RA (1996) Phosphorylation of the cytoplasmic tail of the PTH/PTHrP receptor. J Bone Miner Res 11:578–586

Brown EM, Wilson RE, Thatcher JG, Marynick SP (1981) Abnormal calcium-regulated PTH release in normal parathyroid tissue from patients with adenoma. Am J Med 71:565–570

Brown EM et al (1993) Cloning and characterization of an extracellular Ca^{2+}-sensing receptor from bovine parathyroid. Nature 366:575–580

Burtis WJ et al (1990) Immunochemical characterization of circulating parathyroid hormone-related protein in patients with humoral hypercalcemia of cancer. N Engl J Med 322:1106–1112

Castro M, Dicker F, Vilardaga JP, Krasel C, Bernhardt M, Lohse MJ (2002) Dual regulation of the parathyroid hormone (PTH)/PTH-related peptide receptor signaling by PKC and ß-arrestins. Endocrinology 143:3854–3865

Cheloha RW, Gellman SH, Vilardaga JP, Gardella TJ (2015) PTH receptor-1 signalling-mechanistic insights and therapeutic prospects. Nat Rev Endocrinol 11:712–724

Civitelli R, Martin TJ, Fausto A, Gunsten SL, Hruska KA, Avioli LV (1989) Parathyroid hormone-related peptide transiently increases cytosolic calcium in osteoblast-like cells: comparison with parathyroid hormone. Endocrinology 125:1204–1210

Civitelli R, Hruska KA, Shen V, Avioli LV (1990) Cyclic AMP-dependent and calcium-dependent signals in parathyroid hormone function. Exp Gerontol 25:223–231

Cole JA (1999) Parathyroid hormone activates mitogen-activated protein kinase in opossum kidney cells. Endocrinology 140:5771–5779

Cornish J, Callon KE, Nicholson GC, Reid IR (1997) Parathyroid hormone-related protein-(107-139) inhibits bone resorption in vivo. Endocrinology 138:1299–1304

Cornish J, Callon KE, Lin C, Xiao C, Moseley JM, Reid IR (1999) Stimulation of osteoblast proliferation by C-terminal fragments of parathyroid hormone-related protein. J Bone Miner Res 14:915–922

Cramer SD, Chen Z, Peehl DM (1996) Prostate specific antigen cleaves parathyroid hormone-related protein in the PTH-like domain: inactivation of PTHrP-stimulated cAMP accumulation in mouse osteoblasts. J Urol 156:526–531

Cupp ME, Nayak SK, Adem AS, Thomsen WJ (2013) Parathyroid hormone (PTH) and PTH-related peptide domains contributing to activation of different PTH receptor-mediated signaling pathways. J Pharmacol Exp Ther 345:404–418

Datta NS, Samra TA, Abou-Samra AB (2012) Parathyroid hormone induces bone formation in phosphorylation-deficient PTHR1 knockin mice. Am J Physiol Endocrinol Metab 302:E1183–E1188

Dean T, Vilardaga JP, Potts JT Jr, Gardella TJ (2008) Altered selectivity of parathyroid hormone (PTH) and PTH-related protein (PTHrP) for distinct conformations of the PTH/PTHrP receptor. Mol Endocrinol 22:156–166

Dicker F, Quitterer U, Winstel R, Honold K, Lohse MJ (1999) Phosphorylation-independent inhibition of parathyroid hormone receptor signaling by G protein-coupled receptor kinases. Proc Natl Acad Sci U S A 96:5476–5481

Diefenbach-Jagger H, Brenner C, Kemp BE, Baron W, McLean J, Martin TJ, Moseley JM (1995) Arg21 is the preferred kexin cleavage site in parathyroid-hormone-related protein. Eur J Biochem 229:91–98

Divieti P, Inomata N, Chapin K, Singh R, Juppner H, Bringhurst FR (2001) Receptors for the carboxyl-terminal region of PTH(1-84) are highly expressed in osteocytic cells. Endocrinology 142:916–925

Duvos C, Scutt A, Mayer H (2006) hPTH-fragments (53-84) and (28-48) antagonize the stimulation of calcium release and repression of alkaline phosphatase activity by hPTH-(1-34) in vitro. FEBS Lett 580:1509–1514

Everhart-Caye M, Inzucchi SE, Guinness-Henry J, Mitnick MA, Stewart AF (1996) Parathyroid hormone (PTH)-related protein(1-36) is equipotent to PTH(1-34) in humans. J Clin Endocrinol Metab 81:199–208

Fenton AJ, Martin TJ, Nicholson GC (1993) Long-term culture of disaggregated rat osteoclasts: inhibition of bone resorption and reduction of osteoclast-like cell number by calcitonin and PTHrP[107-139]. J Cell Physiol 155:1–7

Ferrandon S et al (2009) Sustained cyclic AMP production by parathyroid hormone receptor endocytosis. Nat Chem Biol 5:734–742

Ferrari SL, Bisello A (2001) Cellular distribution of constitutively active mutant parathyroid hormone (PTH)/PTH-related protein receptors and regulation of cyclic adenosine 3′,5-′-monophosphate signaling by ß-arrestin2. Mol Endocrinol 15:149–163

Ferrari SL, Behar V, Chorev M, Rosenblatt M, Bisello A (1999) Endocytosis of ligand-human parathyroid hormone receptor 1 complexes is protein kinase C-dependent and involves ß-arrestin2. Real-time monitoring by fluorescence microscopy. J Biol Chem 274:29968–29975

Ferrari S et al (2005) Bone response to intermittent parathyroid hormone is altered in mice null for ß-arrestin2. Endocrinology 146:1854–1862

Fiaschi-Taesch NM, Stewart AF (2003) Minireview: parathyroid hormone-related protein as an intracrine factor--trafficking mechanisms and functional consequences. Endocrinology 144:407–411

Flannery PJ, Spurney RF (2001) Domains of the parathyroid hormone (PTH) receptor required for regulation by G protein-coupled receptor kinases (GRKs). Biochem Pharmacol 62:1047–1058

Frohner IE, Mudrak I, Kronlachner S, Schuchner S, Ogris E (2020) Antibodies recognizing the C terminus of PP2A catalytic subunit are unsuitable for evaluating PP2A activity and holoenzyme composition. Sci Signal 13

Frolik CA, Black EC, Cain RL, Satterwhite JH, Brown-Augsburger PL, Sato M, Hock JM (2003) Anabolic and catabolic bone effects of human parathyroid hormone (1-34) are predicted by duration of hormone exposure. Bone 33:372–379

Gardella TJ, Vilardaga J-P (2015) International union of basic and clinical pharmacology. XCIII. The parathyroid hormone receptors—family B G protein–coupled receptors. Pharmacol Rev 67:310–337

Gesty-Palmer D et al (2006) Distinct ß-arrestin and G protein dependent pathways for parathyroid hormone receptor stimulated ERK1/2 activation. J Biol Chem 281:10856–10864

Gesty-Palmer D, Flannery P, Yuan L, Corsino L, Spurney R, Lefkowitz RJ, Luttrell LM (2009) A β-arrestin-biased agonist of the parathyroid hormone receptor (PTH1R) promotes bone formation independent of G protein activation. Sci Trans Med 1:1ra1

Grundmann M et al (2018) Lack of β-arrestin signaling in the absence of active G proteins. Nat Commun 9:341

Hendy GN, Bennett HP, Gibbs BF, Lazure C, Day R, Seidah NG (1995) Proparathyroid hormone is preferentially cleaved to parathyroid hormone by the prohormone convertase furin. A mass spectrometric study. J Biol Chem 270:9517–9525

Horwitz MJ et al (2006) Safety and tolerability of subcutaneous PTHrP(1-36) in healthy human volunteers: a dose escalation study. Osteoporos Int 17:225–230

Horwitz MJ et al (2010) Parathyroid hormone-related protein for the treatment of postmenopausal osteoporosis: defining the maximal tolerable dose. J Clin Endocrinol Metab 95:1279–1287

Huan J, Olgaard K, Nielsen LB, Lewin E (2006) Parathyroid hormone 7-84 induces Hypocalcemia and inhibits the parathyroid hormone 1-84 secretory response to Hypocalcemia in rats with intact parathyroid glands. J Am Soc Nephrol 17:1923–1930

Huang Z, Chen Y, Nissenson RA (1995) The cytoplasmic tail of the G-protein-coupled receptor for parathyroid hormone and parathyroid hormone-related protein contains positive and negative signals for endocytosis. J Biol Chem 270:151–156

Jilka RL (2007) Molecular and cellular mechanisms of the anabolic effect of intermittent PTH. Bone 40:1434–1446

Jilka RL, Weinstein RS, Bellido T, Roberson P, Parfitt AM, Manolagas SC (1999) Increased bone formation by prevention of osteoblast apoptosis with parathyroid hormone. J Clin Invest 104:439–446

Johnson RW, Suva LJ (2018) Hallmarks of bone metastasis. Calcif Tissue Int 102:141–151

Jüppner H et al (1991) A G protein-linked receptor for parathyroid hormone and parathyroid hormone-related peptide. Science 254:1024–1026

Kamalakar A et al (2017) PTHrP(12-48) modulates the bone marrow microenvironment and suppresses human osteoclast differentiation and lifespan. J Bone Miner Res 32:1421–1431

Karaplis AC, Luz A, Glowacki J, Bronson RT, Tybulewicz VL, Kronenberg HM, Mulligan RC (1994) Lethal skeletal dysplasia from targeted disruption of the parathyroid hormone-related peptide gene. Genes Dev 8:277–289

Karim Z et al (2008) NHERF1 mutations and responsiveness of renal parathyroid hormone. N Engl J Med 359:1128–1135

Kovacs CS (2017) The skeleton is a storehouse of mineral that is plundered during lactation and (fully?) replenished afterwards. J Bone Miner Res 32:676–680

Kronenberg HM (2006) PTHrP and skeletal development. Ann N Y Acad Sci 1068:1–13

Lanske B et al (1996) PTH/PTHrP receptor in early development and Indian hedgehog-regulated bone growth. Science 273:663–666

Leder BZ et al (2015) Effects of abaloparatide, a human parathyroid hormone-related peptide analog, on bone mineral density in postmenopausal women with osteoporosis. J Clin Endocrinol Metab 100:697–706

Lee S et al (2015) A homozygous [Cys25]PTH(1-84) mutation that impairs PTH/PTHrP receptor activation defines a novel form of hypoparathyroidism. J Bone Miner Res 30:1803–1813

Lee SH, Park YH, Jin YB, Kim SU, Hur JK (2020) CRISPR diagnosis and therapeutics with single base pair precision. Trends Mol Med 26:337–350

Liu L et al (2012) Na^+/H^+-exchanger regulatory factor-1 (NHERF1) directly regulates osteogenesis. J Biol Chem 287:43312–43321

Mahon MJ, Donowitz M, Yun CC, Segre GV (2002) Na^+/H^+ exchanger regulatory factor 2 directs parathyroid hormone 1 receptor signalling. Nature 417:858–861

Malecz N, Bambino T, Bencsik M, Nissenson RA (1998) Identification of phosphorylation sites in the G protein-coupled receptor for parathyroid hormone. Receptor phosphorylation is not required for agonist-induced internalization. Mol Endocrinol 12:1846–1856

Mannstadt M, Juppner H, Gardella TJ (1999) Receptors for PTH and PTHrP: their biological importance and functional properties. Am J Phys 277:F665–F675

McCauley LK, Martin TJ (2012) Twenty-five years of PTHrP progress: from cancer hormone to multifunctional cytokine. J Bone Miner Res 27:1231–1239

Miao D, Tong XK, Chan GK, Panda D, McPherson PS, Goltzman D (2001) Parathyroid hormone-related peptide stimulates osteogenic cell proliferation through PKC activation of the Ras/mitogen-activated protein kinase signaling pathway. J Biol Chem 276:32204–32213

Miao D, Li J, Xue Y, Su H, Karaplis AC, Goltzman D (2004) Parathyroid hormone-related peptide is required for increased trabecular bone volume in parathyroid hormone-null mice. Endocrinology 145:3554–3562

Miao D et al (2005) Osteoblast-derived PTHrP is a potent endogenous bone anabolic agent that modifies the therapeutic efficacy of administered PTH 1-34. J Clin Invest 115:2402–2411

Miles RR et al (2000) ADAMTS-1: a cellular disintegrin and metalloprotease with thrombospondin motifs is a target for parathyroid hormone in bone. Endocrinology 141:4533–4542

Morales FC, Takahashi Y, Kreimann EL, Georgescu MM (2004) Ezrin-radixin-moesin (ERM)-binding phosphoprotein 50 organizes ERM proteins at the apical membrane of polarized epithelia. Proc Natl Acad Sci U S A 101:17705–17710

Neer RM et al (2001) Effect of parathyroid hormone (1-34) on fractures and bone mineral density in postmenopausal women with osteoporosis. N Engl J Med 344:1434–1441

Neville MC, McFadden TB, Forsyth I (2002) Hormonal regulation of mammary differentiation and milk secretion. J Mammary Gland Biol Neoplasia 7:49–66

Orloff JJ, Ganz MB, Ribaudo AE, Burtis WJ, Reiss M, Milstone LM, Stewart AF (1992) Analysis of PTHRP binding and signal transduction mechanisms in benign and malignant squamous cells. Am J Phys 262:E599–E607

Orloff JJ, Reddy D, de Papp AE, Yang KH, Soifer NE, Stewart AF (1994) Parathyroid hormone-related protein as a prohormone: posttranslational processing and receptor interactions. Endocr Rev 15:40–60

Pines M et al (1996) Inositol 1-,4-,5-trisphosphate-dependent Ca^{2+} signaling by the recombinant human PTH/PTHrP receptor stably expressed in a human kidney cell line. Bone 18:381–389

Portal-Nunez S, Lozano D, de Castro LF, de Gortazar AR, Nogues X, Esbrit P (2010) Alterations of the Wnt/beta-catenin pathway and its target genes for the N- and C-terminal domains of parathyroid hormone-related protein in bone from diabetic mice. FEBS Lett 584:3095–3100
Potts JT Jr et al (1971) Synthesis of a biologically active N-terminal tetratriacontapeptide of parathyroid hormone. Proc Natl Acad Sci U S A 68:63–67
Qiu T, Crane JL, Xie L, Xian L, Xie H, Cao X (2018) IGF-I induced phosphorylation of PTH receptor enhances osteoblast to osteocyte transition. Bone Res 6:5
Rey A, Manen D, Rizzoli R, Caverzasio J, Ferrari SL (2006) Proline-rich motifs in the PTH/PTHrP-receptor C-terminus mediate scaffolding of c-Src with ß-arrestin2 for ERK1/2 activation. J Biol Chem:38181–38188
Rodda CP et al (1988) Evidence for a novel parathyroid hormone-related protein in fetal lamb parathyroid glands and sheep placenta: comparisons with a similar protein implicated in humoral hypercalcaemia of malignancy. J Endocrinol 117:261–271
Sakwe AM, Engstrom A, Larsson M, Rask L (2002) Biosynthesis and secretion of parathyroid hormone are sensitive to proteasome inhibitors in dispersed bovine parathyroid cells. J Biol Chem 277:17687–17695
Shenolikar S, Voltz JW, Minkoff CM, Wade JB, Weinman EJ (2002) Targeted disruption of the mouse NHERF-1 gene promotes internalization of proximal tubule sodium-phosphate cotransporter type IIa and renal phosphate wasting. Proc Natl Acad Sci U S A 99:11470–11475
Shenoy SK et al (2006) β-Arrestin-dependent, G protein-independent ERK1/2 activation by the β2 adrenergic receptor. J Biol Chem 281:1261–1273
Shimizu M et al (2016) Pharmacodynamic actions of a long-acting PTH analog (LA-PTH) in thyroparathyroidectomized (tptx) rats and normal monkeys. J Bone Miner Res 31:1405–1412
Slatopolsky E et al (2000) A novel mechanism for skeletal resistance in uremia. Kidney Int 58:753–761
Sneddon WB, Friedman PA (2007) ß-arrestin-dependent parathyroid hormone-stimulated ERK activation and PTH1R internalization. Endocrinology 148:4073–4079
Sneddon WB, Gesek FA, Friedman PA (2000) Obligate MAP kinase activation in parathyroid hormone stimulation of calcium transport but not calcium signaling. Endocrinology 141:4185–4193
Sneddon WB et al (2003) Activation-independent parathyroid hormone receptor internalization is regulated by NHERF1 (EBP50). J Biol Chem 278:43787–43796
Soifer NE et al (1992) Parathyroid hormone-related protein. Evidence for secretion of a novel mid-region fragment by three different cell types. J Biol Chem 267:18236–18243
Spurney RF, Flannery PJ, Garner SC, Athirakul K, Liu S, Guilak F, Quarles LD (2002) Anabolic effects of a G protein-coupled receptor kinase inhibitor expressed in osteoblasts. J Clin Invest 109:1361–1371
Suva LJ et al (1987) A parathyroid hormone-related protein implicated in malignant hypercalcemia: cloning and expression. Science 237:893–896
Swarthout JT, Doggett TA, Lemker JL, Partridge NC (2001) Stimulation of extracellular signal-regulated kinases and proliferation in rat osteoblastic cells by parathyroid hormone is PKC-dependent. J Biol Chem 276:7586–7592
Syme CA, Friedman PA, Bisello A (2005) Parathyroid hormone receptor trafficking contributes to the activation of extracellular signal-regulated kinases but is not required for regulation of cAMP signaling. J Biol Chem 280:11281–11288
Thirunavukkarasu K, Halladay DL, Miles RR, Geringer CD, Onyia JE (2002) Analysis of regulator of G-protein signaling-2 (RGS-2) expression and function in osteoblastic cells. J Cell Biochem 85:837–850
van der Lee MM et al (2013) ß-Arrestin-biased signalling of PTH analogues of the type 1 parathyroid hormone receptor. Cell Signal 25:527–538
VanHouten J, Dann P, McGeoch G, Brown EM, Krapcho K, Neville M, Wysolmerski JJ (2004) The calcium-sensing receptor regulates mammary gland parathyroid hormone-related protein production and calcium transport. J Clin Invest 113:598–608

Vilardaga JP, Frank M, Krasel C, Dees C, Nissenson RA, Lohse MJ (2001) Differential conformational requirements for activation of G proteins and the regulatory proteins arrestin and G protein-coupled receptor kinase in the G protein-coupled receptor for parathyroid hormone (PTH)/PTH-related protein. J Biol Chem 276:33435–33443

Vortkamp A, Lee K, Lanske B, Segre GV, Kronenberg HM, Tabin CJ (1996) Regulation of rate of cartilage differentiation by Indian hedgehog and PTH-related protein. Science 273:613–622

Wang HH, Drugge ED, Yen YC, Blumenthal MR, Pang PK (1984) Effects of synthetic parathyroid hormone on hemodynamics and regional blood flows. Eur J Pharmacol 97:209–215

Wang L, Liu S, Quarles LD, Spurney RF (2005) Targeted overexpression of G protein-coupled receptor kinase-2 in osteoblasts promotes bone loss. Am J Physiol Endocrinol Metab 288:E826–E834

Wang B, Bisello A, Yang Y, Romero GG, Friedman PA (2007) NHERF1 regulates parathyroid hormone receptor membrane retention without affecting recycling. J Biol Chem 282:36214–36222

Wang B, Ardura JA, Romero G, Yang Y, Hall RA, Friedman PA (2010) Na/H exchanger regulatory factors control PTH receptor signaling by differential activation of Gα protein subunits. J Biol Chem 285:26976–26986

Wang B, Yang Y, Liu L, Blair HC, Friedman PA (2013) NHERF1 regulation of PTH-dependent bimodal pi transport in osteoblasts. Bone 52:268–277

Washam CL et al (2013) Identification of PTHrP(12-48) as a plasma biomarker associated with breast cancer bone metastasis. Cancer Epidemiol Biomark Prev 22:972–983

Wehbi VL, Stevenson HP, Feinstein TN, Calero G, Romero G, Vilardaga JP (2013) Noncanonical GPCR signaling arising from a PTH receptor-arrestin-Gβγ complex. Proc Natl Acad Sci U S A 110:1530–1535

Wu TL et al (1996) Structural and physiologic characterization of the mid-region secretory species of parathyroid hormone-related protein. J Biol Chem 271:24371–24381

Zhang Q et al (2018) Site-specific polyubiquitination differentially regulates parathyroid hormone receptor-initiated MAPK signaling and cell proliferation. J Biol Chem 293:5556–5571

Zhao LH et al (2019) Structure and dynamics of the active human parathyroid hormone receptor-1. Science 364:148–153

Vitamin D and Bone

Sylvia Christakos, Shanshan Li, Jessica DeLa Cruz, Lieve Verlinden, and Geert Carmeliet

Contents

Abstract

Vitamin D is a principal factor required for mineral and skeletal homeostasis. Vitamin D deficiency during development causes rickets and in adults can result in osteomalacia and increased risk of fracture. 1,25-Dihydroxyvitamin D_3 ($1,25(OH)_2D_3$), the hormonally active form of vitamin D, is responsible for the biological actions of vitamin D which are mediated by the vitamin D receptor (VDR). Mutations in the VDR result in early-onset rickets and low calcium and phosphate, indicating the essential role of $1,25(OH)_2D_3$/VDR signaling in the regulation of mineral homeostasis and skeletal health. This chapter summarizes our current understanding of the production of the vitamin D endocrine hormone, $1,25(OH)_2D_3$, and the actions of $1,25(OH)_2D_3$ which result in the maintenance of skeletal homeostasis. The primary role of $1,25(OH)_2D_3$ is to increase calcium absorption from the intestine and thus to increase the availability of calcium for

S. Christakos (✉) · S. Li · J. DeLa Cruz
Department of Microbiology, Biochemistry and Molecular Genetics, Rutgers New Jersey Medical School, Newark, NJ, USA
e-mail: christak@njms.rutgers.edu

L. Verlinden · G. Carmeliet
Clinical and Experimental Medicine and Endocrinology, KU Leuven, Leuven, Belgium

P. H. Stern (ed.), *Bone Regulators and Osteoporosis Therapy*,
Handbook of Experimental Pharmacology 262, https://doi.org/10.1007/164_2019_338

bone mineralization. Specific actions of $1,25(OH)_2D_3$ on the intestine, kidney, and bone needed to maintain calcium homeostasis are summarized, and the impact of vitamin D status on bone health is discussed.

Keywords
1,25-Dihydroxyvitamin D3 · Bone · Calcium homeostasis · Intestine · Kidney · Vitamin D metabolism

1 The Vitamin D Endocrine System: Metabolism and Molecular Mechanism of Action

Vitamin D is a principal factor required for the development and maintenance of bone as well as for maintaining normal calcium and phosphorus homeostasis. Vitamin D deficiency during bone development causes rickets, and in adults, vitamin D deficiency, which is common in the elderly, causes secondary hyperparathyroidism that can result in osteomalacia and increased risk of fracture (Weaver et al. 2016; Bouillon and Carmeliet 2018). For vitamin D to affect mineral metabolism, it must first be metabolized to its active form. Vitamin D which is synthesized in the skin from 7-dehydrocholesterol in a reaction catalyzed by ultraviolet irradiation or taken in the diet (few foods, which include fish oils and fortified dairy products, contain appreciable amounts of vitamin D) is first hydroxylated in the liver to 25-hydroxyvitamin D_3 [$25(OH)D_3$], the major circulating form of vitamin D and the most reliable index of vitamin D status (Christakos et al. 2019). CYP2R1, a 25-hydroxylase, is likely the key vitamin D based in part on genetic evidence that patients with a mutation in CYP2R1 are deficient in $25(OH)D_3$ and develop vitamin D-dependent rickets (Cheng et al. 2003, 2004; Thacher et al. 2015). Since in *Cyp2r1* null mice levels of $25(OH)D_3$ are reduced but not abolished, it has been suggested that other hydroxylases are also involved in the conversion in the liver of vitamin D to 25(OH)D (Zhu et al. 2013). The second hydroxylation occurs in the proximal renal tubule through the action of mitochondrial 25(OH)D 1 α hydroxylase (CYP27B1) resulting in the conversion of 25(OH)D to the principal hormonal form of vitamin D, 1,25-dihydroxyvitamin D_3 ($1,25(OH)_2D_3$), which is responsible for the biological actions of vitamin D (Christakos et al. 2016). Vitamin D and its metabolites are transported in the blood by vitamin D binding protein (DBP) (Christakos et al. 2016, 2019). Mutation in the CYP27B1 gene results in vitamin D dependency rickets type 1 (VDDR-1), characterized by decreased mineralization, hypocalcemia, and low circulating $1,25(OH)_2D_3$ (Kitanaka et al. 1998). The activity of renal CYP27B1 is under stringent control. Parathyroid hormone (PTH) induced in response to hypocalcemia stimulates CYP27B1 (Jones et al. 2014). FGF23, which promotes renal phosphate excretion and requires klotho, a transmembrane protein, suppresses the expression of CYP27B1 (Shimada et al. 2004; Hu et al. 2013).

$1{,}25(OH)_2D_3$, as a feedback mechanism, regulates its own production by inhibiting CYP27B1, downregulating PTH synthesis by the parathyroid gland and upregulating FGF23 production in bone (Jones et al. 2014; Christakos et al. 2016). Recent mouse genomic studies by Meyer et al. (2017, 2019a) identified a kidney specific enhancer module that mediates basal and PTH-induced expression of *Cyp27b1* and FGF23 and $1{,}25(OH)_2D_3$-mediated repression. Studies using human kidney suggest that a kidney-specific module similar to that observed in the mouse exists in humans (Meyer et al. 2019a). These findings represent an important advance in the vitamin D field since they provide insight for the first time at the genomic level on the mechanisms that control *Cyp27b1* expression. Additional factors including sex hormones and prolactin have been reported to stimulate CYP27B1 (Tanaka et al. 1978; Ajibade et al. 2010). In addition to the kidney, CYP27B1 is also expressed in the placenta (Zehnder et al. 2002) and in small amounts in a number of different tissues, including bone (Bikle 2010). However, the role of CYP27B1 under normal physiological conditions in tissues other than kidney and placenta remains to be determined.

As an additional autoregulatory mechanism, $1{,}25(OH)_2D_3$ induces CYP24A1 (25-hydroxyvitamin D_3 24-hydroxylase) which accelerates the catabolism of $1{,}25(OH)_2D_3$ by catalyzing the conversion of $1{,}25(OH)_2D_3$ into 24-hydroxylated products targeted for excretion (Jones et al. 2014; Christakos et al. 2016). Mutation in the *CYP24A1* gene results in hypercalcemia, hypercalciuria, decreased PTH, and normal to high $1{,}25(OH)_2D_3$ levels (Schlingmann et al. 2011). The levels of 25(OH) D_3 have also been reported to be regulated through the catabolic activity of renal CYP24A1 (Jones et al. 2014; Christakos et al. 2016; Meyer et al. 2019b). CYP24A1 is present not only in kidney but also in all cells that contain VDR (Jones et al. 2014; Christakos et al. 2016). Thus, CYP24A1 not only regulates circulating $1{,}25(OH)_2D_3$ protecting against hypercalcemia but also may modulate the amount of $1{,}25(OH)_2D_3$ in target cells and control the cellular response. In the kidney CYP24A1 is reciprocally regulated when compared to CYP27B1 (suppressed by low calcium and PTH and induced by FGF23 and $1{,}25(OH)_2D_3$). This reciprocal regulation occurs only in the kidney. In nonrenal target cells, CYP24A1 is solely regulated by $1{,}25(OH)_2D_3$ (Meyer et al. 2019b). Recent studies have identified genomic mechanisms resulting in differential regulation of *Cyp24a1* in the kidney and nonrenal target tissues (e.g., intestine and bone) (Meyer et al. 2019b).

The biological activities of $1{,}25(OH)_2D_3$ are mediated by the vitamin D receptor (VDR), a nuclear receptor which is a member of the steroid receptor family. $1{,}25(OH)_2D_3$-occupied VDR heterodimerizes with the retinoid X receptor (RXR) and together with chromatin active co-regulatory proteins interacts with specific DNA sequences (vitamin D response elements) in and around target genes resulting in activation or repression of transcription (Christakos et al. 2016; Pike and Christakos 2017). VDR-binding sites are located at proximal promoters and also many kilobases upstream and downstream and in intronic and exonic sites. Mutations in VDR result in early-onset rickets, low calcium and phosphate, and high PTH, indicating the essential role of VDR in mediating $1{,}25(OH)_2D_3$ regulation of mineral homeostasis and skeletal health (Malloy et al. 2014).

2 Effect of Vitamin D on Calcium Homeostasis

2.1 Intestine

The principal action of vitamin D in maintenance of calcium homeostasis is increased intestinal calcium absorption and thus increased availability of calcium for mineralization of bone. This conclusion is based in part on studies in *Vdr* null mice. Feeding *Vdr* null mice a diet which includes high calcium prevents rickets and osteomalacia and results in the normalization of serum calcium and PTH (Amling et al. 1999; Masuyama et al. 2003). In addition, in humans with a mutation in VDR characterized by resistance to $1,25(OH)_2D_3$ (hereditary vitamin D-resistant rickets; HVDRR), administration of calcium alone has been reported to normalize bone and result in normal mineralization (al-Aqeel et al. 1993). Active intestinal calcium absorption occurs when there is an increased need for calcium (during growth, pregnancy, lactation, and under low dietary calcium conditions). Active calcium absorption is mediated at least in part by a transcellular process involving calcium entry via the apical epithelial calcium channel TRPV6, calcium binding by the intracellular calcium-binding protein calbindin-D_{9k}, and calcium extrusion via the basolateral membrane calcium ATPase PMCA1b (Christakos et al. 2014). TRPV6, calbindin, and PMCA1b are regulated by $1,25(OH)_2D_3$/VDR (Van Cromphaut et al. 2001; Song et al. 2003; Lee et al. 2015). Abnormalities in the *Vdr* null mice develop only after weaning, consistent with studies showing that intestinal VDR, calbindin, and TRPV6 are induced at weaning (Yoshizawa et al. 1997; Song et al. 2003). Overexpression of TRPV6 in the intestine results in hypercalcemia and soft tissue calcification indicating a significant role for TRPV6 in intestinal calcium absorption and suggesting that the inability to transport calcium into the enterocyte may be a primary defect in VDR-dependent rickets (Cui et al. 2012). Direct evidence for a critical role of $1,25(OH)_2D_3$/VDR-mediated intestinal calcium absorption in bone homeostasis was shown in studies in which VDR expression specifically in the intestine of *Vdr* null mice prevented rickets and normalized serum calcium (Xue and Fleet 2009). These findings indicate that intestinal VDR is essential for controlling bone formation. In addition, when VDR is deleted specifically from mouse intestine, there is calcium malabsorption, bone resorption, increased bone fractures, and normal serum calcium (Lieben et al. 2012). Thus when calcium homeostasis cannot be maintained by $1,25(OH)_2D_3$/VDR-mediated intestinal calcium absorption, serum calcium will be maintained at the expense of skeletal integrity.

2.2 Kidney

When serum calcium cannot be maintained by intestinal calcium absorption, in addition to stimulation of osteoclastogenesis by PTH and $1,25(OH)_2D_3$ (see Sect. 2.3), calcium reabsorption occurs in the distal convoluted tubule and collecting tubule and is regulated by $1,25(OH)_2D_3$ and PTH (Christakos et al. 2016). Similar to the intestine, calcium is reabsorbed in the distal tubule by a transcellular process involving calcium entry through calcium channel TRPV5 (75% sequence homology with TRPV6), calcium binding to the calcium-binding protein calbindin [calbindin-D_{9k} (9,000 Mr; only in mouse kidney) and calbindin-D_{28k} (28,000 Mr; in mouse, rat, and human kidney)], and calcium extrusion via the calcium ATPase PMCA1b and the Na^+/Ca^{2+} exchanger (NCX1 or SLC8A1). $1,25(OH)_2D_3$ induces the expression in the kidney of TRPV5 and the calbindins (Song et al. 2003). PTH has been reported to activate TRPV5 via protein kinase A phosphorylation (de Groot et al. 2009). Studies in *Trpv5* null mice show that inactivation of *Trpv5* results in diminished calcium reabsorption in the distal tubule, severe hypercalciuria, and significant changes in bone structure (Hoenderop et al. 2003). It has been suggested that calcium uptake by TRPV5 is a rate-limiting step in renal calcium reabsorption and thus in the maintenance of calcium and bone homeostasis (Hoenderop et al. 2003). The kidney is also a major site of production of $1,25(OH)_2D_3$ and its regulation (Christakos et al. 2010, 2016; Pike and Christakos 2017).

2.3 Bone

As outlined above, loss-of-function mutations in the human VDR lead to resistance toward the actions of $1,25(OH)_2D_3$ and result in the development of hereditary vitamin D-resistant rickets type II (Malloy et al. 2014). This condition is mimicked in systemic *Vdr* null mice that develop hypocalcemia, hypophosphatemia, secondary hyperparathyroidism, and rickets (Li et al. 1997; Yoshizawa et al. 1997; Van Cromphaut et al. 2001). The rickets phenotype is characterized by progressive widening of the epiphyseal growth plates, due to a significant enlargement and disorganization of the zone of hypertrophic chondrocytes. In addition, systemic *Vdr* null mice develop hyperosteoidosis and osteomalacia, due to a delay in bone mineralization that develops when *Vdr* null mice become hypocalcemic after weaning (Li et al. 1997; Yoshizawa et al. 1997; Van Cromphaut et al. 2001). This increase in unmineralized osteoid is accompanied by an increased number of osteoblasts, whereas osteoclast numbers are not altered (Amling et al. 1999). Importantly, the bone and mineral phenotype of systemic *Vdr* null mice is fully corrected by supplementation with a high-calcium/high-lactose diet, underscoring the importance of VDR-mediated intestinal calcium absorption (Li et al. 1998; Amling et al. 1999; Van Cromphaut et al. 2001). Yet, multiple cell types within bone express the VDR, and the highest levels of VDR are present in osteoblasts and osteocytes, which are thus considered as the main mediators of $1,25(OH)_2D_3$ action in bone homeostasis. VDR expression is also present in chondrocytes and in osteoclasts, although much less abundant. As bone homeostasis is regulated by timely controlled interactions between osteoblasts and osteoclasts (Wang et al. 2014; Nakamichi et al. 2017) and between chondrocytes and osteoclasts (Masuyama et al. 2006), VDR expression in all these cell types enables $1,25(OH)_2D_3$ to affect bone development and remodeling. Here, we aim to delineate the role of $1,25(OH)_2D_3$ in bone based on the knowledge obtained from different transgenic mouse models in which *Vdr* expression is either deleted or overexpressed in one of the different bone cell types. First, we will discuss the bone effects of $1,25(OH)_2D_3$ in conditions of a positive calcium balance, where the amount of (re)absorbed calcium equals or exceeds fecal and renal calcium losses. In this condition, serum calcium levels remain normal and allow calcium deposition in bone. Thereafter, we will focus on the effects of $1,25(OH)_2D_3$ in bone when the calcium balance is negative, as caused, for example, by insufficient intestinal calcium absorption.

2.3.1 $1,25(OH)_2D_3$ and Bone Metabolism During a Positive Calcium Balance

When dietary calcium intake is normal, $1,25(OH)_2D_3$ indirectly regulates bone homeostasis and mineralization by guaranteeing adequate calcium supply through stimulation of intestinal and renal calcium (re)absorption (Christakos et al. 2016; Goltzman 2018). In addition, VDR expression in different osteogenic cells enables $1,25(OH)_2D_3$ to directly and locally impact on bone metabolism in a paracrine or autocrine manner. Yet, these latter effects are still not fully elucidated and are likely dependent on the differentiation stage of the osteogenic cells.

Stage-Dependent Inactivation of *Vdr* Expression in Osteoblast-Lineage Cells Points to a Minor Role of Osteoblastic *Vdr* Expression in Bone Homeostasis

Osteoblasts together with chondrocytes differentiate from a common skeletal progenitor cell. Differentiation along the osteoblast lineage is governed by multiple transcription factors including Runx2 and Osterix (Osx) (Huang et al. 2007). When osteoprogenitor cells differentiate to immature osteoblasts, they start to express genes that encode for proteins of the extracellular matrix (ECM) such as type 1 collagen α1 (Col1a1), whereas mature osteoblasts are typified by the production of osteocalcin, a secreted protein with numerous endocrine functions including the regulation of glucose and energy metabolism (Tangseefa et al. 2018; Dirckx et al. 2019). Once osteoblasts become embedded in the bone matrix, they differentiate toward osteocytes, which fulfill a function in coordinating bone formation and resorption, and these cells are characterized by a high expression of dentin matrix protein 1 (Dmp-1) and the Wnt inhibitor sclerostin (Atkins and Findlay 2012).

During the last decades, several transgenic mouse models have been established in which *Vdr* expression was either deleted or overexpressed at a specific stage of osteoblast differentiation (Gardiner et al. 2000; Lieben et al. 2012; Yamamoto et al. 2013; Triliana et al. 2016; Nakamichi et al. 2017) (Fig. 1). All studies were performed in mice that received adequate supply of dietary calcium (ranging from 0.8 to 1.18% dietary calcium). Deletion of *Vdr* expression in osteoblast precursors, by *Osx*-driven Cre recombination (*Osx-Vdr*-cKO mice), does not affect calcium or bone homeostasis (Nakamichi et al. 2017) (Fig. 2). Indeed, serum concentrations of calcium, phosphate, PTH, and $1,25(OH)_2D_3$ are normal in 14-week-old *Osx-Vdr*-cKO mice, while serum FGF23 levels are slightly reduced. Trabecular bone mass in *Osx-Vdr*-cKO mice is similar to control littermates with no overt changes in bone resorption or in bone mineralization. When *Vdr* expression is deleted in immature osteoblasts, under the control of the *Col1a1* promoter (*Col1a1-Vdr*-cKO mice), no differences in bone mass are observed in 4- and 9-week-old animals (Yamamoto

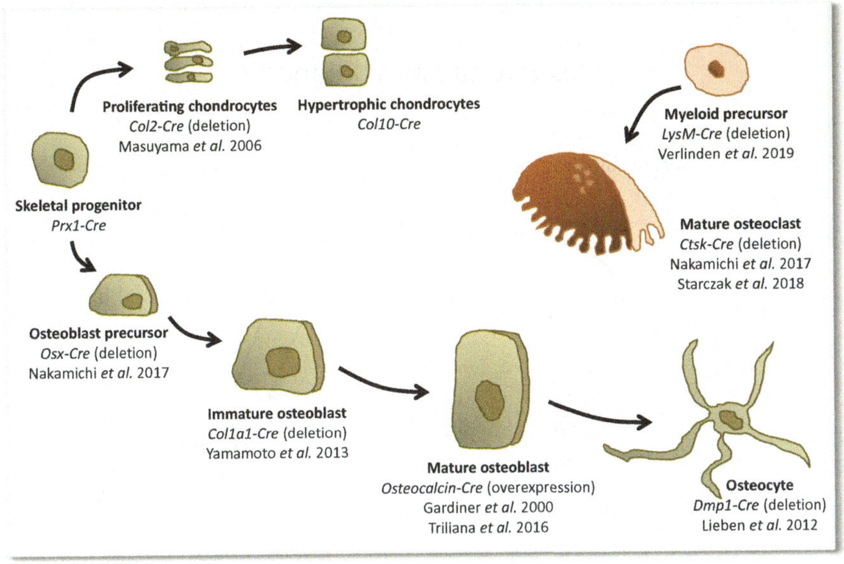

Fig. 1 Overview of different bone cell types in which *Vdr* expression was targeted

et al. 2013). However, in 16-week-old animals, a small – but significant – increase in trabecular bone mass is detected, while no significant changes in cortical bone are present. This increased bone mass is attributed to decreased bone resorption rather than to increased bone formation and is accompanied by reduced expression of the receptor activator of nuclear factor κβ ligand (RANKL), a major stimulator of osteoclastic differentiation. This bone phenotype fits with the in vitro observations showing that $1,25(OH)_2D_3$ signaling increases RANKL expression in osteoblasts (Yasuda et al. 1998), although it is not clear why the increased bone mass is only observed in older mice. Serum levels of calcium and phosphate as well as those of PTH and $1,25(OH)_2D_3$ are normal in *Col1a1-Vdr*-cKO mice.

Deletion of *Vdr* expression in mature osteoblasts and osteocytes under control of the *Dmp-1* promoter (*Dmp1-Vdr*-cKO mice) showed that VDR signaling in mature osteoblasts and osteocytes is redundant for bone homeostasis (Lieben et al. 2012). Trabecular and cortical bone mass are indistinguishable between 8-week-old *Dmp1-Vdr*-cKO mice and their wild-type littermates, as are bone resorption and mineralization. Together, these three different conditional knockout models, in which *Vdr* expression is deleted at different stages of osteoblastic differentiation, suggest that the vitamin D system does not have major direct effects on bone homeostasis in conditions of a positive calcium balance.

On the other hand, studies in mice in which *Vdr* is overexpressed in mature osteoblasts, under the control of the *osteocalcin* promoter (*osteocalcin-Vdr*-cOE mice), point toward a positive effect of VDR signaling on bone mass (Gardiner et al. 2000). Indeed, 4- and 9-month-old *osteocalcin-Vdr*-cOE mice have increased cortical and trabecular bone mass, which is accompanied by enhanced bone

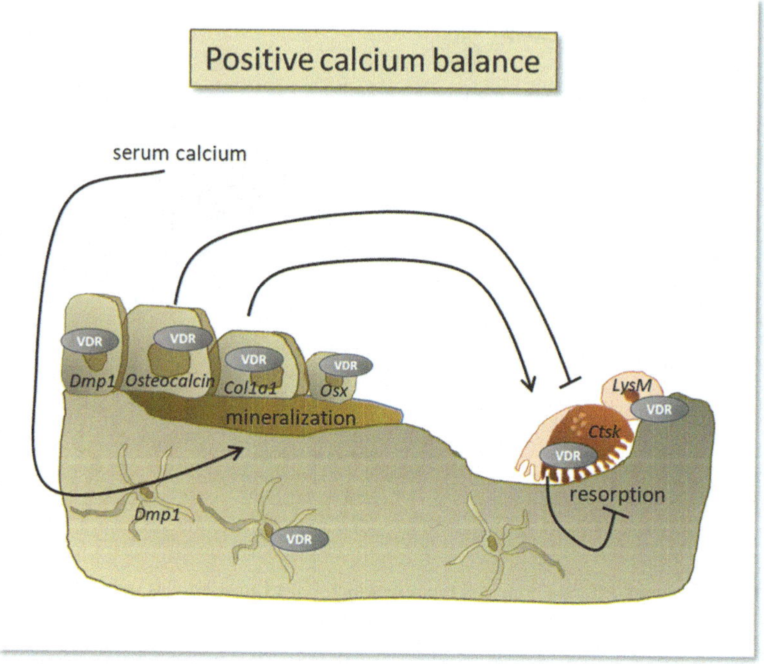

Fig. 2 Cell- and stage-specific effects of *Vdr* expression in osteogenic cells on bone homeostasis during a positive calcium balance. During a positive serum balance, normal serum calcium levels allow calcium deposition in bone and proper mineralization. *Vdr* expression in osteoprogenitors (characterized by *Osx* expression (Nakamichi et al. 2017)) or in mature osteoblasts and osteocytes (*Dmp1* (Lieben et al. 2012)) does not affect bone homeostasis. However, *Vdr* expression in immature osteoblasts (*Col1a1* (Yamamoto et al. 2013)) and in mature osteoblasts (*osteocalcin* (Gardiner et al. 2000; Triliana et al. 2016)) is reported to induce or inhibit bone resorption, respectively. *Vdr* expression in osteoclast progenitors (*LysM* (Verlinden et al. 2019)) does not influence bone resorption, whereas *Vdr* signaling in mature osteoclasts (*Ctsk*) is shown to have no (Nakamichi et al. 2017) or inhibitory effects (Starczak et al. 2018) on osteoclast activity

formation and reduced bone resorption. Mechanistically, the decrease in bone resorption in *osteocalcin-Vdr*-cOE mice is attributed to reduced RANKL and increased OPG, a decoy receptor for RANKL (Misof et al. 2003; Baldock et al. 2006). This decrease in RANKL expression is rather surprising and not fully understood, as VDR signaling has been described to stimulate RANKL expression in osteogenic cells (Yasuda et al. 1998). To investigate whether the observed bone phenotype was dependent on the FVB/N background, these *osteocalcin-Vdr*-cOE mice were backcrossed and studied in a C57Bl6 background (Triliana et al. 2016). In parallel to the previous study, an increase in cortical and trabecular bone mass is observed in 9-week-old male and female mice and is associated with increased bone formation and reduced bone resorption. However, whereas this phenotype is recapitulated in 20-week-old male mice, no differences are present between 20-week-old female *osteocalcin-Vdr*-cOE mice and their wild-type littermates.

Taken together, changes in *Vdr* signaling in osteogenic cells do not manifestly affect bone homeostasis, although some mutant *Vdr* mice develop a bone phenotype. However, both inactivation and overexpression of *Vdr* result in an elevated bone mass (Gardiner et al. 2000; Yamamoto et al. 2013; Triliana et al. 2016), and it is at present not clear how to reconcile these apparently contradictory findings. A possible explanation is that the divergent experimental settings, such as diet, age, genetic background of the animals, and gene dosage effects (overexpression versus deletion), intervene with the effect of VDR on bone homeostasis. Alternatively, the effect of 1,25$(OH)_2D_3$ on bone metabolism might depend on the osteoblastic differentiation stage, as is suggested by the differential response of in vitro cultured osteoblasts and osteocytes to 1,25$(OH)_2D_3$ (St John et al. 2014). Therefore, a direct comparison under controlled and identical circumstances (diet, age, background, analysis method) of different transgenic models with *Vdr* inactivation at specific stages of osteoblast differentiation is required to enhance our understanding of the direct VDR effects in osteoblasts under a positive calcium balance.

Vdr Signaling in Growth Plate Chondrocytes Transiently Regulates Bone and Phosphate Homeostasis

As outlined above, chondrocytes differentiate from the same skeletal progenitor cells as osteoblasts and express low levels of *Vdr*. Systemic *Vdr* null mice have expanded epiphyseal growth plates with a widened zone of hypertrophic chondrocytes, due to a reduced apoptosis rate. This impaired induction of caspase-mediated apoptosis of hypertrophic chondrocytes is caused by the low serum phosphate levels in the systemic *Vdr* null mice (Donohue and Demay 2002; Sabbagh et al. 2005). Correction of the hypophosphatemia by a high-calcium/high-lactose diet leads to normalization of the growth plate phenotype (Li et al. 1998; Amling et al. 1999). Accordingly, inactivation of *Vdr* expression by *Col2*-Cre-driven excision does not affect chondrocyte development (*Col2-Vdr*-cKO mice) (Masuyama et al. 2006). However, early in life (postnatal days 3 and 15), *Col2-Vdr*-cKO mice have transiently reduced trabecular bone mass, which is associated with decreased vascular invasion and reduced osteoclast number at the growth plate. Mechanistically, chondrocytic VDR signaling can stimulate osteoclast formation directly by induction of RANKL expression as evidenced in chondrocyte/splenocyte cocultures. Surprisingly, serum phosphate and 1,25$(OH)_2D_3$ levels are increased in young *Col2-Vdr*-cKO mice, whereas calcium and PTH levels are normal. These changes can be explained by the lower serum levels of FGF23 in mutant mice, and in vitro cultures confirmed that *Vdr* inactivation in chondrocytes results in reduced osteoblastic expression of *Fgf23*. These decreased circulating FGF23 levels in *Col2-Vdr*-cKO mice lead to elevated renal expression of *Cyp27b1* and of the sodium phosphate cotransporter type IIa, which then cause increased serum levels of 1,25$(OH)_2D_3$ and phosphate. Together these data demonstrate that VDR signaling in chondrocytes has endocrine actions and is able to affect bone mass.

Stage-Specific Deletion of Osteoclastic *Vdr* Expression Does Not Manifestly Affect Bone Homeostasis

It is well established that 1,25(OH)$_2$D$_3$ can enhance osteoclast formation in vitro in cocultures of osteoblasts and hematopoietic cells by inducing osteoblastic RANKL production (Nakamichi et al. 2018). However, as osteoclasts express *Cyp27b1* as well as low levels of *Vdr*, it is plausible that osteoclastic VDR expression affects bone resorption in an autocrine or paracrine manner. Recent studies addressed this question by deleting *Vdr* expression either in myeloid cells by use of M lysozyme-driven Cre expression (*LysM-Vdr*-cKO mice) (Verlinden et al. 2019) or in mature osteoclasts by *Cathepsin K*-driven Cre recombination (*Ctsk-Vdr*-cKO mice) (Fig. 2) (Nakamichi et al. 2017; Starczak et al. 2018). VDR inactivation in myeloid cells results in reduced osteoclastic *Vdr* expression and lower induction of *Cyp24a1* in response to 1,25(OH)$_2$D$_3$. However, calcium and bone metabolism are normal in 8-week-old *LysM-Vdr*-cKO mice, and no changes in bone resorption parameters are observed. Correspondingly, in vitro osteoclast formation with hematopoietic cells from *LysM-Vdr*-cKO mice occurs normally, suggesting that osteoclastic *Vdr* expression does not affect osteoclast formation and function (Verlinden et al. 2019). These findings are in agreement with observations in mature osteoclasts. Indeed, Nakamichi et al. reported that 14-week-old *Ctsk-Vdr*-cKO mice have a normal trabecular bone mass and that osteoclast surface is not different from control littermates (Nakamichi et al. 2017). In addition, osteoblast surface and dynamic bone parameters are unaltered arguing against a paracrine role of osteoclastic VDR expression on osteoblast function. However, Starczak et al. described that trabecular bone mass is slightly, but significantly, reduced in 6-week-old *Ctsk-Vdr*-cKO mice (Starczak et al. 2018). Yet, no differences are observed in osteoclast surface, number, or size, and dynamic bone parameters are also similar between *Ctsk-Vdr*-cKO mice and control littermates. After ovariectomy, *Ctsk-Vdr*-cKO mice experience an exacerbated bone loss, which is associated with enhanced osteoclastic activity but not with increased osteoclast formation. Collectively, their data suggest that osteoclastic VDR expression is a positive determinant of bone mass. This apparent discrepancy between the latter studies may point toward a transient role of osteoclastic VDR expression as Starczak et al. studied the bone phenotype of 6-week-old mice, whereas Nakamichi et al. used 14-week-old animals. Alternatively, the observed difference may be gender-related as Starczak et al. used female mice versus male mice in the study of Nakamichi et al. In analogy with the osteoblast-specific *Vdr* knockout models, controlled side-by-side analysis of transgenic models with osteoclastic *Vdr* inactivation will be required to definitively exclude a role for osteoclastic VDR expression in bone homeostasis.

2.3.2 1,25(OH)$_2$D$_3$ and Bone Metabolism During a Negative Calcium Balance

A negative calcium balance occurs when intestinal calcium absorption does not meet the daily calcium demand to maintain stable serum calcium concentrations and to ensure proper calcium deposition within bone. An interesting model to mimic such a negative calcium balance is the intestine-specific *Vdr* null model where intestinal

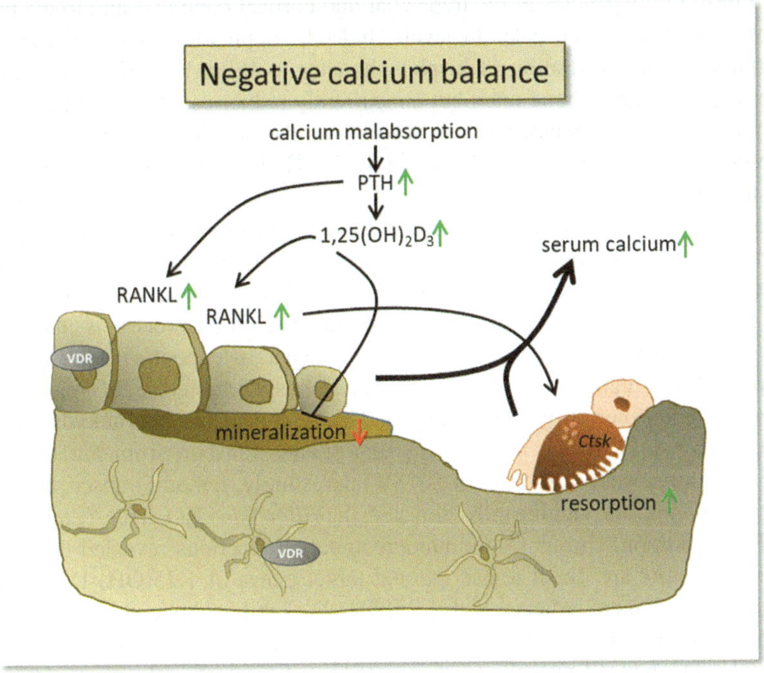

Fig. 3 Effects of a negative calcium balance on bone and the contribution of *Vdr* expression in osteoblasts. In case of intestinal calcium malabsorption, serum levels of PTH and $1{,}25(OH)_2D_3$ increase and stimulate the osteoblastic production of RANKL, which on its turn enhances osteoclastic bone resorption. In addition, high circulating levels of $1{,}25(OH)_2D_3$ decrease bone mineralization by transcriptional induction of mineralization inhibitors through *Vdr*-mediated signaling in late osteoblasts and osteocytes. Elevated bone resorption and decreased mineralization contribute both to the transfer of calcium from the bone to the blood and ensure stable serum calcium concentrations

Vdr expression is inactivated by villin-driven Cre expression (*Villin-Vdr*-cKO mice) (Lieben et al. 2012). Intestinal calcium absorption is decreased in *Villin-Vdr*-cKO mice, and in response serum levels of PTH and $1{,}25(OH)_2D_3$ increase, but calcium and phosphate levels remain within the normal range (Fig. 3). The normal serum calcium levels are in contrast with the hypocalcemia that develops in systemic *Vdr* null mice, suggesting that in *Villin-Vdr*-cKO mice, compensation mechanisms are installed in the kidney and bone to ensure normal serum calcium levels. Indeed, urinary calcium loss is decreased in *Villin-Vdr*-cKO mice due to elevated renal calcium reabsorption. More importantly, *Villin-Vdr*-cKO mice are characterized by a major reduction in bone mass, by enhanced cortical thinning and porosity, by a manifest increase in the amount of unmineralized matrix (hyperosteoidosis) and, finally, by a reduced mineral content of the mineralized bone (hypomineralization). Together these findings indicate a mass transfer of calcium from bone to serum. Increased bone resorption, in response to the high circulating levels of PTH and

$1,25(OH)_2D_3$, contributes to the trabecular and cortical bone loss and to the preservation of normal serum calcium levels. Indeed, suppression of osteoclastic bone resorption by administration of bisphosphonates to *Villin-Vdr*-cKO mice leads to better preservation of bone mass but concurrently reduces serum calcium levels.

In addition to the enhanced bone resorption, also impaired bone mineralization in response to the high serum levels of $1,25(OH)_2D_3$ contributes to this calcium transfer. Osteoblastic bone matrix mineralization is a multistep process, characterized by an initial formation of hydroxyapatite crystals in small extracellular matrix vesicles and subsequent deposition of hydroxyapatite minerals outside the vesicles and accumulation of minerals in the extracellular matrix (Van Driel and Van Leeuwen 2017; Goltzman 2018). This mineralization process is inhibited by high pyrophosphate levels and matrix proteins such as osteopontin (Opn). Pyrophosphate levels are regulated by pyrophosphatase phosphodiesterase enzymes (*Enpp1*, *Enpp3*), which generate pyrophosphate, and by the transmembrane ankylosis protein (encoded by *Ank*), which mediates pyrophosphate transport from intracellular to the extracellular matrix (Goltzman 2018). Interestingly, femoral transcript levels of *Ennp3* and *Ank* are significantly elevated in *Villin-Vdr*-cKO mice (Lieben et al. 2012). In addition, chromatin immunoprecipitation reactions revealed that *Enpp1*, *Enpp3*, and *Ank* are direct transcriptional target genes of $1,25(OH)_2D_3$ signaling. Together these data indicate that $1,25(OH)_2D_3$ inhibits bone mineralization by elevating pyrophosphate levels. In accordance with these findings, lowering the $1,25(OH)_2D_3$-induced pyrophosphate levels by cotreatment with tissue-nonspecific alkaline phosphatase, which reduces pyrophosphate levels, restores mineralization in the presence of $1,25(OH)_2D_3$. In addition, *Opn* gene expression is increased in *Villin-Vdr*-cKO mice, and it is known that $1,25(OH)_2D_3$ induces *Opn* gene expression via VDR-mediated transactivation of the *Opn* gene (Staal et al. 1996). Further study in *Dmp1-Vdr*-cKO mice, with *Vdr* inactivation in mature osteoblasts and osteocytes, revealed that osteocytic *Vdr* expression is involved in the inhibitory effect of $1,25(OH)_2D_3$ on matrix mineralization. Indeed, *Dmp1-Vdr*-cKO mice treated with high doses of $1,25(OH)_2D_3$ do not develop hyperosteoidosis, and the hypercalcemia is less pronounced in comparison with wild-type littermates. In conclusion, these findings revealed that during a negative calcium balance, normocalcemia is maintained at the expense of skeletal integrity. Transfer of bone calcium to serum is ensured by enhanced bone resorption as well as by reduced bone mineralization. Elevated bone resorption in response to PTH and $1,25(OH)_2D_3$ is likely due to enhanced osteoblastic expression of RANKL and not to VDR-mediated signaling in osteoclasts. Indeed, bone loss in *LysM-Vdr*-cKO mice during a negative calcium balance is not different from bone loss in control littermates and suggests that osteoclastic *Vdr* expression does not play an important role in bone homeostasis (Verlinden et al. 2019).

3 Vitamin D and Bone Health

Vitamin D deficiency, as defined by low serum 25(OH)D levels, is common in the elderly and causes secondary hyperparathyroidism which can result in decreased bone density, accelerated bone loss, osteoporosis, and increased risk of fracture (Carmeliet et al. 2015). Risk factors for vitamin D deficiency include inadequate exposure to sunshine, obesity, dark skin tone, and older age (Holick et al. 2011). With older age there is a decline in the ability of the intestine to absorb calcium, a decline in the ability of the kidney to synthesize $1,25(OH)_2D_3$, and an increase in the catabolism of $1,25(OH)_2D_3$ by CYP24A1 which may contribute to the age-related bone loss (Veldurthy et al. 2016). Although controversy exists between the association of vitamin D supplementation and protection against fracture, it is generally recognized that there is a positive relationship between vitamin D sufficiency and a reduction in the risk of fracture (Bouillon and Carmeliet 2018). Several meta-analyses of randomized controlled clinical trials have reported that vitamin D supplementation in conjunction with sufficient calcium intake has clear benefit in protection against fracture, particularly in the elderly who are vitamin D deficient (Chapuy et al. 2002; Larsen et al. 2004; Gallagher et al. 2012; Weaver et al. 2016; Bouillon and Carmeliet 2018; Macdonald et al. 2018). Vitamin D deficiency is defined by the National Institute of Medicine [now known as the National Academy of Medicine (NAM)] as serum 25(OH)D levels below 50 nM (20 ng/mL) (Ross et al. 2011). However, the Endocrine Society guidelines suggest that a threshold of 75 nM (30 ng/mL) is necessary to maintain bone health (Gallagher et al. 2012). It is not clear, however, that optimal vitamin D levels are the same for Caucasians, Black Africans, and Asians. One problem has been that assays for measuring 25(OH)D vary. Efforts are currently being made to standardize results from different laboratories and more laboratories are implementing liquid chromatography/mass spectrometry (LC/MS) to measure vitamin D metabolites (Christakos et al. 2019). Thus, in the future there may be more of a consensus for optimal 25(OH)D levels for different groups. For vitamin D supplementation, the NAM recommends 600 IU/day for ages 1–70 and 800 IU/day for those over 70 as well as 700–1,300 mg calcium/day for adults (Ross et al. 2011). A combination of vitamin D and calcium together with pharmacological intervention [bisphosphonates and RANKL inhibitor (antiresorptive compounds) and PTH (anabolic drug, teriparatide)] has been recommended for the treatment of osteoporotic patients (Rizzoli 2018). Thus correcting vitamin D deficiency together with sufficient calcium intake may optimize pharmacological treatment of osteoporosis.

4 Conclusion

Vitamin D-regulated intestinal calcium absorption is essential in order to maintain calcium homeostasis and skeletal integrity. When calcium cannot be maintained by intestinal calcium absorption (e.g., under conditions of inadequate calcium intake, vitamin D deficiency, or diminished absorption), $1,25(OH)_2D_3$ together with PTH can stimulate osteoclastogenesis and mobilize calcium from bone and can also enhance calcium reabsorption from the renal distal tubule in order to maintain circulating calcium levels and bone homeostasis. These findings emphasize the need to use a combination of calcium and vitamin D in order to prevent or treat osteoporosis, since vitamin D alone may negatively affect bone during a negative calcium balance. During a positive calcium balance, although the mechanisms are not clearly defined, 1,25$(OH)_2D_3$ can modulate bone formation by acting on certain stages of osteoblasts. Thus, through direct and indirect effects, 1,25$(OH)_2D_3$ is a key factor in bone mineralization. Although the mechanisms by which vitamin D deficiency contributes to osteoporosis remain to be defined, it is generally recognized that vitamin D supplementation together with sufficient calcium intake has clear benefit in protection against fracture, particularly in the elderly who are vitamin D deficient. Future studies related to mechanisms by which inadequate vitamin D contributes to osteoporosis and the identification of novel targets of 1,25$(OH)_2D_3$ action in intestine, kidney and bone involved in calcium homeostasis and changes that occur with aging will provide an increased understanding of age-related dysregulation of calcium homeostasis and may suggest candidates for targeted therapies to sustain calcium balance in the elderly.

References

Ajibade DV, Dhawan P, Fechner AJ et al (2010) Evidence for a role of prolactin in calcium homeostasis: regulation of intestinal transient receptor potential vanilloid type 6, intestinal calcium absorption, and the 25-hydroxyvitamin D(3) 1alpha hydroxylase gene by prolactin. Endocrinology 151(7):2974–2984

Al-Aqeel A, Ozand P, Sobki S et al (1993) The combined use of intravenous and oral calcium for the treatment of vitamin D dependent rickets type II (VDDRII). Clin Endocrinol 39(2):229–237

Amling M, Priemel M, Holzmann T et al (1999) Rescue of the skeletal phenotype of vitamin D receptor-ablated mice in the setting of normal mineral ion homeostasis: formal histomorphometric and biomechanical analyses. Endocrinology 140(11):4982–4987

Atkins GJ, Findlay DM (2012) Osteocyte regulation of bone mineral: a little give and take. Osteoporos Int 23(8):2067–2079

Baldock PA, Thomas GP, Hodge JM et al (2006) Vitamin D action and regulation of bone remodeling: suppression of osteoclastogenesis by the mature osteoblast. J Bone Miner Res 21(10):1618–1626

Bikle DD (2010) Extrarenal synthesis of 1,25-dihydroxyvitamin D and its health implications. In: Holick MF (ed) Vitamin D: physiology, molecular biology, and clinical applications. Humana Press, New York, pp 277–295

Bouillon R, Carmeliet G (2018) Vitamin D insufficiency: definition, diagnosis and management. Best Pract Res Clin Endocrinol Metab 32(5):669–684

Carmeliet G, Dermauw V, Bouillon R (2015) Vitamin D signaling in calcium and bone homeostasis: a delicate balance. Best Pract Res Clin Endocrinol Metab 29(4):621–631

Chapuy MC, Pamphile R, Paris E et al (2002) Combined calcium and vitamin D3 supplementation in elderly women: confirmation of reversal of secondary hyperparathyroidism and hip fracture risk: the Decalyos II study. Osteoporos Int 13(3):257–264

Cheng JB, Motola DL, Mangelsdorf DJ et al (2003) De-orphanization of cytochrome P450 2R1 - a microsomal vitamin D 25-hydroxylase. J Biol Chem 278(39):38084–38093

Cheng JB, Levine MA, Bell NH et al (2004) Genetic evidence that the human CYP2R1 enzyme is a key vitamin D 25-hydroxylase. Proc Natl Acad Sci U S A 101(20):7711–7715

Christakos S, Ajibade DV, Dhawan P et al (2010) Vitamin D: metabolism. Endocrinol Metab Clin N Am 39(2):243–253. table of contents

Christakos S, Lieben L, Masuyama R et al (2014) Vitamin D endocrine system and the intestine. Bonekey Rep 3:496

Christakos S, Dhawan P, Verstuyf A et al (2016) Vitamin D: metabolism, molecular mechanism of action, and pleiotropic effects. Physiol Rev 96(1):365–408

Christakos S, Li S, De La Cruz J et al (2019) New developments in our understanding of vitamin metabolism, action and treatment. Metabolism 98:112–120

Cui M, Li Q, Johnson R et al (2012) Villin promoter-mediated transgenic expression of transient receptor potential cation channel, subfamily V, member 6 (TRPV6) increases intestinal calcium absorption in wild-type and vitamin D receptor knockout mice. J Bone Miner Res 27 (10):2097–2107

De Groot T, Lee K, Langeslag M et al (2009) Parathyroid hormone activates TRPV5 via PKA-dependent phosphorylation. J Am Soc Nephrol 20(8):1693–1704

Dirckx N, Moorer MC, Clemens TL et al (2019) The role of osteoblasts in energy homeostasis. Nat Rev Endocrinol 15:651–665

Donohue MM, Demay MB (2002) Rickets in VDR null mice is secondary to decreased apoptosis of hypertrophic chondrocytes. Endocrinology 143(9):3691–3694

Gallagher JC, Sai A, Templin T 2nd et al (2012) Dose response to vitamin D supplementation in postmenopausal women: a randomized trial. Ann Intern Med 156(6):425–437

Gardiner EM, Baldock PA, Thomas GP et al (2000) Increased formation and decreased resorption of bone in mice with elevated vitamin D receptor in mature cells of the osteoblastic lineage. FASEB J 14(13):1908–1916

Goltzman D (2018) Functions of vitamin D in bone. Histochem Cell Biol 149(4):305–312

Hoenderop JG, Van Leeuwen JP, Van Der Eerden BC et al (2003) Renal Ca2+ wasting, hyperabsorption, and reduced bone thickness in mice lacking TRPV5. J Clin Invest 112(12):1906–1914

Holick MF, Binkley NC, Bischoff-Ferrari HA et al (2011) Evaluation, treatment, and prevention of vitamin D deficiency: an Endocrine Society clinical practice guideline. J Clin Endocrinol Metab 96(7):1911–1930

Hu MC, Shiizaki K, Kuro-O M et al (2013) Fibroblast growth factor 23 and Klotho: physiology and pathophysiology of an endocrine network of mineral metabolism. Annu Rev Physiol 75:503–533

Huang W, Yang S, Shao J et al (2007) Signaling and transcriptional regulation in osteoblast commitment and differentiation. Front Biosci 12:3068–3092

Jones G, Prosser DE, Kaufmann M (2014) Cytochrome P450-mediated metabolism of vitamin D. J Lipid Res 55(1):13–31

Kitanaka S, Takeyama K, Murayama A et al (1998) Inactivating mutations in the 25-hydroxyvitamin D3 1alpha-hydroxylase gene in patients with pseudovitamin D-deficiency rickets. N Engl J Med 338(10):653–661

Larsen ER, Mosekilde L, Foldspang A (2004) Vitamin D and calcium supplementation prevents osteoporotic fractures in elderly community dwelling residents: a pragmatic population-based 3-year intervention study. J Bone Miner Res 19(3):370–378

Lee SM, Riley EM, Meyer MB et al (2015) 1,25-Dihydroxyvitamin D3 controls a cohort of vitamin D receptor target genes in the proximal intestine that is enriched for calcium-regulating components. J Biol Chem 290(29):18199–18215

Li YC, Pirro AE, Amling M et al (1997) Targeted ablation of the vitamin D receptor: an animal model of vitamin D-dependent rickets type II with alopecia. Proc Natl Acad Sci U S A 94 (18):9831–9835

Li YC, Amling M, Pirro AE et al (1998) Normalization of mineral ion homeostasis by dietary means prevents hyperparathyroidism, rickets, and osteomalacia, but not alopecia in vitamin D receptor-ablated mice. Endocrinology 139(10):4391–4396

Lieben L, Masuyama R, Torrekens S et al (2012) Normocalcemia is maintained in mice under conditions of calcium malabsorption by vitamin D-induced inhibition of bone mineralization. J Clin Invest 122(5):1803–1815
Macdonald HM, Reid IR, Gamble GD et al (2018) 25-Hydroxyvitamin D threshold for the effects of vitamin D supplements on bone density: secondary analysis of a randomized controlled trial. J Bone Miner Res 33(8):1464–1469
Malloy PJ, Tasic V, Taha D et al (2014) Vitamin D receptor mutations in patients with hereditary 1,25-dihydroxyvitamin D-resistant rickets. Mol Genet Metab 111(1):33–40
Masuyama R, Nakaya Y, Katsumata S et al (2003) Dietary calcium and phosphorus ratio regulates bone mineralization and turnover in vitamin D receptor knockout mice by affecting intestinal calcium and phosphorus absorption. J Bone Miner Res 18(7):1217–1226
Masuyama R, Stockmans I, Torrekens S et al (2006) Vitamin D receptor in chondrocytes promotes osteoclastogenesis and regulates FGF23 production in osteoblasts. J Clin Invest 116(12):3150–3159
Meyer MB, Benkusky NA, Kaufmann M et al (2017) A kidney-specific genetic control module in mice governs endocrine regulation of the cytochrome P450 gene Cyp27b1 essential for vitamin D3 activation. J Biol Chem 292(42):17541–17558
Meyer MB, Benkusky NA, Kaufmann M et al (2019a) Targeted genomic deletions identify diverse enhancer functions and generate a kidney-specific, endocrine-deficient Cyp27b1 pseudo-null mouse. J Biol Chem 294(24):9518–9535
Meyer MB, Lee SM, Carlson AH et al (2019b) A chromatin-based mechanism controls differential regulation of the cytochrome P450 gene Cyp24a1 in renal and non-renal tissues. J Biol Chem. https://doi.org/10.1074/jbc.RA119.010173
Misof BM, Roschger P, Tesch W et al (2003) Targeted overexpression of vitamin D receptor in osteoblasts increases calcium concentration without affecting structural properties of bone mineral crystals. Calcif Tissue Int 73(3):251–257
Nakamichi Y, Udagawa N, Horibe K et al (2017) VDR in osteoblast-lineage cells primarily mediates vitamin D treatment-induced increase in bone mass by suppressing bone resorption. J Bone Miner Res 32(6):1297–1308
Nakamichi Y, Udagawa N, Suda T et al (2018) Mechanisms involved in bone resorption regulated by vitamin D. J Steroid Biochem Mol Biol 177:70–76
Pike JW, Christakos S (2017) Biology and mechanisms of action of the vitamin D hormone. Endocrinol Metab Clin N Am 46(4):815–843
Rizzoli R (2018) Postmenopausal osteoporosis: assessment and management. Best Pract Res Clin Endocrinol Metab 32(5):739–757
Ross AC, Manson JE, Abrams SA et al (2011) The 2011 report on dietary reference intakes for calcium and vitamin D from the Institute of Medicine: what clinicians need to know. J Clin Endocrinol Metab 96(1):53–58
Sabbagh Y, Carpenter TO, Demay MB (2005) Hypophosphatemia leads to rickets by impairing caspase-mediated apoptosis of hypertrophic chondrocytes. Proc Natl Acad Sci U S A 102 (27):9637–9642
Schlingmann KP, Kaufmann M, Weber S et al (2011) Mutations in CYP24A1 and idiopathic infantile hypercalcemia. N Engl J Med 365(5):410–421
Shimada T, Hasegawa H, Yamazaki Y et al (2004) FGF-23 is a potent regulator of vitamin D metabolism and phosphate homeostasis. J Bone Miner Res 19(3):429–435
Song Y, Peng X, Porta A et al (2003) Calcium transporter 1 and epithelial calcium channel messenger ribonucleic acid are differentially regulated by 1,25 dihydroxyvitamin D3 in the intestine and kidney of mice. Endocrinology 144(9):3885–3894
St John HC, Bishop KA, Meyer MB et al (2014) The osteoblast to osteocyte transition: epigenetic changes and response to the vitamin D3 hormone. Mol Endocrinol 28(7):1150–1165
Staal A, Van Wijnen AJ, Birkenhäger JC et al (1996) Distinct conformations of vitamin D receptor/ retinoid X receptor-alpha heterodimers are specified by dinucleotide differences in the vitamin D-responsive elements of the osteocalcin and osteopontin genes. Mol Endocrinol 10(11):1444–1456

Starczak Y, Reinke DC, Barratt KR et al (2018) Absence of vitamin D receptor in mature osteoclasts results in altered osteoclastic activity and bone loss. J Steroid Biochem Mol Biol 177:77–82

Tanaka Y, Castillo L, Wineland MJ et al (1978) Synergistic effect of progesterone, testosterone, and estradiol in the stimulation of chick renal 25-hydroxyvitamin D3-1alpha-hydroxylase. Endocrinology 103(6):2035–2039

Tangseefa P, Martin SK, Fitter S et al (2018) Osteocalcin-dependent regulation of glucose metabolism and fertility: skeletal implications for the development of insulin resistance. J Cell Physiol 233(5):3769–3783

Thacher TD, Fischer PR, Singh RJ et al (2015) CYP2R1 mutations impair generation of 25-hydroxyvitamin D and cause an atypical form of vitamin D deficiency. J Clin Endocrinol Metab 100(7):E1005–E1013

Triliana R, Lam NN, Sawyer RK et al (2016) Skeletal characterization of an osteoblast-specific vitamin D receptor transgenic (ObVDR-B6) mouse model. J Steroid Biochem Mol Biol 164:331–336

Van Cromphaut SJ, Dewerchin M, Hoenderop JG et al (2001) Duodenal calcium absorption in vitamin D receptor-knockout mice: functional and molecular aspects. Proc Natl Acad Sci U S A 98(23):13324–13329

Van Driel M, Van Leeuwen JPTM (2017) Vitamin D endocrinology of bone mineralization. Mol Cell Endocrinol 453:46–51

Veldurthy V, Wei R, Oz L et al (2016) Vitamin D, calcium homeostasis and aging. Bone Res 4:16041

Verlinden L, Janssens I, Doms S et al (2019) Vdr expression in osteoclast precursors is not critical in bone homeostasis. J Steroid Biochem Mol Biol 195:105478

Wang Y, Zhu J, Deluca HF (2014) Identification of the vitamin D receptor in osteoblasts and chondrocytes but not osteoclasts in mouse bone. J Bone Miner Res 29(3):685–692

Weaver CM, Alexander DD, Boushey CJ et al (2016) Calcium plus vitamin D supplementation and risk of fractures: an updated meta-analysis from the National Osteoporosis Foundation. Osteoporos Int 27(1):367–376

Xue Y, Fleet JC (2009) Intestinal vitamin D receptor is required for normal calcium and bone metabolism in mice. Gastroenterology 136(4):1317–1327. e1311–1312

Yamamoto Y, Yoshizawa T, Fukuda T et al (2013) Vitamin D receptor in osteoblasts is a negative regulator of bone mass control. Endocrinology 154(3):1008–1020

Yasuda H, Shima N, Nakagawa N et al (1998) Osteoclast differentiation factor is a ligand for osteoprotegerin/osteoclastogenesis-inhibitory factor and is identical to TRANCE/RANKL. Proc Natl Acad Sci U S A 95(7):3597–3602

Yoshizawa T, Handa Y, Uematsu Y et al (1997) Mice lacking the vitamin D receptor exhibit impaired bone formation, uterine hypoplasia and growth retardation after weaning. Nat Genet 16(4):391–396

Zehnder D, Evans KN, Kilby MD et al (2002) The ontogeny of 25-hydroxyvitamin D3 1α-hydroxylase expression in human placenta and decidua. Am J Pathol 161(1):105–114

Zhu JG, Ochalek JT, Kaufmann M et al (2013) CYP2R1 is a major, but not exclusive, contributor to 25-hydroxyvitamin D production in vivo. Proc Natl Acad Sci U S A 110(39):15650–15655

Gonadal Hormones and Bone

Tomohiko Yoshida, Jun Wang, and Paula H. Stern

Contents

T. Yoshida (✉)
Department of Diabetes, Metabolism and Endocrinology, International University of Health and Welfare School of Medicine, Chiba, Japan
e-mail: tomoyoshida@iuhw.ac.jp

J. Wang
Division of Cancer Epidemiology and Prevention, Department of Preventive Medicine, Northwestern University Feinberg School of Medicine, Chicago, IL, USA

P. H. Stern
Department of Pharmacology, Northwestern University Feinberg School of Medicine, Chicago, IL, USA

P. H. Stern (ed.), *Bone Regulators and Osteoporosis Therapy*,
Handbook of Experimental Pharmacology 262, https://doi.org/10.1007/164_2019_327

Abstract

In both sexes, estrogen is one of the most essential hormones for maintaining bone integrity. Also, especially in men, androgen has beneficial effects on bone independent of estrogen. However, estrogen replacement therapy for postmenopausal women increases the risk of developing breast cancer and endometrial cancer, and androgen replacement therapy for partial androgen deficiency of the aging male increases the risk of developing prostate cancer. Various mechanisms have been proposed on the effects of gonadal hormones on bone, such as effects through cytokines including IL-6 and effects on the OPG/RANKL ratio. In addition, large amounts of new information deriving from high-throughput gene expression analysis raise the possibility of multiple other effects on bone cells. Both estrogen and androgen exert their effects via the estrogen receptor (ER) or the androgen receptor (AR), which belongs to the nuclear receptor superfamily. Compounds such as selective estrogen receptor modulators (SERMs) and selective androgen receptor modulators (SARMs) also bind ER and AR, respectively. However, SERMs and SARMs alter the ER or AR structure differently from estrogen or androgen, resulting in other downstream gene responses. As a result they can exert favorable effects on bone while suppressing the undesirable actions of estrogen and androgen. Elucidation of ER and AR ligand-specific and tissue-specific gene regulation mechanisms will also provide information on the signal transduction mechanisms of other nuclear receptors and will be valuable for the development of new therapeutic agents.

Keywords

Androgen · Estrogen · Male osteoporosis · Postmenopausal osteoporosis · SARMs · SERMs

1 Overview of the Effects of Estrogens and Androgens on Bone

1.1 Postmenopausal Osteoporosis

Estrogen deficiency has been recognized to cause bone fragility for many years. Women in their 50s, an age at which menopause commonly occurs, rapidly lose trabecular bone when determined by dual X-ray absorptiometry (DXA) (Khosla et al. 1998). While DXA imaging only provides areal bone mineral density (aBMD), quantitative computed tomography (QCT) is able to determine the three-dimensional bone mineral density (volumetric BMD, vBMD) and differentiate cortical and trabecular components (Cann 1988). Population-based, cross-sectional QCT studies held in Olmsted County in Minnesota revealed that trabecular vBMD started to decrease before midlife and continued throughout life, and this reduction in vBMD in women could be up to 55% (Riggs et al. 2004). In contrast, a reduction in cortical vBMD was found to begin in midlife, a little later than that of trabecular vBMD. The cortical vBMD decreases have been reported to be up to 25% and are smaller than the reduction in trabecular vBMD. Longitudinal studies have confirmed the findings (Riggs et al. 2008). The decline in trabecular vBMD in women began before age 50, and there was a 37% total trabecular bone loss over a lifetime. However, considerable cortical bone loss started in middle life in women, and the reduction in bone mineral was reported to reach 6% when determined by QCT. Although these structural changes do not always coincide with changes in estrogen levels, it is clear that estrogen has a significant effect on bone strength.

These structural changes before and after menopause appear to increase the incidence of bone fracture. A Chinese population-based study of postmenopausal women revealed that the prevalence of radiographic vertebral fracture increased from 13% in women aged 50 to 59 years to more than 50% in those over 80 years (Cui et al. 2017). Since 1995, oral bisphosphonates have been widely used in clinical practice, and it is necessary to rely on earlier data to determine the pure effects of age and gonadal hormones on bone. Amin et al. examined 1989–1991 statistical data from Olmsted County, Minnesota residents and reported that the incidence of vertebral fracture began to rise at menopause and continued to increase throughout life (Amin et al. 2014). The prevalence of distal forearm fracture increased immediately after menopause but plateaued afterward. These results suggest that the effect of estrogen on bone may differ between cancellous and cortical bone.

1.2 Male Osteoporosis

Estrogens have a critical role in female bone metabolism, but also have a substantial effect on male bone. It was reported that a heterozygous mutation in the estrogen receptor gene in a man caused osteoporosis and delayed skeletal maturation with tall stature (Smith et al. 1994). Aromatase is a key enzyme that catalyzes the conversion of androgens to estrogens, and a man who had a homozygous mutation in the aromatase gene was first reported in 1995 (Morishima et al. 1995). He presented with severe low bone mineral density and delayed bone maturation. Plasma concentrations of estrogen and estrone were below the detection limit, whereas testosterone, 5α-dihydrotestosterone, and androstenedione levels were elevated. The bone turnover markers osteocalcin, pyridinoline, and deoxypyridinoline were also increased. Subsequent reports described a man with aromatase deficiency with different mutations (Miedlich et al. 2016). Unlike the previous case, this patient had no abnormal lipid metabolism or insulin resistance but had a marked decrease in bone density and a delay in bone age. Estradiol replacement successfully induced growth plate closure and increased bone mass.

Both cross-sectional (Slemenda et al. 1997; Szulc et al. 2001) and longitudinal (Khosla et al. 2001) studies have confirmed the critical role of estrogen in male osteoporosis. Plasma estradiol levels have been reported to be more closely correlated with male bone density than testosterone levels. These reports suggest that estrogen has a major impact on male bone metabolism, but it is unclear whether testosterone or estrogen is more important for male bone. Falahati-Nini et al. performed some research to address this question (Falahati-Nini et al. 2000). Fifty-nine elderly men (mean age 68.4) were first treated with a long-acting GnRH agonist to suppress endogenous testosterone (T) and estrogen (E) production. They also took an aromatase inhibitor to suppress the conversion from testosterone to estrogen. Pharmacological T and E inhibition caused a significant increase in bone resorption markers. However, no changes in bone resorption markers were seen in the group receiving T- and E-patches to maintain physiological gonadal hormone levels. Statistical analysis showed that E was the most critical factor in preventing the increase in bone resorption markers, while T involvement was not significant. On the other hand, serum osteocalcin, a bone formation marker, decreased after the suppression of gonadal hormones but increased with the administration of either E or T, and further increased over baseline with simultaneous administration of E and T. Collectively, the authors conclude that E is dominant in the inhibitory effect on bone resorption, but both E and T are involved in promoting bone formation in males.

2 Gonadal Hormones and Bone Cells

2.1 Estrogen Receptors and Bone Cells

The presence of high-affinity receptors for estrogen in osteoblasts was first described in 1988 by Komm et al. in rat ROS 17/2.8 and human HOS TE85 cells (Komm et al. 1988) and by Eriksen and coworkers in human osteoblastic cells from surgical specimens (Eriksen et al. 1988). The estrogen receptors and the genes for the receptors were subsequently reported in other osteoblastic cells (Davis et al. 1994; Etienne et al. 1990; Masuyama et al. 1992), normal osteoblasts, pre-osteoblast-like cells (Grandien et al. 1995; Ikegami et al. 1993; Ohashi et al. 1991; Zhang et al. 1995), osteoclasts, as well as pre-osteoclasts (Fiorelli et al. 1995; Oursler et al. 1991; Oursler et al. 1994; Pensler et al. 1990; Westerlind et al. 1995). It has also been reported in osteocytes (Braidman et al. 1995) and bone endothelial cells (Brandi et al. 1993).

In 1996, Kuiper et al. cloned a second estrogen receptor in rat prostate and ovary (Kuiper et al. 1996), and later studies reported its presence in humans (Mosselman et al. 1996) and mice (Tremblay et al. 1997), thereby establishing two types of ERs, namely, ERα and ERβ. ERα and ERβ reportedly send opposite signals through AP-1 transcription factors (Paech et al. 1997). ERα is considered to be the major receptor for bone metabolism (Sims et al. 2003). As described above, researchers have observed estrogen receptor expression in a variety of cells in bone and bone marrow. However, it is unclear as to what cell is most critical for mediating the effects of estrogen on the bone (Vanderschueren et al. 2014). In a recent review, Hewitt and Korach described four distinct mechanisms in ER-mediated signaling: (1) the genomic mechanism through interaction between ER and the estrogen-responsive element (ERE); (2) the "tethered" mechanism mediated through indirect signal transduction from ER to the AP-1 DNA motif via AP-1 transcription factors, such as FOS/JUN dimer; (3) the non-genomic signaling mediated by membrane-bound ER, or by GPER, a G-protein-coupled receptor to exert rapid effects of estrogen; and

(4) the ligand (estrogen)-independent signaling through the ER-ERE complex activated by a signaling cascade, such as mitogen-activated protein kinase (MAPK), triggered by extracellular growth factor (GF) via a cell membrane GF receptor (GFR) (Hewitt and Korach 2018). It is unclear how these four signaling mechanisms are involved in bone metabolism, and further studies are needed to elucidate this.

2.2 Effects of Estrogens on Osteoblasts, Osteocytes, and Bone Formation

Various mechanisms have been proposed to explain how estrogen deficiency can cause osteoporosis. Chow et al. showed that 17β-estradiol augmented rat vertebral bone mass in a dose-dependent manner when bone resorption was suppressed by bisphosphonate, suggesting a crucial role of estrogens on bone formation (Chow et al. 1992). Ernst et al. reported that 17β-estradiol (E2) stimulated cell number and [^{3}H]thymidine incorporation in rat osteoblastic cells (Ernst et al. 1989). Also, E2 significantly increased mRNA expression of *Col1A1* (procollagen type I alpha 1 chain gene) in primary cultures of rat calvarial osteoblasts, indicating that estrogens have a critical role in the proliferation and function of osteoblasts.

Sclerostin is known as to be a potent inhibitor of osteoblast function and bone formation. According to a cross-sectional study conducted in Connecticut (Mirza et al. 2010), serum sclerostin levels in postmenopausal women were significantly higher than those in premenopausal women, while there was no significant difference in PTH, 25-hydroxy or 1,25-dihydroxy vitamin D levels between two groups. Moreover, the sclerostin levels inversely correlated with estrogens. Wang et al. reported that SOST vaccination caused the elevation of the serum anti-SOST antibody levels in mice, and that serum from SOST vaccinated mice mitigated the reduction in osteogenic gene (*RUNX2*, *Osterix*, and *ALP*) expression in murine osteoblastic MC3T3-E1 cells (Wang et al. 2018). Blockade of sclerostin (SOST) signaling in mice induced by SOST vaccination attenuated the rise in bone resorption markers and the reduction in bone mineral density due to ovariectomy. Lipoprotein receptor 5 (LRP5) is a co-receptor of the canonical Wnt signaling pathway, and gain-of-function mutations in the *LRP5* gene cause resistance to sclerostin, resulting in high bone mass (Little et al. 2002; Roetzer et al. 2018). Mice with a gene mutation in the *LRP5* coding region showed resistance not only to bone loss induced by sclerostin but also to osteopenic effects due to ovariectomy (Niziolek et al. 2015). These data suggest that estrogen may inhibit sclerostin production and stimulate the canonical Wnt signaling pathway, thereby promoting osteoblast proliferation and differentiation. However, it is still unclear how estrogen regulates sclerostin secretion and bone formation in the physiological environment.

2.3 Effects of Estrogens on Osteoclasts and Bone Resorption

Estrogens are thought to have beneficial effects not only on bone formation but also on bone resorption. Pacifici et al. reported the secretion of bone-resorbing cytokines of interleukin 1 (IL-1) and tumor necrosis factor α (TNF-α) from blood mononuclear cells were increased in postmenopausal women or women who underwent ovariectomy, compared to those in premenopausal women (Pacifici et al. 1989, 1991). Interleukin-6 (IL-6) is also known to be a strong stimulator of osteoclastogenesis (Kurihara et al. 1991), but 17β-estradiol significantly inhibited IL-6 production induced by cotreatment of IL-1 and TNF-α by bone marrow-derived stromal cells, human bone-derived cells, and osteoblastic cell lines from mice and rats (Girasole et al. 1992; Kassem et al. 1996). In our own organ culture studies, it was found that parathyroid hormone increased bone resorption and IL-6 production of the neonatal mouse vertebral bone, and that 17β-estradiol blocked both effects (Stern et al. 1997). In a study performed with IL-6 deficient mice, neither bone mass nor bone remodeling rates were changed after ovariectomy (Poli et al. 1994), suggesting a critical role of IL-6 in the bone loss caused by estrogen deficiency. We have previously investigated whether estrogen levels affect IL-6 expression in postmenopausal women. When serum IL-6 values in pre- and post-menopausal women were compared, the values were significantly increased in healthy postmenopausal women (Lakatos et al. 1997). Other groups have subsequently confirmed this and have found that in addition to serum levels of IL-6, soluble IL-6 receptor (sIL-6R) is also elevated in women after ovariectomy (Girasole et al. 1999). Recent reports suggested that a signal mediated through complex formation of soluble IL-6 receptor (sIL-6R) and glycoprotein 130 kDa (gp130) is important for IL-6 (named as "*trans*-signaling"), in addition to conventional cell membrane receptor-derived signaling (Scheller et al. 2014). Although most soluble receptors act as antagonists, the sIL-6R was shown to have agonistic properties and activate gp130 expressing target cells together with IL-6. Studies using a specific antibody to block IL-6 *trans*-signaling indicate that restriction of IL-6 *trans*-signaling prevents trabecular bone loss in ovariectomized mice, but does not inhibit cortical bone loss, suggesting a critical role of IL-6 *trans*-signaling on trabecular bone loss in estrogen deficiency (Lazzaro et al. 2018).

Receptor activator of NF-κB ligand (RANKL) is essential for osteoclastogenesis, and several studies have revealed that estrogens inhibit the RANKL-RANK signaling pathway. Eghbali-Fatourechi et al. reported that RANKL expression was higher in bone marrow mononuclear cells (MNCs) of early postmenopausal women than in cells from premenopausal women (Eghbali-Fatourechi et al. 2003). Moreover, estrogen replacement therapy decreased the RANKL expression in MNCs from postmenopausal women. Also, these changes in the expression of RANKL in the MNCs associated with estrogen levels correlated with the expression of bone resorption markers. Osteoprotegerin (OPG) is a decoy receptor for RANKL and antagonizes the RANKL-RANK signaling pathway. In experiments using human osteoblast-like cells, Hofbauer et al. showed that 17β-estradiol increased OPG mRNA and protein expression in a dose- and time-dependent manner, whereas antiestrogens impaired the effects (Hofbauer et al. 1999). Likewise, 17β-estradiol dose-dependently induced estrogen receptor expression per osteoblastic cell. Subsequent studies have reported that ERα is essential for the expression of OPG and for the increased OPG/RANKL ratio induced by estrogen, but ERβ was not involved in the effect (Lindberg et al. 2001; Saika et al. 2001). Bord et al. reported 17β-estradiol stimulated both RANKL and OPG protein expression in human osteoblasts at 24 h, but at 48 h, the RANKL protein levels had returned their original values, while OPG expression remained elevated, suggesting that estrogen could alter the OPG/RANKL ratio to reduce osteoclastogenesis (Bord et al. 2003). Recent studies have shown that bone lining cells may be the bone cells most responsible for the estrogen-mediated suppression of RANKL production, and that ERα is mainly responsible for these effects (Streicher et al. 2017).

2.4 Effects of Androgens on Bone Cells

Androgens testosterone and 5α-dihydrotestosterone (DHT) exert their actions mainly via the androgen receptor (AR), which is widely expressed in various tissues. Both testosterone and DHT, which has a higher affinity for the androgen receptor than testosterone, are known to be indispensable for exerting full androgen action (Avila et al. 2001). In bone, AR expression in osteoblasts (Colvard et al. 1989), bone marrow stromal cells (Bellido et al. 1995), osteocytes (Abu et al. 1997), hypertrophic chondrocytes (Abu et al. 1997), as well as osteoclasts (Mizuno et al. 1994) have been reported. As mentioned above, estrogen is an essential hormone affecting male bone metabolism, and the action of testosterone on bone is considered to be primarily mediated by estrogen production through the aromatization of testosterone in peripheral tissues. However, Falahati-Nini and coworkers described that both testosterone and estrogens are critical for bone formation in males (Falahati-Nini et al. 2000). Chen and colleagues reported that testosterone inhibited parathyroid hormone (PTH)-induced osteoclastogenesis in a dose-dependent manner (Chen et al. 2001). Moreover, the estrogen receptor antagonist ICI 182780 and the aromatase inhibitor 4-androsten-4-ol-3,17-dione failed to antagonize the suppressive effect of testosterone on PTH-dependent osteoclastogenesis. These results suggest that the inhibitory effect of testosterone on PTH-dependent osteoclastogenesis is mainly mediated by the androgen receptor, although it is uncertain whether this has a physiological impact on bone.

3 Genetic Effects of Gonadal Hormones

3.1 Genetic Effects of Estrogens and Androgens

Sex steroid hormones act on their target cells by binding to members of the nuclear hormone receptor superfamily: estrogens bind to estrogen receptor (ER) α or ERβ, and androgens bind to the androgen receptor (AR). Several reviews have summarized the complexity of classical and nonclassical estrogen signaling mediated by ERα or ERβ and their splice variants (Börjesson et al. 2013; Heldring et al. 2007; Imai et al. 2009a; Manalogas et al. 2013; McDevitt et al. 2008) and the role for direct androgen receptor-mediated and indirect estradiol-mediated effects of androgen on bone (Ebeling 2010; Frenkel et al. 2010; Khosla 2010; Nicks et al. 2010). In addition, studies of targeted deletion of ER or AR in specific bone cells or their precursors (such as osteoblast progenitors, differentiated osteoblasts, osteocytes, and osteoclasts) have provided a better understanding of the mechanism of sex steroid hormones in different bone cell types and cell stages (Rooney and van der Meulen 2017).

3.2 The Role of Gonadal Hormones on Osteoblastic Gene Expression

3.2.1 Estrogen-Regulated Genes via ERα in Osteoblasts (Table 1)

The protective effects of estrogen in bone are due to many mechanisms (Khalid and Krum 2016). Comparison of gene expression patterns in cortical bone from ER-α-deficient (ERKO) mice and nonclassical ER knock-in (NERKI) mice (ER functioning only by nonclassical pathways) revealed that 28% (210 of 763) of the genes differentially expressed in ERKO mice were altered in NERKI mice, suggesting ERE-dependent regulation of these genes in bone. The majority (72%) of the genes regulated in ERKO mice were unique (i.e., not altered in NERKI mice), suggesting that these are regulated by nonclassical mechanisms. Genes involved in bone formation, senescence, apoptosis, and autophagy were significantly regulated (Chokalingam et al. 2012; Syed et al. 2010, 2011). Gene expression analyses of cortical bone from NERKI mice with an osteoporotic phenotype revealed suppression of lymphoid enhancer factor-1 (Lef1), suggesting that the nonclassical pathways are involved in Lef1-mediated Wnt signaling through both the stimulation of secreted Wnt inhibitors and/or disruption of normal β-catenin function (Mödder et al. 2012).

Using cell-specific promoters, ERα has been specifically deleted from osteoblastic cells at different stages: in mesenchymal progenitors via the Prx1 promoter (Prx1-ERαKO) or via the Sp7 promoter (Sp7-ERαKO) (Almeida et al. 2013), in osteoblast progenitors via the Osx1 promoter (Osx1-ERαKO) (Almeida et al. 2013), in early osteoblasts via the Col1a1 promoter (Col1a1-ERαKO) (Almeida et al. 2013), in mature osteoblasts via the OC promoter (OC-ERαKO) (Määttä et al. 2013a; Melville et al. 2014, 2015) or via the Runx2 promoter (Runx2-ERαKO) (Seitz et al. 2012), and in osteocytes via the Dmp1 promoter (Dmp1-ERαKO) (Kondoh et al. 2014; Windahl et al. 2013), as well as in chondrocytes via the Col2α1 promoter (Col2α1-ERαKO) (Börjesson et al. 2010) or via the OC promoter (OC-ERαKO) (Määttä et al. 2013a). Gene expression studies of osteocytes derived from Dmp1-ERαKO mice found that the expression of Mdk and Sostdc1, Wnt inhibitors, was significantly increased, but osteoblast/osteocyte markers in bone were reduced (Runx2, Sp7 and Dmp1) or unchanged (Sost and β-catenin), suggesting the osteoprotective functions

Table 1 Genes regulated by estrogen through ER in osteoblasts

Gene	References
Mediated by ERα	
BSP	(Takai et al. 2014)
FASLG, ALPL	(Krum et al. 2008a, b)
CSE	(Lambertini et al. 2008, 2017)
IGFBP4 (among 46 genes)	(Denger et al. 2008)
LC3, BECN1, ULK1	(Yang et al. 2013)
LRP6, LRP5, LRP4, APOER2	(Gui et al. 2016)
NDUFA10, UQCRC1, COX8A, COX6A2, COX8C, COX6C, COX6B2, COX412, ATP12A	(Lin et al. 2017)
SOST, SOSTDC1	(Fujita et al. 2014)
SVEP1	(Glait-Santar and Benayahu 2012)
miR-145	(Jia et al. 2017)
miR-199a-3p	(Fu et al. 2018)
miR-9	(Fang et al. 2015)
4,353 genes (RNA-sequencing)	(Roforth et al. 2014)
5,403 genes (microarray)	(Shang et al. 2014)
Mediated by ERβ	
SOST	(Galea et al. 2013; Lu et al. 2017)

BSP bone sialoprotein, *FASLG* Fas ligand, *ALPL* alkaline phosphatase, *CSE* cystathione gamma-lyase, *IGFBP4* insulin-like growth factor binding protein 4, *LC3(MAP 1LC3A)* microtubule-associated protein 1 light chain 3 alpha, *BECN1* beclin-1, *ULK1* unc-51 like autophagy activating kinase 1, *LRP* low-density lipoprotein, *APOER2* apolipoprotein E receptor 2, *NDUFA10* NADH: ubiquinone oxidoreductase subunit A10, *UQCRC1* ubiquinol-cytochrome c reductase core protein 1, *COX* cytochrome c oxidase, *ATP12A* ATPase H^+/K^+ transporting non-gastric alpha2 subunit, *SOST* sclerostin, *SOSTDC1* sclerostin domain containing 1, *SVEP1* sushi, von Willebrand factor type A, EGF and pentraxin domain containing 1, *miR* microRNA

of ERα in osteocytes in trabecular bone formation may act through regulating expression of Wnt antagonists (Kondoh et al. 2014; Windahl et al. 2013).

Many genes have been reported to be regulated by estradiol in osteoblasts through ERα, including genes related to apoptosis (FasL, Bcl-2), proliferation (IGFBP4, TGFβ3, Rbbp1), osteoblast differentiation (ALPL, BMP-2, BMP-4, BMP-6), and osteoclastogenesis (OPG) (Krum 2011). Some of the genes, including FasL, ALPL, IGFBP4, ESR1, and Hydrogen sulfide (H2S), were confirmed as direct transcriptional target genes of ERα in osteoblasts by ChIP assays (Denger et al. 2008; Krum et al. 2008a, b; Lambertini et al. 2008, 2017). Krum et al. revealed that ERα controls expression of key osteoblast genes (Alpl, Nfatc1, and Fasl), nuclear receptors (Gper1/Gpr30 and Nr5a2), and other transcription factors (Pax7, Sox5, Foxo1, and Foxo4) (Krum et al. 2008a, b). Svep1 was also found to be directly regulated by ERα in a preosteoblast cell line in collaboration with other transcription factors (transcription factor IIB, nuclear factor-kB, and specificity protein 1) (Glait-Santar and Benayahu 2012). GATA4 may serve as a critical pioneer factor for ERα recruitment to osteoblast-specific enhancers. GATA4 and ERα were both recruited to ERα-binding sites near genes that are specifically expressed in osteoblasts and control osteoblast differentiation (Miranda-Carboni et al. 2011). In addition, ERα also interacted with other pathways to regulate gene expression, such as C/EBPβ and IL6 in the insulin-like growth factor (IGF)-I pathway, TβRI and Runx2 in the transforming growth factor pathway and GSK3 activity, and β-catenin, LRP-6, and TCF-4 in the Wnt signaling pathway (Centrella and McCarthy 2012).

Several high-throughput studies revealed additional gene targets of estrogen in osteoblasts. Denger et al. (2008) identified 46 genes that were direct targets of estrogen in primary human osteoblasts. The majority of upregulated genes were involved in signal transduction and regulation of transcription, and many of the downregulated genes were immune and inflammatory response genes. Fujita et al. (2014) studied associations between estrogen status and expression of genes associated with bone mineral density and risk of fracture which were identified in a GWAS study in human bone samples (Estrada et al. 2012). They found 8 of the 70 genes were modulated by estrogen, including decreases in Sost and Sostdc1, two key inhibitors of Wnt/BMP signaling. Roforth et al. (2014) performed RNA-sequencing in human fetal osteoblasts that overexpressed ERα. They identified 4,353 genes upregulated by estrogen. By using mutant ERα constructs with nuclear only (NOER) functions or nonclassical ERα knock in (NERKI) mutations, they determined that 45% of the genes were nuclear ERE-independent, 27% were nuclear ERE-dependent, and 28% were extra-nuclear. Pathway and gene ontology analyses revealed that genes regulated through the nuclear ERE and nuclear non-ERE pathways were largely involved in transcriptional regulation, whereas genes regulated through extra-nuclear mechanisms were involved in cytoplasmic signaling transduction pathways. Interestingly, they found that 25 of 70 genes linked to bone density and fracture risk from the GWAS study (Estrada et al. 2012) were regulated by estrogen. Shang et al. (2014) treated MC3T3-E1 cells with 17β-estradiol for 5 days and using microarray analysis detected 5,403 differentially expressed genes, of which 1,996 genes were upregulated and 3,407 genes were downregulated. Among them, the expression of genes related to proliferation, differentiation, collagens, and transforming growth factor beta (TGF-β)-related cytokines increased, while the expression of genes related to apoptosis and osteoclast differentiation decreased.

Recent in vitro studies identified more estrogen-targeted genes and miRNAs in different functional pathways related to apoptosis, differentiation, and mineralization. Estrogen was found to protect against apoptosis via promotion of autophagy through the ER-ERK-mTOR pathway by increasing the expression of LC3, beclin1, and ULK1 in MC3T3-E1 osteoblastic cells (Yang et al. 2013) and by suppressing the inhibitory effects of miR-199a-3p on IGF-1 and mTOR pathway in osteocyte-like MLO-Y4 cells (Fu et al. 2018). In human osteoblast-like MG63 cells, estrogen suppressed miR145 expression, which is associated with osteoblast differentiation, which in turn posttranscriptionally suppressed OPG expression (Jia et al. 2017). In primary osteoblasts derived from the calvaria of newborn mice, estrogen induced osteoblast differentiation by upregulating LRP6 and downregulating LRP5, LRP4, and Apoer2, suggesting that estrogen promoted osteoblast differentiation in an early stage (Gui et al. 2016). Estrogen also stimulated bone sialoprotein (BSP) gene transcription in a ligand-independent manner by targeting the CRE and AP1/GRE elements in the rat BSP gene promoter in ROS17/2.8 cells (Takai et al. 2014). In MG-63 cells, a high dose of estrogen upregulated miR-9 and in turn downregulated expression of long noncoding RNA metastasis associated lung adenocarcinoma transcript 1 (MALAT-1) (Fang et al. 2015). In human osteoblast-like U2OS cells, estradiol exposure led to increases in mitochondrial energy metabolism and subsequent osteoblast mineralization. Expressions of the genomic respiratory chain complex NDUFA10, UQCRC1, cytochrome c oxidase (COX)8A, COX6A2, COX8C, COX6C, COX6B2, COX412, and ATP12A genes were upregulated (Lin et al. 2017).

3.2.2 Estrogen-Regulated Genes via ERβ in Osteoblasts

Monroe et al. (2003) treated U2OS cells that overexpressed inducible ERα or ERβ with estrogen for 24 h and detected only a 21% overlap between estrogen-regulated genes in U2OS-ERα and U2OS-ERβ cells, demonstrating that the two ER receptors had different functions in osteoblast-like cells. In a mouse model with deletion of ERβ specifically in early osteoprogenitor cells using the Prx1-Cre promoter, 128 genes in 16 pre-specified pathways were significantly downregulated (Nicks et al. 2016).

ERβ may antagonize ERα action. With shRNA-based depletion of ERα in MG-63 osteoblasts, expression of OPG, MBP2, TGF-β, RUNX2, IGF-1 was significantly reduced, while expression of RANKL was drastically increased. By contrast, ERβ played an opposite role in regulating gene expression of OPG, MBP2, TGF-β, RUNX2, IGF-1, and RANKL. However, double depletion of ERα and ERβ could not rescue the gene expression of these factors in vitro (Wang et al. 2016). In another study, Lu et al. (2017) found that activation of ERβ significantly rescued cell cycle arrest induced by the downregulation of ERα. The synergic effects of ERα and ERβ deletion were mediated via upregulation of SOST gene expression and the subsequent inhibition of OPG and Runx2 gene expression. Thus, ERβ may serve the function of balancing osteoblast viability and differentiation induced by ERα.

ERβ also was involved in estrogen-induced osteoblast proliferation and differentiation via Wnt/β-catenin signaling. In human Saos-2 cells and in mouse long bone-derived osteoblastic cells, ERα and ERβ differentially regulated SOST, which is involved in the osteoblast response to strain. SOST downregulation by ERβ activation led to enhanced Wnt/β-catenin signaling and osteoblast differentiation (Galea et al. 2013). Estrogen induced the accumulation of β-catenin protein in the nucleus which led to interaction with T-cell-specific transcription factor/lymphoid enhancer binding factor (TCF/LEF) transcription factors via ERβ/GSK-3β-dependent Wnt/β-catenin signaling pathway (Yin et al. 2015).

Table 2 Genes regulated by androgen through AR in osteoblasts

Gene	References
OPG	(Chen et al. 2004)
MYBL2, HOXD11, ADCYAP1R1 (among 430 genes)	(Miki et al. 2007)
SOST	(Di Nisio et al. 2015)

OPG osteoprotegerin, *MYBL2* MYB proto-oncogene like 2, *HOXD11* homeobox D11, *ADCYAP1R1* ADCYAP receptor type 1, *SOST* sclerostin

3.2.3 Androgen-Regulated Genes via AR in Osteoblasts (Table 2)

Androgens include aromatizable and non-aromatizable types. Aromatizable androgens, such as testosterone, potentially activate both AR and ER, while DHT (a non-aromatizable form of testosterone) only activates AR. In the nucleus, AR binds as dimers to specific androgen response elements (AREs), interacts with other transcription factors, and recruits coregulatory proteins, as classical AR signaling. AR has been also found to exert nongenomic effects via components of the SRC, ERK, PI3K, and AKT pathways (Vanderschueren et al. 2004). Wang et al. (2012) found that in MG63 osteoblastic cells overexpressing ERβ, DHEA activated the pERK1/2-MAPK signaling pathway, but not p38 and JNK, via a dominant ERβ receptor. Like estrogen, androgen also interacted with other factors, such as TGFβ, IGFs, and IL-6, to regulate gene expression (Krum 2011; Vanderschueren et al. 2014). McCarthy and Centrella (2015) confirmed the complex interactions of local and systemic bone growth regulators with DHT, showing that DHT suppressed the prostaglandin E2 (PGE2) and TGF-β induced IGF-I gene promoter and affected other aspects of TGF-β activity by synergistically increasing Smad-dependent gene promoter activity in osteoblasts.

Testosterone increased OPG mRNA expression in both mouse bone cell cultures and MC3T3-E1 cells (Chen et al. 2004). A microarray analysis of human fetal osteoblasts (hFOB) revealed 430 genes regulated by DHT. Among them, MYBL2, HOXD11, and ADCYAP1R1 were validated in qPCR assays (Miki et al. 2007). Di Nisio et al. (2015) observed serum T levels were negatively correlated with sclerostin. Stimulation of cultured human osteocytes with DHT decreased sclerostin expression in a time- and dose-dependent manner.

Wiren et al. (2010) examined RNA expression from cortical bone of mice with osteoblast-specific AR-overexpressing via Col1a1 promoter: 3.6-kb and 2.3-kb fragments (AR-transgeneOb;3.6-Col1a1 and AR-transgeneOb;2.3-Col1a1) (Wiren et al. 2004, 2008) and observed that TGFβ superfamily and BMP signaling were significantly altered by androgen. Tnfrs11b (osteoprotegerin), Runx2, Tgfb2, Fos, and Jun were upregulated, and Il6, Il1a, and Tnfrsf11a (encoding RANK) were downregulated. Bioinformatic analyses indicated proliferation, osteoblast differentiation, and mineralization as major biological processes affected, suggesting that BMP signaling contributed to androgen inhibition of osteoblast differentiation and mineralization.

AR has been deleted specifically in osteoprogenitor cells using the Prx1-Cre promoter (Ucer et al. 2015), in osteoblasts via Col1a1 promoter (Notini et al. 2007) or via OC promoter (Chiang et al. 2009; Määttä et al. 2013b), and in osteocytes via the Dmp1 promoter (Sinnesael et al. 2012). A microarray study by Russell et al. (2012) using whole bones isolated from OC promotor-driven ARKO mice identified biological processes affected by AR, including skeletal and muscular system development and carbohydrate metabolism. The osteoblast genes Col1a1 and Bglap and the osteoclast genes Ctsk and RANKL (Tnfs11) were upregulated. Genes involved in carbohydrate metabolism (adiponectin and Dpp4) and in growth and development (GH, Tgfb2, Wnt4) were identified as potential targets of androgen action via AR in mineralizing osteoblasts. Huang et al. (2013) found that loss of AR in bone marrow stromal cells (BMSCs) suppressed osteogenesis-related genes, inhibiting osteoblast differentiation from BMSCs, but promoted adipogenesis of BMSCs via Akt activation through IGFBP3-mediated IGF signaling.

3.3 The Effects of Gonadal Hormones on Osteoclastic Gene Expression

3.3.1 Estrogen-Regulated Genes in Osteoclasts (Table 3)

Multiple mechanisms may account for osteoclast regulation by estrogen including repression of pro-osteoclastic cytokines, induction of apoptosis in bone-resorbing osteoclasts, and regulation of the RANKL/OPG ratio (Khalid and Krum 2016). Estrogen-induced apoptosis of osteoclasts may be mediated by MMP3-induced cleavage and solubilization of osteoblast-expressed Fas ligand. In human and murine primary osteoblasts and in calvarial organ cultures, estrogen treatment upregulated MMP3 protein expression, which was co-localized with the osteoblast-specific RUNX2. The soluble FasL induced osteoclast apoptosis (Garcia et al. 2013). A study from Sugatani and Hruska (2013) further revealed that estrogen downregulated miR-21 biogenesis and in turn posttranscriptionally increased protein levels of FasL, one of the targets of miR-21, and induced osteoclastic apoptosis. By using an in vitro co-culture system and an in vivo bone marrow-derived mesenchymal stem cell (BMMSCs) transplantation assay, Shao et al. (2015) also found that estrogen did not increase FasL gene transcription but elevated FasL protein accumulation through microRNA-mediated posttranscriptional regulation. Estrogen downregulated expression of

Table 3 Genes regulated by estrogen through ER in osteoclasts

Gene	References
c-fos, c-Jun, CTSB, CTSL, CTSK, ACP5	(Krum 2011)
SPARC, FN1, ITGA1, CASP2, FASLG, CTNNB1, NFATC1	(Wang and Stern 2011)

c-fos c-fos protein, *c-Jun* Jun proto-oncogene, *CTSB* cathepsin B, *CTSL* cathepsin L, *CTSK* cathepsin K, *ACP5* tartrate-resistant acid phosphatase type 5, *SPARC* secreted protein acidic and cysteine rich, *FN1* fibronectin 1, *ITGA1* integrin subunit alpha 1, *CASP2* caspase 2, *FASLG* Fas ligand, *CTNNB1* catenin beta 1, *NFATC1* nuclear factor of activated T cells 1 miR-181a, a negative modulator of FasL by targeting the 3′-UTR of FasL mRNA, and promoted the apoptosis of CD4 + T lymphocytes (Shao et al. 2018).

ERα has been specifically inactivated in osteoclast precursors via the LysM promoter (LysM-ERαKO) (Martin-Millan et al. 2010; Seitz et al. 2012) and in mature osteoclasts via the Ctsk promoter (Ctsk-ERαKO) (Nakamura et al. 2007). In vivo estrogens directly regulated the life span of mature osteoclasts by inducing expression of pro-apoptotic Fas ligand and decreasing expression of osteoclast-specific genes that control bone resorption activity (Imai et al. 2009b).

Some genes have been found to be regulated by estrogen using cultured mammalian osteoclasts, including genes related to proliferation (c-fos, c-jun), osteoclast resorption (cathepsin B, cathepsin L, cathepsin K, TRAP), and inflammatory cytokines (IL-1RI, IL-1RII) (Krum 2011). To compare sex-specific actions of estrogen on osteoclasts, we studied gene expression profiles of 94 target genes in osteoclasts derived from male or female donor human peripheral blood mononuclear precursor cells treated with 17β-estradiol for 24 h. Expression of SPARC, FN1, ITGA1, CASP2, FASLG, and CTTNB1 was upregulated in the female-derived cells, while CTTNB1 was decreased by estrogen in the male-derived osteoclasts. NFATC1, a transcription factor that effects differentiation, fusion, and activity of osteoclasts, was significantly increased by estrogen in the male-derived cells, but not in the female-derived cells (Wang and Stern 2011).

3.3.2 Androgen-Regulated Genes in Osteoclasts (Table 4)

Androgens decreases osteoclastogenesis by a mechanism distinct from that of estrogen, with a direct inhibitory effect on osteoclast formation. When CD14+ human monocytes were treated with testosterone, the differentiated osteoclasts did not form as many TRAP-positive osteoclasts, whereas estrogen failed to do this in the absence of osteoblasts (Krum et al. 2008a; Michael et al. 2005). In an early in vitro study (Pederson et al. 1999), 24-h treatment of avian osteoclasts with DHT upregulated genes for TGF-β and downregulated cathepsin B and tartrate-resistant acid phosphatase.

In our study to assess sex-specific actions of androgen on osteoclasts, we treated osteoclasts derived from male or female donor human peripheral blood mononuclear precursor cells with testosterone for 24 h. Testosterone increased a number of genes in the cells from the female donors. Two of the genes, LRP5 and MAP 2K4, were also increased by 17β-estradiol, and thus may have been activated as a result of conversion of testosterone to estrogen through aromatase activity. Four genes

Table 4 Genes regulated by androgen through AR in osteoclasts

Gene	References
TGFB1, CTSB, ACP5	(Pederson et al. 1999)
TGFB3, PTEN, ICAM1, TNFα	(Wang and Stern 2011)

TGFB1 transforming growth factor beta 1, *CTSB* cathepsin B, *ACP5* tartrate-resistant acid phosphatase type 5, *TGFB3* transforming growth factor beta 3, *PTEN* phosphatase and tensin homolog, *ICAM1* intracellular adhesion molecule 1, *TNFα*, tumor necrosis factor alpha

(TGFβ3, PTEN, ICAM1, and TNF) that were regulated by testosterone in the female-derived cells were not affected by 17β-estradiol in either male or female-derived cells, and thus these effects of testosterone were not due to its conversion to estrogen (Wang and Stern 2011).

AR deletion specifically in osteoclast precursors by LysM-Cre (Sinnesael et al. 2015) or in mature osteoclasts by Ctsk-Cre (Ucer et al. 2015) had no effect on bone mass or osteoclast number, indicating that the AR in osteoclasts is not critical for bone maintenance. The levels of AR were very low in osteoclast-enriched cultures derived from bone marrow and undetectable in osteoclasts generated from spleen precursors. In co-culture experiments, bone marrow stromal cells (BMSCs) were essential for the suppressive action of AR on osteoclastogenesis and osteoclast activity. Runx2 expression in BMSCs was higher in ARKO mice, suggesting that ARKO mice may more readily commit osteoprogenitor cells to osteoblastogenesis. Thus, AR did not suppress bone resorption through direct actions on osteoclasts, while BMSCs may represent an alternative AR target in the bone marrow milieu (Sinnesael et al. 2015).

4 Pharmacologic Approach to Estrogen and Androgen Deficiency on Bone

4.1 Overview of Osteoporosis Treatment in Elderly Patients with Gonadal Hormone Deficiency

In general, the principle of treatment for patients with hormone insufficiency is the replacement of the deficient hormone. However, this principle does not necessarily apply to the treatment of postmenopausal osteoporosis and osteoporosis in older men associated with low androgen levels. The Women's Health Initiative (WHI) and other trials suggest that estrogen plus progesterone therapy increases the risk of heart failure, breast cancer, stroke, and pulmonary embolism, although it can reduce fracture risk (Rossouw et al. 2002). Estrogen monotherapy also reduces the risk of fractures (Cauley et al. 1995), but unopposed estrogen without suppression of endometrial proliferation by progesterone increases the risk of endometrial cancer (Sulak 1997). Based on these facts, therapies other than estrogen or estrogen plus progesterone are recommended as the first-line agents for the treatment of postmenopausal osteoporosis (see "17. Osteoporosis therapeutics 2019") (Qaseem et al. 2017). Regarding male osteoporosis, there is less evidence than for women, but studies have suggested that the bisphosphonates alendronate (Sawka et al. 2005; Orwoll et al. 2000) and zoledronate (Boonen et al. 2012) increase the vertebral bone mineral density, and significantly reduce the risk of vertebral fractures compared to placebo. The American College of Physicians' clinical guideline in 2017 recommends bisphosphonates to reduce the risk of vertebral fracture in men (Qaseem et al. 2017).

4.2 Selective Androgen Receptor Modulators (SARMs)

Testosterone, along with estrogen, is critical for male bone metabolism. Katznelson et al. investigated the effect of testosterone on the bone in adult men with acquired hypogonadism (Katznelson et al. 1996). Men with acquired hypogonadism had a significantly reduced spine bone mineral density compared to age-matched eugonadal men (1.006 ± 0.024 *vs.* 1.109 ± 0.028 g/cm^2). Testosterone replacement at a dose of 100 mg/week raised spinal and trabecular bone density by $5 \pm 1\%$ and $14 \pm 3\%$, respectively. Serum bone-specific alkaline phosphatase and urinary deoxypyridinoline excretion decreased significantly over 18 months of testosterone treatment. This testosterone action on bone may be partly due to the aromatization to estrogen, but as mentioned above, the expression of AR has been confirmed in various bone cells. Thus, there might be androgen-specific effects on bone.

Androgen is likely to have an anti-osteoporotic effect, but there are concerns about the increased risks of prostate hypertrophy, prostate cancer, and cardiovascular events, although recent studies have revealed testosterone therapy may decrease the risk of prostate cancer through androgen receptor saturation (Kaplan et al. 2016). Also, there are reports that testosterone did not raise the risk of cardiovascular disease (Gagliano-Jucá and Basaria 2019). Long after the discovery of selective estrogen receptor modulators (SERMs), Dalton et al. reported the first selective androgen receptor modulators (SARMs) (Dalton et al. 1998), and later studies developed other compounds that activate AR in a tissue-dependent manner.

Like the ER, the AR belongs to the nuclear receptor superfamily, composed of three major functional domains: (1) the N-terminal domain (NTD), followed by (2) the DNA binding domain (DBD), and (3) the C-terminal ligand-binding domain (LBD) (Tan et al. 2015). When testosterone or DHT binds to AR in the cytoplasm, it further binds to importin-α and is transported into the nucleus. In the nucleus, the DBD of the AR dimer binds to androgen response elements (AREs) of target genes as also recruits many transcriptional cofactors, resulting in induction of androgen-dependent gene transcription. The activation function 1 (AF1) in the NTD is constitutively active, whereas AF2 in LBD is activated in a ligand-dependent manner. The size of the ligand-binding pocket (LBP) in the LBD is variable, which enables binding to various ligands, resulting in various ligand-dependent transcriptional activities.

Kearbey and colleagues investigated the effects of S-4, an arylpropionamide-derived SARM on bone (Kearbey et al. 2007). S-4 treatment at doses greater than 0.1 mg/day significantly inhibited bone loss due to ovariectomy in female rats. Moreover, it dose-dependently reduced the fat mass of ovariectomized and intact rats. The biochemical strength of the femur evaluated by three-point bending analysis did not decrease in S-4-treated ovariectomized animals. S-4 stimulated the differentiation of bone marrow stromal cells toward osteoblast in a dose-dependent manner. These results suggest that S-4, a SARM, could prevent not only male osteoporosis but also postmenopausal osteoporosis.

Unfortunately, as far as we know, there are currently no available SARMs for the treatment of osteoporosis in the clinical setting. Human application of SARMs may depend on the following: (1) how much we could strengthen the favorable effects of androgen in muscle and bone; (2) how much we could reduce the unfavorable effects, such as prostate cancer and cardiovascular events.

4.3 Estrogen Replacement Therapy and the Women's Health Initiative

Since postmenopausal women have an increased risk of coronary artery disease (CHD) compared to premenopausal women, estrogen was expected to have an anti-atherogenic effect but might increase the incidence of breast cancer. Therefore, the WHI started what was a planned 8.5-year study to assess the risks and benefits of estrogen plus progestin therapy for breast cancer and CHD (Rossouw et al. 2002). Unexpectedly, estrogen plus progestin therapy increased CHD by 29% compared to placebo. Moreover, it increased breast cancer incidence by 26%, stroke by 41%, and pulmonary embolism by 113%. Although it reduced hip fractures by 34%, vertebral fractures by 34%, total fractures by 24%, colorectal cancer by 37%, and endometrial cancer by 17%, it worsened the global index summarizing risks and benefits, with an increased risk of 19 events per 10,000 person-years. As a result, the WHI, initially planned for 8.5 years, was forced to cease with an average follow-up period of 5.2 years.

A later WHI trial examined the effects of estrogen monotherapy with conjugated equine estrogen (CEE) in postmenopausal women who had undergone a total hysterectomy (Anderson et al. 2004). The CEE monotherapy increased the risk of stroke but reduced the hip fracture risk, and did not alter the incidence of CHD. From these results, the WHI steering committee recommended against the use of CEE for chronic disease prevention in postmenopausal women.

Subsequently, in the re-evaluation of these two WHI trials, CEE monotherapy (Manson et al. 2013), especially in younger women aged from 50 to 59, showed a 30% reduction in overall mortality, a 45% reduction in myocardial infarction, and a 16% reduction in the global index. From these favorable results, some researchers recommend CEE monotherapy for women aged 50–59 or within 10 years after menopause. However, considering the increased risk of pulmonary embolism of the CEE monotherapy and the efficacy of other antiresorptive agents, such as bisphosphonates, it has been established that first-line therapy for postmenopausal osteoporosis should be other than estrogen monotherapy.

4.4 Selective Estrogen Receptor Modulators (SERMs)

The WHI trials pointed out the unfavorable effects of estrogen, including increased risks of CHD, stroke, and breast cancer, but they reconfirmed the effects of estrogen on preventing fractures. Selective estrogen receptor modulators (SERMs) are compounds that exhibit an estrogenic action or an antiestrogenic action in a tissue-specific manner. SERMs include raloxifene, bazedoxifene, and lasofoxifene used for osteoporosis and tamoxifen used for breast cancer treatment (Fisher et al. 1998; Cummings et al. 1999, 2010; Ettinger et al. 1999; Villiers et al. 2011; Silverman et al. 2012). Similar to estrogen, SERMs suppress fractures while reducing the risk of developing breast cancer, an effect not seen with estrogen. Interestingly, tamoxifen and lasofoxifene stimulate endometrial proliferation associated with an elevated risk of endometrial cancer, whereas raloxifene and bazedoxifene have little effect on the endometrium. Thus, individual SERM have tissue-specific actions.

The Multiple Outcomes of Raloxifene Evaluation (MORE) study is a multicenter, randomized, blinded, placebo-controlled trial that was conducted to investigate the effects of raloxifene on preventing vertebral fractures and non-vertebral fractures (Ettinger et al. 1999). In the MORE study, 7,705 women who had osteoporosis, determined by low bone density or radiographic vertebral fracture, registered at 180 centers in 25 countries. They were more than 2-year postmenopausal, and had no severe or long-term disabilities, and were assigned to receive placebo, raloxifene 60 mg/day, or 120 mg/day for 36 months. All subjects took 500 mg per day of calcium and 400 to 600 units per day of cholecalciferol (vitamin D3). New vertebral fractures were observed in 7.4% of subjects during the observation period. In subjects without previous fractures, new fractures in the placebo group were 4.5%, compared to 2.3% in the raloxifene 60 mg/day group and 2.8% in the 120 mg/day group, suggesting a 40–50% fracture risk reduction by raloxifene. Furthermore, raloxifene also achieved a significant fracture risk reduction in subjects with existing fractures at the beginning, with 21.2% of new fractures in the placebo group compared to 14.7% in the 60 mg/day group (relative risk [RR], 0.7) and 10.7% in the 120 mg/day group (RR, 0.5). Raloxifene augmented bone mineral density in the femoral neck by 2.1% (60 mg) and 2.4% (120 mg) and in the spine by 2.6% (60 mg) and 2.7% (120 mg) compared to placebo. The subjects receiving raloxifene had a higher risk of venous thromboembolism vs. placebo (RR, 3.1). However, raloxifene did not cause genital bleeding or breast pain and also reduced the incidence of breast cancer.

A newer SERM, bazedoxifene (available in Europe and Japan and in the United States available as a combination with conjugated equine estrogen), also reduced the risk of vertebral fractures, with a 6.8% incidence of new vertebral fractures in the placebo group versus 4.5% in the bazedoxifene 20 mg group and 3.9% in the 40/20 mg group, resulting in a 35% or 40% relative risk reduction, respectively (Silverman et al. 2012). There were no differences between the groups for non-vertebral fractures. However, in high-risk groups with femoral neck T-score < -3.0 and/or one or more moderate or severe vertebral fractures, or more than two mild vertebral fractures, bazedoxifene significantly reduced non-vertebral fracture risk by 37%. Bazedoxifene significantly increased bone mineral density and decreased bone markers, serum CTx, and osteocalcin, compared to the placebo group. The incidence of coronary artery disease, stroke, pulmonary thromboembolism, and retinal vein thrombosis was similar between groups, but deep vein thrombosis was higher in the bazedoxifene group (0.5% and 0.6% for bazedoxifene 20 and 40/20 mg, respectively) than in the placebo group (0.2%).

The other SERM, lasofoxifene (available in some countries), also reduced the risk of non-vertebral and vertebral fractures with 0.5 mg, as well as the risk of ER-positive breast cancer, coronary artery disease, and stroke (Cummings et al. 2010). However, like other SERMs, lasofoxifene increased the risk of venous thrombosis.

Studies reported that there was a significant reduction in the incidence of fracture, but also there was an increase in the risk of undesirable side effects, such as venous thrombosis and endometrial hypertrophy when using SERMs. Therefore, bisphosphonates may be the first-line for the treatment of postmenopausal osteoporosis because of their excellent safety and efficacy. Among the SERMs, raloxifene may currently be preferred because the long-term safety data are insufficient for bazedoxifene and lasofoxifene.

There have been various proposals on why SERMs act as agonists or antagonists to estrogen receptors (ERs) in a tissue-dependent manner. When ER binds to estrogen, it binds to an estrogen-responsive element (ERE) and regulates the expression of downstream genes. Similarly, SERMs also regulate transcription of downstream genes via ER, but ligand-dependent ER conformational changes differ between estrogen and SERMs. Also, this difference in conformational change between estrogen and SERM results in different affinities for transcriptional cofactors, resulting in specific downstream gene regulation. In particular, the position of helix 12 (H12) that constitutes the activator function-2 (AF-2) domain in the ligand-binding domain (LBD) region is altered significantly by each ligand, which modifies chromatin accessibility and RNA transcription frequency (Brzozowski et al. 1997; Shiau et al. 1998; Pike et al. 1999; Hewitt and Korach 2018).

Shang and colleagues reported that differences in transcriptional activation of SERMs are dependent on the intracellular environment (Shang and Brown 2002). In their studies, tamoxifen acted as an antagonist in the breast cancer cell line MCF-7 but acted as a partial agonist in Ishikawa cells, an endometrial cancer cell line. This effect correlated with the expression level of c-Myc and IGF-I mRNA upon stimulation with tamoxifen. In Ishikawa cells, c-Myc and IGF-I mRNA were increased by tamoxifen stimulation, whereas there was no difference in their mRNA levels in MCF-7 cells. Also, steroid receptor coactivator-1 (SRC-1) expression was high in Ishikawa cells, but only a slight expression was observed in MCF-7 cells. The findings suggest that the transcriptional activity of SERMs may be influenced by tissue differences in the expression levels of the coactivators such as SRC-1. Both c-Myc and IGF-I are genes that do not have an estrogen response element (ERE) at the transcriptional regulatory site, and IGF-I transcriptional regulation is known to be achieved via the AP-1 DNA motif. This suggests that the action of a SERM may be exerted by a mechanism mediated by other transcriptional regulators (such as AP-1 DNA motif) and not by ERE. They also described differences in the recruitment of transcriptional cofactors between raloxifene and tamoxifen. In Ishikawa cells, Tamoxifen caused the ER-transcription complex formation with SRC-1, while raloxifene attracted co-repressors such as NCoR and SMRT. This suggests that each SERM exerts distinct ER-transactivation through different transcriptional cofactors. Lonard et al. showed that the steady-state levels of SRC-1 and SRC-3 were increased in an ER-dependent manner by the administration of tamoxifen and raloxifene (Lonard et al. 2004). The authors suggested that this increase in SRC-1 and SRC-3 may affect other nuclear receptors that interact with SRC-1 or SRC-3, resulting in unexpected effects of SERMs.

Collectively, the mechanisms of downstream gene regulation by SERMs are complicated, but advances in the understanding of transcriptional regulation by SERMs should facilitate future drug discovery.

References

Abu EO, Horner A, Kusec V et al (1997) The localization of androgen receptors in human bone. J Clin Endocrinol Metab 82(10):3493–3497

Almeida M, Iyer S, Martin-Millan M et al (2013) Estrogen receptor-α signaling in osteoblast progenitors stimulates cortical bone accrual. J Clin Invest 123(1):394–404

Amin S, Achenbach SJ, Atkinson EJ et al (2014) Trends in fracture incidence: a population-based study over 20 years. J Bone Miner Res 29(3):581–589

Anderson GL, Limacher M, Assaf AR et al (2004) Effects of conjugated equine estrogen in postmenopausal women with hysterectomy: the Women's Health Initiative randomized controlled trial. JAMA 291(14):1701–1712

Avila DM, Zoppi S, McPhaul MJ (2001) The androgen receptor (AR) in syndromes of androgen insensitivity and in prostate cancer. J Steroid Biochem Mol Biol 76(1–5):135–142

Bellido T, Jilka RL, Boyce BF et al (1995) Regulation of interleukin-6, osteoclastogenesis, and bone mass by androgens. The role of the androgen receptor. J Clin Invest 95(6):2886–2895

Boonen S, Reginster JY, Kaufman JM et al (2012) Fracture risk and zoledronic acid therapy in men with osteoporosis. N Engl J Med 367(18):1714–1723

Bord S, Ireland DC, Beavan SR et al (2003) The effects of estrogen on osteoprotegerin, RANKL, and estrogen receptor expression in human osteoblasts. Bone 32(2):136–141

Börjesson AE, Lagerquist MK, Liu C et al (2010) The role of estrogen receptor α in growth plate cartilage for longitudinal bone growth. J Bone Miner Res 25(12):2690–2700

Börjesson AE, Lagerquist MK, Windahl S et al (2013) The role of estrogen receptor α in the regulation of bone and growth plate cartilage. Cell Mol Life Sci 70(21):4023–4037

Braidman IP, Davenport LK, Carter HD et al (1995) Preliminary in situ identification of estrogen target cells in bone. J Bone Miner Res 10(1):74–80

Brandi ML, Crescioli C, Tanini A et al (1993) Bone endothelial cells as estrogen targets. Calcif Tissue Int 53(5):312–317

Brzozowski AM, Pike ACW, Dauter Z et al (1997) Molecular basis of agonism and antagonism in the oestrogen receptor. Nature 389(6652):753–758

Cann CE (1988) Quantitative CT for determination of bone mineral density: a review. Radiology 166(2):509–522

Cauley JA, Seeley DG, Ensrud K et al (1995) Estrogen replacement therapy and fractures in older women. Study of Osteoporotic Fractures Research Group. Ann Intern Med 122(1):9–16

Centrella M, McCarthy TL (2012) Estrogen receptor dependent gene expression by osteoblasts – direct, indirect, circumspect, and speculative effects. Steroids 77(3):174–184

Chen Q, Kaji H, Sugimoto T et al (2001) Testosterone inhibits osteoclast formation stimulated by parathyroid hormone through androgen receptor. FEBS Lett 491(1–2):91–93

Chen Q, Kaji H, Kanatani M et al (2004) Testosterone increases osteoprotegerin mRNA expression in mouse osteoblast cells. Horm Metab Res 36(10):674–678

Chiang C, Chiu M, Moore AJ et al (2009) Mineralization and bone resorption are regulated by the androgen receptor in male mice. J Bone Miner Res 24(4):621–631

Chokalingam K, Roforth MM, Nicks KM et al (2012) Examination of ERα signaling pathways in bone of mutant mouse models reveals the importance of ERE-dependent signaling. Endocrinology 153(11):5325–5333

Chow J, Tobias JH, Colston KW et al (1992) Estrogen maintains trabecular bone volume in rats not only by suppression of bone resorption but also by stimulation of bone formation. J Clin Invest 89(1):74–78

Colvard DS, Eriksen EF, Keeting PE et al (1989) Identification of androgen receptors in normal human osteoblast-like cells. Proc Natl Acad Sci U S A 86(3):854–857

Cui L, Chen L, Xia W et al (2017) Vertebral fracture in postmenopausal Chinese women: a population-based study. Osteoporos Int 28(9):2583–2590

Cummings SR, Eckert S, Krueger KA et al (1999) The effect of raloxifene on risk of breast cancer in postmenopausal women: results from the MORE randomized trial. Multiple outcomes of raloxifene evaluation. JAMA 281(23):2189–2197

Cummings SR, Ensrud K, Delmas PD et al (2010) Lasofoxifene in postmenopausal women with osteoporosis. N Engl J Med 362(8):686–696

Dalton JT, Mukherjee A, Zhu Z et al (1998) Discovery of nonsteroidal androgens. Biochem Biophys Res Commun 244(1):1–4

Davis VL, Couse JF, Gray TK et al (1994) Correlation between low levels of estrogen receptors and estrogen responsiveness in two rat osteoblast-like cell lines. J Bone Miner Res 9(7):983–991

Denger S, Bähr-Ivacevic T, Brand H et al (2008) Transcriptome profiling of estrogen-regulated genes in human primary osteoblasts reveals an osteoblast-specific regulation of the insulin-like growth factor binding protein 4 gene. Mol Endocrinol 22(2):361–379

Di Nisio A, De Toni L, Speltra E et al (2015) Regulation of sclerostin production in human male osteocytes by androgens: experimental and clinical evidence. Endocrinology 156(12):4534–4544

Ebeling PR (2010) Androgens and osteoporosis. Curr Opin Endocrinol Diabetes Obes 17 (3):284–292

Eghbali-Fatourechi G, Khosla S, Sanyal A et al (2003) Role of RANK ligand in mediating increased bone resorption in early postmenopausal women. J Clin Invest 111(8):1221–1230

Eriksen EF, Colvard DS, Berg NJ et al (1988) Evidence of estrogen receptors in normal human osteoblast-like cells. Science 241(4861):84–86

Ernst M, Heath JK, Rodan GA (1989) Estradiol effects on proliferation, messenger ribonucleic acid for collagen and insulin-like growth factor-I, and parathyroid hormone-stimulated adenylate cyclase activity in osteoblastic cells from calvariae and long bones. Endocrinology 125 (2):825–833

Estrada K, Styrkarsdottir U, Evangelou E et al (2012) Genome-wide meta-analysis identifies 56 bone mineral density loci and reveals 14 loci associated with risk of fracture. Nat Genet 44 (5):491–501

Etienne MC, Fischel JL, Milano G et al (1990) Steroid receptors in human osteoblast-like cells. Eur J Cancer 26(7):807–810

Ettinger B, Black DM, Mitlak BH et al (1999) Reduction of vertebral fracture risk in postmenopausal women with osteoporosis treated with raloxifene: results from a 3-year randomized clinical trial. JAMA 282(7):637–645

Falahati-Nini A, Riggs BL, Atkinson EJ et al (2000) Relative contributions of testosterone and estrogen in regulating bone resorption and formation in normal elderly men. J Clin Invest 106 (12):1553–1560

Fang D, Yang H, Lin J et al (2015) 17β-estradiol regulates cell proliferation, colony formation, migration, invasion and promotes apoptosis by upregulating miR-9 and thus degrades MALAT-1 in osteosarcoma cell MG-63 in an estrogen receptor-independent manner. Biochem Biophys Res Commun 457(4):500–506

Fiorelli G, Gori F, Petilli M et al (1995) Functional estrogen receptors in a human preosteoclastic cell line. Proc Natl Acad Sci U S A 92(7):2672–2676

Fisher B, Costantino JP, Wickerham DL et al (1998) Tamoxifen for prevention of breast cancer: report of the National Surgical Adjuvant Breast and Bowel Project P-1 study. J Natl Cancer Inst 90(18):1371–1388

Frenkel B, Hong A, Baniwal SK et al (2010) Regulation of adult bone turnover by sex steroids. J Cell Physiol 224(2):305–310

Fu J, Hao L, Tian Y et al (2018) miR-199a-3p is involved in estrogen-mediated autophagy through the IGF-1/mTOR pathway in osteocyte-like MLO-Y4 cells. J Cell Physiol 233(3):2292–2303

Fujita K, Roforth MM, Demaray S et al (2014) Effects of estrogen on bone mRNA levels of sclerostin and other genes relevant to bone metabolism in postmenopausal women. J Clin Endocrinol Metab 99(1):E81–E88

Gagliano-Jucá T, Basaria S (2019) Testosterone replacement therapy and cardiovascular risk. Nat Rev Cardiol 16(9):555–574

Galea GL, Meakin LB, Sugiyama T et al (2013) Estrogen receptor α mediates proliferation of osteoblastic cells stimulated by estrogen and mechanical strain, but their acute down-regulation of the Wnt antagonist Sost is mediated by estrogen receptor β. J Biol Chem 288(13):9035–9048

Garcia AJ, Tom C, Guemes M et al (2013) ERα signaling regulates MMP3 expression to induce FasL cleavage and osteoclast apoptosis. J Bone Miner Res 28(2):283–290

Girasole G, Jilka RL, Passeri G et al (1992) 17 beta-estradiol inhibits interleukin-6 production by bone marrow-derived stromal cells and osteoblasts in vitro: a potential mechanism for the antiosteoporotic effect of estrogens. J Clin Invest 89(3):883–891

Girasole G, Giuliani N, Modena AB et al (1999) Oestrogens prevent the increase of human serum soluble interleukin-6 receptor induced by ovariectomy in vivo and decrease its release in human osteoblastic cells in vitro. Clin Endocrinol (Oxf) 51(6):801–807

Glait-Santar C, Benayahu D (2012) Regulation of SVEP1 gene expression by 17β-estradiol and TNFα in pre-osteoblastic and mammary adenocarcinoma cells. J Steroid Biochem Mol Biol 130 (1–2):36–44

Grandien K, Bäckdahl M, Ljunggren O et al (1995) Estrogen target tissue determines alternative promoter utilization of the human estrogen receptor gene in osteoblasts and tumor cell lines. Endocrinology 136(5):2223–2229

Gui Y, Duan Z, Qiu X et al (2016) Multifarious effects of 17-β-estradiol on apolipoprotein E receptors gene expression during osteoblast differentiation in vitro. Biosci Trends 10(1):54–66

Heldring N, Pike A, Andersson S et al (2007) Estrogen receptors: how do they signal and what are their targets. Physiol Rev 87(3):905–931

Hewitt SC, Korach KS (2018) Estrogen receptors: new directions in the new millennium. Endocr Rev 39:664–675

Hofbauer LC, Khosla S, Dunstan CR et al (1999) Estrogen stimulates gene expression and protein production of osteoprotegerin in human osteoblastic cells. Endocrinology 140(9):4367–4370

Huang CK, Lai KP, Luo J et al (2013) Loss of androgen receptor promotes adipogenesis but suppresses osteogenesis in bone marrow stromal cells. Stem Cell Res 11(2):938–950

Ikegami A, Inoue S, Hosoi T et al (1993) Immunohistochemical detection and northern blot analysis of estrogen receptor in osteoblastic cells. J Bone Miner Res 8(9):1103–1109

Imai Y, Kondoh S, Kouzmenko A et al (2009a) Regulation of bone metabolism by nuclear receptors. Mol Cell Endocrinol 310(1–2):3–10

Imai Y, Youn MY, Kondoh S et al (2009b) Estrogens maintain bone mass by regulating expression of genes controlling function and life span in mature osteoclasts. Ann N Y Acad Sci 1173(Suppl 1):E31–E39

Jia J, Zhou H, Zeng X et al (2017) Estrogen stimulates osteoprotegerin expression via the suppression of miR-145 expression in MG-63 cells. Mol Med Rep 15(4):1539–1546

Kaplan AL, Hu JC, Morgentaler A et al (2016) Testosterone therapy in men with prostate cancer. Eur Urol 69(5):894–903

Kassem M, Harris SA, Spelsberg TC et al (1996) Estrogen inhibits interleukin-6 production and gene expression in a human osteoblastic cell line with high levels of estrogen receptors. J Bone Miner Res 11(2):193–199

Katznelson L, Finkelstein JS, Schoenfeld DA et al (1996) Increase in bone density and lean body mass during testosterone administration in men with acquired hypogonadism. J Clin Endocrinol Metab 81(12):4358–4365

Kearbey JD, Gao W, Narayanan R et al (2007) Selective androgen receptor modulator (SARM) treatment prevents bone loss and reduces body fat in ovariectomized rats. Pham Res 24 (2):328–335

Khalid AB, Krum SA (2016) Estrogen receptors alpha and beta in bone. Bone 87:130–135

Khosla S (2010) Update in male osteoporosis. J Clin Endocrinol Metab 95(1):3–10

Khosla S, Melton LJ, Atkinson EJ et al (1998) Relationship of serum sex steroid levels and bone turnover markers with bone mineral density in men and women: a key role for bioavailable estrogen. J Clin Endocrinol Metab 83(7):2266–2274

Khosla S, Melton LJ, Atkinson EJ et al (2001) Relationship of serum sex steroid levels to longitudinal changes in bone density in young versus elderly men. J Clin Endocrinol Metab 86(8):3555–3561

Komm BS, Terpening CM, Benz DJ et al (1988) Estrogen binding, receptor mRNA, and biologic response in osteoblast-like osteosarcoma cells. Science 241(4861):81–84

Kondoh S, Inoue K, Igarashi K et al (2014) Estrogen receptor α in osteocytes regulates trabecular bone formation in female mice. Bone 60:68–77

Krum SA (2011) Direct transcriptional targets of sex steroid hormones in bone. J Cell Biochem 112 (2):401–408

Krum SA, Miranda-Carboni GA, Hauschka PV et al (2008a) Estrogen protects bone by inducing Fas ligand in osteoblasts to regulate osteoclast survival. EMBO J 27(3):535–545

Krum SA, Miranda-Carboni GA, Lupien M et al (2008b) Unique ERalpha cistromes control cell type-specific gene regulation. Mol Endocrinol 22(11):2393–2406

Kuiper GG, Enmark E, Pelto-Huikko M et al (1996) Cloning of a novel receptor expressed in rat prostate and ovary. Proc Natl Acad Sci U S A 93(12):5925–5930

Kurihara N, Civin C, Roodman GD (1991) Osteotropic factor responsiveness of highly purified populations of early and late precursors for human multinucleated cells expressing the osteoclast phenotype. J Bone Miner Res 6(3):257–261

Lakatos P, Foldes J, Horvath C et al (1997) Serum interleukin-6 and bone metabolism in patients with thyroid function disorders. J Clin Endocrinol Metab 82(1):78–81

Lambertini E, Tavanti E, Torreggiani E et al (2008) ERalpha and AP-1 interact in vivo with a specific sequence of the F promoter of the human ERalpha gene in osteoblasts. J Cell Physiol 216(1):101–110

Lambertini E, Penolazzi L, Angelozzi M et al (2017) The expression of cystathionine gamma-lyase is regulated by estrogen receptor alpha in human osteoblasts. Oncotarget 8(60):101686–101696

Lazzaro L, Tonkin BA, Poulton IJ et al (2018) IL-6 trans-signaling mediates trabecular, but not cortical, bone loss after ovariectomy. Bone 112:120–127

Lin PI, Tai YT, Chan WP et al (2017) Estrogen/ERα signaling axis participates in osteoblast maturation via upregulating chromosomal and mitochondrial complex gene expressions. Oncotarget 9(1):1169–1186

Lindberg MK, Erlandsson M, Alatalo SL et al (2001) Estrogen receptor. J Endocrinol 171 (3):425–433

Little RD, Carulli JP, Del Mastro RG et al (2002) A mutation in the LDL receptor-related protein 5 gene results in the autosomal dominant high-bone-mass trait. Am J Hum Genet 70(1):11–19

Lonard DM, Tsai SY, O'Malley BW (2004) Selective estrogen receptor modulators 4-hydroxytamoxifen and raloxifene impact the stability and function of SRC-1 and SRC-3 coactivator proteins. Mol Cell Biol 24(1):14–24

Lu S, Sun C, Miao C et al (2017) ERβ compensates for the absence of ERα function to promote osteoblast viability by inhibition of SOST signaling. Exp Ther Med 14(4):3387–3392

Määttä JA, Büki KG, Gu G et al (2013a) Inactivation of estrogen receptor α in bone-forming cells induces bone loss in female mice. FASEB J 27(2):478–488

Määttä JA, Büki KG, Ivaska KK et al (2013b) Inactivation of the androgen receptor in bone-forming cells leads to trabecular bone loss in adult female mice. Bonekey Rep 2:440
Manalogas SC, O'Brien CA, Almeida M (2013) The role of estrogen and androgen receptors in bone health and disease. Nat Rev Endocrinol 9(12):699–712
Manson JE, Chlebowski RT, Stefanick ML et al (2013) Menopausal hormone therapy and health outcomes during the intervention and extended poststopping phases of the Women's Health Initiative randomized trials. JAMA 310(13):1353–1368
Martin-Millan M, Almeida M, Ambrogini E et al (2010) The estrogen receptor-alpha in osteoclasts mediates the protective effects of estrogens on cancellous but not cortical bone. Mol Endocrinol 24(2):323–334
Masuyama A, Ouchi Y, Sato F et al (1992) Characteristics of steroid hormone receptors in cultured MC3T3-E1 osteoblastic cells and effect of steroid hormones on cell proliferation. Calcif Tissue Int 51(5):376–381
McCarthy TL, Centrella M (2015) Androgen receptor activation integrates complex transcriptional effects in osteoblasts, involving the growth factors TGF-β and IGF-I, and transcription factor C/EBPδ. Gene 573(1):129–140
McDevitt MA, Glidewell-Kenney C, Jimenez MA et al (2008) New insights into the classical and non-classical actions of estrogen: evidence from estrogen receptor knock-out and knock-in mice. Mol Cell Endocrinol 290(1–2):24–30
Melville K, Kelly NH, Khan SA et al (2014) Female mice lacking estrogen receptor-alpha in osteoblasts have compromised bone mass and strength. J Bone Miner Res 29(2):370–379
Melville KM, Kelly NH, Surita G et al (2015) Effects of deletion of ERα in osteoblast-lineage cells on bone mass and adaptation to mechanical loading differ in female and male mice. J Bone Miner Res 30(8):1468–1480
Michael H, Härkönen PL, Väänänen HK et al (2005) Estrogen and testosterone use different cellular pathways to inhibit osteoclastogenesis and bone resorption. J Bone Miner Res 20(12):2224–2232
Miedlich SU, Karamooz N, Hammes SR (2016) Aromatase deficiency in a male patient-case report and review of the literature. Bone 93:181–186
Miki Y, Suzuki T, Hatori M et al (2007) Effects of aromatase inhibitors on human osteoblast and osteoblast-like cells: a possible androgenic bone protective effects induced by exemestane. Bone 40(4):876–887
Miranda-Carboni GA, Guemes M, Bailey S et al (2011) GATA4 regulates estrogen receptor-alpha-mediated osteoblast transcription. Mol Endocrinol 25(7):1126–1136
Mirza FS, Padhi ID, Raisz LG et al (2010) Serum sclerostin levels negatively correlate with parathyroid hormone levels and free estrogen index in postmenopausal women. J Clin Endocrinol Metab 95(4):1991–1997
Mizuno Y, Hosoi T, Inoue S et al (1994) Immunocytochemical identification of androgen receptor in mouse osteoclast-like multinucleated cells. Calcif Tissue Int 54(4):325–326
Mödder UI, Rudnik V, Liu G et al (2012) A DNA binding mutation in estrogen receptor-α leads to suppression of Wnt signaling via β-catenin destabilization in osteoblasts. J Cell Biochem 113 (7):2248–2255
Monroe DG, Getz BJ, Johnsen SA et al (2003) Estrogen receptor isoform-specific regulation of endogenous gene expression in human osteoblastic cell lines expressing either ERalpha or ERbeta. J Cell Biochem 90(2):315–326
Morishima A, Grumbach MM, Simpson ER et al (1995) Aromatase deficiency in male and female siblings caused by a novel mutation and the physiological role of estrogens. J Clin Endocrinol Metab 80(12):3689–3698
Mosselman S, Polman J, Dijkema R (1996) ERβ: identification and characterization of a novel human estrogen receptor. FEBS Lett 392(1):49–53
Nakamura T, Imai Y, Matsumoto T et al (2007) Estrogen prevents bone loss via estrogen receptor alpha and induction of Fas ligand in osteoclasts. Cell 130(5):811–823
Nicks KM, Fowler TW, Gaddy D (2010) Reproductive hormones and bone. Curr Osteoporos Rep 8 (2):60–67

Nicks KM, Fujita K, Fraser D et al (2016) Deletion of estrogen receptor beta in osteoprogenitor cells increases trabecular but not cortical bone mass in female mice. J Bone Miner Res 31(3):606–614
Niziolek PJ, Bullock W, Warman ML et al (2015) Missense mutations in LRP5 associated with high bone mass protect the mouse skeleton from disuse-and ovariectomy-induced osteopenia. PLoS One 10(11):e0140775
Notini AJ, McManus JF, Moore A et al (2007) Osteoblast deletion of exon 3 of the androgen receptor gene results in trabecular bone loss in adult male mice. J Bone Miner Res 22 (3):347–356
Ohashi T, Kusuhara S, Ishida K (1991) Estrogen target cells during the early stage of medullary bone osteogenesis: immunohistochemical detection of estrogen receptors in osteogenic cells of estrogen-treated male Japanese quail. Calcif Tissue Int 49(2):124–127
Orwoll E, Ettinger M, Weiss S et al (2000) Alendronate for the treatment of osteoporosis in men. N Engl J Med 343(9):604–610
Oursler MJ, Osdoby P, Pyfferoen J et al (1991) Avian osteoclasts as estrogen target cells. Proc Natl Acad Sci U S A 88(15):6613–6617
Oursler MJ, Pederson L, Fitzpatrick L et al (1994) Human giant cell tumors of the bone (osteoclastomas) are estrogen target cells. Proc Natl Acad Sci U S A 91(12):5227–5231
Pacifici R, Rifas L, McCracken R et al (1989) Ovarian steroid treatment blocks a postmenopausal increase in blood monocyte interleukin 1 release. Proc Natl Acad Sci U S A 86(7):2398–2402
Pacifici R, Brown C, Puscheck E et al (1991) Effect of surgical menopause and estrogen replacement on cytokine release from human blood mononuclear cells. Proc Natl Acad Sci U S A 88 (12):5134–5138
Paech K, Webb P, Kuiper GG et al (1997) Differential ligand activation of estrogen receptors ERα and ERβ at AP1 sites. Science 277(5331):1508–1510
Pederson L, Kremer M, Judd J et al (1999) Androgens regulate bone resorption activity of isolated osteoclasts in vitro. Proc Natl Acad Sci U S A 96(2):505–510
Pensler JM, Radosevich JA, Higbee R et al (1990) Osteoclasts isolated from membranous bone in children exhibit nuclear estrogen and progesterone receptors. J Bone Miner Res 5(8):797–802
Pike ACW, Brzozowski AM, Hubbard RE et al (1999) Structure of the ligand-binding domain of oestrogen receptor beta in the presence of a partial agonist and a full antagonist. EMBO J 18 (7):4608–4618
Poli V, Balena R, Fattori E et al (1994) Interleukin-6 deficient mice are protected from bone loss caused by estrogen depletion. EMBO J 13(5):1189–1196
Qaseem A, Forciea MA, McLean RM et al (2017) Treatment of low bone density or osteoporosis to prevent fractures in men and women: a clinical practice guideline update from the American College of Physicians. Ann Intern Med 166(11):818–839
Riggs BL, Melton LJ, Robb RA et al (2004) Population-based study of age and sex differences in bone volumetric density, size, geometry, and structure at different skeletal sites. J Bone Miner Res 19(12):1945–1954
Riggs BL, ML J, Robb RA et al (2008) A population-based assessment of rates of bone loss at multiple skeletal sites: evidence for substantial trabecular bone loss in young adult women and men. J Bone Miner Res 23(2):205–214
Roetzer KM, Uyanik G, Brehm A et al (2018) Novel familial mutation of LRP5 causing high bone mass: genetic analysis, clinical presentation, and characterization of bone matrix mineralization. Bone 107:154–160
Roforth MM, Atkinson EJ, Levin ER et al (2014) Dissection of estrogen receptor alpha signaling pathways in osteoblasts using RNA-sequencing. PLoS One 9(4):e95987
Rooney AM, van der Meulen MCH (2017) Mouse models to evaluate the role of estrogen receptor α in skeletal maintenance and adaptation. Ann N Y Acad Sci 1410(1):85–92
Rossouw JE, Anderson GL, Prentice RL et al (2002) Risks and benefits of estrogen plus progestin in healthy postmenopausal women: principal results from the Women's Health Initiative randomized controlled trial. JAMA 288(3):321–333

Russell PK, Clarke MV, Skinner JP et al (2012) Identification of gene pathways altered by deletion of the androgen receptor specifically in mineralizing osteoblasts and osteocytes in mice. J Mol Endocrinol 49(1):1–10

Saika M, Inoue D, Kido S et al (2001) 17β-estradiol stimulates expression of osteoprotegerin by a mouse stromal cell line, ST-2, via estrogen receptor-α. Endocrinology 142(6):2205–2212

Sawka AM, Papaioannou A, Adachi JD et al (2005) Does alendronate reduce the risk of fracture in men? A meta-analysis incorporating prior knowledge of anti-fracture efficacy in women. BMC Musculoskelet Disord 6(39)

Scheller J, Garbers C, Rose-John S (2014) Interleukin-6: from basic biology to selective blockade of pro-inflammatory activities. Semin Immunol 26(1):2–12

Seitz S, Keller J, Schilling AF et al (2012) Pharmacological estrogen administration causes a FSH-independent osteo-anabolic effect requiring ER alpha in osteoblasts. PLoS One 7(11): e50301

Shang Y, Brown M (2002) Molecular determinants for the tissue specificity of SERMs. Science 295 (5564):2465–2468

Shang ZZ, Li X, Sun HQ et al (2014) Differentially expressed genes and signalling pathways are involved in mouse osteoblast-like MC3T3-E1 cells exposed to 17-β estradiol. Int J Oral Sci 6 (3):142–149

Shao B, Liao L, Yu Y et al (2015) Estrogen preserves Fas ligand levels by inhibiting microRNA-181a in bone marrow-derived mesenchymal stem cells to maintain bone remodeling balance. FASEB J 29(9):3935–3944

Shao B, Fu X, Yu Y et al (2018) Regulatory effects of miRNA-181a on FasL expression in bone marrow mesenchymal stem cells and its effect on CD4+T lymphocyte apoptosis in estrogen deficiency-induced osteoporosis. Mol Med Rep 18(1):920–930

Shiau AK, Barstad D, Loria PM et al (1998) The structural basis of estrogen receptor/coactivator recognition and the antagonism of this interaction by tamoxifen. Cell 95(7):927–937

Silverman SL, Chines AA, Kendler DL et al (2012) Sustained efficacy and safety of bazedoxifene in preventing fractures in postmenopausal women with osteoporosis: results of a 5-year, randomized, placebo-controlled study. Osteoporos Int 23(1):351–363

Sims NA, Clément-Lacroix P, Minet D et al (2003) A functional androgen receptor is not sufficient to allow estradiol to protect bone after gonadectomy in estradiol receptor-deficient mice. J Clin Invest 111(9):1319–1327

Sinnesael M, Claessens F, Laurent M et al (2012) Androgen receptor (AR) in osteocytes is important for the maintenance of male skeletal integrity: evidence from targeted AR disruption in mouse osteocytes. J Bone Miner Res 27(12):2535–2543

Sinnesael M, Jardi F, Deboel L et al (2015) The androgen receptor has no direct antiresorptive actions in mouse osteoclasts. Mol Cell Endocrinol 411:198–206

Slemenda CW, Longcope C, Zhou L et al (1997) Sex steroids and bone mass in older men. Positive associations with serum estrogens and negative associations with androgens. J Clin Invest 100 (7):1755–1759

Smith EP, Boyd J, Frank GR et al (1994) Estrogen resistance caused by a mutation in the estrogen-receptor gene in a man. N Engl J Med 331(16):1056–1061

Stern PH, Tarjan G, Sanders J (1997) Estrogens and bone cells. Progress in the management of the menopause: the proceedings of the 8th international congress on the menopause, Sydney, Australia, November 1996, pp 120–125

Streicher C, Heyny A, Andrukhova O et al (2017) Estrogen regulates bone turnover by targeting RANKL expression in bone lining cells. Sci Rep 7(1):6460

Sugatani T, Hruska KA (2013) Down-regulation of miR-21 biogenesis by estrogen action contributes to osteoclastic apoptosis. J Cell Biochem 114(6):1217–1222

Sulak PJ (1997) Endometrial cancer and hormone replacement therapy. Appropriate use of progestins to oppose endogenous and exogenous estrogen. Endocrinol Metab Clin North Am 26(2):399–412

Syed FA, Mödder UI, Roforth M et al (2010) Effects of chronic estrogen treatment on modulating age-related bone loss in female mice. J Bone Miner Res 25(11):2438–2446
Syed FA, Fraser DG, Monroe DG et al (2011) Distinct effects of loss of classical estrogen receptor signaling versus complete deletion of estrogen receptor alpha on bone. Bone 49(2):208–216
Szulc P, Munoz F, Claustrat B et al (2001) Bioavailable estradiol may be an important determinant of osteoporosis in men: the MINOS study. J Clin Endocrinol Metab 86(1):192–199
Takai H, Matsumura H, Matsui S et al (2014) Unliganded estrogen receptor α stimulates bone sialoprotein gene expression. Gene 539(1):50–57
Tan MH, Li J, Xu HE et al (2015) Androgen receptor: structure, role in prostate cancer and drug discovery. Acta Pharmacol Sin 36(1):3–23
Tremblay GB, Tremblay A, Copeland NG et al (1997) Cloning, chromosomal localization, and functional analysis of the murine estrogen receptor β. Mol Endocrinol 11(3):353–365
Ucer S, Iyer S, Bartell SM et al (2015) The effects of androgens on murine cortical bone do not require AR or ERα signaling in osteoblasts and osteoclasts. J Bone Miner Res 30(7):1138–1149
Vanderschueren D, Vandenput L, Boonen S et al (2004) Androgens and bone. Endocr Rev 25 (3):389–425
Vanderschueren D, Laurent MR, Claessens F et al (2014) Sex steroid actions in male bone. Endocr Rev 35(6):906–960
Villiers DTJ, Chines AA, Palacios S et al (2011) Safety and tolerability of bazedoxifene in postmenopausal women with osteoporosis: results of a 5-year, randomized, placebo-controlled phase 3 trial. Osteoporos Int 22(2):567–576
Wang J, Stern PH (2011) Sex-specific effects of estrogen and androgen on gene expression in human monocyte-derived osteoclasts. J Cell Biochem 112(12):3714–3721
Wang YD, Tao MF, Wang L et al (2012) Selective regulation of osteoblastic OPG and RANKL by dehydroepiandrosterone through activation of the estrogen receptor β-mediated MAPK signaling pathway. Horm Metab Res 44(7):494–500
Wang YX, Li M, Zhang HQ et al (2016) Opposite function of ERα and ERβ in controlling 17β-estradiol-mediated osteogenesis in osteoblasts. Arch Med Res 47(4):255–261
Wang FS, Wu RW, Lain WS et al (2018) Sclerostin vaccination mitigates estrogen deficiency induction of bone mass loss and microstructure deterioration. Bone 112:24–34
Westerlind KC, Sarkar G, Bolander ME et al (1995) Estrogen receptor mRNA is expressed in vivo in rat calvarial periosteum. Steroids 60(7):484–487
Windahl SH, Börjesson AE, Farman HH et al (2013) Estrogen receptor-α in osteocytes is important for trabecular bone formation in male mice. Proc Natl Acad Sci U S A 110(6):2294–2299
Wiren KM, Zhang XW, Toombs AR et al (2004) Targeted overexpression of androgen receptor in osteoblasts: unexpected complex bone phenotype in growing animals. Endocrinology 145 (7):3507–3522
Wiren KM, Semirale AA, Zhang XW et al (2008) Targeting of androgen receptor in bone reveals a lack of androgen anabolic action and inhibition of osteogenesis: a model for compartment-specific androgen action in the skeleton. Bone 43(3):440–451
Wiren KM, Semirale AA, Hashimoto JG et al (2010) Signaling pathways implicated in androgen regulation of endocortical bone. Bone 46(3):710–723
Yang YH, Chen K, Li B et al (2013) Estradiol inhibits osteoblast apoptosis via promotion of autophagy through the ER-ERK-mTOR pathway. Apoptosis 18(11):1363–1375
Yin X, Wang X, Hu X et al (2015) ERβ induces the differentiation of cultured osteoblasts by both Wnt/β-catenin signaling pathway and estrogen signaling pathways. Exp Cell Res 335 (1):107–114
Zhang RW, Supowit SC, Xu X et al (1995) Expression of selected osteogenic markers in the fibroblast-like cells of rat marrow stroma. Calcif Tissue Int 56(4):283–291

Thyroid Hormones, Glucocorticoids, Insulin, and Bone

Peter Lakatos, Balazs Szili, Bence Bakos, Istvan Takacs, Zsuzsanna Putz, and Ildiko Istenes

Contents

Abstract

Several endocrine systems have important effects on bone tissue. Thyroid hormones are essential for normal growth and development. Excess of these hormones will result in clinically significant changes that may require intervention. Glucocorticoids also have a marked effect on bone metabolism by several pathways. Their endogenous or exogenous excess will induce pathological processes that might elevate the risk of fractures. Insulin and the carbohydrate metabolism elicit a physiological effect on bone; however, the lack of insulin (type 1 diabetes) or insulin resistance (type 2 diabetes) have deleterious influence on bone tissue.

P. Lakatos (✉) · B. Szili · B. Bakos · I. Takacs · Z. Putz · I. Istenes
1st Department of Medicine, Semmelweis University, Budapest, Hungary
e-mail: lakatos.peter@med.semmelweis-univ.hu

P. H. Stern (ed.), *Bone Regulators and Osteoporosis Therapy*,
Handbook of Experimental Pharmacology 262, https://doi.org/10.1007/164_2019_314

Keywords
Glucocorticoid-induced osteoporosis · Glucocorticoids · Hyperthyroidism · Hypothyroidism · Insulin · Insulin resistance · Suppression of TSH · Thyroid hormones · Type 1 diabetes · Type 2 diabetes

1 Thyroid Hormones and Bone

Balazs Szili, Bence Bakos, Istvan Takacs, and Peter Lakatos

1.1 Cellular and Nuclear Actions of Thyroid Hormones

Thyroid hormones are produced in the thyroid gland mainly as thyroxin (T_4). Triiodothyronine (T_3) is the biologically active form of thyroid hormone and is predominantly synthesized from T_4 by deiodinases in peripheral tissues. Based on their structure, thyroid hormones are poorly soluble in water and therefore are mostly bound to carrier proteins: thyroxine-binding globulin (TBG), albumin, and transthyretin (Tata 1961). Less than 1% is the free, biologically available form. Thyroid hormones are transported to cells via specific carrier-mediated mechanisms (thyroid hormone transporters) (Visser 2000) and act on their specific nuclear receptors.

Thyroid hormone receptors (TR) are members of the nuclear receptor superfamily (Germain et al. 2006). The common features of these receptors are (1) having a DNA-binding domain containing two zinc fingers, (2) the A/B domain on amino-terminal that is responsible for transcription modulation, and (3) a ligand-binding domain on the carboxy-terminal that is also responsible for receptor dimerization and interplay with corepressors and coactivators. Thyroid hormone binding changes the receptor conformation and allows the activation or repression of particular genes. TR can form heterodimers with other nuclear receptor members (e.g., retinoid X receptor, vitamin D receptor) which can modulate its effects (Williams et al. 1994).

The activated TRs bind to specific regions of the genome (thyroid response elements – TRE) which are located in the promoter regions of the target genes and increase specific gene transcription. Thyroid hormone receptors can also function as repressors on the TRE in the absence of T_3 binding. During this process other nuclear corepressors interact with TRs (e.g., silencing mediator of retinoid and thyroid receptor, Sin3, histone deacetylase) (Shibata et al. 1997).

There are two classes of thyroid receptors, TRα and TRβ, which are products of two genes located on different chromosomes (Chassande et al. 1997; Williams 2000). Both have distinct isoforms with different T_3 binding affinities (Milne et al. 1999). Both TRα and TRβ mRNA are found in skeletal cells and cell lines (Williams et al. 1994). Basically bone effects of thyroid hormones are mediated through the TRα1 isotype (in bone cells its concentration is tenfold higher than TRβ) (O'Shea et al. 2003), but there is also evidence for a role of TRβ (Monfoulet et al. 2011).

In developing bone, thyroid receptors are found in the reserve zone and in proliferating chondrocytes (mature chondrocytes do not express TRs) (Ballock et al. 1999; Stevens et al. 2000). Among bone cells, mainly osteoblasts show TR expression. The direct action of thyroid hormones and the presence of thyroid receptors in osteocytes and osteoclasts are still unclear.

In vitro in osteoblasts, the response to T_3 varies by the applied dose and exposure time and also depends on the species and cell line type (Stevens et al. 2000). Results predominantly show osteoblast proliferation with thyroid hormone treatment (Luegmayr et al. 2000). Adding T_3 to cell cultures increases the levels of osteoblast activity markers, e.g., osteocalcin, alkaline phosphatase, type I collagen, and osteoprotegerin (Banovac and Koren 2000; Varga et al. 2004, 2010). IGF-I levels are also higher when cells are treated with T_3. IGF-I has remarkable osteoanabolic effects, increasing cell proliferation and protein synthesis. IGF-I is a key secondary mediator of T_3 bone anabolic effects (Huang et al. 2000). This process can be modulated by IGF-binding proteins; however the exact mechanism needs further investigation.

Studies that investigated whether bone resorption due to T_3 treatment is mediated by a direct osteoclast effect or secondary to alterations of osteoblast or osteocyte function – or even interaction with bone marrow cells – are controversial (Klaushofer et al. 1995). In vitro isolated osteoclasts do not respond to T_3 therapy (Allain et al. 1992).

The resorptive effect of T_3 is slower and weaker than the effect of parathyroid hormone treatment (Kawaguchi et al. 1994). The interactions of PTH and T_3 with TGFβ are also dissimilar: the early effect of T3 is inhibited by TGFβ, while the later response is amplified (Lakatos and Stern 1992). In contrast TGFβ upregulates PTH expression, and in turn, PTH increases the concentration of TGFβ in rat bone. Osteoclast activation in vivo might also be mediated by T_3-induced interleukin-6 and prostaglandin synthesis (Lakatos et al. 1997).

Remodeling is crucial in maintaining bone health. The majority of in vitro studies have focused on the influence on either resorption or formation. An interesting model was designed for evaluating the effects on remodeling of T_3 in mouse fetal radius and ulna (Soskolne et al. 1990). At lower concentrations, T_3 caused increased growth and positive calcium balance. When higher doses were applied, calcium loss was observed.

Thyroid hormones are essential for bone development and ossification of epiphyseal cartilage. T_3 interacts with several other metabolic pathways during growth plate development. T_3 inhibits chondrocyte proliferation, but it causes their hypertrophic differentiation (Miura et al. 2002). There are several mouse models with or without nonfunctioning TSH receptors (TRα1- and TRα2-deficient mice; all TRα- and TRβ1- and TRβ2-deficient mice) (Gauthier et al. 1999; Bassett et al. 2006). Observations on these models suggest that the cartilage matrix and especially heparan sulfate and proteoglycans are important mediators of the skeletal response to thyroid hormones. These studies have also shown that TRβ is essential for normal bone development. In Pax8 knockout mice, follicular cells are lacking because this transcription factor is crucial in follicular cell development (Mansouri et al. 1998). The common feature in these intrauterine hypothyroid mice is growth retardation, evidence for impaired mineralization and bone formation.

1.2 Thyroid Hormone Resistance Syndromes

TRβ2 is the predominant isoform among thyrotropin receptors in hypothalamo-hypophyseal negative feedback loops. The syndrome of the resistance for thyroid hormone (RTH) is the result of mutations in this receptor (Refetoff et al. 1993). It leads to elevated TSH and T3 levels. Similar mutations of TRα gene causes RTHα which was first described in 2012 (Bochukova et al. 2012).

Studying patients with RTHβ has contributed to understanding the skeletal effects of thyroid hormones. The severity of the symptoms varies among the patients. Common features are short stature, goiter, cognitive deficit, and tachycardia. Bone manifestations include delayed bone age and stippled epiphyses like in hypothyroidism. In some other patients, advanced bone age, premature epiphyseal closure, and early chondrocyte maturation were observed as in hyperthyroidism (Behr et al. 1997). The phenotypic differences might be explained by the distinct target sites where resistance is manifested.

1.3 Bone Disease in Thyroid Dysfunction Syndromes

1.3.1 Hyperthyroidism

One of the first bone consequences of untreated overt hyperthyroidism was reported by von Recklinghausen in 1891 in a young female who died from hyperthyroidism: her bones looked like "worm-eaten" (Recklinghausen 1891). Since the availability of radioiodine treatment and thyrostatic drugs, this severe appearance of thyroid bone disease is a rarity in the literature. Nowadays, owing to an aging population, subclinical and overt hyperthyroidism have become more frequent. Bone mineral density measurements show bone loss, and epidemiological data show increased fracture risk among these patients.

Hyperthyroidism in children and adolescents is a rare entity. In infants, thyroid hormone excess causes accelerated bone development. Advanced bone age can be observed in these patients, and the early closure of growth plates can lead to permanent short stature and even craniosynostosis (Polak et al. 2006).

Bone loss can be observed in every adult patient with overt hyperthyroidism (Vestergaard and Mosekilde 2003). The extent of the bone mineral density reduction can reach 20%. In a large Danish cohort, more than 9,000 patients with suppressed TSH level were observed for a median of 7.5 years (Abrahamsen et al. 2014). A single decreased TSH level increased the risk of hip fractures (HR 1.16). The longer duration of thyrotoxicosis increased the risk, every 6 months with subnormal TSH raised the hip fracture risk by a factor of 1.07 (Abrahamsen et al. 2014). The increased fracture risk was more pronounced in postmenopausal women.

In subclinical hyperthyroidism, the free T_4 and T_3 levels are within the reference range, while the TSH level is suppressed below the lower limit of the normal. This can occur with endogenous thyroid hormone overproduction or excessive thyroid hormone therapy. Clinically manifested bone disease is not a trait of subclinical hyperthyroidism, but observational data provided strong evidence of unfavorable osteological outcomes. The extent of the effect varies in the different studies, but it is difficult to compare them because of the different anatomical locations of BMD measurements, different methodologies, and the diverse populations. The most robust evidence on BMD reduction and increased fracture risk in subclinical hyperthyroidism is available among postmenopausal women. The relative risk for fractures varied between 1.25 and 5 in the different studies (Vadiveloo et al. 2011; Lee et al. 2010).

The cause of the bone loss in hyperthyroidism is the increased turnover and the subsequent imbalance between bone resorption and formation. Histomorphometrical analysis showed an increased number of osteoclasts, larger resorbing areas, and thinning of trabecular bone (Mosekilde and Melsen 1978). In contrast, in young adult patients who were treated with suppressive doses of levothyroxine because of thyroid cancer, the bone HR-pQCT microstructure parameters did not differ significantly from controls (Mendonca Monteiro de Barros et al. 2016). Bone biochemical markers provided further evidence for the increased turnover. Both resorption (collagen-derived pyridinium cross-links and urinary N-telopeptide of type I collagen) (Harvey et al. 1991; Mora et al. 1999; Loviselli et al. 1997) and formation markers (bone specific alkaline phosphatase, osteocalcin) are usually elevated (Loviselli et al. 1997). The increased level of turnover markers (mainly formation markers) remains high for months after treating hyperthyroidism (Pantazi and Papapetrou 2000).

Mineral metabolism is also altered in hyperthyroidism. The increased resorption leads to hypercalcemia in some patients (Mosekilde et al. 1990). The elevated calcium level suppresses parathyroid hormone (PTH) synthesis that will cause increased renal calcium excretion. Suppressed PTH also reduces the 1.25 $(OH)_2$ D-hormone production (Jastrup et al. 1982). This, with the increased gut motility, leads to decreased intestinal calcium absorption. Due to these, calcium balance is negative, so bone mineral content decreases (Pantazi and Papapetrou 2000).

The reversibility of bone loss in hyperthyroid patients after treatment is inconsistent. Observational data (Vestergaard and Mosekilde 2003; Karga et al. 2004) showed decreased BMD 1 year after recovery from hyperthyroidism. A cross-sectional study showed a higher prevalence of osteoporosis (measured at the ultradistal radius) among women with previous self-reported hyperthyroidism (Svare et al. 2009). Data are heterogeneous with respect to the BMD measurement sites and techniques, the vitamin D status, and calcium supplementation. In general, we can conclude that bone loss is partially irreversible in some patients, but there is a need for prospective studies among recovered hyperthyroid patients with adequate calcium and vitamin D supplementation, standardized BMD measurements, and fracture risk assessment.

1.3.2 Hypothyroidism

In contrast to hyperthyroidism, the skeletal effects of hypothyroidism are less pronounced. Reduced thyroid hormone levels cause decreased bone turnover, delayed bone maturation, and bone age among infants with consequent short stature.

Decreased bone turnover and a reduced remodeling rate can also be observed in adulthood. There are only a few data on bone turnover biochemical tests in hypothyroidism. Those show decreased formation (Lakatos et al. 2000) and resorption (Nakamura et al. 1996) markers. Long-term fracture risk and BMD data are also lacking, as there are no patients who remain untreated after establishing the diagnosis of overt hypothyroidism. Among the scarce data, decreased turnover results in no change or a small increase in BMD (Vestergaard and Mosekilde 2002). Overt or subclinical hypothyroidism and elevated TSH do not seem to increase fracture risk.

2 Glucocorticoids and Bone

Bence Bakos, Balazs Szili, Istvan Takacs, and Peter Lakatos

2.1 Introduction

The detrimental effects of systemic glucocorticoid treatment and endogenous glucocorticoid excess on skeletal tissue have long been established. Though corticosteroid-induced osteoporosis was first described almost 90 years ago, it still remains an everyday clinical challenge. In recent years it became clear that even short courses and/or low doses of exogenous corticosteroids have a harmful effect on bone quality, and the significance of endogenous subclinical hypercortisolism in osteoporosis has also been elucidated. Furthermore the role of the skeleton has been implied in some of the systemic metabolic effects of corticosteroid excess. With the improving understanding of the underlying molecular mechanisms of physiologic and pathologic corticosteroid action, guidelines for alternative immunosuppressive strategies and the prevention and treatment of corticosteroid-induced osteoporosis are constantly evolving.

2.2 Subcellular Actions

Glucocorticoids (GCs) exert their effects primarily via genomic actions through nuclear glucocorticoid receptors (GRs) (Evans 1988). GRα and GRβ are members of the nuclear receptor family and the splice variants of the same gene located on chromosome 5 (Baschant and Tuckermann 2010). Although in recent decades membrane-bound and cytoplasmic receptor mechanisms have been elucidated (Song and Buttgereit 2006), the role of these in bone metabolism and bone pathology is currently unclear.

Most known effects of glucocorticoids on bone and other cells are thought to be primarily mediated by GRα (Hartmann et al. 2016). The unligated receptor is located in the cytoplasm as part of a large multiprotein complex consisting primarily of chaperones, heat-shock proteins, and mitogen-activated protein kinases. Ligand binding results in the translocation of the receptor into the nucleus. Although monomeric actions are also established, this is usually followed by homo- or heterodimerization of the receptors and transactivation and transrepression of target genes containing positive or negative GC response elements, respectively. A large number of interactions with other transcription factors, coactivators, and corepressors further complicate the picture, making, for example, the isolation of specific glucocorticoid actions especially challenging. GRβ is considered to have few if any independent effects on its own. It is primarily thought to alter GRα function by blocking its actions. While only a few details have been elucidated regarding its role and regulation in skeletal tissue, GRβ expression seems to affect individual GC sensitivity as its expression negatively correlates with the effectiveness of corticosteroids, for example, in inflammatory conditions (Honda et al. 2000).

The primary regulator of the magnitude of glucocorticoid effect was previously thought to be their serum concentration. Large differences between the sensitivity of different tissues to GC action however led to the recognition of prereceptor GC metabolism (Chapman et al. 2013). The expression of membrane transport proteins (Nixon et al. 2016), GC-activating and GC-inactivating enzymes, shows large variation between cell types and also changes with disease states. Among these proteins 11β-HSD (11β-hydroxysteroid dehydrogenase) isoforms are of most clinical interest in regard to bone metabolism (Gathercole et al. 2013). The isoform 11β-HSD2 is responsible for the transformation and inactivation of cortisol into cortisone and is mostly expressed in mineralocorticoid-sensitive tissues such as the kidneys and colon where it prevents mineralocorticoid receptor overactivation. 11β-HSD1 catalyzes the reverse reaction and is present in GC target tissues such as the liver, muscle, fat, and skin. Osteoblasts, osteocytes, and to a smaller degree osteoclasts all express 11β-HSD1 (Cooper et al. 2000), while the loss of 11β-HSD1 and the presence of 11β-HSD2 were shown in vitro in malignant bone cell lines (Patel et al. 2012). Systemic inflammation was shown to increase 11β-HSD1 expression in bone cells which is suggested to be part of the defense against inflammation-mediated bone loss. This mechanism however may very well contribute to steroid-induced bone loss in case of systemic GC treatment in inflammatory conditions (Hansen et al. 1996).

2.3 Cellular and In Vitro Effects of Corticosteroids

Glucocorticoids have a marked effect on all cell types in the bone. Osteoblastogenesis, osteoblast function and life span, as well as osteocyte longevity are all disturbed by corticosteroid excess, while osteoclast survival is promoted, resulting in a decline in bone formation rate, decreased trabecular thickness, and a net bone loss over time.

2.3.1 Osteoblasts

Endogenous GCs are necessary for proper osteoblastogenesis (Sher et al. 2004) which they primarily induce via the relative downregulation of PPARγ2 (peroxisome proliferator-activated receptor γ2) and upregulation of Runx2 (runt-related transcription factor 2) in mesenchymal stem cells. This process is mediated, among others, by the proteins FHL2 (four and a half LIM domains protein 2) and GILZ (glucocorticoid-induced leucine zipper) (Hamidouche et al. 2008; Zhang et al. 2008). In vitro GCs enhance commitment and differentiation of mesenchymal precursors toward osteoblasts and also stimulate osteoblast-specific gene expression, although high doses seem to have an inhibitory effect on osteoblast proliferation (Hartmann et al. 2016). While physiological concentrations of GCs are necessary for in vivo osteoblastogenesis, glucocorticoid excess upregulates PPARγ2 and diverts mesenchymal stem cell differentiation from osteoblasts, increasing adipogenesis instead (Shi et al. 2000). This might be the physiological basis behind the accumulation of bone marrow fat accompanying bone loss in patients with long-standing corticosteroid treatment (Sui et al. 2016).

Osteoblast maturation and extracellular matrix production are also impaired in the presence of supraphysiological GC concentrations. Decreased levels of anabolic signaling proteins such as IGF-1 (Bennett et al. 1984; McCarthy et al. 1990), TGFβ (Centrella et al. 1991), and PDGF, bone morphogenetic proteins have been demonstrated (Canalis et al. 2007). The in vitro inhibition of intracellular Wnt signaling and the increased synthesis of paracrine Wnt inhibitors like DKK_1 and sclerostin by osteoblasts and osteocytes have also been reported in response to high-dose GCs (Wang et al. 2008). Decreased expression of key protein components of the extracellular matrix like type I collagen and osteocalcin and increased production of mineralization inhibitors in response to GC treatment have been established.

GCs have a marked proapoptotic effect on osteoblasts and osteocytes which is hypothesized to be the most significant among their pro-osteoporotic actions (Weinstein et al. 1998). Multiple mechanisms are involved, but the most notable mediators of these effects are the Fas/FasL death receptor pathway and the Bim (bcl-2 interacting mediator of cell death) gene, the activation of which results in the downstream activation of caspase-3, caspase-7, and caspase-8 (O'Brien et al. 2004). Knockdown of the Bim gene protects osteoblasts from GC-induced apoptosis in cell cultures (Chen et al. 2014).

2.3.2 Osteoclasts

GCs exert both direct and indirect effects on osteoclasts (Sivagurunathan et al. 2005). Among the former the most well-known are the inhibition of caspase-3-dependent apoptosis and the prolongation of osteoclast survival (Jia et al. 2006). Increased resorption activity and pit formation was also demonstrated in response to GC treatment (Sivagurunathan et al. 2005). In several in vitro studies however, similar to the dual effects seen with osteoblasts, larger doses of GCs exerted an inhibitory effect on both osteoclast differentiation and activation. Indirect effects of glucocorticoids, mediated by osteoblast- and osteocyte-derived secondary agents, seem to play an even more significant role in osteoclast stimulation. Of these, the inhibition of OPG (osteoprotegerin) and stimulation of RANKL synthesis and the increase of the RANKL/OPG ratio are the most well studied (Conaway et al. 2016). Though the levels of some stimulatory cytokines like IL-1 and IL-6 also decrease in response the glucocorticoid treatment, at supraphysiological GC doses this effect cannot compensate for the loss of OPG action (Hofbauer et al. 1999).

2.3.3 Osteocytes

Our view of osteocytes as passive remnants of once metabolically active osteoblasts has changed dramatically in recent decades (Dallas et al. 2013). The crucial role osteocytes play in orchestrating modeling, remodeling, and the intricate response of skeletal tissue to mechanical stress is increasingly understood. Osteocyte-derived paracrine mediators are important regulators of bone remodeling. GC excess as previously mentioned shifts the balance of both RANKL/OPG system and the Wnt pathway toward an increase in resorption. GC treatment also results in decreased fluid flow in the osteocyte canaliculi and an increase in osteocyte lacunar size (Lane et al. 2006). Osteocytic osteolysis, a rapid, but reversible decrease in bone mineral content in the osteocyte-adjacent skeletal tissue, is also demonstrated. Osteocyte apoptosis also seems play a role in GC-induced bone loss, though it is only observed in certain animal models and seems to have a dose-dependent threshold (O'Brien et al. 2004; Jia et al. 2011). As bone loss develops even in its absence, it is by no means necessary for the development of glucocorticoid-induced osteoporosis (GIOP). Osteocyte apoptosis is thought to be potentially irreversible as these long-lived cells are hardly truly replaced. The fact that skeletal effects of GCs diminish rapidly after the discontinuation of treatment questions the hypothesis that osteocyte apoptosis plays a crucial role in GIOP. Some authors suggest that extensive and irreversible osteocyte apoptosis in response to GC treatment results in osteonecrosis (Weinstein et al. 2000).

2.4 Animal Studies and Tissue-Level Effects

The interpretation of results from animal studies investigating GC-induced osteoporosis (GIOP) poses a number of challenges. Although the bone loss with steroid treatment is almost universal, results and the primary mechanism of action is largely influenced by the strain, sex and age of mice, and the type of steroid analogue used (Wood et al. 2018). It is also largely dependent on the dose, the route of administration, and the examined bone site. Most animal studies fail to model the underlying immunological disorder for which steroid treatment in human subjects is given, which also a potentially significant factor in bone loss.

The first study that demonstrated GC-induced osteoblast and osteocyte apoptosis in vivo in the Swiss Webster mouse strain was published more than 20 years ago (Weinstein et al. 1998). Several mouse models have since been generated with tissue-level resistance to GC action in the bone (Rauch et al. 2010). This is generally achieved either by the introduction of 11β-HSD2 (Woitge et al. 2001) expression in osteoblasts that is normally lacking or the selective inactivation of GC receptors in bone cells. These animals while resistant to GC-induced osteoporosis generally show a basal phenotype of slightly impaired bone formation with decreased vertebral density, delayed cranial ossification, and reduced periosteal apposition which underscores the abovementioned physiologic role of GCs in osteoblast formation (Sher et al. 2004). In recent years the potential role of mineralocorticoid receptors (MR) in GC-induced osteoporosis has also been suggested (Beavan et al. 2001). Expression of MRs in mouse osteoblasts has been demonstrated, and the pharmacologic blockade of these receptors was shown to decrease the extent of glucocorticoid-induced osteoporosis (GIOP) (Fumoto et al. 2014).

Mouse models also support the in vitro findings of increased osteoclast activity and survival in response to the initiation of GC treatment (Jia et al. 2006). This effect was also shown to be ameliorated by the knock-in expression of 11β-HSD2 in osteoclasts. With prolonged corticosteroid expression, osteoclast activity decreases, possibly due to the inhibition of differentiation. In one study, deletion of osteoclast GC receptors in dexamethasone-treated mice prevented not only the increased resorption but also some of the expected decrease in bone formation (Kim et al. 2006). Though no underlying pathway is clearly identified, this finding raises the probability that certain effects of GCs on osteoblasts are partly osteoclast-mediated. Other similarly designed studies however have so far failed to replicate this result.

Based on animal studies, the involvement of skeletal tissue in systemic side effects of GCs like insulin resistance has been proposed in recent years (Lee et al. 2007a, b; Brennan-Speranza et al. 2012). The improvement of insulin sensitivity and insulin secretion in response to bone-derived uncarboxylated osteocalcin has been noted (Pi et al. 2016), though the receptor for this protein is still unknown. The levels of serum osteocalcin drop dramatically in response to GC treatment, while the induction of 11β-HSD2 in bone cells has been shown to prevent this reduction. These mice also have a better metabolic profile than their wild-type siblings. Conversely, while 11β-HSD1-deficient mice are protected from GC-induced metabolic dysregulation, those with the isolated deletion of the enzyme from the liver, fat, and muscle tissues are not. This also supports the contribution of other tissue types that are generally not considered, in mediating these effects. Results in humans are scant but favorable (Mazziotti et al. 2014; Yeap et al. 2015). A summary of the effects of glucocorticoids can be seen on Fig. 1.

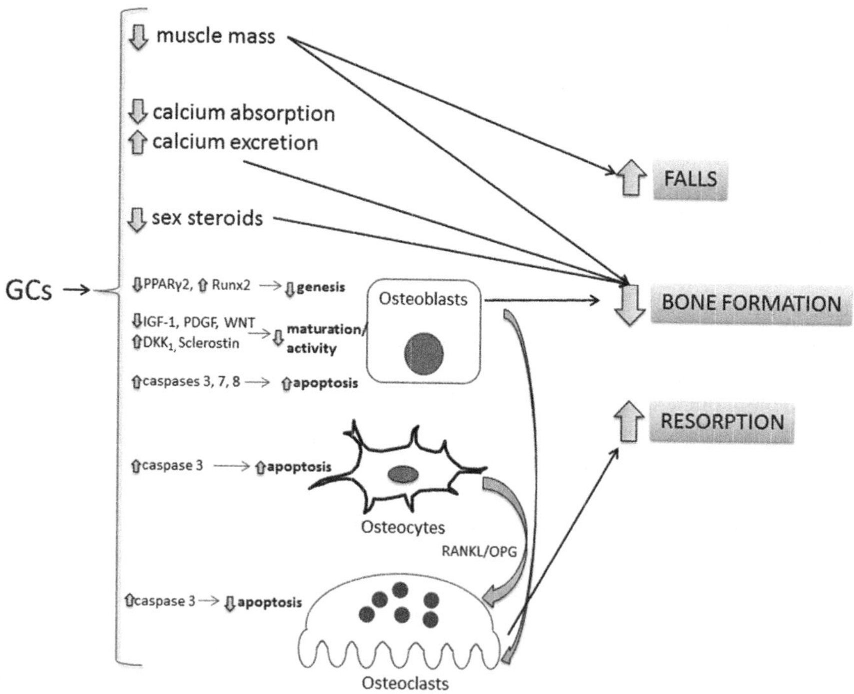

Fig. 1 Effects of glucocorticoids (GCs) on bone

2.5 Clinical Data

While GC-induced osteoporosis has first been described as part of the syndrome of endogenous hypercortisolism (Cushing 1932), with the advent of therapeutic immunosuppression, it is much more often seen as a result of medical treatment. The most clinically relevant changes are decreased formation at trabecular sites and increased resorption on endosteal surfaces (Henneicke et al. 2011). With a significant decrease in bone mineral density, the disorganization of bone structure, and an increase in fracture risk, GIOP is recognized as one of the leading causes of secondary osteoporosis and a distinct metabolic bone disease (Walsh et al. 1996). As previously seen, a number of complex mechanisms are involved in the development of GIOP, with all cell types of the bone being involved. Although no single process seems to be paramount, current understanding places the most emphasis on osteocyte dysfunction (Xiong et al. 2015). This is further complicated by the fact that, while results from in vitro and animal studies emphasize the direct effects of GCs on bone cells, GC-induced osteoporosis (GIOP) is not related solely to actions on the skeletal tissue. Corticosteroid excess is characterized by a net calcium deficit (Ritz et al. 1984) related to decreased intestinal calcium absorption and increased renal calcium excretion. Sexual steroid (Chrousos et al. 1998) and growth hormone production and muscle strength (Tomas et al. 1979) are also impaired, all contributing not only to bone loss but an increased fall and fracture risk. Increased insulin resistance, obesity, and diabetes related to glucocorticoid excess are also known to mediate complex secondary effects on bone metabolism (Shanbhogue et al. 2016).

2.5.1 Endogenous Hypercortisolism

The deleterious effects of endogenous glucocorticoid excess have been first described by Cushing along with the original description of the disease. The severity of bone disease depends on the etiology of glucocorticoid excess (pituitary, adrenal, ectopic) and the time spent untreated. Ectopic ACTH production usually results in more severe disease due to higher cortisol levels. Changes in the levels of most bone formation and resorption markers are not typical and generally do not correlate highly with disease severity. The only exception is serum osteocalcin, the level of which decrease in a reliable fashion in response to cortisol excess and the exposure of skeletal tissue to GCs (Tóth and Grossman 2013). Low serum osteocalcin has even been proposed as a diagnostic marker for Cushing's disease (Belaya et al. 2016). In moderate- to high-risk individuals, this test has been shown to have a sensitivity and specificity of 74 and 94%, respectively, by a preliminary study.

Generally at least half of the patients sustain a clinical fracture during disease course (Vestergaard et al. 2002). Incidence of clinical fractures was shown to be sixfold higher in the 2 years preceding the diagnosis in patients with Cushing's disease. Given their silent nature (Kendler et al. 2016), the rate of vertebral fractures is even higher. In one study 76% of patients had one or multiple vertebral fractures, three-quarter of which were silent. In another report only 5% of the reported 150 fractures in 182 patients were non-vertebral (Belaya et al. 2015). These results emphasize the role of the spine as the most important site of fracture and treatment assessment in these patients. Bone mineral density (BMD) and trabecular bone score (TBS) values though usually decrease in proportion to the severity of the glucocorticoid excess (Chiodini et al. 1998; Belaya et al. 2015) have limited predictive value estimating fracture risk in patients with Cushing's disease.

With successful treatment of hypercortisolism, BMD decline is usually reversible, especially in younger patients (Fütő et al. 2008; Kawamata et al. 2008). Serum osteocalcin levels also increase rapidly (Szappanos et al. 2010). Whether this translates to the frank normalization of fracture risk is still debated. Clinical fracture rates seem to return to baseline after cure. However the subgroup of Cushing's patients who develop adrenal insufficiency and require long-term GC replacement after surgery has been shown to have a reduced BMD. Postmenopausal women are especially high risk in this regard, thus corticosteroid replacement warrants especially careful monitoring in this population (Barahona et al. 2009). Long-term corticosteroid replacement was also associated with osteoporosis and increased fracture risk in patients with congenital adrenal hyperplasia (Falhammar et al. 2007) and primary adrenal insufficiency though results are somewhat contradictory (Koetz et al. 2012; Schulz et al. 2016; Camozzi et al. 2018). The role of steroid replacement is also hard to discern from the role of the underlying disease in these patients. No comparative studies are available regarding the effectiveness of targeted antiporotic treatment in patients with endogenous hypercortisolism. Agents used in iatrogenic GIOP are generally considered to be safe and effective in patients with persistently low BMD.

2.5.2 Autonomous Cortisol Secretion

With the increasing availability of high-resolution imaging techniques, the incidence of adrenal incidentalomas is on the rise. The prevalence of silent adrenal nodules is reported to be between 3 and 10% in elderly patients. Approximately one third of these incidentalomas can be shown to have some degree of autonomous cortisol secretion without the typical signs and symptoms of Cushing's disease (Fassnacht et al. 2016). The most commonly used diagnostic test in this regard is the low-dose dexamethasone suppression test (Goddard et al. 2015). Decreased BMD and osteocalcin levels and increased fracture risk are also established with these patients (Chiodini et al. 2009; Eller-Vainicher et al. 2013). The primary site of bone involvement is the spine as the prevalence of vertebral fractures is reported to be fourfold higher in this population (Morelli et al. 2016). Incidentalomas without any detectable cortisol excess were also shown to have a slightly increased vertebral fracture risk, which raises the possibility of even more subclinical cortisol excess in these cases. The optimal screening for this condition and the subsequent treatment of individuals with adrenal incidentalomas, autonomous cortisol excess, and vertebral fractures are currently unclear. Surgical adrenalectomy was shown to decrease fracture risk (Salcuni et al. 2016); however no randomized trials or guidelines on medical therapy are currently available.

2.5.3 Endogenous Differences in GC Metabolism

A number of studies were aimed to assess the potential effect of physiological differences in endogenous cortisol production on bone health. Methodologies, end points, and patient populations differ substantially between these studies; however even without an underlying illness, there seems to be an association between higher levels of endogenous cortisol and osteoporosis (Heshmati et al. 1998; Raff et al. 1999; Dennison et al. 1999; Cooper et al. 2005; Shi et al. 2015). This relationship once again seems to be strongest in postmenopausal women. The potential future diagnostic and therapeutic consequences of these studies remain to be seen.

Genetic differences of GRs (Huizenga et al. 1998), proteins of glucocorticoid transport and metabolism, have also been examined in relationship to osteoporosis and fracture risk. A few small studies have shown an increased risk of osteoporosis in association with 11β-HSD1 polymorphisms (Hwang et al. 2009; Feldman et al. 2012) although in large-scale studies no SNP has been shown to have a strong effect on the bone. These polymorphisms, however, do seem to play a role in influencing the susceptibility to endogenous or exogenous GC excess and might explain the large individual variations seen in the risk of GIOP development (Szappanos et al. 2009; Koetz et al. 2013; Morgan et al. 2014). 11β-HSD1 catalyzes the activation of not only cortisone but also that of prednisone. In healthy volunteers differences in 11β-HSD1 activity were shown to affect prednisone-induced bone loss irrespective of serum prednisone or prednisolone levels (Cooper et al. 2003).

2.5.4 Exogenous Glucocorticoid Excess

Iatrogenic GIOP is generally associated with prolonged glucocorticoid use in systemic autoimmune conditions. It is important to note, however, that systemic inflammation in and of itself is a major cause for bone loss. This means that in practice it is often hard to separate drug side effect from the impact from the underlying disease. Thus it is possible for GC to offer relative protection against inflammation-induced bone loss under certain conditions (Kirwan et al. 1995; Landewé et al. 2002).

The prevalence of systemic GC use is reported worldwide to be between 1 and 5% (Fardet et al. 2011; Laugesen et al. 2017). This number generally increases with age, and certain studies suggest that it might be increasing over time despite the advances made with much more targeted biological agents. Osteoporosis and fracture risk associated with GC use depends largely on age, sex, dose, and duration of treatment (Majumdar et al. 2013). With prolonged use osteoporosis develops in at least 50% of patients (van Staa et al. 2000a, b). Fracture risk increases rapidly after the initiation of treatment but also decreases promptly after discontinuation. Bone mineral content, though sometimes follows a similar trend, changes much slower indicating that fracture risk is more strongly associated with poor bone quality rather than BMD in these patients. Multiple studies have reported a predominant increase in cortical porosity (Zhu et al. 2015) and at spinal sites a decreased TBS (Paggiosi et al. 2015), with a relatively intact BMD in both premenopausal and postmenopausal women. Increased fall risk as mentioned previously might also be a significant underlying determinant of fractures. With prolonged GC use, the relative risk of vertebral fractures is reported to be around 2.6, while the risk for hip and non-vertebral fractures increases to a lesser degree (RR: ~1,5) (van Staa et al. 2000a, b). Even with 5 mg/day prednisolone, the risk of fractures increased by approximately 20%. These rates are much less daunting in premenopausal women and young men: 3–5%/year risk for vertebral and 2.5–3%/year for non-vertebral fractures (Amiche et al. 2016). This is generally explained by the protective characteristics of sexual steroids. Short-term (<30 days) intermittent glucocorticoid use was previously thought to be relatively safe; recent data indicates increased fracture risk even in this population (De Vries et al. 2007). This correlates with result showing a decrease in bone strength only a few weeks after initiation of treatment (Mellibovsky et al. 2015). No increase is associated with inhaled corticosteroids except at daily dosages above 7.5 mg of prednisolone equivalents (Vestergaard et al. 2005a, b). No increase in fracture risk is associated with other forms of topical corticosteroids (Vestergaard et al. 2005a, b).

Treatment decisions in GIOP are generally based on clinical risk assessment tools as, for the reasons mentioned above, changes in BMD are generally not consistent with fracture risk (Buckley et al. 2017). FRAX is the most widely available such tool and the only one where GC treatment is an independent item of assessment. Additional guidance is available for further correction of FRAX-based fracture risk with GC dose (Kanis et al. 2011). FRAX is generally deemed to be insensitive to the increase of vertebral fracture risk in younger patients undergoing GC treatment. In this setting low spinal BMD (T-score < -1.5) and previous fractures may also guide treatment.

In patients with high fracture risk, bone protective treatment should ideally be started with or shortly after glucocorticoids. Bisphosphonates (Loviselli et al. 1997, Saag et al. 1998, Adachi et al. 2001, Reid et al. 2009), teriparatide (Gluer et al. 2013), and denosumab (Ishiguro et al. 2017; Sawamura et al. 2017; Saag et al. 2018) were all found to effectively increase BMD in this setting; however, data regarding fracture risk reduction is scarce. Recent retrospective analyses show a relative risk reduction of 30–60% with antiosteoporotic treatment in patients undergoing GC treatment. The currently available results suggest, however, the superiority of teriparatide both in fracture risk reduction and BMD improvement (Saag et al. 2007, 2016). Vitamin D and calcium supplementation in itself has minor therapeutic effects, but optimal vitamin D and calcium availability is considered a prerequisite of any successful antiporotic treatment. The use of locally acting steroid formulations if available and an early switch to glucocorticoid-sparing agents (Strehl et al. 2016) if possible are essential prerequisites of successful treatment and prevention. The treatment of concomitant comorbidities such as malabsorption, renal failure, or hypogonadism is also an essential adjunct to antiporotic therapy.

3 Insulin and Bone

Zsuzsanna Putz, Ildiko Istenes, Istvan Takacs, and Peter Lakatos

3.1 The Relationship Between Glucose and Bone Metabolism at the Cellular Level

Significant correlation has been found between calcium and glucose metabolism at many levels recently. Cytosolic calcium concentration ([Ca2+]c) is essential for the regulation of many cellular processes such as insulin secretion, metabolism, proliferation, muscle contraction, and cell signaling. The sarco-/endoplasmic reticulum (ER/SR) Ca2+ ATPase (SERCA) transports Ca2+ from the cytosol to the ER or SR lumen. Three different genes (ATP2A1, ATP2A2, and ATP2A3) encode 11 different isoforms of SERCA pumps. The ATP2A1 gene encodes SERCA1a and 1b isoforms, which are expressed in skeletal muscle, while the ATP2A2 gene encodes SERCA2a, b, and c. The SERCA2a isoform is mainly expressed in cardiomyocytes and slow-twitch skeletal muscle, the SERCA2b is expressed in smooth muscle and non-muscle tissues, while SERCA2c and SERCA3 isoforms are present in hematopoietic cell lines.

It is known that the expression of SERCA2b and SERCA3 is reduced in beta cells of the diabetic pancreas, which leads to impaired intracellular Ca2+ homeostasis (the intracellular Ca2+ level increases but the extracellular Ca2+ level slightly decreases) and diminished insulin secretion as a consequence (Zarain-Herzberg et al. 2014). Furthermore, there is an extracellular Ca-sensing receptor on the surface of the beta cells which plays an important role in its physiological function: activation of the Ca-sensing receptor directly alters the expression and function of various potassium and voltage-dependent calcium channels of the beta cell (Squires et al. 2014).

Obesity can result in dysregulation of cytosolic and organelle Ca2+ fluxes which disrupts equilibrium in metabolic tissues and immune cells and changes organelle homeostasis, signaling pathways, as well as autophagy. Excessive energy intake stimulates de novo lipogenesis in the liver, leading to the accumulation of lipid droplets, as well as the balance of phosphatidylethanolamine (PE) and phosphatidylcholine (PC) synthesis, which are the main phospholipids of organelle membranes. Increased PE/PC ratio of the ER membrane which reduces the function of SERCA2b results in decreased Ca2+ accumulation in the ER lumen.

These alterations at the cellular and organelle levels translate directly into effects on hepatic glucose production, lipogenesis, inflammation, and other processes that affect systemic metabolism and overall metabolic health. Consequently, Ca2+ handling could be considered a critical problem in metabolic diseases. Based on these data, diabetes could actually be also considered as a disorder of calcium homeostasis of beta cells (Arruda and Hotamisligil 2015).

The forkhead box O1 (FOXO1) is considered today as one of the central pathogenic factors in type 1 and type 2 diabetes. FOXO1 plays a significant role in glucose homeostasis since increased FOXO1 activity leads to insulin resistance

(Pajvani and Accili 2015; Lee and Donh 2017). Furthermore, based on preclinical studies with vitamin D receptor null mice, it was found that vitamin D deficiency leads to increased FOXO1 expression, causing decreased insulin sensitivity and impaired glucose tolerance as a consequence (Chen et al. 2016).

The metabolic and endocrine alterations in diabetes affect bone metabolism and, thus, the amount and quality of bone of these patients as well. Type 1 and type 2 diabetics may show various disorders of calcium metabolism, such as impaired calcium absorption and loss of calcium from bone, which may result in osteopenia with decreased osteocalcin levels. Bone tissue can regulate metabolic pathways of glucose tolerance and insulin signaling as well. Osteoblasts and osteoclasts express insulin receptors on their surface. Insulin administration increases osteoblast proliferation and osteoblast and osteoclast differentiation, in favor of bone formation. On the other hand, hyperglycemia has negative effects on osteoblastogenesis since high glucose concentrations may reduce the viability of mesenchymal stem cells. Furthermore, hyperglycemia increases osteoclast activity and results in impaired bone resorption. In addition to that, advanced glycation end products (AGEs) can suppress endoplasmic reticulum function, which is essential for osteoblast differentiation. AGE-modified collagen fibers affect the proliferation and function of both osteoblasts and osteoclasts and the accumulation of AGEs within bone collagen and leads to the stiffness of the collagen network as well (Palermo et al. 2017).

Several studies have shown that there is an inverse relationship between the amount of osteocalcin produced by osteoblasts and HbA1c. It has been suggested that hyperglycemia decreases vitamin D-induced osteocalcin secretion of the osteoblasts (Palermo et al. 2017; Adami 2009; Inaba et al. 1999). Osteocalcin also plays an important role in glucose metabolism beyond bone metabolism. It has been shown that osteocalcin-deficient mice have reduced number of beta cells and decreased insulin production leading to increased insulin resistance and impaired glucose tolerance, and this has been verified in humans as well (Lee et al. 2007a, b; Buday et al. 2008; Guo et al. 2017). The interrelationship between bone, fat, and glucose metabolism can be seen in Fig. 2 (Rosen 2009). The production of osteocalcin is stimulated by serotonin originating from the brainstem and is inhibited by serotonin produced in the gut. Osteocalcin increases insulin secretion of the beta cells in the pancreas. In return, insulin increases leptin synthesis of the adipocytes which increases serotonin secretion in the ventromedial nuclei of the brainstem.

3.2 The Relationship Between Glucose and Bone Metabolism at Clinical Level

In clinical studies, it has been confirmed that the condition of the bone in both type 1 and type 2 diabetes depends on the quality of diabetes treatment. Hyperglycemia seems to contribute to low bone turnover in both type 1 and type 2 diabetics. Hyperglycemia inhibits bone formation by modulating osteoblast phenotype, function, and bone resorption as well (Adami 2009; Campos Pastor et al. 2000). In recent years, data have accumulated suggesting that the metabolic and endocrine alterations

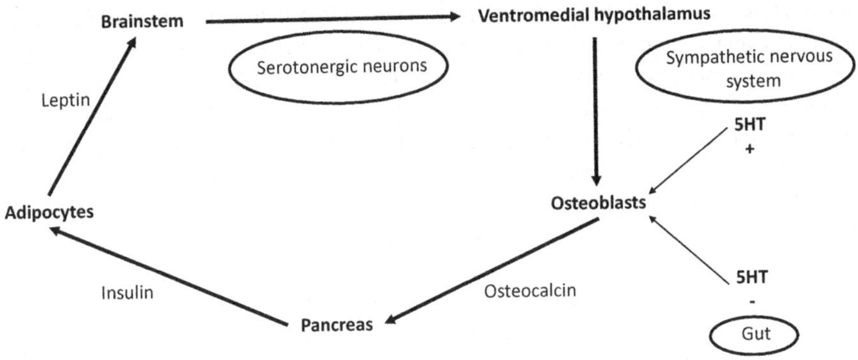

Fig. 2 The interrelationship between bone, fat, and glucose metabolism. Modified from: Rosen, C (2009). Bone: serotonin, leptin and the central control of bone remodeling. Nat. Rev. Rheumatol 5: 667–668

of diabetes (as discussed previously) affect bone quantity and quality, resulting in significantly increased risk of fracture.

In some studies of type 1 diabetes, increased bone resorption and decreased bone formation were observed. Bone mass decreases in these patients due to the lack of insulin and insulin-like growth factor, dysregulation of adipokines, and proliferation of pro-inflammatory cytokines. Therefore, the risk of fracture in type 1 diabetes is 1.5–2 times in each age group (Christensen and Svendsen 1999; Weber and Schwartz 2016).

Type 2 diabetes is characterized by low bone turnover with decreased bone formation and resorption. In type 2 diabetics, insulin resistance and high insulin levels result in higher bone mass but poor bone quality due to the presence of glycated collagen products and reduced cortical thickness, which manifest in an increased risk of bone fracture as a consequence (Grey 2009). In another meta-analysis, the risk of fracture was sixfold higher in type 1 diabetic patients and twofold higher in type 2 diabetes compared to the general population (Strotmeyer and Cauley 2007). Fracture risk in these patients was also associated with an increased tendency to fall due to retinopathy, neuropathy, hypo- and hyperglycemia, and sometimes medication used during treatment. Hypo- and hyperglycemia may cause visual impairment, muscle weakness, and unfavorable coordination, thus increasing the risk of falls (Valderrábano and Linares 2018). Data suggest increased circulating sclerostin concentrations in subjects with type 2 diabetes mellitus (Wang et al. 2018) which may also add to the increased fracture risk in these patients.

Many studies have investigated the effects of antihyperglycemic drugs on the risk of fracture in diabetes. In clinical studies, the most commonly used insulin-sensitizing drug metformin reduced the incidence of all types of fractures in type 2 diabetics by stimulating proliferation and differentiation of osteoblast-like cell lines (Chandran 2017). Nowadays, the use of sulfonylureas has decreased due to their hypoglycemic effects. Information about fracture risk associated with sulfonylurea use is mixed. There have been studies showing that sulfonylurea treatment increases the risk of bone fractures due to the higher incidence of hypoglycemic events and risk of falls (Vestergaard et al. 2005a, b). However, many studies have reported that sulphonylureas have a neutral effect on the risk of fractures in diabetics (Monarni et al. 2008). The insulin sensitizer thiazolidinediones act on the mesenchymal stem cells through the activation of peroxisome proliferator-activated receptor (PPARγ) and increase stem cell differentiation toward adipocyte instead of differentiating to osteoblasts (Adami 2009). Thus, thiazolidinediones increase the number of adipocytes and decrease the number of bone-building cells which results in a decreased bone mass with an increased risk of fracture as a consequence. In fact, based on the results of several randomized and observational studies, glitazones were found to actually double fractures in type 2 diabetics (Loke et al. 2009; Jones et al. 2009; Dormuth et al. 2009; Meier et al. 2008).

Clinical data suggest that the glucagon-like peptide-1 (GLP-1) receptor agonist effect may be useful in preventing bone loss caused by glucose toxicity. They increase bone mass and bone quality by preventing the adverse effects of insulin resistance and hyperglycemia on bone metabolism. Nevertheless, the potential anti-resorptive effect of GLP-1 receptor agonists is not fully understood. They may induce osteogenic differentiation in bone and increase trabecular bone mass and the expression of osteoblast markers (Palermo et al. 2017). The DPP-4 inhibitors also reduce bone fragility, according to a recent meta-analysis (Chandran 2017).

Sodium glucose co-transporter-2 (SGLT-2) inhibitors such as dapagliflozin do not have an impact on bone turnover markers and bone density (Bolinder et al. 2014; Chandran 2017). Among the SGLT2 inhibitors, there are only data on canagliflozin effects on bone, and these reveal a dose-dependent decrease in bone density after 1 year of canagliflozin treatment (Bilezikian et al. 2016). When analyzing the risk of fractures, a significant increase was demonstrated, especially in elderly patients with increased cardiovascular risk, reduced GFR, and diuretic therapy (Watts et al. 2016). These results suggest that the use of SGLT-2 inhibitors in this population requires increased caution.

The effect of insulin on the bone is extremely complex. It has no particular effect on bone mass, but according to the results of several studies, it increases the risk of fractures; however, it has not been confirmed by other studies (Adami 2009). It is not clear whether it is due to the direct effect of insulin or whether this effect is mediated through hypoglycemia and increased risk of falls (Chandran 2017). Hyperinsulinemia caused by insulin resistance negatively affects bone health: It increases BMD in patients with type 2 diabetes; however, it deteriorates bone quality (Baerrett-Connor and Kritz-Silverstein 1996; Stolk et al. 1996).

Diabetics are more likely to take drugs that increase the risk of falls compared to nondiabetics such as sedatives and antidepressants drugs. Commonly used drugs like beta-blockers (Zofková and Matucha 2014), statins (An et al. 2017), thiazides (Aung and Htay 2011), and ACE inhibitors (Gebru et al. 2013) have a beneficial effect on bone, while proton pump inhibitors increase both bone density and fracture risk (Lee et al. 2013). As a conclusion, it can be stated that diabetes causes metabolic and endocrine changes which can lead to higher risk of bone fractures.

References

Abrahamsen B, Jorgensen HL, Laulund AS, Nybo M, Brix TH, Hegedus L (2014) Low serum thyrotropin level and duration of suppression as a predictor of major osteoporotic fractures-the OPENTHYRO register cohort. J Bone Miner Res 29(9):2040–2050. https://doi.org/10.1002/jbmr.2244

Adachi JD, Saag KG, Delmas PD et al (2001) Two-year effects of alendronate on bone mineral density and vertebral fracture in patients receiving glucocorticoids: a randomized, double-blind, placebo controlled extension trial. Arthritis Rheum 44(1):202–211

Adami S (2009) Bone health in diabetes: considerations for clinical management. Curr Med Res Opin 25(5):1057–1072

Allain TJ, Chambers TJ, Flanagan AM, McGregor AM (1992) Tri-iodothyronine stimulates rat osteoclastic bone resorption by an indirect effect. J Endocrinol 133(3):327–331

Amiche MA, Albaum JM, Tadrous M et al (2016) Fracture risk in oral glucocorticoid users: a Bayesian metaregression leveraging control arms of osteoporosis clinical trials. Osteoporos Int 27(5):1709–1718

An T, Hao J, Sun S, Li R, Yang M, Cheng G, You M (2017) Efficacy of statins for osteoporosis: a systematic review and meta-analysis. Osteoporos Int 28(1):47–57

Arruda AP, Hotamisligil GS (2015) Calcium homeostasis and organelle function in the pathogenesis of obesity and diabetes. Cell Metab 22(3):381–397

Aung K, Htay T (2011) Thiazide diuretics and the risk of hip fracture. Cochrane Database Syst Rev. https://doi.org/10.1002/14651858.CD005185.pub2

Baerrett-Connor E, Kritz-Silverstein D (1996) Does hyperinsulinemia preserve bone? Diabetes Care 19:1388–1392

Ballock R, Mita BC, Zhou X, Chen DH, Mink LM (1999) Expression of thyroid hormone receptor isoforms in rat growth plate cartilage in vivo. J Bone Miner Res 14(9):1550–1556. https://doi.org/10.1359/jbmr.1999.14.9.1550

Banovac K, Koren E (2000) Triiodothyronine stimulates the release of membrane-bound alkaline phosphatase in osteoblastic cells. Calcif Tissue Int 67(6):460–465

Barahona MJ, Sucunza N, Resmini E et al (2009) Deleterious effects of glucocorticoid replacement on bone in women after long-term remission of Cushing's syndrome. J Bone Miner Res 24 (11):1841–1846

Baschant U, Tuckermann J (2010) The role of the glucocorticoid receptor in inflammation and immunity. J Steroid Biochem Mol Biol 120(2-3):69–75

Bassett JH, Swinhoe R, Chassande O, Samarut J, Williams GR (2006) Thyroid hormone regulates heparan sulfate proteoglycan expression in the growth plate. Endocrinology 147(1):295–305. https://doi.org/10.1210/en.2005-0485

Beavan S, Horner A, Bord S et al (2001) Colocalization of glucocorticoid and mineralocorticoid receptors in human bone. J Bone Miner Res 16(8):1496–1504

Behr M, Ramsden DB, Loos U (1997) Deoxyribonucleic acid binding and transcriptional silencing by a truncated c-erbA beta 1 thyroid hormone receptor identified in a severely retarded patient with resistance to thyroid hormone. J Clin Endocrinol Metab 82(4):1081–1087. https://doi.org/10.1210/jcem.82.4.3853

Belaya ZE, Hans D, Rozhinskaya LY et al (2015) The risk factors for fractures and trabecular bone-score value in patients with endogenous Cushing's syndrome. Arch Osteoporos 10(1):44

Belaya ZE, Iljin AV, Melnichenko GA et al (2016) Diagnostic performance of osteocalcin measurements in patients with endogenous Cushing's syndrome. BoneKey Rep 5:815

Bennett A, Chen T, Feldman D et al (1984) Characterization of insulin-like growth factor I receptors on cultured rat bone cells: regulation of receptor concentration by glucocorticoids. Endocrinology 115(4):1577–1583

Bilezikian JP, Watts NB, Usiskin K, Polidori D, Fung A, Sullivan D, Rosenthal N (2016) Evaluation of bone mineral density and bone biomarkers in patients with type 2 diabetes treated with canagliflozin. J Clin Endocrinol Metab 101(1):44–51

Bochukova E, Schoenmakers N, Agostini M, Schoenmakers E, Rajanayagam O, Keogh JM, Henning E, Reinemund J, Gevers E, Sarri M, Downes K, Offiah A, Albanese A, Halsall D, Schwabe JW, Bain M, Lindley K, Muntoni F, Vargha-Khadem F, Dattani M, Farooqi IS, Gurnell M, Chatterjee K (2012) A mutation in the thyroid hormone receptor alpha gene. N Engl J Med 366(3):243–249. https://doi.org/10.1056/NEJMoa1110296

Bolinder J, Ljunggren O, Johansson L, Wilding J, Langkilde AM, Sjostrom CD et al (2014) Dapagliflozin maintains glycaemic control while reducing weight and body fat mass over2 years in patients with type 2 diabetes mellitus inadequately controlled on metformin. Diabetes Obes Metab 16(2):159–169

Brennan-Speranza TC, Henneicke H, Gasparini SJ et al (2012) Osteoblasts mediate the adverse effects of glucocorticoids on fuel metabolism. J Clin Invest 122(11):4172–4189

Buckley L, Guyatt G, Fink HA et al (2017) 2017 American College of Rheumatology guideline for the prevention and treatment of glucocorticoid induced osteoporosis. Arthritis Rheumatol 69 (8):1521–1537

Buday B, Kulcsár E, Literáti Nagy B, Horváth T, Vitai M, Vecsei I, Bezzegh K, Kiss J, Péterfai E, Koltay L, Korányi L (2008) The role of osteocalcin in the connection of bone and glucose metabolism in humans. Orv Hetil 149(52):2453–2461

Camozzi V, Betterle C, Frigo AC et al (2018) Vertebral fractures assessed with dual-energy X-ray absorptiometry in patients with Addison's disease on glucocorticoid and mineralocorticoid replacement therapy. Endocrine 59(2):319–329

Campos Pastor MM, López-Ibarra PJ, Escobar-Jiménez F, Serrano Pardo MD, García-Cervigón AG (2000) Intensive insulin therapy and bone mineral density in type 1 diabetes mellitus: a prospective study. Osteoporos Int 11(5):455–459

Canalis E, Mazziotti G, Giustina A et al (2007) Glucocorticoid-induced osteoporosis: pathophysiology and therapy. Osteoporos Int 18:1319–1328

Centrella M, McCarthy TL, Canalis E (1991) Glucocorticoid regulation of transforming growth factor beta 1 activity and binding in osteoblast-enriched cultures from fetal rat bone. Mol Cell Biol 11(9):4490–4496

Chandran M (2017) Diabetes drug effects on the skeleton. Calcif Tissue Int 100(2):133–149

Chapman K, Holmes M, Seckl J (2013) 11b-Hydroxysteroid dehydrogenases: intracellular gatekeepers of tissue glucocorticoid action. Physiol Rev 93(3):1139–1206

Chassande O, Fraichard A, Gauthier K, Flamant F, Legrand C, Savatier P, Laudet V, Samarut J (1997) Identification of transcripts initiated from an internal promoter in the c-erbA alpha locus that encode inhibitors of retinoic acid receptor-alpha and triiodothyronine receptor activities. Mol Endocrinol 11(9):1278–1290. https://doi.org/10.1210/mend.11.9.9972

Chen F, Zhang L, OuYang Y (2014) Glucocorticoid induced osteoblast apoptosis by increasing E4BP4 expression via up-regulation of Bim. Calcif Tissue Int 94(6):640–647

Chen S, Villalta SA, Agrawal DK (2016) FOXO1 mediates vitamin D deficiency-induced insulin resistance in skeletal muscle. J Bone Miner Res 31(3):585–595

Chiodini I, Carnevale V, Torlontano M et al (1998) Alterations of bone turnover and bone mass at different skeletal sites due to pure glucocorticoid excess: study in eumenorrheic patients with Cushing's syndrome. J Clin Endocrinol Metab 83(6):1863–1867

Chiodini I, Morelli V, Masserini B et al (2009) Bone mineral density, prevalence of vertebral fractures, and bone quality in patients with adrenal incidentalomas with and without subclinical hypercortisolism: an Italian multicenter study. J Clin Endocrinol Metab 94(9):3207–3214

Christensen JO, Svendsen OL (1999) Bone mineral in pre- and postmenopausal women with insulin-dependent and non-insulin-dependent diabetes mellitus. Osteoporos Int 10(4):307–311

Chrousos GP, Torpy DJ, Gold PW (1998) Interactions between the hypothalamic-pituitary-adrenal axis and the female reproductive system: clinical implications. Ann Intern Med 129(3):229–240

Conaway HH, Henning P, Lie A, Tuckermann J et al (2016) Activation of dimeric glucocorticoid receptors in osteoclast progenitors potentiates RANKL induced mature osteoclast bone resorbing activity. Bone 93:43–54

Cooper MS, Walker EA, Bland R et al (2000) Expression and functional consequences of 11b-hydroxysteroid dehydrogenase activity in human bone. Bone 27(3):375–381
Cooper MS, Blumsohn A, Goddard PE et al (2003) 11b-Hydroxysteroid dehydrogenase type 1 activity predicts the effects of glucocorticoids on bone. J Clin Endocrinol Metab 88 (8):3874–3877
Cooper MS, Syddall HE, Fall CH et al (2005) Circulating cortisone levels are associated with biochemical markers of bone formation and lumbar spine BMD: the Hertfordshire Cohort Study. Clin Endocrinol (Oxf) 62(6):692–697
Cushing HW (1932) The basophil adenomas of the pituitary body and their clinical manifestations (pituitary basophilism). Bull Johns Hopkins Hosp 50:137–195
Dallas SL, Prideaux M, Bonewald LF (2013) The osteocyte: an endocrine cell ... and more. Endocr Rev 34(5):658–690
de Vries F, Bracke M, Leufkens HG et al (2007) Fracture risk with intermittent high-dose oral glucocorticoid therapy. Arthritis Rheum 56(1):208–214
Dennison E, Hindmarsh P, Fall C et al (1999) Profiles of endogenous circulating cortisol and bone mineral density in healthy elderly men. J Clin Endocrinol Metab 84(9):3058–3063
Dormuth CR, Carney G, Carleton B, Bassett K, Wright JM (2009) Thiazolidinediones and fractures in men and women. Arch Intern Med 169(15):1395–1402
Eller-Vainicher C, Morelli V, Ulivieri FM et al (2013) Bone quality, as measured by trabecular bone score in patients with adrenal incidentalomas with and without subclinical hypercortisolism. J Bone Miner Res 27(10):2223–2230
Evans RM (1988) The steroid and thyroid hormone receptor superfamily. Science 240:889–895
Falhammar H, Filipsson H, Holmdahl G et al (2007) Fractures and bone mineral density in adult women with 21-hydroxylase deficiency. J Clin Endocrinol Metab 92(12):4643–4649
Fardet L, Petersen I, Nazareth I (2011) Prevalence of longterm oral glucocorticoid prescriptions in the UK over the past 20 years. Rheumatology (Oxford) 50(11):1982–1990
Fassnacht M, Arlt W, Bancos I et al (2016) Management of adrenal incidentalomas: European Society of endocrinology clinical practice guideline in collaboration with the European network for the study of adrenal tumors. Eur J Endocrinol 175(2):G1–G34
Feldman K, Szappanos A, Butz H et al (2012) The rs4844880 polymorphism in the promoter region of the HSD11B1 gene associates with bone mineral density in healthy and postmenopausal osteoporotic women. Steroids 77(13):1345–1351
Fumoto T, Ishii KA, Ito M et al (2014) Mineralocorticoid receptor function in bone metabolism and its role in glucocorticoid-induced osteopenia. Biochem Biophys Res Commun 447(3):407–412
Fütő L, Toke J, Patócs A et al (2008) Skeletal differences in bone mineral area and content before and after cure of endogenous Cushing's syndrome. Osteoporos Int 19(7):941–949
Gathercole LL, Lavery GG, Morgan SA et al (2013) 11b-hydroxysteroid dehydrogenase 1: translational and therapeutic aspects. Endocr Rev 34(4):525–555
Gauthier K, Chassande O, Plateroti M, Roux JP, Legrand C, Pain B, Rousset B, Weiss R, Trouillas J, Samarut J (1999) Different functions for the thyroid hormone receptors TRalpha and TRbeta in the control of thyroid hormone production and post-natal development. EMBO J 18(3):623–631. https://doi.org/10.1093/emboj/18.3.623
Gebru Y, Diao TY, Pan H, Mukwaya E, Zhang Y (2013) Potential of RAS inhibition to improve metabolic bone disorders. Biomed Res Int. https://doi.org/10.1155/2013/932691
Germain P, Staels B, Dacquet C, Spedding M, Laudet V (2006) Overview of nomenclature of nuclear receptors. Pharmacol Rev 58(4):685–704. https://doi.org/10.1124/pr.58.4.2
Gluer CC, Marin F, Ringe JD et al (2013) Comparative effects of teriparatide and risedronate in glucocorticoid-induced osteoporosis in men: 18-month results of the EuroGIOPs trial. J Bone Miner Res 28(6):1355–1368
Goddard GM, Ravikumar A, Levine AC (2015) Adrenal mild hypercortisolism. Endocrinol Metab Clin North Am 44(2):371–379
Grey A (2009) Thiazolidinedione-induced skeletal fragility--mechanisms and implications. Diabetes Obes Metab 11(4):275–284

Guo Q, Li H, Xu L, Wu S, Sun H, Zhou B (2017) Undercarboxylated osteocalcin reverts insulin resistance induced by endoplasmic reticulum stress in human umbilical vein endothelial cells. Sci Rep 7(1):46

Hamidouche Z, Hay¨ E, Vaudin P et al (2008) FHL2 mediates dexamethasone-induced mesenchymal cell differentiation into osteoblasts by activating Wnt/ β-catenin signaling-dependent Runx2 expression. FASEB J 22:3813–3822

Hansen M, Florescu A, Stoltenberg M (1996) Bone loss in rheumatoid arthritis. Influence of disease activity, duration of the disease, functional capacity, and corticosteroid treatment. Scand J Rheumatol 25(6):367–376

Hartmann K, Koenen M, Schauer S et al (2016) Molecular actions of glucocorticoids in cartilage and bone during health, disease, and steroid therapy. Physiol Rev 96(2):409–447

Harvey RD, McHardy KC, Reid IW, Paterson F, Bewsher PD, Duncan A, Robins SP (1991) Measurement of bone collagen degradation in hyperthyroidism and during thyroxine replacement therapy using pyridinium cross-links as specific urinary markers. J Clin Endocrinol Metab 72(6):1189–1194. https://doi.org/10.1210/jcem-72-6-1189

Henneicke H, Herrmann M, Kalak R et al (2011) Corticosterone selectively targets endocortical surfaces by an osteoblast-dependent mechanism. Bone 49(4):733–742

Heshmati HM, Riggs BL, Burritt MF et al (1998) Effects of the circadian variation in serum cortisol on markers of bone turnover and calcium homeostasis in normal postmenopausal women. J Clin Endocrinol Metab 83(3):751–756

Hofbauer LC, Gori F, Riggs BL et al (1999) Stimulation of osteoprotegerin ligand and inhibition of osteoprotegerin production by glucocorticoids in human osteoblastic lineage cells: potential paracrine mechanisms of glucocorticoid-induced osteoporosis. Endocrinology 140:4382–4389

Honda M, Orii F, Ayabe T et al (2000) Expression of glucocorticoid receptor β in lymphocytes of patients with glucocorticoid-resistant ulcerative colitis. Gastroenterology 118:859–866

Huang BK, Golden LA, Tarjan G, Madison LD, Stern PH (2000) Insulin-like growth factor I production is essential for anabolic effects of thyroid hormone in osteoblasts. J Bone Miner Res 15(2):188–197. https://doi.org/10.1359/jbmr.2000.15.2.188

Huizenga NA, Koper JW, de Lange P et al (1998) A polymorphism in the glucocorticoid receptor gene may be associated with and increased sensitivity to glucocorticoids in vivo. J Clin Endocrinol Metab 83(1):144–151

Hwang JY, Lee SH, Kim GS et al (2009) HSD11B1 polymorphisms predicted bone mineral density and fracture risk in postmenopausal women without a clinically apparent hypercortisolemia. Bone 45(6):1098–1103

Inaba M, Nishizawa Y, Mita K, Kumeda Y, Emoto M, Kawagishi T, Ishimura E, Nakatsuka K, Shioi A, Morii H (1999) Poor glycemic control impairs the response of biochemical parameters of bone formation and resorption to exogenous 1,25-dihydroxyvitamin D3 in patients with type 2 diabetes. Osteoporos Int 9(6):525–531

Ishiguro S, Ito K, Nakagawa S et al (2017) The clinical benefits of denosumab for prophylaxis of steroid-induced osteoporosis in patients with pulmonary disease. Arch Osteoporos 12(1):44

Jastrup B, Mosekilde L, Melsen F, Lund B, Lund B, Sorensen OH (1982) Serum levels of vitamin D metabolites and bone remodelling in hyperthyroidism. Metabolism 31(2):126–132

Jia D, O'Brien CA, Stewart SA (2006) Glucocorticoids act directly on osteoclasts to increase their life span and reduce bone density. Endocrinology 147:5592–5599

Jia J, Yao W, Guan M et al (2011) Glucocorticoid dose determines osteocyte cell fate. FASEB J 25 (10):3366–3376

Jones SG, Momin SR, Good MW, Shea TK, Patric K (2009) Distal upper and lower limb fractures associated with thiazolidinedione use. Am J Manag Care 15(8):491–496

Kanis JA, Johansson H, Oden A et al (2011) Guidance for the adjustment of FRAX according to the dose of glucocorticoids. Osteoporos Int 22(3):809–816

Karga H, Papapetrou PD, Korakovouni A, Papandroulaki F, Polymeris A, Pampouras G (2004) Bone mineral density in hyperthyroidism. Clin Endocrinol (Oxf) 61(4):466–472. https://doi.org/10.1111/j.1365-2265.2004.02110.x

Kawaguchi H, Pilbeam CC, Woodiel FN, Raisz LG (1994) Comparison of the effects of 3,5,3′-triiodothyroacetic acid and triiodothyronine on bone resorption in cultured fetal rat long bones and neonatal mouse calvariae. J Bone Miner Res 9(2):247–253. https://doi.org/10.1002/jbmr.5650090214

Kawamata A, Iihara M, Okamoto T et al (2008) Bone mineral density before and after surgical cure of Cushing's syndrome due to adrenocortical adenoma: prospective study. World J Surg 32 (5):890–896

Kendler DL, Bauer DC, Davison KS et al (2016) Vertebral fractures: clinical importance and management. Am J Med 129(2):221.e1–221.10

Kim HJ, Zhao H, Kitaura H et al (2006) Glucocorticoids suppress bone formation via the osteoclast. J Clin Invest 116:2152–2160

Kirwan JR et al (1995) The effect of glucocorticoids on joint destruction in rheumatoid arthritis. N Engl J Med 3:142–147

Klaushofer K, Varga F, Glantschnig H, Fratzl-Zelman N, Czerwenka E, Leis HJ, Koller K, Peterlik M (1995) The regulatory role of thyroid hormones in bone cell growth and differentiation. J Nutr 125(7 Suppl):1996S–2003S. https://doi.org/10.1093/jn/125.suppl_7.1996S

Koetz KR, Ventz M, Diederich S et al (2012) Bone mineral density is not significantly reduced in adult patients on low-dose glucocorticoid replacement therapy. J Clin Endocrinol Metab 97 (1):85–92

Koetz KR, van Rossum EF, Ventz M et al (2013) BclI polymorphism of the glucocorticoid receptor gene is associated with increased bone resorption in patients on glucocorticoid replacement therapy. Clin Endocrinol (Oxf) 78(6):831–837

Lakatos P, Stern PH (1992) Effects of cyclosporins and transforming growth factor beta 1 on thyroid hormone action in cultured fetal rat limb bones. Calcif Tissue Int 50(2):123–128

Lakatos P, Foldes J, Horvath C, Kiss L, Tatrai A, Takacs I, Tarjan G, Stern PH (1997) Serum interleukin-6 and bone metabolism in patients with thyroid function disorders. J Clin Endocrinol Metab 82(1):78–81. https://doi.org/10.1210/jcem.82.1.3641

Lakatos P, Foldes J, Nagy Z, Takacs I, Speer G, Horvath C, Mohan S, Baylink DJ, Stern PH (2000) Serum insulin-like growth factor-I, insulin-like growth factor binding proteins, and bone mineral content in hyperthyroidism. Thyroid 10(5):417–423. https://doi.org/10.1089/thy.2000.10.417

Landewé RB, Boers M, Verhoeven AC et al (2002) COBRA combination therapy in patients with early rheumatoid arthritis: long-term structural benefits of a brief intervention. Arthritis Rheum 46(2):347–356

Lane NE, Yao W, Balooch M et al (2006) Glucocorticoid treated mice have localized changes in trabecular bone material properties and osteocyte lacunar size that are not observed in placebo-treated or estrogen-deficient mice. J Bone Miner Res 21(3):466–476

Laugesen K, Jørgensen JOL, Sørensen HT et al (2017) Systemic glucocorticoid use in Denmark: a population-based prevalence study. BMJ Open 7(5):e015237

Lee S, Donh HH (2017) FoxO integration of insulin signaling with glucose and lipid metabolism. J Endocrinol 233(2):R67–R79

Lee NK, Sowa H, Hinoi E, Ferron M et al (2007a) Endocrine regulation of of energy metabolism by the skeleton. Cell 130(3):456–469

Lee NK, Sowa H, Hinoi E, Ferron M, Ahn JD, Confavreux C, Dacquin R, Mee PJ, McKee MD, Jung DY, Zhang Z, Kim JK, Mauvais-Jarvis F, Ducy P, Karsenty G (2007b) Endocrine regulation of energy metabolism by the skeleton. Cell 130(3):456–469

Lee JS, Buzkova P, Fink HA, Vu J, Carbone L, Chen Z, Cauley J, Bauer DC, Cappola AR, Robbins J (2010) Subclinical thyroid dysfunction and incident hip fracture in older adults. Arch Intern Med 170(21):1876–1883. https://doi.org/10.1001/archinternmed.2010.424

Lee J, Youn K, Choi NK, Lee JH, Kang D, Song HJ, Park BJ (2013) A population-based case-control study: proton pump inhibition and risk of hip fracture by use of bisphosphonate. J Gastroenterol 48(9):1016–1022

Loke YK, Singh S, Furberg C (2009) Long-term use of thiazolidinediones and fractures in type 2 diabetes: a meta-analysis. CMAJ 180(1):32–39

Loviselli A, Mastinu R, Rizzolo E, Massa GM, Velluzzi F, Sammartano L, Mela Q, Mariotti S (1997) Circulating telopeptide type I is a peripheral marker of thyroid hormone action in hyperthyroidism and during levothyroxine suppressive therapy. Thyroid 7(4):561–566. https://doi.org/10.1089/thy.1997.7.561

Luegmayr E, Glantschnig H, Varga F, Klaushofer K (2000) The organization of adherens junctions in mouse osteoblast-like cells (MC3T3-E1) and their modulation by triiodothyronine and 1,25-dihydroxyvitamin D3. Histochem Cell Biol 113(6):467–478

Majumdar SR, Morin SN, Lix LM et al (2013) Influence of recency and duration of glucocorticoid use on bone mineral density and risk of fractures: population-based cohort study. Osteoporos Int 24(9):2493–2498

Mansouri A, Chowdhury K, Gruss P (1998) Follicular cells of the thyroid gland require Pax8 gene function. Nat Genet 19(1):87–90. https://doi.org/10.1038/ng0598-87

Mazziotti G, Maffezzoni F, Doga M et al (2014) Outcome of glucose homeostasis in patients with glucocorticoid-induced osteoporosis undergoing treatment with bone active-drugs. Bone 67:175–180

McCarthy TL, Centrella M, Canalis E (1990) Cortisol inhibits the synthesis of insulin-like growth factor-I in skeletal cells. Endocrinology 126(3):1569–1575

Meier C, Kraenzlin ME, Bodmer M, Jick SS, Jick H, Meier CR (2008) Use of thiazolidinediones and fracture risk. Arch Intern Med 168(8):820–825

Mellibovsky L, Prieto-Alhambra D, Mellibovsky F et al (2015) Bone tissue properties measurement by reference point indentation in glucocorticoid-induced osteoporosis. J Bone Miner Res 30 (9):1651–1656

Mendonca Monteiro de Barros G, Madeira M, Vieira Neto L, de Paula Paranhos Neto F, Carvalho Mendonca LM, Correa Barbosa Lima I, Corbo R, Fleiuss Farias ML (2016) Bone mineral density and bone microarchitecture after long-term suppressive levothyroxine treatment of differentiated thyroid carcinoma in young adult patients. J Bone Miner Metab 34(4):417–421. https://doi.org/10.1007/s00774-015-0680-4

Milne M, Kang MI, Cardona G, Quail JM, Braverman LE, Chin WW, Baran DT (1999) Expression of multiple thyroid hormone receptor isoforms in rat femoral and vertebral bone and in bone marrow osteogenic cultures. J Cell Biochem 74(4):684–693

Miura M, Tanaka K, Komatsu Y, Suda M, Yasoda A, Sakuma Y, Ozasa A, Nakao K (2002) Thyroid hormones promote chondrocyte differentiation in mouse ATDC5 cells and stimulate endochondral ossification in fetal mouse tibias through iodothyronine deiodinases in the growth plate. J Bone Miner Res 17(3):443–454. https://doi.org/10.1359/jbmr.2002.17.3.443

Monarni M, Cresci B, Coloumbini A, Pala L, Balzi D, Gori F, Chiasserini V, Marchionni N, Rotella CM, Mannucci E (2008) Bone fractures and hypoglycemic treatment in type 2 diabetic patients: a case-control study. Diabetes Care 31:199–203

Monfoulet LE, Rabier B, Dacquin R, Anginot A, Photsavang J, Jurdic P, Vico L, Malaval L, Chassande O (2011) Thyroid hormone receptor beta mediates thyroid hormone effects on bone remodeling and bone mass. J Bone Miner Res 26(9):2036–2044. https://doi.org/10.1002/jbmr.432

Mora S, Weber G, Marenzi K, Signorini E, Rovelli R, Proverbio MC, Chiumello G (1999) Longitudinal changes of bone density and bone resorption in hyperthyroid girls during treatment. J Bone Miner Res 14(11):1971–1977. https://doi.org/10.1359/jbmr.1999.14.11.1971

Morelli V, Eller-Vainicher C, Palmieri S et al (2016) Prediction of vertebral fractures in patients with monolateral adrenal incidentalomas. J Clin Endocrinol Metab 101(7):2768–2775

Morgan SA, McCabe EL, Gathercole LL et al (2014) 11b-HSD1 is the major regulator of the tissue-specific effects of circulating glucocorticoid excess. Proc Natl Acad Sci U S A 111(24):E2482–E2491

Mosekilde L, Melsen F (1978) Morphometric and dynamic studies of bone changes in hypothyroidism. Acta Pathol Microbiol Scand A 86(1):56–62

Mosekilde L, Eriksen EF, Charles P (1990) Effects of thyroid hormones on bone and mineral metabolism. Endocrinol Metab Clin North Am 19(1):35–63

Nakamura H, Mori T, Genma R, Suzuki Y, Natsume H, Andoh S, Kitahara R, Nagasawa S, Nishiyama K, Yoshimi T (1996) Urinary excretion of pyridinoline and deoxypyridinoline measured by immunoassay in hypothyroidism. Clin Endocrinol (Oxf) 44(4):447–451

Nixon M, Mackenzie SD, Taylor AI et al (2016) ABCC1 confers tissue-specific sensitivity to cortisol versus corticosterone: a rationale for safer glucocorticoid replacement therapy. Sci Transl Med 8(352):352ra109

O'Brien CA, Jia D, Plotkin LI (2004) Glucocorticoids act directly on osteoblasts and osteocytes to induce their apoptosis and reduce bone formation and strength. Endocrinology 145:1835–1841

O'Shea PJ, Harvey CB, Suzuki H, Kaneshige M, Kaneshige K, Cheng SY, Williams GR (2003) A thyrotoxic skeletal phenotype of advanced bone formation in mice with resistance to thyroid hormone. Mol Endocrinol 17(7):1410–1424. https://doi.org/10.1210/me.2002-0296

Paggiosi MA, Peel NF, Eastell R (2015) The impact of glucocorticoid therapy on trabecular bone score in older women. Osteoporos Int 26(6):1773–1780

Pajvani UB, Accili D (2015) The new biology of diabetes. Diabetologia 58(11):2459–2468

Palermo A, D'Onofrio L, Buzzetti R, Manfrini S, Napoli N (2017) Pathophysiology of bone fragility in patients with diabetes. Calcif Tissue Int 100(2):122–132

Pantazi H, Papapetrou PD (2000) Changes in parameters of bone and mineral metabolism during therapy for hyperthyroidism. J Clin Endocrinol Metab 85(3):1099–1106. https://doi.org/10.1210/jcem.85.3.6457

Patel P, Hardy R, Sumathi V et al (2012) Expression of 11b-hydroxysteroid dehydrogenase enzymes in human osteosarcoma: potential role in pathogenesis and as targets for treatments. Endocr Relat Cancer 19(4):589–598

Pi M, Kapoor K, Ye R et al (2016) Evidence for osteocalcin binding and activation of GPRC6A in b-cells. Endocrinology 157(5):1866–1880

Polak M, Legac I, Vuillard E, Guibourdenche J, Castanet M, Luton D (2006) Congenital hyperthyroidism: the fetus as a patient. Horm Res 65(5):235–242. https://doi.org/10.1159/000092454

Raff H, Raff JL, Duthie EH et al (1999) Elevated salivary cortisol in the evening in healthy elderly men and women: correlation with bone mineral density. J Gerontol A Biol Sci Med Sci 54(9): M479–M483

Rauch A, Seitz S, Baschant U et al (2010) Glucocorticoids suppress bone formation by attenuating osteoblast differentiation via the monomeric glucocorticoid receptor. Cell Metab 11(6):517–531

Recklinghausen FDV (1891) Die Fibröse oder deformierende Ostitis, die Osteomalazie und die osteoplastische Carzinose in ihren gegenseitigen Beziehungen. Festchrift Rudolf Virchow (German) Berlin: George Reimer, p 1

Refetoff S, Weiss RE, Usala SJ (1993) The syndromes of resistance to thyroid hormone. Endocr Rev 14(3):348–399. https://doi.org/10.1210/edrv-14-3-348

Reid DM, Devogelaer JP, Saag K et al (2009) Zoledronic acid and risedronate in the prevention and treatment of glucocorticoid induced osteoporosis (HORIZON): a multicentre, double-blind, double-dummy, randomised controlled trial. Lancet 373(9671):1253–1263

Ritz E, Kreusser W, Rambausek M (1984) Effects of glucocorticoids on calcium and phosphate excretion. Adv Exp Med Biol 171:381–397

Rosen C (2009) Bone: serotonin, leptin and the central control of bone remodeling. Nat Rev Rheumatol 5:667–668

Saag KG, Emkey R, Schnitzer TJ et al (1998) Glucocorticoid-Induced Osteoporosis Intervention Study Group. Alendronate for the prevention and treatment of glucocorticoid-induced osteoporosis. N Engl J Med 339(5):292–299

Saag KG, Shane E, Boonen S et al (2007) Teriparatide or alendronate in glucocorticoid-induced osteoporosis. N Engl J Med 357(20):2028–2039

Saag KG, Agnusdei D, Hans D et al (2016) Trabecular bone score in patients with chronic glucocorticoid therapy induced osteoporosis treated with alendronate or teriparatide. Arthritis Rheumatol 68(9):2122–2128

Saag KG, Wagman RB, Geusens P et al (2018) Denosumab versus risedronate in glucocorticoid-induced osteoporosis: a multicentre, randomised, double-blind, active controlled, double-dummy, non-inferiority study. Lancet Diabetes Endocrinol 6(6):445–454

Salcuni AS, Morelli V, Eller Vainicher C et al (2016) Adrenalectomy reduces the risk of vertebral fractures in patients with monolateral adrenal incidentalomas and subclinical hypercortisolism. Eur J Endocrinol 174(3):261–269

Sawamura M, Komatsuda A, Togashi M et al (2017) Effects of denosumab on bone metabolic markers and bone mineral density in patients treated with glucocorticoids. Intern Med 56 (6):631–636

Schulz J, Frey KR, Cooper MS et al (2016) Reduction in daily hydrocortisone dose improves bone health in primary adrenal insufficiency. Eur J Endocrinol 174(4):531–538

Shanbhogue VV, Mitchell DM, Rosen CJ et al (2016) Type 2 diabetes and the skeleton: new insights into sweet bones. Lancet Diabetes Endocrinol 4(2):159–173

Sher LB, Woitge HW, Adams DJ et al (2004) Transgenic expression of 11β -hydroxysteroid dehydrogenase type 2 in osteoblasts reveals an anabolic role for endogenous glucocorticoids in bone. Endocrinology 145:922–929

Shi XM, Blair HC, Yang X et al (2000) Tandem repeat of C/EBP binding sites mediates PPAR2 gene transcription in glucocorticoid-induced adipocyte differentiation. J Cell Biochem 76:518–527

Shi L, Sanchez-Guijo A, Hartmann MF et al (2015) Higher glucocorticoid secretion in the physiological range is associated with lower bone strength at the proximal radius in healthy children: importance of protein intake adjustment. J Bone Miner Res 30(2):240–248

Shibata H, Spencer TE, Onate SA, Jenster G, Tsai SY, Tsai MJ, O'Malley BW (1997) Role of co-activators and co-repressors in the mechanism of steroid/thyroid receptor action. Recent Prog Horm Res 52:141–164; discussion 164–165

Sivagurunathan S, Muir MM, Brennan TC et al (2005) Influence of glucocorticoids on human osteoclast generation and activity. J Bone Miner Res 20(3):390–398

Song IH, Buttgereit F (2006) Non-genomic glucocorticoid effects to provide the basis for new drug developments. Mol Cell Endocrinol 246:142–146

Soskolne WA, Schwartz Z, Goldstein M, Ornoy A (1990) The biphasic effect of triiodothyronine compared to bone resorbing effect of PTH on bone modelling of mouse long bone in vitro. Bone 11(5):301–307

Squires PE, Jones PM, Younis MY, Hills CE (2014) The calcium-sensing receptor and beta-cell function. Vitam Horm 95:249–267

Stevens DA, Hasserjian RP, Robson H, Siebler T, Shalet SM, Williams GR (2000) Thyroid hormones regulate hypertrophic chondrocyte differentiation and expression of parathyroid hormone-related peptide and its receptor during endochondral bone formation. J Bone Miner Res 15(12):2431–2442. https://doi.org/10.1359/jbmr.2000.15.12.2431

Stolk RP, van Daele PL, Pols HA, Burge RH, Hofman A, Birkenhäger JC, Lamberts SW, Grobbee DE (1996) Hyperinsulinemia and bone mineral density in an elderly population: the Rotterdam study. Bone 18:545–549

Strehl C, Bijlsma JW, de Wit M et al (2016) Defining conditions where long-term glucocorticoid treatment has an acceptably low level of harm to facilitate implementation of existing recommendations: viewpoints from an EULAR task force. Ann Rheum Dis 75(6):952–995

Strotmeyer ES, Cauley JA (2007) Diabetes mellitus, bone mineral density, and fracture risk. Curr Opin Endocrinol Diabetes Obes 14:429–435

Sui B, Hu C, Liao L, Chen Y et al (2016) Mesenchymal progenitors in osteopenias of diverse pathologies: differential characteristics in the common shift from osteoblastogenesis to adipogenesis. Sci Rep 6(1):30186

Svare A, Nilsen TI, Bjoro T, Forsmo S, Schei B, Langhammer A (2009) Hyperthyroid levels of TSH correlate with low bone mineral density: the HUNT 2 study. Eur J Endocrinol 161 (5):779–786. https://doi.org/10.1530/EJE-09-0139

Szappanos A, Patócs A, Tõke J et al (2009) BclI polymorphism of the glucocorticoid receptor gene is associated with decreased bone mineral density in patients with endogenous hypercortisolism. Clin Endocrinol (Oxf) 71(5):636–643

Szappanos A, Toke J, Lippai D et al (2010) Bone turnover in patients with endogenous Cushing's syndrome before and after successful treatment. Osteoporos Int 21(4):637–645

Tata JR (1961) The purification of thyroxine-binding globulin and thyroxine-binding prealbumin. Clin Chim Acta 6:819–832

Tomas FM, Munro HN, Young VR (1979) Effect of glucocorticoid administration on the rate of muscle protein breakdown in vivo in rats, as measured by urinary excretion of NT-methylhistidine. Biochem J 178(1):139–146

Tóth M, Grossman A (2013) Glucocorticoid-induced osteoporosis: lessons from Cushing's syndrome. Clin Endocrinol (Oxf) 79(1):1–11

Vadiveloo T, Donnan PT, Cochrane L, Leese GP (2011) The thyroid epidemiology, audit, and research study (TEARS): morbidity in patients with endogenous subclinical hyperthyroidism. J Clin Endocrinol Metab 96(5):1344–1351. https://doi.org/10.1210/jc.2010-2693

Valderrábano RJ, Linares MI (2018) Diabetes mellitus and bone health: epidemiology, etiology and implications for fracture risk stratification. Clin Diabetes Endocrinol 4:9. https://doi.org/10.1186/s40842-018-0060-9

van Staa TP, Leufkens HG, Abenhaim L et al (2000a) Oral corticosteroids and fracture risk: relationship to daily and cumulative doses. Rheumatology (Oxford) 39(12):1383–1389

van Staa TP, Leufkens HG, Abenhaim L et al (2000b) Use of oral corticosteroids and risk of fractures. J Bone Miner Res 15(6):993–1000

Varga F, Spitzer S, Klaushofer K (2004) Triiodothyronine (T3) and 1,25-dihydroxyvitamin D3 (1,25D3) inversely regulate OPG gene expression in dependence of the osteoblastic phenotype. Calcif Tissue Int 74(4):382–387. https://doi.org/10.1007/s00223-003-0033-5

Varga F, Rumpler M, Zoehrer R, Turecek C, Spitzer S, Thaler R, Paschalis EP, Klaushofer K (2010) T3 affects expression of collagen I and collagen cross-linking in bone cell cultures. Biochem Biophys Res Commun 402(2):180–185. https://doi.org/10.1016/j.bbrc.2010.08.022

Vestergaard P, Mosekilde L (2002) Fractures in patients with hyperthyroidism and hypothyroidism: a nationwide follow-up study in 16,249 patients. Thyroid 12(5):411–419. https://doi.org/10.1089/105072502760043503

Vestergaard P, Mosekilde L (2003) Hyperthyroidism, bone mineral, and fracture risk--a meta-analysis. Thyroid 13(6):585–593. https://doi.org/10.1089/105072503322238854

Vestergaard P, Lindholm J, Jørgensen JO et al (2002) Increased risk of osteoporotic fractures in patients with Cushing's syndrome. Eur J Endocrinol 146(1):51–56

Vestergaard P, Rejnmark L, Mosekilde L (2005a) Fracture risk associated with systemic and topical corticosteroids. J Intern Med 257(4):374–384

Vestergaard P, Rejnmark I, Moseklide I (2005b) Relative fracture risk in patients with diabetes mellitus, and the impact of insulin and oral antidiabetic medication on relative fracture risk. Diabetologia 48:1229–1299

Visser TJ (2000) In: Feingold KR, Anawalt B, Boyce A et al (eds) Cellular uptake of thyroid hormones. Endotext, South Dartmouth

Walsh LJ, Wong CA, Pringle M et al (1996) Use of oral corticosteroids in the community and the prevention of secondary osteoporosis: a cross sectional study. BMJ 313(7053):344–346

Wang FS, Ko JY, Yeh DW et al (2008) Modulation of Dickkopf-1 attenuates glucocorticoid induction of osteoblast apoptosis, adipocytic differentiation, and bone mass loss. Endocrinology 149:1793–1801

Wang N, Xue P, Wu X, Ma J, Wang Y, Li Y (2018) Role of sclerostin and dkk1 in bone remodeling in type 2 diabetic patients. Endocr Res 43(1):29–38

Watts NB, Bilezikian JP, Usiskin K, Edwards R, Desai M, Law G, Meininger G (2016) Effects of canagliflozin on fracture risk in patients with type 2 diabetes mellitus. J Clin Endocrinol Metab 101(1):157–166

Weber DR, Schwartz G (2016) Epidemiology of skeletal health in type 1 diabetes. Curr Osteoporos Rep 14(6):327–336

Weinstein RS, Jilka RL, Parfitt AM et al (1998) Inhibition of osteoblastogenesis and promotion of apoptosis of osteoblasts and osteocytes by glucocorticoids. Potential mechanisms of their deleterious effects on bone. J Clin Invest 102(2):274–282

Weinstein RS, Nicholas RW, Manolagas SC (2000) Apoptosis of osteocytes in glucocorticoid-induced osteonecrosis of the hip. J Clin Endocrinol Metab 85(8):2907–2912

Williams GR (2000) Cloning and characterization of two novel thyroid hormone receptor beta isoforms. Mol Cell Biol 20(22):8329–8342. https://doi.org/10.1128/mcb.20.22.8329-8342.2000

Williams GR, Bland R, Sheppard MC (1994) Characterization of thyroid hormone (T3) receptors in three osteosarcoma cell lines of distinct osteoblast phenotype: interactions among T3, vitamin D3, and retinoid signaling. Endocrinology 135(6):2375–2385. https://doi.org/10.1210/endo.135.6.7988420

Woitge H, Harrison J, Ivkosic A (2001) Cloning and in vitro characterization of a 1(I)-collagen 11b-hydroxysteroid dehydrogenase type 2 transgenes as models for osteoblast-selective inactivation of natural glucocorticoids. Endocrinology 142(3):1341–1348

Wood CL, Soucek O, Wong SC et al (2018) Animal models to explore the effects of glucocorticoids on skeletal growth and structure. J Endocrinol 236(1):R69–R91

Xiong J, Piemontese M, Onal M et al (2015) Osteocytes, not osteoblasts or lining cells, are the main source of the RANKL required for osteoclast formation in remodeling bone. PLoS One 10(9): e0138189

Yeap BB, Alfonso H, Chubb SA et al (2015) Higher serum undercarboxylated osteocalcin and other bone turnover markers are associated with reduced diabetes risk and lower estradiol concentrations in older men. J Clin Endocrinol Metab 100(1):63–71

Zarain-Herzberg A, García-Rivas G2, Estrada-Avilés R (2014) Regulation of SERCA pumps expression in diabetes. Cell Calcium 56(5):302–310

Zhang W, Yang N, Shi XM (2008) Regulation of mesenchymal stem cell osteogenic differentiation by glucocorticoid-induced leucine zipper (GILZ). J Biol Chem 283:4723–4729

Zhu TY, Griffith JF, Qin L et al (2015) Cortical thinning and progressive cortical porosity in female patients with systemic lupus erythematosus on long-term glucocorticoids: a 2-year case-control study. Osteoporos Int 26(6):1759–1771

Zofková I, Matucha P (2014) New insights into the physiology of bone regulation: the role of neurohormones. Physiol Res 63(4):421–427

Growth Factors, Carrier Materials, and Bone Repair

Erin L. Hsu and Stuart R. Stock

Contents

E. L. Hsu
Department of Orthopaedic Surgery, Feinberg School of Medicine, Northwestern University, Chicago, IL, USA

Simpson Querrey Institute, Northwestern University, Chicago, IL, USA
e-mail: e-hsu@northwestern.edu

S. R. Stock (✉)
Department of Cell and Developmental Biology, Feinberg School of Medicine, Northwestern University, Chicago, IL, USA

Simpson Querrey Institute, Northwestern University, Chicago, IL, USA
e-mail: s-stock@northwestern.edu

P. H. Stern (ed.), *Bone Regulators and Osteoporosis Therapy*,
Handbook of Experimental Pharmacology 262, https://doi.org/10.1007/164_2020_371

Abstract

This chapter provides an overview of the growth factors active in bone regeneration and healing. Both normal and impaired bone healing are discussed, with a focus on the spatiotemporal activity of the various growth factors known to be involved in the healing response. The review highlights the activities of most important growth factors impacting bone regeneration, with a particular emphasis on those being pursued for clinical translation or which have already been marketed as components of bone regenerative materials. Current approaches the use of bone grafts in clinical settings of bone repair (including bone grafts) are summarized, and carrier systems (scaffolds) for bone tissue engineering via localized growth factor delivery are reviewed. The chapter concludes with a consideration of how bone repair might be improved in the future.

Keywords

3D printing · Bone · Bone regeneration · Bone tissue engineering · Growth factors · Impaired healing · Regenerative medicine · Scaffolds

1 Introduction

Bone healing is a complex process in which various growth factors act in sequence and in concert, producing scar-free fully functional mineralized tissue to replace that which was damaged. The steps in normal bone healing are covered in Sect. 2 of this chapter. We then introduce the most critical of the growth factors involved in this process, and discuss the mechanistic, translational, and – where applicable – clinical data available to date. The spatiotemporal considerations that must be addressed when proposing delivery modalities are also discussed. Finally, we examine the use of scaffolds for growth factor delivery to augment bone healing in preclinical and clinical settings.

Although bone tissue has a remarkable capacity to regenerate after injury, the need to augment this natural process is not uncommon after fracture, trauma, or orthopedic procedures. Impaired healing can occur in "simple" fractures, taking the form of *bony nonunion*. In a recent assessment of the fracture incidence in Scotland over a 5-year period, the overall fracture rate was 11.6 per 1,000 people per year, and the rate of nonunion per fracture in adults was 1.9% (Mills et al. 2017). Similar rates are seen in the USA, where 5–10% of fractures show delayed or impaired healing (Morshed 2014; Tzioupis and Giannoudis 2007), with >100,000/year of these progressing to nonunion (Bishop et al. 2012). Risk factors for nonunion are either patient dependent or patient independent (Hak et al. 2014). The former include advanced age (e.g., osteoporosis), medical comorbidities (diabetes, osteoarthritis with rheumatoid arthritis), use of prescription nonsteroidal anti-inflammatory drugs plus opioids, smoking, and high alcohol consumption while the latter include things like fracture severity, pattern and location, degree of bone loss, presence or absence of infection, etc. (Zura et al. 2016). In addition, some bones are naturally more likely to suffer nonunion than others (Zura et al. 2016), so the clinical picture is complex.

Impaired healing can also occur in orthopedic procedures such as bone or joint fusion (arthrodesis), resulting in a failed fusion, or *pseudarthrosis*. Bone or joint fusions are performed surgically to relieve severe pain and to restore stability, for example, in the foot or ankle or in the spine (Cottrell et al. 2016). Bone grafts or engineered scaffolds are often required for successful arthrodesis, particularly in the spine (Mariscal et al. 2020; Park et al. 2019; Tuchman et al. 2017; Kaiser et al. 2014); even with grafts, impaired bone healing occurs in 5–15% of procedures in the absence of the risk factors listed above, with significantly higher nonunion rates reported in at-risk patient populations (Buza and Einhorn 2016; Hoffmann et al. 2012; Dimar et al. 2009; Andersen et al. 2001; Schmitz et al. 1999; Phan et al. 2018; Adogwa et al. 2013; Raizman et al. 2009; Grabowski and Cornett 2013). Bone grafts/scaffolds also find use in the repair of bone defects which are larger than the critical size which can naturally heal.

The economic impact of impaired bone healing is immense, not only for fractures but also in orthopedic procedures such as spinal fusions (Hak et al. 2014). In addition to indirect costs associated with lost productivity – for which it is difficult to attribute accurate estimates of financial losses – the cost of revision surgery to correct nonunions can be upward of $30,000 per patient (Hak et al. 2014; Adogwa et al. 2013; Kanakaris and Giannoudis 2007). Moreover, revision procedures are themselves not without further medical risk to the patient (Raizman et al. 2009; Ondra and Marzouk 2003). Therefore, the effective management of fractures and augmentation of bone healing are important contemporary challenges.

2 Bone Healing, Growth Factors, and Bone Grafts

The bone healing cascade is a complex and highly orchestrated process encompassing an inflammatory phase with formation of a hematoma, recruitment, and induction of mesenchymal progenitor cells to differentiate into chondroblasts and/or osteoblasts, mineralization, and remodeling (Fig. 1). A number of growth factors and other signaling molecules function to coordinate this complex process, the most well-studied being the bone morphogenetic proteins (BMPs). First identified in 1965 by Dr. Marshall Urist, this family of proteins was and remains the target of enthusiastic research investigation due to the capacity to potently induce bone formation (Urist 1965; Grgurevic et al. 2017). Beyond BMPs, a number of other growth factors are well-known to influence bone growth and repair. Among others, fibroblast growth factors (FGFs), transforming growth factor-beta (TGF-β), platelet-derived growth factor (PDGF), and vascular endothelial growth factor (VEGF) are found in bone and have various effects on cartilage and bone cells, including contributing to BMP activity.

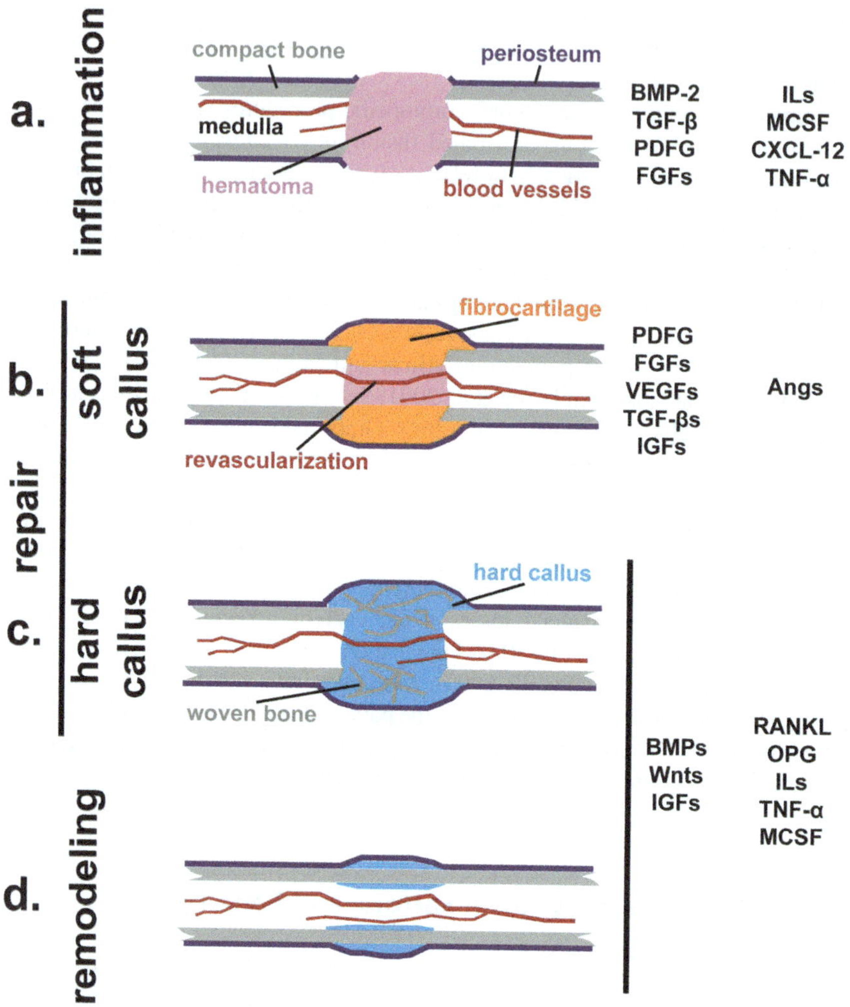

Fig. 1 Stages of bone fracture healing. The main growth factors (GF) active at each stage are listed in the first column to the right of the diagrams; the far right column lists other important factors. (**a**) Inflammation stage. The ILs include IL-1, IL-6, IL-17, and IL-18. (**b**) Repair stage – soft callus formation. FGFs; VEGFs include VEGF A, VEGF B and VEGF C; Angs are Ang-1 and Ang-2 (**c**) Repair stage – hard callus formation. (**d**) Remodeling

2.1 The Bone Healing Cascade

In the case of both acute traumatic injury and surgical intervention, the disruption of blood vessels at the site of injury results in the formation of a hematoma to provide hemostasis. This is followed by a noninfectious inflammatory phase whereby inflammatory cells such as lymphocytes, macrophages, eosinophils, and neutrophils infiltrate the defect site (Claes et al. 2012). Cytokines mediating this chemotactic response include tumor necrosis factor-α (TNF-α), interleukin-1 (IL-1), interleukin-6 (IL-6), interleukin-11 (IL-11), and (interleukin-18 (IL-18) (Einhorn and Gerstenfeld 2015).

Following this inflammatory phase, a fibrovascular response ensues, involving recruitment of blood vessels and mesenchymal progenitor/stem cells (MSCs) (Hankenson et al. 2015). Although these MSCs are derived from several niches, including the bone marrow, adipose tissue, muscle, and potentially the circulating blood, the greatest contributor to the chondroblastic/osteoblastic lineage and ultimately to bone repair appears to be the periosteum (Einhorn and Gerstenfeld 2015; Wang et al. 2016, 2017; Abou-Khalil et al. 2015). These recruited progenitor cells initially undergo a proliferative period for several days, and then respond to extracellular signaling molecules by differentiating into chondroblasts and/or osteoblasts (Wang et al. 1988; Wozney et al. 1988). Endochondral bone formation – in which a cartilage intermediate (i.e., a soft callus) forms prior to ossification – is the mechanism that ensues directly at the site of vascular disruption, due to low oxygen tension (Loiselle and Zuscik 2019). Terminally differentiated hypertrophic chondrocytes within the callus contribute to mineralization, resulting in a calcified cartilage that serves as template for bone formation, mediated by osteoblasts (Einhorn and Gerstenfeld 2015). Osteoclasts are also recruited in response to macrophage colony-stimulating factor (M-CSF), receptor activator of nuclear factor-κB ligand (RANKL), and osteoprotegerin (OPG), which contribute to early remodeling by removal of the cartilaginous callus (El-Jawhari et al. 2016).

More distal to the fracture site where the vasculature experiences less dramatic disruption, intramembranous ossification ensues, with no evidence of callus formation. Fracture stability is also a major contributor to the bone healing mechanism, whereas micromotion in non-stabilized fractures contributes to chondrogenic activity and callus formation and stabilized fractures generally heal without the cartilage intermediate. In intramembranous ossification, periosteal-derived cells undergo osteoblastic differentiation, and the differentiated osteoblasts proceed to lay down new mineral (Loiselle and Zuscik 2019; Morgan et al. 2010).

Initially, the matrix is composed of immature woven bone, which is secondarily remodeled by osteoclasts and replaced with lamellar bone. In this secondary remodeling stage, osteoclast activity is primarily mediated by IL-1, TNF-α, and RANKL (El-Jawhari et al. 2016). Cortical bone at the fracture site – where necrosis ensued due to vascular disruption – is also replaced and remodeled by osteoclasts. These remodeling processes are critical for the restoration of both structure and function of the bone (Hankenson et al. 2015). The end goal of assisted fracture healing is the formation of bridging bone at the fracture site, with re-establishment of biomechanical stability to a degree that matches the pre-fracture state (Augat et al. 2014).

2.2 Grafts for Surgical Augmentation of Bony Healing

Widely recognized as the gold standard approach for augmentation of bone healing in a wide variety of surgical settings, autograft bone is typically harvested from the iliac crest, proximal tibia, or the calcaneus. Autogenous bone contains the necessary components to promote bone repair: It is osteoinductive (inducing progenitor cells to differentiate down the osteoblastic lineage), is osteoconductive (supporting bone ingrowth), and is osteogenic (contains progenitor cells) (Sohn and Oh 2019; Ghodasra et al. 2014; De Long et al. 2007). However, the graft harvest procedure requires a longer operative time and is associated with related complications in up to 20% of patients, including greater intraoperative blood loss, an increased risk of infection, seroma formation, fracture, scarring, and longer-term donor site pain (Myeroff and Archdeacon 2011; Goulet et al. 1997; Pollock et al. 2008; Summers and Eisenstein 1989; Seiler and Johnson 2000; St John et al. 2003; Geideman et al. 2004). Even despite the well-known capacity for autograft bone graft to promote healing of bone defects, nonunions still occur at rates ranging from 5 to 40%, depending on the surgical setting and patient-specific conditions (Haddad et al. 2007; Kitaoka 1999). As such, alternatives to autograft bone which either reduce the volume of bone graft to be harvested (bone graft *extenders*) or entirely obviate the need for bone graft harvest (bone graft *substitutes*) have long been pursued by scientists and clinicians.

Growth factor delivery represents an attractive strategy for surgical augmentation of bone healing. In particular, exogenous growth factor delivery is useful in anatomic sites where the risk of nonunion is high. For example, growth factor delivery is common in the setting of tibial fractures, where the risk of nonunion is high due to limited vascularization at that anatomic site (Tarkin et al. 2010). This contrasts with femoral fractures, where the risk of nonunion is lower, due to a greater degree of vascularization. Growth factors have a well-established track record for osteoinductive, mitogenic, and/or chemotactic activity, demonstrating the capacity to promote effective bone regeneration. However, the mode of delivery has a major impact on both safety and efficacy of the growth factor within a given clinical indication. As discussed in the following sections, several FDA-approved products exist for augmentation of bone repair, but the carriers used to deliver those growth factors are relatively unsophisticated and do not allow for temporal control of local growth factor delivery. With inefficient delivery modalities, high doses of growth factor may be required to achieve the desired potent osteoinductivity; however, supraphysiologic local concentrations of growth factor is associated with adverse effects, such as seroma formation, ectopic, and heterotopic bone formation, etc. (Shi et al. 2017; Fu et al. 2013; Ekrol et al. 2008; Laursen et al. 1999). Therefore, more advanced scaffold delivery systems would enable greater spatiotemporal control over growth factor release, thereby reducing the therapeutic dose while avoiding off-target effects.

3 Growth Factors Involved in Bone Regeneration

Although a plethora of growth factors and other signaling molecules play established roles in bone regeneration and healing, only a relatively small subset of these have been or are being pursued for clinical translation. The most well-known of these are BMPs-2 and -7, but FGF2 and PDGF are also FDA-approved for clinical use, and BMP-6 approval is being pursued currently. These are discussed in detail below, followed by a limited description of other growth factors and signaling molecules that have shown significant promise in preclinical studies. Although the latter are not yet clinically available, they show significant promise in preclinical testing, making them good candidates for eventual evaluation in clinical studies.

3.1 BMP-2

Of the BMPs with known osteochondral function (Even et al. 2012), BMP-2 is the most well-studied. BMP-2 is a disulfide-linked homodimer with the capacity to induce both chondrocyte and osteoblast differentiation (Gothard et al. 2014); as such, it can promote both endochondral and intramembranous bone formation (Hankenson et al. 2015). The cloning of BMP-2 enabled the generation of conditional knockout mouse models, which showed early and severe skeletal abnormalities. Conditional knockout of BMP-2 in the limb bud mesenchyme caused delayed ossification, producing spontaneous forelimb fractures by 13 weeks and hind limb fractures by 23 weeks of age (Tsuji et al. 2006). Histological analysis showed the presence of undifferentiated MSCs at the fracture site with no evidence of chondrogenesis. These results stand in contrast to the deletion of other BMPs in the same setting, where there appears to be some compensation for loss of activity (Tsuji et al. 2008, 2010).

With the availability of recombinant human (rh) BMP-2 came a flood of in vivo studies exploring the utility of the growth factor for bone regenerative purposes. A large number of these studies demonstrated the capacity for direct delivery of rhBMP-2 to repair bone defects in both small (Yasko et al. 1992; Kamal et al. 2019) and large animals (Gothard et al. 2014; He et al. 2010). This includes the repair of segmental defects (both endochondral and intramembranous healing models) (Boerckel et al. 2012; Kirker-Head et al. 1998), lumbar fusions (Akamaru et al. 2003; Fu et al. 2009), and calvarial defects (He et al. 2010), in addition to many studies demonstrating ectopic bone formation in subcutaneous and intramuscular models (Fu et al. 2010; Kimura et al. 2010; Saito and Takaoka 2003; Luca et al. 2010, 2011). A number of other groups have employed gene therapy approaches for indirect delivery of the growth factor (Gothard et al. 2014).

Some clinical studies have suggested that rhBMP-2 use can obviate the need for autogenous bone graft, achieving similar or even superior outcomes relative to autograft bone (Mulconrey et al. 2008; Boden et al. 2000, 2002). The efficacy of rhBMP-2 has been reported in a variety of clinical settings, including open tibial shaft and other traumatic extremity fractures (Govender et al. 2002; Hagen et al.

2012), as well as interbody and posterolateral spine fusions (Kaiser et al. 2014; Hoffmann et al. 2012; Fu et al. 2013; Mulconrey et al. 2008; Boden et al. 2000, 2002; Burkus et al. 2002, 2003; Kraiwattanapong et al. 2005). However, a number of studies also suggest that initial reports of BMP-2 efficacy were overzealous, and adverse events were underreported (Fu et al. 2013; Aro et al. 2011; Lyon et al. 2013). Complications such as heterotopic and ectopic bone formation, inappropriate inflammation, bone resorption, urogenital complications, dysphagia (in the setting of cervical spine), and even induction of cancer have raised serious concerns from patients and clinicians alike over the use of rhBMP-2 (Fu et al. 2013; Garrison et al. 2010). Due to the association of these side effects with supraphysiologic dosing, surgeons have in recent years moved toward using rhBMP-2 more judiciously. Although improved carrier technologies exist and continue to be developed, they are not yet clinically available. FDA approval of a more efficient carrier for rhBMP-2 would likely reduce the dose necessary to achieve high rates of union, while simultaneously reducing complications associated with supraphysiologic use.

3.2 BMP-7

Unlike the conditional limb bud knockout for BMP-2, a similar strategy for BMP-7 deletion showed a less severe phenotype, with no spontaneous fractures or reduced bone mineral density. Further work showed that bone healing in these animals was not impaired (Tsuji et al. 2010), suggesting that BMP-7 is not absolutely required for normal bone healing and that other BMPs can compensate for the lack of this isoform. Even so, its capacity for osteoinductivity is well-established.

BMP-7 has been administered both directly and indirectly in a number of bone regenerative models, using both small and large animals. After showing significant osteoinductivity in small animal ectopic implantation models (Sampath et al. 1992), the growth factor was investigated extensively for its utility in preclinical fracture healing and spine fusion applications. BMP-7 has shown the capacity to promote bone healing in rats, rabbits, dogs, goats, sheep, and nonhuman primates, using collagen matrices or other similar carriers (Ripamonti et al. 2001a, b; Bright et al. 2006; Barrack et al. 2003; Blattert et al. 2002; Blokhuis et al. 2001; Cook et al. 1994, 2002; Cunningham et al. 1999, 2002; den Boer et al. 2002, 2003; Jenis et al. 2002; Lietman et al. 2005; Magin and Delling 2001; Salkeld et al. 2001). Interestingly, although BMP-7 delivered to closed tibial fractures in goats showed some capacity to accelerate fracture healing, delivery of the growth factor on a collagen matrix did not prove beneficial relative to direct delivery of BMP-7 in aqueous solution (Blokhuis et al. 2001; den Boer et al. 2002). Nonetheless, the large body of literature demonstrating positive effects of BMP-7 on bone regeneration, with outcomes in some cases surpassing autogenous bone grafting (den Boer et al. 2003; Cook et al. 1995), prompted the eventual translation and adoption for clinical use (den Boer et al. 2003; Cook et al. 1995).

The implementation of BMP-7 in human patients was first reported for the treatment of tibial osteotomy. Geesink et al. reported in 1999 the direct delivery of rhBMP-7 using a type 1 collagen matrix for enhanced bone formation (Geesink et al. 1999). The first prospective randomized controlled trial compared rhBMP-7, known clinically as OP-1 Device, with autogenous bone graft in the ability to promote healing of established tibial nonunions (Friedlaender et al. 2001). In that study, 3.5 mg of rhOP-1 was delivered to the defect using a bovine type 1 collagen matrix, where clinical outcomes were comparable to the autograft cohort, while obviating the need for autograft harvest and the associated morbidities. The growth factor has since been delivered using other relatively simple modalities, including with hydroxyapatite (hAp) and as an adjunct to allograft bone, with encouraging results (Lietman et al. 2005; Caterini et al. 2016). Notably, rhBMP-7 is not approved for spinal indications, and its use in that setting requires a humanitarian device exemption. However, its use as an adjunct to local autograft has been recommended as an alternative to iliac crest bone grafts for single-level, instrumented spine fusions (Kaiser et al. 2014). On the other hand, conflicting evidence for its efficacy without concurrent delivery of autograft bone – i.e., delivery of rhBMP-7 using an absorbable collagen sponge alone – has dictated that no recommendation be made for its use as a bone graft substitute for spine fusion procedures (Kaiser et al. 2014). Moreover, concerns similar to those associated with supraphysiologic rhBMP-2 exist, including reports of bone resorption with rhBMP-7 use (Ekrol et al. 2008; Laursen et al. 1999).

3.3 BMP-6

After the discovery that rhBMP-6 was osteoinductive in vitro (Yang et al. 2003), its capacity for promoting bone formation and healing was evaluated in preclinical models. In a rabbit model of posterolateral spine fusion, rhBMP-6-induced BMSCs were shown to enhance bone formation and healing (Valdes et al. 2007). MSCs transfected with rhBMP-6 also showed the capacity to produce robust bone formation in the same model (Sheyn et al. 2008). There is some indication that BMP-6 may act on bone via IGF-1 and epidermal growth factor (EGF). When given systemically to enhance bone formation in mice, rhBMP-6 induced expression of IGF-1 and EGF (Grasser et al. 2007). Similar effects were also seen in primary human osteoblasts treated with rhBMP-6, with upregulation of both IGF-1 and EGF and concurrent increases in osteoinductivity markers. When co-delivered with IGF-1 in a rat ectopic bone formation model, bone regeneration and mineralization were enhanced relative to BMP-6 alone (Rico-Llanos et al. 2017). Studies comparing that the osteoinductivity and bone forming capacity of rhBMP-6 with that of either rhBMP-7 or rhBMP-2 have shown conflicting results (Rico-Llanos et al. 2017; Taipaleenmaki et al. 2008; Mizrahi et al. 2013; Vukicevic et al. 2014a). Despite this, continued investigations exploring the utility of rhBMP-6 in combination with a variety of carriers for bone regenerative medicine are ongoing (Gümüşderelioğlu et al. 2019; Li et al. 2019; Vukicevic et al. 2020; Grgurevic et al. 2019).

RhBMP-6 is currently under development for clinical bone regenerative applications as a component of a device (OSTEOGROW). The rhBMP-6 carrier is an autologous whole blood-derived coagulum, which since taken from a patient's peripheral blood is biocompatible and therefore less inflammatory, The impetus for its development has in part been as a response to the challenges noted with the clinical use of rhBMP-2 and rhBMP-7, which are attributed to the burst release resulting from inefficient binding of the supraphysiologic doses of the growth factors to their carriers (Vukicevic et al. 2014b). OSTEOGROW reportedly employs lower doses of the growth factor due to its greater affinity for its carrier and appears to stimulate MSC differentiation and uncouple bone formation from bone resorption (Vukicevic et al. 2014a). The higher carrier affinity could reduce or eliminate the burst release seen with rhBMP-2 and rhBMP-7 products currently on the market, potentially enabling a lower therapeutic dose (Grgurevic 2016). Currently, clinical studies are evaluating the device for acute radius fracture and tibial nonunion indications; however, the literature on this product is still limited, and both safety and efficacy benchmarks must be met before FDA approval and clinical adoption.

3.4 FGFs

Fibroblast growth factors (FGFs) are a large group of structurally related proteins that are involved in a diverse array of biological processes, including tissue repair. Several FGFs, including FGFs 1, 2, 9, and 18 are expressed in bone and can influence bone healing (Lin et al. 2009). FGF23 is a soluble (hormonal) factor that regulates phosphate and vitamin D metabolism and plays an intimate role in bone mineralization (Shimada et al. 2004). Other than the intracellular (FGF11–14) and endocrine (FGFs 19, 21, and 23) members, all FGFs bind to FGF receptors (FGFRs 1-4) to exert their function.

FGFs have mitogenic effects on many different cell types relevant to bone healing, including endothelial cells, MSCs, and differentiated osteoblasts (Globus et al. 1989; Itoh and Ornitz 2004). Although the precise mechanisms by which FGF/FGFR signaling impacts bone regeneration are still unclear, the temporal expression pattern during normal tibial fracture healing in mice has been reported. In that study, FGF9 was induced in the early phase of bone healing, whereas FGFs 16 and 18 were induced in the late stage (Schmid et al. 2009). FGFs 2, 5, and 6 were induced throughout the healing process. Interestingly, FGF receptor expression was greatest in late stages of healing. Of the FGFs, FGF2 is the most abundant in bone, has the greatest expression during bone formation, and is the most relevant for augmentation of fracture healing (Khan et al. 2000).

Genetic studies confirm the importance of FGF2 in bone formation, as *FGF2*-null mice have reduced bone mass, and bone marrow stromal cells from these mice have a reduced capacity for osteoblast differentiation (Montero et al. 2000), possibly by modulation of Wnt signaling (Fei et al. 2011). However, nonspecific overexpression of FGF2 in mice results in impaired bone mineralization and a dwarf phenotype, suggesting a potential negative regulatory role and underscoring the need for a targeted approach in leveraging the growth factor for therapeutic purposes (Coffin et al. 1995).

A number of preclinical studies have evaluated the utility of FGF2 for enhancing bone healing. Local delivery of FGF2 has been investigated in the treatment of tibial shaft fractures, where bone union (as determined by radiography) was significantly and dose-dependently higher with FGF2 treatment relative to placebo control. In the rat calvarial defect model, FGF2 has been delivered using various carriers, where it has enhanced bone regeneration and healing (Yoshida et al. 2015; Hong et al. 2010; Tanaka et al. 2006). FGF2 has also been evaluated as a means to enhance healing of ulnar fractures in nonhuman primates, where the growth factor promoted successful union in 100% of FGF2-treated animals, whereas vehicle-treated animals achieved union at a rate of 40% (Kawaguchi et al. 2001). Bone composition and strength were also enhanced with FGF2 treatment in those animals.

Clinical studies involving FGF2 have been conducted primarily in the arena of periodontal reconstruction (Kitamura et al. 2016; Cochran et al. 2016; de Santana and de Santana 2015; Kitamura et al. 2011; Kitamura et al. 2008). A recent randomized controlled trial evaluating rhFGF2 delivered using a β-tricalcium phosphate (TCP) carrier for the treatment of periodontal defects, the growth factor dose-dependently enhanced healing, with no reported adverse events (Cochran et al. 2016). Similar success was seen with delivery using a hyaluronic acid carrier, with improved regeneration and clinical outcomes (de Santana and de Santana 2015). In orthopedic applications, rhFGF2 was evaluated in a randomized controlled trial in patients undergoing surgical treatment for fresh tibial shaft fracture. Delivered using a gelatin hydrogel, both doses of rhFGF2 evaluated induced significantly higher union rates relative to placebo controls, with no adverse events attributed to rhFGF2 treatment (Kawaguchi et al. 2010). In another study of osteoarthritis patients undergoing high tibial osteotomy (HTO), rhFGF2 dose-dependently enhanced healing, again without any adverse events noted (Kawaguchi et al. 2007).

3.5 PDGF

PDGF is considered a key factor in bone repair, with actions as a mitogen, chemoattractant, and angiogenic mediator (Hollinger et al. 2008a). The growth factor was originally identified in platelets, but subsequently found in other tissues, including bone. It is composed of two polypeptide chains, which form a bioactive homodimer or heterodimer (PDGF-AA, BB, AB, CC, or DD dimers). The dimers signal through two different cell surface receptors, PDGFR-α and PDGF-β; these form both homodimers and heterodimers, and they have different binding affinities for the various isoforms (Seifert et al. 1989). PDGF-BB has the ability to bind all receptor types and is considered the universal PDGF isoform (Hollinger et al. 2008a; Alvarez et al. 2006); as such, it has received the most attention for its potential for translation to clinical applications (Hankenson et al. 2015).

PDGF is an early player in the bone healing cascade. Upon initial injury or surgical intervention, the inflammatory phase is initiated, with cells and pro-inflammatory cytokines flooding the defect site (Caplan and Correa 2011). At formation of the initial hematoma, platelets aggregate and release their PDGF-containing granules into the clot. This attracts and activates neutrophils and macrophages, which themselves then serve as a continued source of PDGF and other growth factors and chemoattractants (Hollinger et al. 2008a). PDGF binds to cell surface receptors on MSCs and pre-osteoblasts and activates phosphoinositide-3 kinase (PI3K)-, AKT-, Stat3-, Grb2-, and Rho/Rac-mediated signaling pathways (Missbach et al. 1999), which are involved in cell proliferation and migration, thereby exerting mitogenic and chemotactic activities (Hankenson et al. 2015; Hollinger et al. 2008a; Tanaka and Liang 1995). BMPs, hedgehog proteins, Wnts, and other factors then act to on these progenitors to induce differentiation into chondrocytes or osteoblasts (Cho et al. 2002; Kugimiya et al. 2005; Murakami and Noda 2000; Hadjiargyrou et al. 2002). However, PDGFs can also induce the BMP inhibitor, gremlin, and thereby influence the degree to which osteoblasts respond to BMPs (Pereira et al. 2000). In another recent study, PDGF signaling was shown to promote negatively regulate osteogenic differentiation of periosteal-derived progenitor cells via inhibition of canonical BMP-2 signaling (Wang et al. 2019). Beyond influencing cell migration, infiltration, and osteoinductivity, PDGF also induces VEGF expression, promoting neovascularization at the bone defect site (Bouletreau et al. 2002; Sato et al. 1993; Guo et al. 2003). The multifaceted biological effects of PDGF imply a need for careful consideration of spatiotemporal factors in delivery of the growth factor to promote bone healing.

In support of mechanistic results indicating a role for PDGF in the bone healing cascade, preclinical studies have shown promise for use of PDGF-BB to promote bone repair. The utility of rhPDGF-BB has been investigated in both rat and rabbit tibial fracture models, where it promoted defect healing, with new bone showing equivalent biomechanical strength to nonoperative controls (Nash et al. 1994; Hollinger et al. 2008b). In a rat calvarial defect, BMSCs overexpressing PDGF-BB were superior to control BMSCs in promoting osteogenesis and angiogenesis (Zhang et al. 2018). In combination with BMSCs and β-TCP, rhPDGF also showed significant potential to enhance bone regeneration and osteointegration in a canine mandibular model (Xu et al. 2016).

The PDGF-BB isoform has now been extensively investigated and is regarded as safe for clinical use in the setting of foot and ankle (Solchaga et al. 2012). To date, three prospective randomized controlled clinical trials have been performed evaluating its utility in this setting in comparison to autogenous bone graft (Daniels et al. 2015; Digiovanni et al. 2011, 2013). A recent meta-analysis of these studies confirms that although longer-term follow-up studies are called for, radiological and clinical outcomes appear similar for rhPDGF-BB treatment vs. autogenous bone graft (Sun et al. 2017).

3.6 Other Growth Factors with Clinical Potential

The growth factors described above are the most well-studied as modulators of bone formation. However, beyond these clinically approved growth factor-based products, a plethora of other growth factors and signaling molecules have also shown significant promise for enhancing bone healing in preclinical studies. One example is vascular endothelial growth factor (VEGF), which appears to have multiple functions in the bone regenerative process (Hankenson et al. 2015). A lack of vascularization of the callus tissue is one of the primary risk factors for nonunion. Certainly, its role as an inducer of new blood vessel formation is critical to the formation of vascularized bone tissue and for delivery of nutrients and oxygen to the healing bone and surrounding soft tissue. A functional vasculature also acts as a conduit for progenitor cells and osteoblasts (Hankenson et al. 2015). However, the growth factor also has direct effects on osteoblasts via BMP production in endothelial cells (Hu and Olsen 2016a). In a mouse femur fracture model, intraoperative application of exogenous VEGF increased both callus ossification and vascularity of adjacent soft tissue (Street et al. 2002). The same group found that when applied in a rabbit segmental defect model, subcutaneous injection of VEGF for 7 days postoperatively significantly increased both callus volume and calcification. Numerous other studies have shown promise for VEGF to potentiate BMP-2-mediated osteogenic activity (Hu and Olsen 2016b; Xiao et al. 2011; Schorn et al. 2017), as discussed [in the next section]. Despite this promise, the short half-life of VEGF (6–8 h) combined with the notion that excessive VEGF could increase the risk of malignancy, means that its delivery for bone regenerative purposes must be rationally controlled (Ennett et al. 2006; Kaigler et al. 2006; Leach et al. 2006; Kasten et al. 2012).

Other examples of growth factors and signaling molecules which have demonstrated the capacity to enhance bone healing in preclinical models include members of the transforming growth factor-β (TGF-β) family (Kandziora et al. 2002a, b; Schmidmaier et al. 2002, 2003a, b), insulin-like growth factor (IGF) (Kandziora et al. 2002a, b; Schmidmaier et al. 2002, 2003a, b; Meinel et al. 2003), stromal cell-derived factor-1 (SDF-1), or CXCL12 (Ho et al. 2015; Herberg et al. 2014, 2015), Indian hedgehog (IHH) (Kim et al. 2013; Zou et al. 2014), and modulators of Wnt signaling (Chen et al. 2007; Komatsu et al. 2010; McGee-Lawrence et al. 2013; Agholme et al. 2010; Minear et al. 2010), among others.

3.7 Combination Growth Factor Delivery for Bone Regeneration

Although no FDA-approved combination therapies that incorporate multiple growth factors currently exist, this approach has been proposed by many groups in recent years. The complexity of the bone healing cascade, which is subject to many spatiotemporal-influencing factors, makes multi-growth factor delivery a promising approach for surgical augmentation (Hadjiargyrou and O'Keefe 2014). The discovery that FGF-2 up-regulates PDGF receptors prompted co-administration of PDGF-BB with VEGF and FGF-2, which enhanced corneal and ischemic limb revascularization (Richardson et al. 2001; Cao et al. 2003). The ability to enhance both osteoinductivity and vascularization could provide additive or synergistic benefits, as could the enhancement of chemotaxis for progenitors, or the growth of those cells. Indeed, VEGF co-delivery with BMP-2 has shown the most promise for promoting robust bone formation, likely attributable to the capacity of the former to both promote angiogenesis and potentiate BMP-2-mediated osteoinductivity (Gothard et al. 2014; Xiao et al. 2011; Schorn et al. 2017; Deckers et al. 2002; Fu et al. 2015; Hou et al. 2009; Huang et al. 2018; Peng et al. 2002; Sukul et al. 2015; Wang et al. 2018, 2020).

Unsurprisingly, studies have also investigated the utility of delivering multiple BMPs for potentiation of osteoinductivity, most notably BMPs 2 and 7 (Sun et al. 2012; Wang et al. 2012; Koh et al. 2008). Although co-delivery of BMPs + TGF-β has shown mixed results (Ripamonti et al. 2001b; Thorey et al. 2011), another series of studies found that co-delivery of TGF-β and IGF-1 showed promise for promoting enhanced osteogenic differentiation and bone healing relative to either growth factor alone (Kandziora et al. 2002a, b; Schmidmaier et al. 2002, 2003a, b). The coordination of bone formation and bone resorption has also been discussed in the context of BMP-2 and TGF-β1 delivery (Tang et al. 2009).

Synergism between FGFs and BMPs has also been proposed. FGF18 appears to regulate BMP-2 expression (Nagayama et al. 2013; Reinhold et al. 2004) and may in this way act to enhance BMP action in vivo (Fujioka-Kobayashi et al. 2012). Interestingly, in addition to regulating PDGF activity, it has also been proposed that FGF2 is an upstream regulator of BMP-2, showing inhibitory effects on the BMP-2 antagonist, noggin (Fakhry et al. 2005). FGF-2 overexpression in MSCs enhanced chondrogenic markers, but downregulated Runx2, Col1, and ALP expression, the latter of which was postulated to be caused by upregulation of MMP-13 (Cucchiarini et al. 2011). In a mandibular co-delivery study with rhBMP-2 and FGF2, the addition of FGF2 resulted in poorer outcomes relative to rhBMP-2-treatment alone (Springer et al. 2008). These studies highlight the need for careful tuning of delivery to achieve appropriate temporal expression of each growth factor for the desired regenerative result.

3.8 Spatiotemporal Growth Factor Delivery Considerations

An important consideration for co-delivery of growth factors is the spatiotemporal nature of their native expression patterns. The temporal and regional gene expression patterns during the healing process have been clarified to some degree, and this information could potentially be used as a benchmark to facilitate augmentation of bone healing where needed. Moreover, monitoring of these benchmarks could be used to identify cases of delayed healing, especially in high-risk patients, such as the aging or diabetic populations. Although only a small number of growth factors have been approved for clinical use, a far greater number have been investigated for their roles in the bone regenerative process. Eventually, this information could enable a more sophisticated application of bioactive molecule delivery for further enhancement of healing.

The temporal expression pattern of TGF-β superfamily members during fracture healing in mice over a 28-day period revealed that despite the close structure and functional relationship of this family of growth factors, each appears to have a distinct role in the healing process. Generally, the mRNA and protein expression levels correlated with the expected pattern of healing, reflecting the known stages within the bone healing cascade. For example, inflammatory cytokines were induced very early in the process, with peak expression of IL-1β and IL-6 at 1 day post-injury (Cho et al. 2002). Coincident with this phase was BMP-2 expression, peaking at 1 day postoperative, then tapering off before rising again at day 7, after which it stayed high until day 21. As reported elsewhere, chondrogenic markers followed the inflammatory phase, with COL1A1 showing gradual expression beginning at the time of injury and peaking at 1–2 weeks, with continued expression out to day 28. TGF-β2 and TGF-β3 were maximally expressed on day 7, although TGF-β1 was constitutively expressed from days 3–21. As expected, osteoblastic markers reflective of bone mineralization rose continuously from around day 7 until days 21–28. During this phase, BMP-4 and -7 showed peaks around day 14, whereas BMP-6 peaked at day 7, tapered off at day 14, and peaked again at day 21 postoperative.

Preclinical studies have highlighted the importance of the spatiotemporal considerations in delivering growth factors for bone regeneration. For example, PDGF has clear mitogenic, chemotactic, and angiogenic influences on various cell types, and it might stand to reason that co-delivery of PDGF with a potent osteoinductive growth factor such as BMP-2 could further boost the regenerative response. However, recent studies suggesting that PDGF may negatively regulate BMP signaling and underscore the need to tightly control delivery such that the growth factors do not unintentionally counteract their respective actions (Wang et al. 2019). In another study which also underscores this notion, brief or sequential treatment of FGFs 2 and 9 enhanced osteogenic markers in primary calvarial osteoblasts, whereas continuous treatment inhibited these markers (Fakhry et al. 2005).

Clinical studies also support the need for temporal control in growth factor delivery. Hara et al. quantified expression levels of TGF-β1 and BMP-2 in the plasma of patients after fracture, and correlated expression to union and nonunion status. Maximal peak expression for both growth factors was delayed by 1 week in the patients who experienced a nonunion relative to those showing normal healing (Hara et al. 2017). Nonetheless, a thorough understanding of the temporal orchestration of growth factor expression during fracture healing, combined with more advanced delivery modalities, presents the opportunity to enhance the healing process using rationally designed approaches for multiple growth factor delivery.

4 Scaffolds for Bone Regeneration

In broad terms, use of scaffolds in bone healing contexts is a subset of tissue engineering. Research into scaffolds for bone applications is a very active field; one biomedical materials journal published a dozen such papers in the first quarter of 2019 (J Biomed Mater Res A, Vol 107, Nos. 1–3). Similar to all other tissue engineering applications, engineered solutions for bone healing ideally include (a) a biochemically and biomechanically compatible structure or matrix that mimics important characteristics of the extra cellular matrix (ECM), (b) morphogenic signals to recruit and direct osteogenic cells, and (c) constituents or structures which allow vascularization to supply nutrients to the new tissue. A simple form of tissue engineering is the immobilization of adjoining bone segments by pins and plates, something encountered in sports medicine. Here, (a)–(c) are intrinsic to the underlying tissue and can heal the fracture, provided that the pieces of bone can be held in close proximity. However, if bone damage is extensive, if the bone gap is too large (i.e., greater than the critical defect size), or if underlying conditions interfere with proper bone healing, scaffolds may be required and will presumably perform best if (a)–(c) are characteristics of the scaffold.

Scaffold properties fall into three general categories: biological, mechanical, and structural. The different size scales range from centimeter to the scale of individual collagen microfibrils and mineral nanoplatelets, i.e., below 100 nm. Some aspects of each grouping are absolutely essential, and others may be desirable.

Biologically, a scaffold for bone healing must be biocompatible and nontoxic, both systemically and locally. For example, poly(lactic-co-glycolic acid (PLGA) is an FDA-approved material for implantation (Makadia and Siegel 2011); its degradation, however, may produce a locally acidic environment which may be inhospitable for osteoprogenitor cells. The scaffold must allow cells to invade it and to produce new matrix that subsequently mineralizes. Biocompatibility with surrounding tissues is essential for clinical success. Ideally, the scaffold should also be biodegradable so that it has been completely replaced by functional bone when healing is complete.

In terms of mechanical requirements for bone scaffolds, it is important to differentiate between nonstructural vs structural scaffolds/grafts. Examples of the former are paste applied to site where bone formation is required, e.g., demineralized bone matrix (DBM) (Morris et al. 2018) or an amorphous collagen sponge loaded with BMP-2 (Burkus et al. 2002) commercially distributed as Infuse™; these add no mechanical rigidity and are strictly delivery platforms. Structural scaffolds are typically designed to have mechanical properties that match those of the tissue into which they will integrate. Scaffolds that are more rigid than bone are typically avoided, because the "excess" stiffness may lead to deleterious effects such as stress-shielding and concomitant localized osteopenia and/or scaffold detachment from the surrounding bone (Denard et al. 2018). The mechanical properties of central interest are the Young's modulus (E) and compressive strength ($\sigma_{y\text{-}c}$). Scaffolds are often tailored to match E and $\sigma_{y\text{-}c}$ of human cancellous bone (2–12 MPa and 0.1–5 GPa, respectively (Wu et al. 2014)) rather than compact bone, which is in excess of 10 times the strength and greater than twice the modulus (Currey 2002). For quantities like modulus, it is important to differentiate between stiffness of the structure as an entity and the stiffness of the material of which it is made.

The structural characteristics of a scaffold should ideally recapitulate the hierarchical structural levels of bone. One can differentiate at least nine levels of structure within bone (Reznikov et al. 2014), and, as discussed below, current scaffolds are simpler, and their designers focus on only a few hierarchical levels. At least two levels of porosity are essential: interconnected microporosity at the 100–1,000 μm level and nanoporosity/nanopatterning of surfaces. Interconnecting microporosity allows (a) blood vessels to grow into the structure, thereby promoting biological integration with the surrounding tissue, (b) counterflow of nutrients and metabolic waste and (potentially) scaffold degradation products or growth factors released from the scaffold, (c) bone ingrowth toward the center of the scaffold, and (d) mechanical interlocking of scaffold and regenerated bone, i.e., osteointegration. Nanoporosity/nanotopography appears to influence osteoinductivity (Qasim et al. 2019) and can be introduced as nanopatterns or can take the form of stochastically distributed surface topology or nanopores (Roseti et al. 2017). An additional structural modulation that is receiving increasing attention is the incorporation of carriers for growth factors and other proteins into scaffolds, for improved control over their release rates (De Witte et al. 2018).

Although scaffolds are produced by a wide range of methods (Roseti et al. 2017), recent years have seen a drastic increase in additive manufacturing techniques such as 3D printing. There are many materials systems being developed and tested using 3D printing approaches. These scaffolds can be produced rapidly, and design iterations can be introduced by changing software files rather than through hardware alterations. Another significant advantage of 3D printing is that once a scaffold system is proven effective, patient-specific scaffolds can be produced rapidly, for example, to exactly match the shape of a nonplanar, critical-sized defect. Alternatively, general purpose 3D-printed scaffolds can be produced and trimmed by the surgeon in the OR. Challenges do still exist in deploying these scaffolds in a minimally invasive surgery setting.

4.1 Metals

Metals have historically been used as orthopedic materials, mainly because of the high strength and fracture resistance needed for load-bearing applications. Titanium and stainless steel implants are also inert. The use of FDA-approved base materials is attractive for porous scaffolds because of the reduced regulatory hurdles. Porous metal scaffolds can be designed to have structural elastic moduli close to that of cancellous bone (rather than the much higher material modulus) and to allow vascular infiltration. For example, increasing porosity from 44% to 65% decreases a Ti scaffold modulus from 1.2 to 0.18 GPa (He et al. 2012). Manufacturing processes include 3D printing followed by sintering (Vlad et al. 2020), conventional casting using a porous mold (Augustin et al. 2020), freeze casting (Yan et al. 2017), or even pressing wires together (He et al. 2012), but additive manufacturing seems to be the current processing focus (Putra et al. 2020). Metallic implants for skull and other craniofacial applications (Park et al. 2016) or for salvage of severely damaged feet/ankles (Hamid et al. 2016) must match the surrounding skeletal contours; patient-specific additive manufacturing has achieved clinical success, perhaps in part because such closely fitting structures are less likely to loosen and more likely to integrate with surrounding bone. However, it is important to note that if porosity is produced by stochastically arranged pores, a minimum level of porosity is required to produce an adequately interconnected structure. Moreover, if a metal scaffold (e.g., Ti, stainless steel) does not degrade in the body, it may not integrate with the surrounding tissue (De Witte et al. 2018); Mg alloy scaffolds (Augustin et al. 2020) corrode in vivo, releasing ions shown to be beneficial to bone regeneration, but rapid degradation rates may lead to nonfunctional scaffolds or potential toxicity (Yang et al. 2018).

4.2 Ceramics

Ceramic implants and coatings also have a long history in orthopedics. Ceramics can be readily incorporated into the porous structures required to facilitate bone healing. Ceramics provide good compressive strength, even in porous structures. However, they are very brittle, which is undesirable in weight-bearing applications. Some but not all ceramics offer bioactivity (osteoconductivity) and tailorable degradation rates, producing ions important in biomineralization. The most common crystalline materials are calcium phosphates [(hAp) and β-TCP], as well as bioactive glass (Fernandes et al. 2018). At the nanometer level, calcium phosphates present surfaces that are of the same makeup as those of the biomineral carbonated hAp (cAp) found in bone, and they do not possess immunogenicity or toxic side effects (Kolk et al. 2012).

Given their brittleness, ceramic scaffolds find use in non-load-bearing locations such as filling bone defects in the oral cavity or critical-sized calvarial defects (Ma et al. 2018). In a study of a hierarchically structured bioglass ceramic scaffold, considerable bone ingrowth was observed 12 weeks post-implantation in a critical-sized calvarial defect in rabbits (Feng et al. 2017). Coating β-TCP scaffold struts with mesoporous bioactive glass improved bone ingrowth into scaffolds filling rabbit calvarial defects 8 weeks post-implantation (Zhang et al. 2015).

4.3 Polymers

Natural polymers and synthetic polymers both find considerable application in scaffolds for bone regeneration, either as stand-alone materials or as components of composites. A brief summary of these materials follows below, and further details are reviewed elsewhere (De Witte et al. 2018).

Natural polymers used in scaffolds include collagen (Hertweck et al. 2018), popular because type I collagen is a principal component of bone; polysaccharides such as chitosan (Lauritano et al. 2020) – derived from the chitin exoskeletons of shellfish and fungi (Croisier and Jérôme 2013) – and alginate (Venkatesan et al. 2015), a biopolymer found in seaweed; silk fibroin (Sartika et al. 2020), a biopolymer produced by the silkworm *Bombyx mori*; and cellulose. The osteoinductive characteristics of natural polymers make them very attractive as scaffolds and include presentation of a range of ligands or of structural motifs (e.g., β-sheet domains (Donnaloja et al. 2020)) favorable for precursors of bone cells

(Fernandez-Yague et al. 2015). Silk fibroin seeded with adipose-derived mesenchymal stem cells appears to increase expression of osteoblast-related genes (BMP-2, COL1a1, OCN) compared to unseeded silk fibroin (Sartika et al. 2020). Scaffolds of natural polymers can be fabricated in a wide range of structures and by a range of techniques, including additive manufacturing. In minimally invasive procedures to address impaired bone healing, hydrogels composed of 3D networks of natural (and synthetic) polymers are an extremely attractive option for delivering bioactive factors. On the other hand, one drawback of natural polymers is the difficulty in altering or controlling their degradation rates.

Synthetic polymers allow one to tailor scaffold degradation rates and tune mechanical properties with greater control than is possible with the use of natural polymers. Polylactic acid (PLA), copolymers PLGA, and polycaprolactone (PCL) (Donnaloja et al. 2020) are among the commonly used synthetic polymers. Hierarchically structured PLA scaffolds with five spatial scales in the architecture and cell seeding have been developed for applications as bone mimetic implants (Sohling et al. 2020); these scaffolds were used with cell seeding. Synthetic polymers generally have lower bioactivity than their natural counterparts, and they can produce inhospitable cell environments or even adverse tissue reactions, due to acidic degradation products. Hybrid natural-synthetic polymer scaffolds (e.g., PLA and alginate (Ataie et al. 2019)) are interesting variants designed to exploit the strengths of each.

4.4 Composites

Each of the monolithic material types described above (metals, ceramics, polymers) have strengths and shortcomings, and there is a long history of combining different material types to obtain more functional composites which incorporate osteoconductivity and osteoinductivity. For a variety of reasons, ceramic-polymer composites comprise the lion's share of current scaffold materials.

Metal-ceramic composites continue to be investigated (e.g., 3D printed and selective laser sintered of porous titanium scaffolds filled with apatitic bone cement (Vlad et al. 2020)). Composites of readily degradable Mg with bioactive glass or particulate hAp or with various bioactive layers have been reviewed recently (Yang et al. 2018). Implant-associated infection remains a challenge and covering porous titanium scaffolds with calcium-phosphate layers containing immobilized silver nanoparticles has been used in vivo to suppress infection (van Hengel et al. 2017).

Bioactive glass-resin composite scaffolds produced by 3D printing have incorporated multiple hierarchical levels to mimic Haversian bone (Zhang et al. 2020), albeit not the sub-osteon and smaller scales. On the other end of the size scale extreme, nanofibrous scaffolds (PLA-alginate) have been coated with hAp (via in situ precipitation) and were biocompatible and supported stem cell osteogenic differentiation (Ataie et al. 2019). The mineral in natural bone is in the form of cAp nanoplatelets, and biomimetic approaches include polymer-nanofiller composites (nanoparticles, nanofibers, nanoplates), which are a relatively new bone tissue engineering material system with the nanophase providing increased bioactivity compared to conventional fillers (Bharadwaz and Jayasuriya 2020). There are advantages of incorporating hAp nanoparticles into porous polymer scaffolds, and incorporating substantial levels of Mg into the hAp lattice can improve bone cell interactivity (Chen et al. 2019). A systematic review of studies of chitosan-based scaffolds found that chitosan has been combined with many biomaterials (hAp, alginate, PLGA, PCL) for improved periodontal regenerative potential (Lauritano et al. 2020).

Collagen calcium-phosphate composites are very attractive materials for bone scaffolds because collagen and calcium phosphate are the main components of bone. Collagen-calcium phosphate scaffolds have employed a wide variety of collagens and TCP or hAp; the form of these mineral phases range from particles to microspheres and the compositions examined include targeted elemental substitutions (Kolodziejska et al. 2020). Bulk processing methods have been employed to produce the composite scaffolds; one of these was developed as a commercial product (Rebaudi et al. 2003) that was tested against a commercial bovine-derived, hAp-including xenograft (Scabbia and Trombelli 2004). Porous collagen-hAp and porous TCP implants have been compared clinically, where the composite performed better than the TCP (Sotome et al. 2016) and was more "surgically friendly" in that it was flexible enough to fit into irregular spaces.

Collagen is well-known for its ability to be mechanically extruded (Meyer et al. 2010; Nitta Casings 2020), and 3D printing of strut-based collagen scaffolds has been achieved (Lode et al. 2016). The authors are unaware, however, of any studies reporting 3D printing of calcium phosphate-loaded collagen scaffolds. Various synthetic polymers have been used to form composites with hAp, and these have been deployed for bone tissue engineering (Qasim et al. 2019). One example is 3D-printed hAp-PLGA composites and variations on the basic design (Driscoll et al. 2020; Hallman et al. 2020; Jakus et al. 2016), which were tested in a rat model of posterolateral spine fusion (PLF). These scaffolds consist of layers of struts, with each layer rotated relative to the preceding by 45° or 90° (Fig. 2). The struts were printed far enough apart to produce interconnected porosity suitable for angiogenesis and bone infiltration. These composites were printed with a synthetic hAp particle size range of 5–30 μm; use of hAp nanoparticles has thus far led to difficulties in controlling printing (Frølich et al. 2016).

4.5 Bioactive Components and Delivery

In addition to osteomimetic structures and surfaces, growth factors, and other bioactive materials can be incorporated into scaffolds for bone tissue engineering. In this setting, the goal is for the growth factors to provide a superior osteoinductive environment for rapid production of bone matrix and subsequent mineralization. As discussed above, bone tissue regeneration is a dynamic process extending over weeks (Cottrell et al. 2016), and the delivered growth factors (GF) must remain in the target area at therapeutic concentrations over an appropriate time frame.

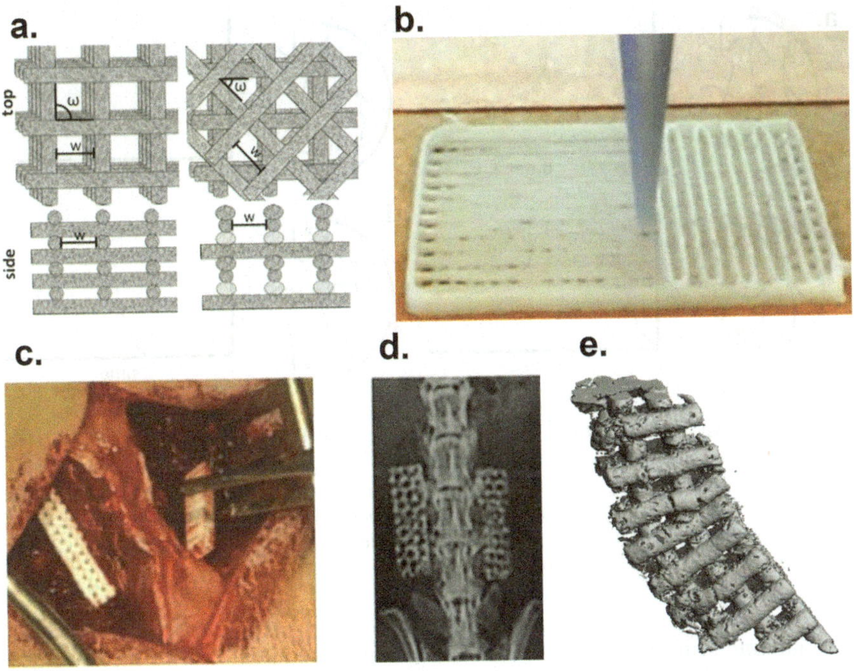

Fig. 2 3D printed scaffolds used in a rat posteriolateral spine fusion model. (**a**) Schematic of struts laid at 0°/90° and 0°/45° relative advancing angles; (**b**) Photograph during printing of a scaffold. (**c**) Surgical placement of the scaffolds in a rat. (**d**) In vivo radiograph of the implanted scaffolds. (**e**) Lab microCT rendering of a scaffold. Images adapted from (Driscoll et al. 2020)

Different designs for GF carriers and their fabrication/synthesis methods are reviewed in detail elsewhere (De Witte et al. 2018; Qasim et al. 2020), but the goal is optimize cellular response by matching biological signaling to the physiological dynamics of bone repair, i.e., through temporally and spatially targeted delivery of physiologically relevant doses of GFs. Not every GF delivery vehicle is compatible with every scaffold because of the scaffold material(s) and their fabrication pathways.

Delivery kinetics of GFs depend on the way they are loaded into/onto the scaffold, and for the purposes of this (review/chapter), what follows focuses on polymer-based scaffolds. Options for GF delivery include direct loading (physical entrapment, adsorption), bonding to the scaffold (covalent or non-covalent), and encapsulation (with biodegradable materials, pressure sensitive biomaterials, micro- or nanospheres) (De Witte et al. 2018; Qasim et al. 2020). Different incorporation methods produce different GF release profiles (Fig. 3) ranging from (a) burst release to (b) prolonged low release rate and (c) intermediate release profiles. Sequences of burst release can also be produced (Fig. 3d). For example, combining delivery methods or using more than one type of microsphere can enable either the sequential release of one GF or of a cascade of different GFs. Adsorption of BMP-2 and other

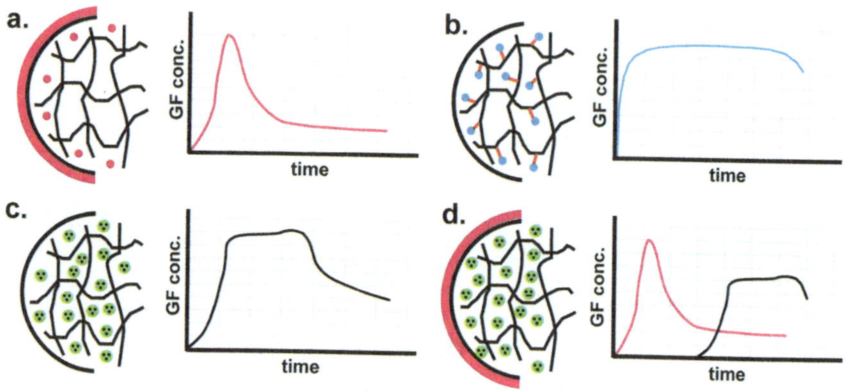

Fig. 3 GF delivery using scaffolds. Schematic of the scaffold and GF incorporation appears on the left of each panel and the GF release profile appears on the right. The local GF concentrations can vary significantly depending on many factors. (**a**) Burst release from a GF applied to the scaffold surface. (**b**) Prolonged low release rate of a covalently bonded GF. (**c**) Intermediate release rate via microencapsulation of a GF. (**d**) Sequence of burst release (e.g. from surface adsorbed GF) and steady release rate (e.g., from microencapsulated GF)

GF onto/into scaffolds (applying GF-containing solution onto the formed scaffold) produces a rapid burst release (Park et al. 2009), which can amount to 60% diffusion out of the scaffold within the first 5 min after implantation (De Witte et al. 2020a). This can be a severe limitation given the short in vivo half-life of GFs, particularly BMP-2 (Azevedo and Pashkuleva 2015; De Witte et al. 2020b) and can lead to untoward biological effects. Nanocarriers for GFs can be immobilized via cross-linking chemistry, and these nanoparticles were shown to persist for up to 4 weeks within a chitosan-based scaffold (De Witte et al. 2020a). Delivery vehicles incorporating components such as gelatin, heparin, hyaluronic acid, or fibronectin provide specific sites for GF immobilization (see (Lee et al. 2011)), and the GF can be attached without harsh chemical processing.

A large number of scaffolds for GF delivery have been reported in recent years, of which only a small number are mentioned here. In one example, a hAp-PLGA composite scaffold carrying a supraphysiologic dose of BMP-2 was tested in a rat PLF model, where the scaffold was completely encapsulated by bone after 8 weeks (Jakus et al. 2016). Alginate has been used as a vehicle for BMP-2 and arginine-glycine-aspartic acid peptides (RGD) delivery (Venkatesan et al. 2015), and a variety of RGD forms can be delivered (Sun and Tan 2013). Chitosan-based composites have been used to deliver growth factors (FGF, BMP) as well as stem cells (Lauritano et al. 2020). In vivo results with BMP-loaded-alginate-chitosan nanocomposite scaffold showed complete closure of a rat calvarial defect after 16 weeks (Florczyk et al. 2013); related rigid, yet injectable hydrogels for minimally invasive surgery (MIS) can have excellent mechanical properties and nanofibrous architecture similar to the natural fibrillar structure of bone (Ghosh et al. 2019).

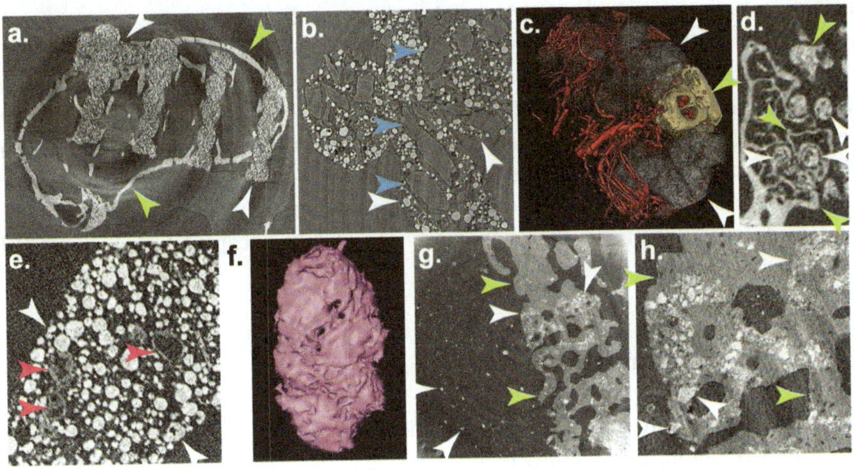

Fig. 4 MicroCT images of 3D printed PLGA-calcium phosphate scaffolds in a rat model of posteriolateral spine fusion (**a–c**, **e–h** synchrotron; **d** laboratory). This information is from studies described in (Nitta Casings 2020; Lode et al. 2016; Driscoll et al. 2020). Arrowheads label different materials: white – calcium phosphate mineral; gold – bone; cyan – DBM; magenta – bone-like spicules. (**a**) BMP-2-coated, PLGA-hAp explanted scaffold imaged 8 weeks postoperatively; (**b**) Portion of an unimplanted PLGA-hAp-DBM scaffold with range of DBM particle shapes visible; (**c**) PLGA-hAp-DBM scaffold 12 weeks post-implantation showing blood vessels perfused with MicroFil® X-ray contrast agent (red), synthetic hAp particles (semitransparent white), and bone (gold); (**d**) PLGA-hAp-DBM scaffold showing bone encapsulating struts; (**e**) Strut of PLGA-hAp-DBM scaffold containing bone-like spicules; (**f**) 3D rendering of a spicule; (**g**) PLGA-β-TCP scaffold showing breakdown of strut structure (wide dispersion of β-TCP particles, left side of image) and accompanying a more rapidly-degrading PLGA copolymer formulation (than in **a–f**). Encapsulation of β-TCP particles is more rapid and complete than with hAp particles (right side of image); (**h**) enlargement of the central portion of panel (**g**). Horizontal field of view are: a – 810 μm; b – 700 μm; d – 2.6 mm; e – 350 μm; g – 1.71 mm; h – 420 μm. The diameter of the reconstructed cylinder in (**c**) is about 4.7 mm, and the length of the spicule in panel (**f**) is 115 μm

Carriers for growth factors need not be nanoscopic. For example, incorporating microscopic particles of milled DBM into hAp-PLGA composites is a simple approach to increase bioactivity of these 3D-printed scaffolds (Driscoll et al. 2020; Hallman et al. 2020) (See Fig. 4b, which depicts a synchrotron microCT cross-sectional image of a strut containing a DBM particle surrounded by hAp particles. When implanted in the rat PLF model, all of the macropores of the DBM-hAp-PLGA composites contained blood vessels by 12 weeks postoperatively (Fig. 4c). Thc DBM-hAp-PLGA composites produced osteointegration superior to DBM-PLGA and somewhat better than hAp-PLGA; Fig. 4d shows bonc growing into the macropores and encapsulating struts of the scaffold. In implanted scaffolds, bone-like spicules grew at the surface of the DBM in cases where the DBM was surrounded by hAp particles, and these spicules had X-ray contrast identical to that of natural bone (Fig. 4e, f). No cells were observed in the vicinity of the bone-like spicules (Driscoll et al. 2020), suggesting that the close local proximity of calcium and phosphate ions into the DBM particles (from nearby hAp) enables renucleation of cAp in the DBM, i.e., it enables renucleation of a bone matrix missing its mineral. This composite of PLGA and synthetic hAp had only begun to be resorbed at 12 weeks postoperative (Driscoll et al. 2020; Hallman et al. 2020), and increasing resorption rates might produce more rapid and complete osteointegration. Changing the copolymer mixture and the calcium phosphate mineral (to β-TCP) leads to more rapid resorption, with strut disintegration dispersing the mineral particles (Fig. 4g), and more extensive bone regrowth more completely encapsulating the mineral (Fig. 4h).

5 Conclusion

Bone healing involves a complex cascade of processes, and impaired healing (fracture nonunion, pseudarthrosis, critical-sized defect) remains a major clinical challenge inadequately addressed by conventional procedures or therapies. In treating refractory bone healing, use of engineered scaffolds and strategically tailored spatiotemporal deployment of growth factors may significantly improve outcomes, but translational (large animal) and clinical studies are still needed for promising designs. In particular, tuned scaffold degradation rates and delivery of multiple growth factors within a single scaffold are approaches worth developing.

General purpose scaffold designs which can be easily adapted in a point-of-care setting (e.g., trimmed to size during surgery) (Driscoll et al. 2020; Hallman et al. 2020; Jakus et al. 2016) will doubtless continue to receive the majority of attention. However, with widespread availability of 3D printing and the increasing pool of expertise in printing scaffolds, it is not unreasonable to expect that production and implantation of patient-specific scaffolds will spread beyond craniofacial reconstructions. Simple-to-use design software will need to be employed to take 3D dimensional inputs from CT scans, but this type of approach has already been pioneered in other areas (e.g., microCT-based finite element analysis of trabecular bone under load). One patient-specific design for healing a refractory nonunion could be a sleeve-like scaffold fitting completely around the circumference of the long bone. The inner surface of the scaffold would match the outer surface of the bone, and the scaffold would necessarily be printed in two halves that would be fit together by the surgeon to surround the bone.

Some types of 3D printing can be performed at ambient temperatures and without harsh solvents, and incorporation of GFs during printing is an attractive prospect. The 3D printing of ceramic-polymer composites incorporating DBM particles (Fig. 4) is one example. Another innovation would be incorporating two GFs into a scaffold, with tailored release rates matching the activity of these GF in the bone healing cascade.

Obtaining approval for clinical use of any new product is a long and involved process. Only a small number of new materials/scaffolds engineered for GF delivery in the setting of impaired bone healing, therefore, are likely to receive the necessary investment for approval. Patient-specific scaffold fabrication via 3D printing is certainly an attractive prospect, but MIS is being used increasingly for many orthopedic procedures. Any materials or scaffolds which are approved for targeted GF delivery must be usable in both conventional surgical and MIS settings, at least in some form, in order to gain adequate clinical acceptance. On the other hand, impaired bone healing is widespread and carries an enormous societal and economic cost; with such high stakes, the difficulties outlined above are certainly worth overcoming.

Acknowledgments We thank Dr. P. Morse (Duke Univ.) for providing the rendering in Fig. 4d. Data in Figs. 2 and 4 was collected at beamline 2-BM of the Advanced Photon Source; use of the Advanced Photon Source was supported by the US Department of Energy, Office of Science, Office of Basic Energy Sciences, under Contract No. DE-AC02-06CH11357. This work was supported by the National Institute of Arthritis, Musculoskeletal, and Skin Diseases, grants R01AR069580 and R01AR072721.

References

Abou-Khalil R et al (2015) Role of muscle stem cells during skeletal regeneration. Stem Cells 33 (5):1501–1511

Adogwa O et al (2013) Revision lumbar surgery in elderly patients with symptomatic pseudarthrosis, adjacent-segment disease, or same-level recurrent stenosis. Part 2. A cost-effectiveness analysis: clinical article. J Neurosurg Spine 18(2):147–153

Agholme F et al (2010) Sclerostin antibody treatment enhances metaphyseal bone healing in rats. J Bone Miner Res 25(11):2412–2418

Akamaru T et al (2003) Simple carrier matrix modifications can enhance delivery of recombinant human bone morphogenetic protein-2 for posterolateral spine fusion. Spine 28(5):429–434

Alvarez RH, Kantarjian HM, Cortes JE (2006) Biology of platelet-derived growth factor and its involvement in disease. Mayo Clin Proc 81(9):1241–1257

Andersen T et al (2001) Smoking as a predictor of negative outcome in lumbar spinal fusion. Spine (Phila Pa 1976) 26(23):2623–2628

Aro HT et al (2011) Recombinant human bone morphogenetic protein-2: a randomized trial in open tibial fractures treated with reamed nail fixation. J Bone Joint Surg Am 93(9):801–808

Ataie M, Shabani I, Seyedjafari E (2019) Surface mineralized hybrid nanofibrous scaffolds based on poly(l-lactide) and alginate enhances osteogenic differentiation of stem cells. J Biomed Mater Res A 107(3):586–596

Augat P et al (2014) Biomechanical methods for the assessment of fracture repair. Injury 45(Suppl 2):S32–S38

Augustin J et al (2020) Comparison of two pore sizes of LAE442 scaffolds and their effect on degradation and osseointegration behavior in the rabbit model. J Biomed Mater Res B Appl Biomater. https://doi.org/10.1002/jbm.b.34607

Azevedo HS, Pashkuleva I (2015) Biomimetic supramolecular designs for the controlled release of growth factors in bone regeneration. Adv Drug Deliv Rev 94:63–76

Barrack RL et al (2003) Induction of bone ingrowth from acetabular defects to a porous surface with OP-1. Clin Orthop 417:41–49

Bharadwaz A, Jayasuriya AC (2020) Recent trends in the application of widely used natural and synthetic polymer nanocomposites in bone tissue regeneration. Korean J Couns Psychother 110:110698

Bishop JA et al (2012) Assessment of compromised fracture healing. J Am Acad Orthop Surg 20 (5):273–282

Blattert TR et al (2002) Successful transpedicular lumbar interbody fusion by means of a composite of osteogenic protein-1 (rhBMP-7) and hydroxyapatite carrier: a comparison with autograft and hydroxyapatite in the sheep spine. Spine 27(23):2697–2705
Blokhuis TJ et al (2001) Biomechanical and histological aspects of fracture healing, stimulated with osteogenic protein-1. Biomaterials 22(7):725–730
Boden SD et al (2000) The use of rhBMP-2 in interbody fusion cages. Definitive evidence of osteoinduction in humans: a preliminary report. Spine 25(3):376–381
Boden SD et al (2002) Use of recombinant human bone morphogenetic protein-2 to achieve posterolateral lumbar spine fusion in humans: a prospective, randomized clinical pilot trial: 2002 Volvo Award in clinical studies. Spine 27(23):2662–2673
Boerckel JD et al (2012) Effects of in vivo mechanical loading on large bone defect regeneration. J Orthop Res 30(7):1067–1075
Bouletreau PJ et al (2002) Factors in the fracture microenvironment induce primary osteoblast angiogenic cytokine production. Plast Reconstr Surg 110(1):139–148
Bright C et al (2006) In vivo evaluation of plasmid DNA encoding OP-1 protein for spine fusion. Spine (Phila Pa 1976) 31(19):2163–2172
Burkus JK et al (2002) Anterior lumbar interbody fusion using rhBMP-2 with tapered interbody cages. J Spinal Disord Tech 15(5):337–349
Burkus JK, Dorchak JD, Sanders DL (2003) Radiographic assessment of interbody fusion using recombinant human bone morphogenetic protein type 2. Spine 28(4):372–377
Buza JA 3rd, Einhorn T (2016) Bone healing in 2016. Clin Cases Miner Bone Metab 13 (2):101–105
Cao R et al (2003) Angiogenic synergism, vascular stability and improvement of hind-limb ischemia by a combination of PDGF-BB and FGF-2. Nat Med 9(5):604–613
Caplan AI, Correa D (2011) PDGF in bone formation and regeneration: new insights into a novel mechanism involving MSCs. J Orthop Res 29(12):1795–1803
Caterini R et al (2016) Treatment of recalcitrant atrophic non-union of the humeral shaft with BMP-7, autologous bone graft and hydroxyapatite pellets. Injury 47(Suppl 4):S71–S77
Chen Y et al (2007) Beta-catenin signaling plays a disparate role in different phases of fracture repair: implications for therapy to improve bone healing. PLoS Med 4(7):e249
Chen S et al (2019) Biomimetic synthesis of Mg-substituted hydroxyapatite nanocomposites and three-dimensional printing of composite scaffolds for bone regeneration. J Biomed Mater Res A 107(11):2512–2521
Cho TJ, Gerstenfeld LC, Einhorn TA (2002) Differential temporal expression of members of the transforming growth factor beta superfamily during murine fracture healing. J Bone Miner Res 17(3):513–520
Claes L, Recknagel S, Ignatius A (2012) Fracture healing under healthy and inflammatory conditions. Nat Rev Rheumatol 8(3):133–143
Cochran DL et al (2016) A randomized clinical trial evaluating rh-FGF-2/beta-TCP in periodontal defects. J Dent Res 95(5):523–530
Coffin JD et al (1995) Abnormal bone growth and selective translational regulation in basic fibroblast growth factor (FGF-2) transgenic mice. Mol Biol Cell 6(12):1861–1873
Cook SD et al (1994) Recombinant human bone morphogenetic protein-7 induces healing in a canine long-bone segmental defect model. Clin Orthop 301:302–312
Cook SD et al (1995) Effect of recombinant human osteogenic protein-1 on healing of segmental defects in non-human primates. J Bone Joint Surg Am 77(5):734–750
Cook SD et al (2002) Healing course of primate ulna segmental defects treated with osteogenic protein-1. J Invest Surg 15(2):69–79
Cottrell JA et al (2016) The biology of bone and ligament healing. Foot Ankle Clin 21(4):739–761
Croisier F, Jérôme C (2013) Chitosan-based biomaterials for tissue engineering. Euro Polmer J 49:780–792

Cucchiarini M et al (2011) Metabolic activities and chondrogenic differentiation of human mesenchymal stem cells following recombinant adeno-associated virus-mediated gene transfer and overexpression of fibroblast growth factor 2. Tissue Eng Part A 17(15–16):1921–1933

Cunningham BW et al (1999) Osteogenic protein versus autologous interbody arthrodesis in the sheep thoracic spine. A comparative endoscopic study using the Bagby and Kuslich interbody fusion device. Spine 24(6):509–518

Cunningham BW et al (2002) Osseointegration of autograft versus osteogenic protein-1 in posterolateral spinal arthrodesis: emphasis on the comparative mechanisms of bone induction. Spine J 2 (1):11–24

Currey JD (2002) Bones: structure and mechanics. Princeton University Press, Princeton

Daniels TR et al (2015) Prospective randomized controlled trial of hindfoot and ankle fusions treated with rhPDGF-BB in combination with a β-TCP-collagen matrix. Foot Ankle Int 36 (7):739–748

De Long WG Jr et al (2007) Bone grafts and bone graft substitutes in orthopaedic trauma surgery. A critical analysis. J Bone Joint Surg Am 89(3):649–658

de Santana RB, de Santana CM (2015) Human intrabony defect regeneration with rhFGF-2 and hyaluronic acid – a randomized controlled clinical trial. J Clin Periodontol 42(7):658–665

De Witte TM et al (2018) Bone tissue engineering via growth factor delivery: from scaffolds to complex matrices. Regen Biomater 5(4):197–211

De Witte TM et al (2020a) Immobilization of nanocarriers within a porous chitosan scaffold for the sustained delivery of growth factors in bone tissue engineering applications. J Biomed Mater Res A 108(5):1122–1135

De Witte TM et al (2020b) Degradable poly(methyl methacrylate)-co-methacrylic acid nanoparticles for controlled delivery of growth factors for bone regeneration. Tissue Eng Part A. https://doi.org/10.1089/ten.tea.2020.0010

Deckers MM et al (2002) Bone morphogenetic proteins stimulate angiogenesis through osteoblast-derived vascular endothelial growth factor A. Endocrinology 143(4):1545–1553

den Boer FC et al (2002) Effect of recombinant human osteogenic protein-1 on the healing of a freshly closed diaphyseal fracture. Bone 31(1):158–164

den Boer FC et al (2003) Healing of segmental bone defects with granular porous hydroxyapatite augmented with recombinant human osteogenic protein-1 or autologous bone marrow. J Orthop Res 21(3):521–528

Denard PJ et al (2018) Stress shielding of the humerus in press-fit anatomic shoulder arthroplasty: review and recommendations for evaluation. J Shoulder Elb Surg 27(6):1139–1147

Digiovanni CW et al (2011) Prospective, randomized, multi-center feasibility trial of rhPDGF-BB versus autologous bone graft in a foot and ankle fusion model. Foot Ankle Int 32(4):344–354

Digiovanni CW et al (2013) Recombinant human platelet-derived growth factor-BB and beta-tricalcium phosphate (rhPDGF-BB/β-TCP): an alternative to autogenous bone graft. J Bone Joint Surg Am 95(13):1184–1192

Dimar JR 2nd et al (2009) Two-year fusion and clinical outcomes in 224 patients treated with a single-level instrumented posterolateral fusion with iliac crest bone graft. Spine J 9(11):880–885

Donnaloja F et al (2020) Natural and synthetic polymers for bone scaffolds optimization. Polymers (Basel) 12(4):E905

Driscoll JA et al (2020) 3D-printed ceramic-demineralized bone matrix Hyperelastic bone composite scaffolds for spinal fusion. Tissue Eng Part A 26(3–4):157–166

Einhorn TA, Gerstenfeld LC (2015) Fracture healing: mechanisms and interventions. Nat Rev Rheumatol 11(1):45–54

Ekrol I et al (2008) A comparison of RhBMP-7 (OP-1) and autogenous graft for metaphyseal defects after osteotomy of the distal radius. Injury 39(Suppl 2):S73–S82

El-Jawhari JJ, Jones E, Giannoudis PV (2016) The roles of immune cells in bone healing; what we know, do not know and future perspectives. Injury 47(11):2399–2406

Ennett AB, Kaigler D, Mooney DJ (2006) Temporally regulated delivery of VEGF in vitro and in vivo. J Biomed Mater Res A 79(1):176–184

Even J, Eskander M, Kang J (2012) Bone morphogenetic protein in spine surgery: current and future uses. J Am Acad Orthop Surg 20(9):547–552

Fakhry A et al (2005) Effects of FGF-2/−9 in calvarial bone cell cultures: differentiation stage-dependent mitogenic effect, inverse regulation of BMP-2 and noggin, and enhancement of osteogenic potential. Bone 36(2):254–266

Fei Y et al (2011) Fibroblast growth factor 2 stimulation of osteoblast differentiation and bone formation is mediated by modulation of the Wnt signaling pathway. J Biol Chem 286 (47):40575–40583

Feng C et al (2017) 3D printing of lotus root-like biomimetic materials for cell delivery and tissue regeneration. Adv Sci (Weinh) 4(12):1700401

Fernandes HR et al (2018) Bioactive glasses and glass-ceramics for healthcare applications in bone regeneration and tissue engineering. Materials (Basel) 11(12):2530

Fernandez-Yague MA et al (2015) Biomimetic approaches in bone tissue engineering: integrating biological and physicomechanical strategies. Adv Drug Deliv Rev 84:1–29

Florczyk SJ et al (2013) Evaluation of three-dimensional porous chitosan-alginate scaffolds in rat calvarial defects for bone regeneration applications. J Biomed Mater Res A 101(10):2974–2983

Friedlaender GE et al (2001) Osteogenic protein-1 (bone morphogenetic protein-7) in the treatment of tibial nonunions. J Bone Joint Surg Am 83-A(Suppl 1 Pt 2):S151–S158

Frølich S et al (2016) Diffraction tomography and Rietveld refinement of a hydroxyapatite bone phantom. J Appl Cryst 49:103–109

Fu TS et al (2009) Enhancement of posterolateral lumbar spine fusion using low-dose rhBMP-2 and cultured marrow stromal cells. J Orthop Res 27(3):380–384

Fu K et al (2010) Prolonged osteogenesis from human mesenchymal stem cells implanted in immunodeficient mice by using coralline hydroxyapatite incorporating rhBMP2 microspheres. J Biomed Mater Res A 92(4):1256–1264

Fu R et al (2013) Effectiveness and harms of recombinant human bone morphogenetic protein-2 in spine fusion: a systematic review and meta-analysis. Ann Intern Med 158(12):890–902

Fu TS et al (2015) Mesenchymal stem cells expressing baculovirus-engineered BMP-2 and VEGF enhance posterolateral spine fusion in a rabbit model. Spine J 15(9):2036–2044

Fujioka-Kobayashi M et al (2012) Cholesteryl group- and acryloyl group-bearing pullulan nanogel to deliver BMP2 and FGF18 for bone tissue engineering. Biomaterials 33(30):7613–7620

Garrison KR et al (2010) Bone morphogenetic protein (BMP) for fracture healing in adults. Cochrane Database Syst Rev 6:CD006950

Geesink RG, Hoefnagels NH, Bulstra SK (1999) Osteogenic activity of OP-1 bone morphogenetic protein (BMP-7) in a human fibular defect. J Bone Joint Surg Br 81(4):710–718

Geideman W, Early JS, Brodsky J (2004) Clinical results of harvesting autogenous cancellous graft from the ipsilateral proximal tibia for use in foot and ankle surgery. Foot Ankle Int 25 (7):451–455

Ghodasra JH et al (2014) Factors influencing arthrodesis rates in a rabbit posterolateral spine model with iliac crest autograft. Eur Spine J 23(2):426–434

Ghosh M et al (2019) Injectable alginate-peptide composite hydrogel as a scaffold for bone tissue regeneration. Nanomaterials (Basel) 9(4):497

Globus RK, Plouet J, Gospodarowicz D (1989) Cultured bovine bone cells synthesize basic fibroblast growth factor and store it in their extracellular matrix. Endocrinology 124 (3):1539–1547

Gothard D et al (2014) Tissue engineered bone using select growth factors: a comprehensive review of animal studies and clinical translation studies in man. Eur Cell Mater 28:166–207. discussion 207-8

Goulet JA et al (1997) Autogenous iliac crest bone graft. Complications and functional assessment. Clin Orthop Relat Res 339:76–81

Govender S et al (2002) Recombinant human bone morphogenetic protein-2 for treatment of open tibial fractures: a prospective, controlled, randomized study of four hundred and fifty patients. J Bone Joint Surg Am 84(12):2123–2134

Grabowski G, Cornett CA (2013) Bone graft and bone graft substitutes in spine surgery: current concepts and controversies. J Am Acad Orthop Surg 21(1):51–60
Grasser WA et al (2007) BMP-6 exerts its osteoinductive effect through activation of IGF-I and EGF pathways. Int Orthop 31(6):759–765
Grgurevic L (2016) Osteogrow development: Bmp6 bone device for enhancing bone healing. Clin Ther 38(10S):e9–e10
Grgurevic L, Pecina M, Vukicevic S, Marshall R (2017) Urist and the discovery of bone morphogenetic proteins. Int Orthop 41(5):1065–1069
Grgurevic L et al (2019) Recombinant human bone morphogenetic protein 6 delivered within autologous blood coagulum restores critical size segmental defects of ulna in rabbits. JBMR Plus 3(5):e10085
Gümüşderelioğlu M et al (2019) Chitosan-based double-faced barrier membrane coated with functional nanostructures and loaded with BMP-6. J Mater Sci Mater Med 31(1):4
Guo P et al (2003) Platelet-derived growth factor-B enhances glioma angiogenesis by stimulating vascular endothelial growth factor expression in tumor endothelia and by promoting pericyte recruitment. Am J Pathol 162(4):1083–1093
Haddad SL et al (2007) Intermediate and long-term outcomes of total ankle arthroplasty and ankle arthrodesis. A systematic review of the literature. J Bone Joint Surg Am 89(9):1899–1905
Hadjiargyrou M, O'Keefe RJ (2014) The convergence of fracture repair and stem cells: interplay of genes, aging, environmental factors and disease. J Bone Miner Res 29(11):2307–2322
Hadjiargyrou M et al (2002) Transcriptional profiling of bone regeneration. Insight into the molecular complexity of wound repair. J Biol Chem 277(33):30177–30182
Hagen A, Gorenoi V, Schonermark MP (2012) Bone graft substitutes for the treatment of traumatic fractures of the extremities. GMS Health Technol Assess 8:Doc04
Hak DJ et al (2014) Delayed union and nonunions: epidemiology, clinical issues, and financial aspects. Injury 45(Suppl 2):S3–S7
Hallman M et al (2020) Influence of geometry and architecture on the in vivo success of 3D-printed scaffolds for spinal fusion. Tissue Eng Part A. https://doi.org/10.1089/ten.TEA.2020.0004
Hamid KS, Parekh SG, Adams SB (2016) Salvage of severe foot and ankle trauma with a 3D printed scaffold. Foot Ankle Int 37(4):433–439
Hankenson KD, Gagne K, Shaughnessy M (2015) Extracellular signaling molecules to promote fracture healing and bone regeneration. Adv Drug Deliv Rev 94:3–12
Hara Y et al (2017) Delayed expression of circulating TGF-beta1 and BMP-2 levels in human nonunion Long bone fracture healing. J Nippon Med Sch 84(1):12–18
He D et al (2010) Effect of recombinant human bone morphogenetic protein-2 on bone regeneration in large defects of the growing canine skull after dura mater replacement with a dura mater substitute. J Neurosurg 112(2):319–328
He G, Liu P, Tan Q (2012) Porous titanium materials with entangled wire structure for load-bearing biomedical applications. J Mech Behav Biomed Mater 5(1):16–31
Herberg S et al (2014) Low-dose bone morphogenetic protein-2/stromal cell-derived factor-1beta cotherapy induces bone regeneration in critical-size rat calvarial defects. Tissue Eng Part A 20 (9–10):1444–1453
Herberg S et al (2015) Mesenchymal stem cell expression of stromal cell-derived factor-1beta augments bone formation in a model of local regenerative therapy. J Orthop Res 33(2):174–184
Hertweck J et al (2018) CD34(+) cells seeded in collagen scaffolds promote bone formation in a mouse calvarial defect model. J Biomed Mater Res B Appl Biomater 106(4):1505–1516
Ho CY et al (2015) Mesenchymal stem cells with increased stromal cell-derived factor 1 expression enhanced fracture healing. Tissue Eng Part A 21(3–4):594–602
Hoffmann MF, Jones CB, Sietsema DL (2012) Adjuncts in posterior lumbar spine fusion: comparison of complications and efficacy. Arch Orthop Trauma Surg 132(8):1105–1110
Hollinger JO et al (2008a) Recombinant human platelet-derived growth factor: biology and clinical applications. J Bone Joint Surg Am 90(Suppl 1):48–54

Hollinger JO et al (2008b) Accelerated fracture healing in the geriatric, osteoporotic rat with recombinant human platelet-derived growth factor-BB and an injectable beta-tricalcium phosphate/collagen matrix. J Orthop Res 26(1):83–90

Hong KS et al (2010) Bone regeneration by bioactive hybrid membrane containing FGF2 within rat calvarium. J Biomed Mater Res A 94(4):1187–1194

Hou H et al (2009) Enhancement of bone formation by genetically-engineered bone marrow stromal cells expressing BMP-2, VEGF and angiopoietin-1. Biotechnol Lett 31(8):1183–1189

Hu K, Olsen BR (2016a) Osteoblast-derived VEGF regulates osteoblast differentiation and bone formation during bone repair. J Clin Invest 126(2):509–526

Hu K, Olsen BR (2016b) The roles of vascular endothelial growth factor in bone repair and regeneration. Bone 91:30–38

Huang B et al (2018) Combination use of BMP2 and VEGF165 promotes osseointegration and stability of titanium implants in irradiated bone. Biomed Res Int 2018:8139424

Itoh N, Ornitz DM (2004) Evolution of the Fgf and Fgfr gene families. Trends Genet 20 (11):563–569

Jakus AE et al (2016) Hyperelastic "bone": a highly versatile, growth factor-free, osteoregenerative, scalable, and surgically friendly biomaterial. Sci Transl Med 8(358):358ra127

Jenis LG et al (2002) The effect of osteogenic protein-1 in instrumented and noninstrumented posterolateral fusion in rabbits. Spine J 2(3):173–178

Kaigler D et al (2006) VEGF scaffolds enhance angiogenesis and bone regeneration in irradiated osseous defects. J Bone Miner Res 21(5):735–744

Kaiser MG et al (2014) Guideline update for the performance of fusion procedures for degenerative disease of the lumbar spine. Part 16: bone graft extenders and substitutes as an adjunct for lumbar fusion. J Neurosurg Spine 21(1):106–132

Kamal AF, Siahaan OSH, Fiolin J (2019) Various dosages of BMP-2 for management of massive bone defect in Sprague Dawley rat. Arch Bone Jt Surg 7(6):498–505

Kanakaris NK, Giannoudis PV (2007) The health economics of the treatment of long-bone non-unions. Injury 38(Suppl 2):S77–S84

Kandziora F et al (2002a) Comparison of BMP-2 and combined IGF-I/TGF-ss1 application in a sheep cervical spine fusion model. Eur Spine J 11(5):482–493

Kandziora F et al (2002b) Experimental fusion of the sheep cervical spine. Part II: effect of growth factors and carrier systems on interbody fusion. Chirurg 73(10):1025–1038

Kasten P et al (2012) Comparison of platelet-rich plasma and VEGF-transfected mesenchymal stem cells on vascularization and bone formation in a critical-size bone defect. Cells Tissues Organs 196(6):523–533

Kawaguchi H et al (2001) Acceleration of fracture healing in nonhuman primates by fibroblast growth factor-2. J Clin Endocrinol Metab 86(2):875–880

Kawaguchi H et al (2007) Local application of recombinant human fibroblast growth factor-2 on bone repair: a dose-escalation prospective trial on patients with osteotomy. J Orthop Res 25 (4):480–487

Kawaguchi H et al (2010) A local application of recombinant human fibroblast growth factor 2 for tibial shaft fractures: a randomized, placebo-controlled trial. J Bone Miner Res 25 (12):2735–2743

Khan SN, Bostrom MP, Lane JM (2000) Bone growth factors. Orthop Clin N Am 31(3):375–388

Kim EJ et al (2013) Ihh and Runx2/Runx3 signaling interact to coordinate early chondrogenesis: a mouse model. PLoS One 8(2):e55296

Kimura Y et al (2010) Controlled release of bone morphogenetic protein-2 enhances recruitment of osteogenic progenitor cells for de novo generation of bone tissue. Tissue Eng Part A 16 (4):1263–1270

Kirker-Head CA et al (1998) Healing bone using recombinant human bone morphogenetic protein 2 and copolymer. Clin Orthop Relat Res 349:205–217

Kitamura M et al (2008) Periodontal tissue regeneration using fibroblast growth factor-2: randomized controlled phase II clinical trial. PLoS One 3(7):e2611

Kitamura M et al (2011) FGF-2 stimulates periodontal regeneration: results of a multi-center randomized clinical trial. J Dent Res 90(1):35–40
Kitamura M et al (2016) Randomized placebo-controlled and controlled non-inferiority phase III trials comparing Trafermin, a recombinant human fibroblast growth factor 2, and enamel matrix derivative in periodontal regeneration in Intrabony defects. J Bone Miner Res 31(4):806–814
Kitaoka HB (1999) Arthrodesis of the ankle: technique, complications, and salvage treatment. Instr Course Lect 48:255–261
Koh JT et al (2008) Combinatorial gene therapy with BMP2/7 enhances cranial bone regeneration. J Dent Res 87(9):845–849
Kolk A et al (2012) Current trends and future perspectives of bone substitute materials – from space holders to innovative biomaterials. J Craniomaxillofac Surg 40(8):706–718
Kolodziejska B, Kaflak A, Kolmas J (2020) Biologically inspired collagen/apatite composite biomaterials for potential use in bone tissue regeneration – a review. Materials (Basel) 13 (7):1748
Komatsu DE et al (2010) Modulation of Wnt signaling influences fracture repair. J Orthop Res 28 (7):928–936
Kraiwattanapong C et al (2005) Comparison of Healos/bone marrow to INFUSE(rhBMP-2/ACS) with a collagen-ceramic sponge bulking agent as graft substitutes for lumbar spine fusion. Spine (Phila Pa 1976) 30(9):1001–1007. discussion 1007
Kugimiya F et al (2005) Involvement of endogenous bone morphogenetic protein (BMP) 2 and BMP6 in bone formation. J Biol Chem 280(42):35704–35712
Lauritano D et al (2020) Nanomaterials for periodontal tissue engineering: chitosan-based scaffolds. A systematic review. Nanomaterials (Basel) 10(4):E605
Laursen M et al (1999) Recombinant bone morphogenetic protein-7 as an intracorporal bone growth stimulator in unstable thoracolumbar burst fractures in humans: preliminary results. Eur Spine J 8(6):485–490
Leach JK et al (2006) Coating of VEGF-releasing scaffolds with bioactive glass for angiogenesis and bone regeneration. Biomaterials 27(17):3249–3255
Lee K, Silva EA, Mooney DJ (2011) Growth factor delivery-based tissue engineering: general approaches and a review of recent developments. J R Soc Interface 8(55):153–170
Li X et al (2019) Synthesis and evaluation of BMMSC-seeded BMP-6/nHAG/GMS scaffolds for bone regeneration. Int J Med Sci 16(7):1007–1017
Lietman SA et al (2005) The effect of recombinant human osteogenic protein-1 on allograft incorporation. J Bone Joint Surg Br 87(9):1292–1297
Lin JM et al (2009) Actions of fibroblast growth factor-8 in bone cells in vitro. Am J Physiol Endocrinol Metab 297(1):E142–E150
Lode A et al (2016) Additive manufacturing of collagen scaffolds by three-dimensional plotting of highly viscous dispersions. Biofabrication 8(1):015015
Loiselle AE, Zuscik MJ (2019) Skeletal healing: cellular and molecular determinants. In: Bilezikian JP (ed) Primer on the metabolic bone diseases and disorders of mineral metabolism. Wiley-Blackwell, Hoboken, pp 101–107
Luca L et al (2010) The effects of carrier nature and pH on rhBMP-2-induced ectopic bone formation. J Control Release 147(1):38–44
Luca L et al (2011) Injectable rhBMP-2-loaded chitosan hydrogel composite: osteoinduction at ectopic site and in segmental long bone defect. J Biomed Mater Res A 96(1):66–74
Lyon T et al (2013) Efficacy and safety of recombinant human bone morphogenetic protein-2/calcium phosphate matrix for closed tibial diaphyseal fracture: a double-blind, randomized, controlled phase-II/III trial. J Bone Joint Surg Am 95(23):2088–2096
Ma H et al (2018) 3D-printed bioceramic scaffolds: from bone tissue engineering to tumor therapy. Acta Biomater 79:37–59
Magin MN, Delling G (2001) Improved lumbar vertebral interbody fusion using rhOP-1: a comparison of autogenous bone graft, bovine hydroxylapatite (bio-Oss), and BMP-7 (rhOP-1) in sheep. Spine 26(5):469–478

Makadia HK, Siegel SJ (2011) Poly lactic-co-glycolic acid (PLGA) as biodegradable controlled drug delivery carrier. Polymers (Basel) 3(3):1377–1397
Mariscal G et al (2020) A meta-analysis of bone morphogenetic protein-2 versus iliac crest bone graft for the posterolateral fusion of the lumbar spine. J Bone Miner Metab 38(1):54–62
McGee-Lawrence ME et al (2013) Sclerostin deficient mice rapidly heal bone defects by activating beta-catenin and increasing intramembranous ossification. Biochem Biophys Res Commun 441 (4):886–890
Meinel L et al (2003) Localized insulin-like growth factor I delivery to enhance new bone formation. Bone 33(4):660–672
Meyer M, Baltzer H, Schwikal K (2010) Collagen fibres by thermoplastic and wet spinning. Mater Sci Eng C 30:1266–1271
Mills LA, Aitken SA, Simpson A (2017) The risk of non-union per fracture: current myths and revised figures from a population of over 4 million adults. Acta Orthop 88(4):434–439
Minear S et al (2010) Wnt proteins promote bone regeneration. Sci Transl Med 2(29):29ra30
Missbach M et al (1999) A novel inhibitor of the tyrosine kinase Src suppresses phosphorylation of its major cellular substrates and reduces bone resorption in vitro and in rodent models in vivo. Bone 24(5):437–449
Mizrahi O et al (2013) BMP-6 is more efficient in bone formation than BMP-2 when overexpressed in mesenchymal stem cells. Gene Ther 20(4):370–377
Montero A et al (2000) Disruption of the fibroblast growth factor-2 gene results in decreased bone mass and bone formation. J Clin Invest 105(8):1085–1093
Morgan EF et al (2010) Correlations between local strains and tissue phenotypes in an experimental model of skeletal healing. J Biomech 43(12):2418–2424
Morris MT, Tarpada SP, Cho W (2018) Bone graft materials for posterolateral fusion made simple: a systematic review. Eur Spine J 27(8):1856–1867
Morshed S (2014) Current options for determining fracture union. Adv Med 2014:708574
Mulconrey DS et al (2008) Bone morphogenetic protein (RhBMP-2) as a substitute for iliac crest bone graft in multilevel adult spinal deformity surgery: minimum two-year evaluation of fusion. Spine 33(20):2153–2159
Murakami S, Noda M (2000) Expression of Indian hedgehog during fracture healing in adult rat femora. Calcif Tissue Int 66(4):272–276
Myeroff C, Archdeacon M (2011) Autogenous bone graft: donor sites and techniques. J Bone Joint Surg Am 93(23):2227–2236
Nagayama T et al (2013) FGF18 accelerates osteoblast differentiation by upregulating Bmp2 expression. Congenit Anom (Kyoto) 53(2):83–88
Nash TJ et al (1994) Effect of platelet-derived growth factor on tibial osteotomies in rabbits. Bone 15(2):203–208
Nitta Casings (2020) Nitta Collagen Gell™. https://www.nittacasings.com/
Ondra SL, Marzouk S (2003) Revision strategies for lumbar pseudarthrosis. Neurosurg Focus 15 (3):E9
Park K-H et al (2009) Bone morphogenic protein-2 (BMP-2) loaded nanoparticles mixed with human mesenchymal stem cell in fibrin hydrogel for bone tissue engineering. J Biosci Bioeng 108(6):530–537
Park EK et al (2016) Cranioplasty enhanced by three-dimensional printing: custom-made three-dimensional-printed titanium implants for skull defects. J Craniofac Surg 27(4):943–949
Park DK et al (2019) Lumbar spine fusion rates with local bone in Posterolateral and combined Posterolateral and Interbody approaches. J Am Acad Orthop Surg Glob Res Rev 3(11):e018
Peng H et al (2002) Synergistic enhancement of bone formation and healing by stem cell-expressed VEGF and bone morphogenetic protein-4. J Clin Invest 110(6):751–759
Pereira RC, Economides AN, Canalis E (2000) Bone morphogenetic proteins induce gremlin, a protein that limits their activity in osteoblasts. Endocrinology 141(12):4558–4563

Phan K et al (2018) Effect of smoking status on successful arthrodesis, clinical outcome, and complications after anterior lumbar Interbody fusion (ALIF). World Neurosurg 110:e998–e1003
Pollock R et al (2008) Donor site morbidity following iliac crest bone harvesting for cervical fusion: a comparison between minimally invasive and open techniques. Eur Spine J 17(6):845–852
Putra NE et al (2020) Multi-material additive manufacturing technologies for Ti-, Mg-, and Fe-based biomaterials for bone substitution. Acta Biomater. S1742-7061(20)30178-1. https://doi.org/10.1016/j.actbio.2020.03.037
Qasim M, Chae DS, Lee NY (2019) Advancements and frontiers in nano-based 3D and 4D scaffolds for bone and cartilage tissue engineering. Int J Nanomedicine 14:4333–4351
Qasim M, Chae DS, Lee NY (2020) Bioengineering strategies for bone and cartilage tissue regeneration using growth factors and stem cells. J Biomed Mater Res A 108(3):394–411
Raizman NM et al (2009) Pseudarthrosis of the spine. J Am Acad Orthop Surg 17(8):494–503
Rebaudi A, Silvestrini P, Trisi P (2003) Use of a resorbable hydroxyapatite-collagen chondroitin sulfate material on immediate postextraction sites: a clinical and histologic study. Int J Periodontics Restorative Dent 23(4):371–379
Reinhold MI et al (2004) FGF18 represses noggin expression and is induced by calcineurin. J Biol Chem 279(37):38209–38219
Reznikov N, Shahar R, Weiner S (2014) Bone hierarchical structure in three dimensions. Acta Biomater 10(9):3815–3826
Richardson TP et al (2001) Polymeric system for dual growth factor delivery. Nat Biotechnol 19 (11):1029–1034
Rico-Llanos GA, Becerra J, Visser R (2017) Insulin-like growth factor-1 (IGF-1) enhances the osteogenic activity of bone morphogenetic protein-6 (BMP-6) in vitro and in vivo, and together have a stronger osteogenic effect than when IGF-1 is combined with BMP-2. J Biomed Mater Res A 105(7):1867–1875
Ripamonti U, Crooks J, Rueger DC (2001a) Induction of bone formation by recombinant human osteogenic protein-1 and sintered porous hydroxyapatite in adult primates. Plast Reconstr Surg 107(4):977–988
Ripamonti U et al (2001b) Bone induction by BMPs/OPs and related family members in primates. J Bone Joint Surg Am 83-A(Suppl 1 Pt 2):S116–S127
Roseti L et al (2017) Scaffolds for bone tissue engineering: state of the art and new perspectives. Mater Sci Eng C Mater Biol Appl 78:1246–1262
Saito N, Takaoka K (2003) New synthetic biodegradable polymers as BMP carriers for bone tissue engineering. Biomaterials 24(13):2287–2293
Salkeld SL et al (2001) The effect of osteogenic protein-1 on the healing of segmental bone defects treated with autograft or allograft bone. J Bone Joint Surg Am 83-A(6):803–816
Sampath TK et al (1992) Recombinant human osteogenic protein-1 (hOP-1) induces new bone formation in vivo with a specific activity comparable with natural bovine osteogenic protein and stimulates osteoblast proliferation and differentiation in vitro. J Biol Chem 267 (28):20352–20362
Sartika D et al (2020) Human adipose-derived mesenchymal stem cells-incorporated silk fibroin as a potential bio-scaffold in guiding bone regeneration. Polymers (Basel) 12(4):E853
Sato N et al (1993) Platelet-derived growth factor indirectly stimulates angiogenesis in vitro. Am J Pathol 142(4):1119–1130
Scabbia A, Trombelli L (2004) A comparative study on the use of a HA/collagen/chondroitin sulphate biomaterial (Biostite) and a bovine-derived HA xenograft (bio-Oss) in the treatment of deep intra-osseous defects. J Clin Periodontol 31(5):348–355
Schmid GJ et al (2009) Fibroblast growth factor expression during skeletal fracture healing in mice. Dev Dyn 238(3):766–774
Schmidmaier G et al (2002) Improvement of fracture healing by systemic administration of growth hormone and local application of insulin-like growth factor-1 and transforming growth factor-beta1. Bone 31(1):165–172

Schmidmaier G et al (2003a) Synergistic effect of IGF-I and TGF-beta1 on fracture healing in rats: single versus combined application of IGF-I and TGF-beta1. Acta Orthop Scand 74(5):604–610
Schmidmaier G et al (2003b) IGF-I and TGF-beta 1 incorporated in a poly(D,L-lactide) implant coating stimulates osteoblast differentiation and collagen-1 production but reduces osteoblast proliferation in cell culture. J Biomed Mater Res B Appl Biomater 65(1):157–162
Schmitz MA et al (1999) Effect of smoking on tibial shaft fracture healing. Clin Orthop Relat Res 365:184–200
Schorn L et al (2017) Vertical bone regeneration using rhBMP-2 and VEGF. Head Face Med 13 (1):11
Seifert RA et al (1989) Two different subunits associate to create isoform-specific platelet-derived growth factor receptors. J Biol Chem 264(15):8771–8778
Seiler JG 3rd, Johnson J (2000) Iliac crest autogenous bone grafting: donor site complications. J South Orthop Assoc 9(2):91–97
Sheyn D et al (2008) Nonvirally engineered porcine adipose tissue-derived stem cells: use in posterior spinal fusion. Stem Cells 26(4):1056–1064
Shi L et al (2017) Heterotopic ossification related to the use of recombinant human BMP-2 in osteonecrosis of femoral head. Medicine 96(27):e7413
Shimada T et al (2004) Targeted ablation of Fgf23 demonstrates an essential physiological role of FGF23 in phosphate and vitamin D metabolism. J Clin Invest 113(4):561–568
Sohling N et al (2020) 3D-printing of hierarchically designed and osteoconductive bone tissue engineering scaffolds. Materials (Basel) 13(8):1836
Sohn HS, Oh JK (2019) Review of bone graft and bone substitutes with an emphasis on fracture surgeries. Biomater Res 23:9
Solchaga LA et al (2012) Safety of recombinant human platelet-derived growth factor-BB in Augment(®) Bone Graft. J Tissue Eng 3(1):2041731412442668
Sotome S et al (2016) Efficacy and safety of porous hydroxyapatite/type 1 collagen composite implantation for bone regeneration: a randomized controlled study. J Orthop Sci 21(3):373–380
Springer IN et al (2008) BMP-2 and bFGF in an irradiated bone model. J Craniomaxillofac Surg 36 (4):210–217
St John TA et al (2003) Physical and monetary costs associated with autogenous bone graft harvesting. Am J Orthop (Belle Mead NJ) 32(1):18–23
Street J et al (2002) Vascular endothelial growth factor stimulates bone repair by promoting angiogenesis and bone turnover. Proc Natl Acad Sci U S A 99(15):9656–9661
Sukul M et al (2015) Effect of local sustainable release of BMP2-VEGF from nano-cellulose loaded in sponge biphasic calcium phosphate on bone regeneration. Tissue Eng Part A 21 (11–12):1822–1836
Summers BN, Eisenstein SM (1989) Donor site pain from the ilium. A complication of lumbar spine fusion. J Bone Joint Surg Br 71(4):677–680
Sun J, Tan H (2013) Alginate-based biomaterials for regenerative medicine applications. Materials (Basel) 6(4):1285–1309
Sun P et al (2012) BMP2/7 heterodimer is a stronger inducer of bone regeneration in peri-implant bone defects model than BMP2 or BMP7 homodimer. Dent Mater J 31(2):239–248
Sun H et al (2017) Recombinant human platelet-derived growth factor-BB versus autologous bone graft in foot and ankle fusion: a systematic review and meta-analysis. J Foot Ankle Surg 23 (1):32–39
Taipaleenmaki H et al (2008) Impact of stromal cell composition on BMP-induced chondrogenic differentiation of mouse bone marrow derived mesenchymal cells. Exp Cell Res 314 (13):2400–2410
Tanaka H, Liang CT (1995) Effect of platelet-derived growth factor on DNA synthesis and gene expression in bone marrow stromal cells derived from adult and old rats. J Cell Physiol 164 (2):367–375
Tanaka E et al (2006) Fibroblast growth factor-2 augments recombinant human bone morphogenetic protein-2-induced osteoinductive activity. Ann Biomed Eng 34(5):717–725

Tang Y et al (2009) TGF-beta1-induced migration of bone mesenchymal stem cells couples bone resorption with formation. Nat Med 15(7):757–765

Tarkin IS, Siska PA, Zelle BA (2010) Soft tissue and biomechanical challenges encountered with the management of distal tibia nonunions. Orthop Clin North Am 41(1):119–126. table of contents

Thorey F et al (2011) Osseointegration by bone morphogenetic protein-2 and transforming growth factor beta2 coated titanium implants in femora of New Zealand white rabbits. Indian J Orthop 45(1):57–62

Tsuji K et al (2006) BMP2 activity, although dispensable for bone formation, is required for the initiation of fracture healing. Nat Genet 38(12):1424–1429

Tsuji K et al (2008) BMP4 is dispensable for skeletogenesis and fracture-healing in the limb. J Bone Joint Surg Am 90(Suppl 1):14–18

Tsuji K et al (2010) Conditional deletion of BMP7 from the limb skeleton does not affect bone formation or fracture repair. J Orthop Res 28(3):384–389

Tuchman A et al (2017) Autograft versus allograft for cervical spinal fusion: a systematic review. Global Spine J 7(1):59–70

Tzioupis C, Giannoudis PV (2007) Prevalence of long-bone non-unions. Injury 38(Suppl 2):S3–S9

Urist MR (1965) Bone: formation by autoinduction. Science 150(698):893–899

Valdes M et al (2007) rhBMP-6 stimulated osteoprogenitor cells enhance posterolateral spinal fusion in the New Zealand white rabbit. Spine Journal 7(3):318–325

van Hengel IAJ et al (2017) Selective laser melting porous metallic implants with immobilized silver nanoparticles kill and prevent biofilm formation by methicillin-resistant *Staphylococcus aureus*. Biomaterials 140:1–15

Venkatesan J et al (2015) Alginate composites for bone tissue engineering: a review. Int J Biol Macromol 72:269–281

Vlad MD et al (2020) Novel titanium-apatite hybrid scaffolds with spongy bone-like micro architecture intended for spinal application: in vitro and in vivo study. Mater Sci Eng C Mater Biol Appl 110:110658

Vukicevic S et al (2014a) The clinical use of bone morphogenetic proteins revisited: a novel biocompatible carrier device OSTEOGROW for bone healing. Int Orthop 38(3):635–647

Vukicevic S et al (2014b) The clinical use of bone morphogenetic proteins revisited: a novel biocompatible carrier device OSTEOGROW for bone healing. Int Orthop 38(3):635–647

Vukicevic S et al (2020) Autologous blood coagulum is a physiological carrier for BMP6 to induce new bone formation and promote posterolateral lumbar spine fusion in rabbits. J Tissue Eng Regen Med 14(1):147–159

Wang EA et al (1988) Purification and characterization of other distinct bone-inducing factors. Proc Natl Acad Sci U S A 85(24):9484–9488

Wang J et al (2012) Low-dose rhBMP2/7 heterodimer to reconstruct peri-implant bone defects: a micro-CT evaluation. J Clin Periodontol 39(1):98–105

Wang C et al (2016) NOTCH signaling in skeletal progenitors is critical for fracture repair. J Clin Invest 126(4):1471–1481

Wang T, Zhang X, Bikle DD (2017) Osteogenic differentiation of periosteal cells during fracture healing. J Cell Physiol 232(5):913–921

Wang T, Guo S, Zhang H (2018) Synergistic effects of controlled-released BMP-2 and VEGF from nHAC/PLGAs scaffold on osteogenesis. Biomed Res Int 2018:3516463

Wang X et al (2019) PDGF modulates BMP2-induced osteogenesis in periosteal progenitor cells. JBMR Plus 3(5):e10127

Wang C, Lu WW, Wang M (2020) Multifunctional fibrous scaffolds for bone regeneration with enhanced vascularization. J Mater Chem B 8(4):636–647

Wozney JM et al (1988) Novel regulators of bone formation: molecular clones and activities. Science 242(4885):1528–1534

Wu S et al (2014) Biomimetic porous scaffolds for bone tissue engineering. Mater Sci Eng R Rep 80:1–36

Xiao C et al (2011) Bone marrow stromal cells with a combined expression of BMP-2 and VEGF-165 enhanced bone regeneration. Biomed Mater 6(1):015013
Xu L et al (2016) Peri-implant bone regeneration using rhPDGF-BB, BMSCs, and beta-TCP in a canine model. Clin Implant Dent Relat Res 18(2):241–252
Yan L et al (2017) Pore structures and mechanical properties of porous titanium scaffolds by bidirectional freeze casting. Korean J Couns Psychother 75:335–340
Yang JH et al (2003) Expression of recombinant human BMP-6 in Escherichia coli and its purification and bioassay in vitro. Sheng Wu Gong Cheng Xue Bao 19(5):556–560
Yang J et al (2018) A review on the exploitation of biodegradable magnesium-based composites for medical applications. Biomed Mater 13(2):022001
Yasko AW et al (1992) The healing of segmental bone defects, induced by recombinant human bone morphogenetic protein (rhBMP-2). A radiographic, histological, and biomechanical study in rats. J Bone Joint Surg Am 74(5):659–670
Yoshida T et al (2015) Bone augmentation using a highly porous PLGA/beta-TCP scaffold containing fibroblast growth factor-2. J Periodontal Res 50(2):265–273
Zhang Y et al (2015) Mesoporous bioactive glass nanolayer-functionalized 3D-printed scaffolds for accelerating osteogenesis and angiogenesis. Nanoscale 7(45):19207–19221
Zhang M et al (2018) The effects of platelet-derived growth factor-BB on bone marrow stromal cell-mediated vascularized bone regeneration. Stem Cells Int 2018:3272098
Zhang M et al (2020) 3D printing of Haversian bone-mimicking scaffolds for multicellular delivery in bone regeneration. Sci Adv 6(12):eaaz6725
Zou S et al (2014) Mesenchymal stem cells overexpressing Ihh promote bone repair. J Orthop Surg Res 9:102
Zura R et al (2016) Epidemiology of fracture nonunion in 18 human bones. JAMA Surg 151(11):e162775

Prostaglandins and Bone

Carol Pilbeam

Contents

Abstract

Prostaglandins (PGs) are highly bioactive fatty acids. PGs, especially prostaglandin E_2 (PGE_2), are abundantly produced by cells of both the bone-forming (osteoblast) lineage and the bone-resorbing (osteoclast) lineage. The inducible cyclooxygenase, COX-2, is largely responsible for most PGE_2 production in bone, and once released, PGE_2 is rapidly degraded in vivo. COX-2 is induced by multiple agonists – hormones, growth factors, and proinflammatory factors – and the resulting PGE_2 may mediate, amplify, or, as we have recently shown for parathyroid hormone (PTH), inhibit responses to these agonists. In vitro, PGE_2 can directly stimulate osteoblast differentiation and, indirectly via stimulation of RANKL in osteoblastic cells, stimulate the differentiation of osteoclasts. The net balance of these two effects of PGE_2 in vivo on bone formation and bone resorption has been hard to predict and, as expected for such a widespread local factor, hard to study. Some of the complexity of PGE_2 actions on bone can be explained by the fact that there are four receptors for PGE_2 (EP1–4). Some of the

C. Pilbeam (✉)
Department of Medicine and Musculoskeletal Institute, UConn Health, Farmington, CT, USA
e-mail: pilbeam@uchc.edu

P. H. Stern (ed.), *Bone Regulators and Osteoporosis Therapy*,
Handbook of Experimental Pharmacology 262, https://doi.org/10.1007/164_2019_332

major actions of PGE_2 in vitro occur via EP2 and EP4, both of which can stimulate cAMP signaling, but there are other distinct signaling pathways, important in other tissues, which have not yet been fully elucidated in bone cells. Giving PGE_2 or agonists of EP2 and EP4 to accelerate bone repair has been examined with positive results. Further studies to clarify the pathways of PGE_2 action in bone may allow us to identify new and more effective ways to deliver the therapeutic benefits of PGE_2 in skeletal disorders.

Keywords
Bone formation · Bone resorption · Cyclooxygenase · EP receptors · NSAIDs · Prostaglandin

1 Introduction

Prostaglandins (PGs) are highly bioactive unsaturated fatty acids. PGs are not stored but are synthesized and released, as needed, and rapidly metabolized. This chapter will focus on prostaglandin E_2 (PGE_2), which is abundantly produced by cells of both the bone-forming and bone-resorbing lineages. Production of PGs depends on the availability of substrate; one of two cyclooxygenases (COXs), constitutively expressed COX-1 or inducible COX-2; and a specific downstream synthase. Most PGE_2 in bone is produced by COX-2. COX-2 is induced by multiple hormones and proinflammatory factors, and the resulting PGE_2 can mediate or modify responses to the agonist, sometimes in very unexpected ways, as we discuss for the COX-2 agonist, parathyroid hormone (PTH). PGE_2 can act on four G-protein-coupled receptors whose signaling pathways have not been fully elucidated in bone cells. It has been 50 years since PGE_2 in bone was first shown to stimulate cyclic AMP (cAMP) production and resorption in bone organ cultures (Klein and Raisz 1970). The early studies in cell and organ cultures led to the conclusion that PGE_2 could increase both bone formation and resorption, and this has led to continuing interest in the potential for therapeutic manipulation of PGE_2 or its receptors.

2 PGE_2 Production

Eicosanoids are bioactive lipids derived from arachidonic acid (AA) and other 20-carbon polyunsaturated fatty acids (PUFAs) (Buczynski et al. 2009; Smith et al. 2011). The term "prostanoid" refers to products of the COX pathway: PGE_2, prostaglandin D_2 (PGD_2), prostaglandin F_{2a} (PGF_{2a}), prostacyclin (PGI_2), and thromboxane (TXA_2). PGs are 20-carbon fatty acids with a cyclopentane ring. Although TXAs have an oxane ring, they are generally discussed under the "PG" heading. The subscript for prostanoids denotes the number of double bonds (e.g., PGE_1, PGE_2, and PGE_3). The two series of prostanoids are the most abundant and the best characterized.

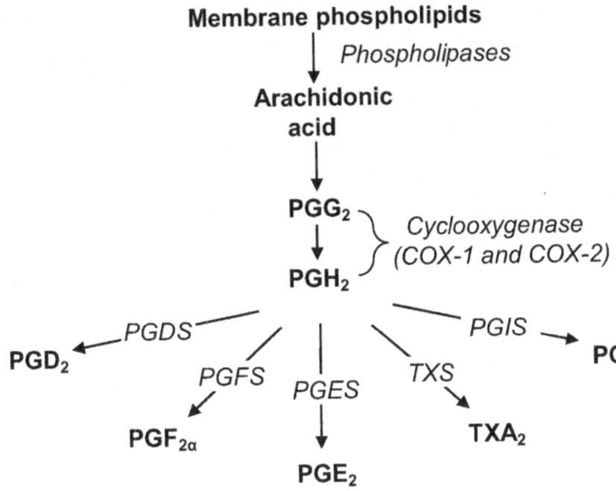

Fig. 1 Major prostanoids generated from arachidonic acid. Free arachidonic acid, released from membrane phospholipids by phospholipases, is converted by a bifunctional enzyme, called cyclooxygenase, to prostaglandin G_2 (PGG_2) in a cyclooxygenase reaction followed by reduction of PGG_2 to prostaglandin H_2 (PGH_2) in a peroxidase reaction. PGH_2 is then converted to specific prostanoids by the terminal synthases

There are three steps in the production of PGs (Fig. 1). The first is the mobilization of AA from membranes by phospholipases. The second step is catalyzed by a bifunctional enzyme that converts free AA to prostaglandin G_2 (PGG_2), first by a cyclooxygenase (hence, the name COX) reaction and then a peroxidase reaction, to prostaglandin H_2 (PGH_2). PGH_2 is then converted by terminal synthases to the various prostanoids.

AA Release Phospholipase A_2 (PLA_2) enzymes catalyze the hydrolysis of membrane phospholipids, from membrane glycerophospholipids, releasing free fatty acids, such as AA. The PLA_2 superfamily has 16 groups and many subgroups (Dennis et al. 2011; Murakami et al. 2011, 2015; Vasquez et al. 2017). The most important PLAs for PG production are probably the Ca^{2+}-independent PLA_2s ($iPLA_2$), the Ca^{2+}-dependent cytosolic PLA_2s ($cPLA_2$), and the secreted PLA_2s ($sPLA_2$). As a general summary, $iPLA_2$ is the primary PLA_2 in cells, producing low levels of free fatty acids, some of which may be AA, needed for daily cellular functions; $cPLA_2$ is the major inducible enzyme hydrolyzing AA-containing phospholipids during infection or inflammation; $sPLA_2$ is also inducible and augments $cPLA_2$ function (Dennis and Norris 2015).

COX Enzymes The bifunctional enzyme is formally named prostaglandin endoperoxide H synthase or prostaglandin G/H synthase (PGHS), and the gene name is *ptgs*. However, because the COX reaction site is inhibited by nonsteroidal anti-inflammatory drugs (NSAIDs), the interest in developing new inhibitors selective for COX-2 led to calling PGHS simply COX and the inhibitors coxibs. The two enzymes for COX, COX-1 and COX-2, are encoded by separate genes (Herschman 1994; Smith et al. 2000). Both COX-1 and COX-2 are N-glycosylated dimeric proteins inserted into the luminal face of the ER and the contiguous inner membrane of the nuclear envelope (Smith et al. 2011).

Despite having similar catalytic mechanisms, COX-1 and COX-2 are independently functioning pathways (Simmons et al. 2004; Smith and Langenbach 2001). COX-2 is rapidly and transiently inducible by multiple factors in many tissues (Kang et al. 2007). COX-1 is expressed at relatively low, stable levels in most tissues and is considered to be "constitutive." The half-life of COX-2 protein is reported to be 2–7 h in various tissues, while the half-life of COX-1 protein is much longer (Kang et al. 2007; Mbonye et al. 2006). However, it is now evident that COX-2 mRNA is also expressed constitutively at low levels in many tissues and cells. A recent study examined signaling pathways underlying constitutive expression of COX-2 in the kidney, gastrointestinal tract, and brain (Kirkby et al. 2016). It is still unclear how much this expression contributes to PG production. Another difference is that COX-2 is much more efficient at using low AA concentrations (below 5 μM) than COX-1 (Swinney et al. 1997). This difference may explain why osteoblasts from COX-2 KO mice make little or no measurable PGE_2 in culture despite the constitutive expression of COX-1 (Chikazu et al. 2005; Choudhary et al. 2003; Okada et al. 2000a; Xu et al. 2007).

It was initially hypothesized that COX-2 was responsible for acute pathological PG responses, while COX-1 produced prostanoids for ongoing "housekeeping" functions, such as maintenance of renal blood flow, platelet aggregation, and gastric cytoprotection. Once highly selective inhibitors of COX-2 activity were available, it became apparent that COX-2 also has physiologic functions and COX-2 selective NSAIDs turned out to have serious adverse side effects, especially on the cardiovascular system, that would limit their use (Grosser et al. 2017a, b).

PG Synthases PGH_2 is converted to each PG by specific PG terminal synthases (Fig. 1), and the synthases may determine the major PG synthesized in a tissue. Prostaglandin E synthase (PGES), which converts PGH_2 to PGE_2, occurs in multiple forms (Hara et al. 2010; Hara 2017). The predominant PGES, mPGES-1, is inducible, located in ER and perinuclear membranes, and regulated similarly to COX-2. mPGES-1 may also be located in the Golgi apparatus, and because COX-2 and $cPLA_2$ may also be located there, the Golgi apparatus may be a dedicated PGE_2 synthesis site (Leslie 2015; Yuan and Smith 2015). mPGES-2 is constitutively expressed and functionally coupled with both COX-1 and COX-2. A third form, cytosolic PGES (cPGES), is preferentially coupled to COX-1 and thought to maintain PGE_2 production for cellular homeostasis (Tanioka et al. 2000). Knockout (KO) mice for the synthases are reviewed in Hara et al. (2010) and Hara (2017). Mice deficient for mPGES-1 have reduced inflammatory and pain responses, and potential inhibitory drugs have been developed (Psarra et al. 2017). Mice deficient for mPGES-2 have no specific phenotype, and deficiency of cPGES in mice is perinatal lethal.

PGE_2 Degradation Degradation of circulating PGE_2 occurs rapidly (Ferreira and Vane 1967), and measurement of PGE_2 metabolites in the urine may be the preferred way to track changes in PGE_2 production in vivo. The first step in degrading PGE_2 is mediated by the enzyme 15-hydroxyprostaglandin dehydrogenase (HPGD, also known as 15-PGDH), which generates metabolites that are orders of magnitude less potent than PGE_2 itself. Many tissues make HPGD, but it is not clear how quickly bone cells degrade PGE_2. It is common practice to follow PGE_2 production in vitro by measuring PGE_2 accumulation in osteoblastic or marrow stromal cell cultures over 2–3 days between medium changes, suggesting shared knowledge of some stability in culture.

In humans, mutations in HPGD are associated with a rare genetic disorder called idiopathic hypertrophic osteoarthropathy, characterized by chronically elevated PGE_2, coarse or thickened skin, and periostosis in bone (Uppal et al. 2008). In mice, inhibition of HPGD increases PGE_2 and potentiates recovery in marrow transplant models and accelerates tissue generation in models of colon and liver injury (Antczak et al. 2017; Desai et al. 2018; Zhang et al. 2015b). Although the known associations of elevated PGE_2 with increased inflammation and tumorigenesis might be thought to limit the usefulness of inhibiting HPGD, initial studies in animal models suggest that a small molecule inhibitor can promote transplant recovery without limiting side effects (Desai et al. 2018).

3 PGE_2 Receptors

There are four G-protein-coupled receptors (GPCRs), called EP1, EP2, EP3, and EP4, mediating actions of PGE_2 (Fig. 1) (Woodward et al. 2011). The EP1 receptor is known to increase Ca^{2+} and may couple to $G\alpha_q$ because studies have reported involvement of the PLC/PKC pathway (Tang et al. 2005). The major signaling pathway for EP3 receptors is $G\alpha_i$-induced adenylate cyclase inhibition. EP2 and EP4 are the receptors most extensively studied in bone. Mice deficient in each EP receptor subtype have been generated, and highly selective agonists for the receptors have been developed (Sugimoto and Narumiya 2007; Woodward et al. 2011).

The first major pathway for PGE_2 signaling identified in bone was cAMP (Klein and Raisz 1970). Both EP2 and EP4 can stimulate $G\alpha_s$ to activate adenylyl cyclases (ACs) and produce cAMP, which can then activate protein kinase A (PKA) or a PKA-independent pathway mediated by EPAC (exchange protein directly activated by cAMP). The PKA pathway is able to crosstalk with other pathways that regulate cell growth, motility, migration, and apoptosis, including the Wnt/β-catenin signaling pathway (Buchanan and DuBois 2006; Castellone et al. 2006; Estus et al. 2016; Hino et al. 2005; Shao et al. 2005). EP2 and EP4 may also transactivate the epidermal growth factor receptor (EGFR) signaling pathway, leading to multiple signaling pathways, including PI3K/Akt/β-catenin. The recruitment of β-arrestin by EP receptors may also activate c-Src, resulting in EGFR transactivation (O'Callaghan and Houston 2015). There was much interest in studying EP receptors in bone in the early days because of the possibility of developing drugs that could target a specific receptor to increase bone mass and fracture healing without the side effect of inhibiting COX. In more recent years, most of the data on EP signaling has come from studies on other tissues.

4 Bone Remodeling and PGE_2

Bone remodeling, the cycle of bone resorption followed by bone formation, occurs throughout our adult life. The net bone balance of these cycles determines whether bone is lost, gained, or maintained. Agents used to prevent or treat osteoporosis, that is, skeletal fragility associated with a high risk of fracture, are aimed at manipulating this cycle. Exogenously applied PGE_2 has been shown to stimulate both resorption and formation in bone, but the role of endogenous PGE_2 in remodeling is still being defined (Blackwell et al. 2010). The potential involvement of endogenous PGE_2 is shown in Fig. 2.

Bone Resorption Early work adding PGs in organ culture showed that exogenous PGs of the E series were potent activators of resorption (Klein and Raisz 1970). Studies with inhibitors of PG production showed that the resorptive effects of many agonists were mediated in part by PGs (Pilbeam et al. 2008). In the early 1990s, the inducible COX-2 was identified (O'Banion et al. 1991; Kujubu et al. 1991; Xie et al. 1991). Subsequently, many resorption agonists were shown to induce COX-2 expression and PGE_2 production in osteoblastic cells that contributed to increasing osteoclasts, including interleukin-1 (IL-1) (Kawaguchi et al. 1994; Sato et al. 1996; Lader and Flanagan 1998; Min et al. 1998), IL-6 (Tai et al. 1997), tumor necrosis

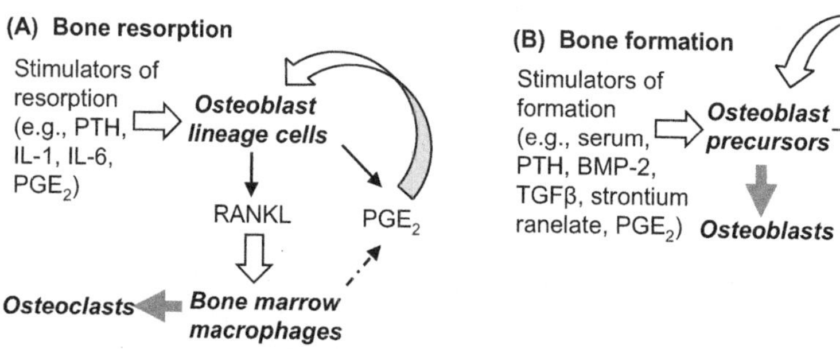

Fig. 2 Potential roles of PGE_2 in bone resorption and formation. (**a**) Agonists of resorption, including PGE_2, act on osteoblastic cells to induce RANKL, which then binds with its receptor, RANK, on bone marrow macrophages (BMMs) and drives them to differentiate into osteoclasts. These agonists are often also inducers of COX-2 expression in osteoblastic cells, as well. The PGE_2 can act back on osteoblasts to induce more RANKL. (RANKL can also induce COX-2 in BMMs.) (**b**) Some stimulators of osteoblastic differentiation also induce COX-2 and produce PGE_2. PGE_2 can then act on the osteoblastic precursors to stimulate osteoblastic differentiation. In some cases, such as serum and strontium ranelate (see text), the major stimulatory effect in cell culture is due to PGE_2

factor-α (TNF-α) (Lader and Flanagan 1998), parathyroid hormone (PTH) (Kawaguchi et al. 1994; Inoue et al. 1995; Okada et al. 2000a), and $1{,}25(OH)_2D_3$ (Okada et al. 2000a). PGE_2 can amplify its own production by inducing COX-2 (Pilbeam et al. 1994). PGE_2 has receptors on both osteoblastic and osteoclastic lineage cells, but stimulates osteoclast differentiation largely indirectly via upregulation of expression of receptor activator of NFκB ligand (RANKL) and inhibition of expression of the decoy RANKL receptor, osteoprotegerin (OPG), in osteoblastic lineage cells (Li et al. 2000) (Fig. 2a). Although both EP2 and EP4 may be involved in PGE_2-stimulated resorption, a number of in vitro studies have suggested that EP4 is the more important receptor for resorption (Miyaura et al. 2000; Suzawa et al. 2000; Raisz and Woodiel 2003; Zhan et al. 2005).

Bone Formation Many studies have shown that PGE_2 stimulates osteoblastic differentiation in murine or rat bone marrow stromal cell and primary osteoblast/calvarial cell cultures (Choudhary et al. 2013; Flanagan and Chambers 1992). In cultured marrow stromal cells or primary osteoblasts from mice with deletion of *ptgs2* or treated with NSAIDs to inhibit COX-2 activity, osteoblastic differentiation is decreased (Choudhary et al. 2013, Okada et al. 2000b, Xu et al. 2007, Zhang et al. 2002). Systemic injections of PGE_2 can increase both periosteal and endosteal bone formation in the rat and produce substantial increases in bone mass (Jee and Ma 1997; Lin et al. 1994; Suponitzky and Weinreb 1998). Systemic administration of PGE_2 in humans (Faye-Petersen et al. 1996; Ueda et al. 1980) and dogs (Norrdin and Shih 1988) has also been shown to increase cortical and cancellous bone mass. Local infusion of PGE_2 has been shown to increase bone in mice (Yoshida et al. 2002).

Similar to resorption agonists, many osteogenic factors induce COX-2 (Fig. 2b), including transforming growth factor β (TGFβ) (Pilbeam et al. 1997), basic fibroblast growth factor (FGF-2) (Kawaguchi et al. 1995), bone morphogenetic protein (BMP-2) (Chikazu et al. 2005), strontium (Choudhary et al. 2007), and fluid shear stress or mechanical loading (Klein-Nulend et al. 1997; Pavalko et al. 1998; Wadhwa et al. 2002). Serum is also a potent inducer of COX-2 expression and PGE_2 production in cultured osteoblasts (Pilbeam et al. 1993). For some osteogenic factors, such as serum, BMP-2, and strontium ranelate, their ability to stimulate osteoblast differentiation in vitro is due largely to their induction of COX-2 produced PGs (Pilbeam et al. 1993; Chikazu et al. 2005; Choudhary et al. 2007).

Both EP2 and EP4 receptors have been positively implicated in the osteogenic and anabolic effects of PGE_2 (Alander and Raisz 2006; Choudhary et al. 2008; Li et al. 2007), while mice deficient in EP1 receptors have enhanced fracture healing, higher trabecular bone volume, increased bone formation, and accelerated osteoblastic differentiation compared to WT mice (Zhang et al. 2011, 2015a). It has been difficult to study EP4 receptors in vivo because EP4 KO murine neonates in a pure C57Bl/6 background die shortly after birth due to patent ductus arteriosus (Segi et al. 1998). Reduced bone mass and impaired fracture healing were found in aged EP4 receptor KO mice compared to WT mice (Li et al. 2005) in contrast to another study which did not find any difference in bone formation between aged WT and EP4 KO mice (Gao et al. 2009). Both of these studies were done in mice with mixed backgrounds to circumvent the patent ductus arteriosus problem, and the difference in phenotypes might be due to variability in backgrounds.

Agonists of EP2 and EP4 can increase fracture healing in animals. Local application of an EP2 agonist and local and systemic application of EP4 agonists have been shown to accelerate bone repair (Li et al. 2003; Paralkar et al. 2003; Tanaka et al. 2004; Yoshida et al. 2002). Clinical studies have not been done with these agonists, perhaps because increased PG can have adverse effects (Markovic et al. 2017).

5 Skeletal Phenotypes of Mice with COX-2 Deficiency

Disruption of genes for PG production has given us some indications of the role of endogenous PGs in vivo. It was clear in early studies that COX-1 KO mice were healthy and survived normally, while COX-2 KO mice had more profound effects (Dinchuk et al. 1995; Langenbach et al. 1995, 1999; Morham et al. 1995). However, one study reported 35% of neonatal COX-2 KO mice died with a patent ductus arteriosus and the mortality increased to 100% when both genes for COX-1 were inactivated, indicating the dependence of COX-2 effects on levels of COX-1 expression (Loftin et al. 2001). Other studies found 20% of COX-2 KO mice dying between 7 and 23 weeks of age secondary to renal dysplasia, despite normal renal development at birth (Morham et al. 1995; Norwood et al. 2000). COX-2 KO female mice were infertile, with multiple failures in female reproductive processes, including ovulation, fertilization, and implantation (Lim et al. 1997). Initial studies of skeletal phenotypes in the C57Bl/6,129 background also reported early death in COX-2 KO mice, gave variable phenotypic results, and suggested that older COX-2 KO mice might have primary hyperparathyroidism (HPTH) (Alam et al. 2005; Xu et al. 2005; Robertson et al. 2006).

Effects of Background Strain Mice for all these initial studies were in the C57Bl/6 inbred strain or mixed C57Bl/6 and 129 inbred strains. Both C57Bl/6 and 129sv mice have been shown to have a natural mutation that results in lack of the gene for the serum phospholipase (GIIA $sPLA_2$) that is important for releasing AA (Kennedy et al. 1995; MacPhee et al. 1995). Hence, COX-2 KOs in these backgrounds are really double KOs, and this may impair the ability of COX-1 to produce PGs and compensate for absent COX-2. The MC3T3-E1 osteoblastic cell line, commonly used to study osteoblastic cells in vitro, was derived from C57Bl/6 mice and may also lack GIIA PLA_2.

Mice in the outbred strain, CD-1, were reported to be heterozygous for the GIIA $sPLA_2$ mutation (Kennedy et al. 1995). We bred COX-2 KO mice into the CD-1 background and found that COX-2 KO mice had no increased mortality and no renal dysfunction and that COX-2 KO females were fertile (Xu et al. 2010). Despite being healthy, 5-month-old male COX-2 KO mice had twofold elevated serum PTH compared to WT mice. COX-2 KO mice also had increased serum markers of bone formation and resorption, decreased femoral BMD by DXA (dual-energy X-ray absorptiometry) and cortical bone thickness by μCT, and small but nonsignificant decreases in trabecular bone volume by μCT or dynamic histomorphometry. We concluded that this bone phenotype could be due to hyperparathyroidism, COX-2 deficiency, or both.

PGE_2 and Sympathetic Activity Recently it was reported that PGE_2 secreted by osteoblastic cells activated EP4 in sensory nerves to increase bone formation by inhibiting sympathetic activity through the central nervous system (Chen et al. 2019). They found that EP4 KO targeted to sensory nerves or COX-2 KO targeted to mature osteoblasts in 3-month-old mice significantly reduced bone volume in adult mice. They postulate that low bone mineral density is sensed by osteoblasts, perhaps because of increased mechanical stresses and stimulated PGE_2 production by osteoblasts. PGE_2 then acts via EP4 on sensory nerves to downregulate sympathetic tone, which then leads osteoblast to increase bone formation. Because so many different genetically engineered mice were subjected to so many different protocols, it is difficult to assess reproducibility. If these results are confirmed, it would mean that targeted COX-2 KO in mature osteoblasts has a larger effect than global COX-2 KO on bone phenotype in mice. Part of the argument for doing this study was the skeletal effects in congenital disorders with insensitivity to pain, disorders caused by several mutations (Nahorski et al. 2015). However, it seems likely that most skeletal problems involving fractures and joint disorders in these patients initially arise from repeated injury due to the lack of pain sensitivity and not from bone loss (Zhang and Haga 2014; Phatarakijnirund et al. 2016; Kayani et al. 2017).

6 COX-2 and PTH: A Special Relationship

PTH is the major hormone responsible for maintenance of calcium homeostasis. It is a major stimulator of bone resorption, acting via a GPCR, which is highly expressed by osteoblast lineage cells and activates both $G\alpha_s$ and $G\alpha_q$ signaling pathways (Vilardaga et al. 2011; Mahon 2012). PTH stimulates bone resorption by increasing RANKL and decreasing osteoprotegerin (OPG) (Boyce and Xing 2008). When PTH is injected intermittently, bone formation is increased more than resorption resulting in bone gain. Intermittent PTH was the first anabolic agent approved for osteoporosis therapy in the USA (Potts and Gardella 2007; Augustine and Horwitz 2013).

Intermittent PTH Both PTH and PGE_2 stimulate cAMP signaling, and both can induce both resorption and formation. We hypothesized that PGE_2 might mediate some of the anabolic effects of PTH but found instead that the anabolic effects of intermittent PTH were increased in COX-2 KO mice (Xu et al. 2010). This led us to consider that PGE_2 might inhibit the osteogenic effects of PTH in vitro. We found that continuous PTH inhibited or had no effect on osteoblastic differentiation in WT marrow stromal cultures but stimulated osteoblastic differentiation in COX-2 KO cultures (Choudhary et al. 2013). The COX-2-dependent inhibition of the osteogenic effects of PTH was shown to be due to a factor secreted by the hematopoietic lineage (bone marrow macrophage) cells in the cultures in response to a combination of

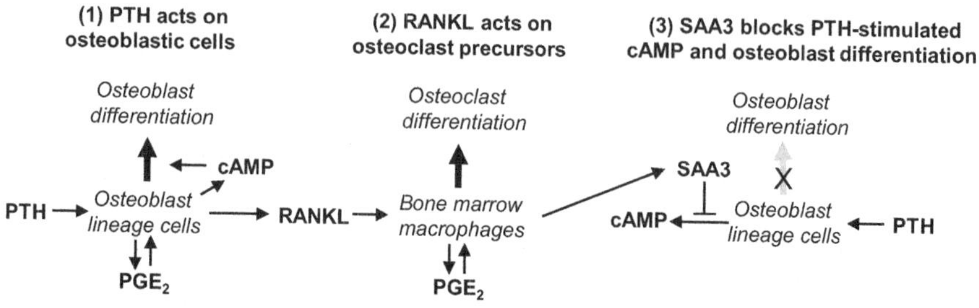

Fig. 3 Role of PGE_2-dependent SAA3 in blocking the osteoblastic response to continuous PTH. (1) PTH acts on osteoblastic precursors to induce cAMP, which causes them to differentiate into mature, bone-forming osteoblasts. PTH also induces COX-2/PGE_2 and RANKL. (2) RANKL acts on bone marrow macrophages (BMMs) to induce them to become osteoclast precursors. RANKL also induces COX-2/PGE_2. RANKL combined with PGE_2 causes the osteoclastic precursors to produce and secrete SAA3. (3) SAA3 acts on the osteoblastic cells to block PTH-stimulated cAMP and suppress osteoblastic differentiation

RANKL, from osteoblastic lineage cells, and PGE_2, produced by either osteoblastic or hematopoietic lineage cells, acting via the EP4 receptor (Choudhary et al. 2013). We subsequently identified the COX-2-dependent secreted inhibitor in vitro as serum amyloid A3 (SAA3) (Choudhary et al. 2016) and showed that secreted SAA3 acted back on osteoblastic cells to inhibit PTH-stimulated cAMP signaling.

Continuous PTH In contrast to intermittent PTH, continuous PTH infusion causes bone loss (Iida-Klein et al. 2005; Robling et al. 2011). It was generally thought that this difference was due to increased resorption when PTH was given continuously. We examined effects of continuous PTH infusion in COX-2 KO mice, which did not express SAA3 in bone marrow macrophage cells when stimulated by PTH-induced RANKL, and in SAA3 KO mice, which have a normal COX-2 response to PTH (Choudhary et al. 2015, 2018). Continuous PTH increased bone formation in both COX-2 KO mice and SAA3 KO mice but suppressed bone formation in WT mice. There was no effect of COX-2 KO or SAA3 KO on PTH-stimulated bone resorption. Because the PTH stimulated increase in bone formation was greater than the PTH stimulated increase in bone resorption, continuous PTH was anabolic in both COX-2 KO and SAA3 KO mice. Hence, our data suggest that the effects of continuous PTH on bone are due to the PGE_2-dependent secretion of SAA3, which suppresses bone formation, and not due to increased bone resorption. Our working hypothesis is shown in Fig. 3.

7 Effects of NSAIDS on Bone

Early studies in animals suggested that NSAIDs impaired fracture healing (Einhorn 2003; Brown et al. 2004; Simon and O'Connor 2007). However, other studies proposed that the effects of NSAIDs on fracture healing were dose and duration dependent and reversible after discontinuation of brief treatment (Gerstenfeld et al. 2007). Recent reviews of animal studies indicate that loss of COX-2 activity primarily affects fracture healing via callus chondrogenesis or endochondral ossification (Geusens et al. 2013; O'Connor et al. 2014; Janssen et al. 2017). Recent reviews of fracture healing in humans conclude that short-term use of NSAIDs does not impair fracture healing (Kurmis et al. 2012; Marquez-Lara et al. 2016; Pountos et al. 2012) or spinal fusion surgery (Sivaganesan et al. 2017). On the other hand, 6 weeks of indomethacin increased risk for non-union after acetabular fracture surgery (Sagi et al. 2014), and chronic use of NSAIDs increased risk for a second hip fracture after hip fracture surgery (Huang et al. 2015). Hence, NSAIDs should probably be used at low dose and for a short duration in situations of bone repair.

It is difficult to obtain data in humans on the effects of NSAIDs on bone loss, given the wide range of NSAID dose, drug adherence, physical activity, inflammatory conditions, and considerable variability at an individual level in the degree of COX-2 inhibition and selectivity attained by selective COX-2 inhibitors (Fries et al. 2006). In a study of men age 65 and older, daily COX-2 inhibitor use in men was associated with lower hip and spine bone mineral density (BMD) compared to nonusers (Richards et al. 2006). In postmenopausal women not on estrogen replacement therapy, it was associated with a higher BMD. There was no effect of COX-2 inhibitor in women on estrogen replacement. A review of the literature on controlled randomized clinical trials with bone remodeling outcomes found some evidence for increased BMD and decreased rate of resorption and no evidence for increased bone formation in NSAID users, but the data were too limited for firm conclusions (Konstantinidis et al. 2013).

8 Summary

PGs are highly bioactive fatty acids, produced by most cells in the body and rapidly released and rapidly degraded. PGE_2 is abundantly produced by both the mesenchymal lineage cells and the hematopoietic cell lineages, which give rise to the bone-forming osteoblasts and the bone-resorbing osteoclasts, respectively, as well as multiple other cells in the bone environment. The production of PGE_2 in bone is highly regulated by multiple factors that induce COX-2. Cell and organ culture studies, as well as in vivo studies of animals given exogenous PGE_2, have demonstrated that PGE_2 can stimulate both bone resorption and bone formation. However, studies of mice with globally absent COX-2 have not shown a major skeletal phenotype. PGE_2 acts at four G-protein-coupled receptors, EP1–4, with distinct signaling pathways. Many of the actions of PGE_2 in bone have been attributed to increasing cAMP via the EP2 and EP4 receptors, and agonists of the EP2 and EP4 receptors have been investigated for their ability to stimulate bone formation and enhance fracture repair. Because COX-2 is induced by multiple hormones, growth factors, and proinflammatory factors, PGE_2 may integrate, amplify, or actually mediate, the responses to these factors, a possibility that is often overlooked. Recent studies have shown that when PTH is given continuously,

PTH-stimulated COX-2/PGE_2 leads to the production of a factor that blocks the osteogenic effects of PTH in vitro and the anabolic effects of PTH in vivo.

As might be expected for a local factor with widespread production, regulated by many factors, and acting at multiple receptors, it has been difficult to define specific functions of PGE_2 in bone as therapeutic targets. As the signaling pathways in bone are better characterized in physiologic and pathologic conditions, it may be possible to identify therapeutic applications of manipulating PGs in skeletal disorders.

Acknowledgments This effort was supported by National Institute of Health Grants NIAMS award AR060286.

References

Alam I, Warden SJ, Robling AG, Turner CH (2005) Mechanotransduction in bone does not require a functional cyclooxygenase-2 (COX-2) gene. J Bone Miner Res 20:438–446
Alander CB, Raisz LG (2006) Effects of selective prostaglandin E2 receptor agonists on cultured calvarial murine osteoblastic cells. Prostaglandins Other Lipid Mediat 81:178–183
Antczak MI, Zhang Y, Wang C, Doran J, Naidoo J, Voruganti S, Williams NS, Markowitz SD, Ready JM (2017) Inhibitors of 15-prostaglandin dehydrogenase to potentiate tissue repair. J Med Chem 60:3979–4001
Augustine M, Horwitz MJ (2013) Parathyroid hormone and parathyroid hormone-related protein analogs as therapies for osteoporosis. Curr Osteoporos Rep 11:400–406
Blackwell KA, Raisz LG, Pilbeam CC (2010) Prostaglandins in bone: good cop? bad cop? Trends Endocrinol Metab 21:294–301
Boyce BF, Xing L (2008) Functions of RANKL/RANK/OPG in bone modeling and remodeling. Arch Biochem Biophys 473:139–146
Brown KM, Saunders MM, Kirsch T, Donahue HJ, Reid JS (2004) Effect of COX-2-specific inhibition on fracture-healing in the rat femur. J Bone Joint Surg Am 86-A:116–123
Buchanan FG, DuBois RN (2006) Connecting COX-2 and Wnt in cancer. Cancer Cell 9:6–8
Buczynski MW, Dumlao DS, Dennis EA (2009) Thematic review series: proteomics. An integrated omics analysis of eicosanoid biology. J Lipid Res 50:1015–1038
Castellone MD, Teramoto H, Gutkind JS (2006) Cyclooxygenase-2 and colorectal cancer chemoprevention: the beta-catenin connection. Cancer Res 66:11085–11088
Chen H, Hu B, Lv X, Zhu S, Zhen G, Wan M, Jain A, Gao B, Chai Y, Yang M, Wang X, Deng R, Wang L, Cao Y, Ni S, Liu S, Yuan W, Chen H, Dong X, Guan Y, Yang H, Cao X (2019) Prostaglandin E2 mediates sensory nerve regulation of bone homeostasis. Nat Commun 10:181
Chikazu D, Li X, Kawaguchi H, Sakuma Y, Voznesensky OS, Adams DJ, Xu M, Hoshi K, Katavic V, Herschman HR, Raisz LG, Pilbeam CC (2005) Bone morphogenetic protein 2 induces cyclooxygenase 2 in osteoblasts via a Cbfa1 binding site: role in effects of bone morphogenetic protein 2 in vitro and in vivo. 2002. J Bone Miner Res 20:1888–1898
Choudhary S, Wadhwa S, Raisz LG, Alander C, Pilbeam CC (2003) Extracellular calcium is a potent inducer of cyclooxygenase-2 in murine osteoblasts through an ERK signaling pathway. J Bone Miner Res 18:1813–1824
Choudhary S, Halbout P, Alander C, Raisz L, Pilbeam C (2007) Strontium ranelate promotes osteoblastic differentiation and mineralization of murine bone marrow stromal cells: involvement of prostaglandins. J Bone Miner Res 22:1002–1010
Choudhary S, Alander C, Zhan P, Gao Q, Pilbeam C, Raisz L (2008) Effect of deletion of the prostaglandin EP2 receptor on the anabolic response to prostaglandin E2 and a selective EP2 receptor agonist. Prostaglandins Other Lipid Mediat 86:35–40

Choudhary S, Blackwell K, Voznesensky O, Deb RA, Pilbeam C (2013) Prostaglandin E2 acts via bone marrow macrophages to block PTH-stimulated osteoblast differentiation in vitro. Bone 56:31–41

Choudhary S, Canalis E, Estus T, Adams D, Pilbeam C (2015) Cyclooxygenase-2 suppresses the anabolic response to PTH infusion in mice. PLoS One 10:e0120164

Choudhary S, Goetjen A, Estus T, Jacome-Galarza CE, Aguila HL, Lorenzo J, Pilbeam C (2016) Serum amyloid A3 secreted by preosteoclasts inhibits parathyroid hormone-stimulated cAMP signaling in murine osteoblasts. J Biol Chem 291:3882–3894

Choudhary S, Santone E, Yee S-P, Lorenzo J, Adams DJ, Goetjen A, McCarthy MB, Mazzocca A, Pilbeam C (2018) Continuous PTH in mice causes bone loss because it induces serum amyloid A (SAA). Endocrinology 159:2759–2776

Dennis EA, Norris PC (2015) Eicosanoid storm in infection and inflammation. Nat Rev Immunol 15:511–523

Dennis EA, Cao J, Hsu YH, Magrioti V, Kokotos G (2011) Phospholipase A2 enzymes: physical structure, biological function, disease implication, chemical inhibition, and therapeutic intervention. Chem Rev 111:6130–6185

Desai A, Zhang Y, Park Y, Dawson DM, Larusch GA, Kasturi L, Wald D, Ready JM, Gerson SL, Markowitz SD (2018) A second-generation 15-PGDH inhibitor promotes bone marrow transplant recovery independent of age, transplant dose, and G-CSF support. Haematologica 103:1054–1064

Dinchuk JE, Car BD, Focht RJ, Johnston JJ, Jaffee BD, Covington MB, Contel NR, Eng VM, Collins RJ, Czerniak PM (1995) Renal abnormalities and an altered inflammatory response in mice lacking cyclooxygenase II. Nature 378:406–409

Einhorn TA (2003) Cox-2: where are we in 2003? the role of cyclooxygenase-2 in bone repair. Arthritis Res Ther 5:5–7

Estus TL, Choudhary S, Pilbeam CC (2016) Prostaglandin-mediated inhibition of PTH-stimulated beta-catenin signaling in osteoblasts by bone marrow macrophages. Bone 85:123–130

Faye-Petersen OM, Johnson WH Jr, Carlo WA, Hedlund GL, Pacifico AD, Blair HC (1996) Prostaglandin E1-induced hyperostosis: clinicopathologic correlations and possible pathogenetic mechanisms. Pediatr Pathol Lab Med 16:489–507

Ferreira SH, Vane JR (1967) Prostaglandins: their disappearance from and release into the circulation. Nature 216:868–873

Flanagan AM, Chambers TJ (1992) Stimulation of bone nodule formation in vitro by prostaglandins E1 and E2. Endocrinology 130:443–448

Fries S, Grosser T, Price TS, Lawson JA, Kapoor S, DeMarco S, Pletcher MT, Wiltshire T, Fitzgerald GA (2006) Marked interindividual variability in the response to selective inhibitors of cyclooxygenase-2. Gastroenterology 130:55–64

Gao Q, Zhan P, Alander CB, Kream BE, Hao C, Breyer MD, Pilbeam CC, Raisz LG (2009) Effects of global or targeted deletion of the EP4 receptor on the response of osteoblasts to prostaglandin in vitro and on bone histomorphometry in aged mice. Bone 45:98–103

Gerstenfeld LC, Al-Ghawas M, Alkhiary YM, Cullinane DM, Krall EA, Fitch JL, Webb EG, Thiede MA, Einhorn TA (2007) Selective and nonselective cyclooxygenase-2 inhibitors and experimental fracture-healing. Reversibility of effects after short-term treatment. J Bone Joint Surg Am 89:114–125

Geusens P, Emans PJ, de Jong JJ, van den Bergh J (2013) NSAIDs and fracture healing. Curr Opin Rheumatol 25:524–531

Grosser T, Ricciotti E, FitzGerald GA (2017a) The cardiovascular pharmacology of nonsteroidal anti-inflammatory drugs. Trends Pharmacol Sci 38:733–748

Grosser T, Theken KN, FitzGerald GA (2017b) Cyclooxygenase inhibition: pain, inflammation, and the cardiovascular system. Clin Pharmacol Ther 102:611–622

Hara S (2017) Prostaglandin terminal synthases as novel therapeutic targets. Proc Jpn Acad Ser B Phys Biol Sci 93:703–723

Hara S, Kamei D, Sasaki Y, Tanemoto A, Nakatani Y, Murakami M (2010) Prostaglandin E synthases: understanding their pathophysiological roles through mouse genetic models. Biochimie 92:651–659

Herschman HR (1994) Regulation of prostaglandin synthase-1 and prostaglandin synthase-2. Cancer Metastasis Rev 13:241–256

Hino S, Tanji C, Nakayama KI, Kikuchi A (2005) Phosphorylation of beta-catenin by cyclic AMP-dependent protein kinase stabilizes beta-catenin through inhibition of its ubiquitination. Mol Cell Biol 25:9063–9072

Huang KC, Huang TW, Yang TY, Lee MS (2015) Chronic NSAIDs use increases the risk of a second hip fracture in patients after hip fracture surgery: evidence from a STROBE-compliant population-based study. Medicine (Baltimore) 94:e1566. https://doi.org/10.1097/MD.0000000000001566

Iida-Klein A, Lu SS, Kapadia R, Burkhart M, Moreno A, Dempster DW, Lindsay R (2005) Short-term continuous infusion of human parathyroid hormone 1-34 fragment is catabolic with decreased trabecular connectivity density accompanied by hypercalcemia in C57BL/J6 mice. J Endocrinol 186:549–557

Inoue H, Tanaka N, Uchiyama C (1995) Parathyroid hormone increases the number of tartrate-resistant acid phosphatase-positive cells through prostaglandin E2 synthesis in adherent cell culture of neonatal rat bones. Endocrinology 136:3648–3656

Janssen MP, Caron MM, van Rietbergen B, Surtel DA, van Rhijn LW, Welting TJ, Emans PJ (2017) Impairment of the chondrogenic phase of endochondral ossification in vivo by inhibition of cyclooxygenase-2. Eur Cell Mater 34:202–216

Jee WS, Ma YF (1997) The in vivo anabolic actions of prostaglandins in bone. Bone 21:297–304

Kang YJ, Mbonye UR, Delong CJ, Wada M, Smith WL (2007) Regulation of intracellular cyclooxygenase levels by gene transcription and protein degradation. Prog Lipid Res 46:108–125

Kawaguchi H, Raisz LG, Voznesensky OS, Alander CB, Hakeda Y, Pilbeam CC (1994) Regulation of the two prostaglandin G/H synthases by parathyroid hormone, interleukin-1, cortisol and prostaglandin E_2 in cultured neonatal mouse calvariae. Endocrinology 135:1157–1164

Kawaguchi H, Pilbeam CC, Gronowicz G, Abreu C, Fletcher BS, Herschman HR, Raisz LG, Hurley MM (1995) Transcriptional induction of prostaglandin G/H synthase-2 by basic fibroblast growth factor. J Clin Invest 96:923–930

Kayani B, Sewell MD, Platinum J, Olivier A, Briggs TWR, Eastwood DM (2017) Orthopaedic manifestations of congenital indifference to pain with anhidrosis (hereditary sensory and autonomic neuropathy type IV). Eur J Paediatr Neurol 21:318–326

Kennedy BP, Payette P, Mudgett J, Vadas P, Pruzanski W, Kwan M, Tang C, Rancourt DE, Cromlish WA (1995) A natural disruption of the secretory group II phospholipase A2 gene in inbred mouse strains. J Biol Chem 270:22378–22385

Kirkby NS, Chan MV, Zaiss AK, Garcia-Vaz E, Jiao J, Berglund LM, Verdu EF, Ahmetaj-Shala B, Wallace JL, Herschman HR, Gomez MF, Mitchell JA (2016) Systematic study of constitutive cyclooxygenase-2 expression: role of NF-kappaB and NFAT transcriptional pathways. Proc Natl Acad Sci U S A 113:434–439

Klein DC, Raisz LG (1970) Prostaglandins: stimulation of bone resorption in tissue culture. Endocrinology 86:1436–1440

Klein-Nulend J, Burger EH, Semeins CM, Raisz LG, Pilbeam CC (1997) Pulsating fluid flow stimulates prostaglandin release and inducible prostaglandin G/H synthase mRNA expression in primary mouse bone cells. J Bone Miner Res 12:45–51

Konstantinidis I, Papageorgiou SN, Kyrgidis A, Tzellos TG, Kouvelas D (2013) Effect of non-steroidal anti-inflammatory drugs on bone turnover: an evidence-based review. Rev Recent Clin Trials 8:48–60

Kujubu DA, Fletcher BS, Varnum BC, Lim RW, Herschman HR (1991) TIS10, a phorbol ester tumor promoter-inducible mRNA from Swiss 3T3 cells, encodes a novel prostaglandin synthase/cyclooxygenase homologue. J Biol Chem 266:12866–12872

Kurmis AP, Kurmis TP, O'Brien JX, Dalen T (2012) The effect of nonsteroidal anti-inflammatory drug administration on acute phase fracture-healing: a review. J Bone Joint Surg Am 94:815–823

Lader CS, Flanagan AM (1998) Prostaglandin E2, interleukin 1alpha, and tumor necrosis factor-alpha increase human osteoclast formation and bone resorption in vitro. Endocrinology 139:3157–3164

Langenbach R, Morham SG, Tiano HF, Loftin CD, Ghanayem BI, Chulada PC, Mahler JF, Lee CA, Goulding EH, Kluckman KD (1995) Prostaglandin synthase 1 gene disruption in mice reduces arachidonic acid-induced inflammation and indomethacin-induced gastric ulceration. Cell 83:483–492

Langenbach R, Loftin C, Lee C, Tiano H (1999) Cyclooxygenase knockout mice: models for elucidating isoform-specific functions. Biochem Pharmacol 58:1237–1246

Leslie CC (2015) Cytosolic phospholipase A(2): physiological function and role in disease. J Lipid Res 56:1386–1402

Li J, Sarosi I, Yan XQ, Morony S, Capparelli C, Tan HL, McCabe S, Elliott R, Scully S, Van G, Kaufman S, Juan SC, Sun Y, Tarpley J, Martin L, Christensen K, McCabe J, Kostenuik P, Hsu H, Fletcher F, Dunstan CR, Lacey DL, Boyle WJ (2000) RANK is the intrinsic hematopoietic cell surface receptor that controls osteoclastogenesis and regulation of bone mass and calcium metabolism. Proc Natl Acad Sci U S A 97:1566–1571

Li M, Ke HZ, Qi H, Healy DR, Li Y, Crawford DT, Paralkar VM, Owen TA, Cameron KO, Lefker BA, Brown TA, Thompson DD (2003) A novel, non-prostanoid EP2 receptor-selective prostaglandin E2 agonist stimulates local bone formation and enhances fracture healing. J Bone Miner Res 18:2033–2042

Li M, Healy DR, Li Y, Simmons HA, Crawford DT, Ke HZ, Pan LC, Brown TA, Thompson DD (2005) Osteopenia and impaired fracture healing in aged EP4 receptor knockout mice. Bone 37:46–54

Li M, Thompson DD, Paralkar VM (2007) Prostaglandin E(2) receptors in bone formation. Int Orthop 31:767–772

Lim H, Paria BC, Das SK, Dinchuk JE, Langenbach R, Trzaskos JM, Dey SK (1997) Multiple female reproductive failures in cyclooxygenase 2-deficient mice. Cell 91:197–208

Lin BY, Jee WSS, Ma YF, Ke HZ, Kimmel DB, Li XJ (1994) Effects of prostaglandin E2 and risedronate administration on cancellous bone in older female rats. Bone 15:489–496

Loftin CD, Trivedi DB, Tiano HF, Clark JA, Lee CA, Epstein JA, Morham SG, Breyer MD, Nguyen M, Hawkins BM, Goulet JL, Smithies O, Koller BH, Langenbach R (2001) Failure of ductus arteriosus closure and remodeling in neonatal mice deficient in cyclooxygenase-1 and cyclooxygenase-2. Proc Natl Acad Sci U S A 98:1059–1064

MacPhee M, Chepenik KP, Liddell RA, Nelson KK, Siracusa LD, Buchberg AM (1995) The secretory phospholipase A2 gene is a candidate for the Mom1 locus, a major modifier of ApcMin-induced intestinal neoplasia. Cell 81:957–966

Mahon MJ (2012) The parathyroid hormone receptorsome and the potential for therapeutic intervention. Curr Drug Targets 13:116–128

Markovic T, Jakopin Z, Dolenc MS, Mlinaric-Rascan I (2017) Structural features of subtype-selective EP receptor modulators. Drug Discov Today 22:57–71

Marquez-Lara A, Hutchinson ID, Nunez F Jr, Smith TL, Miller AN (2016) Nonsteroidal anti-inflammatory drugs and bone-healing: a systematic review of research quality. JBJS Rev. https://doi.org/10.2106/JBJS.RVW.O.00055

Mbonye UR, Wada M, Rieke CJ, Tang HY, DeWitt DL, Smith WL (2006) The 19-amino acid cassette of cyclooxygenase-2 mediates entry of the protein into the endoplasmic reticulum-associated degradation system. J Biol Chem 281:35770–35778

Min YK, Rao Y, Okada Y, Raisz LG, Pilbeam CC (1998) Regulation of prostaglandin G/H synthase-2 expression by interleukin-1 in human osteoblast-like cells. J Bone Miner Res 13:1066–1075

Miyaura C, Inada M, Suzawa T, Sugimoto Y, Ushikubi F, Ichikawa A, Narumiya S, Suda T (2000) Impaired bone resorption to prostaglandin E2 in prostaglandin E receptor EP4-knockout mice. J Biol Chem 275:19819–19823

Morham SG, Langenbach R, Loftin CD, Tiano HF, Vouloumanos N, Jennette JC, Mahler JF, Kluckman KD, Ledford A, Lee CA (1995) Prostaglandin synthase 2 gene disruption causes severe renal pathology in the mouse. Cell 83:473–482

Murakami M, Taketomi Y, Miki Y, Sato H, Hirabayashi T, Yamamoto K (2011) Recent progress in phospholipase A(2) research: from cells to animals to humans. Prog Lipid Res 50:152–192

Murakami M, Sato H, Miki Y, Yamamoto K, Taketomi Y (2015) A new era of secreted phospholipase A(2). J Lipid Res 56:1248–1261

Nahorski MS, Chen YC, Woods CG (2015) New Mendelian disorders of painlessness. Trends Neurosci 38:712–724

Norrdin RW, Shih MS (1988) Systemic effects of prostaglandin E2 on vertebral trabecular remodeling in beagles used in a healing study. Calcif Tissue Int 42:363–368

Norwood VF, Morham SG, Smithies O (2000) Postnatal development and progression of renal dysplasia in cyclooxygenase-2 null mice. Kidney Int 58:2291–2300

O'Banion MK, Sadowski HB, Winn V, Young DA (1991) A serum- and glucorticoid-regulated 4-kilobase mRNA encodes a cyclooxygenase-related protein. J Biol Chem 266:23261–23267

O'Callaghan G, Houston A (2015) Prostaglandin E2 and the EP receptors in malignancy: possible therapeutic targets? Br J Pharmacol 172:5239–5250

O'Connor JP, Manigrasso MB, Kim BD, Subramanian S (2014) Fracture healing and lipid mediators. Bonekey Rep 3:517. https://doi.org/10.1038/bonekey.2014.12

Okada Y, Lorenzo JA, Freeman AM, Tomita M, Morham SG, Raisz LG, Pilbeam CC (2000a) Prostaglandin G/H synthase-2 is required for maximal formation of osteoclast-like cells in culture. J Clin Invest 105:823–832

Okada Y, Tomita M, Gronowicz G, Kawaguchi H, Sohn J, Tanaka Y, Morimoto I, Nakamura T, Raisz L, Pilbeam C (2000b) Effects of cyclooxygenase-2 gene disruption on osteoblastic function. J Bone Miner Res 15:S217

Paralkar VM, Borovecki F, Ke HZ, Cameron KO, Lefker B, Grasser WA, Owen TA, Li M, Silva-Jardine P, Zhou M, Dunn RL, Dumont F, Korsmeyer R, Krasney P, Brown TA, Plowchalk D, Vukicevic S, Thompson DD (2003) An EP2 receptor-selective prostaglandin E2 agonist induces bone healing. Proc Natl Acad Sci U S A 100:6736–6740

Pavalko FM, Chen NX, Turner CH, Burr DB, Atkinson S, Hsieh YF, Qiu J, Duncan RL (1998) Fluid shear-induced mechanical signaling in MC3T3-E1 osteoblasts requires cytoskeleton-integrin interactions. Am J Physiol 275:C1591–C1601

Phatarakijnirund V, Mumm S, McAlister WH, Novack DV, Wenkert D, Clements KL, Whyte MP (2016) Congenital insensitivity to pain: fracturing without apparent skeletal pathobiology caused by an autosomal dominant, second mutation in SCN11A encoding voltage-gated sodium channel 1.9. Bone 84:289–298

Pilbeam CC, Kawaguchi H, Hakeda Y, Voznesensky O, Alander CB, Raisz LG (1993) Differential regulation of inducible and constitutive prostaglandin endoperoxide synthase in osteoblastic MC3T3-E1 cells. J Biol Chem 268:25643–25649

Pilbeam CC, Raisz LG, Voznesensky O, Alander CB, Delman BN, Kawaguchi K (1994) Autoregulation of inducible prostaglandin G/H synthase in osteoblastic cells by prostaglandins. J Bone Miner Res 10:406–414

Pilbeam C, Rao Y, Voznesensky O, Kawaguchi H, Alander C, Raisz LG, Herschman H (1997) Transforming growth factor-β1 regulation of prostaglandin G/H syntase-2 expression in osteoblastic MC3T3-E1 cells. Endocrinology 138:4672–4682

Pilbeam CC, Choudhary S, Blackwell KA, Raisz LG (2008) Prostaglandins and bone metabolism. In: Bilezikian JP, Raisz LG, Martn TJ (eds) Principles of bone biology, 3rd edn. Elsevier/Academic Press, San Diego

Potts JT, Gardella TJ (2007) Progress, paradox, and potential: parathyroid hormone research over five decades. Ann N Y Acad Sci 1117:196–208

Pountos I, Georgouli T, Calori GM, Giannoudis PV (2012) Do nonsteroidal anti-inflammatory drugs affect bone healing? A critical analysis. ScientificWorldJournal 2012:606404. https://doi.org/10.1100/2012/606404
Psarra A, Nikolaou A, Kokotou MG, Limnios D, Kokotos G (2017) Microsomal prostaglandin E2 synthase-1 inhibitors: a patent review. Expert Opin Ther Pat 27:1047–1059
Raisz LG, Woodiel FN (2003) Effects of selective prostaglandin EP2 and EP4 receptor agonists on bone resorption and formation in fetal rat organ cultures. Prostaglandins Other Lipid Mediat 71:287–292
Richards JB, Joseph L, Schwartzman K, Kreiger N, Tenenhouse A, Goltzman D (2006) The effect of cyclooxygenase-2 inhibitors on bone mineral density: results from the Canadian Multicentre Osteoporosis Study. Osteoporos Int 17:1410–1419
Robertson G, Xie C, Chen D, Awad H, Schwarz EM, O'Keefe RJ, Guldberg RE, Zhang X (2006) Alteration of femoral bone morphology and density in COX-2-/- mice. Bone 39:767–772
Robling AG, Kedlaya R, Ellis SN, Childress PJ, Bidwell JP, Bellido T, Turner CH (2011) Anabolic and catabolic regimens of human parathyroid hormone 1-34 elicit bone- and envelope-specific attenuation of skeletal effects in Sost-deficient mice. Endocrinology 152:2963–2975
Sagi HC, Jordan CJ, Barei DP, Serrano-Riera R, Steverson B (2014) Indomethacin prophylaxis for heterotopic ossification after acetabular fracture surgery increases the risk for nonunion of the posterior wall. J Orthop Trauma 28:377–383
Sato T, Morita I, Sakaguchi K, Nakahama KI, Smith WL, DeWitt DL, Murota SI (1996) Involvement of prostaglandin endoperoxide H synthase-2 in osteoclast-like cell formation induced by interleukin-1 beta. J Bone Miner Res 11:392–400
Segi E, Sugimoto Y, Yamasaki A, Aze Y, Oida H, Nishimura T, Murata T, Matsuoka T, Ushikubi F, Hirose M, Tanaka T, Yoshida N, Narumiya S, Ichikawa A (1998) Patent ductus arteriosus and neonatal death in prostaglandin receptor EP4-deficient mice. Biochem Biophys Res Commun 246:7–12
Shao J, Jung C, Liu C, Sheng H (2005) Prostaglandin E2 stimulates the beta-catenin/T cell factor-dependent transcription in colon cancer. J Biol Chem 280:26565–26572
Simmons DL, Botting RM, Hla T (2004) Cyclooxygenase isozymes: the biology of prostaglandin synthesis and inhibition. Pharmacol Rev 56:387–437
Simon AM, O'Connor JP (2007) Dose and time-dependent effects of cyclooxygenase-2 inhibition on fracture-healing. J Bone Joint Surg Am 89:500–511
Sivaganesan A, Chotai S, White-Dzuro G, McGirt MJ, Devin CJ (2017) The effect of NSAIDs on spinal fusion: a cross-disciplinary review of biochemical, animal, and human studies. Eur Spine J 26:2719–2728
Smith WL, Langenbach R (2001) Why there are two cyclooxygenase isozymes. J Clin Invest 107:1491–1495
Smith WL, DeWitt DL, Garavito RM (2000) Cyclooxygenases: structural, cellular, and molecular biology. Annu Rev Biochem 69:145–182
Smith WL, Urade Y, Jakobsson PJ (2011) Enzymes of the cyclooxygenase pathways of prostanoid biosynthesis. Chem Rev 111:5821–5865
Sugimoto Y, Narumiya S (2007) Prostaglandin E receptors. J Biol Chem 282:11613–11617
Suponitzky I, Weinreb M (1998) Differential effects of systemic prostaglandin E2 on bone mass in rat long bones and calvariae. J Endocrinol 156:51–57
Suzawa T, Miyaura C, Inada M, Maruyama T, Sugimoto Y, Ushikubi F, Ichikawa A, Narumiya S, Suda T (2000) The role of prostaglandin E receptor subtypes (EP1, EP2, EP3, and EP4) in bone resorption: an analysis using specific agonists for the respective EPs. Endocrinology 141:1554–1559
Swinney DC, Mak AY, Barnett J, Ramesha CS (1997) Differential allosteric regulation of prostaglandin H synthase 1 and 2 by arachidonic acid. J Biol Chem 272:12393–12398
Tai H, Miyaura C, Pilbeam CC, Tamura T, Ohsugi Y, Koishihara Y, Kubodera N, Kawaguchi H, Raisz LG, Suda T (1997) Transcriptional induction of cyclooxygenase-2 in osteoblasts is involved in interleukin-6-induced osteoclast formation. Endocrinology 138:2372–2379

Tanaka M, Sakai A, Uchida S, Tanaka S, Nagashima M, Katayama T, Yamaguchi K, Nakamura T (2004) Prostaglandin E2 receptor (EP4) selective agonist (ONO-4819.CD) accelerates bone repair of femoral cortex after drill-hole injury associated with local upregulation of bone turnover in mature rats. Bone 34:940–948

Tang CH, Yang RS, Fu WM (2005) Prostaglandin E2 stimulates fibronectin expression through EP1 receptor, phospholipase C, protein kinase Calpha, and c-Src pathway in primary cultured rat osteoblasts. J Biol Chem 280:22907–22916

Tanioka T, Nakatani Y, Semmyo N, Murakami M, Kudo I (2000) Molecular identification of cytosolic prostaglandin E2 synthase that is functionally coupled with cyclooxygenase-1 in immediate prostaglandin E2 biosynthesis. J Biol Chem 275:32775–32782

Ueda K, Saito A, Nakano H, Aoshima M, Yokota M, Muraoka R, Iwaya T (1980) Cortical hyperostosis following long-term administration of prostaglandin E1 in infants with cyanotic congenital heart disease. J Pediatr 97:834–836

Uppal S, Diggle CP, Carr IM, Fishwick CW, Ahmed M, Ibrahim GH, Helliwell PS, Latos-Bielenska A, Phillips SE, Markham AF, Bennett CP, Bonthron DT (2008) Mutations in 15-hydroxyprostaglandin dehydrogenase cause primary hypertrophic osteoarthropathy. Nat Genet 40:789–793

Vasquez AM, Mouchlis VD, Dennis EA (2017) Review of four major distinct types of human phospholipase A2. Adv Biol Regul 10. https://doi.org/10.1016/j.jbior.2017.10.009

Vilardaga JP, Romero G, Friedman PA, Gardella TJ (2011) Molecular basis of parathyroid hormone receptor signaling and trafficking: a family B GPCR paradigm. Cell Mol Life Sci 68:1–13

Wadhwa S, Choudhary S, Voznesensky M, Epstein M, Raisz L, Pilbeam C (2002) Fluid flow induces COX-2 expression in MC3T3-E1 osteoblasts via a PKA signaling pathway. Biochem Biophys Res Commun 297:46–51

Woodward DF, Jones RL, Narumiya S (2011) International Union of Basic and Clinical Pharmacology. LXXXIII: classification of prostanoid receptors, updating 15 years of progress. Pharmacol Rev 63:471–538

Xie WL, Chipman JG, Robertson DL, Erikson RL, Simmons DL (1991) Expression of a mitogen-responsive gene encoding prostaglandin synthase is regulated by mRNA splicing. Proc Natl Acad Sci U S A 88:2692–2696

Xu M, Choudhary S, Goltzman D, Ledgard F, Adams D, Gronowicz G, Koczon-Jaremko B, Raisz L, Pilbeam C (2005) Do cyclooxygenase knockout mice have primary hyperparathyroidism? Endocrinology 146:1843–1853

Xu Z, Choudhary S, Okada Y, Voznesensky O, Alander C, Raisz L, Pilbeam C (2007) Cyclooxygenase-2 gene disruption promotes proliferation of murine calvarial osteoblasts in vitro. Bone 41:68–76

Xu M, Choudhary S, Voznesensky O, Gao Q, Adams D, Diaz-Doran V, Wu Q, Goltzman D, Raisz LG, Pilbeam CC (2010) Basal bone phenotype and increased anabolic responses to intermittent parathyroid hormone in healthy male COX-2 knockout mice. Bone 47:341–352

Yoshida K, Oida H, Kobayashi T, Maruyama T, Tanaka M, Katayama T, Yamaguchi K, Segi E, Tsuboyama T, Matsushita M, Ito K, Ito Y, Sugimoto Y, Ushikubi F, Ohuchida S, Kondo K, Nakamura T, Narumiya S (2002) Stimulation of bone formation and prevention of bone loss by prostaglandin E EP4 receptor activation. Proc Natl Acad Sci U S A 99:4580–4585

Yuan C, Smith WL (2015) A cyclooxygenase-2-dependent prostaglandin E2 biosynthetic system in the Golgi apparatus. J Biol Chem 290:5606–5620

Zhan P, Alander C, Kaneko H, Pilbeam CC, Guan Y, Zhang Y, Breyer MD, Raisz LG (2005) Effect of deletion of the prostaglandin EP4 receptor on stimulation of calcium release from cultured mouse calvariae: impaired responsiveness in heterozygotes. Prostaglandins Other Lipid Mediat 78:19–26

Zhang Y, Haga N (2014) Skeletal complications in congenital insensitivity to pain with anhidrosis: a case series of 14 patients and review of articles published in Japanese. J Orthop Sci 19:827–831

Zhang X, Schwarz EM, Young DA, Puzas JE, Rosier RN, O'Keefe RJ (2002) Cyclooxygenase-2 regulates mesenchymal cell differentiation into the osteoblast lineage and is critically involved in bone repair. J Clin Invest 109:1405–1415

Zhang M, Ho HC, Sheu TJ, Breyer MD, Flick LM, Jonason JH, Awad HA, Schwarz EM, O'Keefe RJ (2011) EP1(−/−) mice have enhanced osteoblast differentiation and accelerated fracture repair. J Bone Miner Res 26:792–802

Zhang M, Feigenson M, Sheu TJ, Awad HA, Schwarz EM, Jonason JH, Loiselle AE, O'Keefe RJ (2015a) Loss of the PGE2 receptor EP1 enhances bone acquisition, which protects against age and ovariectomy-induced impairments in bone strength. Bone 72:92–100

Zhang Y, Desai A, Yang SY, Bae KB, Antczak MI, Fink SP, Tiwari S, Willis JE, Williams NS, Dawson DM, Wald D, Chen WD, Wang Z, Kasturi L, Larusch GA, He L, Cominelli F, Di ML, Djuric Z, Milne GL, Chance M, Sanabria J, Dealwis C, Mikkola D, Naidoo J, Wei S, Tai HH, Gerson SL, Ready JM, Posner B, Willson JK, Markowitz SD (2015b) Inhibition of the prostaglandin-degrading enzyme 15-PGDH potentiates tissue regeneration. Science 348: aaa2340

Cytokines and Bone: Osteoimmunology

Joseph Lorenzo

Contents

J. Lorenzo (✉)
Departments of Medicine and Orthopaedic Surgery, UConn Health, Farmington, CT, USA
e-mail: jlorenzo@uchc.edu

P. H. Stern (ed.), *Bone Regulators and Osteoporosis Therapy*,
Handbook of Experimental Pharmacology 262, https://doi.org/10.1007/164_2019_346

Abstract

Cytokines and hematopoietic growth factors have traditionally been thought of as regulators of the development and function of immune and blood cells. However, an ever-expanding number of these factors have been discovered to have major effects on bone cells and the development of the skeleton in health and disease (Table 1). In addition, several cytokines have been directly linked to the development of osteoporosis in both animal models and in patients. In order to understand the mechanisms regulating bone cells and how this may be dysregulated in disease states, it is necessary to appreciate the diverse effects that cytokines and inflammation have on osteoblasts, osteoclasts, and bone mass. This chapter provides a broad overview of this topic with extensive references so that, if desired, readers can access specific references to delve into individual topics in greater detail.

Keywords

Cytokines · Interleukins · Colony stimulating factors · Osteoblasts · Osteoclasts · Bone mass

1 Receptor Activator of Nuclear Factor-κB Ligand (RANKL), Receptor Activator of Nuclear Factor-κB (RANK), and Osteoprotegerin (OPG)

RANKL, which is also known as TNF superfamily ligand 11 (TNFSF11), is the sole ligand of the critical paracrine system, regulating osteoclast formation. The other members are two receptors: (1) RANK (TNFRSF11A), which is expressed on cell membranes and mediates the downstream signalling events of osteoclastogenesis, and (2) OPG (TNFRSF11B), which is a soluble decoy receptor for RANKL that inhibits RANKL-mediated cell signalling (Suda et al. 1999; Teitelbaum 2000; Walsh et al. 2006).

RANKL potently enhances the formation of osteoclasts from precursor cells and the bone-resorbing activity and life span of mature osteoclasts (Fuller et al. 1998; Lacey et al. 1998; Yasuda et al. 1998). RANKL-deficient mice have significant osteopetrosis and no osteoclasts but a normal number of monocyte/macrophages (Kong et al. 1999). They also fail to erupt teeth, which is a common finding in developmental osteopetrosis. Marrow stromal cells, hypertrophic chondrocytes, osteoblasts, and osteocytes produce RANKL (Nakashima et al. 2011; Xiong et al. 2011). In addition, production of RANKL by B-lymphocytes is important for regulating the enhanced bone resorption and decreased bone mass that occurs after ovariectomy in mice (Onal et al. 2012), while T-lymphocyte-generated RANKL may mediate bone loss in patients with HIV infection (Titanji et al. 2018). Targeted gene deletion of RANKL in the osteocytes and maturing osteoblasts of mice demonstrated the critical nature of RANKL expression by these cells for the maintenance of osteoclastogenesis in post-fetal development (Nakashima et al. 2011; Xiong et al. 2011, 2014). Recently, it was found that membrane-bound

RANKL can bind RANK on vesicles, which are secreted by maturing osteoclasts. These, in turn, can reverse signal into osteoblasts to regulate their activity (Zhang et al. 2017; Ikebuchi et al. 2018).

RANKL exists either in a cell membrane-bound or soluble form (Nakashima et al. 2000). Loss of soluble RANKL increased bone mass and decreased osteoclast number in adult, but not in growing mice. However, deletion of soluble RANKL had no effect on the bone loss that occurs after ovariectomy (Xiong et al. 2018). Most resorption stimulators, including cytokines and hormones, exert their primary osteoclastogenic activity by inducing RANKL expression in mesenchymal lineage cells (Boyle et al. 2003; Suda et al. 1999). Conversely, the shedding of membrane-bound RANKL is a mechanism for inhibiting osteoblast-mediated osteoclast formation by removing RANKL from the cell surface (Hikita et al. 2006). This process depends on expression of matrix metalloproteinase (MMP) 14 (Hikita et al. 2006), since mice lacking this enzyme have increased osteoclast number (Table 1).

OPG is a secreted inhibitor of osteoclast formation that acts as a decoy receptor for RANKL (Lacey et al. 1998; Simonet et al. 1997; Yasuda et al. 1998). It was initially identified as a soluble factor, having inhibitory effects on osteoclastogenesis in vitro (Simonet et al. 1997; Tsuda et al. 1997), and an inducer of osteopetrosis when it was transgenically overexpressed in mice (Simonet et al. 1997). In marrow, it is produced by a variety of cells including stromal cells, B-lymphocytes, and dendritic cells (Yun et al. 1998). Besides RANKL, OPG also binds the TNF-like ligand TRAIL (TNF-related apoptosis-inducing ligand) (Emery et al. 1998). Mice that lack OPG have severe osteoporosis, an increased number of osteoclasts, and arterial calcification (Bucay et al. 1998; Mizuno et al. 1998). The latter finding highlights a potential genetic link between osteoporosis and vascular disease (Boyle et al. 2003). Overexpression of OPG in transgenic mice caused osteopetrosis with decreased osteoclast numbers and extramedullary hematopoiesis (Simonet et al. 1997). Serum levels of OPG increase with age in humans and are directly correlated with whole body bone mass (Coulson et al. 2017). Conversely, serum titers of OPG antibodies are inversely correlated with bone mass (Hauser et al. 2017). Genetic inactivation of OPG is a cause of juvenile Paget's disease (Whyte et al. 2002a), while decreases in OPG are associated with glucocorticoid-induced osteoporosis in animal and in vitro models (Piemontese et al. 2016; Hofbauer et al. 1999a).

The biologically active receptor for RANKL is RANK. It was first identified on dendritic cells (Anderson et al. 1997), but it is also present on osteoclast precursors and mature osteoclasts (Dougall et al. 1999). RANK expression at the RNA level is detected in a variety of cell types and tissues (Anderson et al. 1997). RANK-deficient mice phenocopy the defective osteoclast development of RANKL-deficient mice, confirming the exclusive specificity of RANKL for RANK (Dougall et al. 1999). In humans, gain-of-function mutations in RANK are associated with familial expansile osteolysis, expansible skeletal hyperphosphatasia, and juvenile Paget's disease, all of which are diseases that are associated with excessive formation and activity of osteoclasts (Cecchini et al. 1997; Hayashi et al. 1997; Muguruma and Lee 1998; Suda et al. 1995; Takahashi et al. 1988; Hughes et al. 2000; Whyte and Hughes 2002; Whyte and Mumm 2004; Whyte et al. 2002a, b, 2014).

Table 1 Effects of the major cytokines and cytokine receptors on bone cells

Cytokine	Receptor	Actions on bone cells
RANKL (TNFSF11)	RANK (TNFSFR11A), OPG (TNFSFR11B)	RANKL is the principal stimulator of osteoclast formation, and osteoclast-mediated bone resorption OPG is a soluble decoy receptor for RANKL that inhibits its interaction with its cognate cellular receptor RANK
CSF-1 (M-CSF)	CSF-1R (c-fms)	CSF-1 stimulates the proliferation of the osteoclast precursor and facilitates RANKL-mediated osteoclastogenesis and bone resorption
IL-34	CSF-1R (c-fms)	IL-34 can replicate the actions of CSF-1 on osteoclast precursors and mature osteoclasts
GM-CSF	GM-CSFR	In early myeloid precursors GM-CSF inhibits RANKL-mediated osteoclastogenesis while stimulating their differentiation into dendritic cells. In more mature osteoclast precursors, it has stimulatory effects on maturation and resorptive activity
IL-1 α and β	IL-1R1, IL-1R2	IL-1 is a major proinflammatory cytokine that stimulates osteoclastogenesis and bone resorption through a variety of both direct and indirect actions on osteoclasts and their precursors. IL-1 also inhibits osteoblast-mediated bone formation. IL-1R1 is the major cellular receptor. IL-1R2 is a decoy receptor found on the cell membrane
TNF α and β	TNFR1, TNFR2	Like IL-1, TNF is a major proinflammatory cytokine, which stimulates osteoclastogenesis and bone resorption through a variety of both direct and indirect actions on osteoclasts and their precursors. In addition, both stimulatory and inhibitory effects of TNF on osteoblast lineage cells have been described. Which of these occurs appears dependent on the differentiation stage of the osteoblast-lineage cells. Both TNF receptors are active. Although, many of the effects of TNF on bone cells are mediated by TNFR1
Fas-ligand	Fas	Inhibits osteoblast differentiation and stimulates osteoclastogenesis
TRAIL	TRAIL-R2	Effects on osteoclastogenesis are variable
CD40-ligand	CD40	Accelerate RANKL-induced osteoclastogenesis
IL-4	IL-4R	Inhibits osteoclast and osteoblast activity
IL-6	IL-6R, gp130	Another major proinflammatory cytokine, which has variable effects on bone resorption and formation
IL-11	IL-11R, gb130	Stimulates osteoclastogenesis and osteoblast differentiation
LIF	LIFR, gp130	Variable effects on resorption and formation
Oncostatin-M	OSMR, LIFR, gp130	Inhibits osteoclasts formation and resorption. Stimulates osteoblast differentiation
IL-7	IL-7R, Common γ chain	Variable effects on bone resorption and formation

(continued)

Table 1 (continued)

Cytokine	Receptor	Actions on bone cells
IL-8	CXCR1, CXCR2	Stimulates bone resorption
IL-10	IL-10R	Inhibits osteoclastogenesis and osteoblastogenesis
IL-12	IL-12R	Inhibits osteoclastogenesis
IL-13	IL-13R	Inhibits osteoclast and osteoblast activity
IL-15	IL-15R	Stimulates osteoclastogenesis
IL-17 (A-F)	IL-17R	Variable effects on osteoclastogenesis
IL-18	IL-18R	Inhibits osteoclastogenesis, variable effects on osteoblasts
INF-α and β (type 1)	IFNAR1	Inhibits osteoclasts and osteoblasts
INF-γ (Type II)	IFNGR1, IFNGR2	Variable effects on osteoclasts and osteoblasts
MIF	CD74	Variable effects, including direct inhibition of osteoclastogenesis

While osteoclast-like cells can form in vitro in the absence of RANK or TRAF6 signalling when exposed to a cocktail of cytokines and growth factors (Kim et al. 2014a, 2005b; Kobayashi et al. 2000; Kudo et al. 2002; Hemingway et al. 2011), the significance of this in vitro finding is controversial since osteoclasts are generally absent in RANK-deficient animals (Dougall et al. 1999; Li et al. 2000). In most instances, cytokines and growth factors other than RANKL, which are produced at sites of inflammation or physiologically during bone turnover, act as cofactors that enhance or modulate the response of osteoclasts and their precursors to RANKL-RANK stimulation (Lam et al. 2000; Lee et al. 2005a, b; Takita et al. 2005). However, it was demonstrated that in the absence of the p100 precursor of NF-κB, tumor necrosis factor α (TNF α) was a potent stimulator of osteoclastogenesis (Yao et al. 2009). In addition, in inflammatory arthritis RANKL-independent osteoclastogenesis in vivo has been observed (O'Brien et al. 2016).

2 Colony-Stimulating Factor-1 (CSF-1)

In addition to RANKL, colony-stimulating factor-1 (CSF-1) (also known as macrophage colony stimulating factor or M-CSF) is critical for normal osteoclast formation. This cytokine was originally identified by its ability to regulate macrophage formation (Roth and Stanley 1992). However, it was subsequently shown that a spontaneous mouse mutant (the *op/op* mouse) with a phenotype of absent osteoclasts and defective macrophage/monocyte formation was deficient in CSF-1 (Felix et al. 1990b; Wiktor Jedrzejczak et al. 1990; Yoshida et al. 1990). Injection of CSF-1 into *op/op* mice corrected the defect in osteoclast formation and bone resorption (Felix et al. 1990a), as did expression of CSF-1 protein specifically in osteoblastic cells (Abboud et al. 2002).

Stimulators of bone resorption can increase the production of CSF-1 in bone (Felix et al. 1989; Rubin et al. 1996; Weir et al. 1993), and multiple transcripts of CSF-1 are produced by alternative splicing resulting in either a soluble or membrane bound form of the protein (Cerretti et al. 1988, 1990). In vivo treatment with CSF-1 increased osteoclast number and bone resorption as well as the rate of fracture repair (Sarahrudi et al. 2009). Expression of the membrane-bound form of CSF-1 is regulated by stimulators of resorption and facilitates the differentiation of precursor cells into mature osteoclasts (Rubin et al. 1996; Yao et al. 1998). However, deletion of only the soluble form of CSF-1 in mice prevented the bone loss that occurs with estrogen deficiency (Yao et al. 2017a). This result demonstrated the critical importance of soluble CSF-1 in mediating bone loss in this model of postmenopausal osteoporosis.

CSF-1 has multiple effects on myeloid precursor cells including those that can differentiate into osteoclasts. It stimulates their replication and differentiation (Otero et al. 2009; Arai et al. 1999), and it regulates their motility (Novack and Faccio 2009). In mature osteoclasts, CSF-1 augments RANKL-induced resorptive activity (Hodge et al. 2011). Environmental factors appear to modulate the response of osteoclast precursor cells to CSF-1 since it was found that there were different in vitro responses depending on whether cells were cultured on plastic or bone (De Vries et al. 2015).

Colony-stimulating factor-1 receptor (CSF-1R, also known as c-Fms), which binds CSF-1 to activate intracellular signalling, is a tyrosine kinase (Sherr et al. 1989; Ashmun et al. 1989). Signalling through CSF-1R involves the immunoreceptor tyrosine-based activation motif (ITAM)-containing protein, DAP12, and β-catenin (Otero et al. 2009).

The role of CSF-1 in regulating osteoclast apoptosis has also been examined. Addition of CSF-1 to mature osteoclast cultures prolongs their survival (Fuller et al. 1993; Jimi et al. 1995). This response may be important for the development of the osteopetrotic phenotype in *op/op* mice, since transgenic expression of Bcl-2, which blocks apoptosis in myeloid cells, partially reversed the defects in osteoclast and macrophage development in these animals (Lagasse and Weissman 1997). The effects of CSF-1 on osteoclasts have been linked to activation of a Na/HCO cotransporter (Bouyer et al. 2007). CSF-1 is also a potent stimulator of RANK expression in osteoclast precursor cells (Otero et al. 2009; Arai et al. 1999) and is critical for expanding the osteoclast precursor pool size (Jacquin et al. 2006). Inhibition of CSF-1 signalling is a potential mechanism to treat osteolytic bone diseases (Kumari et al. 2018).

Interleukin (IL)-34 is an additional ligand for CSF-1R (Wei et al. 2010). Like CSF-1, it can be added with RANKL to in vitro cultures to facilitate osteoclastogenesis (Chen et al. 2011). Injection of IL-34 into mice increased the proportion of CD11b-positive cells, which contain osteoclast precursors, and decreased trabecular bone mass (Chen et al. 2011). Production of IL-34 may be responsible for the spontaneous rescue of the osteoclast phenotype that occurs in CSF-1-deficient *op/op* mice with age (Nakamichi et al. 2012). Inability to bind either CSF-1 or IL-34 may also explain why the CSF-1R deficient mouse has a more severe phenotype than the CSF-1-deficient *op/op* mouse (Dai et al. 2002). Differences in the ability of CSF-1R and IL-34 to polarize macrophages into M1 and M2 subsets have also been identified, suggesting additional receptors or cofactors may be involved in this response (Boulakirba et al. 2018).

3 Additional Colony Stimulating Factors

Like CSF-1, the colony stimulating factors, granulocyte macrophage colony-stimulating factor (GM-CSF) and IL-3, affect osteoclast differentiation (Lorenzo et al. 1987b; MacDonald et al. 1986; Shinar et al. 1990). Both have complex actions that are dependent on the lineage stage of the myeloid precursor cells with which they interact. IL-3 is produced by osteoblasts (Birch et al. 1993). It has multiple effects on in vitro osteoclastogenesis (Khapli et al. 2003; Yogesha et al. 2005; Enelow et al. 1992; Fujikawa et al. 2001; Soysa and Alles 2018), which have been associated with both stimulatory and inhibitory responses, depending on the cells that are examined and the culture conditions (Hong et al. 2013). One mechanism by which IL-3 inhibits osteoclastogenesis is through regulation of c-Fos and Id protein expression (Oh et al. 2012). Another is through differential regulation of soluble and membrane-bound RANKL expression in osteoblastic lineage cells (Singh et al. 2018). IL-3 is also reported to inhibit osteoblast differentiation in multiple myeloma (Ehrlich et al. 2005) and to enhance the differentiation of mesenchymal stem cells into osteoblasts (Barhanpurkar et al. 2012).

GM-CSF inhibits RANKL-mediated osteoclastogenesis (Khapli et al. 2003; Udagawa et al. 1997) and enhances the number of osteoclast precursor cells in early multipotential myeloid cells (Kurihara et al. 1990; Takahashi et al. 1991). It does this by directing the common myeloid precursor cell toward the dendritic cell lineage (Alnaeeli et al. 2006; Khapli et al. 2003). One mechanism for this effect is increased shedding of CSF-1R through upregulation of "a disintegrin and metalloproteinase 17" (ADAM 17), which is also known as tumor necrosis factor alpha converting enzyme (TACE) (Hiasa et al. 2009). Interestingly, GM-CSF-expanded dendritic lineage cells can still maintain their capacity to differentiate into osteoclasts and may be a source of osteoclast progenitors in inflammatory conditions (Ruef et al. 2017).

GM-CSF also inhibits expression of monocyte chemotactic protein 1 (MCP-1, CCL2) by osteoclast precursor cells (Kim et al. 2005a). MCP-1 is a chemokine involved in osteoclast motility. Both GM-CSF and IL-3 inhibit expression of TNF receptors on myeloid precursor cells (Yogesha et al. 2005). However, in pre-fusion osteoclast precursors, which are myeloid cells that have been stimulated with RANKL and CSF-1 for 3 days, to a point where they will shortly (within 6 h) fuse into osteoclasts, treatment with GM-CSF + CSF-1 enhanced osteoclastogenesis and mimicked the response to RANKL + CSF-1 (Lee et al. 2009). This increased osteoclastogenesis was mediated by upregulation of the cell fusion protein, dendritic cell-specific transmembrane protein (DC-STAMP) in the pre-fusion osteoclasts (Lee et al. 2009). It has also been shown that if multipotential myeloid precursor cells are cultured sequentially with GM-CSF and then with CSF-1 + RANKL, they form osteoclasts and, in some instances, act as dendritic cells, which present antigen to T-lymphocytes and initiate the adaptive immune response (Lari et al. 2007; Nomura et al. 2008; Alnaeeli and Teng 2009). IL-3 and GM-CSF may also support osteoclast differentiation by stimulating CSF-1 production (Fujikawa et al. 2001).

At relatively high doses G-CSF decreased bone mass in rodents when injected systemically (Soshi et al. 1996; Takamatsu et al. 1998). This response appeared to result from increased osteoclast formation, decreased osteoblast function, and increased osteoblast apoptosis. Similar effects are seen in humans (Takamatsu et al. 1998). G-CSF also mobilized the migration of hematopoietic precursor cells from the bone marrow into the circulation (Lapidot and Petit 2002) and increased the number of circulating osteoclast precursor cells (Purton et al. 1996), which may be related to its ability to increase osteoclast resorptive activity.

In mice, overexpression of G-CSF inhibited the ability of osteoblasts to respond to bone morphogenetic protein (Kuwabara et al. 2001). Short-term treatment of mice with G-CSF decreased endosteal and trabecular osteoblasts by increasing their apoptosis and inhibiting osteoblast precursor cell differentiation (Christopher and Link 2008). Mice overexpressing G-CSF had increased bone resorption. However, in contrast to wild-type mice, mice overexpressing G-CSF did not lose additional bone with ovariectomy (Oda et al. 2005). Targeted deletion of ADAM17 in mice using Cre/Lox technology and the Sox9 promoter to target Cre recombinase to chondroprogenitor cells produced a phenotype of enhanced G-CSF production, osteoporosis, and extramedullary hematopoiesis (Horiuchi et al. 2009). These results suggest that production of ADAM17 on osteoblasts and/or chondroblasts regulates expression of G-CSF.

4 Interleukin-1 (IL-1)

There are two separate IL-1 genes, IL-1α and IL-1β, which have identical activities (Dinarello 1991). IL-1 is a potent peptide stimulator of bone resorption (Lorenzo et al. 1987a; Sabatini et al. 1988). It directly affects RANKL-primed osteoclasts to resorb (Jimi et al. 1999; Jules et al. 2012) by a mechanism dependent on microphthalmia transcription factor (MITF) (Kim et al. 2009). It also has indirect resorptive actions through its ability to stimulate RANKL production (Hofbauer et al. 1999b). In addition, the ability of either RANKL- or 1,25-dihydroxyvitamin D_3 to stimulate osteoclast formation in vitro is mediated, in part, by effects on IL-1 production (Lee et al. 2002a, 2005a, b; Nakamura and Jimi 2006). IL-1 also increases prostaglandin synthesis in the bone (Lorenzo et al. 1987a; Sato et al. 1986), which may further enhance its resorptive activity since prostaglandins are potent resorption stimuli (Klein and Raisz 1970). Direct stimulation of osteoclastogenesis by IL-1 in mixed murine stromal and hematopoietic cell cultures is dependent on RANKL but not TNF α expression in the stromal/osteoblastic cells (Ma et al. 2004).

In mouse models, IL-1 appears to be involved in normal growth plate development (Simsa-Maziel et al. 2013) and bone turnover (Lee et al. 2010). It also may be essential for the systemic bone loss that is seen in some inflammatory conditions due to high TNF production (Polzer et al. 2010). In addition, it mediates some of the effects of estrogen withdrawal on bone loss in both mice (Lorenzo et al. 1998) and humans (Charatcharoenwitthaya et al. 2007). In humans, levels of IL-1 were negatively correlated with bone mass (Ivanova et al. 2012; Zupan et al. 2012).

IL-1 is produced in the bone (Lorenzo et al. 1990b), and its activity is present in bone marrow serum (Kawaguchi et al. 1995; Miyaura et al. 1995). One source of bone cell-derived IL-1 is osteoclast precursor cells, which produce IL-1 when they interact with bone matrix (Yao et al. 2008). There is also a natural inhibitor of IL-1, IL-1 receptor antagonist (IL-1ra), which is an analog of IL-1 that binds, but does not activate, the biologically important type I IL-1 receptor (Arend et al. 1990; Eisenberg et al. 1990; Hannum et al. 1990).

There are two receptors for IL-1: type I and type II (Dinarello 1993). All known biologic responses to IL-1 appear to be mediated exclusively through the type I receptor (Sims et al. 1993). IL-1 receptor type I requires interaction with a second protein, IL-1 receptor accessory protein (IRAcP), to generate post-receptor signals (Huang et al. 1997; Korherr et al. 1997; Wesche et al. 1997). Signalling through type I receptors involves activation of specific TRAFs and NF-κB (Eder 1997; Martin and Falk 1997). IL-1 receptor type II is a decoy receptor that prevents activation of type I receptors (Colotta et al. 1993). One report found a decrease in the bone mass of mice that were deficient in the bioactive type I IL-1 receptor (Bajayo et al. 2005). However, this has not been our experience (Vargas et al. 1996).

Expression of myeloid differentiation factor 88 (MyD88) but not Toll/interleukin-1 receptor domain-containing adaptor inducing interferon-beta (TRIF) was necessary for IL-1 to stimulate RANKL production in osteoblasts and to prolong the survival of osteoclasts (Sato et al. 2004). Survival of osteoclasts by treatment with IL-1 required PI3-kinase/AKT and ERK (Lee et al. 2002b).

The effects of IL-1 on bone in inflammatory states, such as rheumatoid arthritis, are multiple and mediated by both direct and indirect mechanisms. IL-1 stimulates bone resorption and inhibits bone formation through its effects on osteoclasts and osteoblasts, respectively (Boyce et al. 1989; Canalis 1986; Tsuboi et al. 1999). In inflammatory conditions, it is directly involved in the production of a relatively unique population of very active osteoclasts (Shiratori et al. 2018). In addition, it stimulates the production of a variety of secondary factors in the bone microenvironment including prostaglandins and GM-CSF, which have complex effects on bone cells themselves (Niki et al. 2007). IL-1 also has been reported both to inhibit and to stimulate production of OPG in various osteoblastic cell models in vitro (Tanabe et al. 2005; Lambert et al. 2007). IL-1's effects on osteoclast precursor cell migration are mediated by its ability to stimulate production of the chemokine CX_3CL1 (Matsuura et al. 2017; Cao et al. 2016).

IL-1 induces the differentiation of mesenchymal stem cells toward osteoblasts via the noncanonical Wnt-5a/Ror2 pathway (Sonomoto et al. 2012). However, it also inhibits osteoblast migration (Hengartner et al. 2013) and the ability of mesenchymal stem cells to promote tissue regeneration (Martino et al. 2016). IL-1 is involved in the mechanisms of mechanosensing in the bone (Veeriah et al. 2016). Osteocytes regulate osteoclastogenesis by producing a variety of factors including RANKL (Nakashima et al. 2011; Xiong et al. 2011). Treatment of the MLO-Y4 osteocyte-like cell line with IL-1 enhanced RANKL and decreased OPG production, which was reversed by mechanically loading the cells (Kulkarni et al. 2012).

5 Tumor Necrosis Factor (TNF)

Like IL-1, TNF represents a family of two related polypeptides (α and β) that are the products of separate genes (Beutler and Cerami 1986, 1987; Old 1985; Oliff 1988; Paul and Ruddle 1988). TNFα and TNFβ have similar biologic activities and are both potent stimulators of bone resorption (Bertolini et al. 1986; Lorenzo et al. 1987a; Tashjian et al. 1987).

In vivo administration of TNFα was shown to increase the serum calcium of mice (Tashjian et al. 1987) and to stimulate new osteoclast formation and bone resorption (Stashenko et al. 1987). Like IL-1, TNF also enhances the formation of osteoclast-like cells in bone marrow culture (Tashjian et al. 1987) through its ability to increase RANKL production (Hofbauer et al. 1999b). However, TNF stimulates multiple additional cytokines in the bone and many of these enhance the response to RANKL. For example, TNF stimulates osteoclast formation in mixed stromal cell/osteoclast precursor cell cultures by a mechanism that was partially dependent on the production of IL-1 (Wei et al. 2005; Zwerina et al. 2007; Polzer et al. 2010). Likewise, TNF-induced osteolysis is dependent on CSF-1 production (Kitaura et al. 2005).

TNF can also directly stimulate osteoclast formation in vitro, independent of RANK in RANK-deficient mice (Azuma et al. 2000; Kim et al. 2005b; Kobayashi et al. 2000). However, the significance of this in vitro finding is controversial. In vivo administration of TNF to RANK-deficient mice caused only an occasional osteoclast to form (Li et al. 2000). In addition, RANK-deficient mice that also overexpressed TNF had severe osteopetrosis and no osteoclasts (Li et al. 2004a). It was also demonstrated that TNF can stimulate osteoclastogenesis in mice that are deficient in the p100 precursor protein of NF-κB, which is a critical signalling molecule in RANKL-mediated stimulation of osteoclastogenesis and bone resorption (Yao et al. 2009). These results demonstrate that even though both TNF and RANKL are members of the same cytokine superfamily and share multiple overlapping signalling pathways, there are crucial differences between their downstream signalling molecules in osteoclast precursor cells. It has also been shown that in cherubism, a disease with elevated production of TNF due to a gain-of-function mutation in the SRC homology 3 domain binding protein 2 (Sh3bp2) gene, bone resorption can occur through the action of cells that are not classic osteoclasts (Kittaka et al. 2018). There may also be a role for sclerostin, a Wnt signalling inhibitor, in the ability of TNF to induce inflammatory arthritis since deletion or inhibition of sclerostin exacerbated the inflammatory joint destruction seen in TNF transgenic mice (Wehmeyer et al. 2016).

As with IL-1, TNF binds to two cell surface receptors, TNF receptor 1 or p55 and TNF receptor 2 or p75 (Fiers and Sim 1993). However, in contrast to IL-1, both receptors transmit biologic responses. The principal effects on bone cells appear to be mediated through TNF receptor 1 (Zhang et al. 2001; Abbas et al. 2003). Mice deficient in both TNF receptor 1 and TNF receptor 2 have been produced (Erickson et al. 1994; Pfeffer et al. 1993; Rothe et al. 1993). These animals appear healthy and are not reported to have an abnormal bone phenotype. This result implies that TNF is more involved in the bone turnover of inflammatory states and has little effect on bone in homeostasis.

TNF can stimulate the expression of CSF-1R in osteoclast precursor cells (Yao et al. 2006) and, through this mechanism, increase their number (Yao et al. 2006). It also enhances RANK signalling, which activates osteoclasts and their precursor cells (Lam et al. 2000), and it enhances expression of the costimulatory molecule, paired Ig-like receptor A (PIR-A), which enhances nuclear factor of activated T-cells cytoplasmic 1 (NFATc1) activation (Ochi et al. 2007). It was found that RANKL enhances TNF-induced osteoclastogenesis independent of effects on TNF receptor associated factor (TRAF) 6 by degrading TRAF 3 in osteoclast precursor cells (Yao et al. 2017b). Mice that overexpress TNF have an increased number of CD11b high osteoclast precursor cells in their spleen and blood (Li et al. 2004b).

The enhancement of osteoclastogenesis that occurs with TNF signalling involves multiple pathways including phosphoinositide 3-kinase/AKT (PI3K/Akt)-mediated Blimp1 expression (Wu et al. 2017) and recombination signal binding protein for immunoglobulin kappa J region (RBP-J)-mediated regulation of miR-182 (Miller et al. 2016).

TNF has biphasic effects on bone formation and osteoblast function, which are dose and time related (Osta et al. 2014). At lower doses, it stimulated the differentiation of mesenchymal precursor cells into osteoblasts (Osta et al. 2014), while at higher concentrations it inhibited osteoblast function and bone formation (Bertolini et al. 1986; Canalis 1987; Stashenko et al. 1987; Nanes et al. 1989). It was reported to promote fracture repair by enhancing the recruitment of precursor cells to the osteoblastic lineage (Glass et al. 2011). The inhibitory effects of TNF on osteoblasts appear to be direct and mediated by downregulation of the critical transcription factor genes, RUNX2 and osterix (Gilbert et al. 2002, 2004). TNF has many actions on osteoblasts, including inhibiting type 1 collagen (Mori et al. 1996) and osteocalcin synthesis (Gowen et al. 1988; Nanes et al. 1991), which are essential for differentiated osteoblast function. It also stimulated osteoblast apoptosis (Kitajima et al. 1996; Hock et al. 2001), suppressed production of insulin-like growth factor (IGF)-1 (Wang et al. 2018a), and downregulated EphB4 signalling (Scharla et al. 1994). Some of the inhibitory effects of TNF on osteoblast differentiation are mediated by activated transcription factor 3 (ATF3) through a pathway involving c-Jun N-terminal kinase (JNK) (Jeong 2018).

Patients with inflammatory diseases who were treated with anti-TNF therapy were found to have increased bone mineral density (Nigil Haroon et al. 2014; Durnez et al. 2013). However, in patients with inflammatory bowel disease who were treated with anti-TNF therapy, the incidence of fractures was not affected even though bone mass increased (Maldonado-Perez et al. 2018). This result suggests that circulating TNF, which is produced by localized inflammatory pathology, has systemic effects on bone mass but not necessarily on bone strength. Production of TNF by B-lymphocytes may also contribute to the osteoclastogenesis that occurs in periodontitis (Kanzaki et al. 2016). Finally, a polymorphism in the TNF-α gene has been associated with the risk of developing osteoporosis in women (Kotrych et al. 2016).

6 Additional TNF Superfamily Members

6.1 Fas-Ligand (FasL)

FasL, which binds to the receptor Fas on responsive cells, regulates apoptosis and other cellular processes in multiple cell types (Wesche et al. 2005). In osteoblasts, FasL inhibits differentiation through a caspase 8-mediated mechanism (Kovacic et al. 2007). In osteoclasts, addition of FasL to cultures of osteoclast precursor cells, which were also treated with CSF-1 and RANKL, increased osteoclast formation. Osteoclast precursors and mature osteoclasts express Fas and FasL (Park et al. 2005). Expression of Fas was upregulated by RANKL treatment in the RAW 264.7 osteoclast precursor cell line and treatment of mature osteoclasts with Fas-induced apoptosis (Wu et al. 2003). However, in contrast to their similar effects on osteoclastogenesis in cultures of precursor cells, there appears to be divergent roles of RANKL and FasL on mature osteoclast apoptosis. At high concentrations, RANKL inhibited the ability of FasL to induce apoptosis (Wu et al. 2005). The effect that FasL deficiency has on bone mass is controversial. One group has found that bone mass is decreased in FasL-deficient mice (Wu et al. 2003), while another found it to be increased (Katavic et al. 2003). However, the significance of studying bone mass in Fas or FasL-deficient mice is questionable since these models have a generalized lymphoproliferative disorder, which activates a wide variety of immune responses that affect bone and makes it difficult to interpret the results of these studies.

It appears that Fas signalling is involved in the effects of estrogen on the bone (Kovacic et al. 2010). However, there has been controversy about the exact role that FasL has in this response. One group found that stimulation of estrogen receptor α in osteoclasts in mice enhanced FasL production, which, in turn, increased osteoclast apoptosis (Nakamura et al. 2007). In contrast, a second group failed to detect expression of FasL in osteoclasts (Krum et al. 2008). Rather, they found that estrogen enhanced FasL production in osteoblasts. They speculated that estrogen-induced increases in FasL production in osteoblasts regulated osteoclast apoptosis through a paracrine mechanism (Krum et al. 2008). More recently, Fas receptor was shown to be required for estrogen deficiency-induced bone loss (Kovacic et al. 2010). Estrogens, interacting through estrogen receptor alpha (ERα), stimulated metalloproteinase 3 expression on osteoblasts, which induced FasL cleavage from osteoblasts. In turn, this solubilized FasL induced osteoclast apoptosis (Garcia et al. 2012). Fas-Fas ligand interactions may also mediate some effects of interferon (INF) γ on the bone (Kohara et al. 2011) and regulate MMP2 expression in osteoblasts (Svandova et al. 2018).

6.2 TNF-Related Apoptosis-Inducing Ligand (TRAIL)

TRAIL is another TNF-superfamily member that has a wide variety of activities. Its effects on osteoclast function and the bone are also controversial. Some groups have found that treatment of osteoclasts with TRAIL induced apoptosis (Roux et al. 2005) through effects that were mediated by the receptor TRAIL-R2, which is also known as "death receptor 5" (DR5) (Zauli et al. 2010; Colucci et al. 2007). Others have found that TRAIL stimulated osteoclast differentiation through a TRAF-6-dependent mechanism (Yen et al. 2012). In vivo, injection of TRAIL for 8 days in 4-week-old mice induced an increase in bone mass. In vitro, this effect was associated with an increase in the cyclin-dependent kinase inhibitor (CDKI), $p27^{Kip1}$, through effects of TRAIL on the ubiquitin-proteasome pathway (Zauli et al. 2007). TRAIL may also be a factor in the effects that myeloma (Tinhofer et al. 2006) and microgravity have on osteoclasts (Sambandam et al. 2016). However, some groups have failed to find either in vitro or in vivo effects of recombinant TRAIL on osteoclasts or bone mass (Labrinidis et al. 2008).

In cultured human osteoblasts, the ability of TRAIL to induce apoptosis was dependent on the differentiation state. Early cells were more responsive than mature cells (Brunetti et al. 2013). This affect was regulated by differential expression of the active DR5 and decoy DcR2 TRAIL receptors during osteoblast differentiation.

6.3 CD40 Ligand (CD40L)

CD40L is involved in the differentiation of naïve T-lymphocytes into T_H1 effector cells (Loskog and Totterman 2007). In humans, deficiency of CD40L causes X-liked hyper IgM (XHIM) syndrome. Bones of XHIM patients develop spontaneous fractures and are osteopenic (Lopez-Granados et al. 2007). Activated T-lymphocytes from XHIM patients have normal amounts of RANKL but deficient INF-γ production, which may contribute to their decreased bone mass (Lopez-Granados et al. 2007). In addition, expression of CD40L in rheumatoid arthritis synovial cells induced RANKL expression and enhanced the ability of these cells to stimulate osteoclastogenesis. This result suggests that this mechanism is involved in the effects that rheumatoid arthritis has on the bone (Lee et al. 2006a). CD40L was also found to accelerate the osteoclastogenesis that is induced by RANKL and lipopolysaccharide (LPS) (Yokoyama et al. 2011).

The ability of parathyroid hormone (PTH) to stimulate osteoclastogenesis has been reported to involve production of CD40L on T-lymphocytes and the subsequent induction of responses in stromal cells, expressing the receptor CD40 (Gao et al. 2008). CD40L has also been implicated in the bone loss that occurs after ovariectomy in mice (Li et al. 2011) and the anabolic response to intermittent PTH

(Robinson et al. 2015). T-lymphocyte-derived CD40L has been implicated in the ability of osteoclasts to suppress T-lymphocyte activation by a mechanism involving osteoclast-produced indoleamine 2,3-dioxygenase (IDO) (Li et al. 2014). A gene association study found correlation between bone mineral density and single-nucleotide polymorphisms (SNPs) in both CD40 and CD40L (Panach et al. 2016).

7 Interleukin-6 (IL-6)

IL-6, like IL-1 and TNF, has a wide variety of activities on immune cell function and on the replication and differentiation of a number of cell types (Akira et al. 1990; Hirano et al. 1990). Osteoblastic cells (both rodent and human) produce IL-6 and IL-6 receptors (Feyen et al. 1989; Lowik et al. 1989). Another source of IL-6 in the bone microenvironment is bone marrow stromal cells, which can produce IL-6 after they are stimulated with IL-1 and TNF (Girasole et al. 1992). The receptor for IL-6 consists of two parts: a specific IL-6 binding protein (IL-6 receptor), which can be either membrane-bound or soluble, and gp130, an activator protein that is common to a number of IL-6-family cytokine receptors (Kishimoto et al. 1994). Soluble IL-6 receptor binds IL-6, and this complex can then activate cells that contain the gp130 signal peptide (Kishimoto et al. 1994; Tamura et al. 1993). The shedding of IL-6 receptor from osteoblasts is stimulated by IL-1 and TNFα (Franchimont et al. 2005).

The ability of IL-6 to affect bone resorption in vitro is variable and depends on the assay system that is used as both stimulatory and inhibitory effects have been observed (Al-Humidan et al. 1991; Ishimi et al. 1990; Linkhart et al. 1991; Lowik et al. 1989; Yoshitake et al. 2008; Duplomb et al. 2008). It appears that a major effect of IL-6 is to regulate the differentiation of osteoclast progenitor cells into mature osteoclasts (Manolagas and Jilka 1995; Roodman 1992). IL-6 also directly stimulates both RANKL and OPG mRNA production in the bone (Palmqvist et al. 2002) and enhances production of prostaglandins (Liu et al. 2005). In addition, one publication suggested that IL-6 can stimulate osteoclastogenesis in vitro by a RANKL-independent mechanism (Kudo et al. 2003). In contrast, two other publications found IL-6 to directly inhibit RANKL signalling in osteoclast precursor cells and decrease osteoclast formation (Yoshitake et al. 2008; Duplomb et al. 2008).

Examination of IL-6-deficient mice at 8 months of age demonstrated that bone mass was increased, as was tartrate-resistant acid phosphatase (TRAP)-positive osteoclast number and alkaline phosphatase activity in osteoblasts. However, curiously, cathepsin K-positive osteoclasts were decreased in IL-6-deficient mice suggesting that loss of IL-6 inhibited osteoclast maturation and in this way enhanced bone mass. This discrepancy may be related, in part, to differences in the responses of cortical and trabecular bone to membrane-bound and soluble IL-6 receptor (Lazzaro et al. 2018). There was also enhanced apoptosis of osteoclasts in IL-6-deficient mice (Liu et al. 2014). Deletion of IL-6 in mice promoted fracture healing (Huang et al. 2017), and IL-6 appears to influence the bone loss that is seen in mice on a high-fat diet (Wang et al. 2016; Feng et al. 2016).

IL-6 has a variety of effects on mesenchymal cells. It promoted osteogenic differentiation of bone marrow mesenchymal cells (Xie et al. 2018) but inhibited more mature osteoblast differentiation through activation of the Janus kinase/signal transduction and activators of transcription (JAK/STAT), Src homology region 2 domain-containing phosphatase-2/ dual specificity mitogen-activated protein kinase kinase 2 (SHP2/MEK2), and SHP2/AKT pathways (Kaneshiro et al. 2014). It also inhibited osteogenesis in glucocorticoid-treated mice (Li et al. 2016). Osterix, a key transcription factor in early osteoblast differentiation, appears to inhibit IL-6 production since mice heterozygous for osterix deficiency had increased IL-6 production (Baek et al. 2013). In contrast, bone morphogenetic protein, which enhances osteoblast differentiation, was found to enhance IL-6 production in osteoblasts in vitro (Akeel et al. 2012).

IL-6 may mediate some of the bone pathology that is seen with aging (Eriksen et al. 2010) and in a variety of clinical syndromes including Paget's disease (Roodman et al. 1992), hypercalcemia of malignancy (Guise and Mundy 1998), fibrous dysplasia (Yamamoto et al. 1996), giant cell tumors of the bone (Reddy et al. 1994), inflammatory states mediated by TNF or RANKL (Axmann et al. 2009), and Gorham-Stout disease (Devlin et al. 1996). There has been conflicting data about the role of IL-6 in PTH-mediated responses in the bone as some investigators have found it critical (Grey et al. 1999), while others have not (O'Brien et al. 2005). Inhibition of the IL-6 receptor abrogated osteoclastogenesis in vitro and in vivo (Axmann et al. 2009).

In mice, increased IL-6 expression augmented the effects of p62/sequestosome-1 mutations (which is linked to the development of Paget's disease in humans) but did not fully reproduce the Pagetic phenotype, suggesting that additional mechanisms are involved (Teramachi et al. 2014). It was also shown in mice that measles virus nuclear capsid protein, which is also implicated as driving the development of Paget's disease, increased IL-6 expression (Wang et al. 2013).

IL-6 is a mediator of inflammation, and IL-6 inhibition is now used therapeutically to treat inflammatory diseases, like rheumatoid arthritis (Hashizume et al. 2014). Recently, anti-IL-6 therapy was shown to reduce systemic bone loss and osteoclast precursor cell number in a mouse model of rheumatoid arthritis (Tanaka et al. 2014). Furthermore, in humans with rheumatoid arthritis, serum levels of IL-6 inversely correlated with bone mass and directly correlated with disease activity (Abdel Meguid et al. 2013).

8 Additional Interleukin-6 Family Members

IL-6 is a member of a group of cytokines that share the gp130 activator protein in their receptor complex (Manolagas et al. 1995; Romas et al. 1996). Each family member utilizes unique ligand receptors to generate specific binding. Signal transduction through these receptors utilizes the JAK/STAT pathway (Kishimoto et al. 1994).

8.1 Interleukin-11 (IL-11)

IL-11 is produced by bone cells in response to a variety of resorptive stimuli (Elias et al. 1995). It stimulates osteoclast formation in murine bone marrow cultures (Girasole et al. 1994) and bone resorption in a variety of in vitro assays (Hill et al. 1998; Morinaga et al. 1998). Interestingly, it has no effect on isolated mature osteoclasts. In mice deficient in the specific IL-11 receptor, trabecular bone mass is increased. This effect appears to result from decreased bone turnover, which is associated with decreased in vitro osteoclast formation and resorption (Sims et al. 2005). After ovariectomy and subsequent estrogen withdrawal, IL-11 receptor-deficient and wild-type mice lost bone mass at similar rates. This result argues that IL-11 signalling is not involved in the effects of estrogen on bone mass (Sims et al. 2005). However, IL-11 does appear to be involved in the ability of mechanical force to stimulate osteoblast activity in vivo through its effects on Wnt signalling (Kido et al. 2009; Koyama et al. 2008). It has also been shown that IL-11 mediates the stimulatory effects of PTH and mechanical stress on osteoblast differentiation via activator protein 1 (AP-1) and SMAD signalling (Matsumoto et al. 2012).

Mutations in IL-11 receptor or IL-11 have been linked to human disease. The mutation in patients with craniosynostosis was found to generate a defective IL-11 receptor, which, in turn, caused premature closure of cranial sutures (Agthe et al. 2018). In addition, a single-nucleotide polymorphism, which alters the stability of IL11, was linked to short stature (Lokau et al. 2018).

8.2 Leukemia Inhibitory Factor (LIF)

LIF is produced by bone cells in response to a number of resorption stimuli (Greenfield et al. 1996; Marusic et al. 1993; Shiina-Ishimi et al. 1986). The effects of LIF on bone resorption are variable. In a number of in vitro model systems, LIF stimulated resorption by a prostaglandin-dependent mechanism (Reid et al. 1990). However, it also had inhibitory effects in vitro (Lorenzo et al. 1990a; Van Beek et al. 1993). In neonatal murine calvaria cultures, LIF stimulated production of both RANKL and OPG (Palmqvist et al. 2002).

The effects of LIF on osteoblast differentiation in vitro are complex and appear dependent on the dose and the differentiation state of the osteoblastic cultures that are studied (Sims and Johnson 2012). Local injection of LIF in vivo over the calvaria of mice increased both resorption and formation parameters, as well as the thickness of the treated calvaria (Cornish et al. 1993). Expression of LIF was downregulated during osteoblast differentiation (Falconi et al. 2007) by a mechanism that was mediated by micro-RNAs (Oskowitz et al. 2008). In mice that lacked the specific LIF receptor (LIF-R) and, hence, could not respond to LIF, bone volume was reduced and osteoclast number was increased sixfold (Ware et al. 1995). LIF may mediate some of its actions on bone in vivo through inhibition of sclerostin production (Walker et al. 2010). Animals lacking LIF are characterized by giant osteoclasts, which are produced through mechanisms involving Fos-related antigen 2 (Fra-2), hypoxia, hypoxia-induced factor (HIF)1α, and B-cell CLL/Lymphoma 2 (Bcl-2) (Bozec et al. 2008). LIF may also be involved in chondroclast production (Sims and Johnson 2012).

8.3 Oncostatin M (OSM)

Oncostatin M was demonstrated to stimulate multinuclear cell formation in murine and human bone marrow cultures (Heymann et al. 1998; Tamura et al. 1993). However, these cells appeared to be macrophage polykaryons and not osteoclasts (Heymann et al. 1998). In contrast, oncostatin M inhibited osteoclast-like cell formation that was stimulated by 1,25-dihydroxyvitamin D_3 in human marrow cultures (Heymann et al. 1998), and it decreased bone resorption rates in fetal mouse long bone cultures (Jay et al. 1996). In vivo, overexpression of oncostatin M in transgenic mice induced a phenotype of osteopetrosis (Malik et al. 1995). Hence, it appears that oncostatin M is predominantly an inhibitor of osteoclast formation and bone resorption (Mundy 1996). However, oncostatin M can affect cellular responses in bone by binding to either the oncostatin M receptor (OSMR), which produces inhibitory effects on resorption or the LIF-R, which promotes bone formation through inhibition of sclerostin expression (Walker et al. 2010).

Oncostatin M stimulates mesenchymal cells to differentiate toward osteoblasts and osteocytes and inhibits their differentiation toward adipocytes (Song et al. 2007; Brounais et al. 2009), an effect that may be mediated through induction of MCP-1 (Zheng and Guan 2018). Monocytes can produce oncostatin M and drive osteoblast differentiation through a mechanism that is dependent on STAT3 (Guihard et al. 2012; Nicolaidou et al. 2012). Finally, it has also been shown that oncostatin M contributes to the anabolic response of the bone to intermittent administration of PTH (Walker et al. 2012).

The role of all IL-6 family members in osteoclast formation has to be examined in the light of data demonstrating that mice lacking the gp130 activator protein have an increased number of osteoclasts in their bones compared with normal animals (Kawasaki et al. 1997). Since gp130 is an activator of signal transduction for all members of the IL-6 family, this result argues that at least some IL-6 family members have a predominantly inhibitory effect on osteoclast formation and bone resorption.

9 Interleukin-7 (IL-7)

IL-7 is a cytokine that has diverse effects on the hematopoietic and immunologic systems (Namen et al. 1988) and is best known for its non-redundant role in supporting B- and T-lymphopoiesis. Studies have demonstrated that IL-7 also is an important regulator of bone homeostasis (Miyaura et al. 1997; Weitzmann et al. 2002). However, the precise nature of how IL-7 affects osteoclasts and osteoblasts is controversial, since it has a variety of actions in different target cells. Systemic administration of IL-7 upregulated osteoclast formation from human peripheral blood cells by increasing osteoclastogenic cytokine production in T-lymphocytes (Weitzmann et al. 2000). Furthermore, mice with global overexpression of IL-7 had a phenotype of decreased bone mass with increased osteoclast and no change in osteoblast number (Salopek et al. 2008). Significantly, IL-7 did not induce bone resorption and bone loss in T-cell-deficient nude mice in vivo (Toraldo et al. 2003). In addition, treatment of mice with a neutralizing anti-IL-7 antibody inhibited ovariectomy-induced proliferation of early T-cell precursors in the thymus, demonstrating that ovariectomy upregulates T-cell development through IL-7. This latter effect may be a mechanism by which IL-7 regulates ovariectomy-induced bone loss (Ryan et al. 2005). However, the interpretation of results from in vivo IL-7 treatment studies is complicated by secondary effects of IL-7, which result from the production of bone-resorbing cytokines by T-cells in response to activation by IL-7 (Toraldo et al. 2003; Weitzmann et al. 2000; Gendron et al. 2008).

In contrast with previously reported studies (Miyaura et al. 1997; Toraldo et al. 2003; Weitzmann et al. 2000), we found differential effects of IL-7 on osteoclastogenesis (Lee et al. 2003). IL-7 inhibited osteoclast formation in murine bone marrow cells that were cultured for 5 days with CSF-1 and RANKL (Lee et al. 2003). Furthermore, IL-7-deficient mice had markedly increased osteoclast number and decreased trabecular bone mass compared to wild-type controls (Lee et al. 2006b). In addition, we found that trabecular bone loss after ovariectomy was similar in wild-type and IL-7-deficient mice (Lee et al. 2006b). Other investigators have found that IL-7 could stimulate osteoclastogenesis by a mechanism that was independent of RANKL but dependent on STAT5 (Kim et al. 2017), and IL-7 was found to decrease bone mass and increase RANKL through a mechanism dependent on c-Fos/c-Jun (Zhao et al. 2018).

IL-7 mRNA levels in bone increase with ovariectomy and this effect may be linked to alterations in osteoblast function with estrogen withdrawal (Sato et al. 2007; Weitzmann et al. 2002). Treatment of newborn murine calvaria cultures with IL-7 inhibited bone formation, as did injection of IL-7 above the calvaria of mice in vivo (Weitzmann et al. 2002). When IL-7 was overexpressed locally by osteoblasts, trabecular bone mass was increased compared with wild-type mice (Lee et al. 2004). Furthermore, targeted overexpression of IL-7 in IL-7-deficient mice rescued the osteoporotic bone phenotype of the IL-7-deficient mice (Lee et al. 2005b). These studies indicated that the actions of IL-7 on bone cells are dependent on whether IL-7 is delivered systemically or locally. Production of IL-7 by osteoblasts appeared critical for normal B-lymphopoiesis (Wu et al. 2008) and was dependent on mechanistic target of rapamycin complex 1 (mTORC1) signalling in osterix-expressing cells (Wang et al. 2018b; Martin et al. 2018). In sepsis, B-lymphopoiesis decreases through a mechanism in which osteoblastic cells are depleted as is, presumably, IL-7 production by these cells (Terashima et al. 2016). Osteoclast mediated bone resorption can influence B-lymphopoiesis through effects on local IL-7 production in the bone marrow (Mansour et al. 2011).

10 Interleukin-10 (IL-10)

IL-10 is produced by activated T- and B-lymphocytes (Moore et al. 2001). It is a direct inhibitor of osteoclastogenesis (Owens et al. 1996; Xu et al. 1995) and osteoblastogenesis (Van Vlasselaer et al. 1993), which are effects that are associated with increased tyrosine phosphorylation of multiple proteins in osteoclast precursor cells (Hong et al. 2000). The direct effects of IL-10 on RANKL-stimulated osteoclastogenesis include decreases in NFATc1 expression, reduced translocation of this transcription factor into the nucleus (Evans and Fox 2007), and suppressed c-Fos and c-Jun expression (Mohamed et al. 2007). Administration of IL-10 may have utility as a mechanism to control wear-induced osteolysis (Carmody et al. 2002) and the alveolar bone loss of periodontal disease (Zhang et al. 2014). In dental follicle cells, which function to regulate tooth eruption, in vitro treatment with IL-10 inhibited RANKL production and enhanced OPG (Liu et al. 2006). Hence, there appears to also be an indirect effect of IL-10 on osteoclastogenesis that is mediated by its ability to regulate RANKL and OPG production.

Treatment of bone marrow cell cultures with IL-10 suppressed the production of osteoblastic proteins and prevented the onset of mineralization (Van Vlasselaer et al. 1993). IL-10 also inhibited osteoclast formation in bone marrow cultures without affecting macrophage formation or the resorptive activity of mature osteoclasts (Owens and Chambers 1995). This effect appears to involve the production of novel phosphotyrosine proteins in osteoclast precursor cells (Hong et al. 2000). IL-10 also stimulated a novel inducible nitric oxide synthase (Sunyer et al. 1996). IL-10-deficient mice had decreased alveolar bone and decreased indices of osteoblast differentiation (Claudino et al. 2010). A polymorphism in the IL-10 gene was linked to postmenopausal osteoporosis (Kotrych et al. 2016).

4-1BB is an inducible T-cell costimulatory molecule, which interacts with 4-1BB ligand. In vitro treatment of RANKL-stimulated osteoclast precursor cells with 4-1BB ligand enhanced IL-10 production. In addition, expression of IL-10 was greater in RANKL-stimulated wild-type osteoclast precursor cell cultures than in cultured cells from 41-BB-deficient mice (Shin et al. 2006). These results imply that some effects of IL-10 on osteoclasts may be mediated through interactions of 4-1BB with 4-1BB ligand.

11 Interleukin 12 (IL-12), Interleukin 23 (IL-23), Interleukin 27 (IL-27), and Interleukin 35 (IL-35)

IL-12 is a cytokine that is produced by myeloid and other cell types. It induces T_H1 differentiation in T-lymphocytes and the subsequent expression of INF γ (Hsieh et al. 1993). Most authors have found that IL-12 has an inhibitory effect on osteoclastogenesis. However, the mechanisms by which this effect occurs in vitro are controversial. Some authors demonstrated direct inhibitory effects of IL-12 on RANKL-stimulated osteoclastogenesis in purified primary osteoclast precursors and RAW 264.7 cells (Amcheslavsky and Bar-Shavit 2006). This was associated with inhibition of NFATc1 expression in the osteoclast precursor cells. Interestingly, the inhibitory effects of IL-12 on osteoclastogenesis were absent in cells that were pretreated with RANKL (Amcheslavsky and Bar-Shavit 2006). In contrast, others found that the inhibitory effects of IL-12 on osteoclastogenesis were indirect. One group demonstrated that the inhibitory effects of IL-12 was mediated by T-lymphocytes and did not involve production of INF γ (Horwood et al. 2001). A second group disputes this result and found inhibition of osteoclastogenesis by IL-12 in cells from T-lymphocyte depleted cultures and cells from T-lymphocyte-deficient nude mice (Nagata et al. 2003). The latter authors also demonstrated that antibody neutralization of INF γ blocked some of the inhibitory effect of IL-12 on RANKL-stimulated osteoclast formation. In contrast to the majority of reports that IL-12 is an inhibitor of osteoclastogenesis, one group demonstrated that IL-12 induced RANKL expression in human periodontal ligament cells (Issaranggun Na Ayuthaya et al. 2017).

The effects of IL-12 on TNFα-induced osteoclastogenesis have been examined in vivo (Yoshimatsu et al. 2009). It was found that osteoclastogenesis, which was stimulated by injection of TNFα over the calvaria of mice, was decreased when the mice were also treated with IL-12. Furthermore, this effect was not altered by antibody neutralization of T-lymphocytes in the mice. Induction of Fas by TNFα and FasL by IL12 in bone was critical for this response (Kitaura et al. 2002)

IL-23 is an IL-12-related cytokine composed of one subunit of p40, which it shares with IL-12, and one subunit of p19, which is unique (Kastelein et al. 2007). It is critical for the differentiation of the T_H17 subset of T-lymphocytes together with tumor necrosis factor β (TGFβ) and IL-6 (Bettelli et al. 2006). IL-23 appears most important for expanding the population of T_H17 T-lymphocytes. This subset of T-lymphocytes, which produces RANKL, has a high osteoclastogenic potential. The response was mediated by the production of IL-17 (Sato et al. 2006). In an LPS-induced model of inflammatory bone destruction, it was found that there was markedly less bone loss in mice that were deficient either in IL-17 or IL-23 (Sato et al. 2006). Hence, production of both is involved in the bone loss in this model. IL-23 induces RANKL expression in CD4 T-lymphocytes (Ju et al. 2008) and RANK expression in osteoclast precursor cells (Chen et al. 2008). However, the actions of IL-23 on the bone in vivo are controversial. IL-23-deficient mice have decreased bone mass in one report (Quinn et al. 2008) but increased bone mass in another (Adamopoulos et al. 2011). In some studies, IL-23 inhibited osteoclastogenesis through actions that were mediated by CD4 T-lymphocytes (Quinn et al. 2008; Kamiya et al. 2007). In another study IL-23 stimulated osteoclastogenesis in mixed osteoblast-osteoclast precursor cultures (Kang and Zhang 2014). One study identified upregulation of leukotriene B4 as a mechanism by which IL-23 regulated osteoclasts (Bouchareychas et al. 2017).

Curiously, individual neutralization of either IL-17 or IL-23 was found to be more efficacious than their combination as an inhibitor of ovariectomy-induced bone loss (Shukla et al. 2018). IL-23 has also been identified as a critical mediator of the effects of ankylosing spondylitis on the bone (Jo et al. 2017).

Another IL-12-related cytokine, IL-27, was found to have inhibitory effects on osteoclastogenesis in murine bone marrow cultures that were mediated by T-lymphocytes (Kamiya et al. 2007). However, the direct inhibitory effects of IL-27 on RANKL-stimulated osteoclastogenesis were identified as being mediated by inhibition of c-Fos (Furukawa et al. 2009). Recent data has implicated INF γ as a mediator of IL-27's inhibitory effect on osteoclastogenesis (Park et al. 2012), and IL-27 can also inhibit RANKL expression in CD4 T-lymphocytes (Kamiya et al. 2011). In osteoblasts, IL-27 inhibited apoptosis through a mechanism dependent on induction of early growth response -2 gene (Shukla et al. 2017). Bone cells can be a source of IL-27 during inflammation (Larousserie et al. 2017).

Interleukin 35, the most recently discovered IL-12 family member, is a direct inhibitor of osteoclastogenesis (Peng et al. 2018; Yago et al. 2018).

12 Interleukin 15 (IL-15)

IL-15, like IL-7, is a member of the interleukin 2 superfamily and shares many activities with IL-2 including the ability to stimulate lymphocytes. It has been shown to enhance osteoclast progenitor cell number in culture (Ogata et al. 1999). Its receptor is composed of a unique IL-15 receptor α and the β and γ chains of the IL-2 receptor (Budagian et al. 2006). Deletion of IL-15 receptor α in mice produced a phenotype of decreased bone mineralization (Loro et al. 2017). IL-15 production by T-lymphocytes has been linked to the increased osteoclastogenesis and bone destruction seen in rheumatoid arthritis (Miranda-Carus et al. 2006). In animal models of inflammatory bowel disease and staphylococcus aureus sepsis, lack of IL-15 or treatment with an IL-15 inhibitor reduced bone loss and the severity of disease (Brounais-Le Royer et al. 2013; Henningsson et al. 2012). In vitro, IL-15 treatment of mixed murine bone marrow and osteoblast co-cultures demonstrated a role of natural killer (NK) cells in the ability of IL-15 to induce apoptosis in osteoblasts (Takeda et al. 2014). Polymorphisms of the IL-15 gene have been linked to variations in bone mineral density in women (Koh et al. 2009). In rheumatoid arthritis, IL-15 is reported to promote osteoclastogenesis via a pathway that is dependent on phospholipase D1 (Park et al. 2011).

13 Interleukin 17 (IL-17) and Interleukin 25 (IL-25)

IL-17 is a family of related cytokines, which are unique and contain at least six members (A–F) (Weaver et al. 2007). IL-17E is also called interleukin 25 (Fort et al. 2001). These cytokines are central for the development of the adaptive immune response and the products of a subset of CD4 T-lymphocytes with a unique cytokine expression profile, termed T_H17. This contrasts with the more established T-lymphocyte cytokine-expressing subsets T_H1 and T_H2.

IL-17A was initially identified as a stimulator of osteoclastogenesis in mixed cultures of mouse hematopoietic cells and osteoblasts (Kotake et al. 1999). This enhanced resorptive activity was mediated through increased production of prostaglandins and RANKL (Kotake et al. 1999). The direct effects of IL-17A on the differentiation of osteoclast precursor cells is controversial with some investigators finding stimulatory effects (Yago et al. 2009) and others finding it to be inhibitory (Kitami et al. 2010). One report found that low levels of IL-17 regulated osteoclast precursor autophagy (Ke et al. 2018). In another report, IL-17 stimulated rheumatoid synoviocytes to produce RANKL only with 1,25-dihydroxyvitamin D_3 and prostaglandin E_2 (PGE_2) (Ota et al. 2014). Production of IL-17A in rheumatoid arthritis appears to be involved in the production of activated osteoclasts and bone destruction in involved joints (Kotake et al. 1999; Lubberts et al. 2000; Lubberts et al. 2003). Effects of IL-17 on osteoclastogenesis and bone resorption are enhanced by TNFα, which is also produced in the inflamed joints of patients with rheumatoid arthritis (Van bezooijen et al. 1999). Inhibition of IL-17A in an antigen-induced arthritis model reduced the joint and bone destruction that is typically seen and decreased production of RANKL, IL-1 β, and TNF α in the involved lesions (Koenders et al. 2005). Multiple reports have now implicated IL-17 as a critical mediator of the bone loss that occurs in animal models after estrogen withdrawal (Tyagi et al. 2012, 2014; DeSelm et al. 2012). One of these suggested that studies of IL-17 neutralization be initiated to determine its role as a potential therapy to reverse postmenopausal bone loss in humans (Tyagi et al. 2014) and to enhance bone regeneration after fracture (Dixit et al. 2017). Data also implicated T-cell IL-17 production in the ability of PTH to stimulate bone resorption (Neale Weitzmann and Pacifici 2017).

The effects of IL-17 on bone formation are complex. It stimulated mesenchymal cell proliferation (Huang et al. 2009) and the expression of genes associated with early stages of osteoblast differentiation (Wang et al. 2018c), an effect that may be mediated by production of IL-17 by γδ T-cells (Ono et al. 2016). However, it inhibited mature osteoblast differentiation in vitro and the reparative response in a calvarial critical size defect in vivo (Kim et al. 2014b), possibly through effects on Wnt signalling (Wang et al. 2017; Shaw et al. 2016; Uluckan et al. 2016; Mansoori et al. 2017). Paradoxically, one study found that IL-17 augmented the osteogenic response to BMP-2 (Croes et al. 2018).

14 Interleukin 18 (IL-18), Interleukin 33 (IL-33), and Interleukin 37 (IL-37)

IL-18 is similar to IL-1 in its structure and a member of the IL-1 superfamily (Orozco et al. 2007). IL-18 synergizes with IL-12 to induce INF γ production (Okamura et al. 1995), and its levels are increased at sites of inflammation such as rheumatoid arthritis (Yamamura et al. 2001). Osteoblastic cells express IL-18, and its production is induced by treatment with endothelin-1 (Zhong et al. 2014). IL-18 inhibits osteoclast formation through a variety of mechanisms. These include its ability to stimulate GM-CSF (Udagawa et al. 1997), which is produced by T-cells in response to IL-18 treatment (Horwood et al. 1998). It also stimulated INF γ production in vivo in bone (Kawase et al. 2003), and its inhibitory effects on osteoclastogenesis and bone resorption were enhanced by co-treatment with IL-12 (Yamada et al. 2002). IL-18 has been shown to indirectly stimulate osteoclastogenesis through its effects on T-lymphocytes (Dai et al. 2004). Finally, IL-18 is reported to increase production of OPG (Makiishi-Shimobayashi et al. 2001). In IL-18 overexpressing transgenic mice, osteoclasts were decreased, although, curiously, so was bone mass. These results indicate that there may also be effects of IL-18 on bone growth (Kawase et al. 2003). In confirmation of this hypothesis, it was demonstrated that PTH treatment of osteoblasts stimulated IL-18 production. In addition, the anabolic effect of intermittent PTH treatment on trabecular bone mass in IL-18-deficient mice was reduced (Raggatt et al. 2008). IL-18 is also a mitogen for osteoblastic cells in vitro (Cornish et al. 2003).

IL-18 binding protein (IL-18BP) is an antagonist of IL-18 action with anti-inflammatory actions. Treatment of ovariectomized mice with IL1-BP prevented bone loss. It was also found that osteoporotic women had decreased IL-18BP levels and increased amounts of serum IL-18 (Mansoori et al. 2016).

IL-33 is another member of the IL-1 family that has primarily been studied for its effects on T-lymphocytes (Villarreal and Weiner 2014). Its specific receptor is the orphan IL-1 receptor ST2 (also called IL-1R-like 1) (Villarreal and Weiner 2014). IL-33 is expressed by osteoblasts (Schulze et al. 2011; Saleh et al. 2011), and production in these cells is stimulated by PTH and OSM (Saleh et al. 2011). Its effects on bone cells are varied. One report found it to stimulate osteoclastogenesis (Mun et al. 2010), while multiple others found it to be inhibitory (Zaiss et al. 2011; Schulze et al. 2011; Saleh et al. 2011). In addition, there is a report that it had no effects on bone remodelling (Saidi et al. 2011), although these same authors suggested that it may be involved in the development of osteonecrosis of the femoral head, caused by inadequate blood supply (Saidi and Magne 2011). Activation of HIF1α in osteoblasts resulted in increased IL-33 production (Kang et al. 2017). Osteoclastogenesis was decreased in transgenic mice that overexpress IL-33 in osteoblasts (Keller et al. 2011). Mice deficient in IL-33 or ST2 had decreased bone mass (Macari et al. 2018). Interestingly, loss of ST2 in mice prevented ovariectomy-induced bone loss in the maxilla but not in the femur (Macari et al. 2018).

Interleukin 37 (IL-37) is another IL-1 family member that inhibits osteoclast activity (Saeed et al. 2016; Tang et al. 2018). Serum levels of IL-37 were increased in patients with rheumatoid arthritis and ankylosing spondylitis and correlated with disease activity (Yang et al. 2015; Chen et al. 2015).

15 Interferons (INF)

INF γ is a type II interferon with a wide variety of biologic activities. In vitro, INF γ has generally been found to have inhibitory actions on bone resorption (Gowen and Mundy 1986; Peterlik et al. 1985). These appear to be direct and mediated by effects on osteoclast progenitor cells. INF γ inhibits the ability of 1,25-dihydroxyvitamin D_3, PTH, and IL-1 to stimulate osteoclast-like cell formation in cultures of human bone marrow (Takahashi et al. 1986). INF γ also inhibited RANK signalling by accelerating the degradation of TRAF6 through the ubiquitin/proteasome system (Takayanagi et al. 2000), by inhibiting NFATc1 expression, and by activating the NF-κB and JNK pathways (Cheng et al. 2012). Curiously, INF γ is reported to not inhibit resorption in mature osteoclasts (Hattersley et al. 1988). However, INF γ is also reported to have stimulatory effects on resorption through its ability to increase RANKL and TNF-α production in T-lymphocytes (Gao et al. 2007) and through its ability to enhance the fusion of preosteoclasts (Kim et al. 2012). It also appears to mediate the ability of γδ T-lymphocytes and IL-27 to inhibit osteoclastogenesis and resorptive activity (Pappalardo and Thompson 2013; Park et al. 2012).

In osteoblasts INF γ is an inhibitor of proliferation (Cornish et al. 2003; Gowen et al. 1988; Nanes et al. 1989) and has variable effects on differentiation (Gowen et al. 1988; Shen et al. 1988; Smith et al. 1987).

The effects of INF γ on bone in vivo are also variable, as both inhibitory and stimulatory effects have been reported. In mice with collagen-induced arthritis, loss of the INF γ receptor (INF γR) leads to increased bone destruction (Manoury-Schwartz et al. 1997; Vermeire et al. 1997). Similarly, in mice that are injected over their calvaria with bacterial endotoxin, which activates toll-like receptors (TLRs), loss of INF γR resulted in an enhanced resorptive response (Takayanagi et al. 2000). This result is consistent with more recent findings demonstrating that the inhibitory effects of INF γ on osteoclastogenesis are enhanced by activation of TLRs (Ji et al. 2009). The ability of M1 macrophages and Foxp3+ CD8 T-cells to inhibit osteoclastogenesis was dependent on their production of INF γ (Yamaguchi et al. 2016; Shashkova et al. 2016). Finally, in mice that underwent ovariectomy to induce estrogen withdrawal, administration of INF γ enhanced bone mass and prevented the development of the bone loss that occurs in this condition (Duque et al. 2011).

In contrast, intraperitoneal injection of INF γ for 8 days in rats induced osteopenia (Mann et al. 1994). In patients who have osteopetrosis, because they produce defective osteoclasts, administration of INF γ stimulated bone resorption and appeared to partially reverse the disease (Key et al. 1995a). The latter effects are possibly due to the ability of INF γ to stimulate osteoclast superoxide synthesis (Key et al. 1992, 1995b), osteoclast formation in vivo (Vignery et al. 1990), or a generalized immune response (Schoenborn and Wilson 2007).

Type I interferons (INF α and INF β) are typically produced in response to invading pathogens (Takayanagi 2005). Mice deficient in the INF α/β receptor component interferon alpha and beta receptor subunit 1 (IFNAR1) have reduced trabecular bone mass and an increased number of osteoclasts (Takayanagi et al. 2002). RANKL induces INF β in osteoclasts, and INF β, in turn, inhibits RANKL-mediated osteoclastogenesis by decreasing c-Fos expression (Takayanagi et al. 2002) and inducing the production of micro-RNA 155 (miR-155) (Zhang et al. 2012). Osteocytes are a source of INF α (Hayashida et al. 2014). INF β has also been shown to inhibit bone resorption in vitro, although its mechanism of action is not as well studied as that of INF γ and α (Avnet et al. 2007). In vivo, INF α had no effect on bone turnover (Goodman et al. 1999).

16 Additional Cytokines

IL-4 and IL-13 are members of a group of locally acting factors that have been termed "inhibitory cytokines." The effects of IL-4 and IL-13 seem related and appear to affect both osteoblasts and osteoclasts. Transgenic mice that overexpress IL-4 had an osteoporotic phenotype (Lewis et al. 1993). This effect may result from both an inhibition of osteoclast formation and activity (Nakano et al. 1994; Shioi et al. 1991) and an inhibition of bone formation (Okada et al. 1998). IL-13 and IL-4 inhibited IL-1-stimulated bone resorption by decreasing the production of prostaglandins and the activity of cyclooxygenase-2 (Onoe et al. 1996). The direct inhibitory effects of IL-4 on osteoclast precursor cell maturation are more potent than that of IL-13 and involve effects on STAT6, NF-κB, peroxisome proliferator-activated receptor γ1, mitogen-activated protein kinase signalling, Ca^{++} signalling, NFATc1, and c-Fos (Bendixen et al. 2001; Kamel Mohamed et al. 2005; Mangashetti et al. 2005; Moreno et al. 2003; Wei et al. 2001; Yamada et al. 2007). IL-4 along with GM-CSF induces multipotential myeloid cell differentiation toward the dendritic cell lineage (Hiasa et al. 2009).

IL-13 and IL-4 induce cell migration (chemotaxis) in osteoblastic cells (Lind et al. 1995), and they regulate the ability of osteoblasts and vascular endothelial cells to control OPG and RANKL production (Palmqvist et al. 2006; Yamada et al. 2007; Stein et al. 2008; Fujii et al. 2012).

IL-32 is a cytokine that is involved in innate and adaptive immunity. It is produced by T-lymphocytes, natural killer cells, and epithelial cells (Felaco et al. 2009) and has six different splice variants (Nold-Petry et al. 2009). IL-32 stimulated the formation of multinuclear cells that were TRAP- and vitronectin receptor-positive but did not resorb the bone. In addition, it inhibited resorption that was stimulated by RANKL (Mabilleau and Sabokbar 2009). IL-32 and IL-17 can reciprocally stimulate each other in inflamed synovium and influence osteoclastic resorption (Moon et al. 2012). In osteoblasts IL-32γ stimulated bone formation through a mechanism dependent on the micro-RNA, miR-29a (Lee et al. 2017).

Macrophage migration inhibitory factor (MIF) was initially identified as an activity in conditioned medium from activated T-lymphocytes that inhibited macrophage migration in capillary tube assays (Baugh and Bucala 2002). Once purified and cloned (Weiser et al. 1989), it became available for functional studies and was shown to have a variety of activities. In addition to T-lymphocytes, it is produced by pituitary cells and activated macrophages. MIF is a direct inhibitor of osteoclastogenesis in vitro (Jacquin et al. 2009) through its ability to activate Lyn tyrosine kinase (Mun et al. 2014).

In vivo, MIF's effects are complex. Mice that overexpress MIF globally have high turnover osteoporosis (Onodera et al. 2006), while mice deficient in MIF have low turnover osteoporosis with decreased serum indices of bone resorption and bone formation (Jacquin et al. 2009). MIF-deficient mice are also reported to not lose bone mass or increase osteoblast or osteoclast number in bone with ovariectomy (Oshima et al. 2006). Hence, MIF may be another mediator of the effects that estrogen withdrawal has on the bone. Estrogen downregulates MIF expression in activated macrophages (Ashcroft et al. 2003). A similar response may occur in the bone or bone marrow and mediate some of the effects that ovariectomy has on bone mass. MIF also mediates the homing of osteoclast precursor cells to osteolytic sites (Movila et al. 2016).

MIF is made by osteoblasts (Onodera et al. 1996), and its production in these cells was upregulated by a variety of factors including TGF-β, FGF-2, IGF-II, and fetal calf serum (Onodera et al. 1999). In vitro, MIF increased MMP9 and MMP13 expression in osteoblasts (Onodera et al. 2002) and inhibited RANKL-stimulated osteoclastogenesis by decreasing the fusion of precursors, possibly through its ability to inhibit the migration of these cells (Jacquin et al. 2009).

Deletion of CD74, a putative MIF receptor, in mice produced a phenotype of enhanced osteoclastogenesis and decreased bone mass (Mun et al. 2013). In human studies, a polymorphism in the MIF gene was linked to osteoporosis (Ozsoy et al. 2017), and it may mediate some of the pathology of ankylosing spondylosis (Ranganathan et al. 2017).

References

Abbas S, Zhang YH, Clohisy JC, Abu-Amer Y (2003) Tumor necrosis factor-alpha inhibits pre-osteoblast differentiation through its type-1 receptor. Cytokine 22(1–2):33–41

Abboud SL, Woodruff K, Liu C, Shen V, Ghosh-Choudhury N (2002) Rescue of the osteopetrotic defect in op/op mice by osteoblast-specific targeting of soluble colony-stimulating factor-1. Endocrinology 143(5):1942–1949

Abdel Meguid MH, Hamad YH, Swilam RS, Barakat MS (2013) Relation of interleukin-6 in rheumatoid arthritis patients to systemic bone loss and structural bone damage. Rheumatol Int 33(3):697–703

Adamopoulos IE, Tessmer M, Chao CC, Adda S, Gorman D, Petro M, Chou CC, Pierce RH, Yao W, Lane NE, Laface D, Bowman EP (2011) IL-23 is critical for induction of arthritis, osteoclast formation, and maintenance of bone mass. J Immunol 187(2):951–959

Agthe M, Brugge J, Garbers Y, Wandel M, Kespohl B, Arnold P, Flynn CM, Lokau J, Aparicio-Siegmund S, Bretscher C, Rose-John S, Waetzig GH, Putoczki T, Grotzinger J, Garbers C (2018) Mutations in craniosynostosis patients cause defective interleukin-11 receptor maturation and drive craniosynostosis-like disease in mice. Cell Rep 25(1):10–18

Akeel S, El-Awady A, Hussein K, El-Refaey M, Elsalanty M, Sharawy M, Al-Shabrawey M (2012) Recombinant bone morphogenetic protein-2 induces up-regulation of vascular endothelial growth factor and interleukin 6 in human pre-osteoblasts: role of reactive oxygen species. Arch Oral Biol 57(5):445–452

Akira S, Hirano T, Taga T, Kishimoto T (1990) Biology of multifunctional cytokines: IL-6 and related molecules (IL-1 and TNF). FASEB J 4:2860–2867

Al-Humidan A, Ralston SH, Hughes DE, Chapman K, Aarden L, Russell RGG, Gowen M (1991) Interleukin-6 does not stimulate bone resorption in neonatal mouse calvariae. J Bone Miner Res 6:3–7

Alnaeeli M, Teng YT (2009) Dendritic cells: a new player in osteoimmunology. Curr Mol Med 9(7):893–910

Alnaeeli M, Penninger JM, Teng YT (2006) Immune interactions with CD4+ T cells promote the development of functional osteoclasts from murine CD11c+ dendritic cells. J Immunol 177(5):3314–3326

Amcheslavsky A, Bar-Shavit Z (2006) Interleukin (IL)-12 mediates the anti-osteoclastogenic activity of CpG-oligodeoxynucleotides. J Cell Physiol 207(1):244–250

Anderson DM, Maraskovsky E, Billingsley WL, Dougall WC, Tometsko ME, Roux ER, Teepe MC, DuBose RF, Cosman D, Galibert L (1997) A homologue of the TNF receptor and its ligand enhance T-cell growth and dendritic-cell function. Nature 390(6656):175–179

Arai F, Miyamoto T, Ohneda O, Inada T, Sudo T, Brasel K, Miyata T, Anderson DM, Suda T (1999) Commitment and differentiation of osteoclast precursor cells by the sequential expression of c-Fms and receptor activator of nuclear factor kappaB (RANK) receptors. J Exp Med 190(12):1741–1754

Arend WP, Welgus HG, Thompson C, Eisenberg SP (1990) Biological properties of recombinant human monocyte-derived interleukin 1 receptor antagonist. J Clin Invest 85:1694–1697

Ashcroft GS, Mills SJ, Lei K, Gibbons L, Jeong MJ, Taniguchi M, Burow M, Horan MA, Wahl SM, Nakayama T (2003) Estrogen modulates cutaneous wound healing by downregulating macrophage migration inhibitory factor. J Clin Invest 111(9):1309–1318

Ashmun RA, Look AT, Roberts WM, Roussel MF, Seremetis S, Ohtsuka M, Sherr CJ (1989) Monoclonal antibodies to the human CSF-1 receptor (c-fms proto-oncogene product) detect epitopes on normal mononuclear phagocytes and on human myeloid leukemic blast cells. Blood 73(3):827–837

Avnet S, Cenni E, Perut F, Granchi D, Brandi ML, Giunti A, Baldini N (2007) Interferon-alpha inhibits in vitro osteoclast differentiation and renal cell carcinoma-induced angiogenesis. Int J Oncol 30(2):469–476

Axmann R, Bohm C, Kronke G, Zwerina J, Smolen J, Schett G (2009) Inhibition of interleukin-6 receptor directly blocks osteoclast formation in vitro and in vivo. Arthritis Rheum 60 (9):2747–2756

Azuma Y, Kaji K, Katogi R, Takeshita S, Kudo A (2000) Tumor necrosis factor-alpha induces differentiation of and bone resorption by osteoclasts. J Biol Chem 275(7):4858–4864

Baek WY, Park SY, Kim YH, Lee MA, Kwon TH, Park KM, de Crombrugghe B, Kim JE (2013) Persistent low level of osterix accelerates interleukin-6 production and impairs regeneration after tissue injury. PLoS One 8(7):e69859

Bajayo A, Goshen I, Feldman S, Csernus V, Iverfeldt K, Shohami E, Yirmiya R, Bab I (2005) Central IL-1 receptor signaling regulates bone growth and mass. Proc Natl Acad Sci U S A 102 (36):12956–12961

Barhanpurkar AP, Gupta N, Srivastava RK, Tomar GB, Naik SP, Joshi SR, Pote ST, Mishra GC, Wani MR (2012) IL-3 promotes osteoblast differentiation and bone formation in human mesenchymal stem cells. Biochem Biophys Res Commun 18(4):669–675

Baugh JA, Bucala R (2002) Macrophage migration inhibitory factor. Crit Care Med 30(Suppl 1): S27–S35

Bendixen AC, Shevde NK, Dienger KM, Willson TM, Funk CD, Pike JW (2001) IL-4 inhibits osteoclast formation through a direct action on osteoclast precursors via peroxisome proliferator-activated receptor gamma 1. Proc Natl Acad Sci U S A 98(5):2443–2448

Bertolini DR, Nedwin GE, Bringman TS, Smith DD, Mundy GR (1986) Stimulation of bone resorption and inhibition of bone formation in vitro by human tumour necrosis factors. Nature 319:516–518

Bettelli E, Carrier Y, Gao W, Korn T, Strom TB, Oukka M, Weiner HL, Kuchroo VK (2006) Reciprocal developmental pathways for the generation of pathogenic effector TH17 and regulatory T cells. Nature 441(7090):235–238

Beutler B, Cerami A (1986) Cachectin and tumour necrosis factor as two sides of the same biological coin. Nature 320:584–588

Beutler B, Cerami A (1987) Cachectin: more than a tumor necrosis factor. N Engl J Med 316:379–385

Birch MA, Ginty AF, Walsh CA, Fraser WD, Gallagher JA, Bilbe G (1993) PCR detection of cytokines in normal human and pagetic osteoblast-like cells. J Bone Miner Res 8 (10):1155–1162

Bouchareychas L, Grossinger EM, Kang M, Qiu H, Adamopoulos IE (2017) Critical role of LTB4/BLT1 in IL-23-induced synovial inflammation and osteoclastogenesis via NF-kappaB. J Immunol 198(1):452–460
Boulakirba S, Pfeifer A, Mhaidly R, Obba S, Goulard M, Schmitt T, Chaintreuil P, Calleja A, Furstoss N, Orange F, Lacas-Gervais S, Boyer L, Marchetti S, Verhoeyen E, Luciano F, Robert G, Auberger P, Jacquel A (2018) IL-34 and CSF-1 display an equivalent macrophage differentiation ability but a different polarization potential. Sci Rep 8(1):256
Bouyer P, Sakai H, Itokawa T, Kawano T, Fulton CM, Boron WF, Insogna KL (2007) Colony-stimulating factor-1 increases osteoclast intracellular pH and promotes survival via the electroneutral Na/HCO3 cotransporter NBCn1. Endocrinology 148(2):831–840
Boyce BF, Aufdemorte TB, Garrett R, Yates AJP, Mundy GR (1989) Effects of interleukin-1 on bone turnover in normal mice. Endocrinology 125:1142
Boyle WJ, Simonet WS, Lacey DL (2003) Osteoclast differentiation and activation. Nature 423 (6937):337–342
Bozec A, Bakiri L, Hoebertz A, Eferl R, Schilling AF, Komnenovic V, Scheuch H, Priemel M, Stewart CL, Amling M, Wagner EF (2008) Osteoclast size is controlled by Fra-2 through LIF/LIF-receptor signalling and hypoxia. Nature 454(7201):221–225
Brounais B, David E, Chipoy C, Trichet V, Ferre V, Charrier C, Duplomb L, Berreur M, Redini F, Heymann D, Blanchard F (2009) Long term oncostatin M treatment induces an osteocyte-like differentiation on osteosarcoma and calvaria cells. Bone 44(5):830–839
Brounais-Le Royer B, Pierroz DD, Velin D, Frossard C, Zheng XX, Lehr HA, Ferrari-Lacraz S, Ferrari SL (2013) Effects of an interleukin-15 antagonist on systemic and skeletal alterations in mice with DSS-induced colitis. Am J Pathol 182(6):2155–2167
Brunetti G, Oranger A, Carbone C, Mori G, Sardone FR, Mori C, Celi M, Faienza MF, Tarantino U, Zallone A, Grano M, Colucci S (2013) Osteoblasts display different responsiveness to TRAIL-induced apoptosis during their differentiation process. Cell Biochem Biophys 67(3):1127–1136
Bucay N, Sarosi I, Dunstan CR, Morony S, Tarpley J, Capparelli C, Scully S, Tan HL, Xu W, Lacey DL, Boyle WJ, Simonet WS (1998) Osteoprotegerin-deficient mice develop early onset osteoporosis and arterial calcification. Genes Dev 12(9):1260–1268
Budagian V, Bulanova E, Paus R, Bulfone-Paus S (2006) IL-15/IL-15 receptor biology: a guided tour through an expanding universe. Cytokine Growth Factor Rev 17(4):259–280
Canalis E (1986) Interleukin-1 has independent effects on deoxyribonucleic acid and collagen synthesis in cultures of rat calvariae. Endocrinology 118:74–81
Canalis E (1987) Effects of tumor necrosis factor on bone formation in vitro. Endocrinology 121:1596–1604
Cao Y, Jansen ID, Sprangers S, Stap J, Leenen PJ, Everts V, de Vries TJ (2016) IL-1beta differently stimulates proliferation and multinucleation of distinct mouse bone marrow osteoclast precursor subsets. J Leukoc Biol 100(3):513–523
Carmody EE, Schwarz EM, Puzas JE, Rosier RN, O'Keefe RJ (2002) Viral interleukin-10 gene inhibition of inflammation, osteoclastogenesis, and bone resorption in response to titanium particles. Arthritis Rheum 46(5):1298–1308
Cecchini MG, Hofstetter W, Halasy J, Wetterwald A, Felix R (1997) Role of CSF-1 in bone and bone marrow development. Mol Reprod Dev 46(1):75–84
Cerretti DP, Wignall J, Anderson D, Tushinski RJ, Gallis BM, Stya M, Gillis S, Urdal DL, Cosman D (1988) Human macrophage-colony stimulating factor: alternative RNA and protein processing from a single gene. Mol Immunol 25(8):761–770
Cerretti DP, Wignall J, Anderson D, Tushinski RJ, Gallis B, Cosman D (1990) Membrane bound forms of human macrophage colony stimulating factor (M-CSF, CSF-1). Prog Clin Biol Res 352:63–70
Charatcharoenwitthaya N, Khosla S, Atkinson EJ, McCready LK, Riggs BL (2007) Effect of blockade of TNF-alpha and interleukin-1 action on bone resorption in early postmenopausal women. J Bone Miner Res 22(5):724–729

Chen L, Wei XQ, Evans B, Jiang W, Aeschlimann D (2008) IL-23 promotes osteoclast formation by up-regulation of receptor activator of NF-kappaB (RANK) expression in myeloid precursor cells. Eur J Immunol 38(10):2845–2854

Chen Z, Buki K, Vaaraniemi J, Gu G, Vaananen HK (2011) The critical role of IL-34 in osteoclastogenesis. PLoS One 6(4):e18689

Chen B, Huang K, Ye L, Li Y, Zhang J, Zhang J, Fan X, Liu X, Li L, Sun J, Du J, Huang Z (2015) Interleukin-37 is increased in ankylosing spondylitis patients and associated with disease activity. J Transl Med 13:36

Cheng J, Liu J, Shi Z, Jules J, Xu D, Luo S, Wei S, Feng X (2012) Molecular mechanisms of the biphasic effects of interferon-gamma on osteoclastogenesis. J Interferon Cytokine Res 32 (1):34–45

Christopher MJ, Link DC (2008) Granulocyte colony-stimulating factor induces osteoblast apoptosis and inhibits osteoblast differentiation. J Bone Miner Res 23(11):1765–1774

Claudino M, Garlet TP, Cardoso CR, de Assis GF, Taga R, Cunha FQ, Silva JS, Garlet GP (2010) Down-regulation of expression of osteoblast and osteocyte markers in periodontal tissues associated with the spontaneous alveolar bone loss of interleukin-10 knockout mice. Eur J Oral Sci 118(1):19–28

Colotta F, Re F, Muzio M, Bertini R, Polentarutti N, Sironi M, Giri JG, Dower SK, Sims JE, Mantovani A (1993) Interleukin-1 type II receptor: a decoy target for IL-1 that is regulated by IL-4. Science 261:472–475

Colucci S, Brunetti G, Cantatore FP, Oranger A, Mori G, Pignataro P, Tamma R, Grassi FR, Zallone A, Grano M (2007) The death receptor DR5 is involved in TRAIL-mediated human osteoclast apoptosis. Apoptosis 12(9):1623–1632

Cornish J, Callon K, King A, Edgar S, Reid IR (1993) The effect of leukemia inhibitory factor on bone in vivo. Endocrinology 132:1359–1366

Cornish J, Gillespie MT, Callon KE, Horwood NJ, Moseley JM, Reid IR (2003) Interleukin-18 is a novel mitogen of osteogenic and chondrogenic cells. Endocrinology 144(4):1194–1201

Coulson J, Bagley L, Barnouin Y, Bradburn S, Butler-Browne G, Gapeyeva H, Hogrel JY, Maden-Wilkinson T, Maier AB, Meskers C, Murgatroyd C, Narici M, Paasuke M, Sassano L, Sipila S, Al-Shanti N, Stenroth L, Jones DA, McPhee JS (2017) Circulating levels of dickkopf-1, osteoprotegerin and sclerostin are higher in old compared with young men and women and positively associated with whole-body bone mineral density in older adults. Osteoporos Int 28 (9):2683–2689

Croes M, Kruyt MC, Groen WM, van Dorenmalen KMA, Dhert WJA, Oner FC, Alblas J (2018) Interleukin 17 enhances bone morphogenetic protein-2-induced ectopic bone formation. Sci Rep 8(1):7269

Dai XM, Ryan GR, Hapel AJ, Dominguez MG, Russell RG, Kapp S, Sylvestre V, Stanley ER (2002) Targeted disruption of the mouse colony-stimulating factor 1 receptor gene results in osteopetrosis, mononuclear phagocyte deficiency, increased primitive progenitor cell frequencies, and reproductive defects. Blood 99(1):111–120

Dai SM, Nishioka K, Yudoh K (2004) Interleukin (IL) 18 stimulates osteoclast formation through synovial T cells in rheumatoid arthritis: comparison with IL1 beta and tumour necrosis factor alpha. Ann Rheum Dis 63(11):1379–1386

De Vries TJ, Schoenmaker T, Aerts D, Grevers LC, Souza PP, Nazmi K, van de Wiel M, Ylstra B, Lent PL, Leenen PJ, Everts V (2015) M-CSF priming of osteoclast precursors can cause osteoclastogenesis-insensitivity, which can be prevented and overcome on bone. J Cell Physiol 230(1):210–225

DeSelm CJ, Takahata Y, Warren J, Chappel JC, Khan T, Li X, Liu C, Choi Y, Kim YF, Zou W, Teitelbaum SL (2012) IL-17 mediates estrogen-deficient osteoporosis in an Act1-dependent manner. J Cell Biochem 113(9):2895–2902

Devlin RD, Bone HG III, Roodman GD (1996) Interleukin-6: a potential mediator of the massive osteolysis in patients with Gorham-Stout disease. J Clin Endocrinol Metab 81:1893–1897

Dinarello CA (1991) Interleukin-1 and interleukin-1 antagonism. Blood 77:1627–1652

Dinarello CA (1993) Blocking interleukin-1 in disease. Blood Purif 11:118–127
Dixit M, Singh KB, Prakash R, Singh D (2017) Functional block of IL-17 cytokine promotes bone healing by augmenting FOXO1 and ATF4 activity in cortical bone defect model. Osteoporos Int 28(7):2207–2220
Dougall WC, Glaccum M, Charrier K, Rohrbach K, Brasel K, De Smedt T, Daro E, Smith J, Tometsko ME, Maliszewski CR, Armstrong A, Shen V, Bain S, Cosman D, Anderson D, Morrissey PJ, Peschon JJ, Schuh J (1999) RANK is essential for osteoclast and lymph node development. Genes Dev 13(18):2412–2424
Duplomb L, Baud'huin M, Charrier C, Berreur M, Trichet V, Blanchard F, Heymann D (2008) IL-6 inhibits RANKL-induced osteoclastogenesis by diverting cells into the macrophage lineage: key role of Serine727 phosphorylation of STAT3. Endocrinology 149(7):3688–3697
Duque G, Huang DC, Dion N, Macoritto M, Rivas D, Li W, Yang XF, Li J, Liang J, Marino FT, Barralet J, Lascau V, Deschenes C, Ste-Marie LG, Kremer R (2011) Interferon gamma plays a role in bone formation in vivo and rescues osteoporosis in ovariectomized mice. J Bone Miner Res 26(7):1472–1483
Durnez A, Paternotte S, Fechtenbaum J, Landewe RB, Dougados M, Roux C, Briot K (2013) Increase in bone density in patients with spondyloarthritis during anti-tumor necrosis factor therapy: 6-year followup study. J Rheumatol 40(10):1712–1718
Eder J (1997) Tumour necrosis factor alpha and interleukin 1 signalling: do MAPKK kinases connect it all? Trends Pharmacol Sci 18(9):319–322
Ehrlich LA, Chung HY, Ghobrial I, Choi SJ, Morandi F, Colla S, Rizzoli V, Roodman GD, Giuliani N (2005) IL-3 is a potential inhibitor of osteoblast differentiation in multiple myeloma. Blood 106(4):1407–1414
Eisenberg SP, Evans RJ, Arend WP, Verderber E, Brewer MT, Hannum CH, Thompson RC (1990) Primary structure and functional expression from complementary DNA of a human inteìeukin-1 receptor antagonist. Nature 343:341–346
Elias JA, Tang W, Horowitz MC (1995) Cytokine and hormonal stimulation of human osteosarcoma interleukin-11 production. Endocrinology 136:489–498
Emery JG, McDonnell P, Burke MB, Deen KC, Lyn S, Silverman C, Dul E, Appelbaum ER, Eichman C, DiPrinzio R, Dodds RA, James IE, Rosenberg M, Lee JC, Young PR (1998) Osteoprotegerin is a receptor for the cytotoxic ligand TRAIL [In Process Citation]. J Biol Chem 273(23):14363–14367
Enelow RI, Sullivan GW, Carper HT, Mandell GL (1992) Induction of multinucleated giant cell formation from in vitro culture of human monocytes with interleukin-3 and interferon-gamma: comparison with other stimulating factors. Am J Respir Cell Mol Biol 6(1):57–62
Erickson SL, de Sauvage FJ, Kikly K, Carver-Moore K, Pitts-Meek S, Gillett N, Sheehan KC, Schreiber RD, Goeddel DV, Moore MW (1994) Decreased sensitivity to tumour-necrosis factor but normal T-cell development in TNF receptor-2-deficient mice. Nature 372(6506):560–563
Eriksen CG, Olsen H, Husted LB, Sorensen L, Carstens M, Soballe K, Langdahl BL (2010) The expression of IL-6 by osteoblasts is increased in healthy elderly individuals: stimulated proliferation and differentiation are unaffected by age. Calcif Tissue Int 87(5):414–423
Evans KE, Fox SW (2007) Interleukin-10 inhibits osteoclastogenesis by reducing NFATc1 expression and preventing its translocation to the nucleus. BMC Cell Biol 19:4
Falconi D, Oizumi K, Aubin JE (2007) Leukemia inhibitory factor influences the fate choice of mesenchymal progenitor cells. Stem Cells 25(2):305–312
Felaco P, Castellani ML, De Lutiis MA, Felaco M, Pandolfi F, Salini V, De Amicis D, Vecchiet J, Tete S, Ciampoli C, Conti F, Cerulli G, Caraffa A, Antinolfi P, Cuccurullo C, Perrella A, Theoharides TC, Conti P, Toniato E, Kempuraj D, Shaik YB (2009) IL-32: a newly-discovered proinflammatory cytokine. J Biol Regul Homeost Agents 23(3):141–147
Felix R, Fleisch H, Elford PR (1989) Bone-resorbing cytokines enhance release of macrophage colony-stimulating activity by the osteoblastic cell MC3T3- E1. Calcif Tissue Int 44:356–360
Felix R, Cecchini MG, Fleisch H (1990a) Macrophage colony stimulating factor restores in vivo bone resorption in the op/op osteopetrotic mouse. Endocrinology 127:2592–2594

Felix R, Cecchini MG, Hofstetter W, Elford PR, Stutzer A, Fleisch H (1990b) Impairment of macrohhage colony-stimulating factor production and lack of resident bone marrow macrophages in the osteopetrotic op/op mouse. J Bone Miner Res 5:781–789
Feng W, Liu B, Liu D, Hasegawa T, Wang W, Han X, Cui J, Yimin OK, Amizuka N, Li M (2016) Long-term administration of high-fat diet corrects abnormal bone remodeling in the tibiae of interleukin-6-deficient mice. J Histochem Cytochem 64(1):42–53
Feyen JH, Elford P, Di Padova FE, Trechsel U (1989) Interleukin-6 is produced by bone and modulated by parathyroid hormone. J Bone Miner Res 4:633–638
Fiers W, Sim E (1993) Tumor necrosis factor. In: The natural immune system: humoral factors. IRL Press at Oxford University Press, Oxford, p 65
Fort MM, Cheung J, Yen D, Li J, Zurawski SM, Lo S, Menon S, Clifford T, Hunte B, Lesley R, Muchamuel T, Hurst SD, Zurawski G, Leach MW, Gorman DM, Rennick DM (2001) IL-25 induces IL-4, IL-5, and IL-13 and Th2-associated pathologies in vivo. Immunity 15(6):985–995
Franchimont N, Lambert C, Huynen P, Ribbens C, Relic B, Chariot A, Bours V, Piette J, Merville MP, Malaise M (2005) Interleukin-6 receptor shedding is enhanced by interleukin-1beta and tumor necrosis factor alpha and is partially mediated by tumor necrosis factor alpha-converting enzyme in osteoblast-like cells. Arthritis Rheum 52(1):84–93
Fujii T, Kitaura H, Kimura K, Hakami ZW, Takano-Yamamoto T (2012) IL-4 inhibits TNF-alpha-mediated osteoclast formation by inhibition of RANKL expression in TNF-alpha-activated stromal cells and direct inhibition of TNF-alpha-activated osteoclast precursors via a T-cell-independent mechanism in vivo. Bone 51(4):771–780
Fujikawa Y, Sabokbar A, Neale SD, Itonaga I, Torisu T, Athanasou NA (2001) The effect of macrophage-colony stimulating factor and other humoral factors (interleukin-1, -3, -6, and -11, tumor necrosis factor-alpha, and granulocyte macrophage-colony stimulating factor) on human osteoclast formation from circulating cells. Bone 28(3):261–267
Fuller K, Owens JM, Jagger CJ, Wilson A, Moss R, Chambers TJ (1993) Macrophage colony-stimulating factor stimulates survival and chemotactic behavior in isolated osteoclasts. J Exp Med 178:1733–1744
Fuller K, Wong B, Fox S, Choi Y, Chambers TJ (1998) TRANCE is necessary and sufficient for osteoblast-mediated activation of bone resorption in osteoclasts. J Exp Med 188(5):997–1001
Furukawa M, Takaishi H, Takito J, Yoda M, Sakai S, Hikata T, Hakozaki A, Uchikawa S, Matsumoto M, Chiba K, Kimura T, Okada Y, Matsuo K, Yoshida H, Toyama Y (2009) IL-27 abrogates receptor activator of NF-{kappa}B ligand-mediated osteoclastogenesis of human granulocyte-macrophage colony-forming unit cells through STAT1-dependent inhibition of c-Fos. J Immunol 183(4):2397–2406
Gao Y, Grassi F, Ryan MR, Terauchi M, Page K, Yang X, Weitzmann MN, Pacifici R (2007) IFN-gamma stimulates osteoclast formation and bone loss in vivo via antigen-driven T cell activation. J Clin Invest 117(1):122–132
Gao Y, Wu X, Terauchi M, Li JY, Grassi F, Galley S, Yang X, Weitzmann MN, Pacifici R (2008) T cells potentiate PTH-induced cortical bone loss through CD40L signaling. Cell Metab 8 (2):132–145
Garcia AJ, Tom C, Guemes M, Polanco G, Mayorga ME, Wend K, Miranda-Carboni GA, Krum SA (2012) ERalpha signaling regulates MMP3 expression to induce FasL cleavage and osteoclast apoptosis. J Bone Miner Res 28(2):283–290
Gendron S, Boisvert M, Chetoui N, Aoudjit F (2008) Alpha1beta1 integrin and interleukin-7 receptor up-regulate the expression of RANKL in human T cells and enhance their osteoclastogenic function. Immunology 125(3):359–369
Gilbert L, He X, Farmer P, Rubin J, Drissi H, Van Wijnen AJ, Lian JB, Stein GB, Nanes MS (2002) Expression of the osteoblast differentiation factor RUNX2 (Cbfa1/AML3/Pebp2alphaA) is inhibited by tumor necrosis factor-alpha. J Biol Chem 277(4):2695–2701
Gilbert LC, Rubin J, Nanes MS (2004) The p55 TNF receptor mediates TNF inhibition of osteoblast differentiation independent of apoptosis. Am J Physiol Endocrinol Metab 288(5):1011–1018

Girasole G, Jilka RL, Passeri G, Boswell S, Boder G, Williams DC, Manolagas SC (1992) 17á-Estradiol inhibits interleukin-6 production by bone marrow- derived stromal cells and osteoblasts in vitro: a potential mechanism for the antiosteoporotic effect of estrogens. J Clin Invest 89:883–891

Girasole G, Passeri G, Jilka RL, Manolagas SC (1994) Interleukin-11: a new cytokine critical for osteoclast development. J Clin Invest 93:1516–1524

Glass GE, Chan JK, Freidin A, Feldmann M, Horwood NJ, Nanchahal J (2011) TNF-alpha promotes fracture repair by augmenting the recruitment and differentiation of muscle-derived stromal cells. Proc Natl Acad Sci U S A 108(4):1585–1590

Goodman GR, Dissanayake IR, Gorodetsky E, Zhou H, Ma YF, Jee WS, Epstein S (1999) Interferon-alpha, unlike interferon-gamma, does not cause bone loss in the rat. Bone 25 (4):459–463

Gowen M, Mundy GR (1986) Actions of recombinant interleukin 1, interleukin 2, and interferon-gamma on bone resorption in vitro. J Immunol 136:2478–2482

Gowen M, MacDonald BR, Russell GG (1988) Actions of recombinant human gamma-interferon and tumor necrosis factor alpha on the proliferation and osteoblastic characteristics of human trabecular bone cells in vitro. Arthritis Rheum 31:1500–1507

Greenfield EM, Horowitz MC, Lavish SA (1996) Stimulation by parathyroid hormone of interleukin-6 and leukemia inhibitory factor expression in osteoblasts is an immediate-early gene response induced by cAMP signal transduction. J Biol Chem 271:10984–10989

Grey A, Mitnick MA, Masiukiewicz U, Sun BH, Rudikoff S, Jilka RL, Manolagas SC, Insogna K (1999) A role for interleukin-6 in parathyroid hormone-induced bone resorption in vivo. Endocrinology 140(10):4683–4690

Guihard P, Danger Y, Brounais B, David E, Brion R, Delecrin J, Richards CD, Chevalier S, Redini F, Heymann D, Gascan H, Blanchard F (2012) Induction of osteogenesis in mesenchymal stem cells by activated monocytes/macrophages depends on oncostatin M signaling. Stem Cells 30(4):762–772

Guise TA, Mundy GR (1998) Cancer and bone. Endcr Rev 19(1):18–54

Hannum CH, Wilcox CJ, Arend WP, Joslin FG, Dripps DJ, Heimdal PL, Armes LG, Sommer A, Eisenberg SP, Thompson RC (1990) Interleukin-1 receptor antagonist activity of a human interleukin-1 inhibitor. Nature 343:336–340

Hashizume M, Tan SL, Takano J, Ohsawa K, Hasada I, Hanasaki A, Ito I, Mihara M, Nishida K (2014) Tocilizumab, a humanized anti-IL-6R antibody, as an emerging therapeutic option for rheumatoid arthritis: molecular and cellular mechanistic insights. Int Rev Immunol 34 (3):265–279

Hattersley G, Dorey E, Horton MA, Chambers TJ (1988) Human macrophage colony-stimulating factor inhibits bone resorption by osteoclasts disaggregated from rat bone. J Cell Physiol 137:199–203

Hauser B, Zhao S, Visconti MR, Riches PL, Fraser WD, Piec I, Goodson NJ, Ralston SH (2017) Autoantibodies to osteoprotegerin are associated with low hip bone mineral density and history of fractures in axial spondyloarthritis: a cross-sectional observational study. Calcif Tissue Int 101(4):375–383

Hayashi S, Miyamoto A, Yamane T, Kataoka H, Ogawa M, Sugawara S, Nishikawa S, Nishikawa S, Sudo T, Yamazaki H, Kunisada T (1997) Osteoclast precursors in bone marrow and peritoneal cavity. J Cell Physiol 170(3):241–247

Hayashida C, Ito J, Nakayachi M, Okayasu M, Ohyama Y, Hakeda Y, Sato T (2014) Osteocytes produce interferon-beta as a negative regulator of osteoclastogenesis. J Biol Chem 289 (16):11545–11555

Hemingway F, Taylor R, Knowles HJ, Athanasou NA (2011) RANKL-independent human osteoclast formation with APRIL, BAFF, NGF, IGF I and IGF II. Bone 48(4):938–944

Hengartner NE, Fiedler J, Ignatius A, Brenner RE (2013) IL-1beta inhibits human osteoblast migration. Mol Med 19:36–42

Henningsson L, Jirholt P, Bogestal YR, Eneljung T, Adiels M, Lindholm C, McInnes I, Bulfone-Paus S, Lerner UH, Gjertsson I (2012) Interleukin 15 mediates joint destruction in staphylococcus aureus arthritis. J Infect Dis 206(5):687–696

Heymann D, Guicheux J, Gouin F, Cottrel M, Daculsi G (1998) Oncostatin M stimulates macrophage-polykaryon formation in long-term human bone-marrow cultures. Cytokine 10 (2):98–109

Hiasa M, Abe M, Nakano A, Oda A, Amou H, Kido S, Takeuchi K, Kagawa K, Yata K, Hashimoto T, Ozaki S, Asaoka K, Tanaka E, Moriyama K, Matsumoto T (2009) GM-CSF and IL-4 induce dendritic cell differentiation and disrupt osteoclastogenesis through M-CSF receptor shedding by up-regulation of TNF-{alpha} converting enzyme (TACE). Blood 114 (20):4517–4526

Hikita A, Yana I, Wakeyama H, Nakamura M, Kadono Y, Oshima Y, Nakamura K, Seiki M, Tanaka S (2006) Negative regulation of osteoclastogenesis by ectodomain shedding of receptor activator of NF-kappaB ligand. J Biol Chem 281(48):36846–36855

Hill PA, Tumber A, Papaioannou S, Meikle MC (1998) The cellular actions of interleukin-11 on bone resorption in vitro. Endocrinology 139(4):1564–1572

Hirano T, Akira S, Taga T, Kishimoto T (1990) Biological and clinical aspects of interleukin 6. Immunol Today 11:443–449

Hock JM, Krishnan V, Onyia JE, Bidwell JP, Milas J, Stanislaus D (2001) Osteoblast apoptosis and bone turnover. J Bone Miner Res 16(6):975–984

Hodge JM, Collier FM, Pavlos NJ, Kirkland MA, Nicholson GC (2011) M-CSF potently augments RANKL-induced resorption activation in mature human osteoclasts. PLoS One 6(6):e21462

Hofbauer LC, Khosla S, Dunstan CR, Lacey DL, Spelsberg TC, Riggs BL (1999a) Estrogen stimulates gene expression and protein production of osteoprotegerin in human osteoblastic cells. Endocrinology 140(9):4367–4370

Hofbauer LC, Lacey DL, Dunstan CR, Spelsberg TC, Riggs BL, Khosla S (1999b) Interleukin-1beta and tumor necrosis factor-alpha, but not interleukin-6, stimulate osteoprotegerin ligand gene expression in human osteoblastic cells. Bone 25(3):255–259

Hong MH, Williams H, Jin CH, Pike JW (2000) The inhibitory effect of interleukin-10 on mouse osteoclast formation involves novel tyrosine-phosphorylated proteins. J Bone Miner Res 15 (5):911–918

Hong H, Shi Z, Qiao P, Li H, McCoy EM, Mao P, Xu H, Feng X, Wang S (2013) Interleukin-3 plays dual roles in osteoclastogenesis by promoting the development of osteoclast progenitors but inhibiting the osteoclastogenic process. Biochem Biophys Res Commun 440(4):545–550

Horiuchi K, Kimura T, Miyamoto T, Miyamoto K, Akiyama H, Takaishi H, Morioka H, Nakamura T, Okada Y, Blobel CP, Toyama Y (2009) Conditional inactivation of TACE by a Sox9 promoter leads to osteoporosis and increased granulopoiesis via dysregulation of IL-17 and G-CSF. JImmunol 182(4):2093–2101

Horwood NJ, Udagawa N, Elliott J, Grail D, Okamura H, Kurimoto M, Dunn AR, Martin T, Gillespie MT (1998) Interleukin 18 inhibits osteoclast formation via T cell production of granulocyte macrophage colony-stimulating factor. J Clin Invest 101(3):595–603

Horwood NJ, Elliott J, Martin TJ, Gillespie MT (2001) IL-12 alone and in synergy with IL-18 inhibits osteoclast formation in vitro. J Immunol 166(8):4915–4921

Hsieh CS, Macatonia SE, Tripp CS, Wolf SF, O'Garra A, Murphy KM (1993) Development of TH1 CD4+ T cells through IL-12 produced by Listeria-induced macrophages. Science 260 (5107):547–549

Huang J, Gao X, Li S, Cao Z (1997) Recruitment of IRAK to the interleukin 1 receptor complex requires interleukin 1 receptor accessory protein. Proc Natl Acad Sci U S A 94 (24):12829–12832

Huang H, Kim HJ, Chang EJ, Lee ZH, Hwang SJ, Kim HM, Lee Y, Kim HH (2009) IL-17 stimulates the proliferation and differentiation of human mesenchymal stem cells: implications for bone remodeling. Cell Death Differ 16(10):1332–1343

Huang L, Liu S, Song T, Zhang W, Fan J, Liu Y (2017) Blockade of interleukin 6 by rat anti-mouse interleukin 6 receptor antibody promotes fracture healing. Biochem Biokhim 82(10):1193–1199
Hughes AE, Ralston SH, Marken J, Bell C, MacPherson H, Wallace RG, van Hul W, Whyte MP, Nakatsuka K, Hovy L, Anderson DM (2000) Mutations in TNFRSF11A, affecting the signal peptide of RANK, cause familial expansile osteolysis. Nat Genet 24(1):45–48
Ikebuchi Y, Aoki S, Honma M, Hayashi M, Sugamori Y, Khan M, Kariya Y, Kato G, Tabata Y, Penninger JM, Udagawa N, Aoki K, Suzuki H (2018) Coupling of bone resorption and formation by RANKL reverse signalling. Nature 561(7722):195–200
Ishimi Y, Miyaura C, Jin CH, Akatsu T, Abe E, Nakamura Y, Yamaguchi A, Yoshiki S, Matsuda T, Hirano T, Kishimoto T, Suda T (1990) IL-6 is produced by osteoblasts and induces bone resorption. J Immunol 145:3297–3303
Issaranggun Na Ayuthaya B, Everts V, Pavasant P (2017) Interleukin-12 induces receptor activator of nuclear factor-kappa B ligand expression by human periodontal ligament cells. J Periodontol 88(7):e109–e119
Ivanova JT, Boyanov MA, Toshev AK (2012) Polymorphisms of the human IL-1 receptor antagonist gene and forearm bone mineral density in postmenopausal women. Indian J Endocrinol Metab 16(4):580–584
Jacquin C, Gran DE, Lee SK, Lorenzo JA, Aguila HL (2006) Identification of multiple osteoclast precursor populations in murine bone marrow. J Bone Miner Res 21(1):67–77
Jacquin C, Koczon-Jaremko B, Aguila HL, Leng L, Bucala R, Kuchel GA, Lee SK (2009) Macrophage migration inhibitory factor inhibits osteoclastogenesis. Bone 45(4):640–649
Jay PR, Centrella M, Lorenzo J, Bruce AG, Horowitz MC (1996) Oncostatin-M: a new bone active cytokine that activates osteoblasts and inhibits bone resorption. Endocrinology 137:1151–1158
Jeong BC (2018) ATF3 mediates the inhibitory action of TNF-alpha on osteoblast differentiation through the JNK signaling pathway. Biochem Biophys Res Commun 499(3):696–701
Ji JD, Park-Min KH, Shen Z, Fajardo RJ, Goldring SR, McHugh KP, Ivashkiv LB (2009) Inhibition of RANK expression and osteoclastogenesis by TLRs and IFN-{gamma} in human osteoclast precursors. J Immunol 183(11):7223–7233
Jimi E, Shuto T, Koga T (1995) Macrophage colony-stimulating factor and interleukin-1à maintain the survival of osteoclast-like cells. Endocrinology 136:808–811
Jimi E, Nakamura I, Duong LT, Ikebe T, Takahashi N, Rodan GA, Suda T (1999) Interleukin 1 induces multinucleation and bone-resorbing activity of osteoclasts in the absence of osteoblasts/stromal cells. Exp Cell Res 247(1):84–93
Jo S, Koo BS, Lee B, Kwon E, Lee YL, Chung H, Sung IH, Park YS, Kim TH (2017) A novel role for bone-derived cells in ankylosing spondylitis: focus on IL-23. Biochem Biophys Res Commun 491(3):787–793
Ju JH, Cho ML, Moon YM, Oh HJ, Park JS, Jhun JY, Min SY, Cho YG, Park KS, Yoon CH, Min JK, Park SH, Sung YC, Kim HY (2008) IL-23 induces receptor activator of NF-kappaB ligand expression on CD4+ T cells and promotes osteoclastogenesis in an autoimmune arthritis model. J Immunol 181(2):1507–1518
Jules J, Zhang P, Ashley JW, Wei S, Shi Z, Liu J, Michalek SM, Feng X (2012) Molecular basis of requirement of receptor activator of nuclear factor kappaB signaling for interleukin 1-mediated osteoclastogenesis. J Biol Chem 287(19):15728–15738
Kamel Mohamed SG, Sugiyama E, Shinoda K, Hounoki H, Taki H, Maruyama M, Miyahara T, Kobayashi M (2005) Interleukin-4 inhibits RANKL-induced expression of NFATc1 and c-Fos: a possible mechanism for downregulation of osteoclastogenesis. Biochem Biophys Res Commun 329(3):839–845
Kamiya S, Nakamura C, Fukawa T, Ono K, Ohwaki T, Yoshimoto T, Wada S (2007) Effects of IL-23 and IL-27 on osteoblasts and osteoclasts: inhibitory effects on osteoclast differentiation. J Bone Miner Metab 25(5):277–285
Kamiya S, Okumura M, Chiba Y, Fukawa T, Nakamura C, Nimura N, Mizuguchi J, Wada S, Yoshimoto T (2011) IL-27 suppresses RANKL expression in CD4(+) T cells in part through STAT3. Immunol Lett 138(1):47–53

Kaneshiro S, Ebina K, Shi K, Higuchi C, Hirao M, Okamoto M, Koizumi K, Morimoto T, Yoshikawa H, Hashimoto J (2014) IL-6 negatively regulates osteoblast differentiation through the SHP2/MEK2 and SHP2/Akt2 pathways in vitro. J Bone Miner Metab 32(4):378–392

Kang YK, Zhang MC (2014) IL-23 promotes osteoclastogenesis in osteoblast-osteoclast co-culture system. Genet Mol Res 13(2):4673–4679

Kang H, Yang K, Xiao L, Guo L, Guo C, Yan Y, Qi J, Wang F, Ryffel B, Li C, Deng L (2017) Osteoblast hypoxia-inducible factor-1alpha pathway activation restrains osteoclastogenesis via the interleukin-33-MicroRNA-34a-Notch1 pathway. Front Immunol 8:1312

Kanzaki H, Makihira S, Suzuki M, Ishii T, Movila A, Hirschfeld J, Mawardi H, Lin X, Han X, Taubman MA, Kawai T (2016) Soluble RANKL cleaved from activated lymphocytes by TNF-alpha-converting enzyme contributes to osteoclastogenesis in periodontitis. J Immunol 197 (10):3871–3883

Kastelein RA, Hunter CA, Cua DJ (2007) Discovery and biology of IL-23 and IL-27: related but functionally distinct regulators of inflammation. Annu Rev Immunol 25:221–242

Katavic V, Lukic IK, Kovacic N, Grcevic D, Lorenzo JA, Marusic A (2003) Increased bone mass is a part of the generalized lymphoproliferative disorder phenotype in the mouse. J Immunol 170 (3):1540–1547

Kawaguchi H, Pilbeam CC, Vargas SJ, Morse EE, Lorenzo JA, Raisz LG (1995) Ovariectomy enhances and estrogen replacement inhibits the activity of bone marrow factors that stimulate prostaglandin production in cultured mouse calvariae. J Clin Invest 96:539–548

Kawasaki K, Gao YH, Yokose S, Kaji Y, Nakamura T, Suda T, Yoshida K, Taga T, Kishimoto T, Kataoka H, Yuasa T, Norimatsu H, Yamaguchi A (1997) Osteoclasts are present in gp130-deficient mice. Endocrinology 138(11):4959–4965

Kawase Y, Hoshino T, Yokota K, Kuzuhara A, Nakamura M, Maeda Y, Nishiwaki E, Zenmyo M, Hiraoka K, Aizawa H, Yoshino K (2003) Bone malformations in interleukin-18 transgenic mice. J Bone Miner Res 18(6):975–983

Ke D, Fu X, Xue Y, Wu H, Zhang Y, Chen X, Hou J (2018) IL-17A regulates the autophagic activity of osteoclast precursors through RANKL-JNK1 signaling during osteoclastogenesis in vitro. Biochem Biophys Res Commun 497(3):890–896. https://doi.org/10.1016/j.bbrc.2018.02.164

Keller J, Catala-Lehnen P, Wintges K, Schulze J, Bickert T, Ito W, Horst AK, Amling M, Schinke T (2011) Transgenic over-expression of interleukin-33 in osteoblasts results in decreased osteoclastogenesis. Biochem Biophys Res Commun 417(1):217–222

Key LL Jr, Ries WL, Rodriguiz RM, Hatcher HC (1992) Recombinant human interferon gamma therapy for osteopetrosis. J Pediatr 121:119–124

Key LL Jr, Rodriguiz RM, Willi SM, Wright NM, Hatcher HC, Eyre DR, Cure JK, Griffin PP, Ries WL (1995a) Long-term treatment of osteopetrosis with recombinant human interferon gamma. N Engl J Med 332:1594–1599

Key LL Jr, Rodriguiz RM, Willi SM, Wright NM, Hatcher HC, Eyre DR, Cure JK, Griffin PP, Ries WL (1995b) Long-term treatment of osteopetrosis with recombinant human interferon gamma. N Engl J Med 332-9:1594

Khapli SM, Mangashetti LS, Yogesha SD, Wani MR (2003) IL-3 acts directly on osteoclast precursors and irreversibly inhibits receptor activator of NF-kappaB ligand-induced osteoclast differentiation by diverting the cells to macrophage lineage. J Immunol 171(1):142–151

Kido S, Kuriwaka-Kido R, Imamura T, Ito Y, Inoue D, Matsumoto T (2009) Mechanical stress induces interleukin-11 expression to stimulate osteoblast differentiation. Bone 45(6):1125–1132

Kim MS, Day CJ, Selinger CI, Magno CL, Stephens SJ, Morrison NA (2005a) MCP-1 induced human osteoclast-like cells are TRAP, NFATc1 and calcitonin receptor positive but require RANKL for bone resorption. J Biol Chem 281(2):1274–1285

Kim N, Kadono Y, Takami M, Lee J, Lee SH, Okada F, Kim JH, Kobayashi T, Odgren PR, Nakano H, Yeh WC, Lee SK, Lorenzo JA, Choi Y (2005b) Osteoclast differentiation independent of the TRANCE-RANK-TRAF6 axis. J Exp Med 202(5):589–595

Kim JH, Jin HM, Kim K, Song I, Youn BU, Matsuo K, Kim N (2009) The mechanism of osteoclast differentiation induced by IL-1. J Immunol 183(3):1862–1870

Kim JW, Lee MS, Lee CH, Kim HY, Chae SU, Kwak HB, Oh J (2012) Effect of interferon-gamma on the fusion of mononuclear osteoclasts into bone-resorbing osteoclasts. BMB Rep 45 (5):281–286

Kim HR, Kim KW, Kim BM, Jung HG, Cho ML, Lee SH (2014a) Reciprocal activation of CD4+ T cells and synovial fibroblasts by stromal cell-derived factor 1 promotes RANKL expression and osteoclastogenesis in rheumatoid arthritis. Arthritis Rheumatol 66(3):538–548

Kim YG, Park JW, Lee JM, Suh JY, Lee JK, Chang BS, Um HS, Kim JY, Lee Y (2014b) IL-17 inhibits osteoblast differentiation and bone regeneration in rat. Arch Oral Biol 59(9):897–905

Kim JH, Sim JH, Lee S, Seol MA, Ye SK, Shin HM, Lee EB, Lee YJ, Choi YJ, Yoo WH, Kim JH, Kim WU, Lee DS, Kim JH, Kang I, Kang SW, Kim HR (2017) Interleukin-7 induces osteoclast formation via STAT5, independent of receptor activator of NF-kappaB ligand. Front Immunol 8:1376

Kishimoto T, Taga T, Akira S (1994) Cytokine signal transduction. Cell 76:253–262

Kitajima I, Soejima Y, Takasaki I, Beppu H, Tokioka T, Maruyama I (1996) Ceramide-induced nuclear translocation of NF-kappaB is a potential mediator of the apoptotic response to TNF-à in murine clonal osteoblasts. Bone 19:263–270

Kitami S, Tanaka H, Kawato T, Tanabe N, Katono-Tani T, Zhang F, Suzuki N, Yonehara Y, Maeno M (2010) IL-17A suppresses the expression of bone resorption-related proteinases and osteoclast differentiation via IL-17RA or IL-17RC receptors in RAW264.7 cells. Biochimie 92 (4):398–404

Kitaura H, Nagata N, Fujimura Y, Hotokezaka H, Yoshida N, Nakayama K (2002) Effect of IL-12 on TNF-alpha-mediated osteoclast formation in bone marrow cells: apoptosis mediated by Fas/Fas ligand interaction. J Immunol 169(9):4732–4738

Kitaura H, Zhou P, Kim HJ, Novack DV, Ross FP, Teitelbaum SL (2005) M-CSF mediates TNF-induced inflammatory osteolysis. J Clin Invest 115(12):3418–3427

Kittaka M, Mayahara K, Mukai T, Yoshimoto T, Yoshitaka T, Gorski JP, Ueki Y (2018) Cherubism mice also deficient in c-Fos exhibit inflammatory bone destruction executed by macrophages that express MMP14 despite the absence of TRAP+ osteoclasts. J Bone Miner Res 33 (1):167–181

Klein DC, Raisz LG (1970) Prostaglandins: stimulation of bone resorption in tissue culture. Endocrinology 86:1436–1440

Kobayashi K, Takahashi N, Jimi E, Udagawa N, Takami M, Kotake S, Nakagawa N, Kinosaki M, Yamaguchi K, Shima N, Yasuda H, Morinaga T, Higashio K, Martin TJ, Suda T (2000) Tumor necrosis factor alpha stimulates osteoclast differentiation by a mechanism independent of the ODF/RANKL-RANK interaction. J Exp Med 191(2):275–286

Koenders MI, Lubberts E, Oppers-Walgreen B, Van Den BL, Helsen MM, Di Padova FE, Boots AM, Gram H, Joosten LA, Van den Berg WB (2005) Blocking of interleukin-17 during reactivation of experimental arthritis prevents joint inflammation and bone erosion by decreasing RANKL and interleukin-1. Am J Pathol 167(1):141–149

Koh JM, Oh B, Ha MH, Cho KW, Lee JY, Park BL, Shin HD, Bae MA, Kim HJ, Hong JM, Kim TH, Shin HI, Lee SH, Kim GS, Kim SY, Park EK (2009) Association of IL-15 polymorphisms with bone mineral density in postmenopausal korean women. Calcif Tissue Int 85(5):369–378

Kohara H, Kitaura H, Fujimura Y, Yoshimatsu M, Morita Y, Eguchi T, Masuyama R, Yoshida N (2011) IFN-gamma directly inhibits TNF-alpha-induced osteoclastogenesis in vitro and in vivo and induces apoptosis mediated by Fas/Fas ligand interactions. Immunol Lett 137(1–2):53–61

Kong YY, Yoshida H, Sarosi I, Tan HL, Timms E, Capparelli C, Morony S, Oliveira-dos-Santos AJ, Van G, Itie A, Khoo W, Wakeham A, Dunstan CR, Lacey DL, Mak TW, Boyle WJ, Penninger JM (1999) OPGL is a key regulator of osteoclastogenesis, lymphocyte development and lymph-node organogenesis. Nature 397(6717):315–323

Korherr C, Hofmeister R, Wesche H, Falk W (1997) A critical role for interleukin-1 receptor accessory protein in interleukin-1 signaling. Eur J Immunol 27(1):262–267

Kotake S, Udagawa N, Takahashi N, Matsuzaki K, Itoh K, Ishiyama S, Saito S, Inoue K, Kamatani N, Gillespie MT, Martin TJ, Suda T (1999) IL-17 in synovial fluids from patients

with rheumatoid arthritis is a potent stimulator of osteoclastogenesis. J Clin Invest 103 (9):1345–1352
Kotrych D, Dziedziejko V, Safranow K, Sroczynski T, Staniszewska M, Juzyszyn Z, Pawlik A (2016) TNF-alpha and IL10 gene polymorphisms in women with postmenopausal osteoporosis. Eur J Obstet Gynecol Reprod Biol 199:92–95
Kovacic N, Lukic IK, Grcevic D, Katavic V, Croucher P, Marusic A (2007) The Fas/Fas ligand system inhibits differentiation of murine osteoblasts but has a limited role in osteoblast and osteoclast apoptosis. J Immunol 178(6):3379–3389
Kovacic N, Grcevic D, Katavic V, Lukic IK, Grubisic V, Mihovilovic K, Cvija H, Croucher PI, Marusic A (2010) Fas receptor is required for estrogen deficiency-induced bone loss in mice. Lab Invest 90(3):402–413
Koyama Y, Mitsui N, Suzuki N, Yanagisawa M, Sanuki R, Isokawa K, Shimizu N, Maeno M (2008) Effect of compressive force on the expression of inflammatory cytokines and their receptors in osteoblastic Saos-2 cells. Arch Oral Biol 53(5):488–496
Krum SA, Miranda-Carboni GA, Hauschka PV, Carroll JS, Lane TF, Freedman LP, Brown M (2008) Estrogen protects bone by inducing Fas ligand in osteoblasts to regulate osteoclast survival. EMBO J 27(3):535–545
Kudo O, Fujikawa Y, Itonaga I, Sabokbar A, Torisu T, Athanasou NA (2002) Proinflammatory cytokine (TNFalpha/IL-1alpha) induction of human osteoclast formation. J Pathol 198 (2):220–227
Kudo O, Sabokbar A, Pocock A, Itonaga I, Fujikawa Y, Athanasou NA (2003) Interleukin-6 and interleukin-11 support human osteoclast formation by a RANKL-independent mechanism. Bone 32(1):1–7
Kulkarni RN, Bakker AD, Everts V, Klein-Nulend J (2012) Mechanical loading prevents the stimulating effect of IL-1beta on osteocyte-modulated osteoclastogenesis. Biochem Biophys Res Commun 420(1):11–16
Kumari A, Silakari O, Singh RK (2018) Recent advances in colony stimulating factor-1 receptor/c-FMS as an emerging target for various therapeutic implications. Biomed Pharmacother 103:662–679
Kurihara N, Chenu C, Miller M, Civin C, Roodman GD (1990) Identification of committed mononuclear precursors for osteoclast-like cells formed in long term human marrow cultures. Endocrinology 126(5):2733–2741
Kuwabara H, Wada T, Oda T, Yoshikawa H, Sawada N, Kokai Y, Ishii S (2001) Overexpression of the granulocyte colony-stimulating factor gene impairs bone morphogenetic protein responsiveness in mice. Lab Invest 81(8):1133–1141
Labrinidis A, Liapis V, Thai le M, Atkins GJ, Vincent C, Hay S, Sims NA, Zannettino AC, Findlay DM, Evdokiou A (2008) Does Apo2L/TRAIL play any physiologic role in osteoclastogenesis? Blood 111(11):5411–5412
Lacey DL, Timms E, Tan HL, Kelley MJ, Dunstan CR, Burgess T, Elliott R, Colombero A, Elliott G, Scully S, Hsu H, Sullivan J, Hawkins N, Davy E, Capparelli C, Eli A, Qian YX, Kaufman S, Sarosi I, Shalhoub V, Senaldi G, Guo J, Delaney J, Boyle WJ (1998) Osteoprotegerin ligand is a cytokine that regulates osteoclast differentiation and activation. Cell 93(2):165–176
Lagasse E, Weissman IL (1997) Enforced expression of Bcl-2 in monocytes rescues macrophages and partially reverses osteopetrosis in op/op mice. Cell 89(7):1021–1031
Lam J, Takeshita S, Barker JE, Kanagawa O, Ross FP, Teitelbaum SL (2000) TNF-alpha induces osteoclastogenesis by direct stimulation of macrophages exposed to permissive levels of RANK ligand. J Clin Invest 106(12):1481–1488
Lambert C, Oury C, Dejardin E, Chariot A, Piette J, Malaise M, Merville MP, Franchimont N (2007) Further insights in the mechanisms of interleukin-1beta stimulation of osteoprotegerin in osteoblast-like cells. J Bone Miner Res 22(9):1350–1361

Lapidot T, Petit I (2002) Current understanding of stem cell mobilization: the roles of chemokines, proteolytic enzymes, adhesion molecules, cytokines, and stromal cells. Exp Hematol 30 (9):973–981

Lari R, Fleetwood AJ, Kitchener PD, Cook AD, Pavasovic D, Hertzog PJ, Hamilton JA (2007) Macrophage lineage phenotypes and osteoclastogenesis – complexity in the control by GM-CSF and TGF-beta. Bone 40(2):323–336

Larousserie F, Bsiri L, Dumaine V, Dietrich C, Audebourg A, Radenen-Bussiere B, Anract P, Vacher-Lavenu MC, Devergne O (2017) Frontline science: human bone cells as a source of IL-27 under inflammatory conditions: role of TLRs and cytokines. J Leukoc Biol 101 (6):1289–1300

Lazzaro L, Tonkin BA, Poulton IJ, McGregor NE, Ferlin W, Sims NA (2018) IL-6 trans-signalling mediates trabecular, but not cortical, bone loss after ovariectomy. Bone 112:120–127

Lee SK, Kalinowski J, Jastrzebski S, Lorenzo JA (2002a) 1,25 (OH)(2) vitamin D(3)-stimulated osteoclast formation in spleen-osteoblast cocultures is mediated in part by enhanced IL-1alpha and receptor activator of NF-kappaB ligand production in osteoblasts. J Immunol 169 (5):2374–2380

Lee ZH, Lee SE, Kim CW, Lee SH, Kim SW, Kwack K, Walsh K, Kim HH (2002b) IL-1alpha stimulation of osteoclast survival through the PI 3-kinase/Akt and ERK pathways. J Biochem 131(1):161–166

Lee SK, Kalinowski JF, Jastrzebski SL, Puddington L, Lorenzo JA (2003) Interleukin-7 is a direct inhibitor of in vitro osteoclastogenesis. Endocrinology 144(8):3524–3531

Lee S, Kalinowski JF, Adams DJ, Aguila HL, Lorenzo JA (2004) Osteoblast specific overexpression of human interleukin-7 increases femoral trabecular bone mass in female mice and inhibits in vitro osteoclastogenesis. J Bone Miner Res 19:S410

Lee S, Kalinowski J, Adams DJ, Aguila HL, Lorenzo JA (2005a) Osteoblast specific overexpression of human interlukin-7 rescues the bone phenotype of interleukin-7 deficient female mice. J Bone Miner Res 20:S48

Lee SK, Gardner AE, Kalinowski JF, Jastrzebski SL, Lorenzo JA (2005b) RANKL-stimulated osteoclast-like cell formation in vitro is partially dependent on endogenous interleukin-1 production. Bone 38(5):678–685

Lee HY, Jeon HS, Song EK, Han MK, Park SI, Lee SI, Yun HJ, Kim JR, Kim JS, Lee YC, Kim SI, Kim HR, Choi JY, Kang I, Kim HY, Yoo WH (2006a) CD40 ligation of rheumatoid synovial fibroblasts regulates RANKL-medicated osteoclastogenesis: evidence of NF-kappaB-dependent, CD40-mediated bone destruction in rheumatoid arthritis. Arthritis Rheum 54 (6):1747–1758

Lee SK, Kalinowski JF, Jacquin C, Adams DJ, Gronowicz G, Lorenzo JA (2006b) Interleukin-7 influences osteoclast function in vivo but is not a critical factor in ovariectomy-induced bone loss. J Bone Miner Res 21(5):695–702

Lee MS, Kim HS, Yeon JT, Choi SW, Chun CH, Kwak HB, Oh J (2009) Granulocyte-macrophage colony-stimulating factor regulates fusion of mononuclear osteoclasts into bone-resorbing osteoclasts by activating the ras/ERK pathway. J Immunol 188:3390–3399

Lee YM, Fujikado N, Manaka H, Yasuda H, Iwakura Y (2010) IL-1 plays an important role in the bone metabolism under physiological conditions. Int Immunol 22(10):805–816

Lee EJ, Kim SM, Choi B, Kim EY, Chung YH, Lee EJ, Yoo B, Lee CK, Hong S, Kim BJ, Koh JM, Kim SH, Kim YG, Chang EJ (2017) Interleukin-32 gamma stimulates bone formation by increasing miR-29a in osteoblastic cells and prevents the development of osteoporosis. Sci Rep 7:40240

Lewis DB, Liggitt HD, Effmann EL, Motley ST, Teitelbaum SL, Jepsen KJ, Goldstein SA, Bonadio J, Carpenter J, Perlmutter RM (1993) Osteoporosis induced in mice by overproduction of interleukin 4. Proc Natl Acad Sci U S A 90:11618–11622

Li J, Sarosi I, Yan XQ, Morony S, Capparelli C, Tan HL, McCabe S, Elliott R, Scully S, Van G, Kaufman S, Juan SC, Sun Y, Tarpley J, Martin L, Christensen K, McCabe J, Kostenuik P, Hsu H, Fletcher F, Dunstan CR, Lacey DL, Boyle WJ (2000) RANK is the intrinsic hematopoietic cell surface receptor that controls osteoclastogenesis and regulation of bone mass and calcium metabolism. Proc Natl Acad Sci U S A 97(4):1566–1571

Li P, Schwarz EM, O'Keefe RJ, Ma L, Boyce BF, Xing L (2004a) RANK signaling is not required for TNFalpha-mediated increase in CD11(hi) osteoclast precursors but is essential for mature osteoclast formation in TNFalpha-mediated inflammatory arthritis. J Bone Miner Res 19 (2):207–213

Li P, Schwarz EM, O'Keefe RJ, Ma L, Looney RJ, Ritchlin CT, Boyce BF, Xing L (2004b) Systemic tumor necrosis factor alpha mediates an increase in peripheral CD11bhigh osteoclast precursors in tumor necrosis factor alpha-transgenic mice. Arthritis Rheum 50(1):265–276

Li JY, Tawfeek H, Bedi B, Yang X, Adams J, Gao KY, Zayzafoon M, Weitzmann MN, Pacifici R (2011) Ovariectomy disregulates osteoblast and osteoclast formation through the T-cell receptor CD40 ligand. Proc Natl Acad Sci U S A 108(2):768–773

Li H, Lu Y, Qian J, Zheng Y, Zhang M, Bi E, He J, Liu Z, Xu J, Gao JY, Yi Q (2014) Human osteoclasts are inducible immunosuppressive cells in response to T cell-derived IFN-gamma and CD40 ligand in vitro. J Bone Miner Res 29(12):2666–2675

Li X, Zhou ZY, Zhang YY, Yang HL (2016) IL-6 contributes to the defective osteogenesis of bone marrow stromal cells from the vertebral body of the glucocorticoid-induced osteoporotic mouse. PLoS One 11(4):e0154677

Lind M, Deleuran B, Yssel H, Fink-Eriksen E, Thestrup-Pedersen K (1995) IL-4 and IL-13, but not IL-10, are chemotactic factors for human osteoblasts. Cytokine 7(1):78–82

Linkhart TA, Linkhart SG, MacCharles DC, Long DL, Strong DD (1991) Interleukin-6 messenger RNA expression and interleukin-6 protein secretion in cells isolated from normal human bone: regulation by interleukin-1. J Bone Miner Res 6:1285–1294

Liu XH, Kirschenbaum A, Yao S, Levine AC (2005) Cross-talk between the interleukin-6 and prostaglandin E(2) signaling systems results in enhancement of osteoclastogenesis through effects on the osteoprotegerin/receptor activator of nuclear factor-{kappa}B (RANK) ligand/ RANK system. Endocrinology 146(4):1991–1998

Liu D, Yao S, Wise GE (2006) Effect of interleukin-10 on gene expression of osteoclastogenic regulatory molecules in the rat dental follicle. Eur J Oral Sci 114(1):42–49

Liu H, Feng W, Yimin CJ, Lv S, Hasegawa T, Sun B, Li J, Oda K, Amizuka N, Li M (2014) Histological evidence of increased osteoclast cell number and asymmetric bone resorption activity in the tibiae of interleukin-6-deficient mice. J Histochem Cytochem 62(8):556–564

Lokau J, Gottert S, Arnold P, Dusterhoft S, Massa Lopez D, Grotzinger J, Garbers C (2018) The SNP rs4252548 (R112H) which is associated with reduced human height compromises the stability of IL-11. Biochim Biophys Acta Mol Cell Res 1865(3):496–506

Lopez-Granados E, Temmerman ST, Wu L, Reynolds JC, Follmann D, Liu S, Nelson DL, Rauch F, Jain A (2007) Osteopenia in X-linked hyper-IgM syndrome reveals a regulatory role for CD40 ligand in osteoclastogenesis. Proc Natl Acad Sci U S A 104(12):5056–5061

Lorenzo JA, Sousa S, Alander C, Raisz LG, Dinarello CA (1987a) Comparison of the bone-resorbing activity in the supernatants from phytohemagglutinin-stimulated human peripheral blood mononuclear cells with that of cytokines through the use of an antiserum to interleukin 1. Endocrinology 121:1164–1170

Lorenzo JA, Sousa SL, Fonseca JM, Hock JM, Medlock ES (1987b) Colony-stimulating factors regulate the development of multinucleated osteoclasts from recently replicated cells in vitro. J Clin Invest 80(1):160–164

Lorenzo JA, Sousa SL, Leahy CL (1990a) Leukemia inhibitory factor (LIF) inhibits basal bone resorption in fetal rat long bone cultures. Cytokine 2:266–271

Lorenzo JA, Sousa SL, Van Den Brink-Webb SE, Korn JH (1990b) Production of both interleukin-1 alpha and beta by newborn mouse calvaria cultures. J Bone Miner Res 5:77–83

Lorenzo JA, Naprta A, Rao Y, Alander C, Glaccum M, Widmer M, Gronowicz G, Kalinowski J, Pilbeam CC (1998) Mice lacking the type I interleukin-1 receptor do not lose bone mass after ovariectomy. Endocrinology 139(6):3022–3025

Loro E, Ramaswamy G, Chandra A, Tseng WJ, Mishra MK, Shore EM, Khurana TS (2017) IL15RA is required for osteoblast function and bone mineralization. Bone 103:20–30

Loskog A, Totterman TH (2007) CD40L – a multipotent molecule for tumor therapy. Endocr Metab Immune Disord Drug Targets 7(1):23–28

Lowik CW, Van der PG, Bloys H, Hoekman K, Bijvoet OL, Aarden LA, Papapoulos SE (1989) Parathyroid hormone (PTH) and PTH-like protein (PLP) stimulate interleukin-6 production by osteogenic cells: a possible role of interleukin-6 in osteoclastogeneis. Biochem Biophys Res Commun 162:1546–1552

Lubberts E, Joosten LAB, Chabaud M, Van den Bersselaar L, Oppers B, Coenen-de Roo CJJ, Richards CD, Miossec P, Van den Berg WB (2000) IL-4 gene therapy for collagen arthritis suppresses synovial IL-17 and osteoprotegerin ligand and prevents bone erosion. J Clin Invest 105(12):1697–1710

Lubberts E, Van Den BL, Oppers-Walgreen B, Schwarzenberger P, Coenen-De Roo CJ, Kolls JK, Joosten LA, Van den Berg WB (2003) IL-17 promotes bone erosion in murine collagen-induced arthritis through loss of the receptor activator of NF-kappaB ligand/osteoprotegerin balance. J Immunol 170(5):2655–2662

Ma T, Miyanishi K, Suen A, Epstein NJ, Tomita T, Smith RL, Goodman SB (2004) Human interleukin-1-induced murine osteoclastogenesis is dependent on RANKL, but independent of TNF-alpha. Cytokine 26(3):138–144

Mabilleau G, Sabokbar A (2009) Interleukin-32 promotes osteoclast differentiation but not osteoclast activation. PLoS One 4(1):e4173

Macari S, Madeira MFM, Lima ILA, Pereira TSF, Dias GJ, Cirelli JA, de Molon RS, Fukada SY, Szawka RE, Garlet GP, Teixeira MM, Silva TA (2018) ST2 regulates bone loss in a site-dependent and estrogen-dependent manner. J Cell Biochem 119(10):8511–8521

MacDonald BR, Mundy GR, Clark S, Wang EA, Kuehl TJ, Stanley ER, Roodman GD (1986) Effects of human recombinant CSF-GM and highly purified CSF-1 on the formation of multinucleated cells with osteoclast characteristics in long-term bone marrow cultures. J Bone Miner Res 1:227–233

Makiishi-Shimobayashi C, Tsujimura T, Iwasaki T, Yamada N, Sugihara A, Okamura H, Hayashi SS, Terada N (2001) Interleukin-18 up-regulates osteoprotegerin expression in stromal/osteoblastic cells. Biochem Biophys Res Commun 281(2):361–366

Maldonado-Perez MB, Castro-Laria L, Caunedo-Alvarez A, Montoya-Garcia MJ, Giner-Garcia M, Arguelles-Arias F, Romero-Gomez M, Vazquez-Gamez MA (2018) Does the antitumor necrosis factor-alpha therapy decrease the vertebral fractures occurrence in inflammatory bowel disease? J Clin Densitom 22(2):195–202

Malik N, Haugen HS, Modrell B, Shoyab M, Clegg CH (1995) Developmental abnormalities in mice transgenic for bovine oncostatin M. Mol Cell Biol 15:2349–2358

Mangashetti LS, Khapli SM, Wani MR (2005) IL-4 inhibits bone-resorbing activity of mature osteoclasts by affecting NF-kappa B and Ca2+ signaling. J Immunol 175(2):917–925

Mann GN, Jacobs TW, Buchinsky FJ, Armstrong EC, Li M, Ke HZ, Ma YF, Jee WSS, Epstein S (1994) Interferon-gamma causes loss of bone volume in vivo and fails to ameliorate cyclosporin A-induced osteopenia. Endocrinology 135:1077–1083

Manolagas SC, Jilka RL (1995) Mechanisms of disease: bone marrow, cytokines, and bone remodeling – emerging insights into the pathophysiology of osteoporosis. N Engl J Med 332:305–311

Manolagas SC, Bellido T, Jilka RL (1995) New insights into the cellular, biochemical, and molecular basis of postmenopausal and senile osteoporosis: roles of IL-6 and gp130. [Review]. Int J Immunopharmacol 17(2):109–116

Manoury-Schwartz B, Chiocchia G, Bessis N, Abehsira-Amar O, Batteux F, Muller S, Huang S, Boissier MC, Fournier C (1997) High susceptibility to collagen-induced arthritis in mice lacking IFN-gamma receptors. J Immunol 158(11):5501–5506

Mansoori MN, Shukla P, Kakaji M, Tyagi AM, Srivastava K, Shukla M, Dixit M, Kureel J, Gupta S, Singh D (2016) IL-18BP is decreased in osteoporotic women: prevents Inflammasome mediated IL-18 activation and reduces Th17 differentiation. Sci Rep 6:33680

Mansoori MN, Shukla P, Singh D (2017) Combination of PTH (1-34) with anti-IL17 prevents bone loss by inhibiting IL-17/N-cadherin mediated disruption of PTHR1/LRP-6 interaction. Bone 105:226–236

Mansour A, Anginot A, Mancini SJ, Schiff C, Carle GF, Wakkach A, Blin-Wakkach C (2011) Osteoclast activity modulates B-cell development in the bone marrow. Cell Res 21 (7):1102–1115

Martin MU, Falk W (1997) The interleukin-1 receptor complex and interleukin-1 signal transduction. Eur Cytokine Netw 8(1):5–17

Martin SK, Fitter S, El Khawanky N, Grose RH, Walkley CR, Purton LE, Ruegg MA, Hall MN, Gronthos S, Zannettino ACW (2018) mTORC1 plays an important role in osteoblastic regulation of B-lymphopoiesis. Sci Rep 8(1):14501

Martino MM, Maruyama K, Kuhn GA, Satoh T, Takeuchi O, Muller R, Akira S (2016) Inhibition of IL-1R1/MyD88 signalling promotes mesenchymal stem cell-driven tissue regeneration. Nat Commun 7:11051

Marusic A, Kalinowski JF, Jastrzebski S, Lorenzo JA (1993) Production of Leukemia inhibitory factor mRNA and protein by malignant and immortalized bone cells. J Bone Miner Res 8:617–624

Matsumoto T, Kuriwaka-Kido R, Kondo T, Endo I, Kido S (2012) Regulation of osteoblast differentiation by interleukin-11 via AP-1 and Smad signaling. Endocr J 59(2):91–101

Matsuura T, Ichinose S, Akiyama M, Kasahara Y, Tachikawa N, Nakahama KI (2017) Involvement of CX3CL1 in the migration of osteoclast precursors across osteoblast layer stimulated by interleukin-1ss. J Cell Physiol 232(7):1739–1745. https://doi.org/10.1002/jcp.25577

Miller CH, Smith SM, Elguindy M, Zhang T, Xiang JZ, Hu X, Ivashkiv LB, Zhao B (2016) RBP-J-regulated miR-182 promotes TNF-alpha-induced osteoclastogenesis. J Immunol 196 (12):4977–4986

Miranda-Carus ME, Ito-Miguel M, Balsa A, Cobo-Ibanez T, Perez De AC, Pascual-Salcedo D, Martin-Mola E (2006) Peripheral blood T lymphocytes from patients with early rheumatoid arthritis express RANKL and interleukin-15 on the cell surface and promote osteoclastogenesis in autologous monocytes. Arthritis Rheum 54(4):1151–1164

Miyaura C, Kusano K, Masuzawa T, Chaki O, Onoe Y, Aoyagi M, Sasaki T, Tamura T, Koishihara Y, Ohsugi Y, Suda T (1995) Endogenous bone-resorbing factors in estrogen deficiency: cooperative effects of IL-1 and IL-6. J Bone Miner Res 10:1365–1363

Miyaura C, Onoe Y, Inada M, Maki K, Ikuta K, Ito M, Suda T (1997) Increased B-lymphopoiesis by interleukin 7 induces bone loss in mice with intact ovarian function: similarity to estrogen deficiency. Proc Natl Acad Sci U S A 94(17):9360–9365

Mizuno A, Amizuka N, Irie K, Murakami A, Fujise N, Kanno T, Sato Y, Nakagawa N, Yasuda H, Mochizuki S, Gomibuchi T, Yano K, Shima N, Washida N, Tsuda E, Morinaga T, Higashio K, Ozawa H (1998) Severe osteoporosis in mice lacking osteoclastogenesis inhibitory factor/osteoprotegerin. Biochem Biophys Res Commun 247(3):610–615

Mohamed SG, Sugiyama E, Shinoda K, Taki H, Hounoki H, Bdel-Aziz HO, Maruyama M, Kobayashi M, Ogawa H, Miyahara T (2007) Interleukin-10 inhibits RANKL-mediated expression of NFATc1 in part via suppression of c-Fos and c-Jun in RAW264.7 cells and mouse bone marrow cells. Bone 41(4):592–602

Moon YM, Yoon BY, Her YM, Oh HJ, Lee JS, Kim KW, Lee SY, Woo YJ, Park KS, Park SH, Kim HY, Cho ML (2012) IL-32 and IL-17 interact and have the potential to aggravate osteoclastogenesis in rheumatoid arthritis. Arthritis Res Ther 14(6):R246

Moore KW, de Waal MR, Coffman RL, O'Garra A (2001) Interleukin-10 and the interleukin-10 receptor. Annu Rev Immunol 19:683–765

Moreno JL, Kaczmareck M, Keegan AD, Tondravi M (2003) IL-4 suppresses both osteoclast development and mature osteoclast function by a STAT6-dependent mechanism: irreversible inhibition of the differentiation program activated by RANKL. Blood 102(3):1078–1086
Mori L, Iselin S, De Libero G, Lesslauer W (1996) Attenuation of collagen-induced arthritis in 55-kDa TNF receptor type 1 (TNFR1)-IgG1-treated and TNFR1-deficient mice. J Immunol 157:3178–3182
Morinaga Y, Fujita N, Ohishi K, Zhang Y, Tsuruo T (1998) Suppression of interleukin-11-mediated bone resorption by cyclooxygenases inhibitors. J Cell Physiol 175(3):247–254
Movila A, Ishii T, Albassam A, Wisitrasameewong W, Howait M, Yamaguchi T, Ruiz-Torruella M, Bahammam L, Nishimura K, Van Dyke T, Kawai T (2016) Macrophage migration inhibitory factor (MIF) supports homing of osteoclast precursors to peripheral osteolytic lesions. J Bone Miner Res 31(9):1688–1700
Muguruma Y, Lee MY (1998) Isolation and characterization of murine clonogenic osteoclast progenitors by cell surface phenotype analysis. Blood 91(4):1272–1279
Mun SH, Ko NY, Kim HS, Kim JW, Kim DK, Kim AR, Lee SH, Kim YG, Lee CK, Kim BK, Beaven MA, Kim YM, Choi WS (2010) Interleukin-33 stimulates formation of functional osteoclasts from human CD14(+) monocytes. Cell Mol Life Sci 67(22):3883–3892
Mun SH, Won HY, Hernandez P, Aguila HL, Lee SK (2013) Deletion of CD74, a putative MIF receptor, in mice enhances osteoclastogenesis and decreases bone mass. J Bone Miner Res 28 (4):948–959
Mun SH, Oh D, Lee SK (2014) Macrophage migration inhibitory factor down-regulates the RANKL-RANK signaling pathway by activating Lyn tyrosine kinase. Arthritis Rheumatol 66:2482–2493
Mundy GR (1996) An OAF by any other name [editorial; comment]. Endocrinology 137(4):114950
Nagata N, Kitaura H, Yoshida N, Nakayama K (2003) Inhibition of RANKL-induced osteoclast formation in mouse bone marrow cells by IL-12: involvement of IFN-gamma possibly induced from non-T cell population. Bone 33(4):721–732
Nakamichi Y, Mizoguchi T, Arai A, Kobayashi Y, Sato M, Penninger JM, Yasuda H, Kato S, DeLuca HF, Suda T, Udagawa N, Takahashi N (2012) Spleen serves as a reservoir of osteoclast precursors through vitamin D-induced IL-34 expression in osteopetrotic op/op mice. Proc Natl Acad Sci U S A 109(25):10006–10011
Nakamura I, Jimi E (2006) Regulation of osteoclast differentiation and function by interleukin-1. Vitam Horm 74:357–370
Nakamura T, Imai Y, Matsumoto T, Sato S, Takeuchi K, Igarashi K, Harada Y, Azuma Y, Krust A, Yamamoto Y, Nishina H, Takeda S, Takayanagi H, Metzger D, Kanno J, Takaoka K, Martin TJ, Chambon P, Kato S (2007) Estrogen prevents bone loss via estrogen receptor alpha and induction of fas ligand in osteoclasts. Cell 130(5):811–823
Nakano Y, Watanabe K, Morimoto I, Okada Y, Ura K, Sato K, Kasono K, Nakamura T, Eto S (1994) Interleukin-4 inhibits spontaneous and parathyroid hormone-related protein-stimulated osteoclast formation in mice. J Bone Miner Res 9:1533–1539
Nakashima T, Kobayashi Y, Yamasaki S, Kawakami A, Eguchi K, Sasaki H, Sakai H (2000) Protein expression and functional difference of membrane-bound and soluble receptor activator of NF-kappaB ligand: modulation of the expression by osteotropic factors and cytokines. Biochem Biophys Res Commun 275(3):768–775
Nakashima T, Hayashi M, Fukunaga T, Kurata K, Oh-Hora M, Feng JQ, Bonewald LF, Kodama T, Wutz A, Wagner EF, Penninger JM, Takayanagi H (2011) Evidence for osteocyte regulation of bone homeostasis through RANKL expression. Nat Med 17(10):1231–1234
Namen AE, Lupton S, Hjerrild K, Wignall J, Mochizuki DY, Schmierer A, Mosley B, March CJ, Urdal D, Gillis S, Cosman D, Goodwin RG (1988) Stimulation of B-cell progenitors by cloned murine interleukin-7. Nature 333:571–573
Nanes MS, McKoy WM, Marx SJ (1989) Inhibitory effects of tumor necrosis factor-alpha and interferon-gamma on deoxyribonucleic acid and collagen synthesis by rat osteosarcoma cells (ROS 17/2.8). Endocrinology 124:339–345

Nanes MS, Rubin J, Titus L, Hendy GN, Catherwood B (1991) Tumor necrosis factor-alpha inhibits 1,25-dihydroxyvitamin D 3 – stimulated bone Gla protein synthesis in rat osteosarcoma cells (ROS 17/2.8) by a pretranslational mechanism. Endocrinology 128:2577–2582

Neale Weitzmann M, Pacifici R (2017) Parathyroid diseases and T cells. Curr Osteoporos Rep 15 (3):135–141

Nicolaidou V, Wong MM, Redpath AN, Ersek A, Baban DF, Williams LM, Cope AP, Horwood NJ (2012) Monocytes induce STAT3 activation in human mesenchymal stem cells to promote osteoblast formation. PLoS One 7(7):e39871

Nigil Haroon N, Sriganthan J, Al Ghanim N, Inman RD, Cheung AM (2014) Effect of TNF-alpha inhibitor treatment on bone mineral density in patients with ankylosing spondylitis: a systematic review and meta-analysis. Semin Arthritis Rheum 44(2):155–161

Niki Y, Takaishi H, Takito J, Miyamoto T, Kosaki N, Matsumoto H, Toyama Y, Tada N (2007) Administration of cyclooxygenase-2 inhibitor reduces joint inflammation but exacerbates osteopenia in IL-1 alpha transgenic mice due to GM-CSF overproduction. J Immunol 179 (1):639–646

Nold-Petry CA, Nold MF, Zepp JA, Kim SH, Voelkel NF, Dinarello CA (2009) IL-32-dependent effects of IL-1beta on endothelial cell functions. Proc Natl Acad Sci U S A 106(10):3883–3888

Nomura K, Kuroda S, Yoshikawa H, Tomita T (2008) Inflammatory osteoclastogenesis can be induced by GM-CSF and activated under TNF immunity. Biochem Biophys Res Commun 367 (4):881–887

Novack DV, Faccio R (2009) Osteoclast motility: putting the brakes on bone resorption. Ageing Res Rev 10(1):54–61

O'Brien CA, Jilka RL, Fu Q, Stewart S, Weinstein RS, Manolagas SC (2005) IL-6 is not required for parathyroid hormone stimulation of RANKL expression, osteoclast formation, and bone loss in mice. Am J Physiol Endocrinol Metab 289(5):E784–E793

O'Brien W, Fissel BM, Maeda Y, Yan J, Ge X, Gravallese EM, Aliprantis AO, Charles JF (2016) RANK-independent osteoclast formation and bone erosion in inflammatory arthritis. Arthritis Rheumatol 68(12):2889–2900

Ochi S, Shinohara M, Sato K, Gober HJ, Koga T, Kodama T, Takai T, Miyasaka N, Takayanagi H (2007) Pathological role of osteoclast costimulation in arthritis-induced bone loss. Proc Natl Acad Sci U S A 104(27):11394–11399

Oda T, Wada T, Kuwabara H, Sawada N, Yamashita T, Kokai Y (2005) Ovariectomy fails to augment bone resorption and marrow B lymphopoiesis in granulocyte colony-stimulating factor transgenic mice. J Orthop Sci 10(1):70–76

Ogata Y, Kukita A, Kukita T, Komine M, Miyahara A, Miyazaki S, Kohashi O (1999) A novel role of IL-15 in the development of osteoclasts: inability to replace its activity with IL-2. J Immunol 162(5):2754–2760

Oh J, Lee MS, Yeon JT, Choi SW, Kim HS, Shim H, Lee SY, Youn BS, Yokota Y, Kim JH, Kwak HB, Khapli SM, Tomar GB, Barhanpurkar AP, Gupta N, Yogesha SD, Pote ST, Wani MR, Srivastava RK, Kour S, Mishra GC (2012) Inhibitory regulation of osteoclast differentiation by interleukin-3 via regulation of c-Fos and Id protein expression. J Cell Physiol 227(5):1851–1860

Okada Y, Morimoto I, Ura K, Nakano Y, Tanaka Y, Nishida S, Nakamura T, Eto S (1998) Short-term treatment of recombinant murine interleukin-4 rapidly inhibits bone formation in normal and ovariectomized mice. Bone 22(4):361–365

Okamura H, Tsutsui H, Komatsu T, Yutsudo M, Hakura A, Tanimoto T, Torigoe K, Okura T, Nukada Y, Hattori K, Akita K, Namba M, Tanabe F, Konishi K, Fukuda S, Kurimoto M (1995) Cloning of a new cytokine that induces IFN-gamma production by T cells. Nature 378:88–91

Old LJ (1985) Tumor necrosis factor (TNF). Science 230:630–632

Oliff A (1988) The role of tumor necrosis factor (cachectin) in cachexia. Cell 54:141–142

Onal M, Xiong J, Chen X, Thostenson JD, Almeida M, Manolagas SC, O'Brien CA (2012) RANKL expression by B lymphocytes contributes to ovariectomy-induced bone loss. J Biol Chem 287(35):29851–29860

Ono T, Okamoto K, Nakashima T, Nitta T, Hori S, Iwakura Y, Takayanagi H (2016) IL-17-producing gammadelta T cells enhance bone regeneration. Nat Commun 7:10928

Onodera S, Suzuki K, Matsuno T, Kaneda K, Kuriyama T, Nishihira J (1996) Identification of macrophage migration inhibitory factor in murine neonatal calvariae and osteoblasts. Immunology 89(3):430–435

Onodera S, Suzuki K, Kaneda K, Fujinaga M, Nishihira J (1999) Growth factor-induced expression of macrophage migration inhibitory factor in osteoblasts: relevance to the plasminogen activator system. Semin Thromb Hemost 25(6):563–568

Onodera S, Nishihira J, Iwabuchi K, Koyama Y, Yoshida K, Tanaka S, Minami A (2002) Macrophage migration inhibitory factor up-regulates matrix metalloproteinase-9 and -13 in rat osteoblasts. Relevance to intracellular signaling pathways. J Biol Chem 277(10):7865–7874

Onodera S, Sasaki S, Ohshima S, Amizuka N, Li M, Udagawa N, Irie K, Nishihira J, Koyama Y, Shiraishi A, Tohyama H, Yasuda K (2006) Transgenic mice overexpressing macrophage migration inhibitory factor (MIF) exhibit high-turnover osteoporosis. J Bone Miner Res 21 (6):876–885

Onoe Y, Miyaura C, Kaminakayashiki T, Nagai Y, Noguchi K, Chen QR, Seo H, Ohta H, Nozawa S, Kudo I, Suda T (1996) IL-13 and IL-4 inhibit bone resorption by suppressing cyclooxygenase-2-dependent prostaglandin synthesis in osteoblasts. J Immunol 156:758–764

Orozco A, Gemmell E, Bickel M, Seymour GJ (2007) Interleukin 18 and periodontal disease. J Dent Res 86(7):586–593

Oshima S, Onodera S, Amizuka N, Li M, Irie K, Watanabe S, Koyama Y, Nishihira J, Yasuda K, Minami A (2006) Macrophage migration inhibitory factor-deficient mice are resistant to ovariectomy-induced bone loss. FEBS Lett 580(5):1251–1256

Oskowitz AZ, Lu J, Penfornis P, Ylostalo J, McBride J, Flemington EK, Prockop DJ, Pochampally R (2008) Human multipotent stromal cells from bone marrow and microRNA: regulation of differentiation and leukemia inhibitory factor expression. Proc Natl Acad Sci U S A 105 (47):18372–18377

Osta B, Benedetti G, Miossec P (2014) Classical and paradoxical effects of TNF-alpha on bone homeostasis. Front Immunol 5:48

Ota M, Yanagisawa M, Tachibana H, Yokota K, Araki Y, Sato K, Mimura T (2014) A significant induction of neutrophilic chemoattractants but not RANKL in synoviocytes stimulated with interleukin 17. J Bone Miner Metab 33(1):40–47

Otero K, Turnbull IR, Poliani PL, Vermi W, Cerutti E, Aoshi T, Tassi I, Takai T, Stanley SL, Miller M, Shaw AS, Colonna M (2009) Macrophage colony-stimulating factor induces the proliferation and survival of macrophages via a pathway involving DAP12 and beta-catenin. Nat Immunol 10(7):734–743

Owens J, Chambers TJ (1995) Differential regulation of osteoclast formation: interleukin 10 (cytokine synthesis inhibitory factor) suppresses formation of osteoclasts but not macrophages in murine bone marrow cultures. J Bone Miner Res 10(Suppl 1):S220

Owens JM, Gallagher AC, Chambers TJ (1996) IL-1O modulates formation of osteoclasts in murine hemopoietic cultures. J Immunol 157:936–940

Ozsoy AZ, Karakus N, Tural S, Yigit S, Kara N, Alayli G, Tumer MK, Kuru O (2017) Influence of the MIF polymorphism -173G > C on Turkish postmenopausal women with osteoporosis. Z Rheumatol 77(7):629–632

Palmqvist P, Persson E, Conaway HH, Lerner UH (2002) IL-6, leukemia inhibitory factor, and oncostatin M stimulate bone resorption and regulate the expression of receptor activator of NF-kappaB ligand, osteoprotegerin, and receptor activator of NF-kappaB in mouse calvariae. J Immunol 169(6):3353–3362

Palmqvist P, Lundberg P, Persson E, Johansson A, Lundgren I, Lie A, Conaway HH, Lerner UH (2006) Inhibition of hormone and cytokine-stimulated osteoclastogenesis and bone resorption by interleukin-4 and interleukin-13 is associated with increased osteoprotegerin and decreased RANKL and RANK in a STAT6-dependent pathway. J Biol Chem 281(5):2414–2429

Panach L, Pineda B, Mifsut D, Tarin JJ, Cano A, Garcia-Perez MA (2016) The role of CD40 and CD40L in bone mineral density and in osteoporosis risk: a genetic and functional study. Bone 83:94–103

Pappalardo A, Thompson K (2013) Activated gammadelta T cells inhibit osteoclast differentiation and resorptive activity in vitro. Clin Exp Immunol 174(2):281–291

Park H, Jung YK, Park OJ, Lee YJ, Choi JY, Choi Y (2005) Interaction of fas ligand and fas expressed on osteoclast precursors increases osteoclastogenesis. J Immunol 175(11):7193–7201

Park MK, Her YM, Cho ML, Oh HJ, Park EM, Kwok SK, Ju JH, Park KS, Min DS, Kim HY, Park SH (2011) IL-15 promotes osteoclastogenesis via the PLD pathway in rheumatoid arthritis. Immunol Lett 139(1–2):42–51

Park JS, Jung YO, Oh HJ, Park SJ, Heo YJ, Kang CM, Kwok SK, Ju JH, Park KS, Cho ML, Sung YC, Park SH, Kim HY (2012) Interleukin-27 suppresses osteoclastogenesis via induction of interferon-gamma. Immunology 137(4):326–335

Paul NL, Ruddle NH (1988) Lymphotoxin. Annu Rev Immunol 6:407–438

Peng M, Wang Y, Qiang L, Xu Y, Li C, Li T, Zhou X, Xiao M, Wang J (2018) Interleukin-35 inhibits TNF-alpha-induced osteoclastogenesis and promotes apoptosis via shifting the activation from TNF receptor-associated death domain (TRADD)-TRAF2 to TRADD-Fas-associated death domain by JAK1/STAT1. Front Immunol 9:1417

Peterlik M, Hoffmann O, Swetly P, Klaushofer K, Koller K (1985) Recombinant gamma-interferon inhibits prostaglandin-mediated and parathyroid hormone-induced bone resorption in cultured neonatal mouse calvaria. FEBS Lett 185:287–290

Pfeffer K, Matsuyama T, Kundig TM, Wakeham A, Kishihara K, Shahinian A, Wiegmann K, Ohashi PS, Krönke M, Mak TW (1993) Mice deficient for the 55 kd tumor necrosis factor receptor are resistant to endotoxic shock, yet succumb to L. monocytogenes infection. Cell 73:457–467

Piemontese M, Xiong J, Fujiwara Y, Thostenson JD, O'Brien CA (2016) Cortical bone loss caused by glucocorticoid excess requires RANKL production by osteocytes and is associated with reduced OPG expression in mice. Am J Physiol Endocrin Metab 311(3):E587–E593

Polzer K, Joosten L, Gasser J, Distler JH, Ruiz G, Baum W, Redlich K, Bobacz K, Smolen JS, van den Berg W, Schett G, Zwerina J (2010) Interleukin-1 is essential for systemic inflammatory bone loss. Ann Rheum Dis 69(1):284–290

Purton LE, Lee MY, Torok-Storb B (1996) Normal human peripheral blood mononuclear cells mobilized with granulocyte colony-stimulating factor have increased osteoclastogenic potential compared to nonmobilized blood. Blood 87:1802–1808

Quinn JM, Sims NA, Saleh H, Mirosa D, Thompson K, Bouralexis S, Walker EC, Martin TJ, Gillespie MT (2008) IL-23 inhibits osteoclastogenesis indirectly through lymphocytes and is required for the maintenance of bone mass in mice. J Immunol 181(8):5720–5729

Raggatt LJ, Qin L, Tamasi J, Jefcoat SC Jr, Shimizu E, Selvamurugan N, Liew FY, Bevelock L, Feyen JH, Partridge NC (2008) Interleukin-18 is regulated by parathyroid hormone and is required for its bone anabolic actions. J Biol Chem 283(11):6790–6798

Ranganathan V, Ciccia F, Zeng F, Sari I, Guggino G, Muralitharan J, Gracey E, Haroon N (2017) Macrophage migration inhibitory factor induces inflammation and predicts spinal progression in ankylosing spondylitis. Arthritis Rheumatol 69(9):1796–1806

Reddy SV, Takahashi S, Dallas M, Williams RE, Neckers L, Roodman GD (1994) Interleukin-6 antisense deoxyoligonucleotides inhibit bone resorption by giant cells from human giant cell tumors of bone. J Bone Miner Res 9(5):753–757

Reid IR, Lowe C, Cornish J, Skinner SJ, Hilton DJ, Willson TA, Gearing DP, Martin TJ (1990) Leukemia inhibitory factor: a novel bone-active cytokine. Endocinology 126:1416–1420

Robinson JW, Li JY, Walker LD, Tyagi AM, Reott MA, Yu M, Adams J, Weitzmann MN, Pacifici R (2015) T cell-expressed CD40L potentiates the bone anabolic activity of intermittent PTH treatment. J Bone Miner Res 30(4):695–705

Romas E, Udagawa N, Zhou H, Tamura T, Saito M, Taga T, Hilton DJ, Suda T, Ng KW, Martin TJ (1996) The role of gp130-mediated signals in osteoclast development: regulation of interleukin 11 production by osteoblasts and distribution of its receptor in bone marrow cultures. J Exp Med 183:2581–2591

Roodman GD (1992) Interleukin-6: an osteotropic factor? J Bone Miner Res 7:475–478

Roodman GD, Kurihara N, Ohsaki Y, Kukita A, Hosking D, Demulder A, Smith JF, Singer FR (1992) Interleukin 6. A potential autocrine/paracrine factor in Paget's disease of bone. J Clin Invest 89(1):46–52

Roth P, Stanley ER (1992) The biology of CSF-1 and its receptor. Curr Top Microbiol Immunol 181:141–167

Rothe J, Lesslauer W, Lotscher H, Lang Y, Koebel P, Kontgen F, Althage A, Zinkernagel R, Steinmetz M, Bluethmann H (1993) Mice lacking the tumour necrosis factor receptor 1 are resistant to TNF-mediated toxicity but highly susceptible to infection by Listeria monocytogenes. Nature 364:798–802

Roux S, Lambert-Comeau P, Saint-Pierre C, Lepine M, Sawan B, Parent JL (2005) Death receptors, Fas and TRAIL receptors, are involved in human osteoclast apoptosis. Biochem Biophys Res Commun 333(1):42–50

Rubin J, Fan X, Thornton D, Bryant R, Biskobing D (1996) Regulation of murine osteoblast macrophage colony-stimulating factor production by 1,25(OH)2D3. Calcif Tissue Int 59 (4):291–294

Ruef N, Dolder S, Aeberli D, Seitz M, Balani D, Hofstetter W (2017) Granulocyte-macrophage colony-stimulating factor-dependent CD11c-positive cells differentiate into active osteoclasts. Bone 97:267–277

Ryan MR, Shepherd R, Leavey JK, Gao YH, Grassi F, Schnell FJ, Qian WP, Kersh GJ, Weitzmann MN, Pacifici R (2005) An IL-7-dependent rebound in thymic T cell output contributes to the bone loss induced by estrogen deficiency. Proc Natl Acad Sci U S A 102(46):16735–16740

Sabatini M, Boyce B, Aufdemorte T, Bonewald L, Mundy GR (1988) Infusions of recombinant human interleukins 1 alpha and 1 beta cause hypercalcemia in normal mice. Proc Natl Acad Sci U S A 85:5235–5239

Saeed J, Kitaura H, Kimura K, Ishida M, Sugisawa H, Ochi Y, Kishikawa A, Takano-Yamamoto T (2016) IL-37 inhibits lipopolysaccharide-induced osteoclast formation and bone resorption in vivo. Immunol Lett 175:8–15

Saidi S, Magne D (2011) Interleukin-33: a novel player in osteonecrosis of the femoral head? Joint Bone Spine 78(6):550–554

Saidi S, Bouri F, Lencel P, Duplomb L, Baud'huin M, Delplace S, Leterme D, Miellot F, Heymann D, Hardouin P, Palmer G, Magne D (2011) IL-33 is expressed in human osteoblasts, but has no direct effect on bone remodeling. Cytokine 53(3):347–354

Saleh H, Eeles D, Hodge JM, Nicholson GC, Gu R, Pompolo S, Gillespie MT, Quinn JM (2011) Interleukin-33, a target of parathyroid hormone and oncostatin M, increases osteoblastic matrix mineral deposition and inhibits osteoclast formation in vitro. Endocrinology 152(5):1911–1922

Salopek D, Grcevic D, Katavic V, Kovacic N, Lukic IK, Marusic A (2008) Increased bone resorption and osteopenia are a part of the lymphoproliferative phenotype of mice with systemic over-expression of interleukin-7 gene driven by MHC class II promoter. Immunol Lett 121 (2):134–139

Sambandam Y, Baird KL, Stroebel M, Kowal E, Balasubramanian S, Reddy SV (2016) Microgravity induction of TRAIL expression in preosteoclast cells enhances osteoclast differentiation. Sci Rep 6:25143

Sarahrudi K, Mousavi M, Grossschmidt K, Sela N, Konig F, Vecsei V, Aharinejad S (2009) The impact of colony-stimulating factor-1 on fracture healing: an experimental study. J Orthop Res 27(1):36–41

Sato K, Fujii Y, Asano S, Ohtsuki T, Kawakami M, Kasono K, Tsushima T, Shizume K (1986) Recombinant human interleukin 1 alpha and beta stimulate mouse osteoblast-like cells

(MC3T3-E1) to produce macrophage-colony stimulating activity and prostaglandin E2. Biochem Biophys Res Commun 141:285–291
Sato N, Takahashi N, Suda K, Nakamura M, Yamaki M, Ninomiya T, Kobayashi Y, Takada H, Shibata K, Yamamoto M, Takeda K, Akira S, Noguchi T, Udagawa N (2004) MyD88 but not TRIF is essential for osteoclastogenesis induced by lipopolysaccharide, diacyl lipopeptide, and IL-1alpha. J Exp Med 200(5):601–611
Sato K, Suematsu A, Okamoto K, Yamaguchi A, Morishita Y, Kadono Y, Tanaka S, Kodama T, Akira S, Iwakura Y, Cua DJ, Takayanagi H (2006) Th17 functions as an osteoclastogenic helper T cell subset that links T cell activation and bone destruction. J Exp Med 203(12):2673–2682
Sato T, Watanabe K, Masuhara M, Hada N, Hakeda Y (2007) Production of IL-7 is increased in ovariectomized mice, but not RANKL mRNA expression by osteoblasts/stromal cells in bone, and IL-7 enhances generation of osteoclast precursors in vitro. J Bone Miner Res 25(1):19–27
Scharla SH, Strong DD, Mohan S, Chevalley T, Linkhart TA (1994) Effect of tumor necrosis factor-alpha on the expression of insulin-like growth factor I and insulin-like growth factor binding protein 4 in mouse osteoblasts. Eur J Endocrinol 131(3):293–301
Schoenborn JR, Wilson CB (2007) Regulation of interferon-gamma during innate and adaptive immune responses. Adv Immunol 96:41–101
Schulze J, Bickert T, Beil FT, Zaiss MM, Albers J, Wintges K, Streichert T, Klaetschke K, Keller J, Hissnauer TN, Spiro AS, Gessner A, Schett G, Amling M, McKenzie AN, Horst AK, Schinke T (2011) Interleukin-33 is expressed in differentiated osteoblasts and blocks osteoclast formation from bone marrow precursor cells. J Bone Miner Res 26(4):704–717
Shashkova EV, Trivedi J, Cline-Smith AB, Ferris C, Buchwald ZS, Gibbs J, Novack D, Aurora R (2016) Osteoclast-primed Foxp3+ CD8 T cells induce T-bet, eomesodermin, and IFN-gamma to regulate bone resorption. J Immunol 197(3):726–735
Shaw AT, Maeda Y, Gravallese EM (2016) IL-17A deficiency promotes periosteal bone formation in a model of inflammatory arthritis. Arthritis Res Ther 18(1):104
Shen V, Kohler G, Jeffrey JJ, Peck WA (1988) Bone-resorbing agents promote and interferon-gamma inhibits bone cell collagenase production. J Bone Miner Res 3(6):657–666
Sherr CJ, Ashmun RA, Downing JR, Ohtsuka M, Quan SG, Golde DW, Roussel MF (1989) Inhibition of colony-stimulating factor-1 activity by monoclonal antibodies to the human CSF-1 receptor. Blood 73(7):1786–1793
Shiina-Ishimi Y, Abe E, Tanaka H, Suda T (1986) Synthesis of colony-stimulating factor (CSF) and differentiation inducing factor (D-factor) by osteoblastic cells, clone MC3T3-E1. Biochem Biophys Res Commun 134:400–406
Shin HH, Lee JE, Lee EA, Kwon BS, Choi HS (2006) Enhanced osteoclastogenesis in 4-1BB-deficient mice caused by reduced interleukin-10. J Bone Miner Res 21(12):1907–1912
Shinar DM, Sato M, Rodan GA (1990) The effect of hemopoietic growth factors on the generation of osteoclast-like cells in mouse bone marrow cultures. Endocrinology 126:1728–1735
Shioi A, Teitelbaum SL, Ross FP, Welgus HG, Suzuki H, Ohara J, Lacey DL (1991) Interleukin 4 inhibits murine osteoclast formation in vitro. J Cell Biochem 47:272–277
Shiratori T, Kyumoto-Nakamura Y, Kukita A, Uehara N, Zhang J, Koda K, Kamiya M, Badawy T, Tomoda E, Xu X, Yamaza T, Urano Y, Koyano K, Kukita T (2018) IL-1beta induces pathologically activated osteoclasts bearing extremely high levels of resorbing activity: a possible pathological subpopulation of osteoclasts, accompanied by suppressed expression of Kindlin-3 and Talin-1. J Immunol 200(1):218–228
Shukla P, Mansoori MN, Kakaji M, Shukla M, Gupta SK, Singh D (2017) Interleukin 27 (IL-27) alleviates bone loss in estrogen-deficient conditions by induction of early growth response-2 gene. J Biol Chem 292(11):4686–4699
Shukla P, Mansoori MN, Singh D (2018) Efficacy of anti-IL-23 monotherapy versus combination therapy with anti-IL-17 in estrogen deficiency induced bone loss conditions. Bone 110:84–95
Simonet WS, Lacey DL, Dunstan CR, Kelley M, Chang MS, Luthy R, Nguyen HQ, Wooden S, Bennett L, Boone T, Shimamoto G, DeRose M, Elliott R, Colombero A, Tan HL, Trail G, Sullivan J, Davy E, Bucay N, Renshaw-Gegg L, Hughes TM, Hill D, Pattison W, Campbell P,

Sander S, Van G, Tarpley J, Derby P, Lee R, Boyle WJ (1997) Osteoprotegerin: a novel secreted protein involved in the regulation of bone density. Cell 89(2):309–319
Sims NA, Johnson RW (2012) Leukemia inhibitory factor: a paracrine mediator of bone metabolism. Growth Factors 30(2):76–87
Sims JE, Gayle MA, Slack JL, Alderson MR, Bird TA, Giri JG, Colotta F, Re F, Mantovani A, Shanebeck K, Grabstein KH, Dower SK (1993) Interleukin 1 signaling occurs exclusively via the type I receptor. Proc Natl Acad Sci U S A 90:6155–6159
Sims NA, Jenkins BJ, Nakamura A, Quinn JM, Li R, Gillespie MT, Ernst M, Robb L, Martin TJ (2005) Interleukin-11 receptor signaling is required for normal bone remodeling. J Bone Miner Res 20(7):1093–1102
Simsa-Maziel S, Zaretsky J, Reich A, Koren Y, Shahar R, Monsonego-Ornan E (2013) IL-1RI participates in normal growth plate development and bone modeling. Am J Physiol Endocrinol Metab 305(1):E15–E21
Singh K, Piprode V, Mhaske ST, Barhanpurkar-Naik A, Wani MR (2018) IL-3 differentially regulates membrane and soluble RANKL in osteoblasts through metalloproteases and the JAK2/STAT5 pathway and improves the RANKL/OPG ratio in adult mice. J Immunol 200 (2):595–606
Smith DD, Gowen M, Mundy GR (1987) Effects of interferon-gamma and other cytokines on collagen synthesis in fetal rat bone cultures. Endocrinology 120:2494–2499
Song HY, Jeon ES, Kim JI, Jung JS, Kim JH (2007) Oncostatin M promotes osteogenesis and suppresses adipogenic differentiation of human adipose tissue-derived mesenchymal stem cells. J Cell Biochem 101(5):1238–1251
Sonomoto K, Yamaoka K, Oshita K, Fukuyo S, Zhang X, Nakano K, Okada Y, Tanaka Y (2012) Interleukin-1beta induces differentiation of human mesenchymal stem cells into osteoblasts via the Wnt-5a/receptor tyrosine kinase-like orphan receptor 2 pathway. Arthritis Rheum 64 (10):3355–3363
Soshi S, Takahashi HE, Tanizawa T, Endo N, Fujimoto R, Murota K (1996) Effect of recombinant human granulocyte colony-stimulating factor (rh G-CSF) on rat bone: inhibition of bone formation at the endosteal surface of vertebra and tibia. Calcif Tissue Int 58(5):337–340
Soysa NS, Alles N (2018) The role of IL-3 in bone. J Cell Biochem. https://doi.org/10.1002/jcb.27956
Stashenko P, Dewhirst FE, Peros WJ, Kent RL, Ago JM (1987) Synergistic interactions between interleukin 1, tumor necrosis factor, and lymphotoxin in bone resorption. J Immunol 138:1464–1468
Stein NC, Kreutzmann C, Zimmermann SP, Niebergall U, Hellmeyer L, Goettsch C, Schoppet M, Hofbauer LC (2008) Interleukin-4 and interleukin-13 stimulate the osteoclast inhibitor osteoprotegerin by human endothelial cells through the STAT6 pathway. J Bone Miner Res 23(5):750–758
Suda T, Takahashi N, Martin TJ (1995) Modulation of osteoclast differentiation: update 1995. Endcr Rev 13(1):66–80
Suda T, Takahashi N, Udagawa N, Jimi E, Gillespie MT, Martin TJ (1999) Modulation of osteoclast differentiation and function by the new members of the tumor necrosis factor receptor and ligand families. Endocr Rev 20(3):345–357
Sunyer T, Rothe L, Jiang XS, Osdoby P, Collin-Osdoby P (1996) Proinflammatory agents, IL-8 and IL-10, upregulate inducible nitric oxide synthase expression and nitric oxide production in avian osteoclast-like cells. J Cell Biochem 60:469–483
Svandova E, Vesela B, Lesot H, Sadoine J, Poliard A, Matalova E (2018) FasL modulates expression of Mmp2 in osteoblasts. Front Physiol 9:1314
Takahashi N, Mundy GR, Roodman GD (1986) Recombinant human interferon-gamma inhibits formation of human osteoclast-like cells. J Immunol 137:3544–3549
Takahashi N, Akatsu T, Udagawa N, Sasaki T, Yamaguchi A, Moseley JM, Martin TJ, Suda T (1988) Osteoblastic cells are involved in osteoclast formation. Endocrinology 123:2600–2602

Takahashi N, Udagawa N, Akatsu T, Tanaka H, Shionome M, Suda T (1991) Role of colony-stimulating factors in osteoclast development. J Bone Miner Res 6(9):977–985

Takamatsu Y, Simmons PJ, Moore RJ, Morris HA, To LB, Levesque JP (1998) Osteoclast-mediated bone resorption is stimulated during short-term administration of granulocyte colony-stimulating factor but is not responsible for hematopoietic progenitor cell mobilization. Blood 92(9):3465–3473

Takayanagi H (2005) Inflammatory bone destruction and osteoimmunology. J Periodontal Res 40 (4):287–293

Takayanagi H, Ogasawara K, Hida S, Chiba T, Murata S, Sato K, Takaoka A, Yokochi T, Oda H, Tanaka K, Nakamura K, Taniguchi T (2000) T-cell-mediated regulation of osteoclastogenesis by signalling cross-talk between RANKL and IFN-gamma. Nature 408(6812):600–605

Takayanagi H, Kim S, Matsuo K, Suzuki H, Suzuki T, Sato K, Yokochi T, Oda H, Nakamura K, Ida N, Wagner EF, Taniguchi T (2002) RANKL maintains bone homeostasis through c-Fos-dependent induction of interferon-beta. Nature 416(6882):744–749

Takeda H, Kikuchi T, Soboku K, Okabe I, Mizutani H, Mitani A, Ishihara Y, Noguchi T (2014) Effect of IL-15 and natural killer cells on osteoclasts and osteoblasts in a mouse coculture. Inflammation 37(3):657–669

Takita C, Fujikawa Y, Itonaga I, Taira H, Kawashima M, Torisu T (2005) Infliximab acts directly on human osteoclast precursors and enhances osteoclast formation induced by receptor activator of nuclear factor kappaB ligand in vitro. Mod Rheumatol 15(2):97–103

Tamura T, Udagawa N, Takahashi N, Miyaura C, Tanaka S, Yamada Y, Koishihara Y, Ohsugi Y, Kumaki K, Taga T, Kishimoto T, Suda T (1993) Soluble interleukin-6 receptor triggers osteoclast formation by interleukin 6. Proc Natl Acad Sci U S A 90:11924–11928

Tanabe N, Maeno M, Suzuki N, Fujisaki K, Tanaka H, Ogiso B, Ito K (2005) IL-1 alpha stimulates the formation of osteoclast-like cells by increasing M-CSF and PGE2 production and decreasing OPG production by osteoblasts. Life Sci 77(6):615–626

Tanaka K, Hashizume M, Mihara M, Yoshida H, Suzuki M, Matsumoto Y (2014) Anti-interleukin-6 receptor antibody prevents systemic bone mass loss via reducing the number of osteoclast precursors in bone marrow in a collagen-induced arthritis model. Clin Exp Immunol 175 (2):172–180

Tang R, Yi J, Yang J, Chen Y, Luo W, Dong S, Fei J (2018) Interleukin-37 inhibits osteoclastogenesis and alleviates inflammatory bone destruction. J Cell Physiol 234 (5):7645–7658

Tashjian AH Jr, Voelkel EF, Lazzaro M, Goad D, Bosma T, Levine L (1987) Tumor necrosis factor-alpha (Cachectin) stimulates bone resorption in mouse calvaria via a prostaglandin-mediated mechanism. Endocrinology 120:2029–2036

Teitelbaum SL (2000) Bone resorption by osteoclasts. Science 289(5484):1504–1508

Teramachi J, Zhou H, Subler MA, Kitagawa Y, Galson DL, Dempster DW, Windle JJ, Kurihara N, Roodman GD (2014) Increased IL-6 expression in osteoclasts is necessary but not sufficient for the development of Paget's disease of bone. J Bone Miner Res 29(6):1456–1465

Terashima A, Okamoto K, Nakashima T, Akira S, Ikuta K, Takayanagi H (2016) Sepsis-induced osteoblast ablation causes immunodeficiency. Immunity 44(6):1434–1443

Tinhofer I, Biedermann R, Krismer M, Crazzolara R, Greil R (2006) A role of TRAIL in killing osteoblasts by myeloma cells. FASEB J 20(6):759–761

Titanji K, Vunnava A, Foster A, Sheth AN, Lennox JL, Knezevic A, Shenvi N, Easley KA, Ofotokun I, Weitzmann MN (2018) T-cell receptor activator of nuclear factor-kappaB ligand/osteoprotegerin imbalance is associated with HIV-induced bone loss in patients with higher CD4+ T-cell counts. AIDS (London, England) 32(7):885–894

Toraldo G, Roggia C, Qian WP, Pacifici R, Weitzmann MN (2003) IL-7 induces bone loss in vivo by induction of receptor activator of nuclear factor kappa B ligand and tumor necrosis factor alpha from T cells. Proc Natl Acad Sci U S A 100(1):125–130

Tsuboi M, Kawakami A, Nakashima T, Matsuoka N, Urayama S, Kawabe Y, Fujiyama K, Kiriyama T, Aoyagi T, Maeda K, Eguchi K (1999) Tumor necrosis factor-alpha and interleukin-1beta increase the Fas-mediated apoptosis of human osteoblasts. J Lab Clin Med 134(3):222–231

Tsuda E, Goto M, Mochizuki S, Yano K, Kobayashi F, Morinaga T, Higashio K (1997) Isolation of a novel cytokine from human fibroblasts that specifically inhibits osteoclastogenesis. Biochem Biophys Res Commun 234(1):137–142

Tyagi AM, Srivastava K, Mansoori MN, Trivedi R, Chattopadhyay N, Singh D (2012) Estrogen deficiency induces the differentiation of IL-17 secreting Th17 cells: a new candidate in the pathogenesis of osteoporosis. PLoS One 7(9):e44552

Tyagi AM, Mansoori MN, Srivastava K, Khan MP, Kureel J, Dixit M, Shukla P, Trivedi R, Chattopadhyay N, Singh D (2014) Enhanced immunoprotective effects by anti-IL-17 antibody translates to improved skeletal parameters under estrogen deficiency compared with anti-RANKL and anti-TNF-alpha antibodies. J Bone Miner Res 29(9):1981–1992

Udagawa N, Horwood NJ, Elliott J, Mackay A, Owens J, Okamura H, Kurimoto M, Chambers TJ, Martin TJ, Gillespie MT (1997) Interleukin-18 (interferon-gamma-inducing factor) is produced by osteoblasts and acts via granulocyte/macrophage colony-stimulating factor and not via interferon-gamma to inhibit osteoclast formation. J Exp Med 185(6):1005–1012

Uluckan O, Jimenez M, Karbach S, Jeschke A, Grana O, Keller J, Busse B, Croxford AL, Finzel S, Koenders M, van den Berg W, Schinke T, Amling M, Waisman A, Schett G, Wagner EF (2016) Chronic skin inflammation leads to bone loss by IL-17-mediated inhibition of Wnt signaling in osteoblasts. Sci Transl Med 8(330):330ra337

Van Beek E, Van der Wee-Pals L, van de Ruit M, Nijweide P, Papapoulos S, Lowik C (1993) Leukemia inhibitory factor inhibits osteoclastic resorption, growth, mineralization, and alkaline phosphatase activity in fetal mouse metacarpal bones in culture. J Bone Miner Res 8(2):191–198

Van Bezooijen RL, Farih-Sips HC, Papapoulos SE, Lowik CW (1999) Interleukin-17: a new bone acting cytokine in vitro. J Bone Miner Res 14(9):1513–1521

Van Vlasselaer P, Borremans B, Van Der Heuvel R, Van Gorp U, De Waal MR (1993) Interleukin-10 inhibits the osteogenic activity of mouse bone marrow. Blood 82(8):2361–2370

Vargas SJ, Naprta A, Glaccum M, Lee SK, Kalinowski J, Lorenzo JA (1996) Interleukin-6 expression and histomorphometry of bones from mice deficient for receptors for interleukin-1 or tumor necrosis factor. J Bone Miner Res 11:1736–1740

Veeriah V, Zanniti A, Paone R, Chatterjee S, Rucci N, Teti A, Capulli M (2016) Interleukin-1beta, lipocalin 2 and nitric oxide synthase 2 are mechano-responsive mediators of mouse and human endothelial cell-osteoblast crosstalk. Sci Rep 6:29880

Vermeire K, Heremans H, Vandeputte M, Huang S, Billiau A, Matthys P (1997) Accelerated collagen-induced arthritis in IFN-gamma receptor-deficient mice. J Immunol 158 (11):5507–5513

Vignery A, Niven-Fairchild T, Shepard MH (1990) Recombinant murine interferon-gamma inhibits the fusion of mouse alveolar macrophages in vitro but stimulates the formation of osteoclast-like cells on implanted syngenic bone particles in mice in vivo. J Bone Miner Res 5:637–644

Villarreal DO, Weiner DB (2014) Interleukin 33: a switch-hitting cytokine. Curr Opin Immunol 28:102–106

Walker EC, McGregor NE, Poulton IJ, Solano M, Pompolo S, Fernandes TJ, Constable MJ, Nicholson GC, Zhang JG, Nicola NA, Gillespie MT, Martin TJ, Sims NA (2010) Oncostatin M promotes bone formation independently of resorption when signaling through leukemia inhibitory factor receptor in mice. J Clin Invest 120(2):582–592

Walker EC, Poulton IJ, McGregor NE, Ho PW, Allan EH, Quach JM, Martin TJ, Sims NA (2012) Sustained RANKL response to parathyroid hormone in oncostatin M receptor-deficient osteoblasts converts anabolic treatment to a catabolic effect in vivo. J Bone Miner Res 27 (4):902–912

Walsh MC, Kim N, Kadono Y, Rho J, Lee SY, Lorenzo J, Choi Y (2006) Osteoimmunology: interplay between the immune system and bone metabolism. Annu Rev Immunol 24:33–63

Wang FM, Sarmasik A, Hiruma Y, Sun Q, Sammut B, Windle JJ, Roodman GD, Galson DL (2013) Measles virus nucleocapsid protein, a key contributor to Paget's disease, increases IL-6 expression via down-regulation of FoxO3/Sirt1 signaling. Bone 53(1):269–276

Wang C, Tian L, Zhang K, Chen Y, Chen X, Xie Y, Zhao Q, Yu X (2016) Interleukin-6 gene knockout antagonizes high-fat-induced trabecular bone loss. J Mol Endocrinol 57(3):161–170

Wang Z, Jia Y, Du F, Chen M, Dong X, Chen Y, Huang W (2017) IL-17A inhibits osteogenic differentiation of bone mesenchymal stem cells via Wnt signaling pathway. Med Sci Monit 23:4095–4101

Wang LM, Zhao N, Zhang J, Sun QF, Yang CZ, Yang PS (2018a) Tumor necrosis factor-alpha inhibits osteogenic differentiation of pre-osteoblasts by downregulation of EphB4 signaling via activated nuclear factor-kappaB signaling pathway. J Periodontal Res 53(1):66–72

Wang Y, Xiao M, Tao C, Chen J, Wang Z, Yang J, Chen Z, Zou Z, Liu A, Cai D, Jiang Y, Ding C, Li M, Bai X (2018b) Inactivation of mTORC1 signaling in osterix-expressing cells impairs B-cell differentiation. J Bone Miner Res 33(4):732–742

Wang Z, Tan J, Lei L, Sun W, Wu Y, Ding P, Chen L (2018c) The positive effects of secreting cytokines IL-17 and IFN-gamma on the early-stage differentiation and negative effects on the calcification of primary osteoblasts in vitro. Int Immunopharmacol 57:1–10

Ware CB, Horowitz MC, Renshaw BR, Hunt JS, Liggitt D, Koblar SA, Gliniak BC, McKenna HJ, Papayannopoulou T, Thoma B, Cheng L, Donovan PJ, Peschon JJ, Bartlett PF, Willis CR, Wright BD, Carpenter MK, Davison BL, Gearing DP (1995) Targeted disruption of the low-affinity leukemia inhibitory factor receptor gene causes placental, skeletal, neural and metabolic defects and results in perinatal death. Development 121:1283–1299

Weaver CT, Hatton RD, Mangan PR, Harrington LE (2007) IL-17 family cytokines and the expanding diversity of effector T cell lineages. Annu Rev Immunol 25:821–852

Wehmeyer C, Frank S, Beckmann D, Bottcher M, Cromme C, Konig U, Fennen M, Held A, Paruzel P, Hartmann C, Stratis A, Korb-Pap A, Kamradt T, Kramer I, van den Berg W, Kneissel M, Pap T, Dankbar B (2016) Sclerostin inhibition promotes TNF-dependent inflammatory joint destruction. Sci Transl Med 8(330):330ra335

Wei S, Wang MW, Teitelbaum SL, Ross FP (2001) Interleukin-4 reversibly inhibits osteoclastogenesis via inhibition of NF-{kappa}B and MAP kinase signaling. J Biol Chem 277(8):6622–6630

Wei S, Kitaura H, Zhou P, Ross FP, Teitelbaum SL (2005) IL-1 mediates TNF-induced osteoclastogenesis. J Clin Invest 115(2):282–290

Wei S, Nandi S, Chitu V, Yeung YG, Yu W, Huang M, Williams LT, Lin H, Stanley ER (2010) Functional overlap but differential expression of CSF-1 and IL-34 in their CSF-1 receptor-mediated regulation of myeloid cells. J Leukoc Biol 88(3):495–505

Weir EC, Horowitz MC, Baron R, Centrella M, Kacinski BM, Insogna KL (1993) Macrophage colony-stimulating factor release and receptor expression in bone cells. J Bone Miner Res 8 (12):1507–1518

Weiser WY, Temple PA, Witek-Giannotti JS, Remold HG, Clark SC, David JR (1989) Molecular cloning of a cDNA encoding a human macrophage migration inhibitory factor. Proc Natl Acad Sci U S A 86(19):7522–7526

Weitzmann MN, Cenci S, Rifas L, Brown C, Pacifici R (2000) Interleukin-7 stimulates osteoclast formation by up-regulating the T-cell production of soluble osteoclastogenic cytokines. Blood 96(5):1873–1878

Weitzmann MN, Roggia C, Toraldo G, Weitzmann L, Pacifici R (2002) Increased production of IL-7 uncouples bone formation from bone resorption during estrogen deficiency. J Clin Invest 110(11):1643–1650

Wesche H, Korherr C, Kracht M, Falk W, Resch K, Martin MU (1997) The interleukin-1 receptor accessory protein (IL-1RAcP) is essential for IL-1-induced activation of interleukin-1 receptor-associated kinase (IRAK) and stress-activated protein kinases (SAP kinases). J Biol Chem 272 (12):7727–7731

Wesche DE, Lomas-Neira JL, Perl M, Chung CS, Ayala A (2005) Leukocyte apoptosis and its significance in sepsis and shock. J Leukoc Biol 78(2):325–337

Whyte MP, Hughes AE (2002) Expansile skeletal hyperphosphatasia is caused by a 15-base pair tandem duplication in TNFRSF11A encoding RANK and is allelic to familial expansile osteolysis. J Bone Miner Res 17(1):26–29

Whyte MP, Mumm S (2004) Heritable disorders of the RANKL/OPG/RANK signaling pathway. J Musculoskelet Neuronal Interact 4(3):254–267

Whyte MP, Obrecht SE, Finnegan PM, Jones JL, Podgornik MN, McAlister WH, Mumm S (2002a) Osteoprotegerin deficiency and juvenile Paget's disease. N Engl J Med 347(3):175–184
Whyte MP, Reinus WR, Podgornik MN, Mills BG (2002b) Familial expansile osteolysis (excessive RANK effect) in a 5-generation American kindred. Medicine (Baltimore) 81(2):101–121
Whyte MP, Tau C, McAlister WH, Zhang X, Novack DV, Preliasco V, Santini-Araujo E, Mumm S (2014) Juvenile Paget's disease with heterozygous duplication in TNFRSF11A encoding RANK. Bone 68:153–161
Wiktor Jedrzejczak W, Bartocci A, Ferrante AW Jr, Ahmed Ansari A, Sell KW, Pollard JW, Stanley ER (1990) Total absence of colony-stimulating factor 1 in the macrophage-deficient osteopetrotic (op/op) mouse. Proc Natl Acad Sci U S A 87:4828–4832
Wu X, McKenna MA, Feng X, Nagy TR, McDonald JM (2003) Osteoclast apoptosis: the role of Fas in vivo and in vitro. Endocrinology 144(12):5545–5555
Wu X, Pan G, McKenna MA, Zayzafoon M, Xiong WC, McDonald JM (2005) RANKL regulates Fas expression and Fas-mediated apoptosis in osteoclasts. J Bone Miner Res 20(1):107–116
Wu JY, Purton LE, Rodda SJ, Chen M, Weinstein LS, McMahon AP, Scadden DT, Kronenberg HM (2008) Osteoblastic regulation of B lymphopoiesis is mediated by Gs{alpha}-dependent signaling pathways. Proc Natl Acad Sci U S A 105(44):16976–16981
Wu L, Guo Q, Yang J, Ni B (2017) Tumor necrosis factor alpha promotes osteoclast formation via PI3K/Akt pathway-mediated Blimp1 expression upregulation. J Cell Biochem 118 (6):1308–1315
Xie Z, Tang S, Ye G, Wang P, Li J, Liu W, Li M, Wang S, Wu X, Cen S, Zheng G, Ma M, Wu Y, Shen H (2018) Interleukin-6/interleukin-6 receptor complex promotes osteogenic differentiation of bone marrow-derived mesenchymal stem cells. Stem Cell Res Ther 9(1):13
Xiong J, Onal M, Jilka RL, Weinstein RS, Manolagas SC, O'Brien CA (2011) Matrix-embedded cells control osteoclast formation. Nat Med 17(10):1235–1241
Xiong J, Piemontese M, Thostenson JD, Weinstein RS, Manolagas SC, O'Brien CA (2014) Osteocyte-derived RANKL is a critical mediator of the increased bone resorption caused by dietary calcium deficiency. Bone 66:146–154
Xiong J, Cawley K, Piemontese M, Fujiwara Y, Zhao H, Goellner JJ, O'Brien CA (2018) Soluble RANKL contributes to osteoclast formation in adult mice but not ovariectomy-induced bone loss. Nat Commun 9(1):2909
Xu LX, Kukita T, Kukita A, Otsuka T, Niho Y, Iijima T (1995) Interleukin-10 selectively inhibits osteoclastogenesis by inhibiting differentiation of osteoclast progenitors into preosteoclast-like cells in rat bone marrow culture system. J Cell Physiol 165:624–629
Yago T, Nanke Y, Ichikawa N, Kobashigawa T, Mogi M, Kamatani N, Kotake S (2009) IL-17 induces osteoclastogenesis from human monocytes alone in the absence of osteoblasts, which is potently inhibited by anti-TNF-alpha antibody: a novel mechanism of osteoclastogenesis by IL-17. J Cell Biochem 108(4):947–955
Yago T, Nanke Y, Kawamoto M, Kobashigawa T, Yamanaka H, Kotake S (2018) IL-35 inhibits human osteoclastogenesis from monocytes induced by receptor-activator of NF-kappaB ligand. Cent Eur J Immunol 43(2):148–154. https://doi.org/10.5114/ceji.2018.77384
Yamada N, Niwa S, Tsujimura T, Iwasaki T, Sugihara A, Futani H, Hayashi S, Okamura H, Akedo H, Terada N (2002) Interleukin-18 and interleukin-12 synergistically inhibit osteoclastic bone-resorbing activity. Bone 30(6):901–908
Yamada A, Takami M, Kawawa T, Yasuhara R, Zhao B, Mochizuki A, Miyamoto Y, Eto T, Yasuda H, Nakamichi Y, Kim N, Katagiri T, Suda T, Kamijo R (2007) Interleukin-4 inhibition of osteoclast differentiation is stronger than that of interleukin-13 and they are equivalent for induction of osteoprotegerin production from osteoblasts. Immunology 120(4):573–579
Yamaguchi T, Movila A, Kataoka S, Wisitrasameewong W, Ruiz Torruella M, Murakoshi M, Murakami S, Kawai T (2016) Proinflammatory M1 macrophages inhibit RANKL-induced osteoclastogenesis. Infect Immun 84(10):2802–2812
Yamamoto T, Ozono K, Kasayama S, Yoh K, Hiroshima K, Takagi M, Matsumoto S, Michigami T, Yamaoka K, Kishimoto T, Okada S (1996) Increased IL-6-production by cells isolated from the

fibrous bone dysplasia tissues in patients with McCune-Albright syndrome. J Clin Invest 98 (1):30–35
Yamamura M, Kawashima M, Taniai M, Yamauchi H, Tanimoto T, Kurimoto M, Morita Y, Ohmoto Y, Makino H (2001) Interferon-gamma-inducing activity of interleukin-18 in the joint with rheumatoid arthritis. Arthritis Rheum 44(2):275–285
Yang L, Zhang J, Tao J, Lu T (2015) Elevated serum levels of interleukin-37 are associated with inflammatory cytokines and disease activity in rheumatoid arthritis. APMIS 123(12):1025–1031
Yao GQ, Sun BH, Hammond EE, Spencer EN, Horowitz MC, Insogna KL, Weir EC (1998) The cell-surface form of colony-stimulating factor-1 is regulated by osteotropic agents and supports formation of multinucleated osteoclast-like cells. J Biol Chem 273(7):4119–4128
Yao Z, Li P, Zhang Q, Schwarz EM, Keng P, Arbini A, Boyce BF, Xing L (2006) TNF increases circulating osteoclast precursor numbers by promoting their proliferation and differentiation in the bone marrow through up-regulation of c-fms expression. J Biol Chem 281 (17):11846–11855
Yao Z, Xing L, Qin C, Schwarz EM, Boyce BF (2008) Osteoclast precursor interaction with bone matrix induces osteoclast formation directly by an interleukin-1-mediated autocrine mechanism. J Biol Chem 283(15):9917–9924
Yao Z, Xing L, Boyce BF (2009) NF-kappaB p100 limits TNF-induced bone resorption in mice by a TRAF3-dependent mechanism. J Clin Invest 21(38716)
Yao GQ, Troiano N, Simpson CA, Insogna KL (2017a) Selective deletion of the soluble colony-stimulating factor 1 isoform in vivo prevents estrogen-deficiency bone loss in mice. Bone Res 5:17022
Yao Z, Lei W, Duan R, Li Y, Luo L, Boyce BF (2017b) RANKL cytokine enhances TNF-induced osteoclastogenesis independently of TNF receptor associated factor (TRAF) 6 by degrading TRAF3 in osteoclast precursors. J Biol Chem 292(24):10169–10179
Yasuda H, Shima N, Nakagawa N, Yamaguchi K, Kinosaki M, Mochizuki S, Tomoyasu A, Yano K, Goto M, Murakami A, Tsuda E, Morinaga T, Higashio K, Udagawa N, Takahashi N, Suda T (1998) Osteoclast differentiation factor is a ligand for osteoprotegerin osteoclastogenesis-inhibitory factor and is identical to TRANCE/RANKL. Proc Natl Acad Sci U S A 95(7):3597–3602
Yen ML, Hsu PN, Liao HJ, Lee BH, Tsai HF (2012) TRAF-6 dependent signaling pathway is essential for TNF-related apoptosis-inducing ligand (TRAIL) induces osteoclast differentiation. PLoS One 7(6):e38048
Yogesha SD, Khapli SM, Wani MR (2005) Interleukin-3 and granulocyte-macrophage colony-stimulating factor inhibits tumor necrosis factor (TNF)-alpha-induced osteoclast differentiation by down-regulation of expression of TNF receptors 1 and 2. J Biol Chem 280(12):11759–11769
Yokoyama M, Ukai T, Ayon Haro ER, Kishimoto T, Yoshinaga Y, Hara Y (2011) Membrane-bound CD40 ligand on T cells from mice injected with lipopolysaccharide accelerates lipopolysaccharide-induced osteoclastogenesis. J Periodontal Res 46(4):464–474
Yoshida H, Hayashi S, Kunisada T, Ogawa M, Nishikawa S, Okamura H, Sudo T, Shultz LD, Nisikawa S (1990) The murine mutation osteopetrosis is in the coding region of the macrophage colony stimulating factor gene. Nature 345:442–444
Yoshimatsu M, Kitaura H, Fujimura Y, Eguchi T, Kohara H, Morita Y, Yoshida N (2009) IL-12 inhibits TNF-alpha induced osteoclastogenesis via a T cell-independent mechanism in vivo. Bone 45(5):1010–1016
Yoshitake F, Itoh S, Narita H, Ishihara K, Ebisu S (2008) IL-6 directly inhibits osteoclast differentiation by suppressing RANK signaling pathways. J Biol Chem 283(17):11535–11540
Yun TJ, Chaudhary PM, Shu GL, Frazer JK, Ewings MK, Schwartz SM, Pascual V, Hood LE, Clark EA (1998) OPG/FDCR-1, a TNF receptor family member, is expressed in lymphoid cells and is up-regulated by ligating CD40. J Immunol 161(11):6113–6121
Zaiss MM, Kurowska-Stolarska M, Bohm C, Gary R, Scholtysek C, Stolarski B, Reilly J, Kerr S, Millar NL, Kamradt T, McInnes IB, Fallon PG, David JP, Liew FY, Schett G (2011) IL-33 shifts

the balance from osteoclast to alternatively activated macrophage differentiation and protects from TNF-{alpha}-mediated bone loss. J Immunol 186(11):6097–6105
Zauli G, Rimondi E, Stea S, Baruffaldi F, Stebel M, Zerbinati C, Corallini F, Secchiero P (2007) TRAIL inhibits osteoclastic differentiation by counteracting RANKL-dependent p27(Kip1) accumulation in pre-osteoclast precursors. J Cell Physiol 214(1):117–125
Zauli G, Rimondi E, Celeghini C, Milani D, Secchiero P (2010) Dexamethasone counteracts the anti-osteoclastic, but not the anti-leukemic, activity of TNF-related apoptosis inducing ligand (TRAIL). J Cell Physiol 222(2):357–364
Zhang YH, Heulsmann A, Tondravi MM, Mukherjee A, Abu-Amer Y (2001) Tumor necrosis factor-alpha (TNF) stimulates RANKL-induced osteoclastogenesis via coupling of TNF type 1 receptor and RANK signaling pathways. J Biol Chem 276(1):563–568
Zhang J, Zhao H, Chen J, Xia B, Jin Y, Wei W, Shen J, Huang Y (2012) Interferon-beta-induced miR-155 inhibits osteoclast differentiation by targeting SOCS1 and MITF. FEBS Lett 586 (19):3255–3262
Zhang Q, Chen B, Yan F, Guo J, Zhu X, Ma S, Yang W (2014) Interleukin-10 inhibits bone resorption: a potential therapeutic strategy in periodontitis and other bone loss diseases. Biomed Res Int 2014:284836
Zhang S, Wang X, Li G, Chong Y, Zhang J, Guo X, Li B, Bi Z (2017) Osteoclast regulation of osteoblasts via RANKRANKL reverse signal transduction in vitro. Mol Med Rep 16 (4):3994–4000
Zhao JJ, Wu ZF, Yu YH, Wang L, Cheng L (2018) Effects of interleukin-7/interleukin-7 receptor on RANKL-mediated osteoclast differentiation and ovariectomy-induced bone loss by regulating c-Fos/c-Jun pathway. J Cell Physiol 233(9):7182–7194
Zheng W, Guan J (2018) Oncostatin M promotes the osteogenic differentiation of mouse MC3T3E1osteoblasts through the regulation of monocyte chemotactic protein1. Mol Med Rep 18(3):2523–2530
Zhong X, Wang H, Huang S (2014) Endothelin-1 induces interleukin-18 expression in human osteoblasts. Arch Oral Biol 59(3):289–296
Zupan J, Pristovsek N, Mencej-Bedrac S, Komadina R, Prezelj J, Marc J (2012) Interleukin-1alpha gene variants influence bone mineral density and the risk of osteoporotic hip fractures in elderly Slovenian people. Clin Chem Lab Med 50(8):1379–1385
Zwerina J, Redlich K, Polzer K, Joosten L, Kronke G, Distler J, Hess A, Pundt N, Pap T, Hoffmann O, Gasser J, Scheinecker C, Smolen JS, van den Berg W, Schett G (2007) TNF-induced structural joint damage is mediated by IL-1. Proc Natl Acad Sci U S A 104 (28):11742–11747

Chemokines and Bone

Annette Gilchrist

Contents

Abstract

Chemokines are a family of small proteins, subdivided by their conserved cysteine residues and common structural features. Chemokines interact with their cognate G-protein-coupled receptors to elicit downstream signals that result in cell migration, proliferation, and survival. This review presents evidence for how the various CXC and CC subfamily chemokines influence bone hemostasis by acting on osteoclasts, osteoblasts, and progenitor cells. Also discussed are the ways in which chemokines contribute to bone loss as a result of inflammatory diseases such as rheumatoid arthritis, HIV infection, and periodontal infection. Both positive and negative effects of chemokines on bone formation and bone loss are presented. In addition, the role of chemokines in altering the bone microenvironment through effects on angiogenesis and tumor invasion is discussed. Very few therapeutic agents that influence bone formation by targeting chemokines or chemokine receptors are available, although a few are currently being evaluated.

Keywords

Chemokine · Chemokine receptor · G-protein-coupled receptor · Hematopoietic stem cell · Mesenchymal stem cell · Osteoblast · Osteoclast · Osteoimmunology

A. Gilchrist (✉)
Department of Pharmaceutical Sciences, Midwestern University, Downers Grove, IL, USA
e-mail: agilchrist@midwestern.edu

P. H. Stern (ed.), *Bone Regulators and Osteoporosis Therapy*,
Handbook of Experimental Pharmacology 262, https://doi.org/10.1007/164_2020_349

1 Overview of Bone Remodeling

The bone is a metabolically active tissue, and bone remodeling is a dynamic process necessary for skeletal growth and normal bone structure maintenance throughout life. Bone remodeling involves the removal of old or damaged bone by osteoclasts (bone resorption) and subsequent replacement with new bone formed by osteoblasts (bone formation). The balance between bone resorption and bone formation can be disturbed by a variety of factors, including menopause-associated hormonal changes, age-related factors, changes in physical activity, drugs, and secondary diseases.

There are two major types of bone: (a) cortical, which provides a mechanical function and is protective, and (b) trabecular, which provides strength and the majority of the metabolic function. Trabecular bone is the major site of bone remodeling. Bone remodeling is carried out by a functional and anatomic structure known as the basic multicellular unit (BMU). In addition to osteoclasts and osteoblasts, bone remodeling requires the coordinated action of bone lining cells and osteocytes (Hauge et al. 2001). In a quiescent state, the bone surface is covered by a monolayer of bone lining cells (Miller et al. 1989). Bone lining cells are a type of early osteoblast often referred to as spindle-shaped N-cadherin^{+}CD45^{-} osteoblastic cells (SNO) (Zhang et al. 2003). Osteocytes, which play a pivotal role in the initiation of bone remodeling, are embedded within the bone (Bonewald 2007). Mechanical loading is known to play a role in maintaining osteocyte viability. Osteoblasts are derived from mesenchymal stem cells (MSCs) that give rise to osteoprogenitors, which differentiate into preosteoblasts and then mature osteoblasts (Ducy et al. 2000).

MSCs have the capacity of self-renewal. The spindle-shaped, adherent cells are non-hematopoietic-derived stem cells that can differentiate into various mesodermal cell types, including osteoblasts, chondrocytes, and adipocytes (Fig. 1). A key step in bone healing is the localization of MSCs to the site of injury. In the early 1980s, Rodan and Martin (1981) postulated that osteoblasts regulate osteoclast formation and that factors expressed by osteoblasts within the bone are produced in response

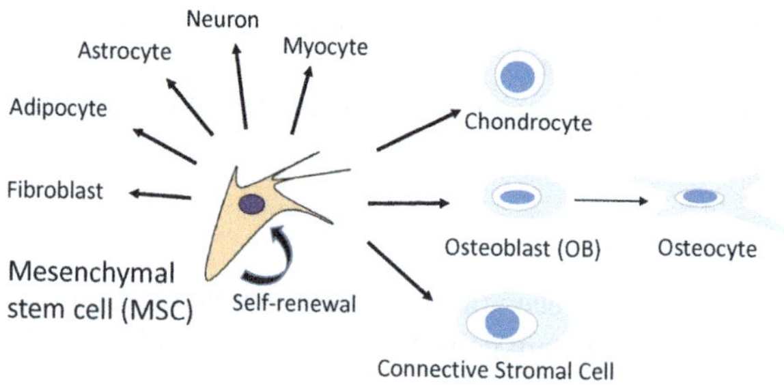

Fig. 1 Differentiation potential of mesenchymal stem cells. Schematic diagram of mesenchymal stem cell (MSC) with multi-lineage potential including chondrocytes, myocytes, neurons, astrocytes, adipocytes, fibroblasts, stromal cells, osteoblasts, and osteocytes

to known stimulators of bone resorption, such as parathyroid hormone (PTH). Another example is the chemokine CXCL12 and its receptor, CXCR4 which have been shown to be important to the recruitment of MSCs during skeletal repair (Kitaori et al. 2009).

While osteocytes and bone lining cells both belong to the osteoblast lineage (Matic et al. 2016), osteoclasts differentiate from mononuclear cells of the monocyte/macrophage lineage derived from hematopoietic stem cells (HSCs), a subset of bone marrow cells that are capable of self-renewal (Boyle et al. 2003). Within the bone marrow (BM) is the "endosteal niche," where HSCs are localized. The endosteal niche is localized in the internal bone shell surface and is close to both the endocortical and trabecular surfaces. SNO-HSC interactions are critical for the regulation of HSCs' fate in the endosteal niche. HSCs give rise to all types of mature blood cells (multipotential) through a process finely controlled by numerous signals emerging from the BM where HSCs reside. Furthermore, as HSCs have the capacity to differentiate into osteoclasts, increased myelopoiesis has been directly linked with increased osteoclastogenesis and bone loss in inflammatory conditions (Charles et al. 2012).

The standard hierarchical model for osteoclast differentiation (Fierro et al. 2017) is that HSC can commit into a multipotent progenitor which gives rise to lymphoid and myeloid lineages but shows very limited if any self-renewal potential. These cells can commit into a common myeloid progenitor cell which can then differentiate in the bone marrow into monoblasts. Monoblasts enter the circulation and can differentiate into immature promonocytes and then mature monocytes. Mature CD11b+/CD14+ RANK+ monocytes can further be differentiated to osteoclasts in response to receptor activator for nuclear factor-kappa-B ligand (RANKL). Tissue resident CD14+ macrophages can also differentiate to osteoclasts. However, various CD11b– and/or CD14– and/or CD11c+ populations have also been shown to differentiate to osteoclasts (Alnaeeli et al. 2006; Charles et al. 2012).

A key development in the late 1980s related to harvesting of large numbers of osteoclast precursors (OCPs) from BM or spleen cells, which could then be cultured in the absence of osteoblast/stromal cells. This advancement rested on the recognition that M-CSF was required for progenitor cells to differentiate into osteoclasts but that M-CSF on its own was unable to complete this process. The ligand for receptor activator of nuclear factor kappa-B (RANK) was identified (Lacey et al. 1998; Yasuda et al. 1998) and turned out to be a member of the TNF ligand family identified in the preceding year (Anderson et al. 1997). Researchers at Amgen found that mice over-expressing osteoprotegerin (OPG) developed marked osteopetrosis as they did not have any osteoclasts present in their bones (Simonet et al. 1997). Subsequent studies showed that B lymphocytes secrete OPG, a potent anti-osteoclastogenic factor that preserves bone mass. Our understanding of the molecular mechanisms that regulate osteoclast formation and activation has advanced rapidly since the discovery of the RANKL/RANK/OPG signaling system. However, this receptor-mediated signaling pathway cannot induce osteoclast formation on its own. Thus, like M-CSF, it is necessary but not sufficient for osteoclastogenesis.

A common problem in bone remodeling occurs when there is an increased number and/or life span of osteoclasts, along with a decrease in the formation and/or life span of osteoblasts. In this circumstance, the rate of resorption exceeds the rate of mineral deposition resulting in an imbalance with subsequent bone loss and deterioration of bone architecture. A decline in the osteoblasts' ability to refill resorption cavities made by osteoclasts can lead to osteoporosis and a reduction of the thickness of the bone packets and thinning of the trabeculae with a subsequent increase for the risk of bone fractures. As the skeleton ages, the quantity of MSCs in the bone marrow decreases. Chen (2004) showed that older mice fail to produce the same number of osteoprogenitor cells when compared to younger mice. Similarly, Shigeno and Ashton (1995) showed a significant decrease in both the number of precursor cells and degree of proliferation starting in the second and third decades of life in humans. Stolzing et al. (2008) also found a decrease in the number and proliferative capacity of MSCs harvested in older humans. Thus, the data indicates that aging decreases the availability and growth potential of MSCs for bone formation.

Bone fractures heal by three partially overlapping phases: the inflammation phase with the initial hematoma and subsequent infiltration of inflammatory cells; the repair phase comprising soft callus formation and intramembranous and endochondral ossification; and the bone remodeling phase, where the initially woven bone is converted to a lamellar bone until the original bone shape is restored (Claes et al. 2012). Both resident and infiltrating cells contribute to all three phases of fracture healing. Formation of a fracture hematoma governs the bone healing process, and the generation of a blood clot is indispensable (Grøgaard et al. 1990). The blood clot acts as temporary scaffold for the infiltration of immune cells, including neutrophils, macrophages, and lymphocytes.

Neutrophils are the first immune cell that influx into the blood clot (Kolar et al. 2010). Patients with an insufficient neutrophil population, or who have neutrophil-specific dysfunctions, suffer from delayed wound healing (Wilgus et al. 2013). Similarly, depletion of neutrophils in the fracture site of a diaphyseal mouse model results in delayed bone regeneration and remodeling. After neutrophil depletion, inflammatory mediators such as cytokines IL-6 and IL-10, as well as chemokines CXCL1 and CCL2, are altered in the fracture hematoma (Kovtun et al. 2016). Conversely, stimulation of neutrophil recruitment by G-CSF supports fracture healing (Ishida et al. 2010; Fukui et al. 2012). However, the detrimental effects of neutrophils are not without controversy as some investigators have found that neutrophil depletion promotes osteogenic differentiation of progenitor cells in a model of growth plate injury (Chung et al. 2006).

The neutrophils are replaced by *macrophages*, and the neutrophils are thought to mediate this switch through their production of CCL2 and IL-6. Macrophages are actively involved throughout all the healing phases (Brancato and Albina 2011), and reducing macrophage numbers or compromising macrophage function results in impaired or delayed bone healing in various animal models (Wu et al. 2013). After the macrophage phase, *T lymphocytes* are recruited into the fracture hematoma and subsequent granulation tissue, whereas few *B lymphocytes* are found at any stage of fracture healing (El Khassawna et al. 2017). Experimental models suggest that lymphocytes are detrimental as mice devoid of T and B lymphocytes have improved healing of diaphyseal fractures (Toben et al. 2011; Könnecke et al. 2014). While the role of B lymphocytes during normal bone remodeling appears minimal, these cells play an important role in many bone diseases. Under physiological conditions, B lymphocytes produce OPG, a potent anti-osteoclastogenic factor, while activated B lymphocytes produce the pro-osteoclastogenic factor RANKL. RANKL/OPG expression by B lymphocytes seems to be driven through activation of mechanistic target of rapamycin complex 1 (mTORC1) (Xu et al. 2016).

It is worth noting that bone tissue is highly vascularized, and the vascular system is critical for bone development and bone remodeling. Additionally, insufficient vascularization is an important challenge in bone tissue engineering. It is becoming increasingly clear that the vascular and skeletal systems are intimately linked (Prisby 2017). Several chemokines are upregulated during osteogenesis (Shaik et al. 2019) suggesting they may play a potential role in differentiation of adipose-derived stem cells (ASCs) for vascular development and integration into bone tissue. Other chemokines, such as CXCL9, may be negative modulators of angiogenesis and osteogenesis (Shen et al. 2019).

2 Osteoimmunology

The term "osteoimmunology" was first used by Arron and Choi (2000) to highlight the cross-regulation and communication that occurs between the immune and skeletal systems. That the immune system plays an important role in bone remodeling became clear with the identification of a role for RANK and RANKL in osteoclast differentiation. That members of the tumor necrosis factor (TNF) superfamily of ligands and receptors would be involved in bone remodeling was a surprise. Initial studies on RANKL signal transduction described it in the context of the immune system where it augmented the ability of dendritic cells to stimulate naive T-cell proliferation and enhance dendritic cell survival. Thus, RANK/RANKL connected immune response to osteoclastic bone resorption. Now, it is readily accepted that bone cells and immune cells share a wide range of molecules, including transcription factors, signaling molecules, and membrane receptors. For example, T and B lymphocytes secrete RANKL and TNF under inflammatory conditions, resulting in increased osteoclast formation and bone resorption in inflammatory states. As a result, inflammatory diseases such as rheumatoid arthritis, HIV infection, periodontal infection, and Crohn's disease are associated with bone loss (Weitzmann 2017). An overwhelming amount of evidence indicates there is a clear interaction between bone remodeling and the immune system. This review will specifically focus on the roles of chemokines in bone remodeling.

3 Chemokines and Chemokine Receptors

Bone remodeling and proper coupling of bone formation to bone resorption are under the control of a multitude of local and systemic factors, including chemokines, a family of immune-related molecules and their cognate G-protein-coupled receptors (GPCRs). Chemokines are a group of small 8–12 kDa proteins, specific to vertebrates, which have key roles in the adaptive immune system through their actions on cell migration, proliferation, activation, differentiation, and survival. All of the chemokines have a similar tertiary structure that includes a disordered N terminus of 6–10 amino acids followed by a long loop (N loop), a 3_{10} helix, a three-stranded β-sheet, and a C-terminal α-helix (Allen et al. 2007). Chemokines are divided into four families based on the spacing of conserved cysteine residues (Bachelerie et al. 2014). There are two major subfamilies, CC (the two cysteines are next to each other) and CXC (the two cysteines are separated by one amino acid), and two minor subfamilies: CX3C (the two cysteines are separated by three amino acids) and XC (the first cysteine is lacking). In all, there are 27 C-C-chemokines, 17 C-X-C-chemokines, 2 XC chemokines, and 1 CX3C chemokine (Table 1). All the chemokines are soluble proteins with two exceptions, CXCL16 (Matloubian et al. 2000) and CXCL1 (Bazan et al. 1997) which can remain tethered to the cell surface by a transmembrane mucin-like stalk.

Chemokines serve as ligands for chemokine receptors. In addition to the 20 conventional chemokine receptors (CCR1–10, CXCR1–6, CXCR8, XCR1, XCR2, and CX_3CR1) in humans, there are four atypical chemokine receptors (ACKRs: ACKR1, ACKR2, ACKR3/CXCR7, and ACKR4). Atypical chemokine receptors lack the ability to engage conventional downstream signaling pathways and instead act by scavenging chemokines. Although some chemokine-chemokine receptor interactions are highly specific such as CXCL16 interacting only with CXCR6, other chemokines are highly promiscuous binding multiple receptors such as observed with CCL7 which binds to CCR1, CCR2, CCR3, CCR5, ACKR1, and ACKR2 (Table 1). All of the chemokine receptors belong to class A GPCRs, and they share the canonical seven-transmembrane helical domain structure. However, chemokine receptors lack a single structural signature and instead display a wide range of amino acid identity (25–80%). There are a group of features that are found more frequently among chemokine receptors than other GPCRs including a length of 340–370 aa; an acidic N-terminal segment; the sequence DRYLAIVHA, or a variation of it, in the second intracellular loop; a short basic third intracellular loop; a cysteine in each of the four extracellular domains; and a tyrosine sulfation motif in the N-terminus.

Our structural knowledge on chemokine receptor has substantially increased over the past decade, with NMR and X-ray crystallography approaches being applied to solve several structures including 6 CC and 1 CXC receptor, as well as the viral chemokine receptor US28 (Wu et al. 2010; Park et al. 2012; Tan et al. 2013; Millard et al. 2014; Burg et al. 2015; Oswald et al. 2016; Zheng et al. 2016; Apel et al. 2019; Jaeger et al. 2019). Within these structures, the co-crystallized

Table 1 Evidence supporting a relationship between individual human chemokines, their associated chemokine receptors, and bone formation or bone resorption

Chemokine	Alternative names	Chemokine receptor	Relationship to bone
CCL1	I-309, TCA-3	CCR8	Gene expression analysis detected upregulation of CCL1 in the development patients with Kashin-Beck disease, a chronic skeletal disorder with osteopenia and osteoporosis (Wen et al. 2016)
CCL2	MCP-1	CCR2 (CCR2A/ CCR2B); CCR4; ACKR1; ACKR2	CCL2 mediates the differentiation of OBs (Molloy et al. 2009), recruits OC precursors, and promotes osteoclastogenesis (Kim et al. 2005; Lu et al. 2007; Miyamoto et al. 2009) RANKL stimulates the formation of osteoclasts in human peripheral blood monocyte cultures, in part, due to an increase in CCL2 production, which was shown by using blocking antibodies to CCL2 (Kim et al. 2005) CCL2 is expressed by resting hepatic stellate cells, the primary hepatic nonparenchymal mesenchymal progenitor of the liver (Chinnadurai et al. 2019) *CCL2*-deficient mice have reduced osteoclast-specific genes (DC-STAMP, NFATc1, and cathepsin K), suggesting impaired osteoclast differentiation (Miyamoto et al. 2009). Mice deficient in *CCL2* also have an elevated bone mass and decreased bone resorption markers (CTX-1 and TRACP 5b) (Sul et al. 2012) Bindarit, an inhibitor of CCL2, protects against bone loss induced by chikungunya virus infection (Chen et al. 2015) CCL2 gene variants are risk factors for osteoporosis and osteopenia (Eraltan et al. 2012), CCL2 is elevated in osteoporosis patients (Fatehi et al. 2017), and synovial tissue and synovial fluid from rheumatoid arthritis patients contain increased concentrations of CCL2 (Iwamoto et al. 2008) CCL2 is expressed in compressed and stretched human periodontal ligament cells during orthodontic tooth movement (Garlet et al. 2008) and may contribute to alveolar bone remodeling as well as root resorption (Asano et al. 2011) Trabecular bone biopsies expressed CCL2, and OB isolated from patients with osteoarthritis, rheumatoid arthritis, and post-trauma showed that CCL2 was constitutively produced (Lisignoli et al. 2002)

(continued)

Table 1 (continued)

Chemokine	Alternative names	Chemokine receptor	Relationship to bone
CCL3	MIP-1α; LD78α	CCR1; CCR5; ACKR2	CCL3 stimulates the recruitment and migration of premature OCs, induces formation of TRAP + multinucleated cells in the absence of RANKL, prolongs the survival of OCs, augments the differentiation of osteoclasts, and inhibits the differentiation of OBs (Yu et al. 2004; Oba et al. 2005; Kim et al. 2006; Vallet et al. 2011) CCL3 is expressed in compressed and stretched human periodontal ligament cells during orthodontic tooth movement (Garlet et al. 2008) and may contribute to alveolar bone remodeling IL-1β and TNF-α stimulated human osteoblast-like cells to express CCL3 mRNA and protein, but these were not detected under basal conditions (Taichman et al. 2000) CCL3 secretion is high in osteoclasts and their precursor cells, compared with osteoblasts (Frisch et al. 2012) CCL3 is elevated in osteoporosis patients (Fatehi et al. 2017), as well as in the synovial tissue and synovial fluid from rheumatoid arthritis patients (Iwamoto et al. 2008) Trabecular bone biopsies expressed CCL3, and OB isolated from patients with osteoarthritis, rheumatoid arthritis, and post-trauma showed that CCL3 was constitutively produced (Lisignoli et al. 2002)
CCL4	MIP-1β	CCR1; CCR3; CCR5; ACKR2	CCL4 is as an important regulator of OC migration via induction of PI3K activation (Xuan et al. 2017) CCL4 is elevated in osteoporosis patients (Fatehi et al. 2017)
CCL5	RANTES	CCR1; CCR2; CCR3; CCR4; CCR5; ACKR1; ACKR2	Knockdown of CCL5 decreased osteogenesis from human mesenchymal stem cells (Liu et al. 2014) CCL5 is secreted from osteoclasts and induces osteoblast chemotaxis (Yano et al. 2005) CCL5 can induce OC migration, resorption activity, adhesion, and survival (Yu et al. 2004) CCL5 is expressed in compressed and stretched human periodontal ligament cells during orthodontic tooth movement (Garlet et al. 2008) and may contribute to alveolar bone remodeling CCL5 is elevated in osteoporosis patients (Fatehi et al. 2017) Trabecular bone biopsies expressed CCL5, and OB isolated from patients with osteoarthritis, rheumatoid arthritis, and post-trauma showed that CCL5 was constitutively produced (Lisignoli et al. 2002)

(continued)

Table 1 (continued)

Chemokine	Alternative names	Chemokine receptor	Relationship to bone
CCL7	MCP-3	CCR1; CCR2; CCR3; CCR5; ACKR1; ACKR2	CCL7 can induce OC migration, resorption activity, adhesion, and survival (Yu et al. 2004) CCL7 is expressed in MLO-Y4 osteocyte-like cells, and expression is increased in osteocytes in response to tooth movement (Kitase et al. 2014) High levels of CCL7 are detected in synovial tissues of individuals suffering from rheumatoid arthritis, osteoarthritis, and reactive arthritis (Haringman et al. 2006)
CCL8	MCP-2	CCR1; CCR2; ACKR2	CCL8 is associated with the severity of periodontitis (Panezai et al. 2017) High levels of CCL8 are detected in synovial tissues of individuals suffering from rheumatoid arthritis, osteoarthritis, and reactive arthritis (Haringman et al. 2006)
CCL9	MIP-1γ	CCR1	CCL9 is expressed by osteoclasts (Lean et al. 2002) RANKL-induced osteoclasts increase production of CCL9, and neutralizing CCL9 antibody reduced RANKL-stimulated osteoclast differentiation (Okamatsu et al. 2004) Microarrays used to assess gene expression during osteoclast differentiation induced by CSF-1 in vivo indicated that CCL9 is strongly induced, and anti-CCL9 antibody inhibited osteoclast differentiation in vitro and suppressed the osteoclast response in CSF-treated rats in vivo (Yang et al. 2006) Neutralizing CCL9 antibody abolished RANKL-stimulated osteoclast formation, and mice treated with MIP-1γ shRNA had less severe osteoarthritis than control mice (Shen et al. 2013)
CCL11	Eotaxin	CCR1; CCR3; ACKR1; ACKR2	CCL11 plays an important role in OC migration, and it is expressed in osteoblasts, and its expression increases during inflammatory conditions (Kindstedt et al. 2017) CCL11 is involved in the bone erosion or damage process (Syversen et al. 2008)
CCL12	MCP-5	CCR2	CCL12 regulates joint formation and limb ossification during development (Longobardi et al. 2012) Low levels of interzone-CCL12 are essential for joint formation and contribute to proper growth plate organization (Longobardi et al. 2012) Upregulation of CCL12 in the perichondrium via conditional knockout of *Ikkβ* inhibited the growth of longitudinal bone (Kobayashi et al. 2015)

(continued)

Table 1 (continued)

Chemokine	Alternative names	Chemokine receptor	Relationship to bone
CCL13	MCP-4	CCR1; CCR2; CCR3; ACKR1; ACKR2	Synovial tissue and synovial fluid from rheumatoid arthritis patients contain increased concentrations of CCL13 (Iwamoto et al. 2008)
CCL15	HCC-2; leukotactin-1 MIP-1γ	CCR1; CCR3	N-terminally truncated CCL15 detected at relatively high levels in synovial fluids from rheumatoid arthritis (Berahovich et al. 2005)
CCL17	TARC	CCR4; ACKR1; ACKR2	CCL17 is abundantly expressed in rheumatoid arthritis synovial tissue compared to osteoarthritis (Miyazaki et al. 2018) Incubation of those cells with titanium significantly upregulated expression of CCL17 by human osteoblasts and in vitro-generated osteoclasts as determined by quantitative real-time PCR and ELISA (Cadosch et al. 2010)
CCL18	PARC; DC-CK1	CCR8	Synovial tissue and synovial fluid from rheumatoid arthritis patients contain increased concentrations of CCL18 (Iwamoto et al. 2008) CCL18 enhances the proliferation and migration of osteosarcoma cell lines MG63 and 143B and upregulated UCA1 through the transcription factor EP300 (Su et al. 2019) Patients undergoing total hip replacement have increased periprosthetic and intraoperative synovial fluid levels of CCL18 versus control patients as quantified via enzyme-linked immunosorbent assay (ELISA) (Trehan et al. 2018) CCL18 is upregulated (RNA and protein) in gingival biopsies from patients with periodontitis (Davanian et al. 2012) CCL18 is a valuable biomarker for Gaucher disease (Raskovalova et al. 2017), which has osteoarticular manifestations including aseptic osteonecrosis, and localized or systemic bone fragility. CCL18 expression correlates with osteonecrosis that occurs despite treatment (Deegan et al. 2011)
CCL19	MIP-3β, ELC	CCR7; CCRL2; ACKR4	CCL19 promotes bone resorption by osteoclasts in an in vivo mice calvarial mode and has been linked to RA pathogenesis (Lee et al. 2017)
CCL20	MIP-3α, LARC	CCR6	CCL20 is expressed in osteoblast progenitors, and its levels increase during osteoblast differentiation. In addition, CCL20 promotes osteoblast survival concordant with activation of the PI3K-AKT pathway, and global deletion of CCL20 results in a reduction in bone mass in mice (Doucet et al. 2016)

(continued)

Table 1 (continued)

Chemokine	Alternative names	Chemokine receptor	Relationship to bone
			CCL20 enhanced osteoblast-mediated osteoclastogenesis, in part through IL-6 production (Pathak et al. 2015) CCL20 was found to increase proliferation of osteoblasts and be an earlier inducer of osteoclast differentiation by increasing the number of pre-osteoclasts, thus favoring cell fusion and MMP-9 release (Lisignoli et al. 2007) CCL20 was found to be abundantly expressed in rheumatoid arthritis synovial tissue compared to osteoarthritis (Feldmann et al. 1996, Lisignoli et al. 2007, Miyazaki et al. 2018)
CCL21	SLC; 6Ckine	CCR7; ACKR4	CCL21 promoted bone resorption by osteoclasts in an in vivo mice calvarial mode and has been linked to RA pathogenesis (Lee et al. 2017) CCL21 triggered angiogenesis indirectly, by activating rheumatoid arthritis fibroblasts and macrophages to secrete proangiogenic factors, such as vascular endothelial growth factor (VEGF), angiopoietin 1, and interleukin (IL)-8 (Pickens et al. 2012)
CCL22	MDC	CCR4; ACKR2	Differentiating osteoclasts constitutively produced CCL22 (Nakamura et al. 2006) Incubation of human osteoblasts and osteoclast cells with titanium significantly upregulated expression of CCL22 in both cell types as determined by quantitative real-time PCR and ELISA assays (Cadosch et al. 2010)
CCL23	MPIF	CCR1	CCL23 was shown to be highly expressed in osteoblasts and chondrocytes in human fetal bone by in situ hybridization and elicited significant chemotactic responses from osteoclast precursors isolated from human osteoclastoma tissue (Votta et al. 2000)
CCL25	TECK	CCR9; CCR11; ACKR4	CCL25 was present in the MG63 osteosarcoma cell line and primary murine osteoblasts (Usui et al. 2016) A micro-computed tomography, histomorphometric and molecular characterization of the alveolar bone healing process in mice following tooth extractions showed an increase in CCL25 at day 7
CCL28	MEC	CCR3, CCR10	CCL28 was abundantly expressed in rheumatoid arthritis synovial tissue compared to osteoarthritis (Miyazaki et al. 2018) and was found at higher levels in subjects with chronic inflammatory diseases than in healthy subjects (Wang et al. 2000) A positive correlation has been observed between clinical periodontal parameters and CCL28 levels (Ertugrul et al. 2013)

(continued)

Table 1 (continued)

Chemokine	Alternative names	Chemokine receptor	Relationship to bone
CXCL1	Gro-α; MGSA	CXCR2	CXCL1 was necessary for LPS-induced OC formation, and antibody blocking studies using anti-CXCL1 antibodies blunted osteoclastogenesis (Valerio et al. 2015; Goto et al. 2016) Trabecular bone biopsies expressed CXCL1, and OB isolated from patients with osteoarthritis, rheumatoid arthritis, and post-trauma showed that CXCL1 was constitutively produced (Lisignoli et al. 2002)
CXCL2	Gro-β	CXCR2	CXCL2 was abundantly expressed in osteoblasts of osteoporotic mice, and CXCL2 neutralization through the use of anti-CXCL2 antibody alleviated bone loss in mice. Similarly, CXCL2 overexpression attenuated proliferation, and differentiation of osteoblasts in vitro and downregulation of CXCL2 promoted osteoblast expansion and differentiation (Yang et al. 2019) CXCL2 was necessary for LPS-induced OC formation, and antibody blocking studies using anti-CXCL2 antibodies blunted osteoclastogenesis (Valerio et al. 2015) Elevated levels of CXCL2 lead to increased recruitment of neutrophils to periodontal ligaments in diabetic mice with induced periodontal disease (Greer et al. 2016; Zheng et al. 2018) CXCL2 was upregulated in primary cultures of osteoblasts isolated from osteoporotic versus non-osteoporotic human bone tissue samples (Trost et al. 2010)
CXCL5	ENA78	CXCR1; CXCR2; ACKR1	CXCL5 was increased in osteosarcoma tissues compared with matched adjacent non-tumor tissues, and levels correlate with advanced clinical stage and metastasis. Moreover, CXCL5 expression levels were increased in osteosarcoma cell lines (Saos-2, MG63, U2OS, SW1353) compared to normal osteoblast hFoB1.19 cells (Dang et al. 2017) Subjects with Paget's disease of the bone have increased CXCL5 mRNA expression in serum as well as bone marrow cells compared to that of normal subjects, and increased levels of CXCL5 may contribute to enhanced levels of RANKL (Sundaram et al. 2013) Cytokine arrays and ELISAs showed that MSC from patients with ankylosing spondylitis, an autoimmune disease characterized by

(continued)

Table 1 (continued)

Chemokine	Alternative names	Chemokine receptor	Relationship to bone
			pathological osteogenesis, secreted more CXCL5 than MSC from healthy donors. In addition, the ability to inhibit osteoclastogenesis is enhanced in MSC from patients with ankylosing spondylitis compared with those from healthy donors, and decreasing CXCL5 reduced this phenomenon (Liu et al. 2019)
CXCL7	NAP-2	CXCR1; CXCR2	The addition of exogenous CXCL7 increased the number of osteoclastic TRAP-positive multinuclear cells in primary mouse bone marrow cultures (Nakao et al. 2009) Antibodies against CXCL7 suppressed OC formation by bone marrow cells (Goto et al. 2016)
CXCL8	IL-8	CXCR1; CXCR2; ACKR1	Human osteoclast precursors were cultured with CXCL8 and then analyzed for multi-nuclei osteoclast formation and activity. CXCL8 enhanced the number of osteoclasts with 3–5 nuclei and >5 nuclei, and the stimulatory effects on osteoclastogenesis were inhibited with IL-6 (Pathak et al. 2015) Trabecular bone biopsies expressed CXCL8, and OB isolated from patients with osteoarthritis, rheumatoid arthritis, and post-trauma showed that CXCL8 was constitutively produced (Lisignoli et al. 2002) CXCL8 is elevated in patients with rheumatoid arthritis (Feldmann et al. 1996)
CXCL9	MIG	CXCR3	Synovial tissue, synovial fluid, or serum from rheumatoid arthritis patients contain increased concentrations of CXCL9 compared to healthy controls (Iwamoto et al. 2008, Kuan et al. 2010)
CXCL10	IP-10	CXCR3	Human osteoblasts express CXCR3, and activation of CXCR3 via CXCL10 increases the expression of alkaline phosphatase and beta-N-acetyl-hexosaminidase in human osteoblasts (Lisignoli et al. 2003a, b) Osteoclast-derived CXCL10 plays a crucial role in bone destruction in a mouse model of rheumatoid arthritis by recruiting CD4+ T cells and promoting them to produce RANKL (Kwak et al. 2008) Augmented production of CXCL10 is required for cancer outgrowth within the bone (Lee et al. 2012) Serum from rheumatoid arthritis patients contains increased concentrations of CXCL10 compared to healthy controls (Kuan et al. 2010)

(continued)

Table 1 (continued)

Chemokine	Alternative names	Chemokine receptor	Relationship to bone
CXCL11	I-TAC	CXCR3; ACKR1; ACKR3	CXCL11 inhibits osteoclastic differentiation of CD14+ monocytes (Coelho et al. 2005)
CXCL12	SDF-1	CXCR4; CXCR7; ACKR3	CXCL12 is important for HSC fate CXCL12 is constitutively produced by BM stromal cells (Pelus et al. 2002) and regulates HSC mobilization in the BM as well as cell adhesion, survival, and cell cycle status (Nervi et al. 2006) CXCL12 expression is found in both immature and mature OB cells (Jung et al. 2006) CXCL12 was important to the recruitment of MSCs during skeletal repair (Kitaori et al. 2009) CXCL12 can induce OC migration, resorption activity, adhesion, and survival (Yu et al. 2003) Antibodies against CXCL12 suppressed OC formation byRAW264.7 cells as well as bone marrow cells (Goto et al. 2016) Synovial tissue and synovial fluid from rheumatoid arthritis patients contain increased concentrations of CXCL12 (Iwamoto et al. 2008) The addition of CXCL12 to collagen implants increases the number of BM-derived osteoprogenitor cells recruited to the site of bone formation, and this effect is blocked by a CXCR4 antibody (Higashino et al. 2011) Plerixafor, a small molecule inhibitor of CXCR4, decreases ulnar loading-induced bone formation (Leucht et al. 2013) Increased osteocalcin expression stimulated by treatment of pluripotential mesenchymal C2C12 cells with rBMP2 is reduced by CXCL12 siRNA or antibodies to CXCL12. Blocking CXCL12 also reduced other markers of osteoblast differentiation, including Runx2 and Osterix and osteoblast transcription factors (Zhu et al. 2007) In a C57BL/6J mouse model, conditional inactivation of CXCR4 results in osteopenia with reduced bone mass, decreased bone mineral density, slowed mineral apposition, and decreased expression of type I collagen (Zhu et al. 2011) Elevated levels of CXCL12 in the synovial and bone tissue of patients with rheumatoid arthritis correlate with pathological bone loss and an increase in the recruitment and activation of osteoclasts at sites of local inflammation (Grassi et al. 2004)

(continued)

Table 1 (continued)

Chemokine	Alternative names	Chemokine receptor	Relationship to bone
CXCL13	BCA-1; BLC	CXCR5	CXCL13 is an important chemokine for functional maintenance of osteoblasts; it was produced by osteoblasts and bone marrow MSC (Lisignoli et al. 2003a, b) and increased the expression of alkaline phosphatase and beta-N-acetyl-hexosaminidase in human osteoblasts (Lisignoli et al. 2003a, b)
CXCL14	BRAK; BMAC; MIP-2γ		CXCL14 expression increased significantly at the root formation stage, compared with the cap or crown formation stage at both transcription and translation levels and has been implicated in the formation of the eruptive pathway of tooth germs via osteoclastogenesis (Yoo et al. 2011)
CXCL16	SR-PSOX	CXCR6	CXCL16 was found to be expressed by human osteocytes in vivo (Hu et al. 2008), and soluble CXCL16 increased migration of osteoclast precursor cells (Li et al. 2012) TGF-β1 increased osteoclast expression of the chemokine CXCL16 to promote osteoblast migration (Ota et al. 2013)
CXCL1			A microarray study carried out on PTHR1-positive osteoblasts (Kusa 4b10 cells) identified CXCL1 as an immediate PTH/PTHrP-responsive gene, and PTH treatment stimulated an increase in secreted CXCL1 protein by Kusa 4b10 cells and calvarial osteoblasts (Onan et al. 2009) CXCL1 is expressed in osteoclast precursors and induces their migration in a dose- dependent manner (Onan et al. 2009) Antibody blocking studies using anti-CXCL1 antibodies blunted osteoclastogenesis in *Dusp1*$^{-/-}$ cells (Valerio et al. 2015)
CXCL2	MIP2α	CXCR2	CXCL2 enhanced the proliferation of OC precursor cells from bone marrow-derived macrophages, and the formation of OCs from bone marrow-derived macrophages was increased with CXCL2 treatment (Ha et al. 2010) Conditioned medium from lipopolysaccharide-treated bone marrow macrophages enhanced migration of osteoclast precursors, and this was blocked with CXCL2-neutralizing antibody (Ha et al. 2011). Blockade of CXCL2 reduced lipopolysaccharide-induced osteoclastogenesis and prevented bone destruction in mice treated with lipopolysaccharide

(continued)

Table 1 (continued)

Chemokine	Alternative names	Chemokine receptor	Relationship to bone
			Antibody blocking studies using anti-CXCL2 antibodies blunted osteoclastogenesis in *Dusp1*$^{-/-}$ cells (Valerio et al. 2015)
CX_3CL1	Fractalkine; neurotactin	CX_3CR1	CX3CL1 has been shown to influence bone remodeling cell types (Gilchrist and Stern 2015) Anti-CX_3CL1 antibody inhibited differentiation of bone marrow-derived OC precursors into osteoclasts during co-culture with osteoblasts (Koizumi et al. 2009) Anti-CX_3CL1 antibody suppressed the migration of OC precursors into the synovium and decreased the number of mature osteoclasts, leading to the inhibition of bone destruction, synovitis, and cartilage destruction (Hoshino-Negishi et al. 2019) Synovial tissue and synovial fluid from rheumatoid arthritis patients contain increased concentrations of CXC3CL1 (Iwamoto et al. 2008), and serum concentrations of CX_3CL1correlate with disease severity

ligands were of a different chemical nature bound to distinct sites, and the receptors themselves were in different conformational states. Yet, all the structures share a broad and open ligand-binding pocket within the transmembrane helical bundle (Zhao et al. 2019). Moreover, the structures suggest that structural changes associated with an "active state" are present, specifically that there is a significant outward shift of the intracellular half of transmembrane (TM) 6, accompanied by an inward movement of the bottom of TM7, and a subtler lateral displacement of the bottom of TM5 (Arimont et al. 2019).

Chemokine receptors activate cells through both G-protein- and non-G-protein-dependent pathways (Gilchrist and Stern 2015). In the case of G-protein-dependent signaling, binding of the chemokine results in GDP/GTP exchange on the Gα subunit. This results in dissociation of the heterotrimeric Gαβγ into an active GTP-bound Gα and a Gβγ dimer that can then go on to activate their downstream signaling pathways such as mobilization of intracellular calcium, activation of Rac, Rho, and cdc42, and inhibition of adenylate cyclase. In addition to signaling through G-proteins, activated chemokine receptors recruit β-arrestin, which can lead to G-protein-independent activation of Akt, p38 MAPK, and ERK1/2.

Several chemokines (Table 1) and their receptors (Table 2) play a critical role in maintaining the balance between bone resorption and bone formation. For example, a number of chemokines can recruit osteoclast precursors or stimulate osteoclastogenesis including CCL2, CCL3, CCL4, CXCL8, and CXCL12 (Silva et al. 2007). Furthermore, CCR1 knockout animals have fewer and thinner trabecular bones, and their osteoblasts show defective differentiation (Hoshino et al. 2010).

Table 2 Individual human chemokine receptors, the chemokines that activate them, and their relationship to bone

Receptor	Relationship to bone
CCR1	CCR1 is expressed by primary osteoblasts and the MC3T3-E1 osteoblast cell line (Yano et al. 2005) Knockdown of CCR1 decreased osteogenesis from human mesenchymal stem cells (Liu et al. 2014) CCR1 is expressed by osteoclasts (Lean et al. 2002), and blocking CCR1 via a pharmalogical inhibitor is sufficient to inhibit osteoclastogenesis (Oba et al. 2005) RANKL-induced osteoclasts increase their production of CCR1 (Okamatsu et al. 2004) Microarrays used to assess gene expression during osteoclast differentiation induced by CSF-1 in vivo indicated that CCR1 is strongly induced (Yang et al. 2006)
CCR2	CCR2 is involved in the chemoattraction of pre-osteoclastic monocytic cells (Garlet et al. 2010) CCR2 is potently induced by RANKL in human osteoclasts (Kim et al. 2006) CCR2 regulates joint formation and limb ossification during development (Longobardi et al. 2012) CCR2 gene variants are risk factors for osteoporosis and osteopenia (Eraltan et al. 2012)
CCR3	CCR3 is expressed by primary osteoblasts and the MC3T3-E1 osteoblast cell line (Yano et al. 2005) CCR3 was significantly upregulated in osteoclasts during their differentiation (Kindstedt et al. 2017)
CCR4	CCR4 is expressed by primary osteoblasts and the MC3T3-E1 osteoblast cell line (Yano et al. 2005) CCR4 is potently induced by RANKL in human osteoclasts (Kim et al. 2006)
CCR5	CCR5 is expressed by primary osteoblasts and the MC3T3-E1 osteoblast cell line (Yano et al. 2005) CCR5 is involved in the chemoattraction of preosteoclastic monocytic cells (Repeke et al. 2011)
CCR6	CCR6 is expressed in osteoblast progenitors, and its levels increase during osteoblast differentiation. In addition, global loss of CCR6 in mice significantly decreases trabecular bone mass coincident with reduced osteoblast numbers (Doucet et al. 2016)
CCR7	CCR7 has been linked to RA pathogenesis (Lee et al. 2017), and its expression in RA monocytes and in vitro differentiated macrophages is closely associated with disease activity score (Van Raemdonck et al. 2019)
CCR8	CCR8 mediates conversion of mesenchymal stem cells to embryoid bodies, three-dimensional aggregates of pluripotent stem cells (Haque et al. 2019)
CCR9	CCR9 is present in the osteoclast RAW 264.7 cell line and murine osteoclast precursor bone marrow macrophages (Usui et al. 2016)
CXCR1	When the expression and functional activity of CXC chemokine receptors was evaluated in human OB obtained post-trauma from old versus young donors, CXCR1 was only expressed by old donors' OB cells (Lisignoli et al. 2003a, b)
CXCR2	Conditioned medium from lipopolysaccharide-treated bone marrow macrophages enhanced migration of osteoclast precursors, and this was blocked with CXCR2 receptor antagonist (Ha et al. 2011)
CXCR3	Activation of CXCR3 via CXCL10 increases the expression of alkaline phosphatase and beta-N-acetyl-hexosaminidase in human osteoblasts (Lisignoli et al. 2003a, b)

(continued)

Table 2 (continued)

Receptor	Relationship to bone
CXCR4	CXCR4 was important to the recruitment of MSCs during skeletal repair (Kitaori et al. 2009) Ablation of CXCR4 in mature osteoblasts in mice resulted in a significant decrease in bone mass and alterations in cancellous bone structure (Shahnazari et al. 2013) When the expression and functional activity of CXC chemokine receptors was evaluated in human OB obtained post-trauma from old versus young donors, CXCR4 was only expressed by old donors' OB cells (Lisignoli et al. 2003a, b) Increased osteocalcin expression stimulated by treatment of pluripotential mesenchymal C2C12 cells with rBMP2 was reduced by antibodies to CXCR4. Blocking CXCR4 also reduced other markers of osteoblast differentiation, including Runx2 and Osterix and osteoblast transcription factors (Zhu et al. 2007)
CXCR5	CXCR5 was reported to be relevant to bone and cartilage repair in human-MSCs grown on plastic or on hyaluronan-based scaffold (Thevenot et al. 2010) and was shown to be highly expressed in osteoblasts and MSCs (Zhang et al. 2015)
CXCR6	CXCR6-positive prostate cancer cells migrated to CXCL16-positive osteocytes, and this resulted in increased bone metastasis (Deng et al. 2010) RANKL decreased CXCR6 in a dose-dependent manner during osteoclastogenesis through the JAK2/STAT3 pathway (Li et al. 2012)
CX_3CR1	CX3CR1 is expressed on osteoclasts and has been shown to play a role in bone-resorbing and inflammatory diseases (Imai and Yasuda 2016) Cx3cr1-deficient mice exhibited slight increases in trabecular and cortical thickness, reduced numbers of osteoclasts, and increased rates of osteoid formation. Cultured Cx3cr1-deficient osteoblastic cells showed reduced calcium deposition, and in vitro studies have suggested a role for CX3CR1 in osteoblast differentiation (Hoshino et al. 2013)
ACKR2	ACKR2−/− mice exhibited increased osteoclast counts and elevated expression of pro-resorptive markers, while the number of osteoblasts and related markers was decreased suggesting ACKR2 may function as a regulator of mechanically induced bone remodeling by affecting the differentiation and activity of bone cells in the periodontal microenvironment (Lima et al. 2017)

In addition, cultured $CCR1^{-/-}$ BM cells generate fewer osteoclasts due to reduced cell fusion, and these cells showed no osteolytic activity. Several chemokines have also been shown to influence skeletal remodeling in physiological and pathological conditions (Brylka and Schinke 2019). As noted earlier, neutrophils are crucial for bone regeneration and remodeling. The impairment of targeted chemotaxis via alterations in chemokine expression has been described in many inflammatory disorders and may in part explain the bone loss noted in rheumatoid arthritis, HIV infection, periodontal infection, and Crohn's disease (Weitzmann 2017).

There are very few clinical therapeutic agents that target chemokines or chemokine receptors. The recent approval of mogamulizumab (Poteligeo™) a monoclonal antibody targeting CCR4 (Kasamon et al. 2019) represents only our third therapeutic targeting chemokine receptors. The other two FDA-approved chemokine-targeted drugs are the CCR5 antagonist maraviroc (Selzentry™/Celsentri™) for HIV infection (FDA 2009) and the CXCR4 inhibitor plerixafor

(Mozobil™) which is used as an immunostimulant to mobilize hematopoietic stem cells in cancer patients into the bloodstream (Brave et al. 2010). There are a number of chemokine and chemokine receptor-directed therapies in clinical trials. For example, phase I and phase II studies have examined the organogermanium compound propagermanium that blocks CCR2 signaling (Yumimoto et al. 2019). PF-04634817, a dual chemokine CCR2/5 receptor antagonist, is being developed for the treatment of diabetic nephropathy (Gale et al. 2018), while BMS-813160, another dual chemokine CCR2/5 receptor antagonist, is being studied in the context of cancer. Several anti-CXCR4 antibodies and small molecule antagonists (burixafor, ulocuplumab, BL-8040, MDX-1338 (BMS-936564), LY2624587, BKT140) have undergone clinical stage testing (Ullah 2019). Similarly, small molecules that target the CXCR4 ligand, namely, CXCL12, such as olaptesed pegol (Ola-PEG), have also been examined. Spiegelmers, synthetic target-binding oligonucleotides built from nonnatural l-nucleotides, bind their targets with high affinity and selectivity. There are two Spiegelmer candidates targeting chemokines, emapticap pegol (NOX-E36; anti-CCL2) and olaptesed pegol (NOX-A12; anti-CXCL12), that are being tested in clinical studies (Vater and Klussmann 2015).

Our understanding of the overlapping roles for the many different chemokines and chemokine receptors has progressed rapidly over the last decade with respect to their impact on bone, including chondrogenesis, osteoblast differentiation, bone formation, bone vascularization, mineral opposition, osteoclastogenesis, and bone resorption. Moreover, it is clear from this review of recent work that chemokines play an important part in many pathological processes related to the bone such as fracture healing, the failure of implants, bone loss due to chronic inflammatory conditions such as rheumatoid arthritis, and the cancer-related metastases. Yet, many questions remain unanswered. Given the preponderance of evidence that chemokines and chemokine receptors play important roles in both normal bone development and pathophysiological disease, it seems surprising not more is being done.

References

Allen SJ, Crown SE, Handel TM (2007) Chemokine: receptor structure, interactions, and antagonism. Annu Rev Immunol 25:787–820

Alnaeeli M, Penninger JM, Teng YT (2006) Immune interactions with CD4+ T cells promote the development of functional osteoclasts from murine CD11c+ dendritic cells. J Immunol 177(5):3314–3326

Anderson DM, Maraskovsky E, Billingsley WL, Dougall WC, Tometsko ME, Roux ER, Teepe MC, DuBose RF, Cosman D, Galibert L (1997) A homologue of the TNF receptor and its ligand enhance T-cell growth and dendritic-cell function. Nature 390(6656):175–179

Apel AK, Cheng RKY, Tautermann CS, Brauchle M, Huang CY, Pautsch A, Hennig M, Nar H, Schnapp G (2019) Crystal structure of CC chemokine receptor 2A in complex with an orthosteric antagonist provides insights for the design of selective antagonists. Structure 27(3):427–438.e425

Arimont M, Hoffmann C, de Graaf C, Leurs R (2019) Chemokine receptor crystal structures: what can be learnt from them? Mol Pharmacol 96(6):765–777
Arron JR, Choi Y (2000) Bone versus immune system. Nature 408(6812):535–536
Asano M, Yamaguchi M, Nakajima R, Fujita S, Utsunomiya T, Yamamoto H, Kasai K (2011) IL-8 and MCP-1 induced by excessive orthodontic force mediates odontoclastogenesis in periodontal tissues. Oral Dis 17(5):489–498
Bachelerie F, Ben-Baruch A, Burkhardt AM, Combadiere C, Farber JM, Graham GJ, Horuk R, Sparre-Ulrich AH, Locati M, Luster AD, Mantovani A, Matsushima K, Murphy PM, Nibbs R, Nomiyama H, Power CA, Proudfoot AE, Rosenkilde MM, Rot A, Sozzani S, Thelen M, Yoshie O, Zlotnik A (2014) International Union of Basic and Clinical Pharmacology. [corrected]. LXXXIX. Update on the extended family of chemokine receptors and introducing a new nomenclature for atypical chemokine receptors. Pharmacol Rev 66(1):1–79
Bazan JF, Bacon KB, Hardiman G, Wang W, Soo K, Rossi D, Greaves DR, Zlotnik A, Schall TJ (1997) A new class of membrane-bound chemokine with a CX3C motif. Nature 385(6617):640–644
Berahovich RD, Miao Z, Wang Y, Premack B, Howard MC, Schall TJ (2005) Proteolytic activation of alternative CCR1 ligands in inflammation. J Immunol 174(11):7341–7351
Bonewald LF (2007) Osteocytes as dynamic multifunctional cells. Ann N Y Acad Sci 1116:281–290
Boyle WJ, Simonet WS, Lacey DL (2003) Osteoclast differentiation and activation. Nature 423(6937):337–342
Brancato SK, Albina JE (2011) Wound macrophages as key regulators of repair: origin, phenotype, and function. Am J Pathol 178(1):19–25
Brave M, Farrell A, Ching Lin S, Ocheltree T, Pope Miksinski S, Lee SL, Saber H, Fourie J, Tornoe C, Booth B, Yuan W, He K, Justice R, Pazdur R (2010) FDA review summary: mozobil in combination with granulocyte colony-stimulating factor to mobilize hematopoietic stem cells to the peripheral blood for collection and subsequent autologous transplantation. Oncology 78(3–4):282–288
Brylka LJ, Schinke T (2019) Chemokines in physiological and pathological bone remodeling. Front Immunol 10:2182
Burg JS, Ingram JR, Venkatakrishnan AJ, Jude KM, Dukkipati A, Feinberg EN, Angelini A, Waghray D, Dror RO, Ploegh HL, Garcia KC (2015) Structural biology. Structural basis for chemokine recognition and activation of a viral G protein-coupled receptor. Science 347(6226):1113–1117
Cadosch D, Gautschi OP, Chan E, Simmen HP, Filgueira L (2010) Titanium induced production of chemokines CCL17/TARC and CCL22/MDC in human osteoclasts and osteoblasts. J Biomed Mater Res A 92(2):475–483
Charles JF, Hsu LY, Niemi EC, Weiss A, Aliprantis AO, Nakamura MC (2012) Inflammatory arthritis increases mouse osteoclast precursors with myeloid suppressor function. J Clin Invest 122(12):4592–4605
Chen TL (2004) Inhibition of growth and differentiation of osteoprogenitors in mouse bone marrow stromal cell cultures by increased donor age and glucocorticoid treatment. Bone 35(1):83–95
Chen W, Foo SS, Taylor A, Lulla A, Merits A, Hueston L, Forwood MR, Walsh NC, Sims NA, Herrero LJ, Mahalingam S (2015) Bindarit, an inhibitor of monocyte chemotactic protein synthesis, protects against bone loss induced by chikungunya virus infection. J Virol 89(1):581–593
Chinnadurai R, Sands J, Rajan D, Liu X, Arafat D, Das R, Anania FA, Gibson G, Kisseleva T, Galipeau J (2019) Molecular genetic and immune functional responses distinguish bone marrow mesenchymal stromal cells from hepatic stellate cells. Stem Cells 37(8):1075–1082
Chung R, Cool JC, Scherer MA, Foster BK, Xian CJ (2006) Roles of neutrophil-mediated inflammatory response in the bony repair of injured growth plate cartilage in young rats. J Leukoc Biol 80(6):1272–1280

Claes L, Recknagel S, Ignatius A (2012) Fracture healing under healthy and inflammatory conditions. Nat Rev Rheumatol 8(3):133–143
Coelho LF, de Freitas Almeida GM, Mennechet FJ, Blangy A, Uzé G (2005) Interferon-alpha and -beta differentially regulate osteoclastogenesis: role of differential induction of chemokine CXCL11 expression. Proc Natl Acad Sci U S A 102(33):11917–11922
Dang H, Wu W, Wang B, Cui C, Niu J, Chen J, Chen Z, Liu Y (2017) CXCL5 plays a promoting role in osteosarcoma cell migration and invasion in autocrine- and paracrine-dependent manners. Oncol Res 25(2):177–186
Davanian H, Stranneheim H, Båge T, Lagervall M, Jansson L, Lundeberg J, Yucel-Lindberg T (2012) Gene expression profiles in paired gingival biopsies from periodontitis-affected and healthy tissues revealed by massively parallel sequencing. PLoS One 7(9):e46440
Deegan PB, Pavlova E, Tindall J, Stein PE, Bearcroft P, Mehta A, Hughes D, Wraith JE, Cox TM (2011) Osseous manifestations of adult Gaucher disease in the era of enzyme replacement therapy. Medicine 90(1):52–60
Deng L, Chen N, Li Y, Zheng H, Lei Q (2010) CXCR6/CXCL16 functions as a regulator in metastasis and progression of cancer. Biochim Biophys Acta 1806(1):42–49
Doucet M, Jayaraman S, Swenson E, Tusing B, Weber KL, Kominsky SL (2016) CCL20/CCR6 signaling regulates bone mass accrual in mice. J Bone Miner Res 31(7):1381–1390
Ducy P, Schinke T, Karsenty G (2000) The osteoblast: a sophisticated fibroblast under central surveillance. Science 289(5484):1501–1504
El Khassawna T, Serra A, Bucher CH, Petersen A, Schlundt C, Könnecke I, Malhan D, Wendler S, Schell H, Volk HD, Schmidt-Bleek K, Duda GN (2017) T lymphocytes influence the mineralization process of bone. Front Immunol 8:562
Eraltan H, Cacina C, Kahraman OT, Kurt O, Aydogan HY, Uyar M, Can A, Cakmakoğlu B (2012) MCP-1 and CCR2 gene variants and the risk for osteoporosis and osteopenia. Genet Test Mol Biomarkers 16(4):229–233
Ertugrul AS, Sahin H, Dikilitas A, Alpaslan N, Bozoglan A (2013) Comparison of CCL28, interleukin-8, interleukin-1β and tumor necrosis factor-alpha in subjects with gingivitis, chronic periodontitis and generalized aggressive periodontitis. J Periodontal Res 48(1):44–51
Fatehi F, Mollahosseini M, Hassanshahi G, Khanamani Falahati-Pour S, Khorramdelazad H, Ahmadi Z, Noroozi Karimabad M, Farahmand H (2017) CC chemokines CCL2, CCL3, CCL4 and CCL5 are elevated in osteoporosis patients. J Biomed Res 31(5):468–470
FDA (2009) FDA advisory committee approves Selzentry. AIDS Patient Care STDs 23(11):987
Feldmann M, Brennan FM, Maini RN (1996) Role of cytokines in rheumatoid arthritis. Annu Rev Immunol 14:397–440
Fierro FA, Nolta JA, Adamopoulos IE (2017) Concise review: stem cells in osteoimmunology. Stem Cells 35(6):1461–1467
Frisch BJ, Ashton JM, Xing L, Becker MW, Jordan CT, Calvi LM (2012) Functional inhibition of osteoblastic cells in an in vivo mouse model of myeloid leukemia. Blood 119(2):540–550
Fukui T, Matsumoto T, Mifune Y, Shoji T, Kuroda T, Kawakami Y, Kawamoto A, Ii M, Kawamata S, Kurosaka M, Asahara T, Kuroda R (2012) Local transplantation of granulocyte colony-stimulating factor-mobilized human peripheral blood mononuclear cells for unhealing bone fractures. Cell Transplant 21(4):707–721
Gale JD, Gilbert S, Blumenthal S, Elliott T, Pergola PE, Goteti K, Scheele W, Perros-Huguet C (2018) Effect of PF-04634817, an oral CCR2/5 chemokine receptor antagonist, on albuminuria in adults with overt diabetic nephropathy. Kidney Int Rep 3(6):1316–1327
Garlet TP, Coelho U, Repeke CE, Silva JS, Cunha Fde Q, Garlet GP (2008) Differential expression of osteoblast and osteoclast chemmoatractants in compression and tension sides during orthodontic movement. Cytokine 42(3):330–335
Garlet TP, Fukada SY, Saconato IF, Avila-Campos MJ, da Silva TA, Garlet GP, Cunha Fde Q (2010) CCR2 deficiency results in increased osteolysis in experimental periapical lesions in mice. J Endod 36(2):244–250
Gilchrist A, Stern P (2015) Chemokines and bone. Clin Rev Bone Miner Metab 13:61–82

Goto Y, Aoyama M, Sekiya T, Kakita H, Waguri-Nagaya Y, Miyazawa K, Asai K, Goto S (2016) CXCR4+ CD45- cells are niche forming for osteoclastogenesis via the SDF-1, CXCL7, and CX3CL1 signaling pathways in bone marrow. Stem Cells 34(11):2733–2743

Grassi F, Cristino S, Toneguzzi S, Piacentini A, Facchini A, Lisignoli G (2004) CXCL12 chemokine up-regulates bone resorption and MMP-9 release by human osteoclasts: CXCL12 levels are increased in synovial and bone tissue of rheumatoid arthritis patients. J Cell Physiol 199(2):244–251

Greer A, Irie K, Hashim A, Leroux BG, Chang AM, Curtis MA, Darveau RP (2016) Site-specific neutrophil migration and CXCL2 expression in periodontal tissue. J Dent Res 95(8):946–952

Grøgaard B, Gerdin B, Reikerås O (1990) The polymorphonuclear leukocyte: has it a role in fracture healing? Arch Orthop Trauma Surg 109(5):268–271

Ha J, Choi HS, Lee Y, Kwon HJ, Song YW, Kim HH (2010) CXC chemokine ligand 2 induced by receptor activator of NF-kappa B ligand enhances osteoclastogenesis. J Immunol 184(9):4717–4724

Ha J, Lee Y, Kim HH (2011) CXCL2 mediates lipopolysaccharide-induced osteoclastogenesis in RANKL-primed precursors. Cytokine 55(1):48–55

Haque NS, Tuteja A, Haque N (2019) CC chemokine CCL1 receptor CCR8 mediates conversion of mesenchymal stem cells to embryoid bodies expressing FOXP3+CCR8+ regulatory T cells. PLoS One 14(7):e0218944

Haringman JJ, Smeets TJ, Reinders-Blankert P, Tak PP (2006) Chemokine and chemokine receptor expression in paired peripheral blood mononuclear cells and synovial tissue of patients with rheumatoid arthritis, osteoarthritis, and reactive arthritis. Ann Rheum Dis 65(3):294–300

Hauge EM, Qvesel D, Eriksen EF, Mosekilde L, Melsen F (2001) Cancellous bone remodeling occurs in specialized compartments lined by cells expressing osteoblastic markers. J Bone Miner Res 16(9):1575–1582

Higashino K, Viggeswarapu M, Bargouti M, Liu H, Titus L, Boden SD (2011) Stromal cell-derived factor-1 potentiates bone morphogenetic protein-2 induced bone formation. Tissue Eng Part A 17(3–4):523–530

Hoshino A, Iimura T, Ueha S, Hanada S, Maruoka Y, Mayahara M, Suzuki K, Imai T, Ito M, Manome Y, Yasuhara M, Kirino T, Yamaguchi A, Matsushima K, Yamamoto K (2010) Deficiency of chemokine receptor CCR1 causes osteopenia due to impaired functions of osteoclasts and osteoblasts. J Biol Chem 285(37):28826–28837

Hoshino A, Ueha S, Hanada S, Imai T, Ito M, Yamamoto K, Matsushima K, Yamaguchi A, Iimura T (2013) Roles of chemokine receptor CX3CR1 in maintaining murine bone homeostasis through the regulation of both osteoblasts and osteoclasts. J Cell Sci 126(Pt 4):1032–1045

Hoshino-Negishi K, Ohkuro M, Nakatani T, Kuboi Y, Nishimura M, Ida Y, Kakuta J, Hamaguchi A, Kumai M, Kamisako T, Sugiyama F, Ikeda W, Ishii N, Yasuda N, Imai T (2019) Role of anti-fractalkine antibody in suppression of joint destruction by inhibiting migration of osteoclast precursors to the synovium in experimental arthritis. Arthritis Rheumatol 71(2):222–231

Hu W, Zhen X, Xiong B, Wang B, Zhang W, Zhou W (2008) CXCR6 is expressed in human prostate cancer in vivo and is involved in the in vitro invasion of PC3 and LNCap cells. Cancer Sci 99(7):1362–1369

Imai T, Yasuda N (2016) Therapeutic intervention of inflammatory/immune diseases by inhibition of the fractalkine (CX3CL1)-CX3CR1 pathway. Inflamm Regen 36:9

Ishida K, Matsumoto T, Sasaki K, Mifune Y, Tei K, Kubo S, Matsushita T, Takayama K, Akisue T, Tabata Y, Kurosaka M, Kuroda R (2010) Bone regeneration properties of granulocyte colony-stimulating factor via neovascularization and osteogenesis. Tissue Eng Part A 16(10):3271–3284

Iwamoto T, Okamoto H, Toyama Y, Momohara S (2008) Molecular aspects of rheumatoid arthritis: chemokines in the joints of patients. FEBS J 275(18):4448–4455

Jaeger K, Bruenle S, Weinert T, Guba W, Muehle J, Miyazaki T, Weber M, Furrer A, Haenggi N, Tetaz T, Huang CY, Mattle D, Vonach JM, Gast A, Kuglstatter A, Rudolph MG, Nogly P, Benz J, Dawson RJP, Standfuss J (2019) Structural basis for allosteric ligand recognition in the human CC chemokine receptor 7. Cell 178(5):1222–1230.e1210

Jung Y, Wang J, Schneider A, Sun YX, Koh-Paige AJ, Osman NI, McCauley LK, Taichman RS (2006) Regulation of SDF-1 (CXCL12) production by osteoblasts; a possible mechanism for stem cell homing. Bone 38(4):497–508

Kasamon YL, Chen H, de Claro RA, Nie L, Ye J, Blumenthal GM, Farrell AT, Pazdur R (2019) FDA approval summary: mogamulizumab-kpkc for mycosis fungoides and sézary syndrome. Clin Cancer Res 25(24):7275–7280

Kim MS, Day CJ, Morrison NA (2005) MCP-1 is induced by receptor activator of nuclear factor-{kappa}B ligand, promotes human osteoclast fusion, and rescues granulocyte macrophage colony-stimulating factor suppression of osteoclast formation. J Biol Chem 280(16):16163–16169

Kim MS, Magno CL, Day CJ, Morrison NA (2006) Induction of chemokines and chemokine receptors CCR2b and CCR4 in authentic human osteoclasts differentiated with RANKL and osteoclast like cells differentiated by MCP-1 and RANTES. J Cell Biochem 97(3):512–518

Kindstedt E, Holm CK, Sulniute R, Martinez-Carrasco I, Lundmark R, Lundberg P (2017) CCL11, a novel mediator of inflammatory bone resorption. Sci Rep 7(1):5334

Kitaori T, Ito H, Schwarz EM, Tsutsumi R, Yoshitomi H, Oishi S, Nakano M, Fujii N, Nagasawa T, Nakamura T (2009) Stromal cell-derived factor 1/CXCR4 signaling is critical for the recruitment of mesenchymal stem cells to the fracture site during skeletal repair in a mouse model. Arthritis Rheum 60(3):813–823

Kitase Y, Lee S, Gluhak-Heinrich J, Johnson ML, Harris SE, Bonewald LF (2014) CCL7 is a protective factor secreted by mechanically loaded osteocytes. J Dent Res 93(11):1108–1115

Kobayashi K, Toguchida J, Karin M, Kato T (2015) IKKβ in postnatal perichondrium remotely controls endochondral ossification of the growth plate through downregulation of MCP-5. Cell Death Differ 22(5):852–861

Koizumi K, Saitoh Y, Minami T, Takeno N, Tsuneyama K, Miyahara T, Nakayama T, Sakurai H, Takano Y, Nishimura M, Imai T, Yoshie O, Saiki I (2009) Role of CX3CL1/fractalkine in osteoclast differentiation and bone resorption. J Immunol 183(12):7825–7831

Kolar P, Schmidt-Bleek K, Schell H, Gaber T, Toben D, Schmidmaier G, Perka C, Buttgereit F, Duda GN (2010) The early fracture hematoma and its potential role in fracture healing. Tissue Eng Part B Rev 16(4):427–434

Könnecke I, Serra A, El Khassawna T, Schlundt C, Schell H, Hauser A, Ellinghaus A, Volk HD, Radbruch A, Duda GN, Schmidt-Bleek K (2014) T and B cells participate in bone repair by infiltrating the fracture callus in a two-wave fashion. Bone 64:155–165

Kovtun A, Bergdolt S, Wiegner R, Radermacher P, Huber-Lang M, Ignatius A (2016) The crucial role of neutrophil granulocytes in bone fracture healing. Eur Cell Mater 32:152–162

Kuan WP, Tam LS, Wong CK, Ko FW, Li T, Zhu T, Li EK (2010) CXCL 9 and CXCL 10 as sensitive markers of disease activity in patients with rheumatoid arthritis. J Rheumatol 37(2):257–264

Kwak HB, Ha H, Kim HN, Lee JH, Kim HS, Lee S, Kim HM, Kim JY, Kim HH, Song YW, Lee ZH (2008) Reciprocal cross-talk between RANKL and interferon-gamma-inducible protein 10 is responsible for bone-erosive experimental arthritis. Arthritis Rheum 58(5):1332–1342

Lacey DL, Timms E, Tan HL, Kelley MJ, Dunstan CR, Burgess T, Elliott R, Colombero A, Elliott G, Scully S, Hsu H, Sullivan J, Hawkins N, Davy E, Capparelli C, Eli A, Qian YX, Kaufman S, Sarosi I, Shalhoub V, Senaldi G, Guo J, Delaney J, Boyle WJ (1998) Osteoprotegerin ligand is a cytokine that regulates osteoclast differentiation and activation. Cell 93(2):165–176

Lean JM, Murphy C, Fuller K, Chambers TJ (2002) CCL9/MIP-1gamma and its receptor CCR1 are the major chemokine ligand/receptor species expressed by osteoclasts. J Cell Biochem 87(4):386–393

Lee JH, Kim HN, Kim KO, Jin WJ, Lee S, Kim HH, Ha H, Lee ZH (2012) CXCL10 promotes osteolytic bone metastasis by enhancing cancer outgrowth and osteoclastogenesis. Cancer Res 72(13):3175–3186

Lee J, Park C, Kim HJ, Lee YD, Lee ZH, Song YW, Kim HH (2017) Stimulation of osteoclast migration and bone resorption by C-C chemokine ligands 19 and 21. Exp Mol Med 49(7):e358

Leucht P, Temiyasathit S, Russell A, Arguello JF, Jacobs CR, Helms JA, Castillo AB (2013) CXCR4 antagonism attenuates load-induced periosteal bone formation in mice. J Orthop Res 31(11):1828–1838

Li C, Zhao J, Sun L, Yao Z, Liu R, Huang J, Liu X (2012) RANKL downregulates cell surface CXCR6 expression through JAK2/STAT3 signaling pathway during osteoclastogenesis. Biochem Biophys Res Commun 429(3–4):156–162

Lima ILA, Silva JMD, Rodrigues LFD, Madureira DF, Fonseca AC, Garlet GP, Teixeira MM, Russo RC, Fukada SY, Silva TAD (2017) Contribution of atypical chemokine receptor 2/ackr2 in bone remodeling. Bone 101:113–122

Lisignoli G, Toneguzzi S, Grassi F, Piacentini A, Tschon M, Cristino S, Gualtieri G, Facchini A (2002) Different chemokines are expressed in human arthritic bone biopsies: IFN-gamma and IL-6 differently modulate IL-8, MCP-1 and rantes production by arthritic osteoblasts. Cytokine 20(5):231–238

Lisignoli G, Piacentini A, Toneguzzi S, Grassi F, Tschon M, Cristino S, Facchini A, Mariani E (2003a) Age-associated changes in functional response to CXCR3 and CXCR5 chemokine receptors in human osteoblasts. Biogerontology 4(5):309–317

Lisignoli G, Toneguzzi S, Piacentini A, Cattini L, Lenti A, Tschon M, Cristino S, Grassi F, Facchini A (2003b) Human osteoblasts express functional CXC chemokine receptors 3 and 5: activation by their ligands, CXCL10 and CXCL13, significantly induces alkaline phosphatase and beta-N-acetylhexosaminidase release. J Cell Physiol 194(1):71–79

Lisignoli G, Piacentini A, Cristino S, Grassi F, Cavallo C, Cattini L, Tonnarelli B, Manferdini C, Facchini A (2007) CCL20 chemokine induces both osteoblast proliferation and osteoclast differentiation: increased levels of CCL20 are expressed in subchondral bone tissue of rheumatoid arthritis patients. J Cell Physiol 210(3):798–806

Liu YC, Kao YT, Huang WK, Lin KY, Wu SC, Hsu SC, Schuyler SC, Li LY, Leigh Lu F, Lu J (2014) CCL5/RANTES is important for inducing osteogenesis of human mesenchymal stem cells and is regulated by dexamethasone. Biosci Trends 8(3):138–143

Liu W, Wang P, Xie Z, Wang S, Ma M, Li J, Li M, Cen S, Tang S, Zheng G, Ye G, Wu X, Wu Y, Shen H (2019) Abnormal inhibition of osteoclastogenesis by mesenchymal stem cells through the miR-4284/CXCL5 axis in ankylosing spondylitis. Cell Death Dis 10(3):188

Longobardi L, Li T, Myers TJ, Rear L, Ozkan H, Li Y, Contaldo C, Spagnoli A (2012) TGF-β type II receptor/MCP-5 axis: at the crossroad between joint and growth plate development. Dev Cell 23(1):71–81

Lu Y, Cai Z, Xiao G, Keller ET, Mizokami A, Yao Z, Roodman GD, Zhang J (2007) Monocyte chemotactic protein-1 mediates prostate cancer-induced bone resorption. Cancer Res 67(8):3646–3653

Matic I, Matthews BG, Wang X, Dyment NA, Worthley DL, Rowe DW, Grcevic D, Kalajzic I (2016) Quiescent bone lining cells are a major source of osteoblasts during adulthood. Stem Cells 34(12):2930–2942

Matloubian M, David A, Engel S, Ryan JE, Cyster JG (2000) A transmembrane CXC chemokine is a ligand for HIV-coreceptor Bonzo. Nat Immunol 1(4):298–304

Millard CJ, Ludeman JP, Canals M, Bridgford JL, Hinds MG, Clayton DJ, Christopoulos A, Payne RJ, Stone MJ (2014) Structural basis of receptor sulfotyrosine recognition by a CC chemokine: the N-terminal region of CCR3 bound to CCL11/eotaxin-1. Structure 22(11):1571–1581

Miller SC, de Saint-Georges L, Bowman BM, Jee WS (1989) Bone lining cells: structure and function. Scanning Microsc 3(3):953–960; discussion 960

Miyamoto K, Ninomiya K, Sonoda KH, Miyauchi Y, Hoshi H, Iwasaki R, Miyamoto H, Yoshida S, Sato Y, Morioka H, Chiba K, Egashira K, Suda T, Toyama Y, Miyamoto T (2009) MCP-1

expressed by osteoclasts stimulates osteoclastogenesis in an autocrine/paracrine manner. Biochem Biophys Res Commun 383(3):373–377
Miyazaki Y, Nakayamada S, Kubo S, Nakano K, Iwata S, Miyagawa I, Ma X, Trimova G, Sakata K, Tanaka Y (2018) Th22 cells promote osteoclast differentiation via production of IL-22 in rheumatoid arthritis. Front Immunol 9:2901
Molloy AP, Martin FT, Dwyer RM, Griffin TP, Murphy M, Barry FP, Brien T, Kerin MJ (2009) Mesenchymal stem cell secretion of chemokines during differentiation into osteoblasts, and their potential role in mediating interactions with breast cancer cells. Int J Cancer 124(2):326–332
Nakamura ES, Koizumi K, Kobayashi M, Saitoh Y, Arita Y, Nakayama T, Sakurai H, Yoshie O, Saiki I (2006) RANKL-induced CCL22/macrophage-derived chemokine produced from osteoclasts potentially promotes the bone metastasis of lung cancer expressing its receptor CCR4. Clin Exp Metastasis 23(1):9–18
Nakao K, Aoyama M, Fukuoka H, Fujita M, Miyazawa K, Asai K, Goto S (2009) IGF2 modulates the microenvironment for osteoclastogenesis. Biochem Biophys Res Commun 378(3):462–466
Nervi B, Link DC, DiPersio JF (2006) Cytokines and hematopoietic stem cell mobilization. J Cell Biochem 99(3):690–705
Oba Y, Lee JW, Ehrlich LA, Chung HY, Jelinek DF, Callander NS, Horuk R, Choi SJ, Roodman GD (2005) MIP-1alpha utilizes both CCR1 and CCR5 to induce osteoclast formation and increase adhesion of myeloma cells to marrow stromal cells. Exp Hematol 33(3):272–278
Okamatsu Y, Kim D, Battaglino R, Sasaki H, Späte U, Stashenko P (2004) MIP-1 gamma promotes receptor-activator-of-NF-kappa-B-ligand-induced osteoclast formation and survival. J Immunol 173(3):2084–2090
Onan D, Allan EH, Quinn JM, Gooi JH, Pompolo S, Sims NA, Gillespie MT, Martin TJ (2009) The chemokine Cxcl1 is a novel target gene of parathyroid hormone (PTH)/PTH-related protein in committed osteoblasts. Endocrinology 150(5):2244–2253
Oswald C, Rappas M, Kean J, Doré AS, Errey JC, Bennett K, Deflorian F, Christopher JA, Jazayeri A, Mason JS, Congreve M, Cooke RM, Marshall FH (2016) Intracellular allosteric antagonism of the CCR9 receptor. Nature 540(7633):462–465
Ota K, Quint P, Weivoda MM, Ruan M, Pederson L, Westendorf JJ, Khosla S, Oursler MJ (2013) Transforming growth factor beta 1 induces CXCL16 and leukemia inhibitory factor expression in osteoclasts to modulate migration of osteoblast progenitors. Bone 57(1):68–75
Panezai J, Ghaffar A, Altamash M, Sundqvist KG, Engström PE, Larsson A (2017) Correlation of serum cytokines, chemokines, growth factors and enzymes with periodontal disease parameters. PLoS One 12(11):e0188945
Park SH, Das BB, Casagrande F, Tian Y, Nothnagel HJ, Chu M, Kiefer H, Maier K, de Angelis AA, Marassi FM, Opella SJ (2012) Structure of the chemokine receptor CXCR1 in phospholipid bilayers. Nature 491(7426):779–783
Pathak JL, Bakker AD, Verschueren P, Lems WF, Luyten FP, Klein-Nulend J, Bravenboer N (2015) CXCL8 and CCL20 enhance osteoclastogenesis via modulation of cytokine production by human primary osteoblasts. PLoS One 10(6):e0131041
Pelus LM, Horowitz D, Cooper SC, King AG (2002) Peripheral blood stem cell mobilization. A role for CXC chemokines. Crit Rev Oncol Hematol 43(3):257–275
Pickens SR, Chamberlain ND, Volin MV, Pope RM, Talarico NE, Mandelin AM, Shahrara S (2012) Role of the CCL21 and CCR7 pathways in rheumatoid arthritis angiogenesis. Arthritis Rheum 64(8):2471–2481
Prisby RD (2017) Mechanical, hormonal and metabolic influences on blood vessels, blood flow and bone. J Endocrinol 235(3):R77–R100
Raskovalova T, Deegan PB, Yang R, Pavlova E, Stirnemann J, Labarère J, Zimran A, Mistry PK, Berger M (2017) Plasma chitotriosidase activity versus CCL18 level for assessing type I Gaucher disease severity: protocol for a systematic review with meta-analysis of individual participant data. Syst Rev 6(1):87
Repeke CE, Ferreira SB, Vieira AE, Silveira EM, Avila-Campos MJ, da Silva JS, Santos CF, Campanelli AP, Trombone AP, Garlet GP (2011) Dose-response met-RANTES treatment of

experimental periodontitis: a narrow edge between the disease severity attenuation and infection control. PLoS One 6(7):e22526
Rodan GA, Martin TJ (1981) Role of osteoblasts in hormonal control of bone resorption – a hypothesis. Calcif Tissue Int 33(4):349–351
Shahnazari M, Chu V, Wronski TJ, Nissenson RA, Halloran BP (2013) CXCL12/CXCR4 signaling in the osteoblast regulates the mesenchymal stem cell and osteoclast lineage populations. FASEB J 27(9):3505–3513
Shaik S, Martin EC, Hayes DJ, Gimble JM, Devireddy RV (2019) Transcriptomic profiling of adipose derived stem cells undergoing osteogenesis by RNA-Seq. Sci Rep 9(1):11800
Shen PC, Lu CS, Shiau AL, Lee CH, Jou IM, Hsieh JL (2013) Lentiviral small hairpin RNA knockdown of macrophage inflammatory protein-1γ ameliorates experimentally induced osteoarthritis in mice. Hum Gene Ther 24(10):871–882
Shen Q, Fan X, Jiang M, Ye Z, Zhou Y, Tan WS (2019) Inhibiting expression of Cxcl9 promotes angiogenesis in MSCs-HUVECs co-culture. Arch Biochem Biophys 675:108108
Shigeno Y, Ashton BA (1995) Human bone-cell proliferation in vitro decreases with human donor age. J Bone Joint Surg Br 77(1):139–142
Silva TA, Garlet GP, Fukada SY, Silva JS, Cunha FQ (2007) Chemokines in oral inflammatory diseases: apical periodontitis and periodontal disease. J Dent Res 86(4):306–319
Simonet WS, Lacey DL, Dunstan CR, Kelley M, Chang MS, Lüthy R, Nguyen HQ, Wooden S, Bennett L, Boone T, Shimamoto G, DeRose M, Elliott R, Colombero A, Tan HL, Trail G, Sullivan J, Davy E, Bucay N, Renshaw-Gegg L, Hughes TM, Hill D, Pattison W, Campbell P, Sander S, Van G, Tarpley J, Derby P, Lee R, Boyle WJ (1997) Osteoprotegerin: a novel secreted protein involved in the regulation of bone density. Cell 89(2):309–319
Stolzing A, Jones E, McGonagle D, Scutt A (2008) Age-related changes in human bone marrow-derived mesenchymal stem cells: consequences for cell therapies. Mech Ageing Dev 129(3):163–173
Su Y, Zhou Y, Sun YJ, Wang YL, Yin JY, Huang YJ, Zhang JJ, He AN, Han K, Zhang HZ, Yao Y, Lv XB, Hu HY (2019) Macrophage-derived CCL18 promotes osteosarcoma proliferation and migration by upregulating the expression of UCA1. J Mol Med 97(1):49–61
Sul OJ, Ke K, Kim WK, Kim SH, Lee SC, Kim HJ, Kim SY, Suh JH, Choi HS (2012) Absence of MCP-1 leads to elevated bone mass via impaired actin ring formation. J Cell Physiol 227(4):1619–1627
Sundaram K, Rao DS, Ries WL, Reddy SV (2013) CXCL5 stimulation of RANK ligand expression in Paget's disease of bone. Lab Investig 93(4):472–479
Syversen SW, Goll GL, Haavardsholm EA, Bøyesen P, Lea T, Kvien TK (2008) A high serum level of eotaxin (CCL 11) is associated with less radiographic progression in early rheumatoid arthritis patients. Arthritis Res Ther 10(2):R28
Taichman RS, Reilly MJ, Matthews LS (2000) Human osteoblast-like cells and osteosarcoma cell lines synthesize macrophage inhibitory protein 1alpha in response to interleukin 1beta and tumour necrosis factor alpha stimulation in vitro. Br J Haematol 108(2):275–283
Tan Q, Zhu Y, Li J, Chen Z, Han GW, Kufareva I, Li T, Ma L, Fenalti G, Li J, Zhang W, Xie X, Yang H, Jiang H, Cherezov V, Liu H, Stevens RC, Zhao Q, Wu B (2013) Structure of the CCR5 chemokine receptor-HIV entry inhibitor maraviroc complex. Science 341(6152):1387–1390
Thevenot PT, Nair AM, Shen J, Lotfi P, Ko CY, Tang L (2010) The effect of incorporation of SDF-1alpha into PLGA scaffolds on stem cell recruitment and the inflammatory response. Biomaterials 31(14):3997–4008
Toben D, Schroeder I, El Khassawna T, Mehta M, Hoffmann JE, Frisch JT, Schell H, Lienau J, Serra A, Radbruch A, Duda GN (2011) Fracture healing is accelerated in the absence of the adaptive immune system. J Bone Miner Res 26(1):113–124
Trehan SK, Zambrana L, Jo JE, Purdue E, Karamitros A, Nguyen JT, Lane JM (2018) An alternative macrophage activation pathway regulator, CHIT1, may provide a serum and synovial fluid biomarker of periprosthetic osteolysis. HSS J 14(2):148–152
Trost Z, Trebse R, Prezelj J, Komadina R, Logar DB, Marc J (2010) A microarray based identification of osteoporosis-related genes in primary culture of human osteoblasts. Bone 46(1):72–80

Ullah TR (2019) The role of CXCR4 in multiple myeloma: cells' journey from bone marrow to beyond. J Bone Oncol 17:100253
Usui M, Okamatsu Y, Sato T, Hanatani T, Moritani Y, Sano K, Yamamoto M, Nakashima K (2016) Thymus-expressed chemokine enhances Porphyromonas gingivalis LPS-induced osteoclast formation via NFATc1 activation. Arch Oral Biol 66:77–85
Valerio MS, Herbert BA, Basilakos DS, Browne C, Yu H, Kirkwood KL (2015) Critical role of MKP-1 in lipopolysaccharide-induced osteoclast formation through CXCL1 and CXCL2. Cytokine 71(1):71–80
Vallet S, Pozzi S, Patel K, Vaghela N, Fulciniti MT, Veiby P, Hideshima T, Santo L, Cirstea D, Scadden DT, Anderson KC, Raje N (2011) A novel role for CCL3 (MIP-1α) in myeloma-induced bone disease via osteocalcin downregulation and inhibition of osteoblast function. Leukemia 25(7):1174–1181
Van Raemdonck K, Umar S, Palasiewicz K, Volkov S, Volin MV, Arami S, Chang HJ, Zanotti B, Sweiss N, Shahrara S (2019) CCL21/CCR7 signaling in macrophages promotes joint inflammation and Th17-mediated osteoclast formation in rheumatoid arthritis. Cell Mol Life Sci. https://doi.org/10.1007/s00018-019-03235-w
Vater A, Klussmann S (2015) Turning mirror-image oligonucleotides into drugs: the evolution of Spiegelmer(®) therapeutics. Drug Discov Today 20(1):147–155
Votta BJ, White JR, Dodds RA, James IE, Connor JR, Lee-Rykaczewski E, Eichman CF, Kumar S, Lark MW, Gowen M (2000) CKbeta-8 [CCL23], a novel CC chemokine, is chemotactic for human osteoclast precursors and is expressed in bone tissues. J Cell Physiol 183(2):196–207
Wang W, Soto H, Oldham ER, Buchanan ME, Homey B, Catron D, Jenkins N, Copeland NG, Gilbert DJ, Nguyen N, Abrams J, Kershenovich D, Smith K, McClanahan T, Vicari AP, Zlotnik A (2000) Identification of a novel chemokine (CCL28), which binds CCR10 (GPR2). J Biol Chem 275(29):22313–22323
Weitzmann MN (2017) Bone and the immune system. Toxicol Pathol 45(7):911–924
Wen Y, Guo X, Hao J, Xiao X, Wang W, Wu C, Wang S, Yang T, Shen H, Chen X, Tan L, Tian Q, Deng HW, Zhang F (2016) Integrative analysis of genome-wide association studies and gene expression profiles identified candidate genes for osteoporosis in Kashin-Beck disease patients. Osteoporos Int 27(3):1041–1046
Wilgus TA, Roy S, McDaniel JC (2013) Neutrophils and wound repair: positive actions and negative reactions. Adv Wound Care 2(7):379–388
Wu B, Chien EY, Mol CD, Fenalti G, Liu W, Katritch V, Abagyan R, Brooun A, Wells P, Bi FC, Hamel DJ, Kuhn P, Handel TM, Cherezov V, Stevens RC (2010) Structures of the CXCR4 chemokine GPCR with small-molecule and cyclic peptide antagonists. Science 330(6007):1066–1071
Wu AC, Raggatt LJ, Alexander KA, Pettit AR (2013) Unraveling macrophage contributions to bone repair. Bonekey Rep 2:373
Xu S, Zhang Y, Liu B, Li K, Huang B, Yan B, Zhang Z, Liang K, Jia C, Lin J, Zeng C, Cai D, Jin D, Jiang Y, Bai X (2016) Activation of mTORC1 in B lymphocytes promotes osteoclast formation via regulation of β-catenin and RANKL/OPG. J Bone Miner Res 31(7):1320–1333
Xuan W, Feng X, Qian C, Peng L, Shi Y, Xu L, Wang F, Tan W (2017) Osteoclast differentiation gene expression profiling reveals chemokine CCL4 mediates RANKL-induced osteoclast migration and invasion via PI3K pathway. Cell Biochem Funct 35(3):171–177
Yang M, Mailhot G, MacKay CA, Mason-Savas A, Aubin J, Odgren PR (2006) Chemokine and chemokine receptor expression during colony stimulating factor-1-induced osteoclast differentiation in the toothless osteopetrotic rat: a key role for CCL9 (MIP-1gamma) in osteoclastogenesis in vivo and in vitro. Blood 107(6):2262–2270
Yang Y, Zhou X, Li Y, Chen A, Liang W, Liang G, Huang B, Li Q, Jin D (2019) CXCL2 attenuates osteoblast differentiation by inhibiting the ERK1/2 signaling pathway. J Cell Sci 132(16). pii: jcs230490

Yano S, Mentaverri R, Kanuparthi D, Bandyopadhyay S, Rivera A, Brown EM, Chattopadhyay N (2005) Functional expression of beta-chemokine receptors in osteoblasts: role of regulated upon activation, normal T cell expressed and secreted (RANTES) in osteoblasts and regulation of its secretion by osteoblasts and osteoclasts. Endocrinology 146(5):2324–2335

Yasuda H, Shima N, Nakagawa N, Yamaguchi K, Kinosaki M, Mochizuki S, Tomoyasu A, Yano K, Goto M, Murakami A, Tsuda E, Morinaga T, Higashio K, Udagawa N, Takahashi N, Suda T (1998) Osteoclast differentiation factor is a ligand for osteoprotegerin/osteoclastogenesis-inhibitory factor and is identical to TRANCE/RANKL. Proc Natl Acad Sci U S A 95(7):3597–3602

Yoo HI, Kang JH, Yang SY, Yong JH, Moon JS, Kim MS, Jung JY, Koh JT, Kim WJ, Oh WM, Lee EJ, Kim SH (2011) Differential expression of cxcl-14 during eruptive movement of rat molar germs. J Exp Zool Part B Mol Dev Evol 316(6):418–426

Yu X, Huang Y, Collin-Osdoby P, Osdoby P (2003) Stromal cell-derived factor-1 (SDF-1) recruits osteoclast precursors by inducing chemotaxis, matrix metalloproteinase-9 (MMP-9) activity, and collagen transmigration. J Bone Miner Res 18(8):1404–1418

Yu X, Huang Y, Collin-Osdoby P, Osdoby P (2004) CCR1 chemokines promote the chemotactic recruitment, RANKL development, and motility of osteoclasts and are induced by inflammatory cytokines in osteoblasts. J Bone Miner Res 19(12):2065–2077

Yumimoto K, Sugiyama S, Mimori K, Nakayama KI (2019) Potentials of C-C motif chemokine 2-C-C chemokine receptor type 2 blockers including propagermanium as anticancer agents. Cancer Sci 110(7):2090–2099

Zhang J, Niu C, Ye L, Huang H, He X, Tong WG, Ross J, Haug J, Johnson T, Feng JQ, Harris S, Wiedemann LM, Mishina Y, Li L (2003) Identification of the haematopoietic stem cell niche and control of the niche size. Nature 425(6960):836–841

Zhang Y, Ma C, Yu Y, Liu M, Yi C (2015) Are CXCL13/CXCR5/FAK critical regulators of MSCs migration and differentiation? Med Hypotheses 84(3):213–215

Zhao S, Wu B, Stevens RC (2019) Advancing chemokine GPCR structure based drug discovery. Structure 27(3):405–408

Zheng Y, Qin L, Zacarías NV, de Vries H, Han GW, Gustavsson M, Dabros M, Zhao C, Cherney RJ, Carter P, Stamos D, Abagyan R, Cherezov V, Stevens RC, IJzerman AP, Heitman LH, Tebben A, Kufareva I, Handel TM (2016) Structure of CC chemokine receptor 2 with orthosteric and allosteric antagonists. Nature 540(7633):458–461

Zheng Y, Ley SH, Hu FB (2018) Global aetiology and epidemiology of type 2 diabetes mellitus and its complications. Nat Rev Endocrinol 14(2):88–98

Zhu W, Boachie-Adjei O, Rawlins BA, Frenkel B, Boskey AL, Ivashkiv LB, Blobel CP (2007) A novel regulatory role for stromal-derived factor-1 signaling in bone morphogenic protein-2 osteogenic differentiation of mesenchymal C2C12 cells. J Biol Chem 282(26):18676–18685

Zhu W, Liang G, Huang Z, Doty SB, Boskey AL (2011) Conditional inactivation of the CXCR4 receptor in osteoprecursors reduces postnatal bone formation due to impaired osteoblast development. J Biol Chem 286(30):26794–26805

Calcium and Bone

Ian R. Reid and Sarah M. Bristow

Contents

I. R. Reid (✉)
Department of Medicine, Faculty of Medical and Health Sciences, University of Auckland, Auckland, New Zealand

Auckland District Health Board, Auckland, New Zealand
e-mail: i.reid@auckland.ac.nz

S. M. Bristow
Department of Medicine, Faculty of Medical and Health Sciences, University of Auckland, Auckland, New Zealand
e-mail: s.bristow@auckland.ac.nz

P. H. Stern (ed.), *Bone Regulators and Osteoporosis Therapy*,
Handbook of Experimental Pharmacology 262, https://doi.org/10.1007/164_2019_324

Abstract

The maintenance of extracellular calcium levels within a narrow range is necessary for normal function of the nervous system, muscle, and coagulation, to maintain mineralization of the skeleton but to avoid calcification of soft tissues. Accordingly, absorption and excretion of calcium is closely regulated, and adult humans can adapt to a wide range of calcium intakes from 300 to 2,000 mg/day. The evidence that low calcium intakes contribute to osteoporosis development is weak, as is evidence that increasing these intakes significantly changes fracture risk. Consistent with this view, the United States Preventive Services Task Force does not support the use of calcium supplements in healthy community-dwelling adults. While some groups continue to recommend that supplements of calcium and vitamin D are given with drug treatments for osteoporosis, this view is not supported by clinical trials which demonstrate anti-fracture efficacy of estrogens and bisphosphonates in the absence of such supplementation. Thus, calcium supplements have only a minor place in contemporary medical practice.

Keywords

Bone · Bone density · Calcium · Calcium balance · Fracture · Osteoporosis · Vitamin D

1 Introduction

Calcium is an element present in biological systems as a positively charged ion and is a major part of the mineral component of bone. It also plays critical roles in intracellular signaling, coagulation, and the function of nerves and muscles (including cardiac muscle), so maintenance of stable extracellular concentrations is a homeostatic priority. Calcium and phosphate are the principal components of the hydroxyapatite crystal, which is the basic building block of bone mineral. Mineral is critical to the compressive strength of bone, but it is the collagen scaffold within which the hydroxyapatite crystals sit that determines bone size and structure (Fig. 1) (Grandfield et al. 2018). This scaffold is laid down by osteoblasts, with mineralization taking place subsequently if extracellular fluid concentrations of calcium and phosphate are normal, and if no mineralization inhibitors are present. Bone is subsequently remodeled by osteoclasts, acting under the direct control of osteocytes, which are in turn influenced by skeletal load, cytokine production, and hormonal influences.

Control of mineralization is a critical homeostatic function – the skeleton needs to be adequately mineralized for optimal strength, but soft tissues must *not* become mineralized since this is associated with cell dysfunction and cell death. Tight control of circulating levels of both calcium and phosphate is critical to maintaining this balance, as is the regulated activity of mineralization inhibitors including pyrophosphate and mineralization-inhibiting proteins such as fetuin-A. Both calcium overload and calcium deficiency can have severe adverse impacts on health.

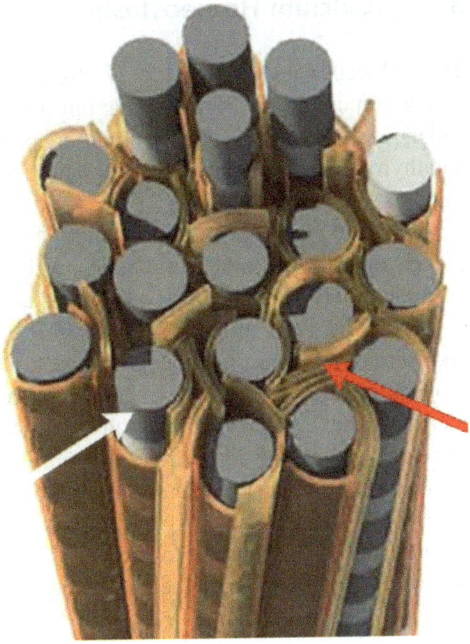

Fig. 1 Model of ultrastructure of human cortical bone, based on studies with scanning transmission electron microscopy. Collagen fibrils are dark gray, as pinpointed by the white arrow. They are about 50 nm in diameter. Most of the mineral of cortical human osteonal bone is in the form of long, thin, polycrystalline plates, known as mineral lamellae (shown in orange) which are curved to wrap closely around the collagen fibrils. In addition, stacks of closely packed uncurved lamellae occur between fibrils (red arrow). From Grandfield et al., Calcified Tissue International 2018, used with permission

2 Calcium Distribution

The adult human body contains about a kilogram of calcium, 99% in the skeleton in the form of hydroxyapatite. The remaining 1% is outside of bone and under tight homeostatic control.

Calcium circulates in plasma in a concentration of 2.2–2.5 mM. Slightly less than half is ionized, about the same proportion is bound to albumin, and the remaining 10% is complexed with anions including phosphate and citrate. It is the ionized or free calcium which is physiologically active and hormonally regulated, and this equilibrates between blood and the extracellular fluid.

Calcium is principally an extracellular cation, intracellular concentrations being almost 10,000-fold lower at about 100 nM. To maintain this gradient, calcium is actively extruded from cells by ATP-dependent calcium pumps, calcium channels, and sodium-calcium exchangers. Maintenance of this steep calcium gradient across the cell membrane is a critical part of maintaining the electrical polarity of the membrane, on which the function of the nervous system and muscle is dependent. The low intracellular calcium concentration allows the ion to act as an intracellular signal, with calcium entry through the cell membrane or its release from intracellular stores acting as a signal to regulate cell functions, such as hormone release.

3 Calcium Homeostasis

This subject is reviewed in detail elsewhere (Vautour and Goltzman 2019). The organs principally involved in calcium fluxes in the body are the gut, the kidneys, and the skeleton (Fig. 2). These fluxes are primarily regulated by the chief cells of the parathyroid glands, where ambient calcium concentrations are sensed by cell surface calcium-sensing receptors.

A reduced plasma calcium concentration leads to reduced binding of calcium ions to their receptor on parathyroid cells, resulting in increased secretion of parathyroid hormone (PTH), which has two principal targets – osteoblasts and the renal tubule. In osteoblasts, PTH stimulates release of RANKL which leads to osteoclastogenesis and increased bone resorption. Osteoblast numbers are also increased, but in physiological circumstances bone resorption usually predominates. In the kidneys, PTH increases the tubular reabsorption of calcium from the glomerular filtrate and increases activity of 25-hydroxyvitamin D 1α-hydroxylase, resulting in increased formation of the most potent vitamin D metabolite, 1,25-dihydroxyvitamin D (1,25 $(OH)_2D$). This, in turn, acts on the proximal small intestine to increase intestinal calcium absorption, but it also stimulates RANKL production by osteoblasts, so can increase bone resorption. The combined direct and indirect effects of PTH on the

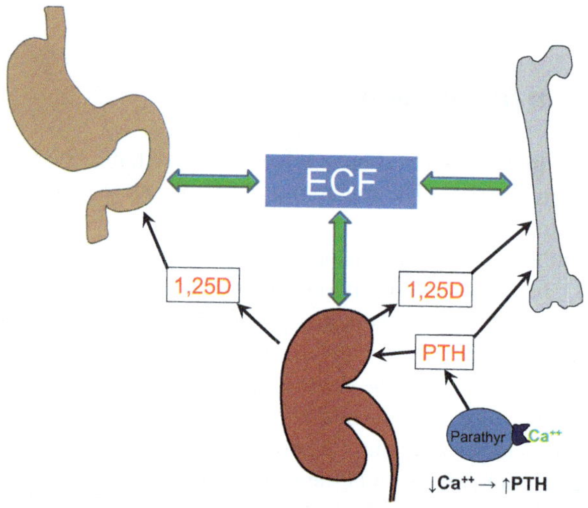

Fig. 2 A simplified schema of calcium homeostasis. Three organs exchange calcium with the extracellular fluid (ECF): the gastrointestinal tract, the bone, and the kidneys. These fluxes are primarily regulated by the parathyroid glands, which secrete parathyroid hormone (PTH) in response to reduced ambient levels of calcium. Calcium-binding to calcium-sensing receptors on the surface of parathyroid chief cells reduces PTH secretion. PTH acts directly on bone to stimulate bone resorption and on the kidneys, to promote renal reabsorption of calcium and the production of 1,25-dihydroxyvitamin D (1,25D). 1,25-Dihydroxyvitamin D primarily regulates calcium absorption in the proximal small intestine but also stimulates bone resorption. Thus, PTH and 1,25-dihydroxyvitamin D act in concert to maintain circulating calcium concentrations. Copyright IR Reid, used with permission

bone, kidney, and gut lead to increased calcium flux into the extracellular fluid, restoring the concentration of ionized calcium to normal. When plasma calcium levels oscillate above normal, the opposite cascade of events occurs, and ionized calcium is again brought back to its set point. There are calcium-sensing receptors in part of the renal tubule which act to directly reduce calcium reabsorption when plasma calcium levels are elevated. PTH also stimulates secretion of the phosphaturic hormone fibroblast growth factor 23 (FGF23) from osteoblasts and osteocytes, which in turn reduces 1,25$(OH)_2$D formation and PTH secretion, providing a negative feedback loop to this sequence of events.

4 Intestinal Calcium Absorption

Dietary calcium intake varies widely across communities, from 100 mg/day in some parts of Africa up to more than 2,000 mg/day in those with high intakes of dairy products. In Western countries, typical adult intakes are 500–1,000 mg/day. The portion of dietary calcium intake that is absorbed varies from 10 to 60%, with a typical value in healthy adults taking a Western diet being about 20%. Thus, the mechanisms outlined above vary the efficiency of absorption to account for variations in dietary intake. This is necessary to allow maintenance of circulating calcium levels while preventing tissue overload with calcium, which could result in soft tissue calcification. At low calcium intakes, most absorption is via a cell-mediated pathway regulated by 1,25$(OH)_2$D, involving calcium entry to the enterocyte via an apical calcium channel, followed by calcium transport across the cell facilitated by the protein calbindin-D9k, then its extrusion across the basolateral membrane by a calcium ATPase. When calcium intakes are high, this pathway is downregulated, and most calcium absorption takes place via passive diffusion down a paracellular route.

Measurement of calcium absorption efficiency is not usually clinically relevant. Low absorption in a healthy individual usually indicates that their calcium intake is high. Fractional calcium absorption is adjusted to maintain ionized calcium at its set point and has not been shown to be related to skeletal health.

5 Vitamin D Metabolism

This topic is dealt with in detail elsewhere in this volume but deserves mention here because of the critical role it plays in intestinal calcium absorption. 1,25$(OH)_2$D directly regulates the active absorption of calcium in the upper small bowel. Severe deficiency of the parent vitamin D results in reduced levels of the principal circulating form of vitamin D, 25-hydroxyvitamin D (25OHD), which can limit synthesis of 1,25$(OH)_2$D via substrate deficiency. The role of 25OHD in calcium metabolism is contentious – it does bind to the vitamin D receptor, though with an affinity 1/1,000 of that of 1,25$(OH)_2$D, but circulates in levels 1,000-fold higher than 1,25$(OH)_2$D. Both metabolites circulate bound to vitamin D binding protein, and

25OHD displaces $1,25(OH)_2D$ from that protein, so its levels directly impact on free $1,25(OH)_2D$ concentrations. The vitamin D endocrine system is directed toward the maintenance of serum calcium, not the maintenance of bone. Thus, high concentrations of $1,25(OH)_2D$ not only stimulate bone resorption but also inhibit bone mineralization, emphasizing the homeostatic priority given to maintenance of normocalcemia, with the skeleton acting as a reservoir for this metabolically important ion (Lieben et al. 2012).

6 Dietary Calcium Requirement

Determining the calcium requirement for healthy individuals remains a vexed issue despite almost a century of work on the subject. Most estimates have been based on calcium balance studies, but the advent of dual-energy X-ray absorptiometry (DXA) for bone density measurement has made direct assessment of bone balance possible.

6.1 Calcium Balance Studies

Balance studies involve measurements of both dietary calcium intake and loss of calcium in the urine and feces over periods of days to weeks. Calcium balance studies from the 1930s through to the 1950s concluded that calcium requirements could be met with daily intakes of only a few hundred milligrams (Malm 1958; Nicholls and Nimalasuriya 1939; Pathak 1958; Walker 1956), and as low as 100 mg in men (Hegsted et al. 1952). Based on these findings, the World Health Organization (WHO) and the Food and Agriculture Organization of the United Nations (FAO) (World Health Organization and Food and Agriculture Organization of the United Nations 1962) concluded in 1962 that:

> Most apparently healthy people – throughout the world – develop and live satisfactorily on a dietary intake of calcium which lies between 300 mg and over 1,000 mg a day. There is so far no convincing evidence that, in the absence of nutritional disorders and especially when the vitamin D status is adequate, an intake of calcium even below 300 mg or above 1,000 mg a day is harmful.

Thus, in 1974, these organizations recommended minimal intakes of calcium for adults of 400–500 mg/day (Food and Agriculture Organization of the United Nations and World Health Organization 1974). These findings fit with intakes in Africa and Asia which have frequently been below these levels, and yet not associated with skeletal health problems until they drop below 200 mg/day, when rickets can develop in children (Thandrayen and Pettifor 2018).

In the 1970s, the focus on this issue in the United States and the United Kingdom became more intense, with most of the work still based on calcium balance studies, since that was the only technique available. Such assessments were undertaken with a view to optimizing bone health, so calcium balance was used as a surrogate for bone balance, which was not then able to be measured directly. A key contributor was Robert Heaney, who published two papers describing calcium balance studies in 168 perimenopausal nuns (Heaney et al. 1977, 1978). They analyzed their data by regressing calcium balance against calcium intake, finding positive relationships in both pre- and postmenopausal women. This analysis has a fundamental problem in that balance is the difference between intake and output, so there is a mathematical inevitability that a positive relationship will be found even if there is no biological basis for it. That problem was not recognized at the time, and the message taken from these studies was that intake and balance were tightly linked and that premenopausal women achieved calcium balance when their dietary intake was 990 mg/day and in postmenopausal woman balance occurred at 1,504 mg/day. These findings guided calcium recommendations in North America over the following 30 years, though in Europe and the United Kingdom, the value of calcium in optimizing postmenopausal bone health was not so firmly embraced.

A subsequent study of calcium balances from both men which circumvented the statistical pitfall of the Heaney studies concluded that calcium balance could be achieved with an average calcium intake of 741 mg/day (Hunt and Johnson 2007) and commented:

> . . . that calcium balance was highly resistant to a change in calcium intake across a broad range of typical dietary calcium intakes (415–1,740 mg/day; between the ~25th and >99th percentiles of typical calcium intake for all female children and adults aged ≥9 years). In other words, homeostatic mechanisms for calcium metabolism seem to be functional across a broad range of typical dietary calcium intakes to minimize calcium losses and accumulations.

A recent study of 137 participants from China has found that intakes of 300 mg/day achieve calcium balance in men and women aged 18–60 years (Fang et al. 2016).

Together, these studies suggest that healthy adults can achieve calcium balance on a wide range of dietary intakes, as long as they had been receiving them long-term. Thus, studies from low dietary calcium environments suggest a very low calcium requirement, whereas those from communities where dairy products form a major part of the diet produce much higher estimate of requirement. The likelihood is that healthy adults can adapt to any of these intakes, as long as their vitamin D status is satisfactory and sufficient time is available – studies that take place after recent changes in intake are likely to be misleading. These conclusions are consistent with the review of calcium physiology provided above and indeed with findings with other important nutrients, for which a wide range of intakes is compatible with good health.

The finding in some studies of zero calcium balance in postmenopausal women is intriguing, since we now know from bone density studies that bone loss is ongoing throughout the postmenopausal period. This is evidence of the lack of reliability of calcium balance as a measure of skeletal balance and may also have been contributed to by studies being carried out after abrupt changes in intake and by extrapolation of data beyond the intakes that were actually studied.

6.2 Bone Balance Studies

In the 1980s it became possible to measure bone density in the spine and proximal femur, as well as throughout the entire skeleton. The latter technique permits assessment of total body bone mineral and total body calcium noninvasively and with minimal radiation exposure over periods of many years, in contrast to the intervals of days to weeks over which calcium balance was previously assessed. Measurements are made while participants are taking their habitual diet and without disturbance to their normal lifestyle.

Cross-sectional studies show there is little relationship between bone density and calcium intake (Anderson et al. 2012; Aptel et al. 1999; Bristow et al. 2019; Kroger et al. 1994; Mavroeidi et al. 2009; Reid et al. 1992). Prospective studies allow direct assessment of long-term bone balance in large numbers of subjects. We have now reported 2 studies of 570 and 698 healthy postmenopausal women, studied over periods of 5 and 6 years, respectively (Bristow et al. 2019; Reid et al. 2015). Both studies demonstrated ongoing loss of bone which was completely unrelated to the average dietary calcium intake measured throughout the study period (Fig. 3). Findings were similar in a study of this design in healthy older men (Bristow et al. 2017). These results are in agreement with the majority of other observational studies of changes in bone density over time, where an effect of calcium intake is not found, as long as osteoporosis therapies or hormone treatment are not being co-administered (Hannan et al. 2000; Hansen et al. 1991; Hosking et al. 1998; Nakamura et al. 2012; Reid et al. 1994; Riggs et al. 1987; van Beresteijn et al. 1990). This directly contradicts the balance studies which indicated that bone loss in older women could be completely prevented with adequate calcium intake.

6.3 Determining Requirement: Calcium or Bone Balance?

We need to consider the merits of each assessment technique in order to decide which should inform our estimates of dietary calcium requirement. Calcium balance is the difference between inputs and outputs. Because the daily difference in inputs and outputs is small, even minor errors in the collection of samples or estimation of their calcium content can result in large errors in the calculation of balance. Calcium balance studies assume that a steady state has been reached during the study period, but it is now apparent that adjustment of bone turnover and intestinal calcium absorption to changed intakes can take several years (Kanis 1991; Kanis and Passmore 1989). Also, the equating of calcium balance with bone balance assumes that soft tissue calcium content is stable, which in older people, particularly those with renal and/or vascular disease, might not be correct (Spiegel and Brady 2012). These factors might account for inconsistencies in findings from the two types of studies. There are other examples of calcium balance studies finding effects which would be implausible if equated with bone density changes, such as a balance of +800 mg/day with high calcium intake in renal failure (equivalent to a 29% improvement in bone density over a year) (Spiegel and Brady 2012), and a report

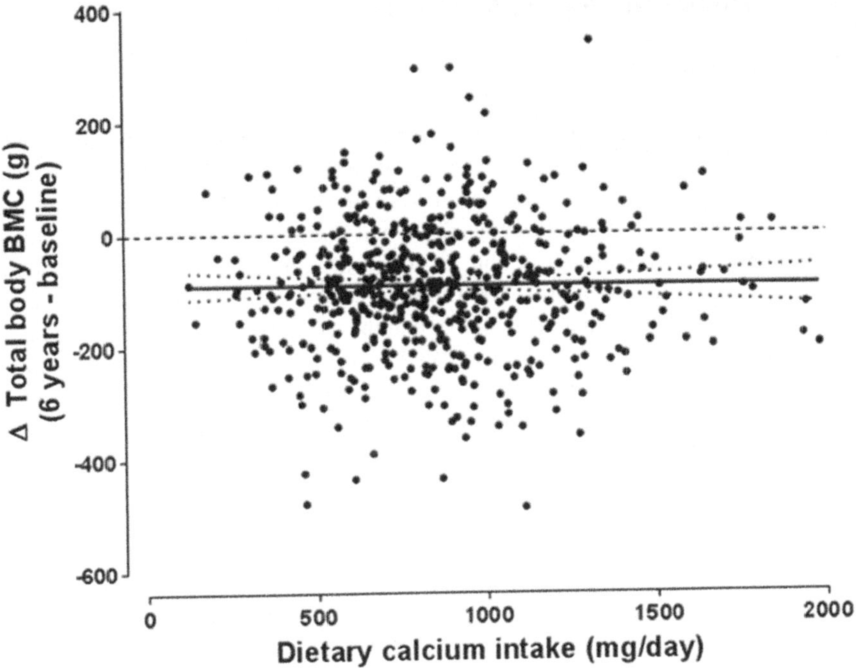

Fig. 3 Absolute change (Δ) in total body bone mineral content (BMC) over 6 years in 698 osteopenic postmenopausal women not receiving bone-active medications, in relation to each woman's average calcium intake assessed at baseline, year 3, and year 6. The regression line (with 95% CIs) for this relationship is shown ($P = 0.99$). From Bristow et al., J Clin Endocrinol Metab, 2019, used with permission

of calcium balance with alkali administration equivalent to an 8%/year improvement in bone density compared with placebo (Moseley et al. 2013), something never demonstrated with direct bone density measurements.

Now with the availability of many randomized controlled trials, it is clear that even calcium intakes approaching 2,000 mg/day are associated with continuing long-term bone loss in postmenopausal women (Reid et al. 2006). This indicates that calcium balance studies are not a satisfactory basis on which to determine optimal calcium intakes and that they have, in some cases, resulted in substantial overestimates of calcium requirements. It must be remembered that calcium balance was originally used as a surrogate for bone balance, which could not be measured at that time. Now bone mass can accurately be estimated with DXA, and it is predictive of fracture, which is not the case for calcium balance. These considerations indicate that calcium requirement should be determined from bone density studies and may be as low as 300 mg/day in vitamin D-replete adults.

7 Studies of Calcium Supplementation

7.1 Biochemical Effects

Men and women on self-selected diets demonstrate an inverse relationship between circulating PTH levels and dietary calcium intake (Bolland et al. 2006; Bristow et al. 2017). Dietary calcium is usually consumed in amounts of <100 mg per serving and has little impact on blood biochemistry beyond that of the meal as a whole, since mixed meals reduce bone resorption. In contrast, high-calcium foods and calcium supplements often contain 500–1,000 mg per serving or dose and increase blood calcium levels toward or above the upper end the normal range, with suppression of PTH and of bone resorption (Bristow et al. 2014). The elevation of serum calcium is present for at least 8 h, but not at 24 h. In individuals taking supplements long-term, PTH and bone formation markers show a sustained reduction of 10–20% (Reid et al. 2008), indicating that these supplements act as weak anti-turnover medications.

7.2 BMD Effects

The numerous studies of the effects of calcium or dairy supplements on bone density have recently been meta-analyzed (Fig. 4). Consistent with the inhibition of bone resorption demonstrated biochemically, there are positive changes in bone density in those randomized to calcium, compared with placebo. At 1 year the between-group advantage is 1.4%, at 2 years 1.3%, and at 3 years and beyond 1.2% from baseline. Thus, this is a noncumulative benefit which is, if anything, gradually diminishing over time. It is consistent with a reduced number of osteoclastic resorption lacunae over the bone surface but suggests no change in long-term bone balance, consistent with the observational data in Fig. 3. A change of this magnitude would not be expected to impact on fractures. The increases in bone density were similar in trials using dietary or supplemental calcium and were independent of co-administration of vitamin D, of calcium dose, and of baseline dietary calcium intake.

7.3 Fracture Effects

A recent comprehensive review of observational studies found that most studies reported no association between calcium intake and fracture (14/22 for total, 17/21 for hip, 7/8 for vertebral, and 5/7 for forearm fracture) (Bolland et al. 2015). The literature describing trials of calcium supplementation on fracture appears to be very confusing. It is made complicated by differences in trial design, specifically whether there was co-administration of vitamin D, and what target population was chosen. Many meta-analyses have been carried out, and these deal with these matters differently, and sometimes not transparently. Also, some meta-analyses only use subgroups from some of the large trials, which can impact substantially on their findings. With the Women's Health Initiative in particular, there have been a number

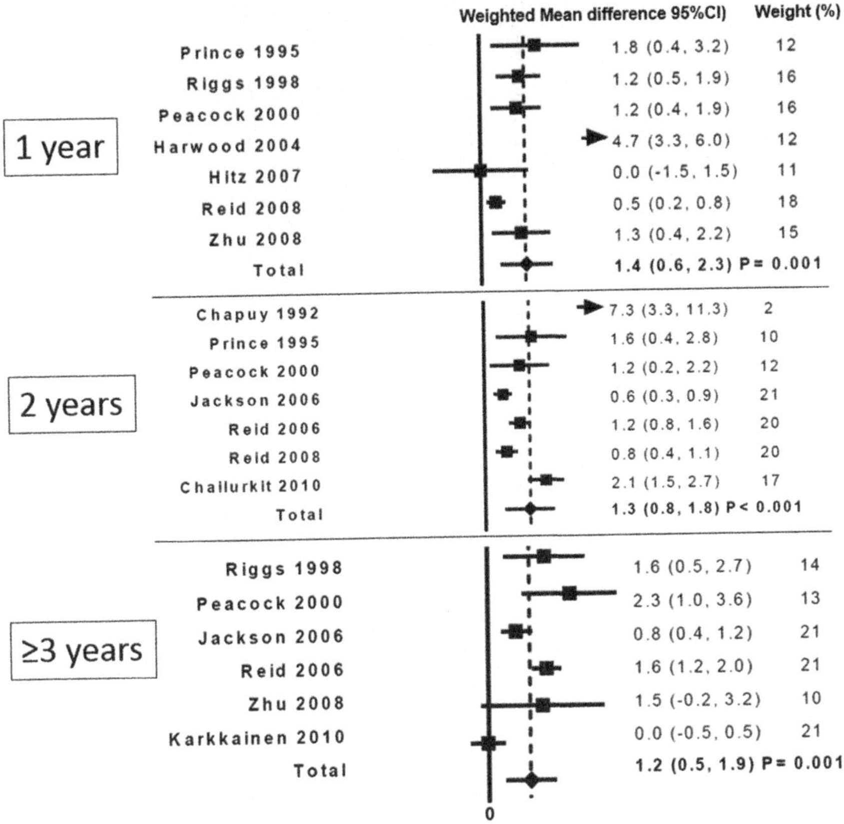

Fig. 4 Random-effects meta-analysis of effect of calcium supplements on percent change in total hip bone mineral density by duration of follow-up. There is a significant positive treatment effect at each time point, but no evidence of greater effects with longer treatment periods. From Tai et al., BMJ, 2015, used with permission

of post hoc analyses published with a wide range of findings. Which of these is used substantially impacts on meta-analytic results as a result of the size of this study. Sometimes their use in meta-analyses has been quite inappropriate (Avenell et al. 2016; Weaver et al. 2016). In contrast, recent meta-analyses in major journals have reached a common conclusion – that calcium supplements do not impact on fracture incidence in community-dwelling adults (Bolland et al. 2015; Zhao et al. 2017) – a position endorsed by other groups such as the United Sates Preventive Services Task Force (Grossman et al. 2018; Kahwati et al. 2018) and the International Osteoporosis Foundation (Harvey et al. 2017).

Formal meta-analysis probably adds little to a general consideration of the major studies. Table 1 sets out the study characteristics and fracture results from those trials which recruited more than 1,000 participants. The Chapuy study randomized 3,000 frail elderly women living in rest homes in France to calcium plus vitamin D, or to

Table 1 Major trials of calcium ± vitamin D on fracture

Study	Setting	N	Age (years)	Calcium (mg/day)	Vitamin D (IU/day)	Duration (months)	Relative risk of fracture	
							Total	Hip
Chapuy 1992,1994	Nursing home	3,270	84 (6)	1,200	800	36	**0.83 (0.71, 0.97)**	**0.77 (0.62, 0.96)**
RECORD 2005	Community	5,292	77 (6)	1,000	800	45	0.93 (0.82, 1.06)	1.10 (0.83, 1.47)
Porthouse 2005	Community	3,314	77 (5)	1,000	800	25	0.96 (0.70, 1.33)	0.71 (0.31, 1.64)
Women's Health Initiative 2006	Community	36,282	62 (7)	1,000	4,000	84	0.97 (0.92, 1.03)	0.88 (0.72, 1.07)
Prince 2006	Community	1,460	75 (3)	1,200	0	60	0.87 (0.69, 1.10)	1.83 (0.68, 4.93)
Reid 2006	Community	1,471	74 (4)	1,000	0	60	0.92 (0.75, 1.14)	**3.43 (1.27, 9.26)**
Salovaara 2010	Community	3,432	67 (2)	1,000	800	36	0.83 (0.62, 1.11)	2.00 (0.37, 10.88)

Age data are mean (SD). Relative risks are given with 95% confidence intervals. Where the confidence intervals do not include 1, data are bolded. Study participants were all female, except in the RECORD study, which was 15% male. Data from Bolland et al. (2015)

placebo (Chapuy et al. 1992, 1994). These women had calcium intakes about 500 mg/day but, more importantly, very low levels of 25OHD, as low as 12 nmol/L in the placebo group during the study [following correction of the assays used to current standards (Lips et al. 1999)]. This study found a dramatic beneficial effect, with a reduction in hip fractures of 23% and of 17% in total fractures. This study remains the clearest indication that calcium and vitamin D impact on fractures and meta-analyses which include it usually produce a positive result. However, it should be remembered that this population was severely vitamin D deficient, to such an extent that many women probably had osteomalacia. Thus, generalization of this result to heathy older people in the community is not appropriate, though it does highlight the nutritional and vitamin D deficiency that can be present in frail institutionalized elderly groups.

The study was followed by a series of large community-based studies, all of which have failed to reproduce the results of the Chapuy study (Jackson et al. 2006; Porthouse et al. 2005; Prince et al. 2006; RECORD Trial Group 2005; Reid et al. 2006; Salovaara et al. 2010). The relative risks of any fracture shown in Table 1 tend to be <1, but none are close to significance in spite of the large size of the studies. In contrast, several studies had relative risks of hip fracture above 1, significantly so in one study. The Women's Health Study appears not to follow this trend, but it has subsequently been reported that there was a significant interaction between randomization to calcium and that to hormone treatment, in terms of the effect on hip fracture (Robbins et al. 2014). In those not assigned to hormone treatment, the hazard ratio for the effect of calcium plus vitamin D on hip fracture was 1.20 (95% confidence interval 0.85–1.69), similar to most of the other studies in Table 1. The Porthouse study was not blinded, and there were a number of fractures for which confirmation was not available. Analysis of all reported fractures found an odds ratio 1.60 (95% confidence interval 0.75–3.40).

While it is now widely accepted that calcium supplements are not useful for primary fracture prevention, there remains some advocacy for calcium co-administration with other osteoporosis medications. There is no evidence that this is necessary, except in situations where there is a risk of hypocalcemia. A formal trial assessing the addition of calcium to alendronate showed no increase in its effects on bone density (Bonnick et al. 2007), and trials of estrogen (Cauley et al. 2003; Lindsay et al. 1980) and the bisphosphonates clodronate and zoledronate (McCloskey et al. 2007; Reid et al. 2018) have found anti-fracture benefits without co-administered calcium comparable to those seen with supplementation.

8 Calcium Intake in Children

The findings in children are very similar to those in adults. Some calcium balance studies suggest that higher intakes are beneficial (Jackman et al. 1997), but neither longitudinal studies of bone density (Lloyd et al. 2000) nor case-control studies of fracture (Händel et al. 2015) show effects of calcium intake. Meta-analyses of trials of calcium supplementation in children do not suggest a meaningful benefit in bone density terms from higher calcium intakes (Winzenberg et al. 2006a, b), a finding supported by a recent trial in 240 peri-pubertal children with low calcium intakes, in whom 18 months of dairy supplements did not affect bone density (Vogel et al. 2017), contradicting previous balance results from the same investigators. With very low baseline calcium intakes, more positive results have been found – increases of 5–6% in forearm bone density were found with 18 months of supplements in Nigerian toddlers aged 12–18 months, but the benefit was lost within 12 months of cessation of supplementation (Umaretiya et al. 2013).

9 Non-bone Effects of Calcium

As noted above, the acute biochemical effects of food calcium and of supplemental calcium are quite distinct, so their non-bone effects are also likely to be different. There is no consistent evidence of adverse effects from dietary calcium intakes up to 2 g/day, though if this is taken as dairy products, then the caloric and fat contents may be inappropriate for some individuals. Much more attention has focused on the non-bone effects of calcium supplements, with concern centering on gastrointestinal, renal, and cardiovascular safety.

9.1 Gastrointestinal

Prescribers of calcium supplements have long been aware of their common gastrointestinal side effects, mainly bloating and constipation. This is a frequent reason for low adherence rates with these supplements (Reid et al. 1993). A formal meta-analysis has shown that such complaints are increased 43% with supplements (Lewis et al. 2012). More concerning, one of the recent calcium trials showed that the risk of admission to hospital with an acute abdominal surgical condition was increased twofold by randomization to the calcium group (Lewis et al. 2012). Thus, the gastrointestinal side effects of calcium are not always just minor discomforts.

9.2 Renal Calculi

Increased calcium intake results in increased urine calcium output as the homeostatic mechanisms that protect the body from calcium overload come into play (Gallagher et al. 2014). Accordingly, the Women's Health Initiative reported a statistically significant 17% increase in renal calculi in the calcium plus vitamin D group (Jackson et al. 2006).

9.3 Cardiovascular Disease

The potential impact of calcium intake on cardiovascular disease has been a focus of attention for many decades, since reports that vascular risk was lower in hard water areas. It has been reviewed in detail recently (Reid et al. 2017). It was demonstrated that calcium supplements improved circulating lipid profiles (Reid et al. 2002) and there is evidence of transient reductions in blood pressure of a few mmHg in trials of calcium supplements (Reid et al. 2005). Accordingly, we pre-specified cardiovascular events as secondary endpoints in the Auckland Calcium Study and found a significant increase in myocardial infarction in those randomized to calcium (Bolland et al. 2008). We subsequently reviewed all other trials of calcium monotherapy from which cardiovascular event data was available (Bolland et al. 2010a). No other study showed a significant increase in events, but an upward trend of about 20% was seen in all the available studies, with the meta-analysis of these trials showing a 27% increase in risk of myocardial infarction.

In contrast, the Women's Health Initiative did not show an increase in risk of myocardial infarction, though in nonobese participants, those investigators reported a relative risk of myocardial infarction of 1.17 (Hsia et al. 2007). This study was unusual in that participants already receiving the trial intervention (i.e., calcium) were recruited and were permitted to continue their own supplements in addition to those provided as the trial intervention. We, therefore, proposed an analysis that focused on participants who were calcium-naive at the time of randomization ($n = 16{,}000$), and this demonstrated a hazard ratio for clinical myocardial infarction of 1.22 (1.00, 1.50) associated with randomization to calcium plus vitamin D (Bolland et al. 2011). Thus, meta-analysis of trials of calcium with or without vitamin D confirmed our original finding with a relative risk of myocardial infarction of 1.24 (1.07, 1.45) and of stroke of 1.15 (1.00, 1.32) (Bolland et al. 2011). We used the calcium-naïve subset of the Women's Health Initiative in this analysis. Subsequent meta-analyses of the effect of calcium alone on myocardial infarction confirm our findings (Lewis et al. 2015; Mao et al. 2013), and calcium plus vitamin D has been confirmed to increase stroke risk by 17% (Khan et al. 2019). When trials of calcium with or without vitamin D are meta-analyzed, the outcome is usually determined by how the Women's Health Initiative data are handled, resulting in contrary findings if the 20,000 women already receiving calcium supplements are included in the analysis (Lewis et al. 2015). Meta-analysis of randomized controlled trials of calcium supplements in patients with chronic kidney disease also demonstrates adverse effects on mortality (Jamal et al. 2013). Observational studies of calcium intake are inconsistent in their findings (Michaëlsson et al. 2013; Xiao et al. 2013), probably as a result of the variable effects of confounding. One widely quoted study showed no effects in its complete cohort, but increases in cardiovascular risk after women using sex hormone treatment were excluded (Harvey et al. 2018).

There are a number of mechanisms by which the acute increase in serum calcium that accompanies supplement ingestion might have adverse cardiovascular effects. Calcium supplements result in blood pressure higher than placebo by >5 mmHg over the 6 h following dosing, and both blood coagulability and the propensity of serum to calcify are also increased (Billington et al. 2017; Bristow et al. 2015, 2016). There is evidence that small increases in serum calcium within the normal range are associated with increased risk of myocardial infarction and death in prospective observational studies (Reid et al. 2016) and this has also been observed in a Mendelian randomization study (Larsson et al. 2017). In contrast, chelation therapy, which acutely reduces serum calcium concentration, reduces incidence of myocardial infarction (Lamas et al. 2013). Higher circulating calcium levels are associated with increased intima-media thickness in the carotid arteries (Ishizaka et al. 2002; Rubin et al. 2007), higher prevalence of aortic calcification (Bolland et al. 2010b), and increased calcified plaque in the coronary arteries (Kwak et al. 2014; Shin et al. 2012). Thus, the adverse effects of calcium supplements on vascular disease are likely to be mediated by the elevation of serum calcium levels which follows each administration, through effects on blood pressure, coagulation, or vessel wall calcification. Interestingly, the closely related ion, strontium, has a similar adverse effect on myocardial infarction (Medicines and Healthcare Products Regulatory Agency (MHRA) 2013).

10 Conclusions

The maintenance of circulating calcium levels within a narrow range is a high homeostatic priority so that mineralization of the skeleton is maintained but calcification of soft tissues is avoided. Accordingly, absorption and excretion of calcium is closely regulated. As a result, humans can adapt to a wide range of calcium intakes, certainly between 300 and 2,000 mg/day, and possibly to higher and lower intakes as well. The evidence that low calcium intakes contribute to osteoporosis development is weak, as is evidence that increasing these intakes significantly changes fracture risk. Consistent with this view, the United States Preventive Services Task Force does not support the use of calcium supplements in healthy community-dwelling adults (Grossman et al. 2018), and the International Osteoporosis Foundation states that "supplementation with calcium alone for fracture reduction is not supported by the literature" (Harvey et al. 2017). While some groups continue to recommend that supplements of calcium and vitamin D should be given with drug treatments for osteoporosis, this view is not supported by clinical trials which demonstrate anti-fracture efficacy of estrogens and bisphosphonates in the absence of such supplementation. Thus, calcium supplements have only a minor place in contemporary medical practice.

References

Anderson JJB, Roggenkamp KJ, Suchindran CM (2012) Calcium intakes and femoral and lumbar bone density of elderly U.S. men and women: National Health and Nutrition Examination Survey 2005-2006 analysis. J Clin Endocrinol Metab 97:4531–4539

Aptel I, Cance-Rouzaud A, Grandjean H (1999) Association between calcium ingested from drinking water and femoral bone density in elderly women: evidence from the EPIDOS cohort. J Bone Miner Res 14:829–833

Avenell A, Bolland MJ, Grey A, Reid IR (2016) Further major uncorrected errors in National Osteoporosis Foundation meta-analyses of calcium and vitamin D supplementation in fracture prevention. Osteoporos Int 28:733–734

Billington EO, Bristow SM, Gamble GD, de Kwant JA, Stewart A, Mihov BV, Horne AM, Reid IR (2017) Acute effects of calcium supplements on blood pressure: randomised, crossover trial in postmenopausal women. Osteoporos Int 28:119–125

Bolland MJ, Grey AB, Ames RW, Horne AM, Gamble GD, Reid IR (2006) Fat mass is an important predictor of parathyroid hormone levels in postmenopausal women. Bone 38:317–321

Bolland MJ, Barber PA, Doughty RN, Mason B, Horne A, Ames R, Gamble GD, Grey A, Reid IR (2008) Vascular events in healthy older women receiving calcium supplementation: randomised controlled trial. BMJ 336:262–266

Bolland MJ, Avenell A, Baron JA, Grey A, MacLennan GS, Gamble GD, Reid IR (2010a) Effect of calcium supplements on risk of myocardial infarction and cardiovascular events: meta-analysis. BMJ 341:C3691

Bolland MJ, Wang TKM, van Pelt NC, Horne AM, Mason BH, Ames RW, Grey AB, Ruygrok PN, Gamble GD, Reid IR (2010b) Abdominal aortic calcification on vertebral morphometry images predicts incident myocardial infarction. J Bone Miner Res 25:505–512

Bolland MJ, Grey A, Avenell A, Gamble GD, Reid IR (2011) Calcium supplements with or without vitamin D and risk of cardiovascular events: reanalysis of the Women's Health Initiative limited access dataset and meta-analysis. BMJ 342:d2040. https://doi.org/10.1136/bmj.d2040

Bolland MJ, Leung W, Tai V, Bastin S, Gamble GD, Grey A, Reid IR (2015) Calcium intake and risk of fracture: systematic review. BMJ 351:h4580

Bonnick S, Broy S, Kaiser F, Teutsch C, Rosenberg E, DeLucca P, Melton M (2007) Treatment with alendronate plus calcium, alendronate alone, or calcium alone for postmenopausal low bone mineral density. Curr Med Res Opin 23:1341–1349

Bristow SM, Gamble GD, Stewart A, Horne L, House ME, Aati O, Mihov B, Horne AM, Reid IR (2014) Acute and 3-month effects of microcrystalline hydroxyapatite, calcium citrate and calcium carbonate on serum calcium and markers of bone turnover: a randomised controlled trial in postmenopausal women. Br J Nutr 112:1611–1620

Bristow SM, Gamble GD, Stewart A, Horne AM, Reid IR (2015) Acute effects of calcium supplements on blood pressure and blood coagulation: secondary analysis of a randomised controlled trial in post-menopausal women. Br J Nutr 114(11):1868–1874

Bristow SM, Gamble GD, Pasch A, O'Neill WC, Stewart A, Horne AM, Reid IR (2016) Acute and 3-month effects of calcium carbonate on the calcification propensity of serum and regulators of vascular calcification: secondary analysis of a randomized controlled trial. Osteoporos Int 27:1209–1216

Bristow SM, Gamble GD, Horne AM, Reid IR (2017) Dietary calcium intake and rate of bone loss in men. Br J Nutr:1–7

Bristow SM, Horne AM, Gamble GD, Mihov B, Stewart A, Reid IR (2019) Dietary calcium intake and bone loss over 6 years in osteopenic postmenopausal women. J Clin Endocrinol Metab 104:3576–3584

Cauley JA, Robbins J, Chen Z, Cummings SR, Jackson RD, LaCroix AZ, LeBoff M, Lewis CE, McGowan J, Neuner J, Pettinger M, Stefanick ML, Wactawski-Wende J, Watts NB (2003) Effects of estrogen plus progestin on risk of fracture and bone mineral density – the Women's Health Initiative randomized trial. JAMA 290:1729–1738

Chapuy MC, Arlot ME, Duboeuf F, Brun J, Crouzet B, Arnaud S, Delmas PD, Meunier PJ (1992) Vitamin D3 and calcium to prevent hip fractures in the elderly women. N Engl J Med 327:1637–1642

Chapuy MC, Arlot ME, Delmas PD, Meunier PJ (1994) Effect of calcium and cholecalciferol treatment for three years on hip fractures in elderly women. BMJ 308:1081–1082

Fang AP, Li KJ, Shi HY, He JJ, Li H (2016) Habitual dietary calcium intakes and calcium metabolism in healthy adults Chinese: a systematic review and meta-analysis. Asia Pac J Clin Nutr 25:776–784

Food and Agriculture Organisation of the United Nations, World Health Organisation (1974) Handbook on human nutritional requirments. World Health Organisation, Geneva

Gallagher JC, Smith LM, Yalamanchili V (2014) Incidence of hypercalciuria and hypercalcemia during vitamin D and calcium supplementation in older women. Menopause 21:1173–1180

Grandfield K, Vuong V, Schwarcz HP (2018) Ultrastructure of bone: hierarchical features from nanometer to micrometer scale revealed in focused ion beam sections in the TEM. Calcif Tissue Int 103:606–616

Grossman DC, Curry SJ, Owens DK, Barry MJ, Caughey AB, Davidson KW, Doubeni CA, Epling JW Jr, Kemper AR, Krist A, Kubik M, Landefeld S, Mangione CM, Silverstein M, Simon MA, Tseng CW (2018) Vitamin D, calcium, OR combined supplementation for the primary prevention of fractures in community-dwelling adults. US Preventive Services Task Force recommendation statement. JAMA 319:1592–1599

Händel MN, Heitmann BL, Abrahamsen B (2015) Nutrient and food intakes in early life and risk of childhood fractures: a systematic review and meta-analysis. Am J Clin Nutr 102:1182–1195

Hannan MT, Felson DT, Dawson-Hughes B, Tucker KL, Cupples LA, Wilson PW, Kiel DP (2000) Risk factors for longitudinal bone loss in elderly men and women: the Framingham Osteoporosis Study. J Bone Miner Res 15:710–720

Hansen MA, Overgaard K, Riis BJ, Christiansen C (1991) Potential risk factors for development of postmenopausal osteoporosis – examined over a 12-year period. Osteoporos Int 1:95–102

Harvey NC, Biver E, Kaufman JM, Bauer J, Branco J, Brandi ML, Bruyere O, Coxam V, Cruz-Jentoft A, Czerwinski E, Dimai H, Fardellone P, Landi F, Reginster JY, Dawson-Hughes B, Kanis JA, Rizzoli R, Cooper C (2017) The role of calcium supplementation in healthy musculoskeletal ageing: an expert consensus meeting of the European Society for Clinical and Economic Aspects of Osteoporosis, Osteoarthritis and Musculoskeletal Diseases (ESCEO) and the International Foundation for Osteoporosis (IOF). Osteoporos Int 28:447–462

Harvey NC, D'Angelo S, Paccou J, Curtis EM, Edwards M, Raisi-Estabragh Z, Walker-Bone K, Petersen SE, Cooper C (2018) Calcium and vitamin D supplementation are not associated with risk of incident ischemic cardiac events or death: findings from the UK Biobank cohort. J Bone Miner Res 33:803–811

Heaney RP, Recker RR, Saville PD (1977) Calcium balance and calcium requirements in middle-aged women. Am J Clin Nutr 30:1603–1611

Heaney RP, Recker RR, Saville PD (1978) Menopausal changes in calcium balance performance. J Lab Clin Med 92:953–963

Hegsted DM, Moscoso I, Collazos Ch C (1952) A study of the minimum calcium requirements of adult men. J Nutr 46:181–201

Hosking DJ, Ross PD, Thompson DE, Wasnich RD, McClung M, Bjarnason NH, Ravn P, Cizza G, Daley M, Yates AJ (1998) Evidence that increased calcium intake does not prevent early postmenopausal bone loss. Clin Ther 20:933–944

Hsia J, Heiss G, Ren H, Allison M, Dolan NC, Greenland P, Heckbert SR, Johnson KC, Manson JE, Sidney S, Trevisan M (2007) Calcium/vitamin D supplementation and cardiovascular events. Circulation 115:846–854

Hunt CD, Johnson LK (2007) Calcium requirements: new estimations for men and women by cross-sectional statistical analyses of calcium balance data from metabolic studies. Am J Clin Nutr 86:1054–1063

Ishizaka N, Ishizaka Y, Takahashi E, Tooda E, Hashimoto H, Nagai R, Yamakado M (2002) Serum calcium concentration and carotid artery plaque: a population-based study. J Cardiol 39:151–157

Jackman LA, Millane SS, Martin BR, Wood OB, McCabe GP, Peacock M, Weaver CM (1997) Calcium retention in relation to calcium intake and postmenarcheal age in adolescent females. Am J Clin Nutr 66:327–333

Jackson RD, LaCroix AZ, Gass M, Wallace RB, Robbins J, Lewis CE, Bassford T, Beresford SAA, Black HR, Blanchette P, Bonds DE, Brunner RL, Brzyski RG, Caan B, Cauley JA, Chlebowski RT, Cummings SR, Granek I, Hays J, Heiss G, Hendrix SL, Howard BV, Hsia J, Hubbell FA, Johnson KC, Judd H, Kotchen JM, Kuller LH, Langer RD, Lasser NL, Limacher MC, Ludlam S, Manson JE, Margolis KL, McGowan J, Ockene JK, O'Sullivan MJ, Phillips L, Prentice RL, Sarto GE, Stefanick ML, Van Horn L, Wactawski-Wende J, Whitlock E, Anderson GL, Assaf AR, Barad D (2006) Calcium plus vitamin D supplementation and the risk of fractures. N Engl J Med 354:669–683

Jamal SA, Vandermeer B, Raggi P, Mendelssohn DC, Chatterley T, Dorgan M, Lok CE, Fitchett D, Tsuyuki RT (2013) Effect of calcium-based versus non-calcium-based phosphate binders on mortality in patients with chronic kidney disease: an updated systematic review and meta-analysis. Lancet 382:1268–1277

Kahwati LC, Weber RP, Pan H, Gourlay M, LeBlanc E, Coker-Schwimmer M, Viswanathan M (2018) Vitamin D, calcium, OR combined supplementation for the primary prevention of fractures in community-dwelling adults evidence report and systematic review for the US Preventive Services Task Force. JAMA 319:1600–1612

Kanis JA (1991) Calcium requirements for optimal skeletal health in women. Calcif Tissue Int 49(suppl):S33–S41

Kanis JA, Passmore R (1989) Calcium supplementation of the diet – I. BMJ 298:137–140

Khan SU, Khan MU, Riaz H, Valavoor S, Zhao D, Vaughan L, Okunrintemi V, Riaz IB, Khan MS, Kaluski E, Murad MH, Blaha MJ, Guallar E, Michos ED (2019) Effects of nutritional supplements and dietary interventions on cardiovascular outcomes: an umbrella review and evidence map effects of supplements and dietary interventions on cardiovascular outcomes. Ann Intern Med 171:190–198

Kroger H, Tuppurainen M, Honkanen R, Alhava E, Saarikoski S (1994) Bone mineral density and risk factors for osteoporosis – a population-based study of 1600 perimenopausal women. Calcif Tissue Int 55:1–7

Kwak SM, Kim JS, Choi Y, Chang Y, Kwon MJ, Jung JG, Jeong C, Ahn J, Kim HS, Shin H, Ryu S (2014) Dietary intake of calcium and phosphorus and serum concentration in relation to the risk of coronary artery calcification in asymptomatic adults. Arterioscler Thromb Vasc Biol 34:1763–1769

Lamas GA, Goertz C, Boineau R, Mark DB, Rozema T, Nahin RL, Lindblad L, Lewis EF, Drisko J, Lee KL, Tact Investigators (2013) Effect of disodium EDTA chelation regimen on cardiovascular events in patients with previous myocardial infarction. The TACT randomized trial. JAMA 309:1241–1250

Larsson SC, Burgess S, Michaëlsson K (2017) Association of genetic variants related to serum calcium levels with coronary artery disease and myocardial infarction. JAMA 318:371–380

Lewis JR, Zhu K, Prince RL (2012) Adverse events from calcium supplementation: relationship to errors in myocardial infarction self-reporting in randomized controlled trials of calcium supplementation. J Bone Miner Res 27:719–722

Lewis JR, Radavelli-Bagatini S, Rejnmark L, Chen JS, Simpson JM, Lappe JM, Mosekilde L, Prentice RL, Prince RL (2015) The effects of calcium supplementation on verified coronary heart disease hospitalization and death in postmenopausal women: a collaborative meta-analysis of randomized controlled trials. J Bone Miner Res 30:165–175

Lieben L, Masuyama R, Torrekens S, Van Looveren R, Schrooten J, Baatsen P, Lafage-Proust MH, Dresselaers T, Feng JQ, Bonewald LF, Meyer MB, Pike JW, Bouillon R, Carmeliet G (2012) Normocalcemia is maintained in mice under conditions of calcium malabsorption by vitamin D-induced inhibition of bone mineralization. J Clin Invest 122:1803–1815

Lindsay R, Hart DM, Forrest C, Baird C (1980) Prevention of spinal osteoporosis in oophorectomised women. Lancet ii:1151–1153

Lips P, Chapuy MC, Dawson-Hughes B, Pols HAP, Holick MF (1999) An international comparison of serum 25-hydroxyvitamin D measurements. Osteoporos Int 9:394–397
Lloyd T, Chinchilli VM, Johnson-Rollings N, Kieselhorst K, Eggli DF, Marcus R (2000) Adult female hip bone density reflects teenage sports-exercise patterns but not teenage calcium intake. Pediatrics 106:40–44
Malm OJ (1958) Calcium requirement and adaptation in adult men. Scand J Clin Lab Invest 10:1–290
Mao P-J, Zhang C, Tang L, Xian Y-Q, Li Y-S, Wang W-D, Zhu X-H, Qiu H-L, He J, Zhou Y-H (2013) Effect of calcium or vitamin D supplementation on vascular outcomes: a meta-analysis of randomized controlled trials. Int J Cardiol 169:106–111
Mavroeidi A, Stewart AD, Reid DM, Macdonald HM (2009) Physical activity and dietary calcium interactions in bone mass in Scottish postmenopausal women. Osteoporos Int 20:409–416
McCloskey EV, Beneton M, Charlesworth D, Kayan K, Detakats D, Dey A, Orgee J, Ashford R, Forster M, Cliffe J, Kersh L, Brazier J, Nichol J, Aropuu S, Jalava T, Kanis JA (2007) Clodronate reduces the incidence of fractures in community-dwelling elderly women unselected for osteoporosis: results of a double-blind, placebo-controlled randomized study. J Bone Miner Res 22:135–141
Medicines and Healthcare Products Regulatory Agency (MHRA) (2013) Strontium ranelate (Protelos): risk of serious cardiac disorders – restricted indications, new contraindications, and warnings. Drug Safety Update 6:S1
Michaëlsson K, Melhus H, Lemming EW, Wolk A, Byberg L (2013) Long term calcium intake and rates of all cause and cardiovascular mortality: community based prospective longitudinal cohort study. BMJ 346:f228
Moseley KF, Weaver CM, Appel L, Sebastian A, Sellmeyer DE (2013) Potassium citrate supplementation results in sustained improvement in calcium balance in older men and women. J Bone Miner Res 28:497–504
Nakamura K, Oyama M, Saito T, Oshiki R, Kobayashi R, Nishiwaki T, Nashimoto M, Tsuchiya Y (2012) Nutritional and biochemical parameters associated with 6-year change in bone mineral density in community-dwelling Japanese women aged 69 years and older: The Muramatsu Study. Nutrition 28:357–361
Nicholls L, Nimalasuriya A (1939) Adaptation to a low calcium intake in reference to the calcium requirements of a tropical population. J Nutr 18:563–577
Pathak CL (1958) Nutritional adaptation to low dietary intakes of calories, proteins, vitamins, and minerals in the tropics. Am J Clin Nutr 6:151–158
Porthouse J, Cockayne S, King C, Saxon L, Steele E, Aspray T, Baverstock M, Birks Y, Dumville J, Francis RM, Iglesias C, Puffer S, Sutcliffe A, Watt I, Torgerson DJ (2005) Randomised controlled trial of supplementation with calcium and cholecalciferol (vitamin D) for prevention of fractures in primary care. BMJ 330:1003–1006
Prince RL, Devine A, Dhaliwal SS, Dick IM (2006) Effects of calcium supplementation on clinical fracture and bone structure – results of a 5-year, double-blind, placebo-controlled trial in elderly women. Arch Intern Med 166:869–875
RECORD Trial Group (2005) Oral vitamin D3 and calcium for secondary prevention of low-trauma fractures in elderly people (randomised evaluation of calcium or vitamin D, RECORD): a randomised placebo-controlled trial. Lancet 365:1621–1628
Reid IR, Ames R, Evans MC, Sharpe S, Gamble G, France JT, Lim TM, Cundy TF (1992) Determinants of total body and regional bone mineral density in normal postmenopausal women – a key role for fat mass. J Clin Endocrinol Metab 75:45–51
Reid IR, Ames RW, Evans MC, Gamble GD, Sharpe SJ (1993) Effect of calcium supplementation on bone loss in postmenopausal women. N Engl J Med 328:460–464
Reid IR, Ames RW, Evans MC, Sharpe SJ, Gamble GD (1994) Determinants of the rate of bone loss in normal postmenopausal women. J Clin Endocrinol Metab 79:950–954

Reid IR, Mason B, Horne A, Ames R, Clearwater J, Bava U, Orr-Walker B, Wu F, Evans MC, Gamble GD (2002) Effects of calcium supplementation on serum lipid concentrations in normal older women: a randomized controlled trial. Am J Med 112:343–347

Reid IR, Horne A, Mason B, Ames R, Bava U, Gamble GD (2005) Effects of calcium supplementation on body weight and blood pressure in normal older women: a randomized controlled trial. J Clin Endocrinol Metab 90:3824–3829

Reid IR, Mason B, Horne A, Ames R, Reid HE, Bava U, Bolland MJ, Gamble GD (2006) Randomized controlled trial of calcium in healthy older women. Am J Med 119:777–785

Reid IR, Ames R, Mason B, Reid HE, Bacon CJ, Bolland MJ, Gamble GD, Grey A, Horne A (2008) Randomized controlled trial of calcium supplementation in healthy, nonosteoporotic, older men. Arch Intern Med 168:2276–2282

Reid IR, Bristow SM, Bolland MJ (2015) Calcium supplements: benefits and risks. J Intern Med 278:354–368

Reid IR, Gamble GD, Bolland MJ (2016) Circulating calcium concentrations, vascular disease and mortality: a systematic review. J Intern Med 279:524–540

Reid IR, Bristow SM, Bolland MJ (2017) Calcium and cardiovascular disease. Endocrinol Metab 32:339–349

Reid IR, Horne AM, Mihov B, Stewart A, Garratt E, Wong S, Wiessing KR, Bolland MJ, Bastin S, Gamble GD (2018) Fracture prevention with zoledronate in older women with osteopenia. N Engl J Med 379:2407–2416

Riggs BL, Wahner HW, Melton LJ 3rd, Richelson LS, Judd HL, O'Fallon WM (1987) Dietary calcium intake and rates of bone loss in women. J Clin Invest 80:979–982

Robbins JA, Aragaki A, Crandall CJ, Manson JE, Carbone L, Jackson R, Lewis CE, Johnson KC, Sarto G, Stefanick ML, Wactawski-Wende J (2014) Women's Health Initiative clinical trials: interaction of calcium and vitamin D with hormone therapy. Menopause 21:116–123

Rubin MR, Rundek T, McMahon DJ, Lee H-S, Sacco RL, Silverberg SJ (2007) Carotid artery plaque thickness is associated with increased serum calcium levels: the Northern Manhattan study. Atherosclerosis 194:426–432

Salovaara K, Tuppurainen M, Karkkainen M, Rikkonen T, Sandini L, Sirola J, Honkanen R, Alhava E, Kroger H (2010) Effect of vitamin D-3 and calcium on fracture risk in 65-to 71-year-old women: a population-based 3-year randomized, controlled trial-the OSTPRE-FPS. J Bone Miner Res 25:1487–1495

Shin S, Kim KJ, Chang HJ, Cho I, Kim YJ, Choi BW, Rhee Y, Lim SK, Yang WI, Shim CY, Ha JW, Jang Y, Chung N (2012) Impact of serum calcium and phosphate on coronary atherosclerosis detected by cardiac computed tomography. Eur Heart J 33:2873–2881

Spiegel DM, Brady K (2012) Calcium balance in normal individuals and in patients with chronic kidney disease on low- and high-calcium diets. Kidney Int 81:1116–1122

Thandrayen K, Pettifor JM (2018) The roles of vitamin D and dietary calcium in nutritional rickets. Bone Reports 8:81–89

Umaretiya PJ, Thacher TD, Fischer PR, Cha SS, Pettifor JM (2013) Bone mineral density in Nigerian children after discontinuation of calcium supplementation. Bone 55:64–68

van Beresteijn EC, van't Hof MA, Schaafsma G, de Waard H, Duursma SA (1990) Habitual dietary calcium intake and cortical bone loss in perimenopausal women: a longitudinal study. Calcif Tissue Int 47:338–344

Vautour L, Goltzman D (2019) Regulation of calcium homeostasis. In: Bilezikian JP (ed) Primer on the metabolic bone diseases and disorders of mineral metabolism, 9th edn. Wiley, Hoboken, pp 165–172

Vogel KA, Martin BR, McCabe LD, Peacock M, Warden SJ, McCabe GP, Weaver CM (2017) The effect of dairy intake on bone mass and body composition in early pubertal girls and boys: a randomized controlled trial. Am J Clin Nutr 105:1214–1229

Walker ARP (1956) Some aspects of nutritional research in South-Africa. Nutr Rev 14:321–324

Weaver CM, Alexander DD, Boushey CJ, Dawson-Hughes B, Lappe JM, LeBoff MS, Liu S, Looker AC, Wallace TC, Wang DD (2016) Calcium plus vitamin D supplementation and risk of fractures: an updated meta-analysis from the National Osteoporosis Foundation. Osteoporos Int 27:367–376

Winzenberg T, Shaw K, Fryer J, Jones G (2006a) Effects of calcium supplementation on bone density in healthy children: meta-analysis of randomised controlled trials. BMJ 333:775

Winzenberg TM, Shaw K, Fryer J, Jones G (2006b) Calcium supplementation for improving bone mineral density in children – art. no. CD005119.pub2. Cochrane Database of Systematic REVIEWS: PUB2-PUB2

World Health Organisation, Food and Agriculture Organisation of the United Nations (1962) Calcium requirements: report of an FAO/WHO Expert Group, Rome, Italy, 23 to 30 May 1961. World Health Organisation, Geneva

Xiao Q, Murphy RA, Houston DK, Harris TB, Chow WH, Park Y (2013) Dietary and supplemental calcium intake and cardiovascular disease mortality. JAMA Intern Med 173:639–646

Zhao J, Zeng X, Wang J, Liu L (2017) Association between calcium or vitamin d supplementation and fracture incidence in community-dwelling older adults: a systematic review and meta-analysis. JAMA 318:2466–2482

FGF23 and Bone and Mineral Metabolism

Seiji Fukumoto

Contents

Abstract
FGF23 is a phosphotropic hormone produced by the bone. FGF23 works by binding to the FGF receptor-Klotho complex. Klotho is expressed in several limited tissues including the kidney and parathyroid glands. This tissue-restricted

S. Fukumoto (✉)
Fujii Memorial Institute of Medical Sciences, Institute of Advanced Medical Sciences, Tokushima University, Tokushima, Japan

P. H. Stern (ed.), *Bone Regulators and Osteoporosis Therapy*,
Handbook of Experimental Pharmacology 262, https://doi.org/10.1007/164_2019_330

expression of Klotho is believed to determine the target organs of FGF23. FGF23 reduces serum phosphate by suppressing the expression of type 2a and 2c sodium-phosphate cotransporters in renal proximal tubules. FGF23 also decreases 1,25-dihydroxyvitamin D levels by regulating the expression of vitamin D-metabolizing enzymes, which results in reduced intestinal phosphate absorption. Excessive actions of FGF23 cause several types of hypophosphatemic rickets/osteomalacia characterized by impaired mineralization of bone matrix. In contrast, deficient actions of FGF23 result in hyperphosphatemic tumoral calcinosis with high 1,25-dihydroxyvitamin D levels. These results indicate that FGF23 is a physiological regulator of phosphate and vitamin D metabolism and indispensable for the maintenance of serum phosphate levels.

Keywords

Hyperphosphatemia · Hypophosphatemia · Klotho · Osteomalacia · Rickets · Tumoral calcinosis

1 Phosphate Metabolism and Rickets/Osteomalacia

Serum phosphate level is regulated by intestinal phosphate absorption, renal phosphate handling, and the dynamic equilibrium between extracellular phosphate and that in the bone or the intracellular space. Renal phosphate handling is believed to be the most important determinant of the serum phosphate level in a chronic state. About 80–90% of phosphate filtered from the glomeruli is reabsorbed in the proximal tubules, and almost all the remaining intraluminal phosphate is excreted into the urine. This proximal tubular phosphate reabsorption is mediated by type 2a and 2c sodium-phosphate cotransporters encoded by *SLC34A1* and *SLC34A3*, respectively, in the brush border membrane of renal proximal tubules (Magagnin et al. 1993; Segawa et al. 2002).

Rickets and osteomalacia are diseases characterized by impaired mineralization of the bone matrix (Fukumoto 2018). Rickets is a disease in childhood. The main problems in patients with rickets are growth retardation and bone deformities such as genu valgus and varum. Osteomalacia develops in subjects with closed epiphyseal growth plates. Patients with osteomalacia complain of muscle weakness and bone pain. There are many causes of rickets and osteomalacia (Table 1). Of these, chronic hypophosphatemia is the most important cause of rickets and osteomalacia.

Chronic hypophosphatemia can be caused by several mechanisms (Table 1). The typical cause of impaired intestinal absorption of phosphate is vitamin D deficiency (Winzenberg and Jones 2013). Vitamin D-dependent rickets type 1 and 2 are caused by inactivating mutations in *CYP27B1* and *vitamin D receptor* (*VDR*), respectively (Kitanaka et al. 1998; Hughes et al. 1988). *CYP27B1* encodes an enzyme responsible for 1,25-dihydroxyvitamin D [1,25(OH)$_2$D] production. In addition, there are several causes of chronic hypophosphatemia due to impaired proximal tubular phosphate reabsorption. Fanconi syndrome is characterized by generalized impairment of proximal tubular reabsorption. In contrast, inactivating mutations in *SLC34A3* result

Table 1 Typical causes for rickets and osteomalacia

Chronic hypophosphatemia
Reduced intestinal phosphate absorption
Vitamin D deficiency
Vitamin D-dependent rickets type 1, 2
Malabsorption (short bowel syndrome, malnutrition, etc.)
Impaired renal phosphate reabsorption
FGF23-related hypophosphatemic rickets/osteomalacia
Hereditary hypophosphatemia rickets with hypercalciuria
Fanconi syndrome
Dent disease
Renal tubular acidosis
Drugs (ifosfamide, adefovir dipivoxil, etc.)
Chronic hypocalcemia
Vitamin D deficiency
Vitamin D-dependent rickets type 1, 2
Impaired mineralization
Hypophosphatasia
Drugs (aluminum, etidronate, etc.)

in hereditary hypophosphatemic rickets with hypercalciuria (HHRH) characterized by selective impairment of phosphate reabsorption (Bergwitz et al. 2006). Chronic hypophosphatemia increases serum 1,25(OH)$_2$D (Tanaka and Deluca 1973) resulting in enhanced intestinal calcium absorption and hypercalciuria in this disease. In contrast to HHRH, there are several hypophosphatemic diseases with relatively low 1,25(OH)$_2$D level such as X-linked hypophosphatemic rickets (XLH) and tumor-induced osteomalacia (Kinoshita and Fukumoto 2018). Patients with these diseases show frank hypophosphatemia. Despite this hypophosphatemia, 1,25(OH)$_2$D usually remains low–low normal indicating that vitamin D metabolism is dysregulated in addition to impaired proximal tubular phosphate reabsorption. XLH is the most frequent cause of genetic hypophosphatemic rickets. There is a murine homologue of XLH called the *Hyp* mouse showing hypophosphatemia and growth retardation (Beck et al. 1997). When *Hyp* mice were parabiosed with wild-type mice, serum phosphate of the wild-type mice decreased indicating that *Hyp* phenotypes are caused by some humoral factor (Meyer et al. 1989). In addition, TIO is a rare paraneoplastic syndrome usually caused by mesenchymal tumors (Minisola et al. 2017). This disease can be cured by complete removal of the responsible tumors again showing that TIO is caused by a humoral mechanism. However, it was unknown whether the humoral factors responsible for XLH/*Hyp* and TIO are the same or not.

2 Discovery and Cloning of FGF23

FGF23 was identified as a responsible gene for autosomal dominant hypophosphatemic rickets (ADHR) by positional cloning (ADHR Consortium 2000). Patients with ADHR show similar clinical features to those with XLH. Patients with ADHR exhibit hypophosphatemia, impaired proximal tubular phosphate reabsorption, and relatively low 1,25(OH)$_2$D levels. *Fgf23* was also cloned by homology to *Fgf15* in mice (Yamashita et al. 2000). Furthermore, FGF23 was identified as a responsible humoral factor for TIO (Shimada et al. 2001). A cDNA library was constructed using mRNA obtained from a tumor which was resected in a patient with TIO and responsible for this disease. By comparison of gene expression with the adjacent bone tissue, several highly expressed genes in the tumor were selected. Then, Chinese hamster ovary cells stably expressing these genes were produced and implanted into nude mice. Of these genes, *FGF23* was identified as a gene whose product induced hypophosphatemia in mice (Shimada et al. 2001).

FGF family members are defined as humoral factors that have a FGF homology region with a β-trefoil structure. There are 22 FGF family members in humans, and these FGF family members are divided into several subfamilies (Itoh and Ornitz 2004). FGF23 belongs to FGF19 subfamily together with FGF19 and FGF21. *FGF19* is an orthologue of murine *Fgf15*. There is no *FGF15* in humans and no *Fgf19* in mice. Human *FGF23* gene encodes a protein with 251 amino acids. The N-terminal 24 amino acids compose a signal peptide, and the secreted full-length FGF23 has 227 amino acids (Fig. 1). A part of the FGF23 protein is proteolytically processed before or during the process of secretion between Arg179 and Ser180 by enzymes that recognize the Arg176-X-X-Arg179 motif (Shimada et al. 2001). FGF23 has a FGF homology region in the N-terminal portion of this processing site. The processed N-terminal and C-terminal fragments of FGF23 do not have the activity to regulate phosphate and vitamin D metabolism (Shimada et al. 2002).

3 Actions of FGF23 in the Kidney

After the cloning of *FGF23*, the actions of FGF23 were examined using recombinant FGF23. When recombinant FGF23 was injected into mice, serum phosphate decreased after several hours (Shimada et al. 2004a). This reduction of serum phosphate was associated with reduced expression of the type 2a sodium-phosphate cotransporter (Shimada et al. 2004a). In addition, *Fgf23* transgenic mice showed reduced expression of *Slc34a1* and *Slc34a3* (Larsson et al. 2004). These results indicated that FGF23 reduces serum phosphate by suppressing the expression of type 2a and 2c sodium-phosphate cotransporters in proximal tubules. Recombinant FGF23 also caused decreased 1,25(OH)$_2$D levels when injected into mice. Vitamins D_2 and D_3 contained in food and absorbed in the intestine, or vitamin D_3 produced in the skin, are first converted to 25-hydroxyvitamin D [25(OH)D] in the liver by 25-hydroxylase. This 25(OH)D is then converted either to 1,25(OH)$_2$D or 24,25-dihycroxyvitamin D [24,25(OH)$_2$D] in the proximal tubules. The conversion to 1,25

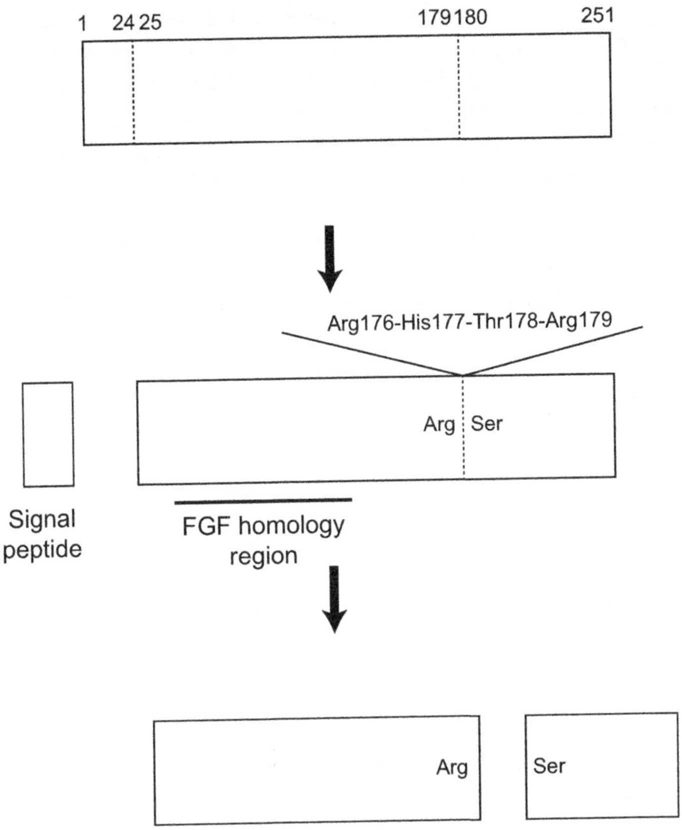

Fig. 1 Structure of FGF23. Human *FGF23* encodes a protein with 251 amino acids. The N-terminal 24 amino acids compose a signal peptide. A part of FGF23 protein is proteolytically processed before or during the process of secretion between Arg179 and Ser180 by enzymes that recognize Arg176-X-X-Arg179 motif. FGF23 has a FGF homology region in the N-terminal portion of this processing site

$(OH)_2D$ is mediated by an enzyme called 25(OH)-1α-hydroxylase encoded by *CYP27B1* (Takeyama et al. 1997). On the other hand, 24-hydroxylase encoded by *CYP24A1* converts 25(OH)D to $24,25(OH)_2D$ (Ohyama et al. 1991). This 24-hydroxylase also converts $1,25(OH)_2D$ to 1,24,25-trihydroxyvitamin D $[1,24,25(OH)_3D]$ for further degradation. Therefore, 25(OH)-1α-hydroxylase works to increase the circulating level of $1,25(OH)_2D$, whereas $1,25(OH)_2D$ levels are decreased by the actions of 24-hydroxylase.

FGF23 was shown to enhance the expression of *Cyp24a1* and suppress that of *Cyp27b1* when injected into mice (Shimada et al. 2004a). By these effects on the expression of vitamin D metabolizing enzymes, FGF23 reduces the serum $1,25(OH)_2D$ level. The reduction in $1,25(OH)_2D$ was observed 3 h after the injection of FGF23. However, the decrease in serum phosphate was first observed 9 h after the administration of FGF23 (Shimada et al. 2004a). The hypophosphatemic effect of

FGF23 develops after its action to decrease 1,25(OH)$_2$D. However, FGF23 decreased serum phosphate in *Vdr* knockout mice indicating that the hypophosphatemic action of FGF23 is not dependent on the reduction in 1,25 (OH)$_2$D (Shimada et al. 2005).

The initial report indicated that FGF23 specifically suppresses renal phosphate reabsorption. When recombinant FGF23 was injected into mice, renal excretion of calcium, sodium, potassium, chloride, and glucose did not change (Shimada et al. 2001). In contrast, there are reports indicating that FGF23 enhances renal calcium and sodium reabsorption (Andrukhova et al. 2014a, b). The different experimental systems used may explain the discrepancy between these studies.

4 FGF Receptors

FGF family members have been shown to bind to and activate FGF receptors (FGFRs) (Itoh and Ornitz 2004). There are four *FGFR* genes (*FGFR1*, *FGFR2*, *FGFR3*, and *FGFR4*). The typical structure of FGFRs is composed of an extracellular domain with three immunoglobulin-like loops, D1, D2, and D3, a membrane-spanning domain, and a cytoplasmic domain with two tyrosine kinase regions. Alternative splicing from *FGFR1*, *FGFR2*, and *FGFR3* produces b and c subtypes of FGFRs (Goetz and Mohammadi 2013). A part of the third immunoglobulin-like loop and the region between this loop and the membrane-spanning domain are different between b and c subtypes. These b and c subtypes of FGFRs have different ligand specificity. Epithelial cells basically produce b subtypes of FGFRs and FGF ligands secreted by mesenchymal cells bind to the subtype. In contrast, mesenchymal cells express c subtypes of FGFRs, and ligands from epithelial cells activate the subtypes (Goetz and Mohammadi 2013).

Prototypical FGF family members such as FGF1 (acidic FGF) and FGF23 (basic FGF) work as local factors, as paracrine or autocrine factors. These FGF family members show high affinity for heparan sulfate (HS). Because of this high affinity for HS, these local FGF family members are trapped and deposited in extracellular matrix around the producing cells. HS also enhances the affinity of FGF ligands for FGFRs (Goetz and Mohammadi 2013).

5 A Receptor for FGF23

Fgf23 is expressed in the liver, heart, and somites during the fetal life (Sitara et al. 2004). After birth, FGF23 is believed to be mainly expressed in the bone, especially osteoblasts/osteocytes (Liu et al. 2003). In contrast, FGF23 works on the kidney as shown above. These results indicate that FGF23 is a systemic humoral factor and secreted into the systemic circulation. In addition to FGF23, both FGF19 and FGF21 work as systemic factors. FGF19 subfamily members have very low or virtually no affinity for HS which explains why these FGF19 subfamily members are systemic humoral factors (Yu et al. 2005). This low affinity for HS also suggested that FGF19 subfamily members require some other molecule for binding to FGFRs.

Because FGF23 is produced by the bone and works in the kidney, there should be a specific receptor for FGF23 in the kidney. Analysis of proteins bound to FGF23 in the kidney identified Klotho as a major one (Urakawa et al. 2006). *Klotho* was identified as a gene whose expression was severely reduced in a mouse line called *Klotho* (Kuro-o et al. 1997). The *Klotho* mice were created in a project to make transgenic mice of rabbit type-I sodium-proton exchanger. Among transgenic mice without the overexpression of the transgene, *Klotho* mice with several phenotypes resembling senescence were identified. *Klotho* mice show growth retardation, short life span, infertility, atrophy of skin, emphysema, and reduced bone mass (Kuro-o et al. 1997). Klotho protein has an extracellular domain with two β-glucosidase-like regions called KL1 and KL2, a single membrane-spanning domain, and a very short intracellular tail. *Klotho* mice were shown to have hyperphosphatemia and high 1,25 $(OH)_2D$ (Kuro-o et al. 1997; Yoshida et al. 2002). These features were also observed in *Fgf23* knockout mice suggesting the functional link between FGF23 and Klotho (Shimada et al. 2004b).

Subsequent studies indicated that Klotho and some FGFRs work as a specific receptor for FGF23 (Kurosu et al. 2006; Urakawa et al. 2006). When Klotho was expressed in several cell lines, FGF23 activated the extracellular signal-regulated kinase (ERK) pathway and induced the expression of *early growth-responsive 1* (*Egr-1*) downstream of ERK. However, this response was not observed in L6 rat myoblast cells which lack endogenous FGFRs expression. When various FGFR subtypes were overexpressed in L6 cells, FGFR1c was shown to mediate the signal from FGF23 in the presence of Klotho. Pull-down assays indicated that the extracellular domain of Klotho binds to FGF23 and FGFR1c (Urakawa et al. 2006). HS was shown to enhance the activity of FGF23 in a Klotho-dependent way. Therefore, it was postulated that FGF23 binds to the FGFR1c-Klotho complex, and HS stabilizes the FGF23-FGFR1c-Klotho complex. Furthermore, another study indicated that FGF23 can bind to FGFR3c and FGFR4 in addition to FGFR1c (Kurosu et al. 2006). In vivo experiments also suggested that FGFR3c and FGFR4 are involved in the actions of FGF23 (Gattineni et al. 2009, 2011; Li et al. 2011). It is possible that several FGFR subtypes can mediate actions of FGF23 together with Klotho.

In vitro studies using anti-FGF23 antibodies suggested that FGFR1c and Klotho bind to N-terminal and C-terminal portion of FGF23, respectively (Yamazaki et al. 2008). This was confirmed by crystallographic analysis of FGF23, ectodomains of FGFR1c, and Klotho (Chen et al. 2018). The C-terminal portion of FGF23 binds to a deep cleft between KL1 and KL2 and a cavity in KL2. A loop of KL2 protrudes from KL2 core and binds to D3 of FGFR1c. The binding pocket in this D3 domain is different between b and c subtypes, and Klotho binds to the c subtypes of FGFRs. The N-terminal portion of FGF23 binds to D2 and D3 domains and the interdomain linker. It was shown that a serine residue present in the D3 domain of FGFR1c-3c and FGFR4 is important for the binding of FGF23. It was proposed that HS induces the dimerization of the FGF23-FGFR1c-Klotho complex making 2:2:2:2: FGF23-FGFR1c-Klotho-HS complex (Chen et al. 2018).

The original study indicated that *Klotho* is expressed in the brain, pituitary gland, kidney, testis, and ovary by RT-PCR. This study showed no expression of *Klotho* in the stomach, lung, skin, and bone. In situ hybridization indicated the expression of *Klotho* in distal convoluted tubules (Kuro-o et al. 1997). From these results, it was postulated that the tissue-restricted expression of *Klotho* determines the target organ of FGF23 because the expression of most FGFRs is not tissue-specific. However, subsequent studies showed the expression of Klotho in other cells such as the renal proximal tubules, heart, and bone (Andrukhova et al. 2012; Takeshita et al. 2018; Dehghani et al. 2018; Komaba et al. 2017). These somewhat discrepant results could be explained by the different sensitivities of the methods employed.

6 Regulation of FGF23 Production

The regulatory mechanisms of FGF23 production have not been fully clarified. FGF23 reduces phosphate and 1,25$(OH)_2$D concentrations. In contrast, phosphate and 1,25$(OH)_2$D have been shown to enhance FGF23 production. FGF23 is undetectable in *VDR* knockout mice, and putative vitamin D response elements were reported in the murine *Fgf23* promoter region (Shimada et al. 2005; Saini et al. 2013). These results indicate that 1,25$(OH)_2$D enhances FGF23 production through the VDR.

High-phosphate diet increased FGF23 levels in both rodent and human (Ferrari et al. 2005; Perwad et al. 2005). However, intravenous administration of sodium phosphate for a few hours did not change FGF23 levels in humans indicating that the regulation of FGF23 level by phosphate is not a rapid response compared to that of PTH secretion by calcium (Ito et al. 2007). High-phosphate diet for 2 weeks increased serum phosphate and FGF23 level in mice (Takashi et al. 2019). While high-phosphate diet did not enhance *Fgf23* expression in the bone, the expression of *polypeptide N-acetylgalactosaminyltransferase 3 (Galnt3)* was enhanced. The *Galnt3* gene product works to prevent the proteolytic processing of FGF23 protein and thus increase FGF23 levels as shown below (Frishberg et al. 2007). Subsequent in vitro analysis using a rat osteoblastic cell line indicated that *Galnt3* is a phosphate-responsive gene. From transcriptomic and proteomic analysis, it was proposed that phosphate activates FGFR1c and the subsequent intracellular signaling through FGF receptor substrate 2α-ERK, its downstream transcription factors enhancing *Galnt3* expression resulting in increased FGF23 levels. However, the detailed mechanism of FGFR1c activation by phosphate is not clear (Takashi et al. 2019).

In addition to 1,25$(OH)_2$D and phosphate, other factors such as inflammation and iron deficiency have been reported to enhance FGF23 production (David et al. 2016). These factors may contribute to the elevation of FGF23 especially in patients with CKD. It is not clear whether these factors have physiological roles in the regulation of FGF23 levels.

7 FGF23 and Hypophosphatemic Diseases

FGF23 was cloned as a responsible gene for ADHR (Consortium 2000). FGF23 was also identified as a responsible humoral factor for TIO (Shimada et al. 2001). ADHR and TIO are characterized by hypophosphatemia and impaired renal tubular phosphate reabsorption. In addition, patients with ADRH or TIO show low to low normal $1,25(OH)_2D$ despite frank hypophosphatemia. These results indicated that ADHR and TIO are caused by excessive actions of FGF23. After the cloning of *FGF23*, several kinds of assays for FGF23 have been created. The intact assay uses two kinds of antibodies that detect the N-terminal and C-terminal region of the processing site of FGF23 and measures only full-length FGF23 with 227 amino acids (Yamazaki et al. 2002). In contrast, two antibodies that recognize the C-terminal region are used in the C-terminal assay (Jonsson et al. 2003). This assay detects both the full-length and the processed C-terminal fragment of FGF23. FGF23 values measured by these two assays usually correlate well (Ito et al. 2005). However, these values can be quite discrepant in some cases as shown below. While FGF23 values measured by the intact assay indicate the amount of biologically active FGF23, those by the C-terminal assay are considered to represent the expression level of *FGF23*.

FGF23 levels measured by both the intact and C-terminal assays were reported to be high in most patients with TIO and XLH (Jonsson et al. 2003; Yamazaki et al. 2002). In contrast, FGF23 levels in patients with other chronic hypophosphatemic diseases such as vitamin D deficiency and Fanconi syndrome were low (Endo et al. 2008). These results indicated that FGF23 measurement is useful for the differential diagnosis of hypophosphatemic diseases. These reports also suggested that XLH is also caused by excessive actions of FGF23. Currently, several FGF23-related hypophosphatemic diseases are considered to be caused by overproduction of FGF23 (Table 2).

7.1 XLH

XLH is the most prevalent cause of genetic hypophosphatemic rickets. Its incidence is estimated to be about 1 in 25,000 births (Haffner et al. 2019). Patients with XLH show hypophosphatemia, impaired proximal tubular phosphate reabsorption, and low–low normal $1,25(OH)_2D$ as in patients with TIO and ADHR. The responsible gene for XLH was identified by positional cloning and named *phosphate-regulating gene with homologies to endopeptidases on the X chromosome* (*PHEX*) (The HYP Consortium 1995). Many mutations in *PHEX* have been reported in patients with XLH. These mutations are considered to be inactivating oncs. *PHEX* encodes a protein with a transmembrane domain. PHEX protein shows a homology to several endopeptidases like neutral endopeptidase and endothelin-converting enzyme 1 (Turner and Tanzawa 1997). However, it is not clear how PHEX works.

Hyp mice were shown to have a deletion of 3′ region of *Phex* gene indicating that these mice were a true genetic model of XLH (Beck et al. 1997). FGF23 was reported to be overexpressed in the bone of *Hyp* mice (Liu et al. 2003). In addition,

Table 2 Diseases caused by deranged actions of FGF23

Excessive actions of FGF23	
Disease	*Responsible gene*
X-linked hypophsophatemic rickets (XLH)	*Phosphate-regulating gene with homologies to endopeptidases on the X chromosome (PHEX)*
Autosomal dominant hypophosphatemic rickets (ADHR)	*FGF23*
Autosomal recessive hypophosphatemic rickets 1 (ARHR1)	*Dentin matrix protein 1 (DMP1)*
Autosomal recessive hypophosphatemic rickets 2 (ARHR2)	*Ectonucleotide pyrophosphatase/ phosphodiesterase 1 (ENPP1)*
Hypophosphatemic dsease with dental anomalies and ectopic calcification	*Family with sequence similarity 20, member C (FAM20C)*
Osteoglophonic dysplasia	*FGFR1*
Jansen-type metaphyseal chondrodysplasia	*Parathyroid hormone 1 receptor (PTH1R)*
McCune-Albright syndrome/fibrous dysplasia	*Guanine nucleotide binding protein, alpha stimulating (GNAS)*
Cutaneous skeletal hypophosphatemia syndrome (CSHS)	*HRA, KRAS, NRAS*
Hypophosphatemic disease with high soluble Klotho	Translocation near *Klotho*
Tumor-induced osteomalacia (TIO)	*Fibronectin 1(FN1)-FGFR1* fusion, *FN1-FGF1* fusion
Intravenous iron polymaltose, saccharated ferrix oxide, ferric carboxymaltose	
Biliary atresia	
Deficient actions of FGF23	
Hyperphosphatemic familial tumoral calcinosis	*Polypeptide N-acetylgalactosaminyltransferase 3 (GALNT3)*
	FGF23
	KLOTHO
Hyperphosphatemic tumoral calcinosis	Autoantibodies to FGF23

Phex inactivation in osteoblasts/osteocytes using *osteocalcin*-Cre transgenic mice caused hypophosphatemia, high FGF23, and enhanced expression of *Fgf23* in the bone (Yuan et al. 2008). Therefore, inactivating mutations in *PHEX* somehow cause excessive production of FGF23 in the bone. However, it is not clear how inactivating mutations of *PHEX* result in enhanced expression of *FGF23*. It was reported that the decreased expression of *Sgne1* is involved in the upregulation of *Fgf23* of *Hyp* mice (Yuan et al. 2013). However, it is not known how inactivating mutations of *Phex* cause altered expression of *Sgne1*.

7.2 ADHR

Mutations in patients with ADHR were observed in codons either for Arg176 or Arg179 just before the processing site between Arg179 and Ser180 (ADHR Consortium 2000). The amino acid sequence before this processing site, Arg176-His177-Thr178-Arg179, matches the consensus R-X-X-R motif recognized by proprotein convertases like furin (Fig. 1). Mutations in patients with ADHR destroy this consensus sequence. In vitro analysis indicated that the mutant FGF23 proteins found in patients with ADHR were actually resistant to the processing (Shimada et al. 2002; White et al. 2001). From these results it was postulated that patients with ADHR show high FGF23 levels because of this resistance to the processing.

However, it was reported that symptoms and biochemical abnormalities in patients with ADHR wax and wane (Imel et al. 2007). Some patients can show normophosphatemia with normal FGF23 levels. FGF23 is high in patients with ADHR when they are hypophosphatemic. This indicates that mutations in *FGF23* do not necessarily cause hypophosphatemia and high FGF23. Even when the mutant FGF23 is resistant to the processing, circulatory FGF23 levels should be kept within the reference range if the regulatory mechanisms of *FGF23* expression are intact. FGF23 levels are low in patients with chronic hypophosphatemia caused by vitamin D deficiency or Fanconi syndrome (Endo et al. 2008). This suggests that FGF23 production in the bone is suppressed in these patients by chronic hypophosphatemia or other associated metabolic changes. It is likely that patients with ADHR show hypophosphatemia and high FGF23 when the expression of *FGF23* in the bone is somehow stimulated. Iron deficiency is proposed to be one of the stimulators (Farrow et al. 2011). However, it is possible that other stimuli that enhance *FGF23* expression contribute to the development of hypophosphatemia in patients with ADHR.

7.3 Autosomal Recessive Hypophosphatemic Rickets (ARHR) 1, 2

These diseases are caused by inactivating mutations in *dentin matrix protein 1* (*DMP1*) (Feng et al. 2006; Lorenz-Depiereux et al. 2006) and *ectonucleotide pyrophosphatase/phosphodiesterase 1* (*ENPP1*) (Levy-Litan et al. 2010; Lorenz-Depiereux et al. 2010). DMP1 is one of the bone matrix proteins. While *Fgf23* was shown to be overexpressed in the bone of *Dmp1* knockout mice (Feng et al. 2006), it is not clear how mutations in *DMP1* cause FGF23 overproduction. ENPP1 is an enzyme that converts pyrophosphate to phosphate. Pyrophosphate is a potent inhibitor of calcification. Inactivating mutations in *ENPP1* had been known to cause generalized arterial calcification of infancy (GACI), often a fatal disease (Ruf et al. 2005). However, hypophosphatemic rickets with high FGF23 was reported among survivors of GACI. While FGF23 is considered to be overexpressed in the bone of these patients, it is not elucidated how inactivating mutations in *ENPP1* cause FGF23 overexpression.

7.4 Hypophosphatemic Disease with Dental Anomalies and Ectopic Calcification

This disease is caused by inactivating mutations in *family with sequence similarity 20, member C* (*FAM20C*) (Rafaelsen et al. 2013). FAM20C is a Golgi kinase that phosphorylates several proteins. FAM20C is known to phosphorylate small integrin-binding ligand N-linked glycoproteins (SIBLINGs) such as osteopontin, bone sialoprotein, and DMP1 which affect bone mineralization (Ishikawa et al. 2012). Inactivating mutations in *ENPP1* were known to cause Raine syndrome characterized by osteosclerosis and periosteal bone formation (Simpson et al. 2007). Raine syndrome is usually a fatal disease. However, hypophosphatemia, high FGF23 levels, dental anomalies, and ectopic calcification were reported in surviving patients (Rafaelsen et al. 2013; Takeyari et al. 2014). *Fam20c* knockout mice were also shown to have hypophosphatemic rickets, high FGF23, and overexpression of FGF23 in the bone (Wang et al. 2012). It is not known how mutations in *FAM20C* result in enhanced expression of FGF23 in the bone.

7.5 Osteoglophonic Dysplasia, Jansen-Type Metaphyseal Chondrodysplasia, and McCune-Albright Syndrome/Fibrous Dysplasia

Osteoglophonic dysplasia is caused by activating mutations in *FGFR1* gene (White et al. 2005). Patients with osteoglophonic dysplasia present with craniosynostosis, prominent supraorbital ridge, depressed nasal bridge, rhizomelic dwarfism, and nonossifying bone lesions. Some patients with osteoglophonic dysplasia show hypophosphatemia with elevated FGF23 levels (White et al. 2005). Jansen-type metaphyseal chondrodysplasia and McCune-Albright syndrome are caused by activating mutations in *parathyroid hormone 1 receptor* (*PTH1R*) and *guanine nucleotide binding protein, alpha stimulating* (*GNAS*), respectively. A patient with Jansen-type metaphyseal chondrodysplasia was reported to show hypophosphatemia and high FGF23 levels (Brown et al. 2009). In addition, some patients with McCune-Albright syndrome/fibrous dysplasia which is caused by postzygotic mosaic mutations show hypophosphatemia with high FGF23 (Riminucci et al. 2003). It was reported that FGF23 levels correlate with the disease burden suggesting that dysplasic bone tissues are the source of FGF23 (Riminucci et al. 2003). These results indicate that signals from FGFR1 and PTH1R-GNAS-cyclic AMP pathway stimulate FGF23 production.

7.6 Other Genetic Hypophosphatemic Diseases

Schimmelpenning-Feuerstein-Mims syndrome (SFMS) is characterized by epidermal nevi and bone dysplasia. This syndrome is caused by postzygotic mosaic activating mutations in *HRAS*, *KRAS*, or *NRAS* (Avitan-Hersh et al. 2014; Lim et al.

2014, 2016). Some patients with SFMS show hypophosphatemia with high FGF23. This condition is called cutaneous skeletal hypophosphatemia syndrome (CSHS). This disease suggests that the RAS-ERK pathway is involved in the regulation of FGF23 production. There was also a report of a hypophosphatemic patient with high FGF23 levels who has a chromosomal translocation near *Klotho* gene (Brownstein et al. 2008). While this patient had a high soluble Klotho level, the cause of high FGF23 is not clear.

7.7 TIO

TIO is one of paraneoplastic syndromes usually caused by mesenchymal tumors called phosphaturic mesenchymal tumor-mixed connective tissue variant (PMTMCT) (Folpe et al. 2004). PMTMCTs can be found in the bone or soft tissue anywhere in the body. Malignant PMTMCTs with metastases have been reported in the literature (Folpe et al. 2004). In addition, some epithelial malignancies such as colon cancer and ovarian cancer were also reported as causes for TIO (Minisola et al. 2017). Tumors responsible for TIO have been shown to secrete several phosphaturic factors including FGF23, secreted frizzled-related protein 4 (sFRP4), matrix extracellular phosphoglycoprotein (MEPE), and FGF7 (Berndt et al. 2003; Rowe et al. 2004; Carpenter et al. 2005). However, elevated levels of these factors except for FGF23 have not been reported in patients with TIO. Tumors responsible for TIO overexpress FGF23. Fusion genes of *fibronectin 1 (FN1)-FGFR1* and *FN1-FGF1* have been reported in some tumors causing TIO (Lee et al. 2015, 2016). In addition, the expression of Klotho has also been reported in these tumors (Kinoshita et al. 2019). Because the expressions of FGFRs are not basically tissue-specific, these results suggest that signals from FGFRs, activated either by the fusion genes or by FGF23 through the FGFR1-Klotho complex, are involved in the overexpression of FGF23 and/or the growth of the causative tumors for TIO.

TIO can be cured by complete removal of the responsible tumor cells by surgery or other methods (Mishra et al. 2019; Tutton et al. 2012). Therefore, it is clinically important to localize the responsible tumors. However, the responsible tumors are often present in the bone and small in size making them difficult to be detected. Somatostatin receptor imaging such as octreotide SPECT/CT, DOTA-TATE PET/CT, and DOTA-TOC PET/CT has been reported to be useful for localizing tumors responsible for TIO (Breer et al. 2014; Paquet et al. 2018). In some cases, venous sampling from draining veins of the tumors was reported to be helpful for the confirmation of the tumor localization (Takeuchi et al. 2004).

7.8 Intravenous Iron Administration

Hypophosphatemic osteomalacia with high FGF23 levels has been reported in patients with iron-deficiency anemia administered with some intravenous iron preparations such as iron polymaltose, saccharated ferric oxide, and ferric carboxymaltose (Schouten et al. 2009; Shimizu et al. 2009; Wolf et al. 2013). Hypophosphatemia was not reported in patients with other intravenous iron preparations including iron maltose and dextrin citrato-iron (III) complex (Shimizu et al. 2009; Wolf et al. 2013). As discussed in the section of ADHR, iron deficiency was shown to enhance *Fgf23* expression. However, iron deficiency does not cause high intact FGF23 levels and hypophosphatemia because the proteolytic processing of FGF23 protein is also stimulated in subjects without *FGF23* mutations. Intravenous iron preparations are considered to inhibit this processing and result in high FGF23 and hypophosphatemia. However, it is not known why only some intravenous preparations of iron prevent the processing and cause high FGF23 levels. Hypophosphatemia and high FGF23 levels improve after stopping these intravenous iron preparations.

7.9 Other FGF23-Related Hypophosphatemic Diseases

Hypophosphatemic disease was reported in several patients with biliary atresia (Wasserman et al. 2016). FGF23 was shown to be expressed in the liver in these patients. Again the detailed mechanisms of FGF23 expression in the liver in these patients have not been elucidated.

8 Treatment of FGF23-Related Hypophosphatemic Diseases

TIO can be cured by the complete removal of responsible tumors for TIO. However, the biochemical abnormalities and symptoms will not improve or recur if there are any residual tumor cells after surgery. It is very important to remove the tumors with sufficient margins of normal tissue if possible. Hypophosphatemia by intravenous iron preparations will improve after the cessation of these drugs. Patients with other hypophosphatemic diseases with high FGF23 levels including those with TIO whose tumors cannot be localized or surgically removed have been treated by neutral phosphate and active vitamin D metabolites (Carpenter et al. 2011). While these medications are effective in improving growth and symptoms of the affected patients, they can be associated with several adverse events such as secondary-tertiary hyperparathyroidism, hypercalciuria, hypercalcemia, nephrocalcinosis, and gastrointestinal symptoms such as diarrhea and abdominal pain. In addition, neutral phosphate needs to be administered three or four times a day creating a compliance problem. Because it has been shown that the excessive actions of FGF23 underlie these hypophosphatemic diseases, the inhibition of FGF23 activities has been considered as a new treatment option.

The isolated C-terminal fragment of FGF23 was shown to inhibit the actions of FGF23 by competing with full-length FGF23 for the binding to the FGFR1c-Klotho complex (Goetz et al. 2010). In addition, a computationally identified compound was shown to bind to FGF23 and interfere with the binding of FGF23 with the FGFR1c-Klotho complex (Xiao et al. 2016). Furthermore, inhibitors of FGFRs and ERK pathway were shown to inhibit the signaling by FGF23 (Wohrle et al. 2013; Zhang et al. 2012). All these molecules have been shown to increase serum phosphate in models of FGF23 excess in vivo. Moreover, anti-FGF23 antibodies were shown to inhibit FGF23 actions both in vitro and in vivo (Aono et al. 2009, 2011; Yamazaki et al. 2008). Administration of anti-FGF23 antibodies increased serum phosphate and 1,25(OH)$_2$D levels and enhanced renal phosphate reabsorption both in wild-type mice and *Hyp* mice. In addition, anti-FGF23 antibodies improved longitudinal growth of long bones, increased bone mineral density, and corrected thickened growth plate and increased osteoid of *Hyp* mice. Furthermore, anti-FGF23 antibodies improved reduced grip power of *Hyp* mice. These results suggested that the inhibition of FGF23 activities by anti-FGF23 antibody improves biochemical, morphological, and clinical features of patients with FGF23-related hypophosphatemic diseases.

From these preclinical results, a monoclonal human anti-FGF23 antibody, burosumab, was developed. The phase I study with 38 adult patients with XLH indicated that single intravenous or subcutaneous administration of burosumab increased serum phosphate, tubular maximum transport of phosphate per glomerular filtration rate (TmP/GFR), and 1,25(OH)$_2$D levels in a dose-dependent manner (Carpenter et al. 2014). Subsequent phase 2 studies with burosumab showed that repeated once every 2 weeks or 4 weeks subcutaneous administration of burosumab increased serum phosphate and improved roentgenological evidence of rickets in child patients with XLH (Carpenter et al. 2018). In addition, once every 2 weeks dosing seemed to be more efficient than once every 4 weeks administration. The phase 3 study showed that once every 2 weeks administration of burosumab was more efficacious than the conventional therapy with neutral phosphate and active vitamin D in the improvement of biochemical abnormalities, roentgenological features of rickets, and growth of child patients with XLH (Imel et al. 2019). Once every 4 weeks, subcutaneous administration of burosumab was also shown to be more effective than placebo in the improvement of biochemical abnormalities, Western Ontario and McMaster Universities Osteoarthritis Index (WOMAC) physical function, WOMAC stiffness, fracture healing, and histomorphometric features of osteomalacia in adult patients with XLH (Insogna et al. 2018, 2019). From these results, burosumab has been approved for clinical use for patients with XLH in several countries.

Theoretically, burosumab seems to be useful for patients with other FGF23-related hypophosphatemic diseases than XLH. Some meeting report suggests that burosumab is also useful for inoperable patients with TIO (https://www.ultragenyx.com/file.cfm/22/docs/Carpenter_2016_ASBMR_Oral_TIO_Final.pdf). However, the efficiency and safety of burosumab for hypophosphatemic patients from other diseases than XLH need to be confirmed by future studies.

9 FGF23 and Hyperphosphatemic Tumoral Calcinosis

In contrast to hypophosphatemic diseases caused by excessive actions of FGF23, deficient actions of FGF23 result in hyperphosphatemic diseases (Table 2). Hyperphosphatemic familial tumoral calcinosis (HFTC) is a rare autosomal recessive disease characterized by hyperphosphatemia, enhanced renal tubular phosphate reabsorption, high 1,25(OH)$_2$D levels, and ectopic calcification especially around large joints (Lyles et al. 1988). Calcified mass in soft tissues sometimes causes pain and whitish discharge from the skin. The biochemical abnormalities in patients with HFTC are similar to those observed in *Klotho* or *Fgf23* knockout mice and just the mirror images of those seen in patients with FGF23-related hypophosphatemic diseases (Kuro-o et al. 1997; Shimada et al. 2004b; Sitara et al. 2004).

Three genes, *GALNT3*, *FGF23*, and *KLOTHO*, have been shown to cause HFTC (Topaz et al. 2004; Araya et al. 2005; Benet-Pages et al. 2005; Ichikawa et al. 2007). *GALNT3* gene encodes an enzyme called polypeptide N-acetylgalactosaminyltransferase 3 (GalNAc-T3) that transfers N-acetylgalactosaine from UDP-N-acetyl-alpha-D-galactosamine to Ser or Thr residue as an initial sugar of mucin-type O-linked glycosylation. There are 20 members of GalNAc-Ts (Bennett et al. 2012). These 20 GalNAc-Ts are considered to have considerable substrate redundancy. No hereditary diseases caused by mutations in *GALNT* genes are known except for *GAlNT3*.

FGF23 protein has three mucin-type O-linked glycans attached to Thr171, Thr178, and Thr200 (Frishberg et al. 2007). While FGF23 protein with three O-linked glycans is resistant to the processing between Arg179 and Ser180, less glycosylated FGF23 protein is cleaved. In addition, Ser180 of FGF23 protein can be phosphorylated by FAM20C (Tagliabracci et al. 2014). This phosphorylation of FGF23 protein was proposed to prevent O-glycosylation of Thr178, thus making FGF23 protein susceptible for the processing. GalNAc-T3 was shown specifically to be able to initiate the O-linked glycosylation of Thr178 (Frishberg et al. 2007; Kato et al. 2006). Therefore, FGF23 protein in patients with biallelic inactivating mutations in *GALNT3* cannot be fully glycosylated and is cleaved into inactive fragments. FGF23 levels in patients with inactivating mutations in *GALNT3* support this theory. Intact and C-terminal assay of FGF23 produce quite discrepant results in these patients. While FGF23 is low by intact assay, it is quite high by C-terminal assay indicating that there is a large amount of processed C-terminal fragment of FGF23 despite the low full-length FGF23 (Frishberg et al. 2007). These results can be explained by the enhanced expression of *FGF23* by hyperphosphatemia or other associated metabolic changes and the susceptibility of the produced FGF23 protein for the processing (Fig. 2).

Patients with HFTC caused by inactivating mutations in *FGF23* also show high- and low-FGF23 levels by C-terminal and intact assays, respectively (Araya et al. 2005). It was shown that the mutant FGF23 protein is trapped in Golgi (Benet-Pages et al. 2005). It is not clear how the C-terminal fragment of FGF23 is secreted. Other report indicated that the mutant full-length FGF23 can be secreted, but cannot activate the FGFR1c-Klotho complex (Shawar et al. 2016). It is possible that mutations in *FGF23* cause HFTC by several mechanisms. A patient with HFTC

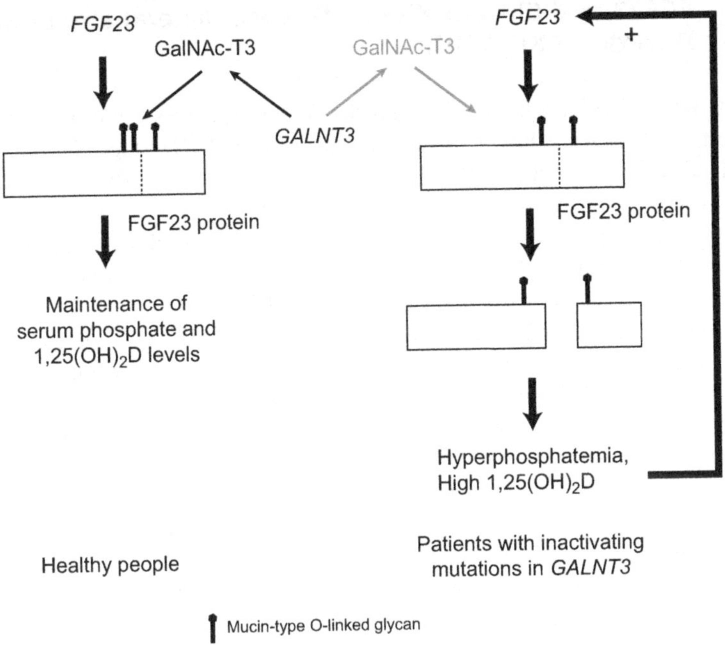

Fig. 2 Mechanism of deficient actions of FGF23 in patients with inactivating mutations in *GALNT3*. GalNAc-T3 encoded by *GALNT3* is specifically involved in the attachment of mucin-type O-linked glycan to Thr178. FGF23 protein with three O-linked glycans are resistant for the processing. This full-length FGF23 is essential for the maintenance of serum phosphate and 1,25 $(OH)_2D$ levels. FGF23 protein cannot be fully glycosylated and is cleaved into inactive fragments in patients with biallelic inactivating mutations in *GALNT3*. The resultant hyperphosphatemia, high $1,25(OH)_2D$, or other associated metabolic changes stimulate *FGF23* expression. Together with the susceptibility for the processing, this enhanced expression of *FGF23* produces high level of C-terminal fragment of FGF23

caused by an inactivating mutation in *Klotho* was reported to show clearly high FGF23 levels by both intact and C-terminal assays (Ichikawa et al. 2007). This HFTC is caused by resistance for FGF23.

In addition to genetic causes, autoantibodies to FGF23 was also shown to cause hyperphosphatemic tumoral calcinosis (Roberts et al. 2018). These results confirm that FGF23 is an essential hormone for the maintenance of serum phosphate and $1,25(OH)_2D$ levels.

Several medications such as phosphate binders and acetazolamide have been used for patients with hyperphosphatemic tumoral calcinosis (Lammoglia and Mericq 2009). In addition, surgical resections of calcified mass may be necessary for symptomatic patients. Recombinant FGF23 is theoretically useful for patients with HFTC caused by mutations in *GALNT3* or *FGF23*. However, recombinant FGF23 has not been clinically developed for this purpose.

10 FGF23 and Chronic Kidney Disease-Mineral and Bone Disorder (CKD-MBD)

Patients with CKD have high FGF23 levels. With the progression of impaired renal function, FGF23 was shown to start to increase in CKD stage 2 when serum phosphate, calcium, and $1,25(OH)_2D$ are all normal (Wolf 2010). It is currently unknown what triggers the increase of FGF23 in this early stage of CKD. This increment of FGF23 is followed by the decrease of $1,25(OH)_2D$, the increase of PTH, and then hyperphosphatemia and hypocalcemia. Patients with end-stage renal disease (ESRD) can show quite high levels of FGF23.

The inhibition of FGF23 actions by antibodies indicated that the increased FGF23 in rat model of early CKD worked to enhance renal phosphate excretion and prevent the development of hyperphosphatemia (Hasegawa et al. 2010). The increased FGF23 also contributes to reduce $1,25(OH)_2D$ levels and thus the development of secondary hyperparathyroidism. FGF23 was reported to suppress the production and secretion of PTH (Ben-Dov et al. 2007). However, chronic actions of FGF23 was also shown to promote the development of secondary hyperparathyroidism (Kawakami et al. 2017). The inhibition of FGF23 actions in a rat model of more advanced CKD resulted in the amelioration of secondary hyperparathyroidism but increased ectopic calcification and mortality (Shalhoub et al. 2012). These results suggest that the increase of FGF23 in patients with CKD is an adaptive response to prevent the development of hyperphosphatemia.

High FGF23 levels were reported to be correlated with worse mortality in patients starting hemodialysis during the subsequent year (Gutierrez et al. 2008). Since then, epidemiological studies mainly conducted in patients with CKD revealed the correlations between high FGF23 levels and various adverse events (Fukumoto and Shimizu 2011). The adverse events reported include low bone mineral density, fractures, progression of CKD, left ventricular hypertrophy, cerebrovascular events, infections, impaired cognitive function, and so on. FGF23 was reported to induce some of these adverse events as shown below. However, the reasons for the correlation are not clear for most of these adverse events.

11 Klotho-Independent Actions of FGF23

FGF23 regulates phosphate and vitamin D metabolism by binding to the FGFRs-Klotho complex. In contract, several Klotho-independent actions of FGF23 have been reported. FGF23 was reported to induce cardiac hypertrophy by directly acting on cardiomyocytes through FGFR4 in a Klotho-independent way (Grabner et al. 2015). This was postulated to explain the correlation between high FGF23 levels and the cardiovascular adverse events including left ventricular hypertrophy and higher mortality in patients with CKD. FGF23 was also shown to enhance the expression of inflammatory cytokines in the liver through FGFR4 (Singh et al. 2016). In addition, FGF23 was reported to impair neutrophil activation through FGFR2 (Rossaint et al. 2016), stimulate tumor necrosis factor α expression possibly via FGR1 in macrophages

(Han et al. 2016), and inhibit alkaline phosphatase expression in osteoblasts through FGFR3 (Murali et al. 2016), all in a Klotho-independent way. Furthermore, FGF23 was shown to suppress PTH secretion both in a Klotho-dependent and Klotho-independent ways (Olauson et al. 2013). These results indicate that FGF23 can activate many subtypes of FGFRs in a Klotho-independent way. However, these findings are in sharp contrast to the enhanced expression of *Egr-1* in response to FGF23 in several restricted tissues expressing Klotho (Urakawa et al. 2006). It is possible that some other molecules than Klotho are involved in the signal transduction from FGF23 in these tissues without Klotho expression. It is also possible that intracellular signals other than ERK and Egr-1 pathway are activated by FGF23 in those cells without Klotho. It is currently unknown how FGF23 activates FGFRs in a Klotho-independent way, and further studies are necessary to establish the significance of these Klotho-independent actions of FGF23.

12 Concluding Remarks

FGF23 has been established as a phosphotropic hormone produced by the bone that regulates phosphate and vitamin D metabolism. Several diseases with abnormal phosphate levels have been shown to be caused by deficient and excessive actions of FGF23 confirming that FGF23 is a physiological humoral factor. However, there still remain many unanswered questions. While several other functions of FGF23 than regulating mineral metabolism and Klotho-independent actions of FGF23 have been reported, it is not clear how FGF23 can activate FGFRs without Klotho. It is not known either how FGF23 can discriminate FGFRs for different actions. Further studies are clearly necessary to understand the physiological and pathophysiological roles of FGF23.

References

Andrukhova O, Zeitz U, Goetz R, Mohammadi M, Lanske B, Erben RG (2012) FGF23 acts directly on renal proximal tubules to induce phosphaturia through activation of the ERK1/2-SGK1 signaling pathway. Bone 51(3):621–628. https://doi.org/10.1016/j.bone.2012.05.015

Andrukhova O, Slavic S, Smorodchenko A, Zeitz U, Shalhoub V, Lanske B, Pohl EE, Erben RG (2014a) FGF23 regulates renal sodium handling and blood pressure. EMBO Mol Med 6 (6):744–759. https://doi.org/10.1002/emmm.201303716

Andrukhova O, Smorodchenko A, Egerbacher M, Streicher C, Zeitz U, Goetz R, Shalhoub V, Mohammadi M, Pohl EE, Lanske B, Erben RG (2014b) FGF23 promotes renal calcium reabsorption through the TRPV5 channel. EMBO J 33(3):229–246. https://doi.org/10.1002/embj.201284188

Aono Y, Yamazaki Y, Yasutake J, Kawata T, Hasegawa H, Urakawa I, Fujita T, Wada M, Yamashita T, Fukumoto S, Shimada T (2009) Therapeutic effects of anti-FGF23 antibodies in hypophosphatemic rickets/osteomalacia. J Bone Miner Res 24(11):1879–1888. https://doi.org/10.1359/jbmr.090509

Aono Y, Hasegawa H, Yamazaki Y, Shimada T, Fujita T, Yamashita T, Fukumoto S (2011) Anti-FGF-23 neutralizing antibodies ameliorate muscle weakness and decreased spontaneous movement of Hyp mice. J Bone Miner Res 26(4):803–810. https://doi.org/10.1002/jbmr.275

Araya K, Fukumoto S, Backenroth R, Takeuchi Y, Nakayama K, Ito N, Yoshii N, Yamazaki Y, Yamashita T, Silver J, Igarashi T, Fujita T (2005) A novel mutation in fibroblast growth factor 23 gene as a cause of tumoral calcinosis. J Clin Endocrinol Metab 90(10):5523–5527. https://doi.org/10.1210/jc.2005-0301

Avitan-Hersh E, Tatur S, Indelman M, Gepstein V, Shreter R, Hershkovitz D, Brick R, Bergman R, Tiosano D (2014) Postzygotic HRAS mutation causing both keratinocytic epidermal nevus and thymoma and associated with bone dysplasia and hypophosphatemia due to elevated FGF23. J Clin Endocrinol Metab 99(1):E132–E136. https://doi.org/10.1210/jc.2013-2813

Beck L, Soumounou Y, Martel J, Krishnamurthy G, Gauthier C, Goodyer CG, Tenenhouse HS (1997) Pex/PEX tissue distribution and evidence for a deletion in the 3′ region of the Pex gene in X-linked hypophosphatemic mice. J Clin Invest 99(6):1200–1209. https://doi.org/10.1172/jci119276

Ben-Dov IZ, Galitzer H, Lavi-Moshayoff V, Goetz R, Kuro-o M, Mohammadi M, Sirkis R, Naveh-Many T, Silver J (2007) The parathyroid is a target organ for FGF23 in rats. J Clin Invest 117 (12):4003–4008. https://doi.org/10.1172/jci32409

Benet-Pages A, Orlik P, Strom TM, Lorenz-Depiereux B (2005) An FGF23 missense mutation causes familial tumoral calcinosis with hyperphosphatemia. Hum Mol Genet 14(3):385–390. https://doi.org/10.1093/hmg/ddi034

Bennett EP, Mandel U, Clausen H, Gerken TA, Fritz TA, Tabak LA (2012) Control of mucin-type O-glycosylation: a classification of the polypeptide GalNAc-transferase gene family. Glycobiology 22(6):736–756. https://doi.org/10.1093/glycob/cwr182

Bergwitz C, Roslin NM, Tieder M, Loredo-Osti JC, Bastepe M, Abu-Zahra H, Frappier D, Burkett K, Carpenter TO, Anderson D, Garabedian M, Sermet I, Fujiwara TM, Morgan K, Tenenhouse HS, Juppner H (2006) SLC34A3 mutations in patients with hereditary hypophosphatemic rickets with hypercalciuria predict a key role for the sodium-phosphate cotransporter NaPi-IIc in maintaining phosphate homeostasis. Am J Hum Genet 78 (2):179–192. https://doi.org/10.1086/499409

Berndt T, Craig TA, Bowe AE, Vassiliadis J, Reczek D, Finnegan R, Jan De Beur SM, Schiavi SC, Kumar R (2003) Secreted frizzled-related protein 4 is a potent tumor-derived phosphaturic agent. J Clin Invest 112(5):785–794. https://doi.org/10.1172/jci18563

Breer S, Brunkhorst T, Beil FT, Peldschus K, Heiland M, Klutmann S, Barvencik F, Zustin J, Gratz KF, Amling M (2014) 68Ga DOTA-TATE PET/CT allows tumor localization in patients with tumor-induced osteomalacia but negative 111In-octreotide SPECT/CT. Bone 64:222–227. https://doi.org/10.1016/j.bone.2014.04.016

Brown WW, Juppner H, Langman CB, Price H, Farrow EG, White KE, McCormick KL (2009) Hypophosphatemia with elevations in serum fibroblast growth factor 23 in a child with Jansen's metaphyseal chondrodysplasia. J Clin Endocrinol Metab 94(1):17–20. https://doi.org/10.1210/jc.2008-022010.1210/jcem.94.2.9988

Brownstein CA, Adler F, Nelson-Williams C, Iijima J, Li P, Imura A, Nabeshima Y, Reyes-Mugica M, Carpenter TO, Lifton RP (2008) A translocation causing increased alpha-klotho level results in hypophosphatemic rickets and hyperparathyroidism. Proc Natl Acad Sci U S A 105(9):3455–3460. https://doi.org/10.1073/pnas.0712361105

Carpenter TO, Ellis BK, Insogna KL, Philbrick WM, Sterpka J, Shimkets R (2005) Fibroblast growth factor 7: an inhibitor of phosphate transport derived from oncogenic osteomalacia-causing tumors. J Clin Endocrinol Metab 90(2):1012–1020. https://doi.org/10.1210/jc.2004-0357

Carpenter TO, Imel EA, Holm IA, Jan de Beur SM, Insogna KL (2011) A clinician's guide to X-linked hypophosphatemia. J Bone Miner Res 26(7):1381–1388. https://doi.org/10.1002/jbmr.340

Carpenter TO, Imel EA, Ruppe MD, Weber TJ, Klausner MA, Wooddell MM, Kawakami T, Ito T, Zhang X, Humphrey J, Insogna KL, Peacock M (2014) Randomized trial of the anti-FGF23 antibody KRN23 in X-linked hypophosphatemia. J Clin Invest 124(4):1587–1597. https://doi.org/10.1172/jci72829

Carpenter TO, Whyte MP, Imel EA, Boot AM, Hogler W, Linglart A, Padidela R, Van't Hoff W, Mao M, Chen CY, Skrinar A, Kakkis E, San Martin J, Portale AA (2018) Burosumab therapy in children with X-linked hypophosphatemia. N Engl J Med 378(21):1987–1998. https://doi.org/10.1056/NEJMoa1714641

Chen G, Liu Y, Goetz R, Fu L, Jayaraman S, Hu MC, Moe OW, Liang G, Li X, Mohammadi M (2018) Alpha-Klotho is a non-enzymatic molecular scaffold for FGF23 hormone signalling. Nature 553(7689):461–466. https://doi.org/10.1038/nature25451

The HYP Consortium (1995) A gene (PEX) with homologies to endopeptidases is mutated in patients with X-linked hypophosphatemic rickets. Nat Genet 11(2):130–136. https://doi.org/10.1038/ng1095-130

ADHR Consortium (2000) Autosomal dominant hypophosphataemic rickets is associated with mutations in FGF23. Nat Genet 26(3):345–348. https://doi.org/10.1038/81664

David V, Martin A, Isakova T, Spaulding C, Qi L, Ramirez V, Zumbrennen-Bullough KB, Sun CC, Lin HY, Babitt JL, Wolf M (2016) Inflammation and functional iron deficiency regulate fibroblast growth factor 23 production. Kidney Int 89(1):135–146. https://doi.org/10.1038/ki.2015.290

Dehghani A, Hafizibarjin Z, Najjari R, Kaseb F, Safari F (2018) Resveratrol and 1,25-dihydroxyvitamin D co-administration protects the heart against D-galactose-induced aging in rats: evaluation of serum and cardiac levels of klotho. Aging Clin Exp Res. https://doi.org/10.1007/s40520-018-1075-x

Endo I, Fukumoto S, Ozono K, Namba N, Tanaka H, Inoue D, Minagawa M, Sugimoto T, Yamauchi M, Michigami T, Matsumoto T (2008) Clinical usefulness of measurement of fibroblast growth factor 23 (FGF23) in hypophosphatemic patients: proposal of diagnostic criteria using FGF23 measurement. Bone 42(6):1235–1239. https://doi.org/10.1016/j.bone.2008.02.014

Farrow EG, Yu X, Summers LJ, Davis SI, Fleet JC, Allen MR, Robling AG, Stayrook KR, Jideonwo V, Magers MJ, Garringer HJ, Vidal R, Chan RJ, Goodwin CB, Hui SL, Peacock M, White KE (2011) Iron deficiency drives an autosomal dominant hypophosphatemic rickets (ADHR) phenotype in fibroblast growth factor-23 (Fgf23) knock-in mice. Proc Natl Acad Sci U S A 108(46):E1146–E1155. https://doi.org/10.1073/pnas.1110905108

Feng JQ, Ward LM, Liu S, Lu Y, Xie Y, Yuan B, Yu X, Rauch F, Davis SI, Zhang S, Rios H, Drezner MK, Quarles LD, Bonewald LF, White KE (2006) Loss of DMP1 causes rickets and osteomalacia and identifies a role for osteocytes in mineral metabolism. Nat Genet 38 (11):1310–1315. https://doi.org/10.1038/ng1905

Ferrari SL, Bonjour JP, Rizzoli R (2005) Fibroblast growth factor-23 relationship to dietary phosphate and renal phosphate handling in healthy young men. J Clin Endocrinol Metab 90 (3):1519–1524. https://doi.org/10.1210/jc.2004-1039

Folpe AL, Fanburg-Smith JC, Billings SD, Bisceglia M, Bertoni F, Cho JY, Econs MJ, Inwards CY, Jan de Beur SM, Mentzel T, Montgomery E, Michal M, Miettinen M, Mills SE, Reith JD, O'Connell JX, Rosenberg AE, Rubin BP, Sweet DE, Vinh TN, Wold LE, Wehrli BM, White KE, Zaino RJ, Weiss SW (2004) Most osteomalacia-associated mesenchymal tumors are a single histopathologic entity: an analysis of 32 cases and a comprehensive review of the literature. Am J Surg Pathol 28(1):1–30

Frishberg Y, Ito N, Rinat C, Yamazaki Y, Feinstein S, Urakawa I, Navon-Elkan P, Becker-Cohen R, Yamashita T, Araya K, Igarashi T, Fujita T, Fukumoto S (2007) Hyperostosis-hyperphosphatemia syndrome: a congenital disorder of O-glycosylation associated with augmented processing of fibroblast growth factor 23. J Bone Miner Res 22(2):235–242. https://doi.org/10.1359/jbmr.061105

Fukumoto S (2018) Targeting fibroblast growth factor 23 signaling with antibodies and inhibitors, is there a rationale? Front Endocrinol 9:48. https://doi.org/10.3389/fendo.2018.00048

Fukumoto S, Shimizu Y (2011) Fibroblast growth factor 23 as a phosphotropic hormone and beyond. J Bone Miner Metab 29(5):507–514. https://doi.org/10.1007/s00774-011-0298-0

Gattineni J, Bates C, Twombley K, Dwarakanath V, Robinson ML, Goetz R, Mohammadi M, Baum M (2009) FGF23 decreases renal NaPi-2a and NaPi-2c expression and induces hypophosphatemia in vivo predominantly via FGF receptor 1. Am J Physiol Renal Physiol 297(2):F282–F291. https://doi.org/10.1152/ajprenal.90742.2008

Gattineni J, Twombley K, Goetz R, Mohammadi M, Baum M (2011) Regulation of serum 1,25 (OH)2 vitamin D3 levels by fibroblast growth factor 23 is mediated by FGF receptors 3 and 4. Am J Physiol Renal Physiol 301(2):F371–F377. https://doi.org/10.1152/ajprenal.00740.2010

Goetz R, Mohammadi M (2013) Exploring mechanisms of FGF signalling through the lens of structural biology. Nat Rev Mol Cell Biol 14(3):166–180. https://doi.org/10.1038/nrm3528

Goetz R, Nakada Y, Hu MC, Kurosu H, Wang L, Nakatani T, Shi M, Eliseenkova AV, Razzaque MS, Moe OW, Kuro-o M, Mohammadi M (2010) Isolated C-terminal tail of FGF23 alleviates hypophosphatemia by inhibiting FGF23-FGFR-Klotho complex formation. Proc Natl Acad Sci U S A 107(1):407–412. https://doi.org/10.1073/pnas.0902006107

Grabner A, Amaral AP, Schramm K, Singh S, Sloan A, Yanucil C, Li J, Shehadeh LA, Hare JM, David V, Martin A, Fornoni A, Di Marco GS, Kentrup D, Reuter S, Mayer AB, Pavenstadt H, Stypmann J, Kuhn C, Hille S, Frey N, Leifheit-Nestler M, Richter B, Haffner D, Abraham R, Bange J, Sperl B, Ullrich A, Brand M, Wolf M, Faul C (2015) Activation of cardiac fibroblast growth factor receptor 4 causes left ventricular hypertrophy. Cell Metab 22(6):1020–1032. https://doi.org/10.1016/j.cmet.2015.09.002

Gutierrez OM, Mannstadt M, Isakova T, Rauh-Hain JA, Tamez H, Shah A, Smith K, Lee H, Thadhani R, Juppner H, Wolf M (2008) Fibroblast growth factor 23 and mortality among patients undergoing hemodialysis. N Engl J Med 359(6):584–592. https://doi.org/10.1056/NEJMoa0706130

Haffner D, Emma F, Eastwood DM, Duplan MB, Bacchetta J, Schnabel D, Wicart P, Bockenhauer D, Santos F, Levtchenko E, Harvengt P, Kirchhoff M, Di Rocco F, Chaussain C, Brandi ML, Savendahl L, Briot K, Kamenicky P, Rejnmark L, Linglart A (2019) Clinical practice recommendations for the diagnosis and management of X-linked hypophosphataemia. Nat Rev Nephrol 15(7):435–455. https://doi.org/10.1038/s41581-019-0152-5

Han X, Li L, Yang J, King G, Xiao Z, Quarles LD (2016) Counter-regulatory paracrine actions of FGF-23 and 1,25(OH)2 D in macrophages. FEBS Lett 590(1):53–67. https://doi.org/10.1002/1873-3468.12040

Hasegawa H, Nagano N, Urakawa I, Yamazaki Y, Iijima K, Fujita T, Yamashita T, Fukumoto S, Shimada T (2010) Direct evidence for a causative role of FGF23 in the abnormal renal phosphate handling and vitamin D metabolism in rats with early-stage chronic kidney disease. Kidney Int 78(10):975–980. https://doi.org/10.1038/ki.2010.313

Hughes MR, Malloy PJ, Kieback DG, Kesterson RA, Pike JW, Feldman D, O'Malley BW (1988) Point mutations in the human vitamin D receptor gene associated with hypocalcemic rickets. Science 242(4886):1702–1705. https://doi.org/10.1126/science.2849209

Ichikawa S, Imel EA, Kreiter ML, Yu X, Mackenzie DS, Sorenson AH, Goetz R, Mohammadi M, White KE, Econs MJ (2007) A homozygous missense mutation in human KLOTHO causes severe tumoral calcinosis. J Clin Invest 117(9):2684–2691. https://doi.org/10.1172/jci31330

Imel EA, Hui SL, Econs MJ (2007) FGF23 concentrations vary with disease status in autosomal dominant hypophosphatemic rickets. J Bone Miner Res 22(4):520–526. https://doi.org/10.1359/jbmr.070107

Imel EA, Glorieux FH, Whyte MP, Munns CF, Ward LM, Nilsson O, Simmons JH, Padidela R, Namba N, Cheong HI, Pitukcheewanont P, Sochett E, Hogler W, Muroya K, Tanaka H, Gottesman GS, Biggin A, Perwad F, Mao M, Chen CY, Skrinar A, San Martin J, Portale AA (2019) Burosumab versus conventional therapy in children with X-linked hypophosphataemia: a randomised, active-controlled, open-label, phase 3 trial. Lancet 393(10189):2416–2427. https://doi.org/10.1016/s0140-6736(19)30654-3

Insogna KL, Briot K, Imel EA, Kamenicky P, Ruppe MD, Portale AA, Weber T, Pitukcheewanont P, Cheong HI, Jan de Beur S, Imanishi Y, Ito N, Lachmann RH, Tanaka H, Perwad F, Zhang L, Chen CY, Theodore-Oklota C, Mealiffe M, San Martin J, Carpenter TO (2018) A randomized, double-blind, placebo-controlled, phase 3 trial evaluating the efficacy of Burosumab, an anti-FGF23 antibody, in adults with X-linked hypophosphatemia: week 24 primary analysis. J Bone Miner Res 33(8):1383–1393. https://doi.org/10.1002/jbmr.3475

Insogna KL, Rauch F, Kamenicky P, Ito N, Kubota T, Nakamura A, Zhang L, Mealiffe M, San Martin J, Portale AA (2019) Burosumab improved histomorphometric measures of osteomalacia in adults with X-linked hypophosphatemia: a phase 3, single-arm, international trial. J Bone Miner Res. https://doi.org/10.1002/jbmr.3843

Ishikawa HO, Xu A, Ogura E, Manning G, Irvine KD (2012) The Raine syndrome protein FAM20C is a Golgi kinase that phosphorylates bio-mineralization proteins. PLoS One 7(8): e42988. https://doi.org/10.1371/journal.pone.0042988

Ito N, Fukumoto S, Takeuchi Y, Yasuda T, Hasegawa Y, Takemoto F, Tajima T, Dobashi K, Yamazaki Y, Yamashita T, Fujita T (2005) Comparison of two assays for fibroblast growth factor (FGF)-23. J Bone Miner Metab 23(6):435–440. https://doi.org/10.1007/s00774-005-0625-4

Ito N, Fukumoto S, Takeuchi Y, Takeda S, Suzuki H, Yamashita T, Fujita T (2007) Effect of acute changes of serum phosphate on fibroblast growth factor (FGF)23 levels in humans. J Bone Miner Metab 25(6):419–422. https://doi.org/10.1007/s00774-007-0779-3

Itoh N, Ornitz DM (2004) Evolution of the Fgf and Fgfr gene families. Trends Genet 20 (11):563–569. https://doi.org/10.1016/j.tig.2004.08.007

Jonsson KB, Zahradnik R, Larsson T, White KE, Sugimoto T, Imanishi Y, Yamamoto T, Hampson G, Koshiyama H, Ljunggren O, Oba K, Yang IM, Miyauchi A, Econs MJ, Lavigne J, Juppner H (2003) Fibroblast growth factor 23 in oncogenic osteomalacia and X-linked hypophosphatemia. N Engl J Med 348(17):1656–1663. https://doi.org/10.1056/NEJMoa020881

Kato K, Jeanneau C, Tarp MA, Benet-Pages A, Lorenz-Depiereux B, Bennett EP, Mandel U, Strom TM, Clausen H (2006) Polypeptide GalNAc-transferase T3 and familial tumoral calcinosis. Secretion of fibroblast growth factor 23 requires O-glycosylation. J Biol Chem 281 (27):18370–18377. https://doi.org/10.1074/jbc.M602469200

Kawakami K, Takeshita A, Furushima K, Miyajima M, Hatamura I, Kuro OM, Furuta Y, Sakaguchi K (2017) Persistent fibroblast growth factor 23 signalling in the parathyroid glands for secondary hyperparathyroidism in mice with chronic kidney disease. Sci Rep 7:40534. https://doi.org/10.1038/srep40534

Kinoshita Y, Fukumoto S (2018) X-linked hypophosphatemia and FGF23-related hypophosphatemic diseases: prospect for new treatment. Endocr Rev 39(3):274–291. https://doi.org/10.1210/er.2017-00220

Kinoshita Y, Takashi Y, Ito N, Ikegawa S, Mano H, Ushiku T, Fukayama M, Nangaku M, Fukumoto S (2019) Ectopic expression of Klotho in fibroblast growth factor 23 (FGF23)-producing tumors that cause tumor-induced rickets/osteomalacia (TIO). Bone Rep 10:100192. https://doi.org/10.1016/j.bonr.2018.100192

Kitanaka S, Takeyama K, Murayama A, Sato T, Okumura K, Nogami M, Hasegawa Y, Niimi H, Yanagisawa J, Tanaka T, Kato S (1998) Inactivating mutations in the 25-hydroxyvitamin D3 1alpha-hydroxylase gene in patients with pseudovitamin D-deficiency rickets. N Engl J Med 338(10):653–661. https://doi.org/10.1056/nejm199803053381004

Komaba H, Kaludjerovic J, Hu DZ, Nagano K, Amano K, Ide N, Sato T, Densmore MJ, Hanai JI, Olauson H, Bellido T, Larsson TE, Baron R, Lanske B (2017) Klotho expression in osteocytes regulates bone metabolism and controls bone formation. Kidney Int 92(3):599–611. https://doi.org/10.1016/j.kint.2017.02.014

Kuro-o M, Matsumura Y, Aizawa H, Kawaguchi H, Suga T, Utsugi T, Ohyama Y, Kurabayashi M, Kaname T, Kume E, Iwasaki H, Iida A, Shiraki-Iida T, Nishikawa S, Nagai R, Nabeshima YI (1997) Mutation of the mouse klotho gene leads to a syndrome resembling ageing. Nature 390 (6655):45–51. https://doi.org/10.1038/36285

Kurosu H, Ogawa Y, Miyoshi M, Yamamoto M, Nandi A, Rosenblatt KP, Baum MG, Schiavi S, Hu MC, Moe OW, Kuro-o M (2006) Regulation of fibroblast growth factor-23 signaling by klotho. J Biol Chem 281(10):6120–6123. https://doi.org/10.1074/jbc.C500457200

Lammoglia JJ, Mericq V (2009) Familial tumoral calcinosis caused by a novel FGF23 mutation: response to induction of tubular renal acidosis with acetazolamide and the non-calcium phosphate binder sevelamer. Horm Res 71(3):178–184. https://doi.org/10.1159/000197876

Larsson T, Marsell R, Schipani E, Ohlsson C, Ljunggren O, Tenenhouse HS, Juppner H, Jonsson KB (2004) Transgenic mice expressing fibroblast growth factor 23 under the control of the alpha1(I) collagen promoter exhibit growth retardation, osteomalacia, and disturbed phosphate homeostasis. Endocrinology 145(7):3087–3094. https://doi.org/10.1210/en.2003-1768

Lee JC, Jeng YM, Su SY, Wu CT, Tsai KS, Lee CH, Lin CY, Carter JM, Huang JW, Chen SH, Shih SR, Marino-Enriquez A, Chen CC, Folpe AL, Chang YL, Liang CW (2015) Identification of a novel FN1-FGFR1 genetic fusion as a frequent event in phosphaturic mesenchymal tumour. J Pathol 235(4):539–545. https://doi.org/10.1002/path.4465

Lee JC, Su SY, Changou CA, Yang RS, Tsai KS, Collins MT, Orwoll ES, Lin CY, Chen SH, Shih SR, Lee CH, Oda Y, Billings SD, Li CF, Nielsen GP, Konishi E, Petersson F, Carpenter TO, Sittampalam K, Huang HY, Folpe AL (2016) Characterization of FN1-FGFR1 and novel FN1-FGF1 fusion genes in a large series of phosphaturic mesenchymal tumors. Mod Pathol 29 (11):1335–1346. https://doi.org/10.1038/modpathol.2016.137

Levy-Litan V, Hershkovitz E, Avizov L, Leventhal N, Bercovich D, Chalifa-Caspi V, Manor E, Buriakovsky S, Hadad Y, Goding J, Parvari R (2010) Autosomal-recessive hypophosphatemic rickets is associated with an inactivation mutation in the ENPP1 gene. Am J Hum Genet 86 (2):273–278. https://doi.org/10.1016/j.ajhg.2010.01.010

Li H, Martin A, David V, Quarles LD (2011) Compound deletion of Fgfr3 and Fgfr4 partially rescues the Hyp mouse phenotype. Am J Physiol Endocrinol Metab 300(3):E508–E517. https://doi.org/10.1152/ajplung.zh5-5845-retr.2011

Lim YH, Ovejero D, Sugarman JS, Deklotz CM, Maruri A, Eichenfield LF, Kelley PK, Juppner H, Gottschalk M, Tifft CJ, Gafni RI, Boyce AM, Cowen EW, Bhattacharyya N, Guthrie LC, Gahl WA, Golas G, Loring EC, Overton JD, Mane SM, Lifton RP, Levy ML, Collins MT, Choate KA (2014) Multilineage somatic activating mutations in HRAS and NRAS cause mosaic cutaneous and skeletal lesions, elevated FGF23 and hypophosphatemia. Hum Mol Genet 23(2):397–407. https://doi.org/10.1093/hmg/ddt429

Lim YH, Ovejero D, Derrick KM, Collins MT, Choate KA (2016) Cutaneous skeletal hypophosphatemia syndrome (CSHS) is a multilineage somatic mosaic RASopathy. J Am Acad Dermatol 75(2):420–427. https://doi.org/10.1016/j.jaad.2015.11.012

Liu S, Guo R, Simpson LG, Xiao ZS, Burnham CE, Quarles LD (2003) Regulation of fibroblastic growth factor 23 expression but not degradation by PHEX. J Biol Chem 278(39):37419–37426. https://doi.org/10.1074/jbc.M304544200

Lorenz-Depiereux B, Bastepe M, Benet-Pages A, Amyere M, Wagenstaller J, Muller-Barth U, Badenhoop K, Kaiser SM, Rittmaster RS, Shlossberg AH, Olivares JL, Loris C, Ramos FJ, Glorieux F, Vikkula M, Juppner H, Strom TM (2006) DMP1 mutations in autosomal recessive hypophosphatemia implicate a bone matrix protein in the regulation of phosphate homeostasis. Nat Genet 38(11):1248–1250. https://doi.org/10.1038/ng1868

Lorenz-Depiereux B, Schnabel D, Tiosano D, Hausler G, Strom TM (2010) Loss-of-function ENPP1 mutations cause both generalized arterial calcification of infancy and autosomal-recessive hypophosphatemic rickets. Am J Hum Genet 86(2):267–272. https://doi.org/10.1016/j.ajhg.2010.01.006

Lyles KW, Halsey DL, Friedman NE, Lobaugh B (1988) Correlations of serum concentrations of 1,25-dihydroxyvitamin D, phosphorus, and parathyroid hormone in tumoral calcinosis. J Clin Endocrinol Metab 67(1):88–92. https://doi.org/10.1210/jcem-67-1-88

Magagnin S, Werner A, Markovich D, Sorribas V, Stange G, Biber J, Murer H (1993) Expression cloning of human and rat renal cortex Na/Pi cotransport. Proc Natl Acad Sci U S A 90 (13):5979–5983. https://doi.org/10.1073/pnas.90.13.5979

Meyer RA Jr, Meyer MH, Gray RW (1989) Parabiosis suggests a humoral factor is involved in X-linked hypophosphatemia in mice. J Bone Miner Res 4(4):493–500. https://doi.org/10.1002/jbmr.5650040407

Minisola S, Peacock M, Fukumoto S, Cipriani C, Pepe J, Tella SH, Collins MT (2017) Tumour-induced osteomalacia. Nat Rev Dis Primers 3:17044. https://doi.org/10.1038/nrdp.2017.44

Mishra SK, Kuchay MS, Sen IB, Garg A, Baijal SS, Mithal A (2019) Successful management of tumor-induced osteomalacia with radiofrequency ablation: a case series. JBMR Plus 3(7): e10178. https://doi.org/10.1002/jbm4.10178

Murali SK, Roschger P, Zeitz U, Klaushofer K, Andrukhova O, Erben RG (2016) FGF23 regulates bone mineralization in a 1,25(OH)2 D3 and Klotho-independent manner. J Bone Miner Res 31 (1):129–142. https://doi.org/10.1002/jbmr.2606

Ohyama Y, Noshiro M, Okuda K (1991) Cloning and expression of cDNA encoding 25-hydroxyvitamin D3 24-hydroxylase. FEBS Lett 278(2):195–198. https://doi.org/10.1016/0014-5793(91)80115-j

Olauson H, Lindberg K, Amin R, Sato T, Jia T, Goetz R, Mohammadi M, Andersson G, Lanske B, Larsson TE (2013) Parathyroid-specific deletion of Klotho unravels a novel calcineurin-dependent FGF23 signaling pathway that regulates PTH secretion. PLoS Genet 9(12): e1003975. https://doi.org/10.1371/journal.pgen.1003975

Paquet M, Gauthe M, Zhang Yin J, Nataf V, Belissant O, Orcel P, Roux C, Talbot JN, Montravers F (2018) Diagnostic performance and impact on patient management of (68)Ga-DOTA-TOC PET/CT for detecting osteomalacia-associated tumours. Eur J Nucl Med Mol Imaging 45 (10):1710–1720. https://doi.org/10.1007/s00259-018-3971-x

Perwad F, Azam N, Zhang MY, Yamashita T, Tenenhouse HS, Portale AA (2005) Dietary and serum phosphorus regulate fibroblast growth factor 23 expression and 1,25-dihydroxyvitamin D metabolism in mice. Endocrinology 146(12):5358–5364. https://doi.org/10.1210/en.2005-0777

Rafaelsen SH, Raeder H, Fagerheim AK, Knappskog P, Carpenter TO, Johansson S, Bjerknes R (2013) Exome sequencing reveals FAM20c mutations associated with fibroblast growth factor 23-related hypophosphatemia, dental anomalies, and ectopic calcification. J Bone Miner Res 28 (6):1378–1385. https://doi.org/10.1002/jbmr.1850

Riminucci M, Collins MT, Fedarko NS, Cherman N, Corsi A, White KE, Waguespack S, Gupta A, Hannon T, Econs MJ, Bianco P, Gehron Robey P (2003) FGF-23 in fibrous dysplasia of bone and its relationship to renal phosphate wasting. J Clin Invest 112(5):683–692. https://doi.org/10.1172/jci18399

Roberts MS, Burbelo PD, Egli-Spichtig D, Perwad F, Romero CJ, Ichikawa S, Farrow E, Econs MJ, Guthrie LC, Collins MT, Gafni RI (2018) Autoimmune hyperphosphatemic tumoral calcinosis in a patient with FGF23 autoantibodies. J Clin Invest 128(12):5368–5373. https://doi.org/10.1172/jci122004

Rossaint J, Oehmichen J, Van Aken H, Reuter S, Pavenstadt HJ, Meersch M, Unruh M, Zarbock A (2016) FGF23 signaling impairs neutrophil recruitment and host defense during CKD. J Clin Invest 126(3):962–974. https://doi.org/10.1172/jci83470

Rowe PS, Kumagai Y, Gutierrez G, Garrett IR, Blacher R, Rosen D, Cundy J, Navvab S, Chen D, Drezner MK, Quarles LD, Mundy GR (2004) MEPE has the properties of an osteoblastic phosphatonin and minhibin. Bone 34(2):303–319. https://doi.org/10.1016/j.bone.2003.10.005

Ruf N, Uhlenberg B, Terkeltaub R, Nurnberg P, Rutsch F (2005) The mutational spectrum of ENPP1 as arising after the analysis of 23 unrelated patients with generalized arterial calcification of infancy (GACI). Hum Mutat 25(1):98. https://doi.org/10.1002/humu.9297

Saini RK, Kaneko I, Jurutka PW, Forster R, Hsieh A, Hsieh JC, Haussler MR, Whitfield GK (2013) 1,25-dihydroxyvitamin D(3) regulation of fibroblast growth factor-23 expression in bone cells: evidence for primary and secondary mechanisms modulated by leptin and interleukin-6. Calcif Tissue Int 92(4):339–353. https://doi.org/10.1007/s00223-012-9683-5

Schouten BJ, Doogue MP, Soule SG, Hunt PJ (2009) Iron polymaltose-induced FGF23 elevation complicated by hypophosphataemic osteomalacia. Ann Clin Biochem 46(Pt 2):167–169. https://doi.org/10.1258/acb.2008.008151

Segawa H, Kaneko I, Takahashi A, Kuwahata M, Ito M, Ohkido I, Tatsumi S, Miyamoto K (2002) Growth-related renal type II Na/Pi cotransporter. J Biol Chem 277(22):19665–19672. https://doi.org/10.1074/jbc.M200943200

Shalhoub V, Shatzen EM, Ward SC, Davis J, Stevens J, Bi V, Renshaw L, Hawkins N, Wang W, Chen C, Tsai MM, Cattley RC, Wronski TJ, Xia X, Li X, Henley C, Eschenberg M, Richards WG (2012) FGF23 neutralization improves chronic kidney disease-associated hyperparathyroidism yet increases mortality. J Clin Invest 122(7):2543–2553. https://doi.org/10.1172/jci61405

Shawar SM, Ramadan AR, Ali BR, Alghamdi MA, John A, Hudaib FM (2016) FGF23-S129F mutant bypasses ER/Golgi to the circulation of hyperphosphatemic familial tumoral calcinosis patients. Bone 93:187–195. https://doi.org/10.1016/j.bone.2015.11.015

Shimada T, Mizutani S, Muto T, Yoneya T, Hino R, Takeda S, Takeuchi Y, Fujita T, Fukumoto S, Yamashita T (2001) Cloning and characterization of FGF23 as a causative factor of tumor-induced osteomalacia. Proc Natl Acad Sci U S A 98(11):6500–6505. https://doi.org/10.1073/pnas.101545198

Shimada T, Muto T, Urakawa I, Yoneya T, Yamazaki Y, Okawa K, Takeuchi Y, Fujita T, Fukumoto S, Yamashita T (2002) Mutant FGF-23 responsible for autosomal dominant hypophosphatemic rickets is resistant to proteolytic cleavage and causes hypophosphatemia in vivo. Endocrinology 143(8):3179–3182. https://doi.org/10.1210/endo.143.8.8795

Shimada T, Hasegawa H, Yamazaki Y, Muto T, Hino R, Takeuchi Y, Fujita T, Nakahara K, Fukumoto S, Yamashita T (2004a) FGF-23 is a potent regulator of vitamin D metabolism and phosphate homeostasis. J Bone Miner Res 19(3):429–435. https://doi.org/10.1359/jbmr.0301264

Shimada T, Kakitani M, Yamazaki Y, Hasegawa H, Takeuchi Y, Fujita T, Fukumoto S, Tomizuka K, Yamashita T (2004b) Targeted ablation of Fgf23 demonstrates an essential physiological role of FGF23 in phosphate and vitamin D metabolism. J Clin Invest 113(4):561–568. https://doi.org/10.1172/jci19081

Shimada T, Yamazaki Y, Takahashi M, Hasegawa H, Urakawa I, Oshima T, Ono K, Kakitani M, Tomizuka K, Fujita T, Fukumoto S, Yamashita T (2005) Vitamin D receptor-independent FGF23 actions in regulating phosphate and vitamin D metabolism. Am J Physiol Renal Physiol 289(5):F1088–F1095. https://doi.org/10.1152/ajprenal.00474.2004

Shimizu Y, Tada Y, Yamauchi M, Okamoto T, Suzuki H, Ito N, Fukumoto S, Sugimoto T, Fujita T (2009) Hypophosphatemia induced by intravenous administration of saccharated ferric oxide: another form of FGF23-related hypophosphatemia. Bone 45(4):814–816. https://doi.org/10.1016/j.bone.2009.06.017

Simpson MA, Hsu R, Keir LS, Hao J, Sivapalan G, Ernst LM, Zackai EH, Al-Gazali LI, Hulskamp G, Kingston HM, Prescott TE, Ion A, Patton MA, Murday V, George A, Crosby AH (2007) Mutations in FAM20C are associated with lethal osteosclerotic bone dysplasia (Raine syndrome), highlighting a crucial molecule in bone development. Am J Hum Genet 81(5):906–912. https://doi.org/10.1086/522240

Singh S, Grabner A, Yanucil C, Schramm K, Czaya B, Krick S, Czaja MJ, Bartz R, Abraham R, Di Marco GS, Brand M, Wolf M, Faul C (2016) Fibroblast growth factor 23 directly targets hepatocytes to promote inflammation in chronic kidney disease. Kidney Int 90(5):985–996. https://doi.org/10.1016/j.kint.2016.05.019

Sitara D, Razzaque MS, Hesse M, Yoganathan S, Taguchi T, Erben RG, Juppner H, Lanske B (2004) Homozygous ablation of fibroblast growth factor-23 results in hyperphosphatemia and impaired skeletogenesis, and reverses hypophosphatemia in Phex-deficient mice. Matrix Biol 23(7):421–432. https://doi.org/10.1016/j.matbio.2004.09.007

Tagliabracci VS, Engel JL, Wiley SE, Xiao J, Gonzalez DJ, Nidumanda Appaiah H, Koller A, Nizet V, White KE, Dixon JE (2014) Dynamic regulation of FGF23 by Fam20C phosphorylation, GalNAc-T3 glycosylation, and furin proteolysis. Proc Natl Acad Sci U S A 111(15):5520–5525. https://doi.org/10.1073/pnas.1402218111

Takashi Y, Kosako H, Sawatsubashi S, Kinoshita Y, Ito N, Tsoumpra MK, Nangaku M, Abe M, Matsuhisa M, Kato S, Matsumoto T, Fukumoto S (2019) Activation of unliganded FGF receptor by extracellular phosphate potentiates proteolytic protection of FGF23 by its O-glycosylation. Proc Natl Acad Sci U S A 116(23):11418–11427. https://doi.org/10.1073/pnas.1815166116

Takeshita A, Kawakami K, Furushima K, Miyajima M, Sakaguchi K (2018) Central role of the proximal tubular αKlotho/FGF receptor complex in FGF23-regulated phosphate and vitamin D metabolism. Sci Rep 8(1):6917. https://doi.org/10.1038/s41598-018-25087-3

Takeuchi Y, Suzuki H, Ogura S, Imai R, Yamazaki Y, Yamashita T, Miyamoto Y, Okazaki H, Nakamura K, Nakahara K, Fukumoto S, Fujita T (2004) Venous sampling for fibroblast growth factor-23 confirms preoperative diagnosis of tumor-induced osteomalacia. J Clin Endocrinol Metab 89(8):3979–3982. https://doi.org/10.1210/jc.2004-0406

Takeyama K, Kitanaka S, Sato T, Kobori M, Yanagisawa J, Kato S (1997) 25-Hydroxyvitamin D3 1alpha-hydroxylase and vitamin D synthesis. Science 277(5333):1827–1830. https://doi.org/10.1126/science.277.5333.1827

Takeyari S, Yamamoto T, Kinoshita Y, Fukumoto S, Glorieux FH, Michigami T, Hasegawa K, Kitaoka T, Kubota T, Imanishi Y, Shimotsuji T, Ozono K (2014) Hypophosphatemic osteomalacia and bone sclerosis caused by a novel homozygous mutation of the FAM20C gene in an elderly man with a mild variant of Raine syndrome. Bone 67:56–62. https://doi.org/10.1016/j.bone.2014.06.026

Tanaka Y, Deluca HF (1973) The control of 25-hydroxyvitamin D metabolism by inorganic phosphorus. Arch Biochem Biophys 154(2):566–574. https://doi.org/10.1016/0003-9861(73)90010-6

Topaz O, Shurman DL, Bergman R, Indelman M, Ratajczak P, Mizrachi M, Khamaysi Z, Behar D, Petronius D, Friedman V, Zelikovic I, Raimer S, Metzker A, Richard G, Sprecher E (2004) Mutations in GALNT3, encoding a protein involved in O-linked glycosylation, cause familial tumoral calcinosis. Nat Genet 36(6):579–581. https://doi.org/10.1038/ng1358

Turner AJ, Tanzawa K (1997) Mammalian membrane metallopeptidases: NEP, ECE, KELL, and PEX. FASEB J 11(5):355–364. https://doi.org/10.1096/fasebj.11.5.9141502

Tutton S, Olson E, King D, Shaker JL (2012) Successful treatment of tumor-induced osteomalacia with CT-guided percutaneous ethanol and cryoablation. J Clin Endocrinol Metab 97(10):3421–3425. https://doi.org/10.1210/jc.2012-1719

Urakawa I, Yamazaki Y, Shimada T, Iijima K, Hasegawa H, Okawa K, Fujita T, Fukumoto S, Yamashita T (2006) Klotho converts canonical FGF receptor into a specific receptor for FGF23. Nature 444(7120):770–774. https://doi.org/10.1038/nature05315

Wang X, Wang S, Li C, Gao T, Liu Y, Rangiani A, Sun Y, Hao J, George A, Lu Y, Groppe J, Yuan B, Feng JQ, Qin C (2012) Inactivation of a novel FGF23 regulator, FAM20C, leads to hypophosphatemic rickets in mice. PLoS Genet 8(5):e1002708. https://doi.org/10.1371/journal.pgen.1002708

Wasserman H, Ikomi C, Hafberg ET, Miethke AG, Bove KE, Backeljauw PF (2016) Two case reports of FGF23-induced hypophosphatemia in childhood biliary atresia. Pediatrics 138(2):e20154453. https://doi.org/10.1542/peds.2015-4453

White KE, Carn G, Lorenz-Depiereux B, Benet-Pages A, Strom TM, Econs MJ (2001) Autosomal-dominant hypophosphatemic rickets (ADHR) mutations stabilize FGF-23. Kidney Int 60(6):2079–2086. https://doi.org/10.1046/j.1523-1755.2001.00064.x

White KE, Cabral JM, Davis SI, Fishburn T, Evans WE, Ichikawa S, Fields J, Yu X, Shaw NJ, McLellan NJ, McKeown C, Fitzpatrick D, Yu K, Ornitz DM, Econs MJ (2005) Mutations that cause osteoglophonic dysplasia define novel roles for FGFR1 in bone elongation. Am J Hum Genet 76(2):361–367. https://doi.org/10.1086/427956

Winzenberg T, Jones G (2013) Vitamin D and bone health in childhood and adolescence. Calcif Tissue Int 92(2):140–150. https://doi.org/10.1007/s00223-012-9615-4

Wohrle S, Henninger C, Bonny O, Thuery A, Beluch N, Hynes NE, Guagnano V, Sellers WR, Hofmann F, Kneissel M, Graus Porta D (2013) Pharmacological inhibition of fibroblast growth factor (FGF) receptor signaling ameliorates FGF23-mediated hypophosphatemic rickets. J Bone Miner Res 28(4):899–911. https://doi.org/10.1002/jbmr.1810

Wolf M (2010) Forging forward with 10 burning questions on FGF23 in kidney disease. J Am Soc Nephrol 21(9):1427–1435. https://doi.org/10.1681/asn.2009121293

Wolf M, Koch TA, Bregman DB (2013) Effects of iron deficiency anemia and its treatment on fibroblast growth factor 23 and phosphate homeostasis in women. J Bone Miner Res 28 (8):1793–1803. https://doi.org/10.1002/jbmr.1923

Xiao Z, Riccardi D, Velazquez HA, Chin AL, Yates CR, Carrick JD, Smith JC, Baudry J, Quarles LD (2016) A computationally identified compound antagonizes excess FGF-23 signaling in renal tubules and a mouse model of hypophosphatemia. Sci Signal 9(455):ra113. https://doi.org/10.1126/scisignal.aaf5034

Yamashita T, Yoshioka M, Itoh N (2000) Identification of a novel fibroblast growth factor, FGF-23, preferentially expressed in the ventrolateral thalamic nucleus of the brain. Biochem Biophys Res Commun 277(2):494–498. https://doi.org/10.1006/bbrc.2000.3696

Yamazaki Y, Okazaki R, Shibata M, Hasegawa Y, Satoh K, Tajima T, Takeuchi Y, Fujita T, Nakahara K, Yamashita T, Fukumoto S (2002) Increased circulatory level of biologically active full-length FGF-23 in patients with hypophosphatemic rickets/osteomalacia. J Clin Endocrinol Metab 87(11):4957–4960. https://doi.org/10.1210/jc.2002-021105

Yamazaki Y, Tamada T, Kasai N, Urakawa I, Aono Y, Hasegawa H, Fujita T, Kuroki R, Yamashita T, Fukumoto S, Shimada T (2008) Anti-FGF23 neutralizing antibodies show the physiological role and structural features of FGF23. J Bone Miner Res 23(9):1509–1518. https://doi.org/10.1359/jbmr.080417

Yoshida T, Fujimori T, Nabeshima Y (2002) Mediation of unusually high concentrations of 1,25-dihydroxyvitamin D in homozygous klotho mutant mice by increased expression of renal 1alpha-hydroxylase gene. Endocrinology 143(2):683–689. https://doi.org/10.1210/endo.143.2.8657

Yu X, Ibrahimi OA, Goetz R, Zhang F, Davis SI, Garringer HJ, Linhardt RJ, Ornitz DM, Mohammadi M, White KE (2005) Analysis of the biochemical mechanisms for the endocrine actions of fibroblast growth factor-23. Endocrinology 146(11):4647–4656. https://doi.org/10.1210/en.2005-0670

Yuan B, Takaiwa M, Clemens TL, Feng JQ, Kumar R, Rowe PS, Xie Y, Drezner MK (2008) Aberrant Phex function in osteoblasts and osteocytes alone underlies murine X-linked hypophosphatemia. J Clin Invest 118(2):722–734. https://doi.org/10.1172/jci32702

Yuan B, Feng JQ, Bowman S, Liu Y, Blank RD, Lindberg I, Drezner MK (2013) Hexa-D-arginine treatment increases 7B2·PC2 activity in hyp-mouse osteoblasts and rescues the HYP phenotype. J Bone Miner Res 28(1):56–72. https://doi.org/10.1002/jbmr.1738

Zhang MY, Ranch D, Pereira RC, Armbrecht HJ, Portale AA, Perwad F (2012) Chronic inhibition of ERK1/2 signaling improves disordered bone and mineral metabolism in hypophosphatemic (Hyp) mice. Endocrinology 153(4):1806–1816. https://doi.org/10.1210/en.2011-1831

The Central Regulation of Bone Mass: Genetic Evidence and Molecular Bases

Gerard Karsenty

Contents

Abstract

The alternation of resorption of preexisting bone by the osteoclasts followed by de novo bone formation by osteoblasts is called bone modeling during childhood and bone remodeling during adulthood. A central question raised by this physiological process that is fundamental to longitudinal growth during childhood and adolescence and that is attacked at the other end of life in the context of osteoporosis is to know how it is regulated. This question was rejuvenated in the late

G. Karsenty (✉)
Departments of Genetics and Development, Vagelos College of Physicians and Surgeons, Columbia University, New York, NY, USA
e-mail: gk2172@columbia.edu

P. H. Stern (ed.), *Bone Regulators and Osteoporosis Therapy*,
Handbook of Experimental Pharmacology 262, https://doi.org/10.1007/164_2020_378

1990s and early 2000s years when the application of mouse genetics made it feasible to test whether there were new endocrine determinants of bone (re)-modeling. Addressing this question, taking into account fundamental cell biology features of bone led to the hypothesis that there should be a coordinated control of bone growth/mass, energy metabolism, and reproduction. Testing genetically and molecularly, this hypothesis revealed that, in vivo, the adipocyte-derived hormone leptin is a powerful inhibitor of bone mass accrual following its signaling in the brain. This chapter details the molecular bases and biological relevance of this regulation of bone mass accrual by leptin.

Keywords

Bone formation · Bone resorption · Brain serotonin · CART · CREB · RANKL · Sympathetic tone

1 Introduction: Why Should Bone Have a Coordinating Role?

A striking cell biological feature of the bone is that it is the only tissue in our body that houses a cell type, the osteoclast, whose only function is to actively destroy the host tissue. This destruction that is energetically demanding occurs daily, from birth to death in hundreds of locations in the skeleton of all bony vertebrates including of course humans. Remarkably, from an energetic vantage point, this destruction is followed, also daily, by de novo bone formation, which is also an energetic taxing process. The alternation of bone resorption followed by de novo bone formation fulfills some critical physiological purposes. Indeed, during childhood, this alternation called then bone modeling is necessary for longitudinal growth (Karsenty and Wagner 2002; Kronenberg 2003). During adulthood, this alternation now called bone remodeling is necessary to repair micro and macro damage, i.e., fractures (Ducy et al. 2000; Teitelbaum 2000; Karsenty and Ferron 2012). It is not an exaggeration to say that bone remodeling has been for mankind the only orthopedic surgeon available for most of the time we lived on this planet.

Taken at face value, each arm of the bone (re)modeling) is energetic costly. It takes energy to destroy anything, especially if one does it in hundreds of locations at a time and everyday. It also takes a lot of energy to rebuild what has just been destroyed. This amount of energy needed for bone (remodeling) is necessarily proportional to the surface covered by the organ involved. Clearly bones cover more surface than most organs in our body. This simple description of a cell biological event intuits to the hypothesis that, maybe, just maybe, energy metabolism and bone (re)modeling could be co-regulated (Holbrook and Barrett-Connor 1993). It is when this hypothetical statement is subjected to the test of clinical medicine that it gained most traction and its real credibility. Indeed, it has been known for centuries that any situation that limits access to food in children, whether it is famine or a disease like anorexia nervosa, simply halts bone modeling and longitudinal growth. By the same token, starvation voluntary or not invariably results in an osteoporosis in adults. At the other end of the spectrum, we also know through modern epidemiological studies that the bone mineral density of individuals overfeeding and overweight is high at virtually all sites tested (Holbrook and Barrett-Connor 1993; Legroux-Gerot et al. 2005; Legroux-Gerot 2007; Misra and Klibanski 2011). No matter what it could be, these two clinical observations that can be viewed as simply the mirror image of each other need a molecular explanation. Certainly, the hypothesis mentioned above would offer, if verified experimentally, a single explanation for these two clinical observations.

If we continue to use clinical medicine for what it is, the best guide to important biological problems and a way to forge hypotheses to study biological problems, we note that gonadal failure invariably causes bone loss and osteoporosis (Riggs et al. 1998). What we did in 2000 was to combine these different and universally accepted clinical observations to come up with a unifying, testable, and potentially medically relevant hypothesis. This hypothesis states that *there may be a coordinated regulation, endocrine in nature since the organs implicated are next to each other of bone mass accrual, energy metabolism, and reproduction* (Ducy 2000). This hypothesis had many inferences; one of them being more actively pursued now, but the most immediate one was to link bone physiology to two other physiologies in the body, something that turned to be very fertile for the field. Another reason that made this hypothesis so attractive and testable at the time we formulated it was that there was then only one regulator that significantly impacts both energy metabolism/appetite and reproduction: leptin (Friedman 2000; Friedman and Halaas 1998; Spiegelman and Flier 1996). This led to a thorough analysis of the bone phenotype leptin-deficient (*ob/ob*) and leptin receptor-deficient (*db/db*) mouse strains (Tartaglia et al. 1995; Zhang et al. 1994). This analysis was conducted without preconceived ideas but taking into account the wealth of knowledge indicating leptin signals in the brain to fulfill its functions by signaling in the brain (Ducy P., Kousteni S. (2015) Leptin and Bone. In: Dagogo-Jack, MD S. (eds) Leptin. Springer, Cham).

2 Leptin Is a Negative Regulator of Bone Mass Accrual

2.1 Bone Mass Increases in the Absence of Leptin Signaling

To state what had been known since the mid-1960s, both the *ob/ob* and *db/db* mice are obese and hypogonadal (Tartaglia et al. 1995; Zhang et al. 1994; Ahima 2004). This obesity has two concurring causes. One is a decrease in energy expenditure, and the other one is a prodigious increase in appetite that can also be seen in patients lacking leptin or its receptor. From a bone biology point of view, the reasonable expectation was that the hypogonadism of the *ob/ob* and mice should result in an osteoporosis. There was another reason to make this prediction, and it is that in both models, circulating glucocorticoid hormones (corticosterone in mice) were elevated, and corticosterone favors bone loss. Nothing of that was observed in mice, or rats or as we will present later in this review, humans lacking leptin signaling (Ducy 2000; Baldock et al. 2005). Instead, X-rays, regular bone histology, histomorphometry, and even microcomputer tomography showed that in the absence of leptin or its receptor, mice demonstrated a marked increase in bone mass, and their long bones were more resistant to fracture upon mechanical loading. This is not an observation limited to the *ob/ob* and *db/db* mice since this high bone mass phenotype is also observed in mice treated with a leptin antagonist, in genetically obese *fa/fa* rats, and in a newly developed model of leptin knockout rats (Vaira et al. 2012; Solomon et al. 2014). Moreover, all lipodystrophic children who have very low circulating leptin levels that were analyzed showed an advanced bone age (see below, Sect. 5).

As one would anticipate given their extremely severe hypogonadism, *ob/ob* and *db/db* mice have an increase in bone resorption (Ducy 2000). It is, however, milder than one would have predicted given their total absence of sex steroids. The molecular bases for this mild increase in bone resorption will be presented later (see below, Sect. 4). The most striking results of the analysis of mice or of other animal models lacking leptin signaling are that despite their high circulating corticosterone levels, these animals had a vast increase in parameters of bone formation (Ducy 2000; Ahima et al. 1999; Reid 1997a, b). Taken at face value and regardless of their molecular bases, those are extremely important observations for our understanding of the control of bone (re)modeling for several reasons. First of all, it shows that bone formation and bone resorption can be decoupled. This increase in bone formation and decrease in bone resorption are a hallmark of the disruption of leptin signaling. Second and simply put, these results indicated that leptin signaling is such a powerful inhibitor of bone formation and bone mass accrual that removing it results in a high bone mass despite gonadal failure and hypercorticism. Until now, no other hormone has been shown to affect bone (re)modeling so dramatically.

2.2 Leptin Action on Bone Mass Accrual Is Independent of its Effect on Body Weight

What could be the cause, the mechanism leading to such an increase in bone formation in the absence of leptin signaling? After ruling out, experimentally, this increase in bone mass was a consequence of an increase in loading on the bones (Suva et al. 2005) (Table 1). Instead many lines of evidence acquired through the

Table 1 Changes in bone mass and remodeling parameters correlate with levels of leptin signaling but not with body weight

	Body weight	Serum leptin	Bone mass	Formation	Resorption
ob/ob	+++	0	+	+	–
db/db	+++	±	+	+	–
Mc4r +/– (–/–)	++	–	–	=	+
Ay/+	++	±	=	ND	ND
Young *ob/ob* on LFD	=	0	+	+	–
ob/+	=	–	+	+	–
A-ZIP transgenic	–	- - -	+	+	–

+ increase, + normal level, ± resistance, – decrease, 0 absence, ND not done histomorphometric studies of the bones of various mouse models indicated that leptin exerts a specific function on bone(re)modeling and that this function is independent of the weight of the animals. Here are some of the many observations that support this contention, mice with obesity not by a lack of leptin signaling either do not display a bone phenotype (Agouti Yellow (Ay) mice) or show, in the case of *Mc4R*-deficient mice a high bone mass caused by a different mechanism (Ducy 2000; Ahn et al. 2006). At the other end of the spectrum, mice with a deficiency in leptin signaling but that are not obese display the same increase in bone formation as first noted in the *ob/ob* mice. For instance, if one looks at very young *ob/ob* mice fed a low fat diet or at *ob/+* mice that have a normal body weight, they both display an increase in bone mass accrual (Ducy 2000). Transgenic mice rendered lipodystrophic through expression of a dominant-negative protein blocking the activity of adipocyte differentiation factors that have a near complete absence of leptin and are lean also exhibit a high bone mass phenotype (Ducy 2000; Moitra et al. 1998). That this phenotype can be fully corrected by increasing the levels of circulating leptin indicating that it is caused by the absence of this hormone but not by the lack of other adipokine (Ducy 2000; Elefteriou et al. 2004).

3 Mechanism of the Regulation of Bone Remodeling by Leptin

3.1 Leptin Regulates Bone Remodeling by a Central Mechanism

A very large body of work both of molecular and genetic nature had shown before we became interested in this hormone that leptin regulates appetite, energy expenditure, and fertility through a central relay (Ahima 2004; Chehab et al. 1996). This raised the then novel hypothesis as to whether there is a central control of bone mass and that leptin orchestrates it. Since then many labs in the world have shown that it was indeed the case, but at the time we performed this experiment, this was viewed, in all fairness, as heretical.

And yet leptin's effect on bone mass accrual relies on a central action. The experimental evidence that supports this claim is too large to be cited. We will cite the most salient ones. First and foremost, delivery of leptin through intracerebroventricular infusion (icv) in *ob/ob* mice, even at a concentration low enough to avoid leakage to the blood stream, results in a decrease in bone mass without any significant loss of weight (Shi et al. 2008). This principal observation was of fundamental importance as it was the first indication ever that there was a central control of bone mass. It also inferred that the threshold of leptin sensitivity was lower for bone mass than for appetite. In support of the first contention, subsequent experiments relying on chemical lesioning showed that the destruction of monosodium glutamate-sensitive neurons in the hypothalamic arcuate nuclei (ARC) affects body weight. In contrast, bone mass accrual is predominantly affected when neurons of the ventromedial hypothalamus nuclei (VMH) are destroyed using gold thioglucose (Takeda et al. 2002). Furthermore, the fact that Ay mice lose bone

Table 2 Dissociation between the central effects of leptin signaling on body weight and bone mass accrual

	Central leptin signaling	Body weight	Bone mass	Formation	Resorption
ob/ob	0	+++	+	+	–
l/l	+	=	–	–	+
ob/ob + leptin icv	+	–	–	–	+
wt + MSG	±	=	=	=	=
wt + GTG	±	–	–	–	+
ob/ob + MSG + leptin icv	±	=	–	–	+
ob/ob + GTG + leptin icv	±	–	+	+	–

+ increase, + normal level, ± partial, – decrease, 0 absence

but not weight when infused icv with leptin, while *ob/ob* mice infused icv with a MC3R/MC4R agonist lose weight but not bone further established that leptin's anorexigenic and anti-osteogenic effects rely on distinct signaling events (Takeda et al. 2002) (Table 2). The notion that the leptin regulation of bone mass is achieved with lower amount of the hormone than the regulation of appetite of energy expenditure was verified with the analysis of the *l/l* mice, a gain-of-function model of the leptin receptor (Bjorbak et al. 2000; Bjornholm et al. 2007). While the *l/l* mice fed a regular chow have a normal appetite, normal energy expenditure, and normal fertility, when it comes to bone mass, they show a mirror image phenotype of the *db/db* mice, i.e., low bone mass, decreased bone formation, and increased bone resorption (Shi et al. 2008; Bjornholm et al. 2007).

The notion that leptin inhibits bone mass accrual is not only an observation made in some rare loss of functions animal models; it is also a pharmacological fact. Specifically, injections or icv infusions leptin also influence bone remodeling. For instance, transgenic mice with a modest, twofold to fourfold, increase in serum concentration of leptin show a lower bone mass than non-transgenic controls (Elefteriou et al. 2004; Takeda et al. 2002). Overexpression of *leptin* in the liver to raise its serum concentration to approximately 1.2 μg/mL (i.e., 250-fold) also induced bone loss, albeit this was of smaller amplitude and at a slower pace than was elicited by icv infusion. This observation suggested that, as it is the case for its control of body weight, leptin's regulation of bone remodeling was inhibited when its serum concentration becomes too high (Könner and Brüning Jens 2012; Mark 2013; Motyl and Rosen 2012). Such a "leptin resistance" mechanism certainly explains the paradoxical beneficial effect on bone mass accrual seen in the case of peripheral injections of massive amounts of leptin (50 μg/day, i.e., 700-fold the amount used in icv studies) (Steppan et al. 2000; Cornish et al. 2002). Indeed, since leptin centrally affects bone mass accrual at a lower threshold than it affects body weight, one would expect that the effect of leptin on bone remodeling would also be more quickly impacted by a leptin resistance mechanism in the case one would "force" the system and deliver massive amount of the hormone.

At the same time the notion that a central control of bone mass existed and was in fact of great importance emerged, rigorous cell biological evidence ruled out that leptin acts locally on bone cells. We will mention here the most important evidence along these lines. First, no technique to assay gene expression has been able yet to detect the expression of the leptin receptor in differentiated, bone-making osteoblasts (Ducy 2000; Ding et al. 2012). One should not rely on negative data. This is why, in a purely functional assay, when cultured, ex vivo osteoblasts derived from wild-type or *db/db* mice proliferate and differentiate similarly. More definitely, neither expression of leptin in osteoblasts or the conditional inactivation of *Lepr* specifically in osteoblasts causes a bone phenotype in mice (Shi et al. 2008; Takeda et al. 2002), whereas this receptor in neurons induces a high bone mass phenotype similar to the one observed in *db/db* mice (Shi et al. 2008; Yadav et al. 2009).

3.2 Leptin Regulates the Bone Through a Serotonin Relay

The hypothalamus has been from the onset of this research, a focal point of leptin biology and yet leptin does not affect neuronal activity in hypothalamic neurons (T. Horvath personal communication). Moreover, deletion of the leptin receptor in neurons of the VMH nucleus in the hypothalamus does not affect bone mass in any measurable way (Yadav et al. 2009; Balthasar et al. 2004). These discrepant observations were explained when it was shown that leptin needs to signal in serotonergic neurons of the raphe to eventually influence, via serotonin the activity of VMH neurons of the hypothalamus (Yadav et al. 2009; Oury and Karsenty 2011; Ducy and Karsenty 2010). Others and we showed that the signaling form of the leptin receptor is expressed in the neurons of the dorsal and median raphe nuclei that synthesize and secrete serotonin. Leptin is an inhibitor of serotonin synthesis and secretion by these neurons, and as a result, the brain content of serotonin is high in *ob/ob* and *db/db* mice and is decreased upon leptin icv infusion in wild-type mice (Yadav et al. 2009). Accordingly, mice lacking Toh2 the gene encoding for the rate-limiting enzyme necessary for serotonin synthesis in the raphe show a low bone mass phenotype mirroring the one observed in *ob/ob* mice, and inactivating one allele of *Tph2* in *ob/ob* mice corrected their high bone mass (Yadav et al. 2009). Conversely, when *LepR* is specifically inactivated in the *Tph2*-expressing neurons of the brainstem, these mice develop a high bone mass phenotype similar to that of *db/db* mice (Yadav et al. 2009). Hence, brain serotonin is a positive regulator of bone mass accrual, and brain serotonin synthesis is inhibited by leptin signaling to the brainstem neurons.

3.3 Leptin Regulation of Bone Mass Accrual Is Mediated by Serotonin Signaling to VMH Neurons Via the Htr2c Receptor

Classical neuron tracing studies showed that serotonergic neurons do project from the brainstem to the VMH nuclei (Yadav et al. 2009) and that serotonin signals to VMH neurons via the Htr2c receptor to favor bone mass accrual. Hence, deleting this receptor in the mouse results in a low bone mass phenotype secondary to a decreased bone formation and an increased in bone resorption, and reactivating the expression of Htr2c specifically in VMH neurons corrects this phenotype (Yadav et al. 2009). Furthermore, the fact that *Tph2+/−; Htr2c+/−* compound heterozygous mice have a similar bone phenotype as *Tph2−/−* mice supports the notion that this receptor is the main serotonin receptors mediating brain serotonin regulation of bone mass accrual (Yadav et al. 2009).

Multiple experimental evidence has shown that Htr2c uses Ca^{2+} as a second messenger and a cascade of CaM kinases (Oury et al. 2010). In situ hybridization revealed that the genes encoding CaMKKβ, CaMKIIa, and CaMKIV are most highly expressed in the SF1-positive neurons of the VMH that mediate leptin-dependent regulation of bone mass accrual (Takeda et al. 2002; Oury et al. 2010). The functional connection between serotonin, CaMKKβ/CaMKIV expression in the VMH, and bone remodeling is demonstrated by the demonstration that mice deficient in *CaMKKβ* (or *CaMKIV*) heterozygous for *Tph2* and *CaMKKβ* (or *CaMKIV*) deficiency in VMH neurons or double develop a similar bone phenotype as the one observed *Tph2−/−* mice. Both in vitro and in vivo evidence showed that the transcription factor CREB mediates the serotonin regulation of bone mass accrual (Oury et al. 2010). Further, the addition of serotonin to wild-type hypothalamic explants, but not to *Htr2c−/−* explants, induces CREB phosphorylation on Ser133, i.e., its activation. In the explant assay, phosphorylation of CREB and CaMKIV co-localizes in the same neurons upon serotonin treatment, and mice lacking *Creb* in VMH neurons only ($Creb_{VHM}-/-$) or compound heterozygous mice for *Creb* and *Htr2c* ($Creb_{VHM}-/-$; *Htr2c+/−*) or for *Creb* and *CaMKIV* ($Creb_{VHM}+/-$; $CaMKIV_{VHM}+/-$) all show a low bone mass phenotype caused by a decrease in bone formation and an increase in bone resorption. Thus, the negative regulation of bone remodeling by leptin is dependent upon inhibition of brainstem serotonergic neurons, which normally activate VMH neurons via an Htr2c→CaMKKβ→CaMKIV→CREB cascade and increase bone formation and decrease the resorption of bone.

4 Establishing the Efferent Mediators of the Central Regulation of Bone Remodeling by Leptin

Once the afferent road map pathway whereby leptin regulates bone mass accrual had been, at least in part, deciphered, it became more important to unravel the molecular mechanisms whereby the VMH regulates the function of bone cells. This could be achieved by inducing the release of a blood-borne factor, relaying the information via the peripheral nervous system, or both. As it turns out leptin uses both mechanisms to regulate bone mass accrual.

4.1 Sympathetic Tone Mediates Regulation of Bone Formation and Resorption by Leptin

The notion that leptin regulation of bone mass accrual is mediated in part by a nervous relay came from a variant of the classical parabiosis experiment revealed that the *ob* gene encoded a circulating factor (Coleman 1973). In our study, rather than connecting an *ob/ob* with a wild-type mouse, we connected it to a second *ob/ob* mouse. In that setting, only one of the two mice received leptin icv at a dose that does not leak into the general circulation (Takeda et al. 2002). As expected the mouse receiving leptin lost bone within a month. This mouse served as a positive control in this experiment. The contralateral and experimental mouse in this system sharing its blood content with the infused mouse did not lose bone. Again, taken at face value, this observation indicated quite clearly that leptin uses a nervous relay as its main effector on bone cells. This was particularly important since the autonomic nervous system is perturbed by the absence of leptin signaling (Bray and York 1998).

Indeed, a major feature of the *ob/ob* and *db/db* mice is a massive decrease of the sympathetic tone (Bray and York 1998). As it was shown in all models engineered to define leptin's central regulation of bone remodeling, changes in the sympathetic tone correlate with the bone phenotype observed (Table 3). A decrease in sympathetic tone was found both in mice without leptin signaling in all neurons or only in brainstem neurons (Shi et al. 2008; Yadav et al. 2009). Conversely, there is increased sympathetic output in mice in which leptin activity is increased directly (*l/l* mice) or indirectly (*Tph2−/−*, *Htr2b−/−* or *Creb* $_{VMH}$*−/−* mice) (Shi et al. 2008; Yadav et al. 2009; Oury et al. 2010). One of the main genes downregulated in the hypothalamus of each of these three mouse strains encodes tyrosine hydroxylase, the enzyme catalyzing the initial step in the synthesis of catecholamines (Oury et al. 2010). Genetic experiments provided direct confirmation that leptin acts on bone remodeling via the sympathetic nervous system.

The enzymatic conversion of dopamine by the enzyme dopamine β-hydroxylase (Dbh) is required for the synthesis of a norepinephrine and epinephrine, the two mediators of sympathetic tone. Remarkably, *Dbh*-deficient mice have increased bone mass is increased and icv leptin is unable to correct this phenotype such as it does in *ob/ob* mice (Takeda et al. 2002). Osteoblasts are the cells in the bone that mediate the sympathetic regulation of bone formation and bone resorption (see below, Part 4). They express the β2 adrenergic receptor (Adrβ2), whereas osteoclasts do not. Inactivation of *Adrβ2* in osteoblasts only or in all cells results in a high bone mass phenotype caused by an increase in bone formation and a decrease in bone resorption that cannot be corrected by icv infusion of leptin (Takeda et al. 2002; Kajimura et al. 2011). Conversely, *ob/ob* mice treated with the β-adrenergic agonist isoproterenol lose bone (Takeda et al. 2002). Similar observations were made in wild-type mice or rats (Takeda et al. 2002; Bonnet et al. 2005, 2007a, b). More

Table 3 Correlation between leptin signaling, brain serotonin, sympathetic output, and bone remodeling

	Central leptin signaling	Brain serotonin	Sympathetic output	Bone mass
ob/ob	0	+	−	+
db/db	0	+	−	+
LepR $_{BS}$*−/−*	±	+	−	+
LepR $_{ARC}$*−/−*	±	−	+	−
LepR $_{VMH}$*−/−*	±	−	+	−
wt + leptin icv	+	−	+	−
LepR $_{BS}$*−/−* + leptin icv	±	+	−	+
Tph2−/−	=	0	+	−
ob/ob; Tph2+/−	0	=	=	=
Htr2c−/−	=	=	+	−
Tph2+/−; Htr2c+/−	=	−	+	−
CaMKKβ $_{VMH}$*−/−*	=	=	+	−
Htr2c+/−; CaMKKβ $_{VMH}$*+/−*	=	=	+	−
CaMKIV $_{VMH}$*−/−*	=	=	+	−
Htr2c+/−; CaMKIV $_{VMH}$*+/−*	=	=	+	−
Creb $_{VMH}$*−/−*	=	=	+	−
Htr2c+/−; Creb $_{VMH}$*+/−*	=	=	+	−
Dbh−/−	=	=	−	+

+ increase, + normal level, ± partial, − decrease, 0 absence
BS, inactivation in brainstem SERT expressing neurons, ARC, inactivation in arcuate POMC expressing neurons, VMH, inactivation in VMH Sf1 expressing neurons

importantly, from a biomedical point of view, wild-type mice or rats treated with the β-blocker propranolol gain bone to such an extent that it can prevent the deleterious consequence of ovariectomy, the classical model of postmenopausal bone loss (Takeda et al. 2002; Bonnet et al. 2006, 2008). Such results were obtained recently in women given a β-blocker (Khosla et al. 2018). Further, several epidemiological studies have reported beneficial effects of β-blockers on BMD and/or fracture risk (Schlienger et al. 2004; Bonnet et al. 2007a, b; Pasco et al. 2005; Turker et al. 2006). More generally, a meta-analysis of eight studies demonstrated that β-blockers reduce hip fracture risk (pooled relative risk 0.72) and any fracture risk (pooled relative risk 0.86) (Wiens et al. 2006), and reflex sympathetic dystrophy, a disease characterized by local sympathetic activation and bone loss, is most often treated with β-blockers (Kurvers 1998; Patel and Elefteriou 2007). The finding of the role of leptin in regulating sympathetic tone through effects on Dbh could provide a molecular explanation for the decreased risk of fracture associated with the treatment of patients with some β-blockers. However, it should be noted that most studies are observational, and prospective randomized clinical trials that take the specificity and dosage of β-blockers into consideration are needed.

4.2 A Second Potential Leptin Mediator: Leptin Regulation of Cart Expression in the Hypothalamus Affects Bone Resorption

A second, less potent and for now far less understood, mode of action of leptin on bone mass accrual involves a neurotransmitter that garnered a lot of interest in the early 2000s cocaine- and amphetamine-regulated transcript (CART). Of note, CART is also present in the general circulation (Thim et al. 1998, 1999; Wierup et al. 2004). *CART* expression in the brain is increased by leptin and is low in the brain of *ob/ob* mice (Vrang et al. 2002; Kristensen et al. 1998; Elias et al. 1998).

We showed that mice lacking *Cart* develop a late-onset osteoporosis characterized by a sole increase in bone resorption thus indicating that CART acts as an inhibitor of bone resorption (Elefteriou et al. 2005). When *Cart−/−* mice are infused icv with leptin, they lose bone faster than wild-type mice do because of a more severe increase in bone resorption (Elefteriou et al. 2005). This is explained by the fact that leptin's positive regulation of bone resorption by the sympathetic nervous system occurs normally but is not mitigated by an increase in CART levels. An increase in *CART* expression and/or circulating levels of CART is the likely cause of the high bone mass bone/low bone resorption phenotype observed in *Mc4R*-deficient mice and that observed in patients heterozygous for inactivating mutations in this gene (Ahn et al. 2006; Elefteriou et al. 2005). There is also evidence of conservation between species of this downstream effector of leptin regulation of bone remodeling. Patients heterozygous for inactivating mutations in *MC4R* develop a late-onset increase in bone mass associated with a decrease in bone resorption (Ahn et al. 2006; Yeo et al. 1998). As it is the case in *Mc4r*-deficient mice, these patients have significantly increased serum levels or CART (Ahn et al. 2006).

5 Leptin Actions on Bone Remodeling as Related to Human Bone Biology

The most compelling indications that leptin's effect to negatively regulate bone mass accrual is conserved from rodents to humans were the findings that a patient with a mutation in the *ob* gene displayed a high bone mass phenotype and the observation that human patients with lipodystrophy display advanced the bone age, an indirect indicator of bone formation (Elefteriou et al. 2005). Beyond these observations, however, it is clear that the situation is undoubtedly more complex. Indeed, positive, negative, or no associations between serum leptin levels and bone mineral density have been reported, confounding the interpretation of leptin's effect on bone mass and the correlation with rodent studies (Iwamoto et al. 2000; Sato et al. 2001; Blum et al. 2003; Rauch et al. 1998; Martini et al. 2001). For example, in obese individuals, body weight can be positively correlated with increased bone mass (Holbrook and Barrett-Connor 1993). This observation can be viewed as a direct contradiction to the negative role of leptin derived from animal studies, since leptin levels are proportional to fat mass. However, obesity is also associated with the development of leptin resistance, which could resolve this discrepancy as it explains the persistence of appetite despite high levels of leptin (Könner and Brüning Jens 2012; Mark 2013; Caro et al. 1996; Schwartz et al. 1996).

Only a limited number of human studies have examined the direct effect of leptin administration on bone mineral density. In one study, there was an increase in bone mass and decrease in body weight after long-term leptin therapy in an obese 9-year-old girl with congenital leptin deficiency (Farooqi et al. 1999). However, it is difficult to make any firm conclusions from these findings because of the patient's age and the effects of ongoing skeletal growth on bone mass. More recently, subcutaneous administration of leptin in two women with lipodystrophy was found to have no significant effect on bone mineral density (Simha et al. 2002). Larger clinical studies are therefore needed in order to clarify the role of leptin in vivo in humans and to assess the contribution of the central and peripheral roles of leptin to the overall maintenance of bone turnover in human beings.

The downstream effectors of leptin regulation of bone remodeling are also conserved between species. Patients heterozygous for inactivating mutations in MC4R develop a late-onset increase in bone mass, associated with decreased bone resorption. These patients have significantly increased levels of CART, similar to Mc4r-deficient mice. Previously discussed studies (see above, Sect. 4) showed corresponding effects of the sympathetic nervous system in mice and humans, suggesting correspondence at the level of the signaling pathway. However, as noted, the human studies on b blockers are largely observational, and more extensive clinical trials are needed.

References

Ahima RS (2004) Body fat, leptin, and hypothalamic amenorrhea. N Engl J Med 351:959–962

Ahima RS, Kelly J, Elmquist JK, Flier JS (1999) Distinct physiologic and neuronal responses to decreased leptin and mild hyperleptinemia. Endocrinology 140:4923–4931

Ahn JD, Dubern B, Lubrano-Berthelier C, Clement K, Karsenty G (2006) Cart overexpression is the only identifiable cause of high bone mass in melanocortin 4 receptor deficiency. Endocrinology 147:3196–3202

Baldock PA et al (2005) Hypothalamic control of bone formation: distinct actions of leptin and y2 receptor pathways. J Bone Miner Res 20:1851–1857

Balthasar N et al (2004) Leptin receptor signaling in POMC neurons is required for normal body weight homeostasis. Neuron 42:983–991

Bjorbak C et al (2000) SOCS3 mediates feedback inhibition of the leptin receptor via Tyr985. J Biol Chem 275:40649–40657

Bjornholm M et al (2007) Mice lacking inhibitory leptin receptor signals are lean with normal endocrine function. J Clin Invest 117:1354–1360

Blum M et al (2003) Leptin, body composition and bone mineral density in premenopausal women. Calcif Tissue Int 73:27–32

Bonnet N et al (2005) Severe bone alterations under β2 agonist treatments: bone mass, microarchitecture and strength analyses in female rats. Bone 37:622–633

Bonnet N et al (2006) Dose effects of propranolol on cancellous and cortical bone in ovariectomized adult rats. J Pharmacol Exp Ther 318:1118–1127

Bonnet N et al (2007a) Doping dose of salbutamol and exercise: deleterious effect on cancellous and cortical bones in adult rats. J Appl Physiol 102:1502–1509

Bonnet N et al (2007b) Protective effect of beta blockers in postmenopausal women: influence on fractures, bone density, micro and macroarchitecture. Bone 40:1209–1216

Bonnet N et al (2008) Low dose beta-blocker prevents ovariectomy-induced bone loss in rats without affecting heart functions. J Cell Physiol 217:819–827

Bray GA, York DA (1998) The MONA LISA hypothesis in the time of leptin. Recent Prog Horm Res 53:95–117

Caro JF et al (1996) Decreased cerebrospinal-fluid/serum leptin ratio in obesity: a possible mechanism for leptin resistance. Lancet 348:159–161

Chehab FF, Lim ME, Lu R (1996) Correction of the sterility defect in homozygous obese female mice by treatment with the human recombinant leptin. Nat Genet 12:318–320

Coleman DL (1973) Effects of parabiosis of obese with diabetes and normal mice. Diabetologia 9:294–298

Cornish J et al (2002) Leptin directly regulates bone cell function in vitro and reduces bone fragility in vivo. J Endocrinol 175:405–415

Ding L, Saunders TL, Enikolopov G, Morrison SJ (2012) Endothelial and perivascular cells maintain haematopoietic stem cells. Nature 481:457–462

Ducy P (2000) Leptin inhibits bone formation through a hypothalamic relay: a central control of bone mass. Cell 100:197–207

Ducy P, Karsenty G (2010) The two faces of serotonin in bone biology. J Cell Biol 191:7–13

Ducy P, Schinke T, Karsenty G (2000) The osteoblast: a sophisticated fibroblast under central surveillance. Science 289:1501–1504

Elefteriou F et al (2004) Serum leptin level is a regulator of bone mass. Proc Natl Acad Sci U S A 101:3258–3263

Elefteriou F et al (2005) Leptin regulation of bone resorption by the sympathetic nervous system and CART. Nature 434:514–520

Elias CF et al (1998) Leptin activates hypothalamic CART neurons projecting to the spinal cord. Neuron 21:1375–1385

Farooqi IS et al (1999) Effects of recombinant leptin therapy in a child with congenital leptin deficiency. N Engl J Med 341:879–884

Friedman JM (2000) Obesity in the new millennium. Nature 404:632–634

Friedman JM, Halaas JL (1998) Leptin and the regulation of body weight in mammals. Nature 395:763–770

Holbrook TL, Barrett-Connor E (1993) The association of lifetime weight and weight control patterns with bone mineral density in an adult community. Bone Miner 20:141–149

Iwamoto I et al (2000) Relationships between serum leptin level and regional bone mineral density, bone metabolic markers in healthy women. Acta Obstet Gynecol Scand 79:1060–1064

Kajimura D et al (2011) Genetic determination of the cellular basis of the sympathetic regulation of bone mass accrual. J Exp Med 208:841–851

Karsenty G, Ferron M (2012) The contribution of bone to whole-organism physiology. Nature 481:314–320

Karsenty G, Wagner EF (2002) Reaching a genetic and molecular understanding of skeletal development. Dev Cell 2:389–406

Khosla S, Drake MT, Volkman TL, Thicke BS, Achenbach SJ, Atkinson EJ, Joyner MJ, Rosen CJ, Monroe DG, Farr JN (2018) Sympathetic β1-adrenergic signaling contributes to regulation of human bone metabolism. J Clin Invest 128:4832–4842

Könner AC, Brüning Jens C (2012) Selective insulin and Leptin resistance in metabolic disorders. Cell Metab 16:144–152

Kristensen P et al (1998) Hypothalamic CART is a new anorectic peptide regulated by leptin. Nature 393:72–76

Kronenberg HM (2003) Developmental regulation of the growth plate. Nature 423:332–336

Kurvers HA (1998) Reflex sympathetic dystrophy: facts and hypotheses. Vasc Med 3:207–214

Legroux-Gerot I (2007) Evaluation of bone loss and its mechanisms in anorexia nervosa. Calcif Tissue Int 81:174–182
Legroux-Gerot I, Vignau J, Collier F, Cortet B (2005) Bone loss associated with anorexia nervosa. Joint Bone Spine 72:489–495
Mark AL (2013) Selective leptin resistance revisited. Am J Physiol Regul Integr Comp Physiol 305: R566–R581
Martini G et al (2001) Influence of insulin-like growth factor-1 and leptin on bone mass in healthy postmenopausal women. Bone 28:113–117
Misra M, Klibanski A (2011) Bone health in anorexia nervosa. Curr Opin Endocrinol Diabetes Obes 18:376–382
Moitra J et al (1998) Life without white fat: a transgenic mouse. Genes Dev 12:3168–3181
Motyl KJ, Rosen CJ (2012) Understanding leptin-dependent regulation of skeletal homeostasis. Biochimie 94:2089–2096
Oury F, Karsenty G (2011) Towards a serotonin-dependent leptin roadmap in the brain. Trends Endocrinol Metab 22:382–387
Oury F et al (2010) CREB mediates brain serotonin regulation of bone mass through its expression in ventromedial hypothalamic neurons. Genes Dev 24:2330–2342
Pasco JA, Henry MJ, Nicholson GC, Schneider HG, Kotowicz MA (2005) Beta-blockers reduce bone resorption marker in early postmenopausal women. Ann Hum Biol 32:738–745
Patel MS, Elefteriou F (2007) The new field of Neuroskeletal biology. Calcif Tissue Int 80:337–347
Rauch F, Blum WF, Klein K, Allolio B, Schonau E (1998) Does leptin have an effect on bone in adult women? Calcif Tissue Int 63:453–455
Reid IR (1997a) Glucocorticoid osteoporosis--mechanisms and management. Eur J Endocrinol 137:209–217
Reid IR (1997b) Steroid-induced osteoporosis. Osteoporosis Int 7(Suppl 3):S213–S216
Riggs BL, Khosla S, Melton LJ (1998) A unitary model for involutional osteoporosis: estrogen deficiency causes both type I and type II osteoporosis in postmenopausal women and contributes to bone loss in aging men. J Bone Miner Res 13:763–773
Sato M et al (2001) Association between serum leptin concentrations and bone mineral density, and biochemical markers of bone turnover in adult men. J Clin Endocrinol Metab 86:5273–5276
Schlienger RG, Kraenzlin ME, Jick SS, Meier CR (2004) Use of beta-blockers and risk of fractures. JAMA 292:1326–1332
Schwartz MW, Peskind E, Raskind M, Boyko EJ, Porte D Jr (1996) Cerebrospinal fluid leptin levels: relationship to plasma levels and to adiposity in humans. Nat Med 2:589–593
Shi Y et al (2008) Dissociation of the neuronal regulation of bone mass and energy metabolism by leptin in vivo. Proc Natl Acad Sci U S A 105:20529
Simha V, Zerwekh JE, Sakhaee K, Garg A (2002) Effect of subcutaneous Leptin replacement therapy on bone metabolism in patients with generalized Lipodystrophy. J Clin Endocrinol Metabol 87:4942–4945
Solomon G, Atkins A, Shahar R, Gertler A, Monsonego-Ornan E (2014) Effect of peripherally administered leptin antagonist on whole body metabolism and bone microarchitecture and biomechanical properties in the mouse. Am J Physiol Endocrinol Metab 306:E14–E27
Spiegelman BM, Flier JS (1996) Adipogenesis and obesity: rounding out the big picture. Cell 87:377–389
Steppan CM, Crawford DT, Chidsey-Frink KL, Ke H, Swick AG (2000) Leptin is a potent stimulator of bone growth in Ob/Ob mice. Regul Pept 92:73–78
Suva LJ, Gaddy D, Perrien DS, Thomas RL, Findlay DM (2005) Regulation of bone mass by mechanical loading: microarchitecture and genetics. Curr Osteoporos Rep 3:46–51
Takeda S et al (2002) Leptin regulates bone formation via the sympathetic nervous system. Cell 111:305–317
Tartaglia LA et al (1995) Identification and expression cloning of a leptin receptor, OB-R. Cell 83:1263–1271
Teitelbaum SL (2000) Bone resorption by osteoclasts. Science 289:1504–1508

Thim L, Kristensen P, Larsen PJ, Wulff BS (1998) CART, a new anorectic peptide. Int J Biochem Cell Biol 30:1281–1284

Thim L, Kristensen P, Nielsen PF, Wulff BS, Clausen JT (1999) Tissue-specific processing of cocaine- and amphetamine-regulated transcript peptides in the rat. Proc Natl Acad Sci 96:2722–2727

Turker S, Karatosun V, Gunal I (2006) Beta-blockers increase bone mineral density. Clin Orthop Relat Res 443:73–74

Vaira S et al (2012) Creation and preliminary characterization of a leptin knockout rat. Endocrinology 153:5622–5628

Vrang N, Kristensen P, Tang-Christensen M, Larsen PJ (2002) Effects of leptin on arcuate pro-opiomelanocortin and cocaine-amphetamine-regulated transcript expression are independent of circulating levels of corticosterone. J Neuroendocrinol 14:880–886

Wiens M, Etminan M, Gill SS, Takkouche B (2006) Effects of antihypertensive drug treatments on fracture outcomes: a meta-analysis of observational studies. J Intern Med 260:350–362

Wierup N et al (2004) Cocaine- and amphetamine-regulated transcript (CART) is expressed in several islet cell types during rat development. J Histochem Cytochem 52:169–177

Yadav VK et al (2009) A serotonin-dependent mechanism explains the leptin regulation of bone mass, appetite, and energy expenditure. Cell 138:976–989

Yeo GS et al (1998) A frameshift mutation in MC4R associated with dominantly inherited human obesity. Nat Genet 20:111–112

Zhang Y et al (1994) Positional cloning of the mouse obese gene and its human homologue. Nature 372:425–432

Genetics of Skeletal Disorders

Fadil M. Hannan, Paul J. Newey, Michael P. Whyte,
and Rajesh V. Thakker

Contents

F. M. Hannan
Nuffield Department of Women's and Reproductive Health, University of Oxford, Oxford, UK

P. J. Newey
Division of Molecular and Clinical Medicine, Ninewells Hospital and Medical School, University of Dundee, Dundee, UK

M. P. Whyte
Center for Metabolic Bone Disease and Molecular Research, Shriners Hospitals for Children – St. Louis, St. Louis, MO, USA

Division of Bone and Mineral Diseases, Department of Internal Medicine, Washington University School of Medicine at Barnes-Jewish Hospital, St. Louis, MO, USA

R. V. Thakker (✉)
Academic Endocrine Unit, Radcliffe Department of Medicine, University of Oxford, Oxford, UK
e-mail: rajesh.thakker@ndm.ox.ac.uk

P. H. Stern (ed.), *Bone Regulators and Osteoporosis Therapy*,
Handbook of Experimental Pharmacology 262, https://doi.org/10.1007/164_2020_350

Abstract

Bone and mineral diseases encompass a variety of conditions that involve altered skeletal homeostasis and are frequently associated with changes in circulating calcium, phosphate, or vitamin D metabolites. These disorders often have a genetic etiology and comprise monogenic disorders caused by a single-gene mutation, which may be germline or somatic, or an oligogenic or polygenic condition involving multiple genetic variants. Single-gene mutations causing Mendelian diseases are usually highly penetrant, whereas the gene variants contributing to oligogenic or polygenic disorders are each associated with smaller effects with additional contributions from environmental factors. The detection of monogenic disorders is clinically important and facilitates timely assessment and management of the patient and their affected relatives. The diagnosis of monogenic metabolic bone disorders requires detailed clinical assessment of the wide variety of symptoms and signs associated with these diseases. Thus, clinicians should undertake a systematic approach commencing with careful history taking and physical examination, followed by appropriate laboratory and skeletal imaging investigations. Finally, clinicians should be familiar with the range of molecular genetic tests available to ensure their appropriate use and interpretation. These considerations are reviewed in this chapter.

Keywords

Bone · Genetic · Next-generation sequencing · Osteoporosis · Variant

1 Introduction

Bone and mineral disorders commonly have a genetic basis (Table 1), which may be due to a pathogenic germline single-gene variant (i.e., Mendelian or monogenic disorder), a somatic single-gene variant (i.e., post-zygotic mosaic disorder), or involve multiple genetic variants (i.e., oligogenic or polygenic disorders) (Hannan et al. 2019; Mortier et al. 2019). Genetic variants causing Mendelian diseases typically have a large effect (i.e., penetrance), whereas oligogenic or polygenic disorders are caused by multiple genetic variants, each of which have smaller effects, in combination with environmental factors (Newey et al. 2018). Although most monogenic disorders arise from rare coding region pathogenic variants, the majority of common genetic variants associated with polygenic traits are located in noncoding regions, usually in proximity to candidate genes implicated in the respective phenotype (Ward and Kellis 2012). Considerable overlap exists between genes causing monogenic bone disorders and those contributing to polygenic skeletal phenotypes.

Table 1 Examples of monogenic metabolic bone disorders, modes of inheritance, and genetic etiology

Mode of inheritance/disease	Gene(s)	Chromosomal location	References
Autosomal dominant			
Osteogenesis imperfecta (OI), types I–IV	*COL1A1, COL1A2*	17q21.33, 7q21.3	Marini et al. (2017)
Osteogenesis imperfecta (OI), type V	*IFITM5*	11p15.5	Cho et al. (2012) and Semler et al. (2012)
Autosomal dominant hypophosphatemic rickets	*FGF-23*	12p13.32	Consortium (2000)
Autosomal dominant high bone mass, type 1	*LRP5*	11q13.2	Little et al. (2002)
Autosomal dominant high bone mass, type 2	*LRP6*	12p13.2	Whyte et al. (2019)
Early-onset osteoporosis	*WNT1*	12q13.12	Laine et al. (2013)
Osteoporosis and skeletal dysplasia	*SGMS2*	4q25	Pekkinen et al. (2019)
Familial hypocalciuric hypercalcemia (FHH), types 1–3	*CASR, GNA11, AP2S1*	3q21.1, 19p13.3, 19q13.3	Hannan et al. (2012) and Nesbit et al. (2013a, b)
Autosomal dominant hypocalcemia (ADH), types 1–2	*CASR, GNA11*	3q21.1, 19p13.3	Nesbit et al. (2013a) and Pearce et al. (1996)
Familial expansile osteolysis	*TNFRSF11A*	18q21.33	Ralston (2008) and Whyte (2018a)
Hypophosphatasia	*TNSALP/ ALPL*	1p36.12	Whyte (2016)
Vitamin D-dependent rickets, type 3	*CYP3A4*	7q22.1	Roizen et al. (2018)
Pseudohypoparathyroidism, type 1a (PHP1a)[a]	*GNAS*	20q13.3	Lemos and Thakker (2015)
Pseudopseudohypoparathyroidism (PPHP)[a]	*GNAS*	20q13.3	Lemos and Thakker (2015)
Pseudohypoparathyroidism, type 1b (PHP1b)[a]	*GNAS, NESP55, STX16*	20q13.3	Lemos and Thakker (2015)
Autosomal recessive			
Osteogenesis imperfecta (OI), type VI	*SERPINF1*	17p13.3	Becker et al. (2011)
Osteogenesis imperfecta (OI), type VII	*CRTAP*	3p22.3	Morello et al. (2006)
Osteogenesis imperfecta (OI), type VIII	*P3H1/ LEPRE1*	1p34.2	Baldridge et al. (2008)
Osteogenesis imperfecta (OI), type XV	*WNT1*	12q13.12	Laine et al. (2013)
Hypophosphatasia	*TNSALP/ ALPL*	1p36.12	Whyte (2016)

(continued)

Table 1 (continued)

Mode of inheritance/disease	Gene(s)	Chromosomal location	References
Neonatal severe hyperparathyroidism (NSHPT)	*CASR*	3q21.1	Hannan et al. (2012)
Vitamin D-dependent rickets, type 1	*CYP27B1*	12q14.1	Carpenter et al. (2017)
Vitamin D-dependent rickets, type 2	*VDR*	12q13.11	Carpenter et al. (2017)
Autosomal recessive hypophosphatemic rickets	*DMP1*, *ENPP1*	4q22.1, 6q23.2	Feng et al. (2006) and Levy-Litan et al. (2010)
Hereditary hypophosphatemic rickets with hypercalciuria	*SLC34A3*	9q34.3	Bergwitz et al. (2006) and Lorenz-Depiereux et al. (2006)
Osteoporosis-pseudoglioma syndrome	*LRP5*	11q13.2	Ai et al. (2005)
Sclerosteosis, type 1	*SOST*	17q21.31	Brunkow et al. (2001)
Sclerosteosis, type 2	*LRP4*	11p11.2	Whyte et al. (2018)
Pyle's disease	*SFRP4*	7p14.1	Kiper et al. (2016)
Juvenile Paget's disease	*TNFRSF11B*	8q24.12	Whyte et al. (2002)
X-linked dominant			
X-linked hypophosphatemia (XLH)	*PHEX*	Xp22.11	Dixon et al. (1998)
X-linked recessive			
X-linked osteoporosis	*PLS3*	Xq23	van Dijk et al. (2013)
Dent disease, type 1	*CLCN5*	Xp11.23	Devuyst and Thakker (2010)
Mitochondrial			
Mitochondrial encephalomyopathy with lactic acidosis and stroke-like episodes (MELAS)	Mitochondrial genome	–	Tengan et al. (1998)
Kearns-Sayre syndrome	Mitochondrial genome	–	Wilichowski et al. (1997)
Mosaicism			
McCune-Albright syndrome (polyostotic fibrous dysplasia)	*GNAS*	20q13.3	Boyce and Collins (2019)
Osteogenesis imperfecta (OI)[b]	*COL1A1/* *COL1A2*	17q21.33, 7q21.3	–

Adapted from *Hannan FM* et al. *Br J Clin Pharmacol 2019; 85: 1147–1160*

[a]Parentally imprinted

[b]Autosomal disorder manifesting as post-zygotic somatic mosaicism in the developing fetus or arising from germline mosaicism in an apparently unaffected parent

The identification of these loci has improved our understanding of the molecular basis of metabolic bone disorders and provided targets for novel therapeutic agents (Karasik et al. 2016; Richards et al. 2012; Rivadeneira and Makitie 2016). In this chapter, the genetics of metabolic bone and mineral disorders, and the clinical and molecular diagnostic approaches required to investigate these typically heritable disorders, are reviewed.

2 Genetics of Bone and Mineral Disorders

2.1 Inheritance

Skeletal diseases are often single-gene monogenic disorders or polygenic complex traits (Mortier et al. 2019; Newey et al. 2018). Monogenic bone and mineral metabolic disturbances may be inherited as one of the six traits (Table 1): autosomal dominant (e.g., osteogenesis imperfecta (OI) due to collagen type I alpha-1 and collagen type I alpha-2 chain (*COL1A1* and *COL1A2)* mutations (Marini et al. 2017); autosomal recessive (e.g., vitamin D-dependent rickets types I and II due to mutations of the renal 1α-hydroxylase (*CYP27B1*) and vitamin D receptor (*VDR*) genes, respectively (Carpenter et al. 2017)); X-linked recessive (e.g., X-linked osteoporosis due to mutations of the plastin 3 (*PLS3*) gene (van Dijk et al. 2013)); X-linked dominant (e.g., X-linked hypophosphatemia (XLH) due to mutations of the phosphate endopeptidase on the X chromosome (*PHEX*) gene (Carpenter et al. 2017; Dixon et al. 1998)); Y-linked (e.g., azoospermia and oligospermia due to deletions of regions of the Y chromosome (Hannan et al. 2019)); and non-Mendelian mitochondrial disorders (e.g., hypoparathyroidism associated with the Kearns-Sayre syndrome and mitochondrial encephalomyopathy with lactic acidosis and stroke-like episodes (MELAS) syndrome (Tengan et al. 1998; Wilichowski et al. 1997)). The pattern of transmission of polygenic disorders may be especially complex because of environmental factors. Examples of polygenic skeletal and mineral disorders include osteoporosis (Estrada et al. 2012; Kemp et al. 2017) and hypercalciuric nephrolithiasis (Stechman et al. 2009). Furthermore, polygenic traits such as osteoporosis can also occur as monogenic disorders, yet be overlooked, e.g., X-linked osteoporosis due to *PLS3* mutations or early-onset osteoporosis due to heterozygous mutations of the Wnt family member 1 (*WNT1*) or sphingomyelin synthase (*SGMS2*) genes (Laine et al. 2013; Pekkinen et al. 2019; van Dijk et al. 2013). In addition to "classical" Mendelian modes of inheritance, some kindreds exhibit apparent inherited disease due to alternate mechanisms. Thus, germline mosaicism (in which a post-zygotic mutation occurs during or prior to gametogenesis in a parent) may cause a seemingly autosomal recessive mode of inheritance with multiple affected children of apparently unaffected parents (Biesecker and Spinner 2013). Disease traits may also be conditioned by epigenetic mechanisms, which cause parent-of-origin effects. For example, germline mutations and epigenetic changes at the *GNAS* complex locus cause distinct pseudohypoparathyroidism phenotypes (Lemos and Thakker 2015). Thus, maternally inherited inactivating coding region mutations in *GNAS* give rise to pseudohypoparathyroidism type 1A (PHP1A), which is characterized by parathyroid hormone (PTH) resistance together with Albright's hereditary osteodystrophy (AHO), while the identical paternally inherited mutations instead give rise to pseudopseudohypoparathyroidism (PPHP), which is characterized by AHO in the absence of PTH resistance (Table 1) (Lemos and Thakker 2015).

2.2 Genetic Heterogeneity

Clinical assessment, investigation, and treatment of patients with bone and mineral disorders require familiarity with the diversity of phenotypes caused by underlying hereditary disease. Thus, establishing the correct diagnosis requires careful clinical work-up as well as appropriate laboratory investigation and genetic testing. While the diagnosis of some skeletal disorders may be apparent from characteristic clinical or radiographic features, many diseases share overlapping phenotypes such as decreased bone mineral density (BMD), skeletal fragility, and altered mineralization (Newey et al. 2018). Similarly, while certain disorders are caused by mutation(s) in a single culpable gene, other diseases are a consequence of marked genetic heterogeneity with mutation(s) occurring in one of the many candidate genes. Thus, in some situations a broad genetic differential diagnosis persists. For example, although most cases of OI are caused by mutations in the *COL1A1* and *COL1A2* type I collagen genes (Marini et al. 2017), >15 additional genes, which are mainly involved in posttranslational processing of collagen, account for a small percentage of OI cases (Cho et al. 2012; Laine et al. 2013; Morello et al. 2006; Pyott et al. 2013; Semler et al. 2012). Similarly, hypophosphatemic rickets is caused by mutations of genes encoding phosphatonins like fibroblast growth factor-23 (FGF-23) or bone cell proteins that regulate the expression and secretion of FGF-23 (e.g., PHEX, dentin matrix protein 1 (DMP1), and ectonucleotide pyrophosphatase/phosphodiesterase 1 (ENPP1)) (Consortium 2000; Dixon et al. 1998; Feng et al. 2006; Levy-Litan et al. 2010) or by mutations affecting renal sodium phosphate cotransporters (e.g., SLC34A3) (Table 1) (Bergwitz et al. 2006; Lorenz-Depiereux et al. 2006). In addition, familial hypocalciuric hypercalcemia types 1–3 (FHH1-3), which are disorders of calcium homeostasis, have been shown to involve germline loss-of-function mutations affecting the calcium-sensing receptor (CaSR), G-protein subunit-α11 ($G\alpha_{11}$), and adaptor-related protein complex-2 σ-subunit (AP2σ), respectively (Table 1) (Hannan et al. 2012; Nesbit et al. 2013a, b).

In contrast to when phenotypically similar disorders are caused by mutations in one of many different genes, mutations in the same gene may underlie a range of clinical phenotypes, with some considered distinctive diseases. For example, mutations in the signal peptide of receptor activator of NF-κB (RANK) can cause familial expansile osteolysis (FEO), expansile skeletal hyperphosphatasia (ESH), and early-onset familial Paget's disease of bone (PDB)], which arise as a consequence of rapid remodeling (Ralston 2008; Whyte 2018a). Disease severity can also be determined by whether a mutation is harbored in the heterozygous or homozygous state. For example, the severe perinatal and infantile forms of hypophosphatasia, an inborn error of metabolism characterized by low serum alkaline phosphatase (ALP) activity, have an autosomal recessive mode of inheritance, whereas more mild and later-onset forms are usually inherited in an autosomal dominant manner (Table 1) (Whyte 2018b). In addition, some disorders are caused by loss- or gain-of-function mutations affecting the same gene. Thus, loss-of-function CaSR mutations cause FHH1 or neonatal severe hyperparathyroidism (NSHPT), whereas gain-of-function CaSR mutations cause autosomal dominant hypocalcemia type 1 (ADH1) or Bartter syndrome type V (Hannan et al. 2012; Pearce et al. 1996; Watanabe et al. 2002). Given this complex relationship between patient phenotype and the underlying genotype, establishing the genetic diagnosis may be challenging yet crucial for the evaluation of patients and kindreds with skeletal diseases.

2.3 Molecular Insights from Monogenic Disorders and Polygenic Traits

Genetic investigation of monogenic diseases has provided fundamental insights into the molecular regulation of bone biology and mineral homeostasis. Thus, utilization of classical gene discovery approaches, which involve studying affected kindreds for co-segregation with polymorphic genetic markers to define the chromosomal location, followed by DNA sequence analysis of genes located within the candidate region (Newey et al. 2018), has identified several Wnt pathway components, which regulate bone mass (Fig. 1). Indeed, autosomal recessive loss-of-function mutations of the LDL receptor-related protein 5 (*LRP5*) gene, which encodes a key Wnt co-receptor (Fig. 1), have been shown to cause osteoporosis-pseudoglioma syndrome, which is characterized by juvenile osteoporosis and congenital or childhood-onset blindness (Ai et al. 2005), whereas heterozygous gain-of-function mutations in *LRP5* (Little et al. 2002) and *LRP6* (Whyte et al. 2019), which encode the LRP5 and LRP6 cognate co-receptors, respectively, cause autosomal dominant high bone mass. In addition, autosomal recessive loss-of-function mutations of sclerostin (*SOST*), a Wnt-β-catenin inhibitor, develop sclerosteosis type 1, which is associated with progressive bone overgrowth (Brunkow et al. 2001; Whyte et al. 2018), while individuals harboring a homozygous 52 kb deletion containing an enhancer element downstream of the *SOST* gene manifest van Buchem disease, which has a milder but similar skeletal phenotype compared to sclerosteosis type 1 (Balemans et al. 2002; van Lierop et al. 2013). Furthermore, biallelic truncating mutations in secreted frizzled-related protein 4 (*SFRP4*), which encodes a soluble Wnt inhibitor (Fig. 1), have been reported in patients with Pyle's disease, a disorder characterized by cortical bone loss, fracture, and limb deformity (Kiper et al. 2016). Moreover, next-generation sequencing (NGS) approaches (Shendure et al. 2017), which include whole exome sequencing (WES) and whole genome sequencing (WGS), are being increasingly utilized to investigate monogenic diseases. Such approaches have demonstrated that heterozygous loss-of-function mutations of

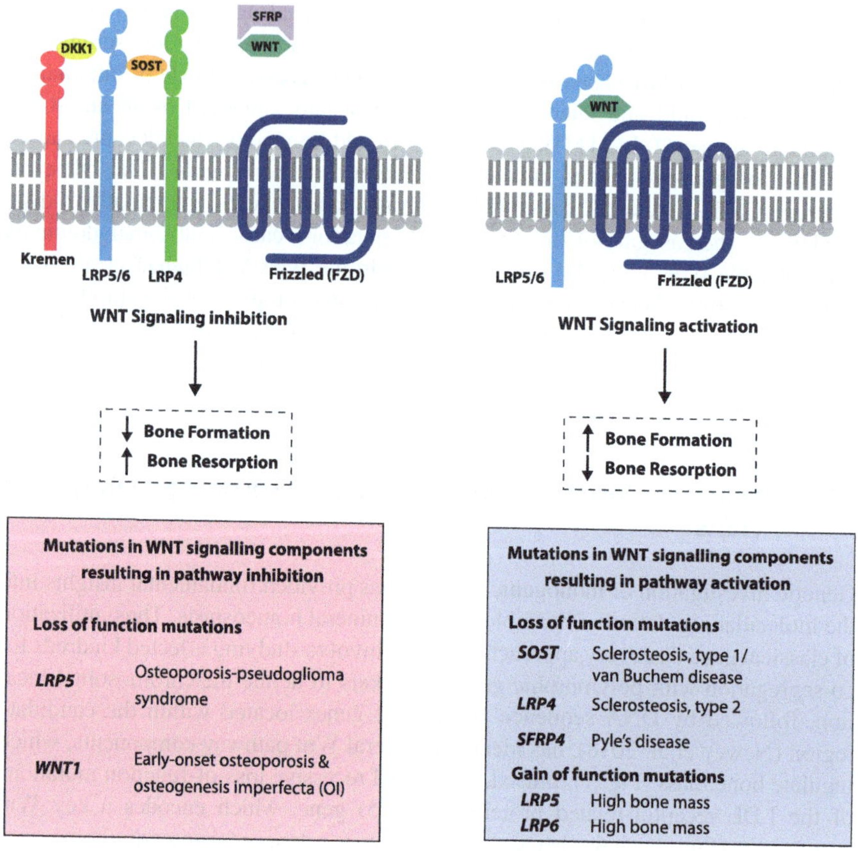

Fig. 1 Schematic representation of Wnt signaling pathway components reported to be mutated in disorders of bone development and skeletal homeostasis. Activation of the canonical Wnt pathway increases bone mass, and this is mediated by the binding of extracellular Wnt ligands (dark green) to a transmembrane receptor complex comprising the Wnt co-receptor LRP5 or LRP6 (LRP5/6, light blue) and a member of the frizzled (FZD) family (dark blue). In contrast, inhibition of the canonical Wnt pathway decreases bone mass (Baron and Kneissel 2013; Krishnan et al. 2006). This inhibition is mediated by extracellular factors such as sclerostin (SOST, orange) and Dickkopf-related protein 1 (DKK1, yellow), which bind to the LRP5/6 co-receptor, thereby preventing activation by Wnt ligands as well as recruiting inhibitory transmembrane proteins such as LRP4, which is a SOST-interacting protein (light green), and the Kremen proteins (pink), which are high-affinity DKK1 receptors that functionally cooperate with DKK1 to decrease Wnt signaling (Mao et al. 2002). Secreted frizzled-related proteins (SFRPs, purple) also inhibit the canonical Wnt pathway by sequestering Wnt ligands. The importance of the canonical Wnt pathway for the regulation of bone mass has been highlighted by loss-of-function mutations affecting SOST and LRP4 and by gain-of-function mutations of LRP5 and LRP6, which lead to the disorder called high bone mass (Balemans et al. 2002; Brunkow et al. 2001; Fijalkowski et al. 2016; Little et al. 2002), and also by loss-of-function mutations of LRP5 and the Wnt1 ligand, which lead to monogenic osteoporosis disorders (Ai et al. 2005; Laine et al. 2013). From *Hannan FM* et al. *Br J Clin Pharmacol 2019; 85: 1147–1160*

WNT1 cause autosomal dominant early-onset osteoporosis, whereas biallelic loss-of-function *WNT1* mutations lead to an autosomal recessive form of OI (Laine et al. 2013). In addition, NGS has revealed that alterations in nonprotein-encoding genes can cause skeletal diseases (Mortier et al. 2019). Thus, a heterozygous germline microRNA mutation has been shown to cause a novel autosomal dominant skeletal dysplasia in two unrelated families (Grigelioniene et al. 2019). This point mutation increased expression of the chondrocyte-specific microRNA-140 (miR-140) and led to derepression of wild-type miR-140 targets while repressing mutant miR-140 targets, consistent with both loss- and gain-of-function effects (Grigelioniene et al. 2019).

Genetic investigation of polygenic disorders such as osteoporosis has utilized genome-wide association studies (GWAS), which require large populations of cases and controls (Estrada et al. 2012; Karasik et al. 2016; Kemp et al. 2017; Richards et al. 2012). Such studies generally involve direct or imputed genotyping of large numbers of common (e.g., minor allele frequency > 5%) and infrequent (e.g., minor allele frequency 1–5%) single nucleotide polymorphisms/variants (SNPs/SNVs) to identify genetic loci enriched for the trait (Kruglyak 2008; Newey et al. 2018). GWAS has further highlighted the pivotal role of Wnt signaling in bone biology and shown that many Wnt pathway components (>15 genes), including *LRP5* and *SOST*, are candidate genes for bone mineral density (BMD) (Estrada et al. 2012; Kemp et al. 2017) and that *WNT16* is a key determinant of cortical bone strength (Moverare-Skrtic et al. 2014; Zheng et al. 2012). Moreover, a GWAS of BMD estimated by quantitative heel ultrasound has reported that the disheveled associated activator of morphogenesis 2 (*DAAM2*) gene, which likely modulates canonical Wnt signaling, has a major influence on bone composition, strength, and mineralization (Morris et al. 2019). Furthermore, osteoporosis GWAS variants have been mapped to target gene promoters using chromatin conformation capture methods (Chesi et al. 2019). Such studies, which involved human mesenchymal stem cell-derived osteoblasts, have identified the inhibitor of growth family member 3 (*ING3*) and ependymin-related 1 (*EPDR1*) genes as potential osteoblastic regulators of BMD (Chesi et al. 2019).

In addition to investigating the influence of common variants in polygenic traits, some studies have highlighted a role for low-frequency and rare variants with larger effect sizes, which provide further insight into skeletal biology. For example, in Icelandic people, a rare truncating variant of the leucine-rich receptor containing G-protein-coupled receptor (*LGR4*) gene was shown to be associated with phenotypic traits that included low BMD and osteoporosis (Styrkarsdottir et al. 2013), while two rare *COL1A2* coding region variants were also detected in this population in association with low BMD (Styrkarsdottir et al. 2013). A further study of individuals of European ancestry reported a low-frequency, noncoding variant in close proximity to the Engrailed homeobox-1 (*EN1*) gene, which was associated with ~fourfold and ~threefold greater effect size on BMD and fracture risk, respectively, compared to reported common variants (Zheng et al. 2015). This and other studies indicate a role for *EN1* in bone biology (Adamska et al. 2004; Mitchell et al. 2016), likely through an interaction with Wnt factors, and regulation of bone turnover, thereby highlighting the utility of these genetic approaches for investigating the molecular pathogenesis of bone and mineral metabolic diseases.

2.4 Genes and Pathways as Therapeutic Targets

A key purpose of characterizing the genetic basis for bone and mineral disorders is the identification of genes and cellular pathways that may be targeted therapeutically. Indeed, many treatments now target key components identified from these studies. For example, the discovery that low skeletal ALP activity causes hypophosphatasia led to the multinational approval in 2015 of the bone-targeted enzyme-replacement biologic asfotase alfa (Whyte 2016). This is further highlighted by the identification of FGF-23 as a key promoter of renal tubular phosphate excretion, which led to the approval in 2018 of burosumab, an anti-FGF-23 monoclonal antibody, for the treatment of XLH (Collins 2018). A phase 3 trial has demonstrated that children with XLH, who received subcutaneous burosumab injections every 2 weeks, achieved normal serum phosphate concentrations, better growth, and reduction in rickets severity compared to those who continued treatment with conventional therapy comprising oral phosphate and active vitamin D (Imel et al. 2019). Understanding the molecular basis of bone cell function has resulted in the development of the monoclonal antibody denosumab for the treatment of osteoporosis. Denosumab blocks RANKL within the OPG/RANKL/RANK/NF-κB signaling pathway to inhibit the formation and activity of osteoclasts, thereby inhibiting bone resorption (Karasik et al. 2016). Treatment with denosumab significantly reduced fracture risk in women with postmenopausal osteoporosis (Cummings et al. 2009). Osteoporosis therapies are now also directed at the Wnt pathway. This includes romosozumab, an anti-sclerostin monoclonal antibody that received FDA approval in 2019 for the treatment of postmenopausal osteoporosis. Romosozumab causes a rapid and marked increase in bone formation, primarily through modeling-based mechanisms, while simultaneously decreasing bone resorption (Ferrari 2018). The Fracture Study in Postmenopausal Women with Osteoporosis (FRAME) clinical trial showed that once-monthly subcutaneous administration of romosozumab over 12 months reduced new vertebral fractures and clinical fractures by >70% and >35%, respectively (Cosman et al. 2016).

3 Approach to the Patient with Genetic Bone and Mineral Disease

3.1 Clinical Approach

Bone and mineral disorders can manifest a variety of clinical symptoms and signs, and an awareness of the range of potential phenotypes is central to appropriate investigation and treatment (Newey et al. 2018). Clinicians have at their disposal many biochemical and radiological tools to aid diagnosis, but they should be used judiciously and following the acquisition of a detailed personal and family history and the undertaking of a careful clinical examination (Newey et al. 2018). As the range and complexity of molecular genetic testing increases, selecting appropriate investigations may be challenging. In addition, the interpretation of genetic test results requires the clinician to recognize and assess potential uncertainties and limitations. Finally, it is important to include the patient in decision-making processes and ensure that appropriate informed consent is acquired before testing (Newey et al. 2018).

3.2 Medical History and Physical Examination

The history of present illness provides key information regarding disease etiology, pathogenesis, and prognosis and is important for guiding diagnosis and therapy. Determining whether signs and symptoms have begun recently or been longstanding can prompt different diagnostic considerations and interventions. It is also important to review prior medical records, radiographs, and laboratory tests to aid diagnosis and prognostication (Newey et al. 2018). Physical examination can reveal a range of findings for diagnosis including skeletal deformities, which are common and perhaps unique in children. Diagnosis may emanate from recognition of a single physical finding, e.g., large café-au-lait spots (McCune-Albright syndrome), premature loss of deciduous teeth (hypophosphatasia), blue or gray sclerae (osteogenesis imperfecta), hallux valgus (fibrodysplasia ossificans progressiva), alopecia (vitamin D-dependent rickets, type II), brachydactyly (PHP1a and PPHP), syndactyly (sclerosteosis types 1 and 2), torus palatinus (high bone mass due to activating *LRP5* or *LRP6* mutations), or numerous surgical scars (multiple endocrine neoplasia (MEN) syndromes) (Carpenter et al. 2017; Collins et al. 2012; Lemos and Thakker 2015; Marini et al. 2017; Pignolo et al. 2011; Thakker et al. 2012; Whyte 2016; Whyte et al. 2018, 2019). For some genetic diseases, an amalgam of physical findings suggests the diagnosis, e.g., rickets with craniotabes at birth, a rachitic rosary (enlargement of the costochondral junctions) developing during the first year of life, and leg deformities and short stature occurring in infancy or childhood (Carpenter et al. 2017). In adults, skeletal deformation arising in childhood can cause much of the morbidity from metabolic bone disease. Bowing of the lower limbs can lead to osteoarthritis, especially in the knees. Without a comprehensive physical examination, these issues may go undetected.

3.3 Family History and Mode of Disease Inheritance

A variety of metabolic bone disorders have a monogenic etiology, and this may be suspected because of an early age of onset, occurrence of clinical features consistent with a syndromic disease, or a family history of the condition (Newey et al. 2018). The family history is required to establish the mode of disease inheritance. Thus, in autosomal dominant disorders such as FHH (Table 1) (Hannan et al. 2018), the proband typically has one affected parent, and the disease affects both sexes and is transmitted by either the father or mother. An autosomal dominant mode of inheritance is therefore disclosed by prior or prospective study of the relatives' affected status. In contrast, in autosomal recessive diseases, which can affect both sexes, the proband is born to parents who may be related and are usually asymptomatic "carriers." Therefore, parental consanguinity, which can be apparent, or inapparent involving endogamy within a specific geographical location and a "founder" mutation, can provide an important clue for autosomal recessive inheritance. In X-linked recessive diseases such as early-onset osteoporosis with fractures due to *PLS3* mutations (Table 1) (van Dijk et al. 2013), only males are typically affected, who are born to often unaffected parents yet the mother is an asymptomatic carrier with affected male relatives. There is no male-to-male transmission. In X-linked dominant diseases such as XLH (Table 1) (Carpenter et al. 2017; Dixon et al. 1998), both males and females are affected. However, females are usually more mildly and variably affected than males, and 50% of children (boys and girls) from an affected woman will have the disease, and all of the daughters, but none of the sons, of an affected man will have the disease (Hannan et al. 2019). In Y-linked diseases such as azoospermia and oligospermia, only males are affected, and unless representing a sporadic case, they have an affected father (patrilineal inheritance) and all sons of an affected man will have the disease (Hannan et al. 2019). In contrast, mitochondrial inherited disorders such as Kearns-Sayre syndrome and MELAS (Table 1) (Tengan et al. 1998; Wilichowski et al. 1997) can affect both sexes, and these disorders are only transmitted by an affected mother (matrilineal inheritance) from her egg mitochondrial DNA (Hannan et al. 2019). These modes of inheritance can be influenced by (1) non-penetrance or variable expression in autosomal dominant disorders such as MEN1 (Thakker et al. 2012); (2) imprinting, whereby expression of an autosomal dominant disorder is conditioned by whether it is maternally or paternally transmitted (e.g., PHP1a versus PPHP) (Lemos and Thakker 2015); (3) anticipation, whereby some dominant disorders become increasingly severe, or have earlier onset, in successive generations; (4) pseudodominant inheritance of autosomal recessive disorders reflecting repeated consanguineous marriages in successive generations; and (5) mosaicism in which an individual has two or more populations of cells with different genotypes because of post-zygotic mutations occurring during their development from a single fertilized egg (e.g., McCune-Albright syndrome) (Boyce and Collins 2019). In the particular circumstance of germline mosaicism arising from somatic mutation during gametogenesis, there may be confusion regarding diagnosis and recurrence risk because of apparently unaffected parents having multiple affected offspring that would suggest autosomal recessive inheritance, but actually reflect an autosomal dominant disorder such as OI type II (Biesecker and Spinner 2013). Hence, characterization of the family history can establish the mode of inheritance, help diagnose a genetic disorder, and also identify individuals at risk (Newey et al. 2018).

4 Overview of Genetic Tests

4.1 Clinical Value of Genetic Testing

Identifying the genetic basis for a patient's bone and mineral disorder facilitates appropriate treatment and assessment of the patient for associated clinical features that may not have been apparent at initial evaluation. A genetic diagnosis can also provide prognostic information and facilitate genetic counseling as well as the testing of relatives who may be asymptomatic "carriers." Furthermore, family members found not at risk of having or transmitting the disease can be reassured. For parents with children affected with severe skeletal disease, the identification of a genetic cause may lead to prenatal counseling and/or preconception genetic testing for future pregnancies. Genetic testing may also aid risk profiling. For example, SNPs associated with osteoporosis have been reported to predict fracture risk in bisphosphonate-treated patients (Lee et al. 2016), and potential genetic markers of anti-resorptive-induced osteonecrosis of the jaw have also been identified (Yang et al. 2018).

4.2 Pretest Considerations

For individuals with a possible genetic metabolic bone disease, several factors require consideration before arranging genetic analysis (Fig. 2). These include the clinical phenotype, mode of inheritance, and availability of additional pedigree members to aid the diagnosis. For example, DNA sequencing of "trios" (i.e., the affected proband and both parents) may facilitate detection of compound heterozygous or de novo mutations (Goldstein et al. 2013). Moreover, the type of underlying genetic abnormality will influence the likelihood of an investigation achieving the correct genetic diagnosis. For example, direct DNA sequencing methods are suitable for detecting coding region single nucleotide variants (SNVs) or small insertions or deletions ("indels") which cause the majority of monogenic bone and mineral disorders. However, these sequencing techniques may not detect whole or partial gene deletions that are associated with some monogenic syndromes or be suitable for identifying large chromosomal abnormalities (e.g., 22q11.2 microdeletion in DiGeorge syndrome), whose detection requires alternative approaches (Fig. 2 and Table 2) (Gijsbers and Ruivenkamp 2011). For some monogenic disorders, gene panel analysis should be considered if genetic heterogeneity is likely (e.g., in OI or FHH) (Hannan et al. 2018; Marini et al. 2017). Thus, genetic testing that fails to detect an abnormality does not necessarily exclude a genetic disease, but instead may reflect (1) the limitations of the utilized genetic methodology (e.g., inadequate resolution or coverage), (2) incorrect assumptions regarding the clinical phenotype or mode of inheritance, or (3) an alternative genetic cause to the one being tested (Newey et al. 2018). Consequently, sequential or simultaneous genetic tests can be required to ensure comprehensive evaluation of the underlying etiology, although such testing may be limited by local availability and cost (Hannan et al. 2019).

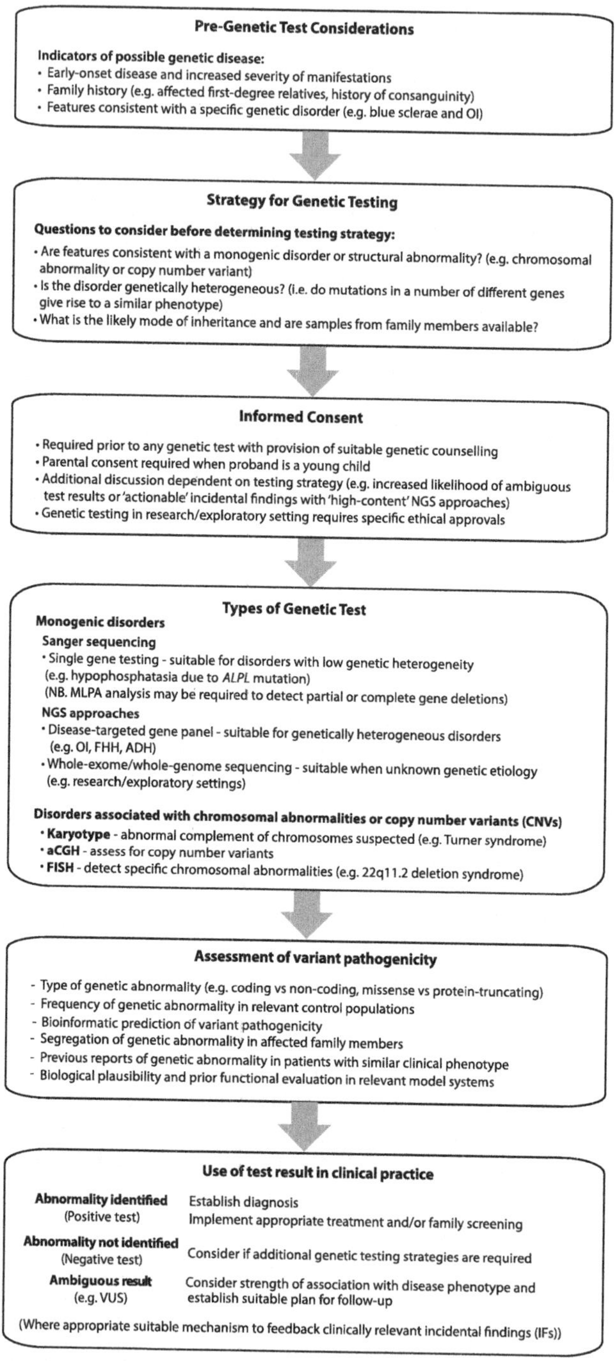

Fig. 2 Flowchart outlining considerations for genetic testing in patients with metabolic bone disease. From *Hannan FM* et al. *Br J Clin Pharmacol 2019; 85: 1147–1160*

Table 2 Examples of genetic tests, their molecular resolution, and utility

Genetic test	Resolution	Abnormalities detected	Additional notes
Detection of chromosomal abnormalities including copy number variations (CNVs)			
Karyotype: G-banding (trypsin-Giemsa staining)	5–10 Mb	Aneuploidy Large chromosomal deletions, duplications, translocations, inversions, insertions	Limited resolution Requirement to study many cells to detect mosaicism
Fluorescence in situ hybridization (FISH)	50 kb to 2 Mb (dependent on size of probes employed)	Structural chromosomal abnormalities (e.g., microdeletions, translocations)	Labor-intensive Low resolution limits its use Unsuitable where unknown genetic etiology
Multiplex ligation probe amplification	Probe dependent 50–70 nucleotides Single exon deletion or duplication possible	Copy number variations (CNVs) including (partial) gene deletions or duplications	Low cost, technically simple method Simultaneous evaluation of multiple genomic regions Not suitable for genome-wide approaches Not suitable for analysis of single cells
Array comparative genomic hybridization (aCGH)	10 kb (high resolution) 1 Mb (low resolution) (dependent on probes set)	Genome-wide copy number variations (CNVs)	Inability to detect balanced translocations Useful for detection of low-level mosaicism
Single nucleotide polymorphism (SNP) array	~50–400 kb (dependent on probe set)	Genome-wide detection of SNP genotypes Copy number variations (CNVs)	Inability to detect balanced translocation Useful for detection of low-level mosaicism Detection of copy number neutral regions or absence of heterozygosity (i.e., due to uniparental disomy)
Detection of monogenic disorders (and copy number variations (CNVs))			
First-generation sequencing (Sanger)			
Single-gene test	Single nucleotide (exonic regions and intron/exon boundaries of candidate gene)	Single nucleotide variants (SNVs) Small insertions or deletions ("indels")	Relative high cost/base May miss large deletions/duplications Unsuitable where unknown genetic etiology
Next-generation sequencing			
Disease-targeted gene panels	Single nucleotide (exonic regions and intron/exon boundaries of candidate genes)	Single nucleotide variants (SNVs) Small insertions or deletions ("indels")	May lack complete coverage of exonic regions (may require Sanger sequencing to fill in "gaps") Increased likelihood of

(continued)

Table 2 (continued)

Genetic test	Resolution	Abnormalities detected	Additional notes
			identifying variants of uncertain significance (VUS) as number of genes increases Unsuitable where unknown genetic etiology
Whole exome sequencing (WES)	Single nucleotide (all exonic regions and intron/exon boundaries)	Single nucleotide variants (SNVs) Small insertions or deletions ("indels") Copy number variations (CNVs)	Not all exons may be covered/captured Difficulties with GC-rich regions and presence of homologous regions/ pseudogenes Small indels may not be captured Bioinformatic expertise required for data analysis High likelihood of incidental findings and VUSs Detection of CNVs requires additional data analysis (i.e., loss of heterozygosity mapping across exonic regions) Suitable for disease associated gene discovery
Whole Genome Sequencing (WGS)	Single nucleotide	Single nucleotide variants (SNVs) Small insertions or deletions ("indels") Copy number variations (CNVs) (translocations/ rearrangements)	Relative high cost Large data sets generated and complex data analysis requiring bioinformatic expertise High likelihood of incidental findings and VUSs CNV analysis possible but may present specific challenges Suitable for disease associated gene discovery

From *Hannan FM* et al. *Br J Clin Pharmacol 2019; 85: 1147–1160*

4.3 Detection of Chromosomal Abnormalities, Copy Number Variations (CNVs), and Mutations Causing Disease

4.3.1 Karyotype

Conventional karyotyping is the initial test for assessing major chromosomal abnormalities such as aneuploidy (i.e., abnormal number of chromosomes) or large insertions, deletions, duplications, inversions, or reciprocal translocations (Table 2) (Dave and Sanger 2007; Gijsbers and Ruivenkamp 2011). Such abnormalities may be suspected in the presence of major congenital abnormalities, marked developmental delay, or features of a specific chromosomal abnormality disorder such as Klinefelter and Turner syndromes, both of which may be associated with osteoporosis (Faienza et al. 2016; Ferlin et al. 2010). Karyotyping is usually performed using peripheral blood leucocytes, which are cultured, prior to evaluation by high-resolution G-banding (Giemsa staining) of at least 20 metaphase nuclei. Evaluation of multiple cells allows a reliable analysis of each chromosome and also facilitates identification of mosaicism (Dave and Sanger 2007; Gijsbers and Ruivenkamp 2011). High-resolution G-band karyotype analysis identifies the most major chromosomal defects; however, its resolution is limited to ~5–10 Mb of DNA and will therefore not identify smaller abnormalities such as CNVs (Table 2) (Dave and Sanger 2007; Gijsbers and Ruivenkamp 2011).

4.3.2 Fluorescence In Situ Hybridization (FISH)

FISH utilizes DNA probes, which hybridize to specific target regions on metaphase chromosomes, thereby enabling visualization by fluorescence microscopy (Dave and Sanger 2007; Gijsbers and Ruivenkamp 2011). A range of chromosomal abnormalities can thus be identified resulting from absence of probe binding (e.g., due to a deletion), additional probe binding (e.g., due to a duplication), or probes binding to an aberrant chromosomal region (e.g., due to a translocation or inversion). Molecular resolution of FISH is typically 50 kb–2 Mb (Table 2). The major limitation is that it does not allow genome-wide analysis, but is limited to detecting abnormalities covered by the probe sets (e.g., detection of 22q11.2 deletion in DiGeorge syndrome). Alternate FISH-based methods that allow the simultaneous evaluation of several regions of interest have been developed, which include whole-chromosome painting probes (termed multiplex FISH (M-FISH) and spectral karyotyping (SKY)), with each chromosome labeled with a different color (Dave and Sanger 2007; Gijsbers and Ruivenkamp 2011). While such techniques can identify interchromosomal abnormalities such as translocations, they do not identify small deletions, duplications, or inversions (Dave and Sanger 2007; Gijsbers and Ruivenkamp 2011).

4.3.3 Multiplex Ligation-Dependent Probe Amplification (MLPA)

MLPA utilizes a pool of custom-designed probes to amplify specific genomic regions and is used to detect small chromosomal abnormalities such as single or partial gene deletions (Gijsbers and Ruivenkamp 2011). Adjacent probes hybridize to each other and undergo ligation followed by polymerase chain reaction (PCR) amplification of the ligated product, with products separated by electrophoresis (Table 2). The probes are generally designed to detect partial or complete gene deletions. In addition, MLPA can evaluate alterations in methylation, for example, pseudohypoparathyroidism 1b (PHP1b), which is associated with deletion of one or more of the four differentially methylated regions (Tafaj and Juppner 2017).

4.3.4 Array-Based Screening

Array comparative genomic hybridization (aCGH) is used for the genome-wide detection of small chromosomal abnormalities (Table 2) and commonly represents first-line investigation for multiple congenital abnormalities, which include skeletal manifestations and/or neurodevelopmental delay (Gandomi et al. 2015; Koczkowska et al. 2017). The aCGH technique involves mixing the patient's DNA sample (labeled green) with a reference DNA sample (labeled red), prior to applying the sample to the array platform for competitive hybridization with a set of immobilized reference DNA fragments. Automated analysis of the array measuring red-green fluorescence facilitates identification of deletions and duplications in the patient's sample (Gijsbers and Ruivenkamp 2011; Kharbanda et al. 2015). The aCGH methodology is frequently used to detect copy number variants (CNVs). However, all individuals harbor small CNVs without apparent adverse health effects, while several potentially pathogenic CNVs have reduced penetrance and do not always cause disease (Newey et al. 2018).

SNP arrays may also detect CNVs and be used for genome-wide genotyping (Table 2). For example, deletions or uniparental disomy spanning several adjacent SNPs on the array can reveal loss of heterozygosity (LOH), while copy number gains (e.g., duplication) may be indicated by increased numbers of different genotypes (Gijsbers and Ruivenkamp 2011). SNP arrays can also be used for homozygosity mapping to localize recessive disorders in the offspring of consanguineous parents (Caparros-Martin et al. 2017).

4.3.5 DNA Sequence Analysis

Sanger DNA sequencing remains the gold standard for detecting DNA sequence variants due to the high fidelity of the DNA polymerase used during DNA amplification (base accuracy of >99.99%) (Falardeau et al. 2017; Lazarus et al. 2014). However, it is labor-intensive and generally reserved for disorders with limited genetic heterogeneity such as single-gene disorders, e.g., hypophosphatasia caused only by *TNSALP/ALPL* mutations (Tenorio et al. 2017). Sanger sequencing is increasingly being replaced by NGS approaches, which allow the massively parallel sequencing of large amounts of genetic material. The three major uses of NGS are WGS, WES, and disease-targeted gene panel sequencing (Table 2). WGS analyzes the entire genome DNA sequence including coding and noncoding regions and can detect SNVs, small insertions or deletions, and CNVs (Newey et al. 2018). In contrast, WES analyzes ~20,000 protein-coding genes (i.e., the "exome"), which comprise 1–2% of the genome, and is expected to harbor most disease-causing mutations (Newey et al. 2018). Consistent with this, WES has represented the mainstay of disease gene discovery studies over the past decade and led to the identification of the genetic cause for a range of metabolic bone diseases (e.g., *AP2S1* mutations in FHH type 3 (Nesbit et al. 2013b); *BMP1* mutations causing increased BMD and recurrent fractures (Asharani et al. 2012); *SFRP4* mutations in Pyle's disease (Kiper et al. 2016); and *CYP3A4* mutations in vitamin D-dependent rickets type 3 (Roizen et al. 2018)). Disease-targeted sequencing represents the principal NGS method used in clinical practice and is used to analyze small or large numbers of genes (e.g., <10 to >150 genes) associated with a specific disease (Falardeau et al. 2017; Lazarus et al. 2014; Rehm 2013). Indeed, NGS disease-targeted panels are established for genetically heterogeneous disorders, which include OI, disorders of calcium sensing, and hypophosphatemic rickets (Arvai et al. 2016; Bardai et al. 2017; Polla et al. 2015).

4.4 Assessment of Variant Pathogenicity

A major challenge arising from increasing utilization of genetic testing is ascribing pathogenicity to individual variants. Several international bodies have published guidelines recommending that variant classification systems utilize the terms "pathogenic," "variant of unknown significance (VUS)," and "not pathogenic or benign" and that evidence for pathogenicity is gathered from multiple sources (Toledo et al. 2017; Richards et al. 2015). The initial analysis of non-synonymous variants should report the frequency of the variant in large-scale population-level databases such as GnomAD and the Exome Aggregation Consortium (ExAC) (GnomAD 2019; Lek et al. 2016). A frequency of $>1\%$ in such databases indicates that the variant is likely a benign polymorphism. However, the presence of variants at low frequencies ($<1\%$) does not exclude their pathogenicity as the databases may contain representative individuals from the disease population (e.g., The Cancer Genome Atlas) (Richards et al. 2015). In addition, clinical variant and disease databases (e.g., ClinVar, Human Gene Mutation Database) should be consulted to determine if the gene has previously been linked with the condition (Johnston and Biesecker 2013). Further investigation of variants involves using in silico algorithms such as SIFT, PolyPhen-2, and MutationTaster to predict whether amino acid changes are likely to affect protein function (Adzhubei et al. 2013; Schwarz et al. 2010; Vaser et al. 2016). In addition, demonstration of evolutionary conservation of the variant residue or location of the variant residue within a critical structural domain of the affected protein, using in silico tools such as Clustal Omega and PyMOL, respectively, can indicate that a variant may be pathogenic (Newey et al. 2018). Once a variant has been assessed by these bioinformatics methods and still remains a plausible candidate, then determining whether the variant segregates only with affected family members is key to attributing disease association (Newey et al. 2018). Functional analysis should also be considered, and the choice of experimental technique will depend on the type of variant (e.g., splice site variants are assessed by measurement of RNA and protein expression, whereas missense mutations may require functional cellular assays), the tissue affected in the disease, and the availability of samples (Newey et al. 2018). Ideally, assessments should be made using patient tissue where possible, for example, in cell lines derived from lymphoblastoids or skin fibroblasts or histological sections of tissues removed by biopsy or surgery. In choosing an in vitro assay, the heritability of the variant should also be considered as overexpression systems may exaggerate the effect of a variant (Newey et al. 2018).

4.5 Special Circumstances for Genetic Testing

4.5.1 Genetic Tests for Mosaicism

Genetic bone disorders can arise from somatic mosaicism (e.g., *GNAS* mutations in McCune-Albright syndrome) (Collins et al. 2012) and may also rarely occur as germline mosaicism arising from somatic mutation during gametogenesis in seemingly unaffected parents (e.g., OI type II). This may cause diagnostic confusion, as in this setting, one parent is carrying the mutation in their gametes, and may give rise to more than one affected offspring, suggesting possible autosomal recessive inheritance, in contrast to the underlying autosomal dominant inheritance pattern (Cohen et al. 2015). Detection of mosaicism has been improved by sensitive genome-wide testing strategies (e.g., aCGH, SNP arrays, and NGS), which can detect low-level mosaicism (e.g., 5% for SNP array) (Biesecker and Spinner 2013). However, selecting the appropriate test depends on the clinical phenotype, the type of mutation suspected (e.g., SNV, CNV, aneuploidy), the likely extent of mosaicism, and its tissue distribution. In general, blood lymphocyte DNA will suffice, but analysis of other affected tissues may be required (e.g., fibroblasts or bone) (Kang et al. 2018; Lindhurst et al. 2011).

4.5.2 Genetic Tests for Prenatal Diagnosis

Genetic testing can be performed at preimplantation or prenatal stages and is used to detect severe disorders such as perinatal lethal OI (Pyott et al. 2011). Preimplantation diagnosis involves a single cell being obtained from the embryo several days after in vitro fertilization to identify single-gene defects or chromosomal abnormalities, thereby allowing selection of unaffected embryos for uterine implantation (Vermeesch et al. 2016), whereas prenatal genetic testing is undertaken once pregnancy is established (Vermeesch et al. 2016). Typically, detection of fetuses at risk from genetic disease involves cells being obtained for DNA analysis using invasive methods such as chorionic villous sampling or amniocentesis (Vermeesch et al. 2016). However, progress made in the detection of cell-free circulating fetal DNA in the maternal circulation (e.g., after ~10 weeks of gestation) now offers the potential for noninvasive prenatal testing (NIPT) (van den Veyver and Eng 2015). Thus, a maternal blood sample may facilitate fetal sex determination, which is important for X-linked disorders, and allow screening for aneuploidy (Breveglieri et al. 2019). Monogenic skeletal disorders such as OI, achondroplasia, and craniosynostosis have also been diagnosed by NIPT (Zhang et al. 2019). However, this approach is limited to autosomal dominant disorders caused by de novo or paternally inherited mutations, whereas maternally inherited or autosomal recessive disorders are more challenging to diagnose given the similarity between maternal DNA and the maternally inherited region of the fetal genome (Breveglieri et al. 2019).

4.6 Informed Consent and Ethical Considerations

Genetic evaluation of skeletal disorders may clearly benefit the patient and family; however, such testing also poses clinical and ethical challenges. For example, NGS approaches frequently identify variants of uncertain significance (VUS) (Richards et al. 2015), and communicating the relevance of such variants to the patient remains a major challenge. It is clear that explaining the possibility of such ambiguous results should comprise part of informed consent prior to genetic testing (Fig. 2) (Newey et al. 2018). Studies have also indicated that variants reported as pathogenic could instead be benign (Manrai et al. 2016) or far less penetrant than previously recognized, and such diagnostic misclassification could potentially lead to inappropriate investigation or treatment. Thus, caution is required in data interpretation and both the clinician and patient must understand the genetic test result and its potential limitations. Genetic testing may also lead to incidental findings, which are abnormalities unrelated to the clinical question but of potential significance to the patient and family. For example, the additional identification of penetrant pathogenic mutations in hereditary cancer genes (e.g., *BRCA1*, *BRCA2*) could result in complex ethical considerations, and plans for managing such findings should be in place beginning with the informed consent pathway (Rigter et al. 2014).

5 Conclusions

Bone and mineral diseases commonly have a genetic basis and represent either a monogenic disorder due a germline or somatic single-gene mutation or an oligogenic or polygenic disorder involving several genetic variants. Recognition of these heritable disorders is fundamental for appropriate investigations and treatment for patients and families. The advent of high-content genetic testing employing NGS approaches has revolutionized investigation and diagnosis of genetic disease. The clinician must now acquire an understanding of these complex tests to combine with his/her fundamental skills of history taking and physical examination to ensure their judicious use to benefit patients.

References

Adamska M, MacDonald BT, Sarmast ZH, Oliver ER, Meisler MH (2004) En1 and Wnt7a interact with Dkk1 during limb development in the mouse. Dev Biol 272:134–144. https://doi.org/10.1016/j.ydbio.2004.04.026

Adzhubei I, Jordan DM, Sunyaev SR (2013) Predicting functional effect of human missense mutations using PolyPhen-2. Curr Protoc Hum Genet 76:7–20. https://doi.org/10.1002/0471142905.hg0720s76

Ai M, Heeger S, Bartels CF, Schelling DK, Osteoporosis-Pseudoglioma Collaborative G (2005) Clinical and molecular findings in osteoporosis-pseudoglioma syndrome. Am J Hum Genet 77:741–753. https://doi.org/10.1086/497706

Arvai K et al (2016) Next-generation sequencing of common osteogenesis imperfecta-related genes in clinical practice. Sci Rep 6:28417. https://doi.org/10.1038/srep28417

Asharani PV et al (2012) Attenuated BMP1 function compromises osteogenesis, leading to bone fragility in humans and zebrafish. Am J Hum Genet 90:661–674. https://doi.org/10.1016/j.ajhg.2012.02.026

Baldridge D et al (2008) CRTAP and LEPRE1 mutations in recessive osteogenesis imperfecta. Hum Mutat 29:1435–1442. https://doi.org/10.1002/humu.20799

Balemans W et al (2002) Identification of a 52 kb deletion downstream of the SOST gene in patients with van Buchem disease. J Med Genet 39:91–97

Bardai G, Ward LM, Trejo P, Moffatt P, Glorieux FH, Rauch F (2017) Molecular diagnosis in children with fractures but no extraskeletal signs of osteogenesis imperfecta. Osteoporos Int 28:2095–2101. https://doi.org/10.1007/s00198-017-4031-2

Baron R, Kneissel M (2013) WNT signaling in bone homeostasis and disease: from human mutations to treatments. Nat Med 19:179–192. https://doi.org/10.1038/nm.3074

Becker J et al (2011) Exome sequencing identifies truncating mutations in human SERPINF1 in autosomal-recessive osteogenesis imperfecta. Am J Hum Genet 88:362–371. https://doi.org/10.1016/j.ajhg.2011.01.015

Bergwitz C et al (2006) SLC34A3 mutations in patients with hereditary hypophosphatemic rickets with hypercalciuria predict a key role for the sodium-phosphate cotransporter NaPi-IIc in maintaining phosphate homeostasis. Am J Hum Genet 78:179–192. https://doi.org/10.1086/499409

Biesecker LG, Spinner NB (2013) A genomic view of mosaicism and human disease. Nat Rev Genet 14:307–320. https://doi.org/10.1038/nrg3424

Boyce AM, Collins MT (2019) Fibrous dysplasia/McCune-Albright syndrome: a rare, mosaic disease of Gαs activation. Endocr Rev. https://doi.org/10.1210/endrev/bnz011

Breveglieri G, D'Aversa E, Finotti A, Borgatti M (2019) Non-invasive prenatal testing using fetal DNA. Mol Diagn Ther 23:291–299. https://doi.org/10.1007/s40291-019-00385-2

Brunkow ME et al (2001) Bone dysplasia sclerosteosis results from loss of the SOST gene product, a novel cystine knot-containing protein. Am J Hum Genet 68:577–589

Caparros-Martin JA et al (2017) Molecular spectrum and differential diagnosis in patients referred with sporadic or autosomal recessive osteogenesis imperfecta. Mol Genet Genomic Med 5:28–39. https://doi.org/10.1002/mgg3.257

Carpenter TO, Shaw NJ, Portale AA, Ward LM, Abrams SA, Pettifor JM (2017) Rickets. Nat Rev Dis Primers 3:17101. https://doi.org/10.1038/nrdp.2017.101

Chesi A et al (2019) Genome-scale capture C promoter interactions implicate effector genes at GWAS loci for bone mineral density. Nat Commun 10:1260. https://doi.org/10.1038/s41467-019-09302-x

Cho TJ et al (2012) A single recurrent mutation in the 5′-UTR of IFITM5 causes osteogenesis imperfecta type V. Am J Hum Genet 91:343–348. https://doi.org/10.1016/j.ajhg.2012.06.005

Cohen AS, Wilson SL, Trinh J, Ye XC (2015) Detecting somatic mosaicism: considerations and clinical implications. Clin Genet 87:554–562. https://doi.org/10.1111/cge.12502

Collins M (2018) Burosumab: at long last, an effective treatment for FGF23-associated hypophosphatemia. J Bone Miner Res 33:1381–1382. https://doi.org/10.1002/jbmr.3544

Collins MT, Singer FR, Eugster E (2012) McCune-Albright syndrome and the extraskeletal manifestations of fibrous dysplasia. Orphanet J Rare Dis 7(Suppl 1):S4. https://doi.org/10.1186/1750-1172-7-S1-S4

ADHR Consortium (2000) Autosomal dominant hypophosphataemic rickets is associated with mutations in FGF23. Nat Genet 26:345–348. https://doi.org/10.1038/81664

Cosman F et al (2016) Romosozumab treatment in postmenopausal women with Osteoporosis. N Engl J Med 375:1532–1543. https://doi.org/10.1056/NEJMoa1607948

Cummings SR et al (2009) Denosumab for prevention of fractures in postmenopausal women with osteoporosis. N Engl J Med 361:756–765. https://doi.org/10.1056/NEJMoa0809493

Dave BJ, Sanger WG (2007) Role of cytogenetics and molecular cytogenetics in the diagnosis of genetic imbalances. Semin Pediatr Neurol 14:2–6. https://doi.org/10.1016/j.spen.2006.11.003
Devuyst O, Thakker RV (2010) Dent's disease. Orphanet J Rare Dis 5:28. https://doi.org/10.1186/1750-1172-5-28
Dixon PH et al (1998) Mutational analysis of PHEX gene in X-linked hypophosphatemia. J Clin Endocrinol Metab 83:3615–3623. https://doi.org/10.1210/jcem.83.10.5180
Estrada K et al (2012) Genome-wide meta-analysis identifies 56 bone mineral density loci and reveals 14 loci associated with risk of fracture. Nat Genet 44:491–501. https://doi.org/10.1038/ng.2249
Faienza MF, Ventura A, Colucci S, Cavallo L, Grano M, Brunetti G (2016) Bone fragility in turner syndrome: mechanisms and prevention strategies. Front Endocrinol 7:34. https://doi.org/10.3389/fendo.2016.00034
Falardeau F, Camurri MV, Campeau PM (2017) Genomic approaches to diagnose rare bone disorders. Bone 102:5–14. https://doi.org/10.1016/j.bone.2016.07.020
Feng JQ et al (2006) Loss of DMP1 causes rickets and osteomalacia and identifies a role for osteocytes in mineral metabolism. Nat Genet 38:1310–1315. https://doi.org/10.1038/ng1905
Ferlin A, Schipilliti M, Di Mambro A, Vinanzi C, Foresta C (2010) Osteoporosis in Klinefelter's syndrome. Mol Hum Reprod 16:402–410. https://doi.org/10.1093/molehr/gaq026
Ferrari SL (2018) Osteoporosis: Romosozumab to rebuild the foundations of bone strength. Nat Rev Rheumatol 14:128. https://doi.org/10.1038/nrrheum.2018.5
Fijalkowski I et al (2016) A novel domain-specific mutation in a sclerosteosis patient suggests a role of LRP4 as an anchor for sclerostin in human bone. J Bone Miner Res 31:874–881. https://doi.org/10.1002/jbmr.2782
Gandomi SK, Parra M, Reeves D, Yap V, Gau CL (2015) Array-CGH is an effective first-tier diagnostic test for EFTUD2-associated congenital mandibulofacial dysostosis with microcephaly. Clin Genet 87:80–84. https://doi.org/10.1111/cge.12328
Gijsbers AC, Ruivenkamp CA (2011) Molecular karyotyping: from microscope to SNP arrays. Horm Res Paediatr 76:208–213. https://doi.org/10.1159/000330406
GnomAD (2019) Genome aggregation database. https://gnomad.broadinstitute.org. Accessed 3 Dec 2019
Goldstein DB, Allen A, Keebler J, Margulies EH, Petrou S, Petrovski S, Sunyaev S (2013) Sequencing studies in human genetics: design and interpretation. Nat Rev Genet 14:460–470. https://doi.org/10.1038/nrg3455
Grigelioniene G et al (2019) Gain-of-function mutation of microRNA-140 in human skeletal dysplasia. Nat Med 25:583–590. https://doi.org/10.1038/s41591-019-0353-2
Hannan FM et al (2012) Identification of 70 calcium-sensing receptor mutations in hyper- and hypo-calcaemic patients: evidence for clustering of extracellular domain mutations at calcium-binding sites. Hum Mol Genet 21:2768–2778. https://doi.org/10.1093/hmg/dds105
Hannan FM, Kallay E, Chang W, Brandi ML, Thakker RV (2018) The calcium-sensing receptor in physiology and in calcitropic and noncalcitropic diseases. Nat Rev Endocrinol 15:33–51. https://doi.org/10.1038/s41574-018-0115-0
Hannan FM, Newey PJ, Whyte MP, Thakker RV (2019) Genetic approaches to metabolic bone diseases. Br J Clin Pharmacol 85:1147–1160. https://doi.org/10.1111/bcp.13803
Imel EA et al (2019) Burosumab versus conventional therapy in children with X-linked hypophosphataemia: a randomised, active-controlled, open-label, phase 3 trial. Lancet 393:2416–2427. https://doi.org/10.1016/S0140-6736(19)30654-3
Johnston JJ, Biesecker LG (2013) Databases of genomic variation and phenotypes: existing resources and future needs. Hum Mol Genet 22:R27–R31. https://doi.org/10.1093/hmg/ddt384
Kang H et al (2018) Somatic activating mutations in MAP 2K1 cause melorheostosis. Nat Commun 9:1390. https://doi.org/10.1038/s41467-018-03720-z
Karasik D, Rivadeneira F, Johnson ML (2016) The genetics of bone mass and susceptibility to bone diseases. Nat Rev Rheumatol 12:323–334. https://doi.org/10.1038/nrrheum.2016.48

Kemp JP et al (2017) Identification of 153 new loci associated with heel bone mineral density and functional involvement of GPC6 in osteoporosis. Nat Genet 49:1468–1475. https://doi.org/10.1038/ng.3949

Kharbanda M, Tolmie J, Joss S (2015) How to use... microarray comparative genomic hybridisation to investigate developmental disorders. Arch Dis Child Educ Pract Ed 100:24–29. https://doi.org/10.1136/archdischild-2014-306022

Kiper POS et al (2016) Cortical-bone fragility--insights from sFRP4 deficiency in Pyle's disease. N Engl J Med 374:2553–2562. https://doi.org/10.1056/NEJMoa1509342

Koczkowska M et al (2017) Genomic findings in patients with clinical suspicion of 22q11.2 deletion syndrome. J Appl Genet 58:93–98. https://doi.org/10.1007/s13353-016-0366-1

Krishnan V, Bryant HU, Macdougald OA (2006) Regulation of bone mass by Wnt signaling. J Clin Investig 116:1202–1209. https://doi.org/10.1172/JCI28551

Kruglyak L (2008) The road to genome-wide association studies. Nat Rev Genet 9:314–318. https://doi.org/10.1038/nrg2316

Laine CM et al (2013) WNT1 mutations in early-onset osteoporosis and osteogenesis imperfecta. N Engl J Med 368:1809–1816. https://doi.org/10.1056/NEJMoa1215458

Lazarus S, Zankl A, Duncan EL (2014) Next-generation sequencing: a frameshift in skeletal dysplasia gene discovery. Osteoporos Int 25:407–422. https://doi.org/10.1007/s00198-013-2443-1

Lee SH et al (2016) Prediction of future osteoporotic fracture occurrence by genetic profiling: a 6-year follow-up observational study. J Clin Endocrinol Metab 101:1215–1224. https://doi.org/10.1210/jc.2015-3972

Lek M et al (2016) Analysis of protein-coding genetic variation in 60,706 humans. Nature 536:285–291. https://doi.org/10.1038/nature19057

Lemos MC, Thakker RV (2015) GNAS mutations in Pseudohypoparathyroidism type 1a and related disorders. Hum Mutat 36:11–19. https://doi.org/10.1002/humu.22696

Levy-Litan V et al (2010) Autosomal-recessive hypophosphatemic rickets is associated with an inactivation mutation in the ENPP1 gene. Am J Hum Genet 86:273–278. https://doi.org/10.1016/j.ajhg.2010.01.010

Lindhurst MJ et al (2011) A mosaic activating mutation in AKT1 associated with the Proteus syndrome. N Engl J Med 365:611–619. https://doi.org/10.1056/NEJMoa1104017

Little RD et al (2002) A mutation in the LDL receptor-related protein 5 gene results in the autosomal dominant high-bone-mass trait. Am J Hum Genet 70:11–19. https://doi.org/10.1086/338450

Lorenz-Depiereux B et al (2006) Hereditary hypophosphatemic rickets with hypercalciuria is caused by mutations in the sodium-phosphate cotransporter gene SLC34A3. Am J Hum Genet 78:193–201. https://doi.org/10.1086/499410

Manrai AK et al (2016) Genetic misdiagnoses and the potential for health disparities. N Engl J Med 375:655–665. https://doi.org/10.1056/NEJMsa1507092

Mao B et al (2002) Kremen proteins are Dickkopf receptors that regulate Wnt/beta-catenin signalling. Nature 417:664–667. https://doi.org/10.1038/nature756

Marini JC et al (2017) Osteogenesis imperfecta. Nat Rev Dis Primers 3:17052. https://doi.org/10.1038/nrdp.2017.52

Mitchell JA et al (2016) Rare EN1 variants and pediatric bone mass. J Bone Miner Res 31:1513–1517. https://doi.org/10.1002/jbmr.2833

Morello R et al (2006) CRTAP is required for prolyl 3- hydroxylation and mutations cause recessive osteogenesis imperfecta. Cell 127:291–304. https://doi.org/10.1016/j.cell.2006.08.039

Morris JA et al (2019) An atlas of genetic influences on osteoporosis in humans and mice. Nat Genet 51:258–266. https://doi.org/10.1038/s41588-018-0302-x

Mortier GR et al (2019) Nosology and classification of genetic skeletal disorders: 2019 revision. Am J Med Genet A 179:2393–2419. https://doi.org/10.1002/ajmg.a.61366

Moverare-Skrtic S et al (2014) Osteoblast-derived WNT16 represses osteoclastogenesis and prevents cortical bone fragility fractures. Nat Med 20:1279–1288. https://doi.org/10.1038/nm.3654

Nesbit MA et al (2013a) Mutations affecting G-protein subunit alpha11 in hypercalcemia and hypocalcemia. N Engl J Med 368:2476–2486. https://doi.org/10.1056/NEJMoa1300253
Nesbit MA et al (2013b) Mutations in AP2S1 cause familial hypocalciuric hypercalcemia type 3. Nat Genet 45:93–97. https://doi.org/10.1038/ng.2492
Newey PJ, Gorvin CM, Whyte MP, Thakker RV (2018) Chapter 1 – introduction to genetics of skeletal and mineral metabolic diseases. In: Thakker RV, Whyte MP, Eisman JA, Igarashi T (eds) Genetics of bone biology and skeletal disease, 2nd edn. Academic Press, Cambridge, pp 1–21. https://doi.org/10.1016/B978-0-12-804182-6.00001-0
Pearce SH et al (1996) A familial syndrome of hypocalcemia with hypercalciuria due to mutations in the calcium-sensing receptor. N Engl J Med 335:1115–1122
Pekkinen M et al (2019) Osteoporosis and skeletal dysplasia caused by pathogenic variants in SGMS2. JCI Insight 4:e126180. https://doi.org/10.1172/jci.insight.126180
Pignolo RJ, Shore EM, Kaplan FS (2011) Fibrodysplasia ossificans progressiva: clinical and genetic aspects. Orphanet J Rare Dis 6:80. https://doi.org/10.1186/1750-1172-6-80
Polla DL et al (2015) Use of targeted exome sequencing for molecular diagnosis of skeletal disorders. PLoS One 10:e0138314. https://doi.org/10.1371/journal.pone.0138314
Pyott SM, Pepin MG, Schwarze U, Yang K, Smith G, Byers PH (2011) Recurrence of perinatal lethal osteogenesis imperfecta in sibships: parsing the risk between parental mosaicism for dominant mutations and autosomal recessive inheritance. Genet Med 13:125–130. https://doi.org/10.1097/GIM.0b013e318202e0f6
Pyott SM et al (2013) WNT1 mutations in families affected by moderately severe and progressive recessive osteogenesis imperfecta. Am J Hum Genet 92:590–597. https://doi.org/10.1016/j.ajhg.2013.02.009
Ralston SH (2008) Juvenile Paget's disease, familial expansile osteolysis and other genetic osteolytic disorders. Best Pract Res Clin Rheumatol 22:101–111. https://doi.org/10.1016/j.berh.2007.11.005
Rehm HL (2013) Disease-targeted sequencing: a cornerstone in the clinic. Nat Rev Genet 14:295–300. https://doi.org/10.1038/nrg3463
Richards JB, Zheng HF, Spector TD (2012) Genetics of osteoporosis from genome-wide association studies: advances and challenges. Nat Rev Genet 13:576–588. https://doi.org/10.1038/nrg3228
Richards S et al (2015) Standards and guidelines for the interpretation of sequence variants: a joint consensus recommendation of the American College of Medical Genetics and Genomics and the Association for Molecular Pathology. Genet Med 17:405–424. https://doi.org/10.1038/gim.2015.30
Rigter T, van Aart CJ, Elting MW, Waisfisz Q, Cornel MC, Henneman L (2014) Informed consent for exome sequencing in diagnostics: exploring first experiences and views of professionals and patients. Clin Genet 85:417–422. https://doi.org/10.1111/cge.12299
Rivadeneira F, Makitie O (2016) Osteoporosis and bone mass disorders: from gene pathways to treatments. Trends Endocrinol Metab 27:262–281. https://doi.org/10.1016/j.tem.2016.03.006
Roizen JD et al (2018) CYP3A4 mutation causes vitamin D-dependent rickets type 3. J Clin Investig 128:1913–1918. https://doi.org/10.1172/JCI98680
Schwarz JM, Rodelsperger C, Schuelke M, Seelow D (2010) MutationTaster evaluates disease-causing potential of sequence alterations. Nat Methods 7:575–576. https://doi.org/10.1038/nmeth0810-575
Semler O et al (2012) A mutation in the 5'-UTR of IFITM5 creates an in-frame start codon and causes autosomal-dominant osteogenesis imperfecta type V with hyperplastic callus. Am J Hum Genet 91:349–357. https://doi.org/10.1016/j.ajhg.2012.06.011
Shendure J, Balasubramanian S, Church GM, Gilbert W, Rogers J, Schloss JA, Waterston RH (2017) DNA sequencing at 40: past, present and future. Nature 550:345–353. https://doi.org/10.1038/nature24286
Stechman MJ, Loh NY, Thakker RV (2009) Genetic causes of hypercalciuric nephrolithiasis. Pediatr Nephrol 24:2321–2332. https://doi.org/10.1007/s00467-008-0807-0

Styrkarsdottir U et al (2013) Nonsense mutation in the LGR4 gene is associated with several human diseases and other traits. Nature 497:517–520. https://doi.org/10.1038/nature12124

Tafaj O, Juppner H (2017) Pseudohypoparathyroidism: one gene, several syndromes. J Endocrinol Invest 40:347–356. https://doi.org/10.1007/s40618-016-0588-4

Tengan CH, Kiyomoto BH, Rocha MS, Tavares VL, Gabbai AA, Moraes CT (1998) Mitochondrial encephalomyopathy and hypoparathyroidism associated with a duplication and a deletion of mitochondrial deoxyribonucleic acid. J Clin Endocrinol Metab 83:125–129. https://doi.org/10.1210/jcem.83.1.4497

Tenorio J et al (2017) Molecular and clinical analysis of ALPL in a cohort of patients with suspicion of Hypophosphatasia. Am J Med Genet A 173:601–610. https://doi.org/10.1002/ajmg.a.37991

Thakker RV et al (2012) Clinical practice guidelines for multiple endocrine neoplasia type 1 (MEN1). J Clin Endocrinol Metab 97:2990–3011. https://doi.org/10.1210/jc.2012-1230

Toledo RA et al (2017) Consensus statement on next-generation-sequencing-based diagnostic testing of hereditary phaeochromocytomas and paragangliomas. Nat Rev Endocrinol 13:233–247. https://doi.org/10.1038/nrendo.2016.185

van den Veyver IB, Eng CM (2015) Genome-wide sequencing for prenatal detection of Fetal single-gene disorders. Cold Spring Harb Perspect Med 5:a023077. https://doi.org/10.1101/cshperspect.a023077

van Dijk FS et al (2013) PLS3 mutations in X-linked osteoporosis with fractures. N Engl J Med 369:1529–1536. https://doi.org/10.1056/NEJMoa1308223

van Lierop AH, Hamdy NA, van Egmond ME, Bakker E, Dikkers FG, Papapoulos SE (2013) Van Buchem disease: clinical, biochemical, and densitometric features of patients and disease carriers. J Bone Miner Res 28:848–854. https://doi.org/10.1002/jbmr.1794

Vaser R, Adusumalli S, Leng SN, Sikic M, Ng PC (2016) SIFT missense predictions for genomes. Nat Protoc 11:1–9. https://doi.org/10.1038/nprot.2015.123

Vermeesch JR, Voet T, Devriendt K (2016) Prenatal and pre-implantation genetic diagnosis. Nat Rev Genet 17:643–656. https://doi.org/10.1038/nrg.2016.97

Ward LD, Kellis M (2012) Interpreting noncoding genetic variation in complex traits and human disease. Nat Biotechnol 30:1095–1106. https://doi.org/10.1038/nbt.2422

Watanabe S et al (2002) Association between activating mutations of calcium-sensing receptor and Bartter's syndrome. Lancet 360:692–694

Whyte MP (2016) Hypophosphatasia – aetiology, nosology, pathogenesis, diagnosis and treatment. Nat Rev Endocrinol 12:233–246. https://doi.org/10.1038/nrendo.2016.14

Whyte MP (2018a) Chapter 26 – Mendelian disorders of RANKL/OPG/RANK/NF-κB signaling. In: Thakker RV, Whyte MP, Eisman JA, Igarashi T (eds) Genetics of bone biology and skeletal disease, 2nd edn. Academic Press, Cambridge, pp 453–468. https://doi.org/10.1016/B978-0-12-804182-6.00026-5

Whyte MP (2018b) Chapter 28 – hypophosphatasia and how alkaline phosphatase promotes mineralization. In: Thakker RV, Whyte MP, Eisman JA, Igarashi T (eds) Genetics of bone biology and skeletal disease, 2nd edn. Academic Press, Cambridge, pp 481–505. https://doi.org/10.1016/B978-0-12-804182-6.00028-9

Whyte MP, Obrecht SE, Finnegan PM, Jones JL, Podgornik MN, McAlister WH, Mumm S (2002) Osteoprotegerin deficiency and juvenile Paget's disease. N Engl J Med 347:175–184. https://doi.org/10.1056/NEJMoa013096

Whyte MP et al (2018) Sclerosteosis: report of type 1 or 2 in three Indian Tamil families and literature review. Bone 116:321–332. https://doi.org/10.1016/j.bone.2018.07.022

Whyte MP et al (2019) New explanation for autosomal dominant high bone mass: mutation of low-density lipoprotein receptor-related protein 6. Bone 127:228–243. https://doi.org/10.1016/j.bone.2019.05.003

Wilichowski E et al (1997) Hypoparathyroidism and deafness associated with pleioplasmic large scale rearrangements of the mitochondrial DNA: a clinical and molecular genetic study of four children with Kearns-Sayre syndrome. Pediatr Res 41:193–200. https://doi.org/10.1203/00006450-199704001-01165

Yang G et al (2018) SIRT1/HERC4 locus associated with bisphosphonate-induced osteonecrosis of the jaw: an exome-wide association analysis. J Bone Miner Res 33:91–98. https://doi.org/10.1002/jbmr.3285

Zhang J et al (2019) Non-invasive prenatal sequencing for multiple Mendelian monogenic disorders using circulating cell-free fetal DNA. Nat Med 25:439–447. https://doi.org/10.1038/s41591-018-0334-x

Zheng HF et al (2012) WNT16 influences bone mineral density, cortical bone thickness, bone strength, and osteoporotic fracture risk. PLoS Genet 8:e1002745. https://doi.org/10.1371/journal.pgen.1002745

Zheng HF et al (2015) Whole-genome sequencing identifies EN1 as a determinant of bone density and fracture. Nature 526:112–117. https://doi.org/10.1038/nature14878

Pathogenesis of Osteoporosis

Ahmed Al Saedi, Nicole Stupka, and Gustavo Duque

Contents

Abstract

Osteoporosis is a condition where bone resorption exceeds bone formation leading to degeneration. With an aging population, the prevalence of osteoporosis is on the rise. Although advances in the field have made progress in targeting the mechanisms of the disease, the efficacy of current treatments remains limited and is complicated by unexpected side effects. Therefore, to overcome this treatment gap, new approaches are needed to identify and elucidate the cellular mechanisms mediating the pathogenesis of osteoporosis, which requires a strong understanding of bone biology. This chapter will focus on bone cells (osteoclasts, osteoblasts, and osteocytes) and their role in the bone turnover process in normal physiology and in pathology. With regard to osteoclast function, the regulators

A. Al Saedi · N. Stupka · G. Duque (✉)
Australian Institute for Musculoskeletal Science (AIMSS), The University of Melbourne and Western Health, St. Albans, VIC, Australia

Department of Medicine, Melbourne Medical School – Western Precinct, The University of Melbourne, St. Albans, VIC, Australia
e-mail: gustavo.duque@unimelb.edu.au

P. H. Stern (ed.), *Bone Regulators and Osteoporosis Therapy*,
Handbook of Experimental Pharmacology 262, https://doi.org/10.1007/164_2020_358

and underpinning signaling pathways leading to bone resorption will be discussed. Decreased osteoblastogenesis also contributes to bone deterioration with aging and osteoporosis; hence the factors and signaling pathways mediating osteoblast formation and function will be examined. Osteocytes are mature osteoblasts embedded in bone matrix and act as endocrine cells; their role in bone health and pathology will also be reviewed. In addition, this chapter will explore the emerging role of adipocytes in bone biology and the implications of increased bone marrow fat infiltration with aging on bone degeneration. In conclusion, a greater understanding of the pathogenesis of osteoporosis is of utmost importance in order to develop more effective treatments for osteoporosis and other bone diseases.

Keywords

Bone remodeling · Marrow fat · Osteoblasts · Osteoclasts · Osteocytes

1 Introduction to Bone Turnover: Cells and Regulators

Bone remodeling is a lifelong process where mature bone tissue is resorbed from the skeleton and new bone tissue is formed. In healthy adults, the bone remodeling process is important for adapting to changes in load-bearing (e.g., increased weight-bearing activity or longer term mobilization) (Jaworski et al. 1980) and repairing the microdamage (Burr et al. 1998). Key factors controlling the rate of bone remodelling are systemic hormones (including parathyroid hormone (PTH), 1,25-dihydroxyvitamin D3 (1,25-D3), sex steroids) and changes in mechanical loading (Hadjidakis and Androulakis 2006).

Various triggers can induce the signaling pathways mediating bone remodeling. Critically important is the cross talk between osteoblasts, osteoclasts, and osteocytes (Fig. 1). Mature osteoclasts are multinucleated bone resorptive cells, originating from myeloid hematopoietic progenitor cells, in contrast to osteoblasts which originate from mesenchymal stem cells. When triggered, both osteoblasts and osteocytes can synthesize and secrete receptor activator of nuclear factor kappa-B ligand (RANKL), a member of the tumor necrosis factor (TNF) ligand superfamily, which then binds to RANKL receptors on osteoclasts and preosteoclasts. This initial step leads the maturation of preosteoclasts to mature osteoclasts and to osteoclast activation to begin the bone resorption process. Once osteoclasts are firmly attached to the bone surface, they begin to remove bone packets resulting in the creation of bone structural units. Attached osteoclasts secrete proteinases (e.g., cathepsin K) to degrade organic bone matrix and acids (e.g., hydrochloric acid) to dissolve bone minerals. Once bone resorption is complete, a monolayer of osteoblasts will cover the newly resorbed surface and deposit a new thin layer of bone collagen matrix, called the cement line, which in turn will be remodeled to mineralized bone. Some osteoblasts will infiltrate this new bone matrix and mature into

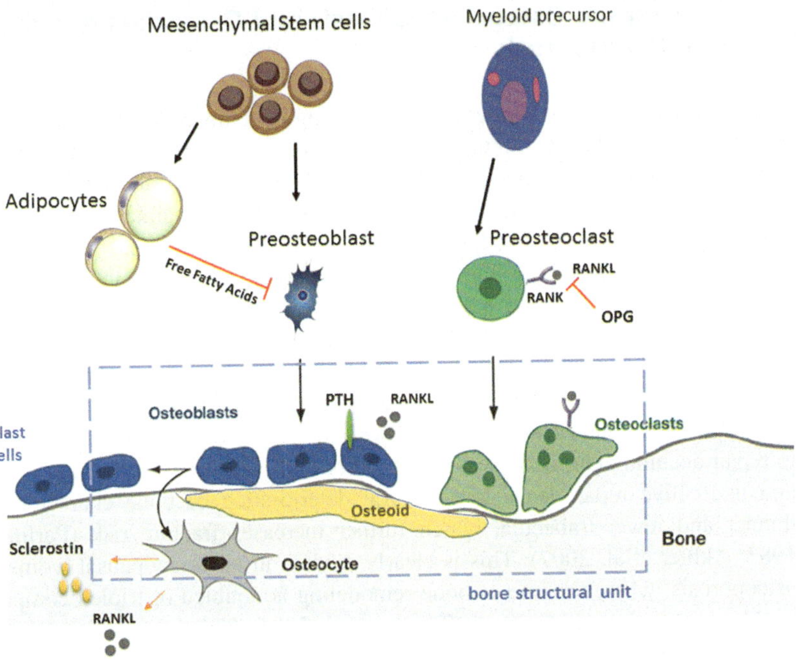

Fig. 1 Bone remodeling cycle. Pre-osteoblasts differentiate from mesenchymal stem cells in the bone marrow, then acquiring the phenotype of osteoblasts. As the osteoid becomes mineralized by osteoblasts, these cells become enclosed in lacunae as osteocytes. Osteoclasts are giant multinucleate cells that differentiate from hematopoietic cells of the monocytes/macrophage lineage in the bone marrow. Bone remodeling starts with phase quiescence then preosteoclast recruitment and osteoclast differentiation, followed by bone resorption and pre-osteoblast recruitment and osteoblast differentiation, and finally finishes with bone formation. Abbreviations are as follows: parathyroid hormone (PTH), osteoprotegerin (OPG), and receptor activator of nuclear factor kappa-B ligand (RANKL)

osteocytes. Within mineralized bone, embedded osteocytes are connected to each other through canaliculi. These are microscopic cytoplasmic canals between the lacunae of osteocytes and ossified bone (Rochefort et al. 2010; Milovanovic and Busse 2019).

Osteocyte connections through these cytoplasmic canals are extremely important in regulating osteoclast and osteoblast function and recognizing changes in mechanical loading and activity. This in turn allows for the tight regulation of bone reabsorption and bone formation; any imbalances in these processes will have negative impacts on bone balance. Negative bone balance is very evident in postmenopausal women where bone resorption exceeds bone formation (Rochefort and Benhamou 2013).

2 Increased Osteoclastogenesis and Osteoclast Activity in Osteoporosis

The remodeling process is essential for repairing bone microdamage and preserving the mechanical integrity of the skeleton; however, it causes transient weakness in the locus of resorptive surface. Bone remodeling is also critically important for maintaining plasma calcium homeostasis. An inability to maintain plasma calcium homeostasis, as is seen in chronic renal disease (Hruska et al. 2017), leads to excessive bone remodeling and compromised bone strength. If there is excessive bone remodeling, trabecular bone is more vulnerable to degeneration compared to cortical bone (Yeh et al. 2017). Where the trabecular bone surface is eradicated due to persistent, excessive remodeling, bone formation is severely compromised at that location. Furthermore, excessive activation of resorption increases the number of bone structural units leading to weakness, more microdamage, and an inability to repair accumulating damage. In osteoporosis, the accumulation of microdamage and ineffective repair leads to structural deterioration of bone characterized by thinner and fewer trabecula, which further increases fracture risk (Parfitt et al. 1983; Akhter et al. 2007). This is clearly evident in postmenopausal women with osteoporosis, where the rate of bone remodeling is doubled or tripled compared to healthy, pre-menopausal women (Recker et al. 2004). This increase in pathological bone remodeling can be assessed noninvasively by bone resorption markers, such as N-terminal telopeptide (NTX) and C-telopeptide of type 1 collagen (CTX). These markers predict fracture risk better than bone mineral density (BMD) (Riggs et al. 1996). Despite numerous population studies showing a correlation between increased remodeling rates, as indicated by NTX and CTX, and fracture risk (Garnero et al. 1996; Milovanovic and Busse 2019), the application of this knowledge to individual patients in the clinic is very limited. There is still too much reliance on bone mineral density to diagnose and manage osteoporosis.

Osteoclasts are the main cells driving the resorption process, which is initiated by the activation of RANK by its ligand (RANKL). When RANKL is bound to RANK on pre-osteoclast and osteoclast, the expression of osteoclast-specific genes is initiated. This drives differentiation, cell survival, migration, and attachment of osteoclasts. RANK signaling is antagonized by osteoprotegrin (OPG), also known as osteoclastogenesis inhibitory factor (OCIF), which as soluble receptor. By binding to RANKL, OPG prevents ligand-receptor interactions (Boyce and Xing 2008) and is critically important in the inhibition of osteoclastogenesis and bone resorption (Simonet et al. 1997). OPG is largely expressed and secreted by osteoblast lineage cells of bone and other cell types (found in the blood vessels, lungs, breast, and skin).

RANK signaling, downstream of ligand binding to receptor, is complex and begins with Tumor necrosis factor receptor (TNFR)-associated factor 6 (TRAF6), which acts as a key connector regulating osteoclastogenic signal transduction pathways (Tan et al. 2017; Park et al. 2017b) (Fig. 2). Highlighting the importance of TRAF6 to osteoclastogenesis, TRAF6 mutations in mice lead to decreased osteoclast activity and result in osteopetrosis (Teti and Econs 2017), a condition which makes bones abnormally dense and prone to breakage (Sobacchi et al. 2013).

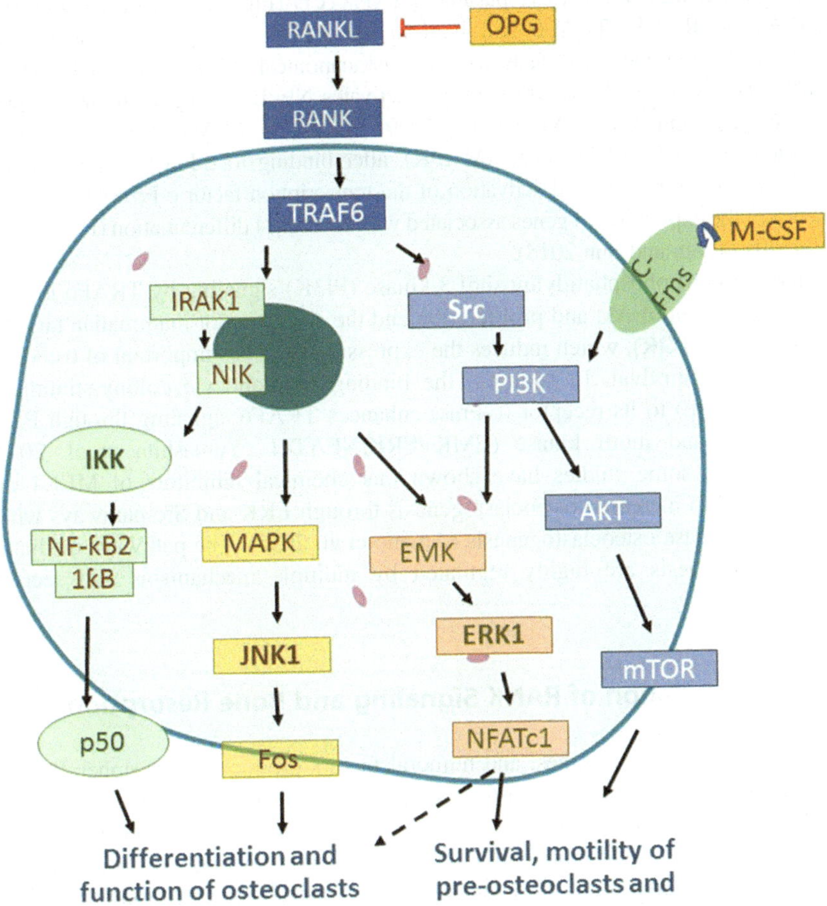

Fig. 2 Osteoclast signaling pathway. Abbreviations are as follows: osteoprotegerin (OPG), receptor activator of nuclear factor kappa-B ligand (RANKL), inhibitor of NF-κB kinase (IKK), c-Jun N-terminal kinase (JNK), extracellular signal-regulated kinase (ERK), proto-oncogene tyrosine-protein kinase (Src), tumor necrosis factor receptor (TNFR), mitogen-activated protein kinase (MAPK), phosphatidylinositol 3-kinase (PI3K), the mammalian target of rapamycin (mTOR), monocyte colony-stimulating factor (M-CSF), interleukin-1 receptor-associated kinase 1 (IRAK1), Tumor necrosis factor receptor (TNFR)-associated factor 6 (TRAF6), ELKL motif kinase (EMK), mitogen-activated protein kinase (MKK)

Many distinct signaling cascades are induced downstream of TRAF6 following RANKL and RANK binding, which are important for the proliferation and survival of preosteoclasts and for osteoclast differentiation and function. The four major signaling cascades include the protein kinase inhibitor of IκB kinase (IKK), c-Jun N-terminal kinase (JNK), extracellular signal-regulated kinase (ERK), and proto-oncogene tyrosine-protein kinase (Src) (Cappariello et al. 2014) (Fig. 2). Osteoclast differentiation is driven by several transcription factors, including p50, Fos, and Nuclear factor of activated T-cells, cytoplasmic 1 (NFATc1) (Fig. 2). Following binding of RANKL to RANK, TRAF6 will bind to interleukin-1 receptor-associated kinase 1 (IRAK1) and then associate with the noncanonical NF-κB signaling pathway (NIK). This activates IKK, which in turn activates NFκB leading to the transcription of osteoclastogenic genes (Wesche et al. 1999). In addition, TRAF6 also signals through mitogen-activated protein kinase (MAPK), after binding to c-Jun N-terminal kinase (JNK) resulting in increased activation of the transcription factor c-Fos, which in turn increases the expression of genes associated with osteoclast differentiation (Fig. 2) (Park et al. 2017a; Shi and Sun 2018).

Induction of phosphatidylinositol 3-kinase (PI3K) signaling by TRAF6 leads to phosphorylation of Akt and proliferation and the activation of mammalian target of rapamycin (mTOR), which induces the expression of genes important of osteoclast motility and survival. In addition, the binding of monocyte colony-stimulating factor (M-CSF) to its receptor (c-Fms) enhances TRAF6 signaling through PI3K/Akt/mTOR and motif kinase (EMK)/ERK/NFATc1 (Yamashita et al. 2012). Furthermore, some studies have shown that chemical inhibitors of MEK1 and mTOR lead to increased osteoclastogenesis through ERK and Src pathways which can also increase osteoclastogenesis (Amano et al. 2015). The pathways mediating osteoclastogenesis are highly regulated by multiple mechanisms and secreted factors.

3 Regulation of RANK Signaling and Bone Resorption

Various hormones, cytokines, and humoral factors produced by peripheral organs and tissues can also influence osteoclastogenesis, and these potentiate or intersect with RANKL signaling pathways. For example, estrogen deficiency postmenopause results in imbalanced remodeling with a significant increase in bone turnover leading to a progressive loss of trabecular and cortical bone (Khosla et al. 2012). With the loss of estrogen postmenopause, there is increased inflammation. Proinflammatory cytokines, such as interleukin-1 (IL-1) and TNFα, are negatively regulated by estrogen, and postmenopause these increase rapidly and lead to the enhanced formation of functional osteoclasts (Pfeilschifter et al. 2002; Shulman 2009). Additional age-related changes driving bone loss include increased inflammation, secondary hyperparathyroidism, decreased testosterone, reduced mechanical loading due to lifestyle related factors, the accumulation of microfractures, changes in the protein content of matrix material, and other factors (Boskey and Imbert 2017).

Inflammation in particular is a powerful modulator of RANK signaling, and this depends on various cell surface receptor and ligand interactions. The best described interactions include those with IL-1, TNF-α, prostaglandin E2 (PGE2), and TGFβ (Li et al. 2019a). The IL-1 receptor and the TNF receptor 1 both signal through TRAF6 and thus have synergistic effects on RANK-mediated TRAF6 activation (Yan et al. 2001). TGFβ regulates the components of the RANK pathway through Smad1 and thus regulating human osteoclast differentiation (Battaglino et al. 2002). However, the effects of inflammatory mediators are complex and may not always stimulate osteoclastogenesis. For example, when interferon-β is secreted by osteoclasts, it will act in an autocrine fashion to downregulate the expression of c-Fos, a critical transcription factor involved in osteoclast development (Takayanagi et al. 2002).

The discovery of RANK signaling cascades in osteoclasts has provided insight into the mechanisms of osteoclastogenesis, the activation of bone resorption, and how bone structure and mass are impacted by hormonal signals. Further investigation of these pathways will provide the molecular insight necessary for the development of therapeutics to treat osteoporosis and other bone disease.

4 Decreased Osteoblastogenesis and Bone Formation in Osteoporosis

In healthy adult bone, bone formation occurs in three stages: the production of an organic, collagen-rich matrix, followed by maturation of this osteoid matrix, and finally matrix mineralization. All three stages occur at the same rate and are regulated by osteoblasts (Hadjidakis and Androulakis 2006). Bone degeneration in osteoporosis can be due to decreased bone formation, a high resorption rate which exceeds tissue capacity for bone formation, and/or an imbalance between organic matrix production and mineralization (Jilka 2003). The latter can be due to a decrease in activity or number of osteoblasts (Manolagas 2000) or a lack of initial minerals (e.g., calcium, magnesium, and phosphate) during new bone formation (Xiao et al. 2016; Tornquist et al. 2019).

Osteoblasts are derived from multipotent mesenchymal stem cells (MSC) (Bianco et al. 2001), which can differentiate not only into osteoblasts but also adipocytes, chondrocytes, myoblasts, and fibroblasts. The differentiation of MSCs into pre-osteoblasts and pre-osteoblasts into osteoblasts is induced by the binding of growth factors, in particular BMP-2, TGFβ, and fibroblast growth factor (FGF3) to their respective receptors. This initiates signaling cascades which culminate in the activation of the osteoblast transcription factors runt-related transcription factor 2 (Runx2), and osterix (Fig. 3). TGFβ superfamily members include not only TGFβ, but also BMPs, and signal via serine/threonine receptor kinases and the phosphorylation of Smad-1, -5, and -8 proteins (Chen et al. 2012). Phosphorylated Smads bind with the common Smad4 in the cytosol and are then translocated to the nucleus, where they induce transcription of osteoblast genes. TGFβ superfamily receptors also signal through mitogen-activated protein kinase (MAPK) pathways. The FGF receptor signals through its tyrosine kinase domains which are activated when receptor-ligand binding induces a conformational change. These tyrosine kinase domains phosphorylate and activate MAPK and protein kinase C (PKC) (Chen et al. 2004; Ebisawa et al. 1999). Both MAPK and PKC are associated with osteoblast differentiation, survival, and motility (Pan et al. 2018; Galea et al. 2014) (Fig. 2).

In mature osteoblasts, FGF (Globus et al. 1989), BMP (Dumic-Cule et al. 2018), and insulin-like growth factor-1 (IGF-1) (Canalis et al. 1993) regulate signaling pathways important for bone formation. Osteoblasts also express receptors for

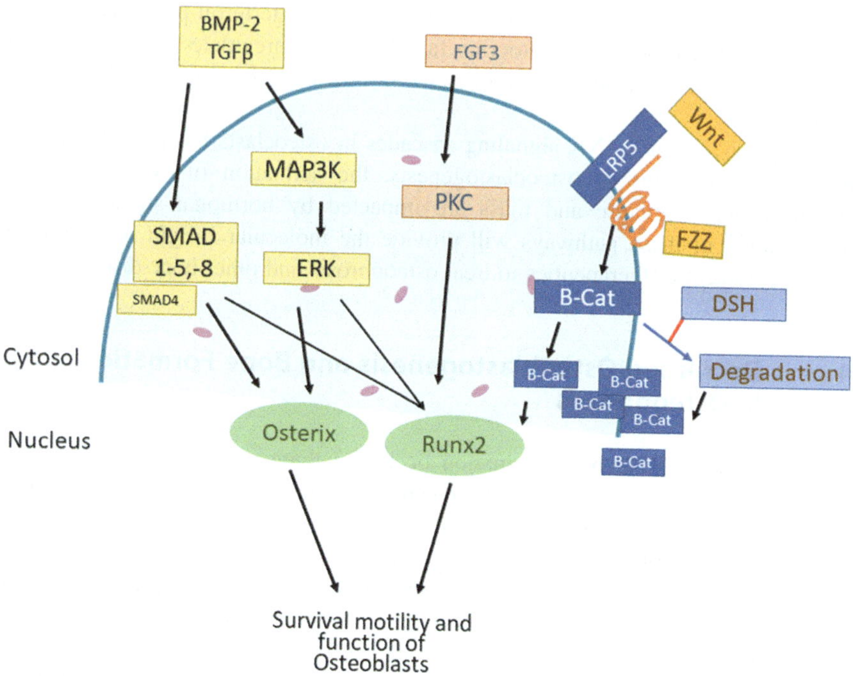

Fig. 3 Osteoblast signaling pathway. Abbreviations are as follows: mesenchymal stem cells (MSC), insulin-like growth factors (IGF), fibroblast growth factor (FGF), bone morphogenetic proteins (BMP), runt-related transcription factor 2 (Runx2), protein kinase C (PKC), frizzled (FZZ), low-density lipoprotein receptor-related protein (LRP), intracellular protein disheveled (DSH), β-catenin (βCat), mitogen-activated protein kinase (MAPK), and extracellular signal-regulated kinases (ERK)

classical hormones, such as PTH, PTH-related protein, thyroid hormone growth hormone, insulin, progesterone, and prolactin, and these all have anabolic effects on bone formation (Zofkova 2018; Pi and Quarles 2013).

Wnt ligands are a family of secreted glycoproteins which also regulate osteoblast expansion and differentiation. Wnt ligands bind to the frizzled (FZZ)/low-density lipoprotein receptor-related protein (LRP) complex, which leads to the activation of the intracellular protein disheveled (DSH). DSH inhibits the activity of a complex protein structure which usually transports β-catenin to the proteasome for degradation. This results in the stabilization and accumulation of β-catenin in the cytosol, which in turn facilitates its translocation to the nucleus. There it binds to the transcription factors T-cell factor/lymphoid enhancer factor (Lef1/Tcf), which then leads to Runx2 transcription (Bodine and Komm 2006; Yavropoulou and Yovos 2007). In osteoblasts, Wnt signaling also regulates noncanonical signaling pathways independent of β-catenin. Osteoblast function and bone formation can be inhibited by natural, extracellular Wnt antagonists, such as dickkopfs and secreted frizzled-related proteins (Westendorf et al. 2004; Pinzone et al. 2009).

Differentiated osteoblasts line the bone surface in specialized regions known as bone structural units, where they participate in bone formation by synthesizing a collagen-rich matrix and regulating its mineralization (Hadjidakis and Androulakis 2006). Following the completion of matrix mineralization, approximately 15% of osteoblasts embedded into the matrix will differentiate into osteocytes; others will become bone lining cells, while the majority will undergo apoptosis (Heino and Hentunen 2008).

Targeting osteoblastogenesis pathways is important for the discovery of new pharmaceutical strategies to promote bone anabolism. This is an important area of research, because most current osteoporosis treatments are anti-resorptive. Pharmacological agents with a dual effect on bone formation and reabsorption will have greater efficacy for the management of osteoporosis.

5 Osteocytes: Bone Endocrine Cells as Emerging Regulators in the Pathogenesis of Osteoporosis

Osteocytes are long-lived cells within the bone matrix, derived from osteoblasts, whose main function is the control of bone remodeling (Jilka and O'Brien 2016). Osteocytes produce many important proteins which modulate bone resorption and formation, the most important of which are RANKL and OPG. RANKL is essential for osteoclast formation and increasing bone resorption (Jilka et al. 2010), while OPG is a soluble receptor that inhibits RANKL action (Bucay et al. 1998). The relative ratios and localization of these proteins determine osteoclast maturation and function. Furthermore, when microdamage occurs, osteocytes undergo apoptosis, and the initiation of remodeling process is stopped due to discontinuation of RANKL secretion (Xiong et al. 2011). This may be relevant to the pathogenesis of osteoporosis where there are high numbers of microfractures due to weakening bones.

Osteocytes also secrete sclerostin, encoded by the *SOST* gene, into the bone microenvironment (Brunkow et al. 2001). Sclerostin binds to Lrp5 and Lrp6 on osteoblasts, and this inhibits Wnt signaling, which is required for osteoblast differentiation and function (Balemans et al. 2002). Inhibiting of sclerostin leads to increased bone formation, even in the absence of bone resorption. Inhibiting sclerostin interactions with Lrp5 and Lrp6 promotes Wnt signaling, which leads to osteoblast commitment and differentiation and increased bone formation (Ominsky et al. 2014). Osteocytes control the rate of bone resorption and the balance between resorption and formation within the bone structural unit by secreting both RANKL, OPG, and sclerostin. This is critically important given their role in mechanosensing and allowing the skeleton to gain mass in response to increased load or lose mass when the load is reduced. Skeletal adaptation studies have focused on the osteocyte network, because of its large surface area generated by the presence of long cytoplasmic connections that extend from the osteocyte body into the bone matrix and maintain contact among osteocytes and other cells.

Osteocytes are important for bone homeostasis. Osteocytes stimulate the loss of bone mass in the context of unloading (e.g., bed rest or space flight) or pathology by increasing RANKL and sclerostin secretion. This stimulates osteoclastogenesis and inhibits osteoblastogenesis, respectively. Hence, in the context of pathology where bone loss is increased, inhibiting osteocyte function could have positive effects on preserving bone mass. Indeed, osteocyte ablation prevented loss of cancellous bone after hind limb unloading in preclinical studies using rats and mice (Tatsumi et al. 2007; Xiao et al. 2011).

In addition to their role in bone remodeling, osteocytes contribute to mineral homeostasis by liberating mineral from the surface of the lacunae in which they reside (Qing et al. 2012). Osteocytes express both PTH receptors and vitamin D receptors (Saini et al. 2013). PTH stimulates bone resorption to release calcium from this depot while also promoting renal calcium retention and production of 1,25-D3 (Watson and Hanley 1993). PTH also stimulates RANKL and decreases OPG production in osteocytes, as well as in osteoblasts (Ma et al. 2001; Onal et al. 2012).

Osteocytes are the main mechanosensing cells in bone; however, it is not known whether or how osteocyte function is impaired with aging or disease. There is limited evidence suggesting lacunae become smaller and more spherical with aging and osteoporosis (Hemmatian et al. 2017). The functional implications of this are not known. Hence, fundamental studies are needed to determine the role of osteocytes in aging-related bone loss.

6 Marrow Fat and Osteoporosis: Marrow Fat Infiltration and Compromised Bone Remodeling

Bone health and remodeling depend not only on osteoclasts, osteoblasts, and osteocytes; in the past decade, there has been increasing recognition of the importance of bone marrow adipocytes. Research has focused on understanding bone marrow adipocyte origin, biology, and their role in regulating the bone marrow niche (Guerra et al. 2018). The inverse correlation between low bone mass and greater bone marrow adiposity is well established in mice, rats, and humans (Shen et al. 2013), and indeed fat infiltration is a hallmark of osteoporosis (Li et al. 2019b). With age and to an even greater extent in osteoporosis, there is an increase in the differentiation of progenitor MSCs to adipocytes rather than osteoblasts (Bianco et al. 2001; Xiong et al. 2011), and adipocytes end up populating most of the bone marrow milieu in osteoporotic bones (Bermeo et al. 2014). What is not yet known is whether the initial increase bone marrow adiposity is a driver of low bone mass or a consequence. What is clear, is that once established, bone marrow fat infiltration will actively contribute to the loss of bone mass.

Bone marrow adipocytes tend to have a distinct metabolic profile and needs compared to other cells in the bone marrow niche. Rather than employing glucose metabolism for energy, as is done by osteoblasts and osteocytes, bone marrow adipocytes depend on lipolysis of their intracellular lipid store to liberate free fatty acids for oxidative metabolism (Bartelt et al. 2017). However, excess production and release of saturated free fatty acids from adipocytes are very harmful to the bone marrow niche (Gasparrini et al. 2009). Adipocytes can also release adipokines; these also disrupt the balance between bone resorption and formation (Li et al. 2018). Adipocytes secrete various fatty acids and lipid metabolites, with the 17 carbon saturated fatty acid palmitate being the most predominant and highly toxic (Elbaz et al. 2010). Saturated fatty acids, and palmitate in particular, have been shown in a number of animal and human cell culture models to be toxic to osteoblasts, compromising their differentiation, function, and survival (Gunaratnam et al. 2013, 2014; Kim et al. 2008). Similarly, this toxic effect was observed in cultured osteocytes (Al Saedi et al. 2019b). The toxic effects of palmitate on osteoblasts and osteocytes include dysfunctional autophagy and the induction of apoptosis (Al Saedi et al. 2019a). This leads to impaired osteoblastogenesis and a reduced capacity for organic matrix deposition and mineralization (Gunaratnam et al. 2013, 2014; Hocking et al. 2012; Al Saedi et al. 2019a). Elucidating the mechanisms which promote adipocyte over osteoblast differentiation in progenitor MSCs and limit fat infiltration of the bone marrow niche should have anabolic effects on bone formation, which would be a significant advance in the management of osteoporosis.

7 Conclusions

Our understanding of the pathophysiology of osteoporosis is ongoing and evolving and now encompasses defects in the microarchitecture and material properties of bone tissue, imbalance in bone remodeling rates, microdamage, and repair recognition. Elucidating the importance of these factors has increased our understanding of bone quality and the mechanisms that underpin it. In addition, there is growing interest in marrow fat and its direct impact on bone quality, microstructure, and bone cells function and survival, providing a novel insights into mechanisms of age-related bone loss and osteoporosis. Current osteoporosis treatments have unexpected side effects and are of limited efficacy as they predominantly target bone reabsorption. Ideally the treatment goal of osteoporosis should be increased bone formation, rather than only stopping resorption. New treatments for this debilitating condition are urgently required, as new effective drugs for the treatment of osteoporosis will not only reduce the economic burden of this condition for society but also greatly improve the quality of life for the globally increasing population with osteoporosis.

References

Akhter MP, Lappe JM, Davies KM, Recker RR (2007) Transmenopausal changes in the trabecular bone structure. Bone 41(1):111–116. https://doi.org/10.1016/j.bone.2007.03.019

Al Saedi A, Goodman CA, Myers DE, Hayes A, Duque G (2019a) Rapamycin affects palmitate-induced lipotoxicity in osteoblasts by modulating apoptosis and autophagy. J Gerontol A 75(1):58–63. https://doi.org/10.1093/gerona/glz149

Al Saedi A, Bermeo S, Plotkin L, Myers DE, Duque G (2019b) Mechanisms of palmitate-induced lipotoxicity in osteocytes. Bone 127:353–359. https://doi.org/10.1016/j.bone.2019.06.016

Amano S, Chang YT, Fukui Y (2015) ERK5 activation is essential for osteoclast differentiation. PLoS One 10(4):e0125054. https://doi.org/10.1371/journal.pone.0125054

Balemans W, Patel N, Ebeling M, Van Hul E, Wuyts W, Lacza C et al (2002) Identification of a 52 kb deletion downstream of the SOST gene in patients with van Buchem disease. J Med Genet 39(2):91–97. https://doi.org/10.1136/jmg.39.2.91

Bartelt A, Koehne T, Todter K, Reimer R, Muller B, Behler-Janbeck F et al (2017) Quantification of bone fatty acid metabolism and its regulation by adipocyte lipoprotein lipase. Int J Mol Sci. 18(6):pii: E1264. https://doi.org/10.3390/ijms18061264

Battaglino R, Kim D, Fu J, Vaage B, Fu XY, Stashenko P (2002) c-myc is required for osteoclast differentiation. J Bone Miner Res 17(5):763–773. https://doi.org/10.1359/jbmr.2002.17.5.763

Bermeo S, Gunaratnam K, Duque G (2014) Fat and bone interactions. Curr Osteoporos Rep 12(2):235–242. https://doi.org/10.1007/s11914-014-0199-y

Bianco P, Riminucci M, Gronthos S, Robey PG (2001) Bone marrow stromal stem cells: nature, biology, and potential applications. Stem Cells 19(3):180–192. https://doi.org/10.1634/stemcells.19-3-180

Bodine PV, Komm BS (2006) Wnt signaling and osteoblastogenesis. Rev Endocr Metab Disord 7(1–2):33–39. https://doi.org/10.1007/s11154-006-9002-4

Boskey AL, Imbert L (2017) Bone quality changes associated with aging and disease: a review. Ann N Y Acad Sci 1410(1):93–106. https://doi.org/10.1111/nyas.13572

Boyce BF, Xing L (2008) Functions of RANKL/RANK/OPG in bone modeling and remodeling. Arch Biochem Biophys 473(2):139–146. https://doi.org/10.1016/j.abb.2008.03.018

Brunkow ME, Gardner JC, Van Ness J, Paeper BW, Kovacevich BR, Proll S et al (2001) Bone dysplasia sclerosteosis results from loss of the SOST gene product, a novel cystine knot-containing protein. Am J Hum Genet 68(3):577–589. https://doi.org/10.1086/318811

Bucay N, Sarosi I, Dunstan CR, Morony S, Tarpley J, Capparelli C et al (1998) osteoprotegerin-deficient mice develop early onset osteoporosis and arterial calcification. Genes Dev 12(9):1260–1268. https://doi.org/10.1101/gad.12.9.1260

Burr DB, Turner CH, Naick P, Forwood MR, Ambrosius W, Hasan MS, Pidaparti R (1998) Does microdamage accumulation affect the mechanical properties of bone? J Biomech 31(4):337–345. https://doi.org/10.1016/s0021-9290(98)00016-5

Canalis E, Pash J, Gabbitas B, Rydziel S, Varghese S (1993) Growth factors regulate the synthesis of insulin-like growth factor-I in bone cell cultures. Endocrinology 133(1):33–38. https://doi.org/10.1210/endo.133.1.8319580

Cappariello A, Maurizi A, Veeriah V, Teti A (2014) The great beauty of the osteoclast. Arch Biochem Biophys 558:70–78. https://doi.org/10.1016/j.abb.2014.06.017

Chen D, Zhao M, Mundy GR (2004) Bone morphogenetic proteins. Growth Factors 22(4):233–241. https://doi.org/10.1080/08977190412331279890

Chen G, Deng C, Li YP (2012) TGF-beta and BMP signaling in osteoblast differentiation and bone formation. Int J Biol Sci 8(2):272–288. https://doi.org/10.7150/ijbs.2929

Dumic-Cule I, Peric M, Kucko L, Grgurevic L, Pecina M, Vukicevic S (2018) Bone morphogenetic proteins in fracture repair. Int Orthop 42(11):2619–2626. https://doi.org/10.1007/s00264-018-4153-y

Ebisawa T, Tada K, Kitajima I, Tojo K, Sampath TK, Kawabata M et al (1999) Characterization of bone morphogenetic protein-6 signaling pathways in osteoblast differentiation. J Cell Sci 112(Pt 20):3519–3527

Elbaz A, Wu X, Rivas D, Gimble JM, Duque G (2010) Inhibition of fatty acid biosynthesis prevents adipocyte lipotoxicity on human osteoblasts in vitro. J Cell Mol Med 14(4):982–991. https://doi.org/10.1111/j.1582-4934.2009.00751.x

Galea GL, Meakin LB, Williams CM, Hulin-Curtis SL, Lanyon LE, Poole AW, Price JS (2014) Protein kinase Calpha (PKCalpha) regulates bone architecture and osteoblast activity. J Biol Chem 289(37):25509–25522. https://doi.org/10.1074/jbc.M114.580365

Garnero P, Sornay-Rendu E, Chapuy MC, Delmas PD (1996) Increased bone turnover in late postmenopausal women is a major determinant of osteoporosis. J Bone Miner Res 11(3):337–349. https://doi.org/10.1002/jbmr.5650110307

Gasparrini M, Rivas D, Elbaz A, Duque G (2009) Differential expression of cytokines in subcutaneous and marrow fat of aging C57BL/6J mice. Exp Gerontol 44(9):613–618. https://doi.org/10.1016/j.exger.2009.05.009

Globus RK, Plouet J, Gospodarowicz D (1989) Cultured bovine bone cells synthesize basic fibroblast growth factor and store it in their extracellular matrix. Endocrinology 124(3):1539–1547. https://doi.org/10.1210/endo-124-3-1539

Guerra DAP, Paiva AE, Sena IFG, Azevedo PO, Batista ML Jr, Mintz A, Birbrair A (2018) Adipocytes role in the bone marrow niche. Cytometry A 93(2):167–171

Gunaratnam K, Vidal C, Boadle R, Thekkedam C, Duque G (2013) Mechanisms of palmitate-induced cell death in human osteoblasts. Biol Open 2(12):1382–1389

Gunaratnam K, Vidal C, Gimble J, Duque G (2014) Mechanisms of palmitate-induced lipotoxicity in human osteoblasts. Gen Endocrinol 155(1):108–116

Hadjidakis DJ, Androulakis II (2006) Bone remodeling. Ann N Y Acad Sci 1092:385–396. https://doi.org/10.1196/annals.1365.035

Heino TJ, Hentunen TA (2008) Differentiation of osteoblasts and osteocytes from mesenchymal stem cells. Curr Stem Cell Res Ther 3(2):131–145. https://doi.org/10.2174/157488808784223032

Hemmatian H, Bakker AD, Klein-Nulend J, Van Lenthe GH (2017) Aging, osteocytes, and mechanotransduction. Curr Osteoporos Rep 15(5):401–411. https://doi.org/10.1007/s11914-017-0402-z

Hocking L, Whitehouse C, Helfrich M (2012) Autophagy: a new player in skeletal maintenance. J Bone Miner Res 27(7):1439–1447

Hruska KA, Sugatani T, Agapova O, Fang Y (2017) The chronic kidney disease - mineral bone disorder (CKD-MBD): advances in pathophysiology. Bone 100:80–86. https://doi.org/10.1016/j.bone.2017.01.023

Jaworski ZF, Liskova-Kiar M, Uhthoff HK (1980) Effect of long-term immobilisation on the pattern of bone loss in older dogs. J Bone Joint Surg Br 62-b(1):104–110

Jilka RL (2003) Biology of the basic multicellular unit and the pathophysiology of osteoporosis. Med Pediatr Oncol 41(3):182–185. https://doi.org/10.1002/mpo.10334

Jilka RL, O'Brien CA (2016) The role of osteocytes in age-related bone loss. Curr Osteoporos Rep 14(1):16–25. https://doi.org/10.1007/s11914-016-0297-0

Jilka RL, O'Brien CA, Bartell SM, Weinstein RS, Manolagas SC (2010) Continuous elevation of PTH increases the number of osteoblasts via both osteoclast-dependent and -independent mechanisms. J Bone Miner Res 25(11):2427–2437. https://doi.org/10.1002/jbmr.145

Khosla S, Oursler MJ, Monroe DG (2012) Estrogen and the skeleton. Trends Endocrinol Metab 23(11):576–581. https://doi.org/10.1016/j.tem.2012.03.008

Kim J-E, Ahn M-W, Baek S-H, Lee IK, Kim Y-W, Kim J-Y et al (2008) AMPK activator, AICAR, inhibits palmitate-induced apoptosis in osteoblast. Bone 43:394–404

Li Z, Hardij J, Bagchi DP, Scheller EL, MacDougald OA (2018) Development, regulation, metabolism and function of bone marrow adipose tissues. Bone 110:134–140. https://doi.org/10.1016/j.bone.2018.01.008

Li J, Ayoub A, Xiu Y, Yin X, Sanders JO, Mesfin A et al (2019a) TGFbeta-induced degradation of TRAF3 in mesenchymal progenitor cells causes age-related osteoporosis. Nat Commun 10(1):2795

Li Y, Meng Y, Yu X (2019b) The unique metabolic characteristics of bone marrow adipose tissue. Front Endocrinol 10:69. https://doi.org/10.3389/fendo.2019.00069

Ma YL, Cain RL, Halladay DL, Yang X, Zeng Q, Miles RR et al (2001) Catabolic effects of continuous human PTH (1--38) in vivo is associated with sustained stimulation of RANKL and inhibition of osteoprotegerin and gene-associated bone formation. Endocrinology 142(9):4047–4054. https://doi.org/10.1210/endo.142.9.8356

Manolagas SC (2000) Birth and death of bone cells: basic regulatory mechanisms and implications for the pathogenesis and treatment of osteoporosis. Endocr Rev 21(2):115–137. https://doi.org/10.1210/edrv.21.2.0395

Milovanovic P, Busse B (2019) Inter-site variability of the human osteocyte lacunar network: implications for bone quality. Curr Osteoporos Rep 17(3):105–115. https://doi.org/10.1007/s11914-019-00508-y

Ominsky MS, Niu QT, Li C, Li X, Ke HZ (2014) Tissue-level mechanisms responsible for the increase in bone formation and bone volume by sclerostin antibody. J Bone Miner Res 29(6):1424–1430

Onal M, Galli C, Fu Q, Xiong J, Weinstein RS, Manolagas SC, O'Brien CA (2012) The RANKL distal control region is required for the increase in RANKL expression, but not the bone loss, associated with hyperparathyroidism or lactation in adult mice. Mol Endocrinol 26(2):341–348. https://doi.org/10.1210/me.2011-1149

Pan BL, Tong ZW, Li SD, Wu L, Liao JL, Yang YX et al (2018) Decreased microRNA-182-5p helps alendronate promote osteoblast proliferation and differentiation in osteoporosis via the Rap1/MAPK pathway. Biosci Rep. 38(6):pii: BSR20180696. https://doi.org/10.1042/bsr20180696

Parfitt AM, Mathews CH, Villanueva AR, Kleerekoper M, Frame B, Rao DS (1983) Relationships between surface, volume, and thickness of iliac trabecular bone in aging and in osteoporosis. Implications for the microanatomic and cellular mechanisms of bone loss. J Clin Invest 72(4):1396–1409. https://doi.org/10.1172/jci111096

Park JH, Lee NK, Lee SY (2017a) Current understanding of RANK signaling in osteoclast differentiation and maturation. Mol Cells 40(10):706–713

Park JH, Lee NK, Lee SY (2017b) Current understanding of RANK signaling in osteoclast differentiation and maturation. Mol Cells 40(10):706–713. https://doi.org/10.14348/molcells.2017.0225

Pfeilschifter J, Koditz R, Pfohl M, Schatz H (2002) Changes in proinflammatory cytokine activity after menopause. Endocr Rev 23(1):90–119. https://doi.org/10.1210/edrv.23.1.0456

Pi M, Quarles LD (2013) Novel bone endocrine networks integrating mineral and energy metabolism. Curr Osteoporos Rep 11(4):391–399. https://doi.org/10.1007/s11914-013-0178-8

Pinzone JJ, Hall BM, Thudi NK, Vonau M, Qiang YW, Rosol TJ, Shaughnessy JD Jr (2009) The role of Dickkopf-1 in bone development, homeostasis, and disease. Blood 113(3):517–525. https://doi.org/10.1182/blood-2008-03-145169

Qing H, Ardeshirpour L, Pajevic PD, Dusevich V, Jahn K, Kato S et al (2012) Demonstration of osteocytic perilacunar/canalicular remodeling in mice during lactation. J Bone Miner Res 27(5):1018–1029. https://doi.org/10.1002/jbmr.1567

Recker R, Lappe J, Davies KM, Heaney R (2004) Bone remodeling increases substantially in the years after menopause and remains increased in older osteoporosis patients. J Bone Miner Res 19(10):1628–1633. https://doi.org/10.1359/jbmr.040710

Riggs BL, Melton LJ 3rd, O'Fallon WM (1996) Drug therapy for vertebral fractures in osteoporosis: evidence that decreases in bone turnover and increases in bone mass both determine antifracture efficacy. Bone 18(3 Suppl):197s–201s

Rochefort GY, Benhamou CL (2013) Osteocytes are not only mechanoreceptive cells. Int J Numer Method Biomed Eng 29(10):1082–1088. https://doi.org/10.1002/cnm.2561

Rochefort GY, Pallu S, Benhamou CL (2010) Osteocyte: the unrecognized side of bone tissue. Osteoporos Int 21(9):1457–1469. https://doi.org/10.1007/s00198-010-1194-5

Saini V, Marengi DA, Barry KJ, Fulzele KS, Heiden E, Liu X et al (2013) Parathyroid hormone (PTH)/PTH-related peptide type 1 receptor (PPR) signaling in osteocytes regulates anabolic and catabolic skeletal responses to PTH. J Biol Chem 288(28):20122–20134. https://doi.org/10.1074/jbc.M112.441360

Shen W, Velasquez G, Chen J, Jin Y, Heymsfield SB, Gallagher D, Pi-Sunyer FX (2013) Comparison of the relationship between bone marrow adipose tissue and volumetric bone mineral density in children and adults. J Clin Densitom 17(1):163–169

Shi JH, Sun SC (2018) Tumor necrosis factor receptor-associated factor regulation of nuclear factor kappaB and mitogen-activated protein kinase pathways. Front Immunol 9:1849

Shulman LP (2009) Androgens and menopause. Minerva Ginecol 61(6):491–497

Simonet WS, Lacey DL, Dunstan CR, Kelley M, Chang MS, Luthy R et al (1997) Osteoprotegerin: a novel secreted protein involved in the regulation of bone density. Cell 89(2):309–319. https://doi.org/10.1016/s0092-8674(00)80209-3

Sobacchi C, Schulz A, Coxon FP, Villa A, Helfrich MH (2013) Osteopetrosis: genetics, treatment and new insights into osteoclast function. Nat Rev Endocrinol 9(9):522–536. https://doi.org/10.1038/nrendo.2013.137

Takayanagi H, Kim S, Matsuo K, Suzuki H, Suzuki T, Sato K et al (2002) RANKL maintains bone homeostasis through c-Fos-dependent induction of interferon-beta. Nature 416(6882):744–749. https://doi.org/10.1038/416744a

Tan EM, Li L, Indran IR, Chew N, Yong EL (2017) TRAF6 mediates suppression of osteoclastogenesis and prevention of ovariectomy-induced bone loss by a novel prenylflavonoid. J Bone Miner Res 32(4):846–860. https://doi.org/10.1002/jbmr.3031

Tatsumi S, Ishii K, Amizuka N, Li M, Kobayashi T, Kohno K et al (2007) Targeted ablation of osteocytes induces osteoporosis with defective mechanotransduction. Cell Metab 5(6):464–475. https://doi.org/10.1016/j.cmet.2007.05.001

Teti A, Econs MJ (2017) Osteopetroses, emphasizing potential approaches to treatment. Bone 102:50–59. https://doi.org/10.1016/j.bone.2017.02.002

Tornquist E, Isaksson H, Turunen MJ (2019) Mineralization of cortical bone during maturation and growth in rabbits. J Bone Miner Metab. https://doi.org/10.1007/s00774-019-01068-y

Watson PH, Hanley DA (1993) Parathyroid hormone: regulation of synthesis and secretion. Clin Invest Med 16(1):58–77

Wesche H, Gao X, Li X, Kirschning CJ, Stark GR, Cao Z (1999) IRAK-M is a novel member of the Pelle/interleukin-1 receptor-associated kinase (IRAK) family. J Biol Chem 274(27):19403–19410. https://doi.org/10.1074/jbc.274.27.19403

Westendorf JJ, Kahler RA, Schroeder TM (2004) Wnt signaling in osteoblasts and bone diseases. Gene 341:19–39. https://doi.org/10.1016/j.gene.2004.06.044

Xiao Z, Dallas M, Qiu N, Nicolella D, Cao L, Johnson M et al (2011) Conditional deletion of Pkd1 in osteocytes disrupts skeletal mechanosensing in mice. FASEB J 25(7):2418–2432. https://doi.org/10.1096/fj.10-180299

Xiao W, Wang Y, Pacios S, Li S, Graves DT (2016) Cellular and molecular aspects of bone remodeling. Front Oral Biol 18:9–16. https://doi.org/10.1159/000351895

Xiong J, Onal M, Jilka RL, Weinstein RS, Manolagas SC, O'Brien CA (2011) Matrix-embedded cells control osteoclast formation. Nat Med 17(10):1235–1241. https://doi.org/10.1038/nm.2448

Yamashita T, Takahashi N, Udagawa N (2012) New roles of osteoblasts involved in osteoclast differentiation. World J Orthop 3(11):175–181. https://doi.org/10.5312/wjo.v3.i11.175

Yan T, Riggs BL, Boyle WJ, Khosla S (2001) Regulation of osteoclastogenesis and RANK expression by TGF-beta1. J Cell Biochem 83(2):320–325

Yavropoulou MP, Yovos JG (2007) The role of the Wnt signaling pathway in osteoblast commitment and differentiation. Hormones (Athens) 6(4):279–294

Yeh PS, Lee YW, Chang WH, Wang W, Wang JL, Liu SH, Chen RM (2017) Biomechanical and tomographic differences in the microarchitecture and strength of trabecular and cortical bone in the early stage of male osteoporosis. PLoS One 14(8):e0219718

Zofkova I (2018) Involvement of bone in systemic endocrine regulation. Physiol Res 67(5):669–677. https://doi.org/10.33549/physiolres.933843

Structural and Metabolic Assessment of Bone

Radhika R. Narla and Susan M. Ott

Contents

Abstract

The assessment of bone structure and metabolism should focus on the bone strength. Many factors are involved, and although bone density is an important component, it is not the same as bone strength. Other aspects of bone quality

R. R. Narla · S. M. Ott (✉)
Department of Medicine, University of Washington, Seattle, WA, USA
e-mail: smott@uw.edu

P. H. Stern (ed.), *Bone Regulators and Osteoporosis Therapy*,
Handbook of Experimental Pharmacology 262, https://doi.org/10.1007/164_2020_376

include bone volume, micro-architecture, material composition, and ability to repair damage. This chapter briefly reviews some of the methods that can be used to assess both density and quality of bone. Non-invasive measurements of density or structure include dual X-ray absorptiometry (DXA), quantitative computed tomography, ultrasound, and magnetic resonance imaging. DXA is most widely used and has advantages of safety and accessibility, but there are limitations in the interpretation of the results, and in clinical practice positioning errors are frequently seen. Invasive methods are used primarily for research. Samples of bone can be used to measure structure by histology as well as micro-computed tomography and infra-red spectroscopy or backscattered electron microscopy. Force can be directly applied to bone samples to measure the bones strength. Impact microindentation is a new minimally invasive technique that measures bone hardness. Metabolic assessment includes blood and urine tests that reflect diseases that cause bone loss, particularly problems with mineral metabolism. Tetracycline-labelled bone biopsies are the standard for measuring bone formation. Non-invasive biochemical tests of bone formation and resorption can evaluate a patient's skeletal physiology.

Keywords

Biochemical markers · Bone density · Bone strength · Dual photon absorptiometry · Histomorphometry

1 Introduction

Increasing bone strength is a major goal for therapies aimed at preventing fractures. Both bone mineral density (BMD) and bone quality contribute to bone strength (Fig. 1) (Ott 2016). We will first review technologies that can be used to quantify and assess BMD and structure and then review methods including laboratory values and markers that can assess bone metabolism.

We define bone as the organic bone matrix, whether mineralized or not. Bone tissue includes bone and also fat, marrow cells, vessels, and extracellular spaces. Bone volume is the proportion of the bone within the bone tissue (Parfitt et al. 1987) (normally about 20% in cancellous bone and >90% in cortical bone). BMD (g/cm^3) is the weight of minerals within a unit volume of bone tissue. The BMD depends on both the bone volume and the degree of mineralization within the bone. Usually greater bone volume results in increased bone strength, but in some cases, such as osteomalacia, fluoride treatment, or osteopetrosis, the bone is weak despite normal or high bone volume, due to poor quality. If the hydroxyapatite crystals are packed more closely, the BMD can increase without an increase in bone volume.

The bone is continuously remodeling, in discrete regions called bone metabolic units (BMU) (Ott 2002). A BMU can be initiated by microdamage or hormonal signals and is directed by osteocytes, the cells embedded within the bone. Next the osteoclasts resorb a volume of the bone; thereafter the osteoblasts secrete new bone matrix which is then mineralized. The relative rates of bone formation and resorption

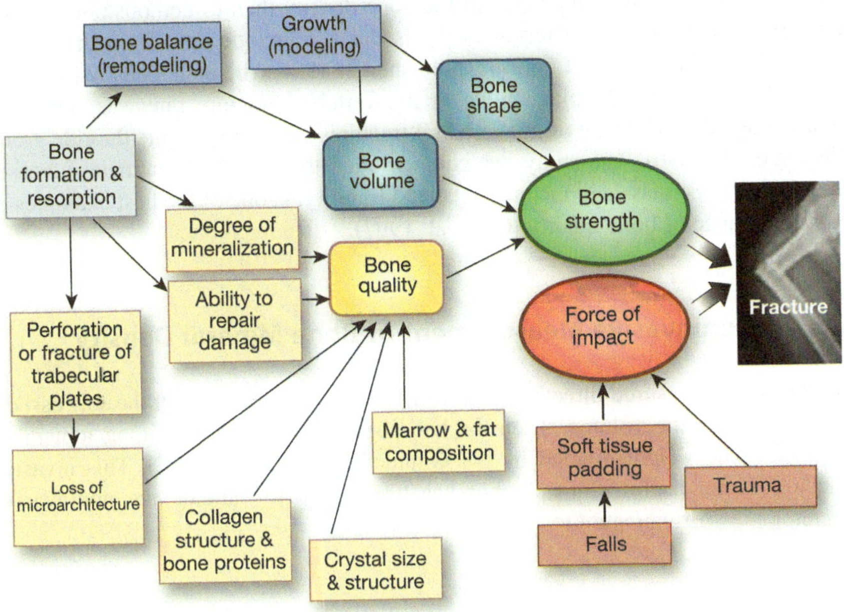

Fig. 1 Some factors that determine fracture. Notice that bone mineral density is "missing," because BMD is determined by both the bone volume and the degree of mineralization (Ott 2016)

will determine the bone volume. Bone volume is lost if resorption is greater than formation, and this will cause decrease in BMD and osteoporosis (porous bone). Inadequate mineralization can also decrease BMD; this is uncommon but seen in osteomalacia (soft bone). Conversely, increased mineralization will increase BMD, and this is seen after treatments with antiresorptive medications such as bisphosphonates (Roschger et al. 1997). BMD is also increased if calcium is replaced by a heavier mineral such as strontium (Rizvi et al. 2016).

Bone shape (geometry) also determines the strength. The bone has trabecular and cortical compartments. The trabecular bone is composed mostly of rods and plates in a sponge-like structure, and it is found in the end of long bones, in vertebrae, and in the pelvis (Seeman 2013; Ott 2018). In long cylindrical bones, which are mainly cortical bone, the strength in bending is proportional to the fourth power of the radius (Szulc et al. 2006). Thus, for the same amount of material, the strength is greater if the diameter is larger, at the expense of a thinner cortex. In humans there is continual expansion at the periosteum of long bones throughout life, which partly compensates for bone loss. This slow expansion does not require prior bone resorption.

In clinical trials of osteoporosis medications, an increase in BMD is usually associated with a decrease in fracture risk (Kanis et al. 2019). Some have suggested that drugs should be approved by government agencies like the US Federal Drug Administration if a drug increases bone density (Bouxsein et al. 2019). This would make it less expensive to bring new medications to market, because fewer subjects are needed to document increases in the bone density than to demonstrate fracture reduction. However, these recent analyses did not include data from the pivotal randomized clinical trial of fluoride, published in 1990. The bone density increased by 35% in the lumbar spine; however the vertebral fracture rate was similar to placebo and the incidence of nonvertebral fractures was higher in the fluoride group (Riggs et al. 1990). The study, which concluded that fluoride treatment increases cancellous-bone mass and increases skeletal fragility, is one of the best examples of the importance of bone quality (Ott et al. 1997).

2 Noninvasive Measurement of Bone Mineral Density

BMD can be measured directly by excising a bone, cleaning off all the soft tissues, and measuring the volume. Then the bone is ashed and the remaining mineral is weighed. The weight divided by the volume is the BMD (g/cm^3). This invasive method is not useful clinically but has been done to validate the noninvasive technologies (Erman and Ott 1988).

2.1 Radiographs

In the 1950s to 1960s, bone density was estimated by measurement of the cortical thickness in radiographs of the hands. The overall patterns of bone loss with aging and menopause were described by these measurements (Garn 1972), but they were not as sensitive as the current measurements. Radiologists can detect osteopenia (low bone density), but this approach is not very sensitive and is not consistently found until the bone density is <0.73 g/cm^2 (Jergas et al. 1994).

2.2 Absorptiometry

Single-photon absorptiometry was a technique that could measure bone density at the distal radius, using a radioactive source which was passed through the bone (Neer 1992). This technique was further developed using two beams with different energies, from a gadolinium source. Dual photon absorptiometry scans were able to measure density at the spine and hip because the second energy could correct for changes in the soft tissue. This technique has been supplanted by dual-energy x-ray absorptiometry (DXA).

Dual-energy x-ray absorptiometry (DXA) was introduced in the mid-1980s and is a quick and safe way to estimate BMD and predict skeletal fracture risk (Sartoris and Resnick 1989). Its radiation dose is minimal, and it is well tolerated by patients. While DXA remains a cornerstone of case detection of osteoporosis, its advantages and limitations must be understood in order to correctly interpret the results. We will highlight some of these major concepts.

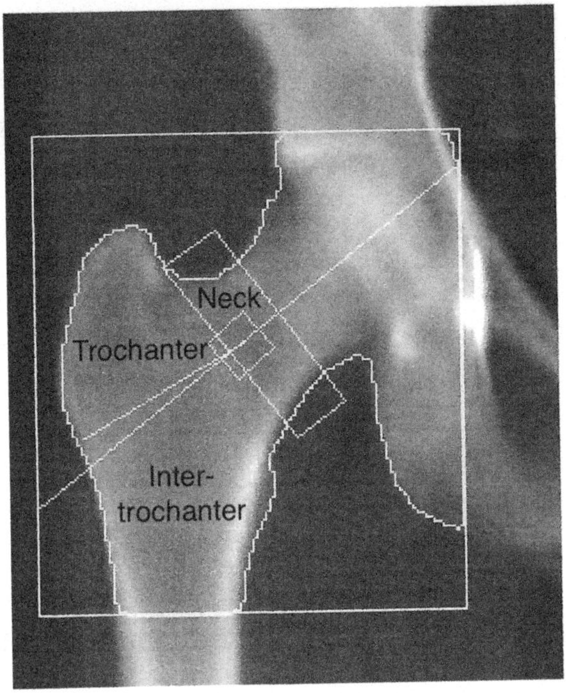

Fig. 2 Image from DXA of the hip, showing placement of the regions of interest on the projected area. The total hip is the sum of the neck, trochanter, and intertrochanter. DXA does not consider the length in the posterior-anterior direction

During the actual acquisition of images, x-ray beams with two energies are passed through bones, and what is not absorbed is detected by sensors on the other side. For each energy, the attenuation of bone and of an average non-bone tissue are known, so differential equations can solve for the density of bone and of non-bone at each pixel scanned. The software formulates an image of the bone and calculates the projected area (Fig. 2). The result is an "areal density" measured in g/cm^2. DXA images are therefore a two-dimensional (vertical and horizontal) condensation of what is a three-dimensional structure (Jain and Vokes 2017; Ott et al. 1997).

Accuracy of DXA is defined as how well the measured value reflects the true of the density of the bone. Cadaveric bone is scanned to measure the bone mineral content (BMC). The bone is then ashed, and this mass is weighed to determine bone mineral value. The accuracy error of DXA bone densitometers is better than 10% (Ott et al. 1997; Baim et al. 2005).

The precision of the machine measures the reproducibility of a bone densitometry technique. When measuring a brick phantom, this is approximately 0.5–1%. Precision is greatly influenced by a range of factors including technologist's skill and training, inconsistent selection of vertebral levels, or site of measurement (Baim et al. 2005; Lekamwasam and Lenora 2003; Wahner et al. 1994). Of mention, the precision in patients depends on accurate positioning. One study examined the effect of leg rotation by 10° both internally and externally from the standard position in a group of 50 women and found a significant change in the femoral neck BMD in 12% of subjects after the internal rotation and 8% after the external rotation

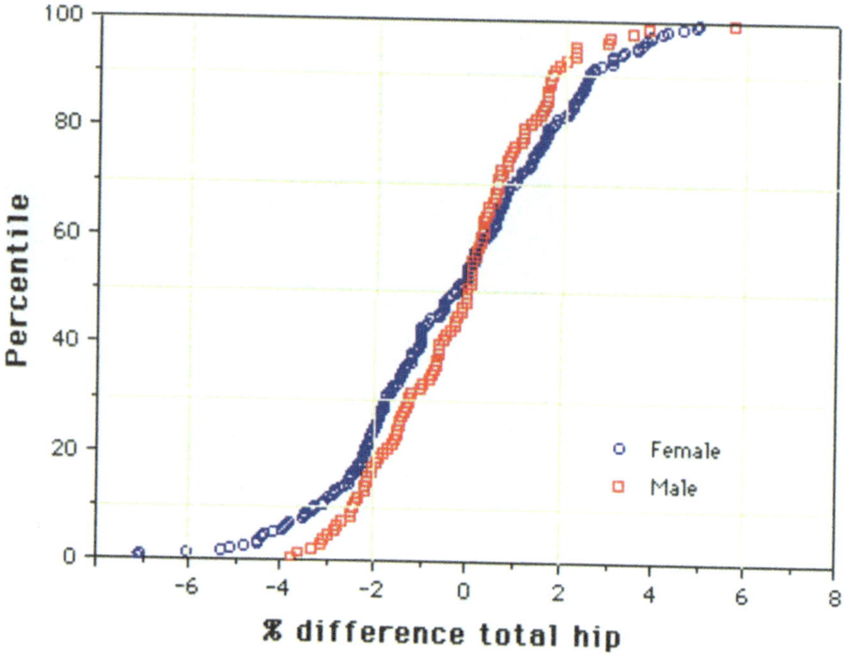

Fig. 3 Reproducibility of hip scans in 300 older subjects measured twice on the same day. The plot shows cumulative incidence of men or women according to the percent difference between scans. For example, 30% of women had "lost" more than 2%

(Lekamwasam and Lenora 2003). In the NHANES survey, the precision at the femur neck was 3.2% (Wahner et al. 1994). We measured 300 subjects older than 65 years with repeat scans on the same day, after they walked around the room, and 50.8% of patients had a difference less than 2%; 34% had a difference between 2% and 4%, and 6.3% were between 4% and 6% (Fig. 3, unpublished data). It is therefore important to determine whether a change in a patient is real or represents a precision error. A statistically significant change in BMD has been defined as a change in excess of 2.77 times of the precision error at each BMD site (Wahner et al. 1994). Furthermore, recognition of artifacts and disease processes that may influence BMD results can be of major importance in the optimal interpretation of DXA scans. For example, degenerative changes in the bone, scoliosis, aortic calcifications, and fractured vertebrae can cause falsely elevated BMD readings. Rotation of the lumbar vertebrae can lead a projected area that is larger than the area in normal rotation, thereby decreasing the areal bone density. Hip measurements rely on the rotational position, which may be difficult when patients have arthritis or are obese. In addition, since it is a 2D measure, larger bones have higher BMD measurements than smaller bones, even if the true three-dimensional BMD is the same (Ott et al. 1997).

The three commercial manufacturers of DXA instruments (Hologic, Lunar, and Norland) do not give standardized results, so it is difficult to compare measurements from one machine to the other. The differences in reported BMDs can be up to 10–15%. For example, the Hologic results are about 6% lower than the Lunar even though both are expressed in g/cm^2. This caused difficulties in clinical practice, and eventually the data from the National Health and Nutrition Examination Survey (NHANES) studies (Looker et al. 1998) were adopted as a reference base for all the bone densitometers, and results were expressed as standard deviations from young normal white females. The data were obtained on Hologic machines, and results from others are adjusted using regression equations (Lu et al. 2001). The T-score is the number of standard deviations from the average Caucasian woman aged 20–29. Note that the T-score is a linear transformation of the bone density. The Z-score is the number of standard deviations from the average of a person with the same age and other characteristics, such as gender, ethnicity, or weight, as the person being measured.

The conventional locations for BMD measurement are the lumbar spine, proximal femur, and forearm. These reference locations were chosen because they are common sites of osteoporotic fractures. Longitudinal studies have shown that the hip BMD gives the best prediction of future osteoporotic fractures (Johnell et al. 2005; Choksi et al. 2018). Evaluation of the BMD of the non-dominant forearm can also be informative in certain situations such as hyperparathyroidism or androgen deprivation therapy where there is preferential loss of cortical bone in the forearm. Finally, morbidly obese patients who are over the limit of the DEXA table can opt for non-dominant forearm measurement. Although the T-scores at any site will be similar for persons at peak bone density, they diverge with aging. For example, an average 80-year-old woman has a T-score of −2.0 at the hip and −3.0 at the radius.

BMD can also be measured for the whole body. This technique can be used for infants as well as animal models. The machines can adjust the size of the pixels during scanning, and programs are available for rodent and primate studies of bone disease.

By convention, osteoporosis is defined as a T-score lower than −2.5 at the proximal femur (Kanis et al. 1994). The question remains whether the T score < -2.5 SD cutoff should be used to diagnose osteoporosis in other groups that are non-Caucasian women or in men. For instance, it is known that in Asia, hip fracture risk is lower than in USA, but BMD is lower as well. Finally, it is critical to remember in a patient with a history of a fragility fracture, it is not necessary to have a T-score lower than −2.5 to make a diagnosis of osteoporosis.

2.3 Fracture Risk Assessment Tool (FRAX)

The Fracture Risk Assessment Tool or FRAX was developed to calculate the 10-year probability of a major osteoporotic fracture or hip fracture taking into account the femoral neck BMD as well as certain clinical risk factors KANIS (Baim et al. 2005; Lekamwasam and Lenora 2003; Wahner et al. 1994). The purpose of the FRAX calculator was to consider risk factors independent of BMD. FRAX was developed using data from 9 prospective population-based cohorts around the world and validated in 11 prospective population-based cohorts (>1 million patient years) by adjusting each calculator for local hip fracture incidence rates (Kanis et al. 2007).

Many population-based cohort studies have shown international consistency for the risk factors used in FRAX which include a prior history of fracture, a parental history of hip fracture, current cigarette smoking, intake greater than 3 daily units of alcohol (in the USA, this is 2 drinks/day), presence of rheumatoid arthritis, and presence of secondary osteoporosis such as hypogonadism or premature menopause, malabsorption, chronic liver or kidney disease, and inflammatory bowel disease (Kanis et al. 2007).

There are limitations to using the FRAX. There are no gradations for exposure to glucocorticoids, alcohol, or cigarettes. History of falls is not included. Not all secondary factors are included (Kanis et al. 2007). The FRAX calculator assumes that the relationship between BMI and mortality in all ethnic backgrounds are similar, but there are no published studies to show this is true for certain groups such as African Americans, Hispanics, and Asians (Kanis et al. 2019). Despite this, a study of the Women's Health Initiative found that additional clinical risk factors did not improve the performance of FRAX (Crandall et al. 2019).

2.4 Trabecular Bone Score (TBS)

A grayscale textural analysis obtained from lumbar spine DXA images can provide an index known as trabecular bone score (TBS). Well-structured or homogenous trabecular bone will receive higher scores than less well-textured bone. Clinically it is useful in patients whose FRAX are borderline. As it can be measured on pre-obtained DXAs, there is growing research to review its effectiveness (Hans et al. 2017).

TBS has been shown in several studies to provide information on osteoporotic fracture risk independent of BMD. In a study using the Manitoba BMD Registry, it was found that for each standard deviation reduction in TBS, there was a 36% increase in major osteoporotic fracture risk (hazard ratio [HR] was 1.36) (Leslie et al. 2014). After modifying for clinical risk factors and femoral BMD, lumbar spine TBS was found to be a small but significant predictor of major osteoporotic fracture. Thus, TBS's greatest utility could be using it in patients who have borderline BMD levels for treatment threshold (Martineau et al. 2017).

In patients with type 2 diabetes mellitus, the BMD tends to be higher than in age-matched non-diabetics, even in the presence of increased fracture risk. Furthermore, FRAX calculator does not take into effect patients with diabetes as a clinical risk factor. However, the same is not true with the TBS score, which provides a potential added advantage to using TBS in diabetes patients. Studies have shown LS-TBS was lower for diabetics, in both unadjusted and adjusted models, and that LS-TBS score can add value to prediction of osteoporotic fractures (Martineau et al. 2017; Leslie et al. 2013). TBS score can be influenced by age, ethnicity, or BMI.

TBS should be used in conjunction with BMD measurements and/or FRAX calculator tool. Further research is ongoing and necessary to elucidate the role of TBS.

2.5 Quantitative Ultrasound

Quantitative ultrasound (QUS) of the bone was introduced in 1984 by Langton et al. as a device for examining bone tissue (Langton et al. 1984). QUS devices measure the transmission and reflectance of ultrasound waves through limb bones, specifically the heel or phalanges. QUS is most often performed in the calcaneus of the heel because it is comprised primarily of trabecular bone and can easily be accessed and measured by a technician for a QUS measurement (Hans and Baim 2017).

The interaction of the bone and the ultrasound wave is analyzed to provide quantitative variables regarding skeletal site properties such as density, structure, or strength. Parameters assessed by QUS are broadband ultrasound attenuation (BUA), speed of sound (SOS), and calculated index of stiffness, often referred to as the quantitative ultrasound index (QUI). BUA, reported as decibels per megahertz (dB/MHz), is a measurement of the loss of attenuation of the sound wave as it passes through the bone. A higher bone density is associated with a higher BUA. SOS has a typical range of 3,000–3,600 meters per second (m/s) with cortical bone and 1,650–2,300 m/s for trabecular bone. A higher bone density is also associated with a higher SOS measurement (Hans et al. 1997). The measurements correlate to bone volume measured by microCT and bone mechanical strength measured by compressive mechanical testing (Qin et al. 2013).

Many clinical studies have found that certain QUS parameters, such as broadband ultrasound attenuation (BUA) and speed of sound (SOS), are significantly associated with fracture risk (Moayyeri et al. 2012; Chan et al. 2012). The EPIDOS study, a large prospective trial of the risk of hip and other fractures in France, demonstrated that the two measured parameters BUA and speed of sound (SOS) can both predict hip fracture risk in elderly women over the age of 75 (Hans et al. 1996). Huopio et al. found that the hazard ratio (HR) for fractures increased by 1.90 (95% CI, 1.25–2.91) per SD decrease in QUS stiffness (Huopio et al. 2004). Based on a pooled meta-analysis of three prospective studies, Moayyeri et al. also concluded that the BUA, SOS, and stiffness were significantly associated with fracture risk (Moayyeri et al. 2012).

Advantages of QUS include the fact that the machines are portable, cost less compared to DXA machines, and do not expose patients to radiation. The potentially could be used on spaceships to monitor bone strength in astronauts (Qin et al. 2013). While the amount of evidence for QUS as an independent marker of fracture risk is increasing, it is still not well standardized (McLeod et al. 2015). However, if DXA is not available, QUS could potentially be used to identify subjects at low or high risk of osteoporotic fracture. The ability of the test to correctly identify those without the disease (true negative rate) is called specificity. The reported specificity values by one study of QUS are 83–87% at the calcaneus (Steiner et al. 2019). Therefore, the role of QUS could be case screening, stratifying high-risk patients and limiting the number of patients who need a DXA scan and BMD measurement.

Limitations of QUS include inability to measure sites which are typically related to osteoporotic fractures (proximal femur, vertebrae, distal radius). Falsely low QUS readings can occur in patients with peripheral edema. Lack of standardization among the different QUS technologies and measurement techniques makes it difficult to compare different QUS devices.

2.6 Quantitative Computed Tomography (QCT) and Peripheral QCT (pQCT)

Quantitative computed tomography (QCT) uses a low-dose CT scan protocol and measures 3D volumetric trabecular and cortical bone density at the spine and hip locations. QCT can differentiate between cortical and trabecular bone. The radiation attenuation is measured in Hounsfield units which can be calibrated to known densities to determine bone mineral density.

With QCT, there is less impact of arthritic changes on BMD compared to DXA (particularly important in older patients) (Engelke 2017). In addition, DXA measures includes both cortical and trabecular bone, whereas QCT can quantify separately the trabecular or cortical bone density. Johannesdottir et al. reviewed this and found in 5 years among cohorts of 100 men and 200 women aged 66–90 years, women lost cortical thickness and cortical BMD more rapidly than men in both regions and unfortunately, this was only weakly reflected in total femoral neck DXA results (Johannesdottir et al. 2013).

Measurements with QCT are true density and not the "areal density" that is measured and calculated by DXA, so no adjustment is necessary due to size. This was shown in a study of patients with growth hormone receptor deficiency, whose DXA measurements were much lower than relatives without the mutation, but the QCT and bone volume (from bone biopsies) were similar (Bachrach et al. 1998).

QCT can also provide additional analyses of the trabecular region, also known as individual trabeculae segmentation (ITS) that assesses the orientation and the ratio of rod and plate trabecular elements as well as cortical thickness and porosity. Finally, QCT data can be further utilized to estimate bone strength using finite element analysis (FEM) (Keaveny 2010). Some studies have shown FEM can calculate and predict vertebral strength (Crawford et al. 2003).

There are several commercial producers of QCT equipment, leading to poor standardization. Radiation dose during QCT is considerably higher than that of DXA. Limited number of longitudinal, fracture prediction studies comparing DXA versus QCTs have been done.

2.7 High-Resolution Peripheral QCT (HRpQCT)

This more recent technology uses the same principles as QCT and measures three-dimensional bone density at the radius or tibia. It allows clearer distinction between cortical and trabecular bone and can measure the cortical dimensions (Fig. 4) (Agarwal et al. 2016; Leonard et al. 2019). Some cross-sectional studies have concluded that poor bone microarchitecture assessed by high-resolution peripheral QCT (HR-pQCT) is associated with higher odds of fracture in postmenopausal women and older men (Szulc et al. 2018). Currently HR-pQCT machines are used in research settings because they are expensive. Other limitations include motion artifacts and long scan times (Bonaretti et al. 2017).

Fig. 4 An example of HRpQCT imaging, showing separation of cortical and trabecular bone in the distal radius. Note this is a three-dimensional image. Courtesy of Dr. Mary Leonard and Kyla Kent, Stanford University

2.8 Magnetic Resonance Imaging (MRI)

Magnetic resonance imaging (MRI) is a noninvasive imaging modality used widely in clinical medicine because it can determine areas of increased metabolic activity. For the bone, it is the most sensitive method of detecting avascular necrosis and can be helpful in detecting malignancies, stress fractures, and other causes of bone marrow edema.

In addition, high-resolution MRI can provide novel information about bone microstructure and picturing trabecular and cortical aspects of the bone. High-resolution MRI using T2-weighted images can show trabecular structure of the tibia and visualize trabecular plate structure and perforations. This "virtual biopsy" requires intensive image processing and thus has been used only in research settings.

Water can exist in two forms, one is bound to the collagen, or bound water, and second, as free water, also known as pore water. The bound water can reflect the collagen matrix measurement, while the pore water can show bone porosity. This distinction has been allowed for use of ultrashort echo time (UTE) sequences as part of the whole-body clinical magnetic resonance (MR) imaging scanners to provide this distinction and offer bone porosity data (Jerban et al. 2020).

Such high-resolution images dramatically demonstrate the increased trabecular perforations that occur with estrogen deficiency (Wehrli et al. 2008), as well as repair of the trabecular microstructure in hypogonadal men treated with testosterone (Benito et al. 2005).

3 Invasive Techniques to Measure Structure: Bone Biopsy

Bone biopsies in patients are performed on the anterior iliac crest because this is a safe location, and there are normative reference data available (Dempster et al. 2013). In animal studies, biopsies are performed in the metaphyseal regions of the long bones, ribs, or vertebra. Sections of bone are directly measured using imaging software, which gives the volume of bone as a percentage of the bone tissue (BV/TV) (Parfitt et al. 1987). Using equations that assume a plate-like structure, the trabecular thickness and separation are calculated from the bone volume and bone surface area. Trabecular thickness can also be measured directly. Methods to assess connectivity on the two-dimensional section include analysis of nodes and star volume (Chappard et al. 2008; Muller et al. 1998; Oleksik et al. 2000).

Bone biopsies are expensive, and the core sample is small relative to the total amount of the bone (Fig. 5). The measurement error on a given sample is small, but the average difference between two biopsies taken from the same patient is over 25%, due to sampling differences (Chappard et al. 2008; de Vernejoul et al. 1981). The standard skeletal location is the iliac crest because that is a safe region to remove the bone, but it is not a weight-bearing bone, and the results may not reflect the bone density in other regions. Bone biopsies are not routinely used to manage osteoporosis patients. They are primarily used in research settings or in patients with complex problems that can't be explained by bone density results (Malluche et al. 2007).

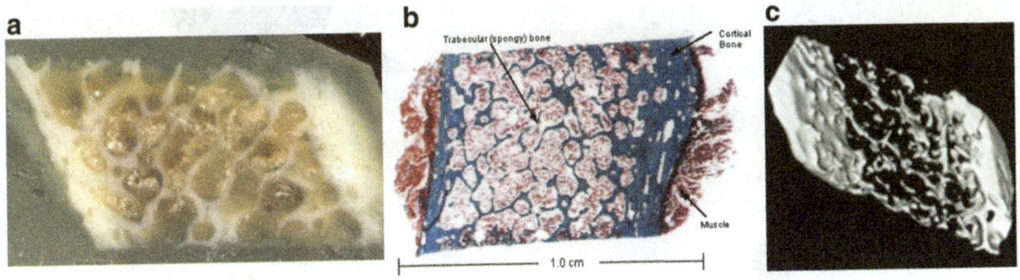

Fig. 5 Human iliac crest bone biopsies, (**a**) embedded sample, (**b**) stained section, (**c**) micro-CT image

A variety of studies can be done to characterize bone from biopsy samples. These techniques are not used clinically but are important in research studies. The three-dimensional structure of the trabecular and cortical bone can be measured using micro-computed tomography (micro-CT) of the core sample (Muller et al. 1998) (Fig. 5c). Bone structural indices related to connectivity of the cancellous bone are generated by the software.

The mineralization density can be measured using back-scattered electron microscopy (Roschger et al. 1997, 2008). With this technique, a beam of electrons is focused on the surface of the sample, and the scattered energies are detected. This allows determination of the density at each pixel of surface (Fig. 6). The average density as well as the range of densities is assessed. Microradiography of a thick section of the bone can also be used to measure the mineralization density (Boivin et al. 2003). Normally bone osteons have a range of ages, and thus the mineralization density is heterologous, because more recently formed bone is less mineralized than older bone. Uniformly high mineralization density can result in brittle bone in which cracks propagate more easily.

The bone strength of a core sample can be measured using biomechanical testing. This directly measures the force required to break the bone, as well as the force that first causes bone deformation. Standard methods test the strength using three-point bending. Force-displacement testing measures stress-strain curves and allows computation of mechanical parameters such as the Young's modulus and the yield and ultimate strength (Turner 2002; Hernandez and van der Meulen 2017; Nishiyama 2019; Varela and Jolette 2018).

MRI with ultrashort echo time predicts bone microstructural properties from harvested human tibial samples (Jerban et al. 2020) and can map bound and pore water as well as collagen protons in cortical bone (Jerban et al. 2019).

On an even smaller scale, methods can assess composition of the bone. Fourier transform infra-red spectroscopy and Raman microscopic imaging measure the size of the bone crystals and degree of mineralization to matrix (Fig. 7) (Faibish et al. 2006; Bala et al. 2012). Increased bone mineral particle size is associated with increased bone fragility; crystals that are too small do not reinforce the bone composite, suggesting there also is an optimal size range for bone mineral crystals.

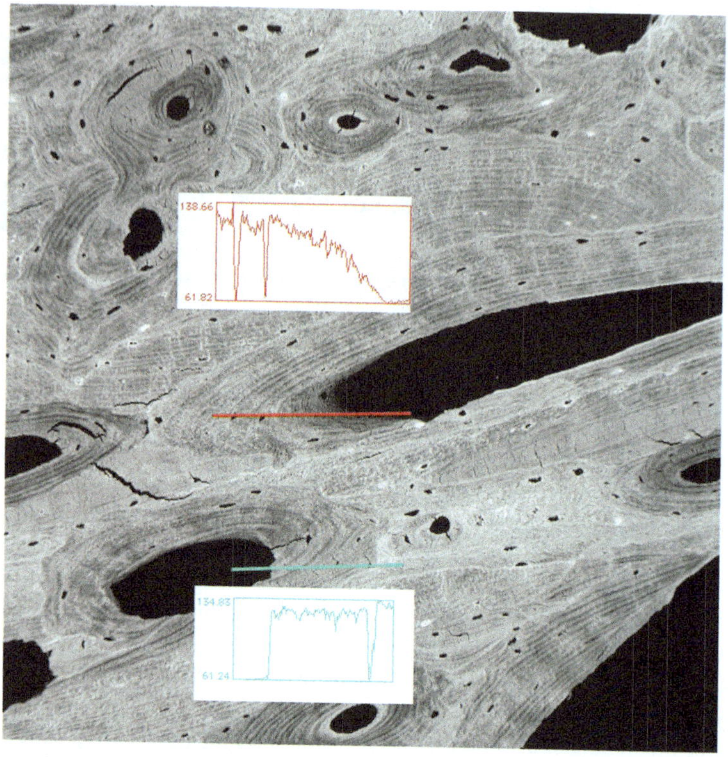

Fig. 6 Backscattered electron micrograph showing degree of mineralization. Lighter shades represent denser mineral. The plots show the gradual increase in mineralization in an osteon where newly formed bone is darker than the older bone (with spikes as the line crosses an osteocyte lacunae)

Other techniques done on samples of bone tissue include atomic force microscopy to analyze mineral properties of the bone (Hassenkam et al. 2004; Tong et al. 2003). Nuclear magnetic imaging measures the water that is bound to collagen molecules. Bound water decreases with aging and is strongly related to bone toughness (Cho et al. 2003; Wu et al. 2003). D-periodicity is measured with electron microscopy, as the distance between collagen molecules. This could be related to formation of micro-cracks (Luo et al. 2014). Nanoindentation measures the hardness of the bone (the force required to make a small indentation with a diamond point) (Bala et al. 2012; Newman et al. 2016).

3.1 Impact Microindentation

Bone indentation is a new minimally invasive technique that can estimate the hardness of the bone (Arnold et al. 2017; Herrera and Diez-Perez 2017; Rozental et al. 2018; Rufus-Membere et al. 2018). An incision is made over the anterior

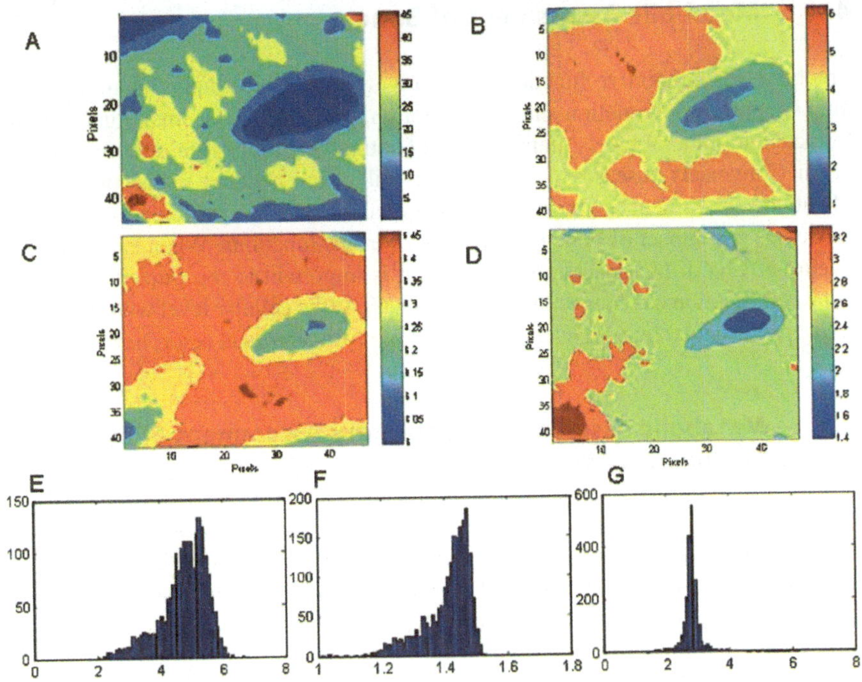

Fig. 7 Infrared imaging spectroscopy. (**a**) The distribution of mineral intensities as revealed by the integrated area under the mineral phosphate band (900–1,200 cm^{-1}) in the raw data. (**b**) The ratio of the integrated areas of the mineral phosphate band to the amide I peak shows the anatomic distribution of the mineral/matrix ratio. The axes are in pixels, where each pixel is 6.25 μm. (**c**) The ratio of the intensity of subbands at 1030 cm^{-1} and 1,020 cm^{-1} shows the distribution of the crystallinity parameter. (**d**) The ratio of intensity of subbands at 1660 and 1,690 cm^{-1} show a parameter related to the ratio of nonreducible and reducible collagen cross-links. (**e**) Pixel distribution for the image of mineral/matrix ratio in Fig. 2b is shown. Note the distribution is skewed to the left. (**f**) Pixel distribution for the image of crystallinity in figure **c** is shown. The distribution is skewed to the left, showing the presence of relatively more small crystallites than larger ones. (**g**) Pixel distribution for the image of collagen cross-link ratio shown in figure **d** is shown. Note the peak is sharp, showing limited variation in collagen maturity (Faibish et al. 2006)

surface of the mid-tibia and the instrument measures the force needed to indent the bone. Multiple indentations are performed with recommendations for at least 11, separated by 2 mm. The result is termed the bone material strength index. This technique requires trained operators, and it can't be done if the tibia has thicker soft tissue covering. The studies related to these measurements to fracture prevalence have inconsistent results. To date, there aren't any large prospective studies of normal ranges or of incident fracture prediction.

4 Metabolic Assessment of Bone

The bone has supportive functions for many systems in the body, including structural support for locomotion and for respiration, protection of internal organs, buffer for acid and mineral levels, transmission of sound, and regulation of energy metabolism. Conversely, diseases of most systems will have a negative impact on the bone. Metabolic assessment can therefore include evaluation of the etiologies of bone loss, as well as the impact of bone disease on its various functions. When evaluation the overall effects of pharmacological agents, it is important to recognize that altering bone cell function can have non-skeletal effects. At the cellular level within the bone tissue, different techniques can quantitate bone formation and resorption.

4.1 Metabolic Assessment: Laboratory Assessment of Systemic Factors Causing Bone Loss

The bone is directly involved in mineral metabolism, and persistent abnormal serum levels will be associated with altered function of the bone. Further clinical evaluation is necessary to determine if abnormal minerals are due to primary bone disease if the bone is reacting to a secondary disease (Diab and Watts 2013).

Calcium, Phosphorous, and Magnesium In the routine chemistry panel, the total calcium is measured. If there are any abnormalities, one should first correct for low albumin and then confirm with a free ionized calcium level. High or low calcium levels require a PTH measurement. Phosphorous gives additional clues to potential disorders causing bone disease. In primary hyperparathyroidism, serum phosphorous is classically low. In osteomalacia, including tumor-induced osteomalacia caused by tumors of mesenchymal origin which secrete FGF-23, there are decreased serum phosphorous levels. Magnesium also plays a role but the exact mechanisms remain unclear. In states of severe magnesium depletion, such as due to nutrition or excessive alcohol intake, there is a phenomenon of decreased PTH release and even PTH resistance leading to hypocalcemia (Hofbauer et al. 2010).

Complete Blood Count (CBC) and Albumin CBC screens for anemia which can be seen in patients with malnutrition, celiac disease, and myeloma. Leukocytosis is seen in leukemia. Albumin is required to correct for any calcium issues, especially states of falsely low calcium due to hypoalbuminemia. It also provides a sense of protein stores and nutritional status of the patient (Gaudio et al. 2020).

24-h Urine Calcium and Creatinine Twenty-four-hour urine for calcium measurement is useful to determine if a patient has adequate calcium intake and absorption because low levels of calcium could suggest malabsorption, such as in patients with gastrointestinal disorders with inflammatory bowel disease and celiac disease or history of undergoing bariatric surgery. Assessment of urinary calcium is also necessary in patients with history of kidney stones. It can detect idiopathic hypercalciuria which very often is accompanied with some degrees of bone demineralization and increased susceptibility to fractures. The 24-h creatinine is needed at the same time to ensure it's an adequate collection of urine. High urine calcium is also a sign of excess bone resorption or excess intestinal absorption, both of which can be caused by vitamin D toxicity.

Serum Creatinine or Cystatin C Chronic kidney disease is invariably associated with bone abnormalities. Creatinine is generated in the muscle, and serum levels are misleading in muscular disorders, including the muscle atrophy that is commonly observed in elderly patients. Cystatin C levels are more accurate determinations of the renal function in these cases, because the source is from white blood cells. Some studies have demonstrated an association between chronic kidney disease and incident hip fracture using cystatin-C as a biomarker of renal function (LaCroix et al. 2008).

Liver Function Tests (LFTs) LFTs would help assess any form of hepatic disease. They would also be indicated if there was any history of significant alcohol intake which is another risk factor for osteoporosis. Patients with liver disease also have immobility, risk for hypogonadism, and vitamin D deficiency because hepatic enzymes are required for 25-hydroxylation of vitamin D (Jeong and Kim 2019).

Alkaline Phosphatase Alkaline phosphatase (ALP) is part of LFTs measurement, but also it is synthesized in the bone. Elevated levels of ALP can be found in patients healing from fractures, Paget's disease or in cases of severe vitamin D deficiency or osteomalacia. If the routine ALP is elevated, a bone-specific alkaline phosphatase can be checked. On the other hand, low alkaline phosphatase is a sign of hypophosphatasia, a genetic disease with variable severity that can be mistaken for osteoporosis. It is important to recognize this because treatment with bisphosphonates or calcium supplements in these patients can cause atypical femur fractures (Shapiro and Lewiecki 2017).

Thyroid Function Hyperthyroidism causes increased bone loss. Osteoblasts have receptors for thyroid hormone. Observational studies suggest screening for asymptomatic hyperthyroidism with thyroid stimulating hormone is cost-effective in the evaluation of osteoporosis (Bauer et al. 2001).

25-Hydroxycholecalciferol (25-OH vitamin D) and its more active metabolite 1,25-dihydroxycholecalciferol (1,25-vitamin D) regulate intestinal calcium and phosphorous absorption and stimulate bone resorption leading to the maintenance of serum calcium concentration. Severe deficiency of 25(OH)D causes osteomalacia. There remains debate about the level of vitamin D which can cause bone disease, but a recent comprehensive study of the bone, including markers, radiographs, and biopsies, failed to find any abnormalities until levels were below 3o nmol/L (12 ng/mL) (Shah et al. 2017). Clinically, vitamin D status can be measured by serum 25(OH)D, which gives the most accurate assessment of vitamin D stores (Bouillon et al. 2013). Serum 25(OH)D is transported in the blood and mostly attached (88%) to the vitamin D-binding protein (VDBP) with remaining small percentage joined to albumin or free or unbound to plasma proteins. Large ethnic differences, such as in African Americans, have been observed that could account for some differences in the affinities of the D-binding proteins. For instance, some studies have shown that while total serum 25(OH)D is generally lower in black than white Americans, free 25(OH)D is almost identical in values (Aloia et al. 2015). However, measuring free 25 (OH)D remains difficult. While there are commercially available immunoassays, the majority of research studies used a monoclonal antibody enzyme linked immunosorbent assay (ELISA, R&D Systems) to measure VDBP. Procedures to ensure accuracy and sampling are still ongoing with the assays (Binkley et al. 2004).

Parathyroid Hormone (PTH) Hypersecretion of PTH hormone leads to cortical bone loss. Parathyroid is measured in serum by radioimmunoassay. Older methods that measured the inactive C-terminal fragments are no longer recommended. The hormone is cleaved in several locations, and assays that measure the "intact" hormone actually do not capture the first seven amino acids. Newer assays do measure the entire ("whole") hormone, but clinically this rarely changes the diagnosis (Chen et al. 2018). As mentioned previously, determination of PTH-mediated or non-PTH-mediated etiologies for calcium abnormalities is warranted. In patients with suspected primary or normocalcemic hyperparathyroidism, one should also order BMD assessment for the distal 1/3 radius which is composed of cortical bone. If a patient has clinical evidence of hyperparathyroidism, either primary or normocalcemic, and have osteoporosis based on DXA testing or history of fragility fracture, they indeed have indications for consideration of operative management with parathyroidectomy (Bilezikian et al. 2014).

Evaluation of bone metabolism in unusual cases may also include tests for celiac sprue (anti-tissue transglutaminase), multiple myeloma (protein electrophoresis and free light chain ratio), mastocytosis (trysin), and Cushing's syndrome (dexamethasone suppression testing).

4.2 Metabolic Assessment of Bone: Invasive Techniques

Much of our understanding of bone physiology at the tissue level has been based on histomorphometry of undecalcified sections of bone biopsies (Parfitt et al. 1987; Ott 2002). On these sections, three kinds of bone cells have been identified: osteoclasts, osteoblasts, and osteocytes. The bone is formed in packets by a team of cells termed the bone metabolic unit (BMU). A BMU originates after signals from the osteocytes which detect microdamage. First the osteoclasts are recruited from the circulating pre-osteoclasts, and then the bone is resorbed, followed by osteoblasts forming new bone to fill in the resorption cavity.

Bone biopsies allow direct visualization of these bone cells and quantitation of the bone formation rates (Fig. 8). Biopsies can also provide estimation of the bone volume and, with new micro-CT scans, can determine trabecular connectivity. In

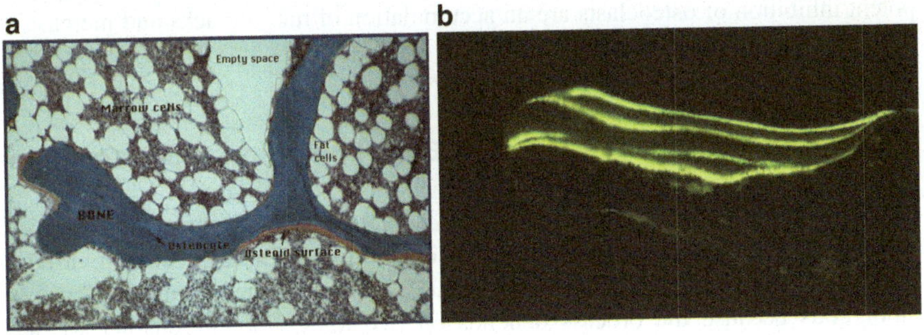

Fig. 8 (**a**) Goldner's stained section of a bone biopsy from a patient with osteoporosis. The bone volume is only 9% the tissue volume (normally it is about 20%). The surface, however, is normal with 13% of the surface covered by osteoid, which is of normal thickness. (**b**) Tetracycle labels show formation rate of bone. Shown here is a biopsy from a patient who was given four separate labels

addition, bone micro-cracks can be observed using fusion-red staining. Goldner's staining differentiates mineralized bone from the unmineralized matrix (osteoid). The majority of the normal bone surface is quiescent. About 2% of the surface is eroded and associated with osteoclasts, whereas 6% is lined with unmineralized osteoid which is covered by a layer of osteoblasts (Recker et al. 2018). When the osteoid becomes mature, mineral is deposited, so in newly formed bone there is a mineralization front that follows the formation of osteoid.

Tetracycline deposits in the bone at the same location as mineral, and these linear deposits can be seen with fluorescent microscopy (Fig. 8). A dose of tetracycline given for 2 or 3 days will deposit in a line wherever there is new bone being mineralized. A second dose given about 8–15 days later will result in another line, and the distance between these lines is used to measure the bone apposition rate. Normally the rate of apposition of new osteoid is the same as the subsequent rate of mineralization. The area formed between tetracycline labels (length of the label times the distance between labels) is measured, giving the bone formation rate. The bone resorption rate is not as clearly defined, because the eroded cavities often have smooth edges and the depth of erosion must be extrapolated from nearby surfaces. The number of osteoclasts (best identified with stains for acid phosphatase) is related to the bone resorption rate.

The bone formation and resorption rates are vital in understanding how medications for osteoporosis can improve the bone strength. Anabolic drugs increase the bone formation rate and can increase the bone volume. Antiresorptive drugs decrease the bone resorption rate. If antiresorptive drugs have no actions on the osteoblasts, there will be a transient increase in the bone volume as previous resorption cavities fill with new bone, but thereafter the bone formation rate will decrease as well. Drugs such as denosumab which strongly inhibit bone resorption also strongly inhibit bone formation (Dempster et al. 2018). Long-term results of potent inhibition of osteoclasts are an accumulation of micro-cracks and increase in mineralization density (Bala et al. 2012).

4.3 Noninvasive Assessment of Bone Formation and Resorption

Bone biochemical markers are noninvasive laboratory tests of either serum or urine that characterize the state of bone formation or resorption. These are often referred to as "bone turnover markers" (Greenblatt et al. 2017), but the term "turnover" can be ambiguous because the process depends on two rates. For this chapter we will specify bone formation markers and bone resorption markers (Chavassieux et al. 2015). In normal bone physiology, these two processes are coupled, since formation follows resorption, so when resorption is high, formation is also high, and therefore either rate would measure bone turnover. Similarly, when both formation and resorption are low, the patient has low turnover. Generally, the marker levels are high during childhood and adolescence, low during adult years, and increase after menopause (Eastell et al. 2012). They decrease following treatment with hormone replacement therapy or antiresorptive drugs. However, in pathological conditions there can be uncoupling, so that a patient may have high bone resorption as well as low bone formation (e.g., with multiple myeloma). It is also possible to have high formation with low resorption (e.g., in patients treated with antibodies against sclerostin), and in these situations the term "bone turnover" has no defined meaning.

4.4 Bone Formation Markers

The most specific markers are based on collagen. To form the bone matrix, osteoblasts secrete pro-peptides of type 1 collagen which undergo proteolytic cleavage. The *N-terminal pro-peptide of type I collagen (P1NP)* and carboxy-terminal pro-peptide of type 1 collagen (PICP) will then circulate. Serum or s-PINP has grown to be the main bone formation marker as there are no influence of diurnal variation and food intake on its laboratory level (Eastell and Szulc 2017; Krege et al. 2014).

A special mention is necessary on the different PINP laboratory assays. The s-PINP assays measure both the trimeric and monomer forms of PINP vs. the intact s-PINP assay that measures the trimeric isomer alone. The trimer is hepatically cleared, while monomeric form is renally cleared. Therefore, intact P1NP assay (IDS-iSYS) is less affected by CKD than total P1NP assay (Roche). In the USA, the intact s-PINP assay manufactured by Orion Diagnostica uses a manual radioimmunoassay. This is in contrast to total s-PINP assay manufactured by Roche Diagnostics and available in Europe (Bauer et al. 2012).

Osteoblasts also contain alkaline phosphatase on their cell membranes, and serum *bone isoform or specific alkaline phosphatase (BAP)* is another bone formation marker that is often used as it's also not affected by chronic kidney disease (CKD) or diurnal variation. The isoforms of alkaline phosphatase including BAP are measured using heat inactivation, but the rate of heating has been difficult to standardize. Other reported methods include electrophoresis, immunoradiometric, immunoassays, and high-performance liquid chromatography (HPLC)-based methods. Among these, immunoassays provide better assay reproducibility and precision, but cross-reactivity with the liver isoform is still present, especially in patients with liver disease. BAP increases with increased bone formation but is not as direct a measure of new bone formed as P1NP because it may increase in other situations where osteoblasts are active but unable to form bone. For example, in osteomalacia the alkaline phosphatase is often high, but the matrix cannot mineralize so bone formation is abnormal (Woitge et al. 1996). This marker may plan another role in chronic kidney disease, because it is associated with vascular calcifications (Nizet et al. 2020).

Lastly, *osteocalcin* is secreted during formation of osteoid, the organic substance that is needed for mineralization. The assay for osteocalcin is sensitive to freeze-thaw cycles, and the results have poor precision, so it is no longer used clinically as a measure of bone formation (Greenblatt et al. 2017). Osteocalcin contains three glutamic acid residues, which are converted to gamma-carboxyglutamate (gla) by vitamin K-dependent posttranslational carboxylation. New research into this protein suggests that it is important in energy metabolism, increases insulin sensitivity in muscles, and is a circulating hormone. Whether the osteoblasts secrete carboxylated or uncarboxylated osteocalcin can depend on both availability of vitamin K and on the status of energy metabolism. When caloric intake is lower than expenditure, the bone formation is regulated downwards and the osteoblasts secrete more uncarboxylated osteocalcin (Zoch et al. 2016).

4.5 Bone Resorption Markers

The most specific bone resorption markers are also based on collagen. After collagen molecules are secreted by the osteoblasts, they line up in specific order, forming staggered layers that are always the same space from each other with C-terminal and N-terminal ends. The alignment causes covalent pyridinium cross-linking between adjacent collagen molecules. These cross-linking regions are called pyridinoline (PYD) and deoxypyridinoline (DPD) (Hanson and Eyre 1996). This makes the collagen fibrils stronger. After this matrix is mineralized, the bone will be stable until it is resorbed by osteoclasts (normally after 5–10 years). The osteoclasts secrete cathepsin K which cleaves the collagen into fragments, but the cross-linking PYD and DPD are stable, enter the circulation, and are excreted in the urine. Assays have been developed to measure these collagen fragments in the blood and in the urine. The ones studied and most often used are the urine *N-telopeptide (NTX)* and the serum *C-telopeptide (CTX)*.

These collagen cross-linking peptides have been studied in virtually every recent clinical trial or observational study related to osteoporosis. They decrease with antiresorptives (estrogen, raloxifene, bisphosphonates, and denosumab). In most large cohort studies the CTX or NTX have been shown to predict future fractures

(Garnero et al. 1996). In randomized trials, they have been positively associated with incident fracture rates (Bauer et al. 2004). They may also be able to play a role in determining dosage of medications (Raje et al. 2016; Patel et al. 2014).

There are some limitations in the use of markers, because they increase following a fracture (Ivaska et al. 2007). There is also diurnal variation that should be taken into account (Greenblatt et al. 2017).

Other serum tests of bone resorption include *tartrate-resistant acid phosphatase (TRACP)*. There are two forms of TRACP in the circulation: TRACP5a and TRACP5b. TRACP5b is derived from osteoclasts and increases when there are more active osteoclasts. TRACP5b is not affected by kidney disease (Vervloet and Brandenburg 2017).

4.6 Potential of Serum Sclerostin

Osteocytes are derived from osteoblasts, and they serve to sense the mechanical forces applied to bone and help regulate bone resorption and formation by producing locally active factors including RANKL, prostaglandins, nitrous oxide, and sclerostin. *Sclerostin* is a protein produced by osteocytes that inhibits the Wnt signaling pathway and prevents osteoblast generation and bone formation. In previous studies sclerostin was found to be significantly higher in postmenopausal women than in premenopausal controls (Mirza et al. 2010). Sclerostin levels may be higher in type 2 diabetes mellitus patients, and this possibly could explain an increased risk of vertebral fractures independent of lumbar spine BMD (Ferrari et al. 2018; Yamamoto et al. 2013). However, serum levels also correlate positively with bone density (Ueland et al. 2019), possibly because the serum levels reflect both the number of osteocytes and their activity rates. More research is underway for sclerostin and its role in assessment of bone metabolism.

5 Conclusions

Identification and diagnosis of metabolic bone disease can be fascinating yet challenging. There are a variety of structural and metabolic bone assessment tools available and emerging to image the skeleton and assess bone mass and bone strength. At the cellular level within the bone tissue, different techniques can quantify and provide a more dynamic reflection of bone formation and bone resorption. We have emphasized some of these modalities and advancements in technology and laboratory testing, their caveats, and how to direct their use and interpretation in clinical practice when trying to identify a cause for bone loss and fractures. Further research is necessary in hopes to continue to better characterize and identify abnormalities with bone remodeling and abnormal bone architecture, especially for this patients at risk of osteoporosis and fracture.

References

Agarwal S, Rosete F, Zhang C, McMahon DJ, Guo XE, Shane E et al (2016) In vivo assessment of bone structure and estimated bone strength by first- and second-generation HR-pQCT. Osteoporos Int 27(10):2955–2966

Aloia J, Mikhail M, Dhaliwal R, Shieh A, Usera G, Stolberg A et al (2015) Free 25(OH)D and the vitamin D paradox in African Americans. J Clin Endocrinol Metab 100(9):3356–3363

Arnold M, Zhao S, Ma S, Giuliani F, Hansen U, Cobb JP et al (2017) Microindentation – a tool for measuring cortical bone stiffness? A systematic review. Bone Joint Res 6(9):542–549

Bachrach LK, Marcus R, Ott SM, Rosenbloom AL, Vasconez O, Martinez V et al (1998) Bone mineral, histomorphometry, and body composition in adults with growth hormone receptor deficiency. J Bone Miner Res 13:415–421

Baim S, Wilson CR, Lewiecki EM, Luckey MM, Downs RW Jr, Lentle BC (2005) Precision assessment and radiation safety for dual-energy X-ray absorptiometry: position paper of the International Society for Clinical Densitometry. J Clin Densitom 8(4):371–378

Bala Y, Depalle B, Farlay D, Douillard T, Meille S, Follet H et al (2012) Bone micromechanical properties are compromised during long-term alendronate therapy independently of mineralization. J Bone Miner Res 27(4):825–834

Bauer DC, Ettinger B, Nevitt MC, Stone KL (2001) Risk for fracture in women with low serum levels of thyroid-stimulating hormone. Ann Intern Med 134(7):561–568

Bauer DC, Black DM, Garnero P, Hochberg M, Ott S, Orloff J et al (2004) Change in bone turnover and hip, non-spine, and vertebral fracture in alendronate-treated women: the fracture intervention trial. J Bone Miner Res 19(8):1250–1258

Bauer D, Krege J, Lane N, Leary E, Libanati C, Miller P et al (2012) National bone health alliance bone turnover marker project: current practices and the need for US harmonization, standardization, and common reference ranges. Osteoporos Int 23(10):2425–2433

Benito M, Vasilic B, Wehrli FW, Bunker B, Wald M, Gomberg B et al (2005) Effect of testosterone replacement on trabecular architecture in hypogonadal men. J Bone Miner Res 20 (10):1785–1791

Bilezikian JP, Brandi ML, Eastell R, Silverberg SJ, Udelsman R, Marcocci C et al (2014) Guidelines for the management of asymptomatic primary hyperparathyroidism: summary statement from the Fourth International Workshop. J Clin Endocrinol Metab 99(10):3561–3569

Binkley N, Krueger D, Cowgill CS, Plum L, Lake E, Hansen KE et al (2004) Assay variation confounds the diagnosis of hypovitaminosis D: a call for standardization. J Clin Endocrinol Metab 89(7):3152–3157

Boivin G, Lips P, Ott SM, Harper KD, Sarkar S, Pinette KV et al (2003) Contribution of raloxifene and calcium and vitamin D3 supplementation to the increase of the degree of mineralization of bone in postmenopausal women. J Clin Endocrinol Metab 88(9):4199–4205

Bonaretti S, Vilayphiou N, Chan CM, Yu A, Nishiyama K, Liu D et al (2017) Operator variability in scan positioning is a major component of HR-pQCT precision error and is reduced by standardized training. Osteoporos Int 28(1):245–257

Bouillon R, Van Schoor NM, Gielen E, Boonen S, Mathieu C, Vanderschueren D et al (2013) Optimal vitamin D status: a critical analysis on the basis of evidence-based medicine. J Clin Endocrinol Metab 98(8):E1283–E1304

Bouxsein ML, Eastell R, Lui LY, Wu LA, de Papp AE, Grauer A et al (2019) Change in bone density and reduction in fracture risk: a meta-regression of published trials. J Bone Miner Res 34 (4):632–642

Chan MY, Nguyen ND, Center JR, Eisman JA, Nguyen TV (2012) Absolute fracture-risk prediction by a combination of calcaneal quantitative ultrasound and bone mineral density. Calcif Tissue Int 90(2):128–136

Chappard C, Marchadier A, Benhamou L (2008) Interindividual and intraspecimen variability of 3-D bone microarchitectural parameters in iliac crest biopsies imaged by conventional micro-computed tomography. J Bone Miner Metab 26(5):506–513

Chavassieux P, Portero-Muzy N, Roux JP, Garnero P, Chapurlat R (2015) Are biochemical markers of bone turnover representative of bone histomorphometry in 370 postmenopausal women? J Clin Endocrinol Metab 100(12):4662–4668
Chen H, Han X, Cui Y, Ye Y, Purrunsing Y, Wang N (2018) Parathyroid hormone fragments: new targets for the diagnosis and treatment of chronic kidney disease-mineral and bone disorder. Biomed Res Int 2018:9619253
Cho G, Wu Y, Ackerman JL (2003) Detection of hydroxyl ions in bone mineral by solid-state NMR spectroscopy. Science 300:1123–1127
Choksi P, Jepsen KJ, Clines GA (2018) The challenges of diagnosing osteoporosis and the limitations of currently available tools. Clin Diabetes Endocrinol 4:12
Crandall CJ, Larson J, Cauley JA, Schousboe JT, LaCroix AZ, Robbins JA et al (2019) Do additional clinical risk factors improve the performance of fracture risk assessment tool (FRAX) among postmenopausal women? Findings from the women's health initiative observational study and clinical trials. JBMR Plus 3(12):e10239
Crawford RP, Cann CE, Keaveny TM (2003) Finite element models predict in vitro vertebral body compressive strength better than quantitative computed tomography. Bone 33(4):744–750
de Vernejoul MC, Kuntz D, Miravet L, Goutallier D, Ryckewaert A (1981) Bone histomorphometric reproducibility in normal patients. Calcif Tissue Int 33(4):369–374
Dempster DW, Compston JE, Drezner MK, Glorieux FH, Kanis JA, Malluche H et al (2013) Standardized nomenclature, symbols, and units for bone histomorphometry: a 2012 update of the report of the ASBMR Histomorphometry Nomenclature Committee. J Bone Miner Res 28 (1):2–17
Dempster DW, Zhou H, Recker RR, Brown JP, Recknor CP, Lewiecki EM et al (2018) Remodeling- and modeling-based bone formation with teriparatide versus denosumab: a longitudinal analysis from baseline to 3 months in the AVA study. J Bone Miner Res 33(2):298–306
Diab D, Watts N (2013) Secondary osteoporosis: differential diagnosis and workup. Clin Obstet Gynecol 56:686–693
Eastell R, Szulc P (2017) Use of bone turnover markers in postmenopausal osteoporosis. Lancet Diabetes Endocrinol 5(11):908–923
Eastell R, Garnero P, Audebert C, Cahall DL (2012) Reference intervals of bone turnover markers in healthy premenopausal women: results from a cross-sectional European study. Bone 50 (5):1141–1147
Engelke K (2017) Quantitative computed tomography-current status and new developments. J Clin Densitom 20(3):309–321
Erman J, Ott SM (1988) Accuracy of dual photon absorptiometry in excised femurs. J Nucl Med 29:1853–1855
Faibish D, Ott S, Boskey A (2006) Mineral changes in osteoporosis: a review. Clin Orthop Relat Res 443:28–38
Ferrari SL, Abrahamsen B, Napoli N, Akesson K, Chandran M, Eastell R et al (2018) Diagnosis and management of bone fragility in diabetes: an emerging challenge. Osteoporos Int 29 (12):2585–2596
Garn S (1972) The course of bone gain and the phases of bone loss. Orthop Clin North Am 3:503–520
Garnero P, Hausherr E, Chapuy MC, Marcelli C, Grandjean H, Muller C et al (1996) Markers of bone resorption predict hip fracture in elderly women: the EPIDOS prospective study. J Bone Miner Res 11(10):1531–1538
Gaudio A, Xourafa A, Rapisarda R, Zanoli L, Signorelli SS, Castellino P (2020) Hematological diseases and osteoporosis. Int J Mol Sci 21(10):4787
Greenblatt MB, Tsai JN, Wein MN (2017) Bone turnover markers in the diagnosis and monitoring of metabolic bone disease. Clin Chem 63(2):464–474
Hans D, Baim S (2017) Quantitative ultrasound (QUS) in the management of osteoporosis and assessment of fracture risk. J Clin Densitom 20:322–333

Hans D, Dargent-Molina P, Schott AM, Sebert JL, Cormier C, Kotzki PO et al (1996) Ultrasonographic heel measurements to predict hip fracture in elderly women: the EPIDOS prospective study. Lancet 348(9026):511–514
Hans D, Fuerst T, Duboeuf F (1997) Quantitative ultrasound bone measurement. Eur Radiol 7 (Suppl 2):S43–S50
Hans D, Šteňová E, Lamy O (2017) The trabecular bone score (TBS) complements DXA and the FRAX as a fracture risk assessment tool in routine clinical practice. Curr Osteoporos Rep 15 (6):521–531
Hanson D, Eyre D (1996) Molecular site specificity of pyridinoline and pyrrole cross-links in type i collagen of human bone. J Biol Chem 271(43):26508–26516
Hassenkam T, Fantner G, Cutroni J, Weaver J, Morse D, Hansma P (2004) High-resolution AFM imaging of intact and fractured trabecular bone. Bone 35:4–10
Hernandez CJ, van der Meulen MC (2017) Understanding bone strength is not enough. J Bone Miner Res 32(6):1157–1162
Herrera S, Diez-Perez A (2017) Clinical experience with microindentation in vivo in humans. Bone 95:175–182
Hofbauer L, Hamann C, Ebeling P (2010) Approach to the patient with secondary osteoporosis. Eur J Endocrinol 162:1009–1020
Huopio J, Kröger H, Honkanen R, Jurvelin J, Saarikoski S, Alhava E (2004) Calcaneal ultrasound predicts early postmenopausal fractures as well as axial BMD. A prospective study of 422 women. Osteoporos Int 15(3):190–195
Ivaska KK, Gerdhem P, Akesson K, Garnero P, Obrant KJ (2007) Effect of fracture on bone turnover markers: a longitudinal study comparing marker levels before and after injury in 113 elderly women. J Bone Miner Res 22(8):1155–1164
Jain RK, Vokes T (2017) Dual-energy X-ray absorptiometry. J Clin Densitom 20(3):291–303
Jeong H, Kim D (2019) Bone diseases in patients with chronic liver disease. Int J Mol Sci 20:4270
Jerban S, Ma Y, Wong JH, Nazaran A, Searleman A, Wan L et al (2019) Ultrashort echo time magnetic resonance imaging (UTE-MRI) of cortical bone correlates well with histomorphometric assessment of bone microstructure. Bone 123:8–17
Jerban S, Ma Y, Jang H, Namiranian B, Le N, Shirazian H et al (2020) Water proton density in human cortical bone obtained from ultrashort echo time (UTE) MRI predicts bone microstructural properties. Magn Reson Imaging 67:85–89
Jergas M, Uffmann M, Escher H, Glüer CC, Young KC, Grampp S et al (1994) Interobserver variation in the detection of osteopenia by radiography and comparison with dual x-ray absorptiometry of the lumbar spine. Skelet Radiol 23(3):195–199
Johannesdottir F, Aspelund T, Reeve J, Poole KE, Sigurdsson S, Harris TB et al (2013) Similarities and differences between sexes in regional loss of cortical and trabecular bone in the mid-femoral neck: the AGES-Reykjavik longitudinal study. J Bone Miner Res 28(10):2165–2176
Johnell O, Kanis JA, Oden A, Johansson H, De Laet C, Delmas P et al (2005) Predictive value of BMD for hip and other fractures. J Bone Miner Res 20(7):1185–1194
Kanis J, Melton J, Christiansen C, Johnston C, Khaltaev N (1994) The diagnosis of osteoporosis. J Bone Mineral Res 9:1137–1141
Kanis JA, Oden A, Johnell O, Johansson H, De Laet C, Brown J et al (2007) The use of clinical risk factors enhances the performance of BMD in the prediction of hip and osteoporotic fractures in men and women. Osteoporos Int 18(8):1033–1046
Kanis JA, Cooper C, Rizzoli R, Reginster JY, Scientific Advisory Board of the European Society for C, Economic Aspects of O et al (2019) European guidance for the diagnosis and management of osteoporosis in postmenopausal women. Osteoporos Int 30(1):3–44
Keaveny TM (2010) Biomechanical computed tomography-noninvasive bone strength analysis using clinical computed tomography scans. Ann N Y Acad Sci 1192:57–65
Krege JH, Lane NE, Harris JM, Miller PD (2014) PINP as a biological response marker during teriparatide treatment for osteoporosis. Osteoporos Int 25(9):2159–2171

LaCroix AZ, Lee JS, Wu L, Cauley JA, Shlipak MG, Ott S et al (2008) Cystatin-C, renal function and incidence of hip fracture in postmenopausal women. J Am Soc Geriatr Soc 56:1434–1441

Langton CM, Palmer SB, Porter RW (1984) The measurement of broadband ultrasonic attenuation in cancellous bone. Eng Med 13(2):89–91

Lekamwasam S, Lenora RS (2003) Effect of leg rotation on hip bone mineral density measurements. J Clin Densitom 6(4):331–336

Leonard MB, Wehrli FW, Ziolkowski SL, Billig E, Long J, Nickolas TL et al (2019) A multi-imaging modality study of bone density, bone structure and the muscle – bone unit in end-stage renal disease. Bone 127:271–279

Leslie WD, Aubry-Rozier B, Lamy O, Hans D (2013) TBS (trabecular bone score) and diabetes-related fracture risk. J Clin Endocrinol Metab 98(2):602–609

Leslie WD, Johansson H, Kanis JA, Lamy O, Oden A, McCloskey EV et al (2014) Lumbar spine texture enhances 10-year fracture probability assessment. Osteoporos Int 25(9):2271–2277

Looker AC, Wahner HW, Dunn WL, Calvo MS, Harris TB, Heyse SP et al (1998) Updated data on proximal femur bone mineral levels of US adults. Osteoporos Int 8(5):468–489

Lu Y, Fuerst T, Hui S, Genant H (2001) Standardization of bone mineral density at femoral neck, trochanter and ward's triangle. Osteoporos Int 12:438–444

Luo Q, Leng H, Wang X, Zhou Y, Rong Q (2014) The role of water and mineral-collagen interfacial bonding on microdamage progression in bone. J Orthop Res 32(2):217–223

Malluche HH, Mawad H, Monier-Faugere MC (2007) Bone biopsy in patients with osteoporosis. Curr Osteoporos Rep 5(4):146–152

Martineau P, Silva BC, Leslie WD (2017) Utility of trabecular bone score in the evaluation of osteoporosis. Curr Opin Endocrinol Diabetes Obes 24(6):402–410

McLeod KM, Johnson S, Rasali D, Verma A (2015) Discriminatory performance of the calcaneal quantitative ultrasound and osteoporosis self-assessment tool to select older women for dual-energy x-ray absorptiometry. J Clin Densitom 18(2):157–164

Mirza FS, Padhi ID, Raisz LG, Lorenzo JA (2010) Serum sclerostin levels negatively correlate with parathyroid hormone levels and free estrogen index in postmenopausal women. J Clin Endocrinol Metab 95(4):1991–1997

Moayyeri A, Adams JE, Adler RA, Krieg MA, Hans D, Compston J et al (2012) Quantitative ultrasound of the heel and fracture risk assessment: an updated meta-analysis. Osteoporos Int 23 (1):143–153

Muller R, Van Campenhout H, Van Damme B, Van Der Perre G, Dequeker J, Hildebrand T et al (1998) Morphometric analysis of human bone biopsies: a quantitative structural comparison of histological sections and micro-computed tomography. Bone 23:59–66

Neer RM (1992) The utility of single-photon absorptiometry and dual-energy x-ray absorptiometry. J Nucl Med 33(1):170–171

Newman C, Creecy A, Granke M, Nyman JS, Tian N, Hammond MA et al (2016) Raloxifene improves skeletal properties in an animal model of cystic chronic kidney disease. Kidney Int 89:95

Nishiyama K (2019) Advanced techniques of bone mass measurements and strength in adults. In: Bilezekien J (ed) Primer on the metabolic bone diseases and disorders of mineral metabolis. Wiley, Hoboken

Nizet A, Cavalier E, Stenvinkel P, Haarhaus M, Magnusson P (2020) Bone alkaline phosphatase: an important biomarker in chronic kidney disease –mineral and bone disorder. Clin Chim Acta 501:198–206

Oleksik A, Ott S, Vedi S, Bravenboer N, Compston J, Lips P (2000) Bone structure in patients with low bone mineral density with or without vertebral fractures. J Bone Mineral Res 15:1368–1375

Ott S (2002) Histomorphometric analysis of bone remodeling. In: Bilezikian J, Raisz L, Rodan GA (eds) Principles of bone biology. Academic Press, San Diego, pp 303–320

Ott SM (2016) Bone strength: more than just bone density. Kidney Int 89(1):16–19

Ott SM (2018) Cortical or trabecular bone: what's the difference? Am J Nephrol 47(6):373–375

Ott SM, O'Hanlan M, Lipkin EW, Newell-Morris L (1997) Evaluation of vertebral volumetric vs. areal bone mineral density during growth. Bone 20(6):553–556

Parfitt AM, Drezner MK, Glorieux FH, Kanis JA, Malluche H, Meunier PJ et al (1987) Bone histomorphometry: standardization of nomenclature, symbols, and units. J Bone Mineral Res 2:595–610

Patel CG, Yee AJ, Scullen TA, Nemani N, Santo L, Richardson PG et al (2014) Biomarkers of bone remodeling in multiple myeloma patients to tailor bisphosphonate therapy. Clin Cancer Res 20 (15):3955–3961

Qin YX, Lin W, Mittra E, Xia Y, Cheng J, Judex S et al (2013) Prediction of trabecular bone qualitative properties using scanning quantitative ultrasound. Acta Astronaut 92(1):79–88

Raje N, Vescio R, Montgomery CW, Badros A, Munshi N, Orlowski R et al (2016) Bone marker-directed dosing of zoledronic acid for the prevention of skeletal complications in patients with multiple myeloma: results of the Z-MARK study. Clin Cancer Res 22(6):1378–1384

Recker RR, Akhter MP, Lappe JM, Watson P (2018) Bone histomorphometry in transiliac biopsies from 48 normal, healthy men. Bone 111:109–115

Riggs B, Hodgson S, O'Fallon M, Chao E, Wahner H, Mums J et al (1990) Fluoride treatment on the fracture rate in postmenopausal women with osteoporosis. New Engl J Med 322:802–809

Rizvi B, Da Silva E, Slatkovska L, Cheung AM, Tavakkoli J, Pejović-Milić A (2016) Technical note: bone mineral density measurements of strontium-rich trabecular bone-mimicking phantoms using quantitative ultrasound. Med Phys 43(11):5817

Roschger P, Fratzl P, Klaushofer K, Rodan G (1997) Mineralization of cancellous bone after alendronate and sodium fluoride treatment: a quantitative backscattered electron imaging study on minipig ribs. Bone 20(5):393–397

Roschger P, Paschalis EP, Fratzl P, Klaushofer K (2008) Bone mineralization density distribution in health and disease. Bone 42(3):456–466

Rozental TD, Walley KC, Demissie S, Caksa S, Martinez-Betancourt A, Parker AM et al (2018) Bone material strength index as measured by impact microindentation in postmenopausal women with distal radius and hip fractures. J Bone Miner Res 33(4):621–626

Rufus-Membere P, Holloway-Kew KL, Diez-Perez A, Kotowicz MA, Pasco JA (2018) Feasibility and tolerability of bone impact microindentation testing: a cross-sectional, population-based study in Australia. BMJ Open 8(12):e023959

Sartoris DJ, Resnick D (1989) Dual-energy radiographic absorptiometry for bone densitometry: current status and perspective. AJR Am J Roentgenol 152(2):241–246

Seeman E (2013) Age- and menopause-related bone loss compromise cortical and trabecular microstructure. J Gerontol A Biol Sci Med Sci 68(10):1218–1225

Shah S, Chiang C, Sikaris K, Lu Z, Bui M, Zebaze R et al (2017) Serum 25-hydroxyvitamin d insufficiency in search of a bone disease. J Clin Endocrinol Metab 102(7):2321–2328

Shapiro JR, Lewiecki EM (2017) Hypophosphatasia in adults: clinical assessment and treatment considerations. J Bone Miner Res 32(10):1977–1980

Steiner B, Dimai HP, Steiner H, Cirar S, Fahrleitner-Pammer A (2019) Prescreening for osteoporosis with quantitative ultrasound in postmenopausal white women. J Ultrasound Med 38 (6):1553–1559

Szulc P, Seeman E, Duboeuf F, Sornay-Rendu E, Delmas PD (2006) Bone fragility: failure of periosteal apposition to compensate for increased endocortical resorption in postmenopausal women. J Bone Miner Res 21(12):1856–1863

Szulc P, Boutroy S, Chapurlat R (2018) Prediction of fractures in men using bone microarchitectural parameters assessed by high-resolution peripheral quantitative computed tomography-the prospective STRAMBO study. J Bone Miner Res 33(8):1470–1479

Tong W, Glimcher MJ, Katz JL, Kuhn L, Eppell SJ (2003) Size and shape of mineralites in young bovine bone measured by atomic force microscopy. Calcif Tissue Int 72:592–598

Turner CH (2002) Biomechanics of bone: determinants of skeletal fragility and bone quality. Osteoporosis Int 13:97–104

Ueland T, Stilgren L, Bollerslev J (2019) Bone matrix levels of dickkopf and sclerostin are positively correlated with bone mass and strength in postmenopausal osteoporosis. Int J Mol Sci 20(12):2896

Varela A, Jolette J (2018) Bone toolbox: biomarkers, imaging tools, biomechanics, and histomorphometry. Toxicol Pathol 46(5):511–529

Vervloet MG, Brandenburg VM (2017) Circulating markers of bone turnover. J Nephrol 30 (5):663–670

Wahner HW, Looker A, Dunn WL, Walters LC, Hauser MF, Novak C (1994) Quality control of bone densitometry in a national health survey (NHANES III) using three mobile examination centers. J Bone Miner Res 9(6):951–960

Wehrli FW, Ladinsky GA, Jones C, Benito M, Magland J, Vasilic B et al (2008) In vivo magnetic resonance detects rapid remodeling changes in the topology of the trabecular bone network after menopause and the protective effect of estradiol. J Bone Miner Res 23(5):730–740

Woitge HW, Seibel MJ, Ziegler R (1996) Comparison of total and bone-specific alkaline phosphatase in patients with nonskeletal disorder or metabolic bone diseases. Clin Chem 42 (11):1796–1804

Wu Y, Ackerman JL, Strawich ES, Rey C, Kim HM, Glimcher MJ (2003) Phosphate ions in bone: identification of a calcium-organic phosphate complex by 31P solid-state NMR spectroscopy at early stages of mineralization. Calcif Tissue Int 72:610–626

Yamamoto M, Yamauchi M, Sugimoto T (2013) Elevated sclerostin levels are associated with vertebral fractures in patients with type 2 diabetes mellitus. J Clin Endocrinol Metab 98 (10):4030–4037

Zoch ML, Clemens TL, Riddle RC (2016) New insights into the biology of osteocalcin. Bone 82:42–49

Osteoporosis Therapeutics 2020

Roland Kocijan, Klaus Klaushofer, and Barbara M. Misof

Contents

Abstract

Numerous safe and efficient drug therapies are currently available to decrease risk of low trauma fractures in patients with osteoporosis including postmenopausal, male, and secondary osteoporosis. In this chapter, we give first an overview of the most important outcomes regarding fracture risk reduction, change in bone mineral density (BMD by DXA) and/or bone markers of the phase III clinical studies of well-established therapies (such as Bisphosphonates, Denosumab or

R. Kocijan · K. Klaushofer · B. M. Misof (✉)
Ludwig Boltzmann Institute of Osteology at Hanusch Hospital of OEGK and AUVA Trauma Centre Meidling, 1st Med. Dept. Hanusch Hospital, Vienna, Austria
e-mail: roland.kocijan@osteologie.lbg.ac.at; klaus.klaushofer@osteologie.at; barbara.misof@osteologie.lbg.ac.at

P. H. Stern (ed.), *Bone Regulators and Osteoporosis Therapy*,
Handbook of Experimental Pharmacology 262, https://doi.org/10.1007/164_2020_373

Teriparatide) and also novel therapies (such as Romosozumab or Abaloparatide) and highlight their mechanisms of action at bone tissue/material level. The latter understanding is not only essential for the choice of drug, duration and discontinuation of treatment but also for the interpretation of the clinical outcomes (in particular of eventual changes in BMD) after drug administration. In the second part of this chapter, we focus on the management of different forms of osteoporosis and give a review of the respective current guidelines for treatment. Adverse effects of treatment such as atypical femoral fractures, osteonecrosis of the jaw or influence of fracture healing are considered also in this context.

Keywords

Anabolic treatment · Antiresorptive treatment · Bone tissue and material quality · Osteoporosis guidelines · Osteoporosis management

Abbreviations

AFFs	Atypical femoral fractures
ALN	Alendronate
BMD	Bone mineral density
BMDD	Bone mineralization density distribution
BPs	Bisphosphonates
BSUs	Bone structural units
CTX	Carboxy-terminal collagen crosslinks (CTX)
DXA	Dual-energy X-ray absorptiometry
FRAX	Fracture Risk Assessment Tool
GCs	Glucocorticoids
GIOP	Glucocorticoid-induced osteoporosis
HRT	Hormone replacement therapy
ONJ	Osteonecrosis of the jaw
PINP	Procollagen I N-terminal propeptide
PTH	Parathyroid hormone
PTHrP	Parathyroid hormone-related protein
RIS	Risedronate
SERMs	Selective estrogen receptor modulators
TBS	Trabecular bone score

1 Introduction

Osteoporosis therapeutics are used to decrease fracture risk in postmenopausal, male, and secondary osteoporosis. The arsenal of anti-osteoporotic drugs has grown recently with the approval of novel treatment options in addition to older well-established therapies.

The gold standard for the diagnosis of osteoporosis is bone mineral density measurement by dual-energy X-ray absorptiometry (DXA). However, the majority of women sustaining osteoporotic fractures have T-scores above −2.5. Treatment might be advisable for patients with high fracture risk but who do not fulfill the DXA criteria for osteoporosis. In contrast, patients with low BMD but without any other risk factors might not benefit from anti-osteoporotic medication. Apart from the osteoporosis therapeutics' specific modes of action and clinical outcomes, agents might also have side effects and might influence fracture healing. Duration and/or sequence of therapies is another essential point to be considered. The use of osteo-anabolic agents, for instance, is limited in duration and repetition due to preclinical concerns regarding osteosarcoma risk. Moreover, a switch from osteo-anabolic to antiresorptive treatment seems mandatory to preserve the newly gained bone mass during anabolic therapy.

Thus, the important questions for successful treatment are: Whom to treat? What agent to use? When to start/stop therapy? How to continue after discontinuation? To help answer these questions, we present an overview on how anti-osteoporotic drugs affect clinical outcomes, bone tissue, and material quality in the first part of this chapter and report on the current treatment guidelines/recommendations in the second part.

2 Osteoporosis Therapeutics: Clinical Outcomes, Effects on Bone Tissue, and Material Quality

2.1 Background of Osteoporosis and the Influence of Osteoporosis Therapeutics on Bone Tissue/Material Quality

Osteoporotic patients suffer from low bone volume which might be associated with altered bone material quality leading to decreased mechanical competence of bone. Consequently, the strategy of osteoporosis therapeutics is to decelerate bone loss or to increase bone volume by either an antiresorptive or anabolic action. Apart from bone volume and microarchitecture, material properties play an important role in the mechanical properties of the material itself and in the overall mechanical integrity of bone (Seeman 2008; Fratzl et al. 2009). Thus, changes in bone tissue and material obtained from transiliac bone biopsy samples are important for establishing the safety of treatments as well as for the understanding of changes in BMD (by DXA) and beneficial effects on fracture risk due to treatment.

Osteoporosis is associated with accelerated bone loss, which is considered to be due to a negative bone balance during remodeling (caused by either decreased bone formation by the osteoblasts or greater resorption depth by the osteoclasts). The latter action is particularly harmful in high turnover postmenopausal osteoporosis (Thomsen et al. 1996). Histomorphometric assessment of the bone microstructure typically shows increases in trabecular separation and cortical porosity, the removal of trabecular elements, and decreases in trabecular bone volume and cortical thickness (Han et al. 1996, 1997). In fluorochrome-labeled biopsy samples, dynamic parameters of bone formation (mineralizing surface and bone formation rate) are increased, indicative of high bone turnover (Recker et al. 2018).

In a high bone turnover state, bone tissue age is on average decreased. The bone packets (bone structural units, BSUs) are on average younger as a higher percentage of bone volume/area undergoes remodeling and formation. Additionally, there is a high probability for the bone packets to be resorbed before they become old. As many of the bone material properties are dependent on tissue age (show a specific change from newly formed to mature bone matrix), they are affected by alterations in bone turnover. One of these material properties is the bone matrix mineralization. Once the mineralization process has started in newly formed osteoid, the calcium concentration increases at first rapidly (primary mineralization phase) and later slowly (secondary mineralization phase) until the final (plateau) level of mineralization is achieved (Fuchs et al. 2008; Akkus et al. 2003; Bala et al. 2013). For interpreting the mineralization pattern (the degree and the distribution of the calcium concentrations in the bone material), the bone mineralization density distribution (BMDD) was introduced as an important bone material quality characteristic (Roschger et al. 2008). In healthy cancellous bone, the BMDD reveals only minor variation with age (from 25 to 97 years), skeletal site, sex, or ethnic origin (Roschger et al. 2003; Boivin and Meunier 2002). This almost constancy enabled the establishment of general reference levels of bone matrix mineralization (Roschger et al. 2003).

It is well established how bone turnover influences the BMDD (Roschger et al. 2008; Boivin et al. 2009). High bone turnover shifts the BMDD toward lower calcium concentrations, while vice versa lower bone turnover causes a shift to higher calcium concentrations. Thus, in most cases postmenopausal osteoporosis is associated with reduced calcium concentrations of the bone material (Roschger et al. 2014).

Apart from the BMDD, the mineral composition and the matrix content and composition may be affected by and contribute to bone fragility in osteoporosis (Paschalis et al. 2016, 2019). We will mention only those key parameters for which treatment effects have been reported (Boskey 2013). In osteoporotic bone, decreased crystallinity, increased carbonate-to-phosphate ratio (Gourion-Arsiquaud et al. 2013), as well as abnormally high collagen cross-link ratios were observed (Gourion-Arsiquaud et al. 2013; Paschalis et al. 2004).

In addition to the modifications due to turnover changes, the dimensions (for instance, the thickness) of the hydroxyapatite particles embedded in the collagen matrix provide information about proper mineralization processes and basic aspects of the bone nanocomposite material (Fratzl et al. 2009). This is especially important when considering safety aspects of treatment. In particular, the altered structure of the bone nanocomposite after sodium fluoride treatment was found to be an unfavorable contributor to the deterioration of the mechanical competence of bone after treatment (Fratzl et al. 1994).

The action of the anti-osteoporotic agents on bone tissue/material is also important for the interpretation of therapy monitoring. It has to be noted that changes in

BMD (by DXA) might be reflecting changes in bone volume or in bone material density or in both (Fratzl et al. 2007). Consequently, BMD (by DXA) increases can be caused by increased bone volume and/or by increments of bone matrix mineralization during therapy. If the therapy agent contains bone-seeking heavy elements (as this is the case for strontium ranelate), the increases in BMD might be also due to the incorporation of this element to bone mineral (Blake and Fogelman 2007).

2.2 Effects of Antiresorptive Agents

2.2.1 Raloxifene and Hormone Replacement Therapy (HRT)

The declining estrogen levels during menopause are an important contributor to the pathogenesis of osteoporosis and predispose women to osteoporosis (Eastell 2006; Levin et al. 2018). Although estrogen therapy remains controversial (Valdes and Bajaj 2019), treatments with estrogen or with selective estrogen receptor modulators (SERMs) are approved for the treatment of postmenopausal osteoporosis and might provide skeletal benefits in postmenopausal osteoporosis (Ishtiaq et al. 2015) but have antagonistic effects on the breast and endometrium (Cranney and Adachi 2005). Previous studies suggested a decline in the risk for vertebral fractures and invasive breast cancer, especially estrogen receptor-positive invasive breast cancers; however, there is also an incline in the risk of fatal strokes and venous thromboembolism (Barrett-Connor et al. 2006). Moreover, a risk reduction in non-vertebral fracture risk was not observed for raloxifene (Ensrud et al. 2008).

Histomorphometric analysis of bone biopsy samples from treated patients showed decreases in dynamic indices of bone formation (Ott et al. 2002; Weinstein et al. 2003). In the few studies where bone matrix mineralization was analyzed in postmenopausal osteoporotic patients after treatment with estrogen or SERMs, an increase in degree of mineralization or mineral/matrix ratio was reported (Boivin and Meunier 2002; Boivin et al. 2005; Paschalis et al. 2003) consistent with the aforementioned histomorphometrically observed decrease in bone turnover. In line with the latter, the collagen cross-link ratio and the mineral crystallinity were also increased after treatment (Paschalis et al. 2003).

2.2.2 Bisphosphonates

Bisphosphonates (BPs) represent the major class of drugs for treatment of bone diseases associated with high bone turnover and have been used for osteoporosis treatment for about 50 years now. BPs reduce bone turnover rapidly and efficiently after initiation of therapy, as can be observed by the decrease in bone turnover markers as early as 1 week up to few weeks depending on the type of bisphosphonate (Naylor et al. 2016). Additionally, BPs lead to an increase in BMD at the hip and spine (Bone et al. 2004). For the oral bisphosphonates alendronate (ALN) and risedronate (RIS) as well as zoledronic acid, a reduction in the risk of vertebral fractures, hip fractures, and non-vertebral fractures was observed (Amiche et al. 2018), while a recent meta-analysis failed to provide evidence for reduction of hip fracture or non-vertebral fracture reduction after ibandronate (Barrionuevo et al. 2019).

The histomorphometric analysis of transiliac bone biopsy samples from BP-treated patients has shown a decrease in dynamic bone formation indices after BP (Bone et al. 1997; Chavassieux et al. 1997), while bone volume was unchanged (Chavassieux et al. 1997) or slightly increased (Recker et al. 2008). However, proof of smaller increases in bone volume by histomorphometry is challenging. Combined information from BMD and mineralization changes from similar patients was used for the detection of small increases in bone volume after RIS (Fratzl et al. 2007). Another important observation regarding BP therapy was that cortical porosity was decreased after ALN (Roschger et al. 2001), RIS (Borah et al. 2010), or ibandronate (Misof et al. 2014).

For short-term therapy with BPs up to 3 years, the frequently observed effect was an increase in the degree and a decrease in the heterogeneity of mineralization (Roschger et al. 2001, 2014; Zoehrer et al. 2006), which occurs in osteoporotic bone with commonly lower degree of mineralization at baseline. Interestingly, the higher the patient's bone turnover and the lower the mineralization at baseline, the larger was the increase in degree of mineralization after BP treatment (Misof et al. 2017). After long-term treatment (5 years and longer), bone matrix mineralization was found to be within reference levels (Roschger et al. 2010; Borah et al. 2006) without any adverse effect on the nanocomposite material (i.e., thickness of the mineral particles) (Roschger et al. 2010). Considering these results, it seems unlikely that the effect on bone matrix mineralization is a contributor to atypical femoral fractures (AFFs), which are considered to be a consequence of long-term bone turnover reduction by BPs (Shane et al. 2014; van der Meulen and Boskey 2012).

The relationship between the changes in bone matrix mineralization and those in bone turnover is well understood (Roschger et al. 2008; Boivin et al. 2009; Ruffoni et al. 2008; Bala et al. 2011; Boskey et al. 2005). Less is known about potential effects of BPs on the mineralization processes. Studies on bone from patients or treated animals suggested no significant effect of ALN or RIS on the temporal course of mineral accumulation in bone (Misof et al. 2020; Fuchs et al. 2011). However, the matrix mineralization in patients treated with zoledronic acid from the Health Outcomes and Reduced Incidence with Zoledronic Acid Once Yearly-Pivotal Fracture Trial (HORIZON-Pivotal Fracture trial) was increased beyond normal reference values (Misof et al. 2013). Together with observed alterations in newly formed bone matrix, these results might point toward an accelerated mineral accumulation during the mineralization processes in zoledronic acid treatment (Gamsjaeger et al. 2013).

After short-term treatment with BP, collagen cross-link ratio also showed on average a reduced heterogeneity (Boskey et al. 2009; Donnelly et al. 2012). Collagen cross-link ratios restricted to newly formed bone indicated differences between different types of BP (Hofstetter et al. 2012). Long-term (10 years) ALN use did not alter the bone material properties, with the exception of transient differences that were seen at actively forming trabecular surfaces exclusively (Hassler et al. 2015). Another study compared key spectroscopic parameters between patients receiving ALN for 10 years and those who stopped ALN after 5 years (Boskey et al. 2018). No differences were observed suggesting that the drug holiday had no influence on the bone material.

2.2.3 Denosumab

Denosumab is an antiresorptive agent that produces a rapid and ongoing suppression of bone turnover markers and an increase of BMD at the lumbar spine and hip (McClung et al. 2013). Denosumab treatment decreases vertebral, non-vertebral, and hip fracture risk in postmenopausal women up to 10 years of treatment (Cummings et al. 2009; Bone et al. 2017). Peripheral fracture risk (including the wrist, forearm, and humerus) decreased after long-term treatment with denosumab, indicating beneficial effects on both trabecular and cortical bone (Bilezikian et al. 2019; Ferrari et al. 2019). Besides improving BMD, denosumab also affects trabecular and cortical bone microstructure by reduction of cortical porosity (Reid et al. 2010; Zebaze et al. 2016), increase in cortical density and thickness (Poole et al. 2015), and increase in trabecular bone score (TBS) (Tsai et al. 2017; McClung et al. 2017; DiGregorio et al. 2015).

Bone histomorphometry and histologic evaluation of transiliac biopsy samples from denosumab-treated patients revealed normal bone microarchitecture; decrease in static and dynamic bone formation, without evidence of adverse effects on mineralization; or the formation of lamellar bone (Reid et al. 2010; Brown et al. 2014; Chavassieux et al. 2019). Cessation of denosumab for about 2 years showed that bone turnover reduction was reversible (Brown et al. 2011). Moreover, the effects on BMDD were found to be similar to those after BP, namely, an increase in the average degree and a decrease in the heterogeneity of mineralization in both cancellous and cortical compartments versus placebo-treated patients (Dempster et al. 2018).

In contrast to bisphosphonates, drug holiday or treatment interruption is not an option with denosumab, as discontinuers were at greater fracture risk than persistent users (Tripto-Shkolnik et al. 2020; Anastasilakis et al. 2017; Brown et al. 2016). An explanation for this might be the rise in bone resorption after cessation of denosumab, which is reflected by the rebound response of bone turnover markers. However, the exact mechanisms of the pathophysiology are unknown. The rebound effect was especially observed in patients without prior exposure to BPs. In BP-pre-treated subjects, as well as in patients after a single denosumab injection, carboxy-terminal collagen crosslinks (CTX) remained in the normal range in most patients (Uebelhart et al. 2017). A single infusion of zoledronic acid could prevent bone loss in denosumab discontinuers (Anastasilakis et al. 2019).

2.3 Effects of Anabolic Agents

2.3.1 Teriparatide and Abaloparatide

Teriparatide, a recombinant, human parathyroid hormone-fragment (rhPTH1–34), and abaloparatide, a parathyroid hormone-related protein (PTHrP), are osteo-anabolic therapeutics. Parathyroid hormone (PTH) 1–84, the pharmaceutical form of parathyroid hormone, was also approved for the therapy of osteoporosis, but is no longer available. While a continuous increase in PTH (e.g., in primary hyperparathyroidism) leads to bone resorption and bone loss, short and intermittent pulses initiate bone formation, explaining the osteo-anabolic effects of PTH, teriparatide, and abaloparatide. When PTH treatment is initiated, bone formation markers, such as procollagen I N-terminal propeptide (PINP), typically increase rapidly, while bone resorption markers, such as CTX, incline more slowly, until a crossover after a few months of therapy, indicating a closure of the bone-forming window. Treatment with teriparatide leads to a fast but reversible increase in BMD (Neer et al. 2001). A rise in BMD and greater improvements in bone strength and stiffness (assessed by high-resolution quantitative computed tomography-based finite element analysis) were also seen in patients with glucocorticoid-induced osteoporosis (GIOP) and teriparatide treatment when compared to RIS-treated subjects (Glüer et al. 2013).

When compared to BPs, greater reduction in the incidence of vertebral fractures was found for teriparatide (Kendler et al. 2018). In a recent meta-analysis including 23 randomized clinical trials, a significant, 56% risk reduction for hip fractures in patients on teriparatide therapy was shown, when compared to controls, while no evidence for a positive effect on humerus, forearm, or wrist fractures was observed (Díez-Pérez et al. 2019). Preclinical studies with PTH treatment on rats have raised concerns of an increased risk of osteosarcoma; however, a causal association could not be proven in humans (Andrews et al. 2012). PTH treatment is still restricted to 24 months of therapy.

As expected from an osteo-anabolic therapy, teriparatide and PTH analogous in general increase bone formation indices and in part also increase trabecular bone volume or cortical width in biopsy samples (Dempster et al. 2001; Lindsay et al. 2007; Cohen et al. 2013; Miki et al. 2004; Jiang et al. 2003; Paschalis et al. 2018). In addition, positive effects on cortical thickness of the hip were also shown in CT studies (Erikson et al. 2014).

As the discontinuation of teriparatide leads to a rapid decrease in BMD (Leder et al. 2009), subsequent antiresorptive therapy is required to preserve BMD and avoid fractures. ALN or denosumab was reported to be effective in maintaining and even increasing BMD after PTH therapy (Black et al. 2005; Leder et al. 2015a). On the other hand, antiresorptive-pre-treated patients might not benefit as well from osteo-anabolic treatment compared to treatment-naïve patients (Leder et al. 2015a; Obermayer-Pietsch et al. 2009; Ettinger et al. 2004). Combined anabolic with concomitant antiresorptive treatment seems to be a therapeutic option in patients with severe osteoporosis at high fracture risk (Cosman 2014; Seeman 2011).

The parathyroid hormone-related protein (PTHrP) abaloparatide has a comparable mode of action like teriparatide with a higher formation-to-resorption ratio reflected by higher BMD gains at the hip (Leder et al. 2015b). A significant reduction of vertebral fracture risk was shown after 18 months of treatment when compared to placebo (Miller et al. 2016). Similar to teriparatide, an antiresorptive treatment should be initiated after stopping abaloparatide. ALN after abaloparatide led to a continuous increase in BMD and fracture risk reduction (Cosman et al. 2017).

On the bone material level, treatment with teriparatide is known to cause a decrease in the average degree and an increase in the heterogeneity of bone mineralization (Paschalis et al. 2005; Misof et al. 2003) which could be fully explained by the increase in the percentage of newly formed, less mineralized bone areas in the treated patients. Clear differences in the effects of anabolic and antiresorptive treatment on bone matrix mineralization were reported (Dempster et al. 2016). Less is known about treatment with parathyroid hormone-related peptide abaloparatide. In the analysis of bone biopsy samples from treated patients, no evidence of adverse effects on histomorphometric indices of bone structure, formation, or resorption were reported (Moreira et al. 2017). Cortical porosity was increased which seems to be a typical, likely transient feature for treatment with PTH(1–84), PTH(1–34), or PTHrP (Moreira et al. 2017).

2.3.2 Romosozumab

Romosozumab blocks sclerostin, a potent inhibitor of bone formation, which is mainly produced by osteocytes and is known to be one of the major inhibitors of the osteoblast differentiation (MacDonald et al. 2009). The effect of lack of sclerostin was revealed in patients with sclerosteosis and van Buchem disease, two very rare autosomal-recessive bone disorders (Brunkow et al. 2001) characterized by high bone mass. The recently introduced romosozumab is an osteo-anabolic, monoclonal antibody, which leads to an increase in bone formation and a decrease in bone resorption (Padhi et al. 2011). Blocking sclerostin by romosozumab leads to osteoblast stimulation and bone formation.

In total, 19 clinical studies in which 14,000 patients were enrolled analyze the efficacy and side effects of romosozumab. Subcutaneously administered treatment led to dose-dependent increases in BMD (Padhi et al. 2011; McClung et al. 2014). Bone formation markers (PINP) increased rapidly. In contrast to teriparatide or abaloparatide, bone resorption markers (CTX) decreased simultaneously (Padhi et al. 2011), while other laboratory parameters such as calcium were unchanged. No significant safety concerns were observed in this study (Padhi et al. 2011). Binding antibodies were found in 20% of romosozumab patients (3% neutralizing antibodies); however, they did not influence pharmacodynamics, pharmacokinetics, or safety (McClung et al. 2014).

In the FRAME (FRActure study in postmenopausal woMen with ostEoporosis) phase III trial, a significant risk reduction of new vertebral fractures of 73% was observed after romosozumab and subsequent denosumab, compared to placebo and subsequent denosumab (Cosman et al. 2016). The study however failed to show a reduction in non-vertebral fracture risk, likely influenced by a low number of fractures in the placebo group (Cosman et al. 2016). In a post hoc analysis, investigating the role of regional background, a risk reduction for non-vertebral fractures was observed in the rest-of-world population, but not in patients from Latin America (Cosman et al. 2018a). The effectiveness of romosozumab in male patients with osteoporosis and fractures (except hip fractures) was shown in the BRIDGE trial (placeBo-contRolled study evaluating the effIcacy anD safety of romosozumab in treatinG mEn with osteoporosis). After 12 months a significantly higher BMD gain was observed for patients receiving romosozumab compared to placebo (Lewiecki et al. 2018). Subsequent alendronate or denosumab following romosozumab had beneficial effects on fracture risk compared to subsequent placebo as reported in the ARCH (Active-contRolled fraCture study in postmenopausal women with osteoporosis at High risk, Phase III) trial and in the FRAME Extension study (Saag et al. 2017; Lewiecki et al. 2019). Moreover, the BMD increase continued under denosumab treatment and did not reach a plateau (Lewiecki et al. 2019; Miyauchi et al. 2019). This indicates that BMD gain in patients receiving romosozumab followed by denosumab is higher than with ongoing denosumab therapy and that the rapid increase in BMD with romosozumab is an advantage when switching to an antiresorptive treatment (Cosman et al. 2018b). As has been described for teriparatide, osteo-anabolic treatment with romosozumab has to be followed by an antiresorptive agent, in order to preserve the acquired bone mass and to maintain a low fracture risk (McClung 2018; Kendler et al. 2019).

It has been noted that the anabolic effect of romosozumab is restricted to the first months of treatment. After 12 months of therapy, the bone-forming effect wanes. Furthermore, as also observed in other osteo-anabolic agents, increases in BMD with romosozumab seem to be smaller in antiresorptive-pre-treated patients compared to treatment-naïve subjects. However, romosozumab caused larger hip BMD increases than teriparatide in bisphosphonate-pre-treated patients (Langdahl et al. 2017).

At this time it is unknown how sclerostin antibody treatment affects bone material quality in human bone; however, for cynomolgus bone, no change in average mineralization was observed despite large increases in bone volume in this animal model (Ross et al. 2014).

3 Management of Osteoporosis 2020

3.1 Review of Osteoporosis Guidelines

Currently, more than 50 national guidelines for the diagnosis and treatment of osteoporosis are available, based on fracture risk or T-score-based thresholds. The thresholds for starting anti-osteoporotic therapy are usually based on the major osteoporotic and hip fracture risk, assessed by FRAX. Moreover, age-dependent intervention thresholds were reported to be practicable and cost-effective (Kanis et al. 2019). Treatment decision, solely based on BMD, is critical, due to low sensitivity of BMD for fracture prediction (Kanis et al. 2018). Nevertheless, osteoporosis therapeutics should be started in patients who have sustained a recent low trauma fracture, in order to prevent further fractures. Especially in women >65 years with an osteoporotic fracture, a further assessment is not necessary. However, in younger postmenopausal women, BMD measurements should be performed (Kanis et al. 2019).

According to the European guidance for the diagnosis and management of osteoporosis in postmenopausal women, oral bisphosphonates (ALN, RIS, and ibandronate) are recommended as first-line anti-osteoporotic drugs in patients, in whom oral bisphosphonates are not contraindicated. As an alternative, intravenous bisphosphonates or denosumab should be administered. Raloxifene and hormone replacement therapy are additional options. Teriparatide is indicated in patients with high fracture risk (Kanis et al. 2019).

Based on the recently published guidelines for the pharmacological management of osteoporosis in postmenopausal women by the Endocrine Society, bisphosphonates are again recommended as the initial treatment in patients at high fracture risk (Eastell et al. 2019). Two intravenous (zoledronic acid, annually, and ibandronate, quarterly) and three oral bisphosphonates (ALN, weekly; RIS, weekly or monthly; and ibandronate, monthly) are available in most countries. Bisphosphonate treatment should be re-evaluated after 3 (intravenous) to 5 (oral) years and after a new fracture (Kanis et al. 2019). The ASBMR Task Force suggests that a drug holiday can be considered in patients with low to moderate fracture risk after the respective treatment periods. However, patients at high risk should continue the bisphosphonate therapy or should switch to an alternative therapy (Adler et al. 2016). In patients receiving a drug holiday, fracture risk and BMD should be re-assessed every 2–4 years (Eastell et al. 2019).

Denosumab, administered twice yearly, is recommended as an alternative therapy in postmenopausal osteoporosis patients at high fracture risk and in osteoporosis patients with renal insufficiency. However, denosumab is not recommended in severe renal insufficiency (CKD-5 or dialysis) patients, due to the risk of severe hypocalcemia (Camacho et al. 2016). Denosumab was approved in 2010 by the FDA and EMA for the treatment of postmenopausal women with osteoporosis at high risk. The human monoclonal RANKL antibody is also approved for the treatment of men with osteoporosis at high risk for fracture and glucocorticoid-induced osteoporosis (GIOP) as well as men at high risk for fracture receiving androgen deprivation therapy for non-metastatic prostate cancer and women at high risk for fracture receiving aromatase inhibitors for breast cancer.

Fracture risk in denosumab users should be re-evaluated after 5–10 years (Eastell et al. 2019). As mentioned above, discontinuation or interruption of denosumab therapy is associated with a rebound of bone turnover and consequently increased vertebral fracture risk. Continuation of denosumab treatment up to 10 years or switching to an alternative antiresorptive therapy is recommended by the ECTS and the Endocrine Society (Eastell et al. 2019; Tsourdie et al. 2017). For those who switch to another therapy, treatment might be started 6 months after the last denosumab injection or when bone turnover markers such as CTX begin to rise (although the optimal time point is currently unclear). Bisphosphonate should be given when the effects of denosumab are no longer evident and bone turnover is increasing again.

Teriparatide was approved by the FDA (Food and Drug Administration) and EMA (European Medicines Agency) for the treatment of postmenopausal women with osteoporosis and men with hypogonadal and primary osteoporosis at high risk for fractures. Moreover, teriparatide is indicated for the therapy of glucocorticoid-induced osteoporosis (GIOP). Abaloparatide is indicated for the treatment of post-menopausal women with osteoporosis at high risk for fracture. It was approved by the FDA in 2017. However, abaloparatide is not available in Europe as in 2018 the Committee for Medicinal Products for Human Use (CHMP) expressed concerns that the main study failed to show the efficacy of abaloparatide in non-vertebral fracture risk reduction.

PTH analogs are especially indicated in patients with a positive history of osteoporotic fractures and multiple risk factors (especially GC use) and in patients who have failed other osteoporosis therapeutics, defined as fractures under antiresorptive therapy for more than 2 years (treatment failure). An osteo-anabolic therapy is limited to a maximum of 2 years. Patients who have completed an osteo-anabolic treatment must receive an antiresorptive osteoporosis therapy (Eastell et al. 2019).

Based on the Practice Guidelines of the American College of Physicians, hormone therapy or raloxifene should not be used for the treatment of osteoporotic women. Both therapies failed to show a significant risk reduction of non-vertebral or hip fracture risk. Moreover, both therapies are related to side adverse events including thromboembolism (Croke 2018). However, this guideline neglects osteo-anabolic treatment and led to discussion in the field of bone research (Kanis et al. 2018).

The American Association of Clinical Endocrinologists and American College of Endocrinology state in their 2016 clinical practice guidelines that oral agents should be used in patients at low to moderate fracture risk. Injectable agents, such as intravenous bisphosphonates, denosumab, or teriparatide, can be considered as initial treatment for patients at highest fracture risk. In contrast to the American College of Physician's (ACP) guideline, raloxifene is suggested as alternative drug for patients at high risk of vertebral, but not hip, fractures (Camacho et al. 2016).

Romosozumab was recently included to the latest guidelines for the management of postmenopausal osteoporosis by the Endocrine Society (Shoback et al. 2020). It has been approved in the USA and Canada for the treatment of osteoporosis in postmenopausal women with high fracture risk and is currently also available in Japan, South Korea, and Australia. In Europe, the concerns regarding cardiovascular side effects have delayed the decision of the EMA. However, romosozumab was approved for the treatment of "severe osteoporosis in postmenopausal women at high risk of fracture and with no history of myocardial infarction or stroke" in December 2019. Although adverse events and serious adverse events were balanced in the placebo-controlled FRAME study (Cosman et al. 2016), a numerical imbalance in cardiovascular serious adverse events has been observed in the alendronate-controlled study with 50 patients in the romosozumab group and 38 in the alendronate group reporting these events (odds ratio, 1.31; 95% CI, 0.85–2.00) (Saag et al. 2017). A potential risk of myocardial infarction as well as stroke and cardiovascular death has been proposed. Sclerostin was found not only in bone but also in the aorta and may act as a negative regulator of vascular calcification (Saag et al. 2017). Therefore, romosozumab should not be administered in patients who have had a serious cardiovascular history.

3.2 Glucocorticoid-Induced Osteoporosis (GIOP)

The development of conventional and biological disease-modifying drugs as well as small molecules has revolutionized the treatment of immunological disorders for many disciplines including rheumatology, neurology, gastroenterology, or dermatology. Nonetheless, 1–2% of the population still receives long-term glucocorticoids (GCs) (Compston 2018; Overman et al. 2013) GCs are known to tremendously influence bone turnover. Osteoclast differentiation is enhanced, leading to increased bone resorption which in particular can be observed within the first weeks of treatment. Mesenchymal stem cells mainly differentiate to adipocytes and not to osteoblasts, reflected by a low bone formation; moreover, GCs induce osteocyte apoptosis (Henneicke et al. 2014). Further risk factors for secondary osteoporosis in GC users are hypogonadism and an altered renal and intestinal calcium handling (Compston 2018). Consequently, fracture risk is significantly and dose-dependently increased in GC users, starting soon after initiation with the highest fracture risk for vertebral fractures (van Staa et al. 2000). According to the European guideline by DVO (Dachverband für Osteologie), anti-osteoporotic treatment should be started in postmenopausal women and men with high-dose GC therapy (Aprednislon $\geq$7.5 mg daily) for $\geq$3 months, if (1) the T-score is ≤ -1.5 at the lumbar spine, femoral neck, or total hip or (2) $\geq$1 vertebral fracture or (3) $\geq$3 peripheral fractures are present (DVO-Leitlinie (DVO-Guidelines) 2017). In case of moderate GC doses (Aprednislon $\geq$2.5 mg–7.5 mg daily) for $\geq$3 months, the DXA threshold for initiation of therapy should be raised for +1.0 T-scores to a maximum of −2.0 T-score. The evidence for inhaled GCs as a risk factor for low traumatic fractures is currently low. However, high-dose inhalative GC therapy should be considered a moderate risk factor for osteoporotic fractures in COPD patients >50 years (DVO-Leitlinie (DVO-Guidelines) 2017). The GIOP guidelines of the ACR (American College of Rheumatology) also include premenopausal women. This guideline is mainly based on the fracture risk, assessed by FRAX (Fracture Risk Assessment Tool) and the presence of osteoporotic fractures. In premenopausal women of childbearing potential, treatment with oral bisphosphonates is recommended, with teriparatide as the second line. Intravenous bisphosphonates and denosumab are suggested as alternative therapies in women at high risk, for whom oral bisphosphonates and teriparatide are inappropriate. For women not of childbearing potential and men, oral bisphosphonates are recommended as the drugs of choice. Other recommended drugs are intravenous bisphosphonates, teriparatide, and denosumab. Raloxifene should only be prescribed for postmenopausal women, if none of the abovementioned therapies are available (Buckley et al. 2017).

3.3 Osteoporosis Therapeutics After Fracture

Osteoporotic fractures are the main risk factors for secondary fractures. More than 25% of patients sustain a secondary fracture within 1 year after a vertebral fracture (Lindsay et al. 2005). As has been reported repeatedly, mortality is also significantly increased after osteoporotic fractures with the highest risk being for hip fractures (Kannegaard et al. 2010). Based on recently published data, more than 85% of patients do not receive a sufficient anti-osteoporotic treatment after hip fracture, despite the fact that appropriate anti-osteoporotic therapy decreases mortality (Behanova et al. 2019) and further fracture risk in these patients. Initiation of antiresorptive therapy was recommended at least 2 weeks after surgical treatment of hip fractures (Lyles et al. 2007). However, the early initiation of oral bisphosphonates after humerus fractures did not influence clinical or radiological outcomes of fracture healing (Seo et al. 2016). In a subanalysis of the FREEDOM trial, delayed fracture healing was not observed in denosumab users (Adami et al. 2012). Bisphosphonates, applied after fracture, increase callus size and mineralization and reduce callus remodeling. Moreover, an improvement in mechanical strength was suggested. Denosumab and raloxifene do not seem to have negative effects on bone repair. In an ESCEO (European Society for Clinical and Economic Aspects of Osteoporosis, Osteoarthritis and Musculoskeletal Diseases) Consensus Paper, no evidence for impaired fracture healing was reported, even after major surgery or when administered immediately after fracture (Goldhahn et al. 2012).

Vertebroplasty is a non-pharmaceutical option for the treatment of acute vertebral fractures (Anderson et al. 2012). Although there has been concern that vertebroplasty might increase secondary vertebral fracture risk due to changes in stiffness of the cemented region, meta-analyses of vertebroplasty revealed no increased secondary fracture risk compared to conservative treatment. Greater pain relief, functional recovery, as well as better life quality have been reported after vertebroplasty (Anderson et al. 2012; Zhang and Zhai 2019). Similar effects including pain relief and improvement in life quality were also shown for teriparatide treatment after vertebral fractures. Clinical outcomes were comparable, but the costs of conservative treatment were lower when compared to vertebroplasty (Ma et al. 2020). Teriparatide has also led to enhanced fracture healing after acute vertebral fracture with less progressive collapse than in untreated or bisphosphonate-treated subjects (Min et al. 2019). Some positive effects on callus formation and union time were suggested for teriparatide-treated patients with complete diaphyseal AFF (atypical femoral fractures) after intramedullary nailing (Shin et al. 2019). Based on the ESCEO Consensus Paper, teriparatide increases callus formation and improves biomechanical strength (Goldhahn et al. 2012). Therefore, teriparatide treatment after acute vertebral fracture or in case of delayed as well as non-union fracture might be an interesting alternative therapeutic approach. However, it has to be noted that the use of teriparatide for fracture healing is off-label, because currently no osteoporosis therapeutics are approved for the treatment of fractures (Roberts and Ke 2018). In case of fracture under ongoing teriparatide treatment, therapy does not have to be stopped. Teriparatide can also be initiated immediately after fracture.

Animal models also showed promising data for romosozumab. A higher bone mineral apposition rate at fracture callus, volumetric bone mineral density (vBMD), and bone volume were found in a mouse osteotomy model when compared to vehicle groups. Moreover, the maximum load of failure at the fracture callus assessed by mechanical testing showed significant benefits in the romosozumab-treated group (Cui et al. 2013).

3.4 Osteonecrosis of the Jaw (ONJ)

Osteonecrosis of the jaw (ONJ) is a severe complication of antiresorptive therapy. As ONJ is related to dose and duration of antiresorptive treatment, it occurs much more often in cancer patients who receive much higher doses of bisphosphonates or denosumab. More than 90% of ONJ cases were found in cancer patients receiving antiresorptive therapy (Khan et al. 2016). Several risk factors have been identified for development of ONJ. Invasive dental interventions such as tooth extraction, dental implants, or periodontal surgery as well as local infections are the main triggers for ONJ (Song 2019). A higher incidence of ONJ was recently found in postmenopausal denosumab users after oral procedures including dental implants, tooth extraction, or jaw surgery (0.68% versus 0.05% in patients without any procedure) (Watts et al. 2019). Other risk factors are diabetes mellitus, GC use, chronic inflammation, and smoking. It has been suggested that the incidence of ONJ can be reduced with preventive dental interventions including dental treatment before initiating antiresorptive drugs, antibiotic treatment before and after surgical procedures, and protection of oral health (Song 2019; Karna et al. 2018). The International Task Force on Osteonecrosis of the Jaw recommends that patients who undergo invasive oral surgery should stop antiresorptive therapy following the procedure. Treatment can be restarted after soft tissue healing has occurred (Khan et al. 2015). It is unclear whether that recommendation is of clinical significance, since bisphosphonates remain in the bone for many years. Based on a position paper, osteoporosis patients receiving oral bisphosphonates for <4 years are at low risk for ONJ and do not have to discontinue treatment prior dental surgery (Svejda et al. 2015).

ONJ has also been reported in patients receiving romosozumab. Routinely oral examination should therefore be performed prior to romosozumab therapy as well. For teriparatide, no increased risk for ONJ was described. Rather, ONJ improved within 6 months of teriparatide therapy and led to partial or complete remission (Ohbayashi et al. 2019).

3.5 Atypical Femoral Fractures (AFFs)

Atypical femoral fractures (AFFs) are another rare but severe complication of antiresorptive agents. The ASBMR Task Force revised the definition of AFFs in 2013 in order to highlight the unusual radiographic findings (Shane et al. 2014). AFFs are stress or insufficiency fractures, located at the femoral diaphysis with characteristic radiographic and clinical features. In addition, at least four of five major criteria are required for the diagnosis of AFF. The incidence of AFF is low and ranges from 50 to 130 cases per 100,000 patient-years (Shane et al. 2014; Starr et al.

2018). In a systematic review including 23 studies, the incidence of AFFs was reported to be even lower (3.0–9.8 per 100,000 person-years) (Khow et al. 2017). AFFs are more common in patients on long-term bisphosphonate treatment. However, the risk may decrease when bisphosphonates are stopped. In the FREEDOM trial on denosumab in postmenopausal osteoporosis, two women developed AFFs (8 per 100,000 participant years) after 3 and 7 years of therapy, respectively (Bone et al. 2017).

In addition to antiresorptive therapy, aGC (Girgis et al. 2010), diabetes mellitus, cardiovascular disease, sarcopenia, chronic kidney disease, and smoking are also associated with AFF (Muschitz et al. 2015). The benefit-risk ratio clearly favors bisphosphonates, especially for treatment durations from 3 to 5 years, because for each AFF, more than 1,200 fractures can be prevented (Black et al. 2019).

3.6 Adherence to Osteoporosis Therapeutics

Adherence of the patient to the osteoporosis therapy is essential for a desirable clinical outcome, especially fracture risk reduction. Higher compliance levels are associated with lower fracture rates (Siris et al. 2006). However, the persistence of daily or weekly orally administered bisphosphonates was reported to be low, with even lower persistence rates for daily administration compared with weekly administration (Downey et al. 2006). The pooled estimate of persistence with oral BP therapy was reported to be 45% after 12 months and 30% after 24 months (Karlsson et al. 2015). Compared with intravenous or oral bisphosphonates, persistence for denosumab seems to be much higher (Hadji et al. 2015). In a retrospective, observational study on more than 2,000 women, persistence with denosumab was 83% at 12 months and 62% at 24 months. A higher persistence was found in patients who had previously received osteoporosis treatment, calcium and vitamin D. A low persistence was found in glucocorticoid users (Karlsson et al. 2015). These data suggest that adequate compliance and persistence are major concerns with oral osteoporosis therapy.

4 Conclusion

As summarized in this chapter, there are a number of safe and efficient drug therapies available to treat osteoporosis and reduce the risk of low trauma fractures in patients with bone diseases. The enormous progress in the development of new therapeutic strategies, however, is still accompanied by severe clinical deficits in the real-world situation. For example, the lack of sensitive diagnostic approaches to detect individual patients with high fracture risk and undertreatment of patients after atraumatic fractures remain important issues. Even in countries with highly developed health systems, where most or all treatment options are available, only a small percentage of fracture patients are appropriately followed with secondary prevention. On the other hand, there are patients, often in early postmenopause, who receive medication although their fracture risk is relatively low and could be further improved by lifestyle modification and physical exercise as well as by calcium and vitamin D substitution. These problems remain major challenges for clinical routine, teaching, and post-graduate education. It is therefore mandatory to keep in mind the fundamentals of good clinical practice, consider the patients' individual pathophysiological mechanisms and risk factors, and then select appropriate, safe, efficient, and cost-effective treatments suitable for the specific situation of each patient.

References

Adami S, Libanati C, Boonen S (2012) Denosumab treatment in postmenopausal women with osteoporosis does not interfere with fracture-healing. J Bone Joint Surg Am 94:1–7

Adler RA, Fuleihan EH, Bauer DC et al (2016) Managing osteoporosis in patients on long-term bisphosphonate treatment: report of a task force of the American Society for Bone and Mineral Research. J Bone Miner Res 31(1):16–35

Akkus O, Polyakova-Akkus A, Adar F, Schaffler MB (2003) Aging of microstructural compartments in human compact bone. J Bone Miner Res 18:1012–1019

Amiche MA, Levesque LE, Gomes T, Adachi JD, Cadarette SM (2018) Effectiveness of oral bisphosphonates in reducing fracture risk among oral glucocorticoid users: three matched cohort analyses. J Bone Miner Res 33:419–429

Anastasilakis AD, Polyzos SA, Makras P, Aubry-Rozier B, Kauris S, Lamy O (2017) Clinical features of 24 patients with rebound-associated vertebral fractures after denosumab discontinuation: systematic review and additional cases. J Bone Miner Res 32(6):1291–1296

Anastasilakis AD, Papapoulos SE, Polyzos SA, Appelman-Dijkstra NM, Makras P (2019) Zoledronate for the prevention of bone loss in women discontinuing denosumab treatment. A prospective 2-year clinical trial. J Bone Miner Res. https://doi.org/10.1002/jbmr.3853

Anderson PA, Froyshteter BA, Tontz WL (2012) Meta-analysis of vertebral augmentation compared with conservative treatment for osteoporotic spinal fractures. J Bone Miner Res 28 (2):372–382

Andrews EB, Gilsenan AWAW, Midkiff K, Sherrill B, Wu Y, Mann BH, Masica D (2012) The US postmarketing surveillance study of adult osteosarcoma and teriparatide: study design and findings from the first 7 years. J Bone Miner Res 27(12):2429–2437

Bala Y, Farlay D, Chapurlat RD, Boivin G (2011) Modifications of bone material properties in postmenopausal osteoporotic women long-term treated with alendronate. Eur J Endocrinol 165:647–655

Bala Y, Farlay D, Boivin G (2013) Bone mineralization: from tissue to crystal in normal and pathological contexts. Osteoporos Int 24:2153–2166. https://doi.org/10.1007/s00198-012-2228-y

Barrett-Connor E, Mosca L, Collins P et al (2006) Raloxifene Use for The Heart (RUTH) Trial Investigators. Effects of raloxifene on cardiovascular events and breast cancer in postmenopausal women. N Engl J Med 355:125–137

Barrionuevo P, Kapoor E, Asi N et al (2019) Efficacy of pharmacological therapies for the prevention of fractures in postmenopausal women: a network meta-analysis. J Clin Endo Crinol Metab 104:1623–1630

Behanova M, Reichardt B, Stamm TA, Zwerina J, Klaushofer K, Kocijan R (2019) Treatment effects of bisphosphonates and denosumab on survival and refracture from real-world data of hip-fractured patients. Calcif Tissue Int 105:630–641

Bilezikian JP, Lin CJF, Brown JP et al (2019) Long-term denosumab treatment restores cortical bone loss and reduces fracture risk at the forearm and humerus: analyses from the FREEDOM extension cross-over group. Osteoporos Int 30:1855–1864

Black DM, Bilezikian JP, Ensrud KE et al (2005) One year of alendronate after one year of parathyroid hormone (1-84) for osteoporosis. N Engl J Med 353(6):555–565

Black DM, Abrahamsen B, Bouxsein ML, Einhorn T, Napoli N (2019) Atypical femur fractures: review of epidemiology, relationship to bisphosphonates, prevention, and clinical management. Endocr Rev 40:333–368

Blake GM, Fogelman I (2007) Effect of bone strontium on BMD measurements. J Clin Densitom 10:34–38

Boivin G, Meunier PJ (2002) The degree of mineralization of bone tissue measured by computerized quantitative contact microradiography. Calcif Tissue Int 70:503–511

Boivin G, Vedi S, Purdie DW, Compston JE, Meunier PJ (2005) Influence of estrogen therapy at conventional and high doses on the degree of mineralization of iliac bone tissue: a quantitative microradiographic analysis in postmenopausal women. Bone 36:562–567

Boivin G, Farlay D, Bala Y, Doublier A, Meunier PJ, Delmas PD (2009) Influence of remodeling on the mineralization of bone tissue. Osteoporos Int 20:1023–1026. https://doi.org/10.1007/s00198-009-0861-x

Bone HG, Downs RW Jr, Tucci JR, Harris ST, Weinstein RS, Licata AA, McClung MR, Kimmel DB, Gertz BJ, Hale E, Polvino WJ (1997) Dose-response relationships for alendronate treatment in osteoporotic elderly women. Alendronate elderly osteoporosis study centers. J Clin Endocrinol Metab 82(1):265–274

Bone HG, Hosking D, Devogelaer JP et al (2004) Ten years' experience with alendronate for osteoporosis in postmenopausal women. N Engl J Med 350:1189–1199

Bone HG, Wagman RB, Brandi ML, Brown JP, Chapurlat R, Cummings SR, Czerwiński E, Fahrleitner-Pammer A, Kendler DL, Lippuner K, Reginster JY, Roux C, Malouf J, Bradley MN, Daizadeh NS, Wang A, Dakin P, Pannacciulli N, Dempster DW, Papapoulos S (2017) 10 years of denosumab treatment in postmenopausal women with osteoporosis: results from the phase 3 randomised FREEDOM trial and open-label extension. Lancet Diabetes Endocrinol 5 (7):513–523. https://doi.org/10.1016/S2213-8587(17)30138-9

Borah B, Dufresne TE, Ritman EL, Jorgensen SM, Liu S, Chmielewski PA, Phipps RJ, Zhou X, Sibonga JD, Turner RT (2006) Long-term risedronate treatment normalizes mineralization and continues to preserve trabecular architecture: sequential triple biopsy studies with micro-computed tomography. Bone 39:345–352

Borah B, Dufresne T, Nurre J, Phipps R, Chmielewski P, Wagner L, Lundy M, Bouxsein M, Zebaze R, Seeman E (2010) Risedronate reduces intracortical porosity in women with osteoporosis. J Bone Miner Res 25(1):41–47

Boskey AL (2013) Bone composition: relationship to bone fragility and antiosteoporotic drug effects. Bonekey Rep 2:447. https://doi.org/10.1038/bonekey.2013.181. eCollection 2013. Review. Erratum in: Bonekey Rep. 2015;4:710

Boskey AL, DiCarlo E, Paschalis E, West P, Mendelsohn R (2005) Comparison of mineral quality and quantity in iliac crest biopsies from high- and low-turnover osteoporosis: an FT-IR microspectroscopic investigation. Osteoporos Int 16(12):2031–2038

Boskey AL, Spevak L, Weinstein RS (2009) Spectroscopic markers of bone quality in alendronate-treated postmenopausal women. Osteoporos Int 20(5):793–800. https://doi.org/10.1007/s00198-008-0725-9

Boskey AL, Spevak L, Ma Y, Wang H, Bauer DC, Black DM, Schwartz AV (2018) Insights into the bisphosphonate holiday: a preliminary FTIRI study. Osteoporos Int 29(3):699–705. https://doi.org/10.1007/s00198-017-4324-5

Brown JP, Dempster DW, Ding B, Dent-Acosta R, San Martin J, Grauer A, Wagman RB, Zanchetta J (2011) Bone remodeling in postmenopausal women who discontinued denosumab treatment: off-treatment biopsy study. J Bone Miner Res 26(11):2737–2744. https://doi.org/10.1002/jbmr.448

Brown JP, Reid IR, Wagman RB, Kendler D, Miller PD, Jensen JE, Bolognese MA, Daizadeh N, Valter I, Zerbini CA, Dempster DW (2014) Effects of up to 5 years of denosumab treatment on bone histology and histomorphometry: the FREEDOM study extension. J Bone Miner Res 29 (9):2051–2056. https://doi.org/10.1002/jbmr.2236

Brown JP, Ferrari S, Gilchrist N et al (2016) Discontinuation of denosumab and associated vertebral fracture incidence: analysis from a phase 3 placebo-controlled study of denosumab and its open-label extension. ACR/ARHP Annual Meeting Abstract Number: 1028

Brunkow ME, Gardner JC, Ness JV et al (2001) Bone dysplasia sclerosteosis results from loss of the SOST gene product, a novel cystine knot–containing protein. Am J Hum Genet 68:577–589

Buckley L, Guyatt G, Fink HA et al (2017) 2017 American college of rheumatology guideline for the prevention and treatment of glucocorticoid-induced osteoporosis. Arthritis Rheumatol 69 (8):1521–1537

Camacho PM, Petak SM, Binkley N et al (2016) American association of clinical endocrinologists and American college of endocrinology clinical practice guidelines for the diagnosis and treatment of postmenopausal osteoporosis. Endocr Pract 22(Suppl 4):1–42

Chavassieux PM, Arlot ME, Reda C, Wei L, Yates AJ, Meunier PJ (1997) Histomorphometric assessment of the long-term effects of alendronate on bone quality and remodeling in patients with osteoporosis. J Clin Invest 100(6):1475–1480

Chavassieux P, Portero-Muzy N, Roux JP, Horlait S, Dempster DW, Wang A, Wagman RB, Chapurlat R (2019) Reduction of cortical bone turnover and erosion depth after 2 and 3 years of denosumab: iliac bone histomorphometry in the FREEDOM trial. J Bone Miner Res 34 (4):626–631

Cohen A, Stein EM, Recker RR, Lappe JM, Dempster DW, Zhou H, Cremers S, McMahon DJ, Nickolas TL, Müller R, Zwahlen A, Young P, Stubby J, Shane E (2013) Teriparatide for idiopathic osteoporosis in premenopausal women: a pilot study. J Clin Endocrinol Metab 98 (5):1971–1981

Compston J (2018) Glucocorticoid-induced osteoporosis: an update. Endocrine 61:7–16

Cosman F (2014) Anabolic and antiresorptive therapy for osteoporosis: combination and sequential approaches. Curr Osteoporos Rep 12:385–395. https://doi.org/10.1007/s11914-014-0237-9

Cosman F, Crittenden DB, Adachi JD et al (2016) Romosozumab treatment in postmenopausal women with osteoporosis. N Engl J Med 375:1532–1543

Cosman F, Miller PD, Williams GC et al (2017) Eighteen months of treatment with subcutaneous abaloparatide followed by 6 months of treatment with alendronate in postmenopausal women with osteoporosis: results of the ACTIVExtend trial. Mayo Clin Proc 92(2):201–210

Cosman F, Crittenden DB, Ferrari S et al (2018a) Romosozumab FRAME study: a post hoc analysis of the role of regional background fracture risk on nonvertebral fracture outcome. J Bone Miner Res 33(8):1407–1416

Cosman F, Crittenden DB, Ferrari S et al (2018b) FRAME study: the foundation effect of building bone with 1 year of romosozumab leads to continued lower fracture risk after transition to denosumab. J Bone Miner Res 33(7):1219–1226

Cranney A, Adachi JD (2005) Benefit-risk assessment of raloxifene in postmenopausal osteoporosis. Drug Saf 28:721–730

Croke LM (2018) Treatment of low BMD and osteoporosis to prevent fractures: updated guideline from the ACP. Am Fam Physician 97(5):352–353

Cui L, Cheng H, Song C, Li C, Simone WS, Ke HZ, Li G (2013) Time-dependent effects of sclerostin antibody on a mouse fracture healing model. J Musculoskelet Neuronal Interact 13 (2):178–184

Cummings SR, Martin JS, McClung MR, FREEDOM Trial et al (2009) Denosumab for prevention of fractures in postmenopausal women with osteoporosis. N Engl J Med 361:756–765

Dempster DW, Cosman F, Kurland ES, Zhou H, Nieves J, Woelfert L, Shane E, Plavetić K, Müller R, Bilezikian J, Lindsay R (2001) Effects of daily treatment with parathyroid hormone on bone microarchitecture and turnover in patients with osteoporosis: a paired biopsy study. J Bone Miner Res 16(10):1846–1853

Dempster DW, Roschger P, Misof BM, Zhou H, Paschalis EP, Alam J, Ruff VA, Klaushofer K, Taylor KA (2016) Differential effects of teriparatide and zoledronic acid on bone mineralization

density distribution at 6 and 24 months in the SHOTZ study. J Bone Miner Res 31:1527–1535. https://doi.org/10.1002/jbmr.2825
Dempster DW, Brown JP, Fahrleitner-Pammer A, Kendler D, Rizzo S, Valter I, Wagman RB, Yin X, Yue SV, Boivin G (2018) Effects of long-term denosumab on bone histomorphometry and mineralization in women with postmenopausal osteoporosis. J Clin Endocrinol Metab. https://doi.org/10.1210/jc.2017-02669
Díez-Pérez A, Marin F, Erikson EF, Kendler DL, Krege JH, Delgado-Rodriguez M (2019) Effects of teriparatide on hip and upper limb fractures in patients with osteoporosis: a systematic review and meta-analysis. Bone 120:1–8
DiGregorio S, Del Rio L, Rodriguez-Tolra J, Bonel E, Garcia M, Winzenrieth R (2015) Comparison between different bone treatments on areal bone mineral density (aBMD) and bone microarchitectural texture as assessed by the trabecular bone score (TBS). Bone 75:138–143
Donnelly E, Meredith DS, Nguyen JT, Gladnick BP, Rebolledo BJ, Shaffer AD, Lorich DG, Lane JM, Boskey AL (2012) Reduced cortical bone compositional heterogeneity with bisphosphonate treatment in postmenopausal women with intertrochanteric and subtrochanteric fractures. J Bone Miner Res 27:672–678. https://doi.org/10.1002/jbmr.560
Downey TW, Foltz SH, Boccuzzi SJ, Omar MA, Kahler KH (2006) Adherence and persistence associated with the pharmacologic treatment of osteoporosis in a managed care setting. South Med J 99(6):570–575
DVO-Leitlinie (DVO-Guidelines) (2017). https://www.dv-osteologie.org/uploads/Leitlinie%202017/Finale%20Version%20Leitlinie%20Osteoporose%202017_end.pdf (in German)
Eastell R (2006) Pathogenesis of postmenopausal osteoporosis. In: Favus MJ (ed) Primer on the metabolic bone diseases and disorders of mineral metabolism. ASBMR, Washington, pp 259–262
Eastell R, Rosen CJ, Black DM, Cheung AM, Murad MH, Shoback D (2019) Pharmacological management of osteoporosis in postmenopausal women: an endocrine society* clinical practice guideline. J Clin Endocrinol Metab 104:1595–1622
Ensrud KE, Stock JL, Barrett-Connor E, Grady D, Mosca L, Khaw KT, Zhao Q, Agnusdei D, Cauley JA (2008) Effects of raloxifene on fracture risk in postmenopausal women: the Raloxifene use for the heart trial. J Bone Miner Res 23:112–120
Erikson EF, Keaveny TM, Gallagher ER, Krege JH (2014) Literature review: the effects of teriparatide therapy at the hip in patients. Bone 67:246–256
Ettinger B, San Martin J, Crans G, Pavo I (2004) Differential effects of teriparatide on BMD after treatment with raloxifene or alendronate. J Bone Miner Res 19:745–751
Ferrari S, Butler PW, Kendler DL et al (2019) Further nonvertebral fracture reduction beyond 3 years for up to 10 years of denosumab treatment. J Clin Endocrinol Metab 104:3450–3461
Fratzl P, Roschger P, Eschberger J, Abendroth B, Klaushofer K (1994) Abnormal bone mineralization after fluoride treatment in osteoporosis: a small-angle x-ray-scattering study. J Bone Miner Res 9:1541–1549
Fratzl P, Roschger P, Fratzl-Zelman N, Paschalis EP, Phipps R, Klaushofer K (2007) Evidence that treatment with risedronate in women with postmenopausal osteoporosis affects bone mineralization and bone volume. Calcif Tissue Int 81:73–80
Fratzl P, Gupta HS, Roschger P, Klaushofer K (2009) Bone nanostructure and its relevance for mechanical performance, disease and treatment. In: Vogel V (ed) Nanotechnology, Nanomedicine, vol 5. Wiley, Weinheim, pp 345–360
Fuchs RK, Allen MR, Ruppel ME, Diab T, Phipps RJ, Miller LM, Burr DB (2008) In situ examination of the time-course for secondary mineralization of Haversian bone using synchrotron Fourier transform infrared microspectroscopy. Matrix Biol 27:34–41
Fuchs RK, Faillace ME, Allen MR, Phipps RJ, Miller LM, Burr DB (2011) Bisphosphonates do not alter the rate of secondary mineralization. Bone 49(4):701–705. https://doi.org/10.1016/j.bone.2011.05.009

Gamsjaeger S, Hofstetter B, Zwettler E, Recker R, Gasser JA, Eriksen EF, Klaushofer K, Pa-schalis EP (2013) Effects of 3 years treatment with once-yearly zoledronic acid on the kinetics of bone matrix maturation in osteoporotic patients. Osteoporos Int 24(1):339–347

Girgis CM, Sher D, Seibel MJ (2010) Atypical femoral fractures and bisphosphonate use. N Engl J Med 362(19):1848–1849

Glüer CC, Marin F, Ringe JD et al (2013) Comparative effects of teriparatide and risedronate in glucocorticoid-induced osteoporosis in men: 18-month results of the EuroGIOPs trial. J Bone Miner Res 28(6):1355–1368

Goldhahn J, Feron JM, Kanis J et al (2012) Implications for fracture healing of current and new osteoporosis treatments: an ESCEO consensus paper. Calcif Tissue Int 90(5):343–353

Gourion-Arsiquaud S, Lukashova L, Power J, Loveridge N, Reeve J, Boskey AL (2013) Fourier transform infrared imaging of femoral neck bone: reduced heterogeneity of mineral-to-matrix and carbonate-to-phosphate and more variable crystallinity in treatment-naive fracture cases compared with fracture-free controls. J Bone Miner Res 28:150–161. https://doi.org/10.1002/jbmr.1724

Hadji P, Papaioannou N, Gielen E et al (2015) Persistence, adherence, and medication-taking behavior in women with postmenopausal osteoporosis receiving denosumab in routine practice in Germany, Austria, Greece, and Belgium: 12-month results from a European non-interventional study. Osteoporos Int 26(10):2479–2489

Han ZH, Palnitkar S, Rao DS, Nelson D, Parfitt AM (1996) Effect of ethnicity and age or menopause on the structure and geometry of iliac bone. J Bone Miner Res 11:1967–1975. Erratum in: J Bone Miner Res 1997;12:867

Han ZH, Palnitkar S, Rao DS, Nelson D, Parfitt AM (1997) Effects of ethnicity and age or menopause on the remodeling and turnover of iliac bone: implications for mechanisms of bone loss. J Bone Miner Res 12:498–508. Erratum in: J Bone Miner Res 1999;14:660

Hassler N, Gamsjaeger S, Hofstetter B, Brozek W, Klaushofer K, Paschalis EP (2015) Effects of long-term alendronate treatment on postmenopausal osteoporosis bone material properties. Osteoporos Int 26(1):339–352. https://doi.org/10.1007/s00198-014-2929-5

Henneicke H, Gasparini SJ, Brennan-Speranza TC, Zhou H, Seibel MJ (2014) Glucocorticoids and bone: local effects and systemic implications. Trends Endocrinol Metab 25(4):197–211

Hofstetter B, Gamsjaeger S, Phipps RJ, Recker RR, Ebetino FH, Klaushofer K, Paschalis EP (2012) Effects of alendronate and risedronate on bone material properties in actively forming trabecular bone surfaces. J Bone Miner Res 27(5):995–1003. https://doi.org/10.1002/jbmr.1572

Ishtiaq S, Fogelman I, Hampson G (2015) Treatment of post-menopausal osteoporosis: beyond bisphosphonates. J Endocrinol Invest 38:13–29. https://doi.org/10.1007/s40618-014-0152-z

Jiang Y, Zhao JJ, Mitlak BH, Wang O, Genant HK, Eriksen EF (2003) Recombinant human parathyroid hormone (1-34) [teriparatide] improves both cortical and cancellous bone structure. J Bone Miner Res 18(11):1932–1941

Kanis JA, Cooper C, Rizzoli R, Reginster JY (2018) Guideline of the American College of Physicians on the treatment of osteoporosis. Osteoporos Int 29(7):1505–1510

Kanis JA, Cooper C, Rizzoli R, Reginster JY (2019) European guidance for the diagnosis and management of osteoporosis in postmenopausal women. Osteoporos Int 30(1):3–44

Kannegaard PN, van der Mark SN, Eiken P, Abrahamsen P (2010) Excess mortality in men compared with women following a hip fracture. National analysis of comedications, comorbidity and survival. Age Aging 39(2):203–209

Karlsson L, Lundkvist J, Psachoulia E, Intorcia M, Ström O (2015) Persistence with denosumab and persistence with oral bisphosphonates for the treatment of postmenopausal osteoporosis: a retrospective, observational study, and a meta-analysis. Osteoporos Int 26(10):2401–2411

Karna H, Gonzalez J, Harveen S, Radia HS, Sedghizadeh PP, Enciso R (2018) Risk-reductive dental strategies for medication related osteonecrosis of the jaw among cancer patients: a systematic review with meta-analyses. Oral Oncol 85:15–23

Kendler DL, Marin F, Zerbini CAF et al (2018) Effects of teriparatide and risedronate on new fractures in post-menopausal women with severe osteoporosis (VERO): a multicentre, double-blind, double-dummy, randomised controlled trial. Lancet 391:230–240

Kendler DL, Bone HG, Massari F et al (2019) Bone mineral density gains with a second 12-month course of romosozumab therapy following placebo or denosumab. Osteoporos Int 30 (12):2437–2448

Khan AA, Morrison A, Hanley DA (2015) Diagnosis and management of osteonecrosis of the jaw: a systematic review and international consensus. J Bone Miner Res 30(1):3–23

Khan A, Morrison A, Cheung A, Hashem W, Compston J (2016) Osteonecrosis of the jaw (ONJ): diagnosis and management in 2015. Osteoporos Int 27:853–859

Khow KS, Shibu P, Yu SCY, Chehade MJ, Visvanathan R (2017) Epidemiology and postoperative outcomes of atypical femoral fractures in older adults: a systematic review. J Nutr Health Aging 21(1):83–91

Langdahl BL, Libanati C, Crittenden DB et al (2017) Romosozumab (sclerostin monoclonal antibody) versus teriparatide in postmenopausal women with osteoporosis transitioning from oral bisphosphonate therapy: a randomised, open-label, phase 3 trial. Lancet 390:1585–1594

Leder BZ, Neer R, Wyland JJ, Lee HW, Burnett SAM, Finkelstein J (2009) Effects of teriparatide treatment and discontinuation in postmenopausal women and eugonadal men with osteoporosis. J Clin Endocrinol Metab 94(8):2915–2921

Leder BZ, Tsai JN, Uihlein AV, Wallaca PM, Lee H, Neer RN (2015a) Denosumab and teriparatide transitions in postmenopausal osteoporosis (the DATA-switch study): extension of a randomised controlled trial. Lancet 386(9999):1147–1155

Leder B, O'Dea LSL, Zanchetta JR, Kumar P, Banks K, McKay K, Lyttle R, Hattersley G et al (2015b) Effects of abaloparatide, a human parathyroid hormone-related peptide analog, on bone mineral density in postmenopausal women with osteoporosis. J Clin Endocrinol Metab 100:697–706

Levin VA, Jiang X, Kagan R (2018) Estrogen therapy for osteoporosis in the modern era. Osteoporos Int 29:1049–1055. https://doi.org/10.1007/s00198-018-4414-z

Lewiecki EM, Blicharski T, Goemaere S et al (2018) A phase III randomized placebo-controlled trial to evaluate efficacy and safety of romosozumab in men with osteoporosis. J Clin Endocrinol Metab 103:3183–3193

Lewiecki EM, Dinavahi RV, Lazaretti-Castro M et al (2019) One year of romosozumab followed by two years of denosumab maintains fracture risk reductions: results of the FRAME extension study. J Bone Miner Res 34(3):419–428

Lindsay R, Burge RT, Strauss DM (2005) One year outcomes and costs following a vertebral fracture. Osteoporos Int 16:78–85

Lindsay R, Zhou H, Cosman F, Nieves J, Dempster DW, Hodsman AB (2007) Effects of a one-month treatment with PTH(1-34) on bone formation on cancellous, endocortical, and periosteal surfaces of the human ilium. J Bone Miner Res 22(4):495–502

Lyles KW, Colón-Emeric CS, Magaziner JS (2007) Zoledronic acid and clinical fractures and mortality after hip fracture. N Engl J Med 357(18):1799–1809

Ma Y, Wua X, Xiaoa X, Maa Y, Fenga L, Yanb W, Chena J, Yan D (2020) Effects of teriparatide versus percutaneous vertebroplasty on pain relief, T quality of life and cost-effectiveness in postmenopausal females with acute osteoporotic vertebral compression fracture: a prospective cohort study. Bone 31. https://doi.org/10.1016/j.bone.2019.115154

MacDonald BT, Tamai K, He X (2009) Wnt/b-catenin signaling: components, mechanisms, and diseases. Dev Cell 17:9–26

McClung MR (2018) Romosozumab for the treatment of osteoporosis. Osteoporosis Sarcopenia 4:11–15

McClung MR, Lippuner K, Brandi ML et al (2013) Effect of denosumab on bone mineral density and biochemical markers of bone turnover: 8-year results of a phase 2 clinical trial. Osteoporos Int 24:227–235

McClung MR, Grauer A, Boonen S et al (2014) Romosozumab in postmenopausal women with low bone mineral density. N Engl J Med 370:412–420

McClung MR, Lippuner K, Brandi ML et al (2017) Effect of denosumab on trabecular bone score in postmenopausal women with osteoporosis. Osteoporos Int 28(10):2967–2973

Miki T, Nakatsuka K, Naka H, Masaki H, Imanishi Y, Ito M, Inaba M, Morii H, Nishizawa Y (2004) Effect and safety of intermittent weekly administration of human parathyroid hormone 1-34 in patients with primary osteoporosis evaluated by histomorphometry and microstructural analysis of iliac trabecular bone before and after 1 year of treatment. J Bone Miner Metab 22 (6):569–576

Miller PD, Hattersley G, Riis BJ et al (2016) Effect of abaloparatide vs placebo on new vertebral fractures in postmenopausal women with osteoporosis: a randomized clinical trial. JAMA 316 (7):722–733

Min HK, Ahn JH, Ha KY et al (2019) Effects of anti-osteoporosis medications on radiological and clinical results after acute osteoporotic spinal fractures: a retrospective analysis of prospectively designed study. Osteoporos Int 30(11):2249–2256

Misof BM, Roschger P, Cosman R, Kurland ES, Tesch W, Messmer P, Dempster DW, Nieves J, Shane E, Fratzl P, Klaushofer K, Bilezikian J, Lindsay R (2003) Effects of intermittent parathyroid hormone administration on bone mineralization density distribution in iliac crest biopsies from patients with osteoporosis: a paired study before and after treatment. J Clin Endocrinol Metab 88:1150–1156

Misof BM, Roschger P, Gabriel D, Paschalis EP, Eriksen EF, Recker RR, Gasser JA, Klaushofer K (2013) Annual intravenous zoledronic acid for three years increased cancellous bone matrix mineralization beyond normal values in the HORIZON biopsy cohort. J Bone Miner Res 28 (3):442–448

Misof BM, Patsch JM, Roschger P, Muschitz C, Gamsjaeger S, Paschalis EP, Prokop E, Klaushofer K, Pietschmann P, Resch H (2014) Intravenous treatment with ibandronate normalizes bone matrix mineralization and reduces cortical porosity after two years in male osteoporosis: a paired biopsy study. J Bone Miner Res 29(2):440–449. https://doi.org/10.1002/jbmr.2035

Misof BM, Blouin S, Lueger S, Paschalis EP, Recker RR, Phipps R, Klaushofer K, Roschger P (2017) Baseline mineralizing surface determines the magnitude of the bisphosphonate effect on cortical bone mineralization in postmenopausal osteoporotic patients. J Musculoskelet Neuronal Interact 17(3):183–191

Misof BM, Roschger P, Zhou H, Nieves JW, Bostrom M, Cosman F, Lindsay R, Klaushofer K, Dempster DW (2020) No evidence for alteration in early secondary mineralization by either alendronate, teriparatide or combination of both in transiliac bone biopsy samples from post-menopausal osteoporotic patients. Bone Rep 12:100253. https://doi.org/10.1016/j.bonr.2020.100253

Miyauchi A, Dinavahi RV, Crittenden DB et al (2019) Increased bone mineral density for 1 year of romosozumab, vs placebo, followed by 2 years of denosumab in the Japanese subgroup of the pivotal FRAME trial and extension. Arch Osteoporos 14(1):59

Moreira CA, Fitzpatrick LA, Wang Y, Recker RR (2017) Effects of abaloparatide-SC (BA058) on bone histology and histomorphometry: the ACTIVE phase 3 trial. Bone 97:314–319. https://doi.org/10.1016/j.bone.2016.11.004

Muschitz C, Thaler HW, Dimai HP et al (2015) Atypical femoral fractures - ongoing and history of bone-specific therapy, concomitant diseases, medications, and survival. J Clin Densitom 19 (3):359–367

Naylor KE, Jacques RM, Paggiosi M, Gossiel F, Peel NF, McCloskey EV, Walsh JS, Eastell R (2016) Response of bone turnover markers to three oral bisphosphonate therapies in postmenopausal osteoporosis: the TRIO study. Osteoporos Int 27:21–31. https://doi.org/10.1007/s00198-015-3145-7

Neer RM, Arnaud CD, Zanchetta JR et al (2001) Effect of parathyroid hormone (1-34) on fractures and bone mineral density in postmenopausal women with osteoporosis. N Engl J Med 344 (19):1434–1441

Obermayer-Pietsch MF, McCloskey EV et al (2009) Effects of two years of daily teriparatide treatment on bmd in postmenopausal women with severe osteoporosis with and without prior antiresorptive treatment. J Bone Miner Res 23:1591–1600
Ohbayashi Y, Iwasaki A, Nakai F, Mashiba T, Miyake M (2019) A comparative effectiveness pilot study of teriparatide for medication-related osteonecrosis of the jaw: daily versus weekly administration. Osteoporos Int. https://doi.org/10.1007/s00198-019-05199-w
Ott SM, Oleksik A, Lu Y, Harper K, Lips P (2002) Bone histomorphometric and biochemical marker results of a 2-year placebo-controlled trial of raloxifene in postmenopausal women. J Bone Miner Res 17:341–348
Overman RA, Yeh JY, Deal C (2013) Prevalence of oral glucocorticoid usage in the United States: a general population perspective. Arthritis Care Res 65(2):294–298
Padhi D, Jang G, Stouch B, Fang L, Posvar E (2011) Single-dose, placebo-controlled, randomized study of AMG 785, a sclerostin monoclonal antibody. J Bone Miner Res 26(1):19–26
Paschalis EP, Boskey AL, Kassem M, Eriksen EF (2003) Effect of hormone replacement therapy on bone quality in early postmenopausal women. J Bone Miner Res 18:955–959
Paschalis EP, Shane E, Lyritis G, Skarantavos G, Mendelsohn R, Boskey AL (2004) Bone fragility and collagen cross-links. J Bone Miner Res 19:2000–2004
Paschalis EP, Glass EV, Donley DW, Eriksen EF (2005) Bone mineral and collagen quality in iliac crest biopsies of patients given teriparatide: new results from the fracture prevention trial. J Clin Endocrinol Metab 90:4644–4649
Paschalis EP, Fratzl P, Gamsjaeger S, Hassler N, Brozek W, Eriksen EF, Rauch F, Glorieux FH, Shane E, Dempster D, Cohen A, Recker R, Klaushofer K (2016) Aging versus postmenopausal osteoporosis: bone composition and maturation kinetics at actively-forming trabecular surfaces of female subjects aged 1 to 84 years. J Bone Miner Res 31:347–357. https://doi.org/10.1002/jbmr.2696
Paschalis EP, Krege JH, Gamsjaeger S et al (2018) Teriparatide treatment increases mineral content and volume in cortical and trabecular bone of iliac crest: a comparison of infrared imaging with X-ray–based bone assessment techniques. J Bone Miner Res 33(12):2230–2235
Paschalis EP, Klaushofer K, Hartmann MA (2019) Material properties and osteoporosis. F1000Res 8. https://doi.org/10.12688/f1000research.18239.1
Poole KE, Treece GM, Gee AH, Brown JP, McClung MR, Wang A, Libanati C (2015) Denosumab rapidly increases cortical bone in key locations of the femur: a 3D bone mapping study in women with osteoporosis. J Bone Miner Res 30(1):46–54
Recker RR, Delmas PD, Halse J, Reid IR, Boonen S, García-Hernandez PA, Supronik J, Lewiecki EM, Ochoa L, Miller P, Hu H, Mesenbrink P, Hartl F, Gasser J, Eriksen EF (2008) Effects of intravenous zoledronic acid once yearly on bone remodeling and bone structure. J Bone Miner Res 23:6–16
Recker RR, Lappe JM, Davies M, Kimmel D (2018) Perimenopausal bone histomorphometry before and after menopause. Bone 108:55–61. https://doi.org/10.1016/j.bone.2017.12.016
Reid IR, Miller PD, Brown JP, Kendler DL, Fahrleitner-Pammer A, Valter I, Maasalu K, Bolognese MA, Woodson G, Bone H, Ding B, Wagman RB, San Martin J, Ominsky MS, Dempster DW (2010) Denosumab phase 3 bone histology study group. Effects of denosumab on bone histomorphometry: the FREEDOM and STAND studies. J Bone Miner Res 25 (10):2256–2265. https://doi.org/10.1002/jbmr.149
Roberts SJ, Ke HZ (2018) Anabolic strategies to augment bone fracture healing. Curr Osteoporos Rep 16:1–2
Roschger P, Rinnerthaler S, Yates J, Rodan GA, Fratzl P, Klaushofer K (2001) Alendronate increases degree and uniformity of mineralization in cancellous bone and decreases the porosity in cortical bone of osteoporotic women. Bone 29(2):185–191
Roschger P, Gupta HS, Berzlanovich A, Ittner G, Dempster DW, Fratzl P, Cosman F, Parisien M, Lindsay R, Nieves JW, Klaushofer K (2003) Constant mineralization density distribution in cancellous human bone. Bone 32:316–323
Roschger P, Paschalis EP, Fratzl P, Klaushofer K (2008) Bone mineralization density distribution in health and disease. Bone 42:456–466

Roschger P, Lombardi A, Misof BM, Maier G, Fratzl-Zelman N, Kimmel D, LaMotta A, Fratzl P, Klaushofer K (2010) Mineralization density distribution of postmenopausal osteoporotic bone is restored to normal after long-term alendronate treatment: qBEI and sSAXS data from the fracture intervention trial long-term extension (FLEX). J Bone Miner Res 25:48–55

Roschger P, Misof B, Paschalis E, Fratzl P, Klaushofer K (2014) Changes in the degree of mineralization with osteoporosis and its treatment. Curr Osteoporos Rep 12:338–350. https://doi.org/10.1007/s11914-014-0218-z

Ross RD, Edwards LH, Acerbo AS, Ominsky MS, Virdi AS, Sena K, Miller LM, Sumner DR (2014) Bone matrix quality after sclerostin antibody treatment. J Bone Miner Res 29 (7):1597–1607. https://doi.org/10.1002/jbmr.2188

Ruffoni D, Fratzl P, Roschger P, Phipps R, Klaushofer K, Weinkamer R (2008) Effect of temporal changes in bone turnover on the bone mineralization density distribution: a computer simulation study. J Bone Miner Res 23(12):1905–1914. https://doi.org/10.1359/jbmr.080711

Saag K, Petersen J, Brandi ML et al (2017) Romosozumab or alendronate for fracture prevention in women with osteoporosis. N Engl J Med 377:1417–1427

Seeman E (2008) Bone quality: the material and structural basis of bone strength. J Bone Miner Metab 26:1–8

Seeman E (2011) Perspectives anabolic plus antiresorptive: is one plus one more or less two? IBMS BoneKey 8:221–228. https://doi.org/10.1138/20110510. http://www.bonekey-ibms.org/cgi/content/full/ibmske;8/5/221

Seo JB, Yoo JS, Ryu JW, Yu KW (2016) Influence of early bisphosphonate administration for fracture healing in patients with osteoporotic proximal humerus fractures. Clin Orthop Surg 8:437–443

Shane E, Burr D, Abrahamsen B, Adler RA, Brown TD, Cheung AM, Cosman F, Curtis JR, Dell R, Dempster DW, Ebeling PR, Einhorn TA, Genant HK, Geusens P, Klaushofer K, Lane JM, McKiernan F, McKinney R, Ng A, Nieves J, O'Keefe R, Papapoulos S, Howe TS, van der Meulen MC, Weinstein RS, Whyte MP (2014) Atypical subtrochanteric and diaphyseal femoral fractures: second report of a task force of the american society for bone and mineral research. J Bone Miner Res 29:1–23

Shin WC, Moon NH, Jang JH, Seo HU, Suh KT (2019) A retrospective bicenter comparative study of surgical outcomes of atypical femoral fracture: potential effect of teriparatide on fracture healing and callus formation. Bone. https://doi.org/10.1016/j.bone.2019.08.006

Shoback D, Rosen CJ, Black DM, Cheung AM, Murad MH, Eastell R (2020) Pharmacological management of osteoporosis in postmenopausal women: an endocrine society guideline update. J Clin Endocrinol Metab 105(3)

Siris ES, Harris ST, Rosen CJ, Barr CE, Arvesen JN, Abbott TA, Silverman S (2006) Adherence to bisphosphonate therapy and fracture rates in osteoporotic women: relationship to vertebral and nonvertebral fractures from 2 US claims database. Mayo Clin Proc 81(8):1013–1022

Song M (2019) Dental care for patients taking antiresorptive drugs: a literature review. Restor Dent Endod 44(4):e42

Starr J, Tay YKD, Shane E (2018) Current understanding of epidemiology, pathophysiology, and management of atypical femur fractures. Curr Osteoporos Rep 16(4):519–529

Svejda B, Muschitz C, Gruber R, Brandtner C, Svejda C, Gasser R, Santler G, Dimai HP (2015) Positionspapier zur medikamentenassoziierten Osteonekrose des Kiefers (MRONJ). Stomatologie. https://doi.org/10.1007/s00715-015-0064-3

Thomsen JS, Mosekilde L, Mosekilde E (1996) Quantification of remodeling parameter sensitivity--assessed by a computer simulation model. Bone 19:505–511

Tripto-Shkolnik L, Fund N, Rouach V et al (2020) Fracture incidence after denosumab discontinuation: real-world data from a large healthcare provider. Bone 130:115150. https://doi.org/10.1016/j.bone.2019.115150

Tsai JN, Jiang LA, Lee H, Hans D, Leder BZ (2017) Effects of teriparatide, denosumab, or both on spine trabecular microarchitecture in DATA-switch: a randomized controlled trial. J Clin Densitom 20(4):507–512

Tsourdie E, Langdahl B, Cohen-Solal M et al (2017) Discontinuation of denosumab therapy for osteoporosis: a systematic review and position statement by ECTS. Bone 105:11–17
Uebelhart B, Rizzoli R, Ferrari SL (2017) Retrospective evaluation of serum CTX levels after denosumab discontinuation in patients with or without prior exposure to bisphosphonates. Osteoporos Int 28(9):2701–2705
Valdes A, Bajaj T (2019) Estrogen therapy. StatPearls. StatPearls Publishing, Treasure Island
van der Meulen MC, Boskey AL (2012) Atypical subtrochanteric femoral shaft fractures: role for mechanics and bone quality. Arthritis Res Ther 14(4):220
van Staa TP, Leufkens HGM, Abenhaim L, Zhang B, Cooper C (2000) Use of oral corticosteroids and risk of fractures. J Bone Miner Res 15(6):993–1000
Watts NB, Grbic JT, Binkley N (2019) Oral procedures and events in postmenopausal women with osteoporosis treated with denosumab for up to 10 years. J Clin Endocrinol Metab 104 (6):2443–2452
Weinstein RS, Parfitt AM, Marcus R, Greenwald M, Crans G, Muchmore DB (2003) Effects of raloxifene, hormone replacement therapy, and placebo on bone turnover in postmenopausal women. Osteoporos Int 14:814–822
Zebaze R, Libanati C, McClung MR et al (2016) Denosumab reduces cortical porosity of the proximal femoral shaft in postmenopausal women with osteoporosis. J Bone Miner Res 31 (10):1827–1834
Zhang L, Zhai P (2019) A comparison of percutaneous vertebroplasty versus conservative treatment in terms of treatment effect for osteoporotic vertebral compression fractures: a meta-analysis. Surg Innov 27(1):19–25
Zoehrer R, Roschger P, Fratzl P, Durchschlag E, Paschalis E, Phipps R, Klaushofer K (2006) Effects of 3- and 5-year treatment with risedronate on the bone mineral density distribution of cancellous bone in human iliac crest biopsies. J Bone Miner Res 21:1106–1112

Reduced Bone Modeling and Unbalanced Bone Remodeling: Targets for Antiresorptive and Anabolic Therapy

Sabashini K. Ramchand and Ego Seeman

We shall not cease from exploration
And the end of all our exploring
Will be to arrive where we started
And know the place for the first time.
T.S. Eliot, Little Gidding, The Four Quartets

Contents

S. K. Ramchand (✉)
Department of Medicine, Endocrine Unit, Massachusetts General Hospital, Harvard University, Boston, MA, USA

Department of Medicine, Endocrine Unit, Austin Hospital, The University of Melbourne, Melbourne, VIC, Australia
e-mail: sramchand@mgh.harvard.edu

E. Seeman
Department of Medicine, Endocrine Unit, Austin Hospital, The University of Melbourne, Melbourne, VIC, Australia

Mary MacKillop Institute for Health Research, Australian Catholic University, Fitzroy, VIC, Australia

P. H. Stern (ed.), *Bone Regulators and Osteoporosis Therapy*,
Handbook of Experimental Pharmacology 262, https://doi.org/10.1007/164_2020_354

Abstract

Bone loss during advancing age is the net result of reduced modeling-based bone formation upon the outer (periosteal) envelope and unbalanced remodeling by basic multicellular units (BMUs) upon the three (intracortical, endocortical, and trabecular) components of the inner (endosteal) bone envelope. Each BMU deposits less bone than resorbed, reducing total bone volume and deteriorating the microstructure of the diminished residual bone volume.

Antiresorptive agents like bisphosphonates reduce, but do not abolish, the rate of bone remodeling – fewer BMUs remodel, "turn over," the volume of bone. Residual unbalanced remodeling continues to slowly reduce total bone volume and deteriorate bone microstructure. By contrast, denosumab virtually abolishes remodeling so the decrease in bone volume and the deterioration in microstructure cease. The less remodeled matrix remains, leaving more time to complete the slow process of secondary mineralization which reduces the heterogeneity of matrix mineralization and allows it to become glycosylated, changes that may make the smaller and microstructurally deteriorated bone volume more brittle. Neither class of antiresorptive restores bone volume or its microstructure, despite increases in bone mineral density misleadingly suggesting otherwise. Nevertheless, these agents reduce vertebral and hip fractures by 50–60% but only reduce nonvertebral fractures by 20–30%.

Restoring bone volume, microstructure, and material composition, "curing" bone fragility, may be partly achieved using anabolic therapy. Teriparatide, and probably abaloparatide, produce mainly remodeling-based bone formation by acting on BMUs existing in their resorption, reversal, or formation phase at the time of treatment and by promoting bone formation in newly initiated BMUs. Romosozumab produces modeling-based bone formation almost exclusively and decreases the surface extent of bone resorption. All three anabolic agents reduce vertebral fracture risk relative to untreated controls; parathyroid hormone 1-34 and romosozumab reduce vertebral fracture risk more greatly than risedronate or alendronate, respectively. Evidence for nonvertebral or hip fracture risk reduction relative to untreated or antiresorptive-treated controls is lacking or inconsistent. Only one study suggests sequential romosozumab followed by alendronate reduces vertebral, nonvertebral, and hip fracture risk compared to continuous alendronate alone. Whether combined antiresorptive and anabolic therapy result in superior fracture risk reduction than monotherapy is untested.

Keywords

Anabolic · Antiresorptives · Bone mineral density · Fracture · Osteoporosis

1 Introduction

Bone remodeling is balanced during young adulthood; equal volumes of bone are resorbed and replaced by each basic multicellular unit (BMU) (Hattner et al. 1965). Around midlife, remodeling becomes unbalanced; each BMU resorbs a smaller volume of bone at discrete points upon the three (intracortical, endocortical, and trabecular) components of the endosteal (inner) surface of the bone, but the volume of new bone subsequently deposited is even less than the volume resorbed (Lips et al. 1978). In women, estrogen deficiency associated with menopause worsens age-related remodeling imbalance because the life span of osteoclasts increases, resulting in an increased volume of bone resorbed, while the life span of osteoblasts decreases further (Manolagas 2000). Less bone is deposited in a larger cavity worsening the focal net volume deficit. Bone loss and microstructural deterioration are amplified by the increased birth rate of BMUs associated with estrogen deficiency; more unbalanced remodeling events occur upon the three components of the endosteal surface, renewing, "turning over," and eroding, a larger proportion of the ever-diminishing total bone volume (Parfitt 1984).

This remodeling imbalance is the necessary and sufficient morphological basis of bone loss (Parfitt 1984). Remodeling imbalance upon the intracortical surfaces of the Haversian canals enlarges them focally producing more surface upon which unbalanced remodeling can be initiated (Bjornerem et al. 2011). Canals traversing the inner cortex (adjacent to the medullary canal) coalesce, cavitating the cortex, a morphological effect that contributes more to cortical thinning than unbalanced endocortical remodeling (which enlarges the medullary canal) (Zebaze et al. 2010). Age-related resorption upon trabeculae causes them to thin. Trabeculae are lost after menopause as the larger resorption cavities perforate and remove them completely (Dempster et al. 1995). Trabeculae become less interconnected and more rod-like (Dempster et al. 1995). Loss of trabeculae with their surfaces results in slowing or cessation of trabecular bone remodeling because there is no longer a surface for remodeling to be initiated upon. Total bone surface may remain unchanged or increase, but it is now located in the cortical compartment (Bjornerem et al. 2011). Intracortical remodeling intensity increases, and bone loss becomes predominantly cortical (Zebaze et al. 2010). There is little trabecular bone remaining, and the more rapid bone loss erodes an ever-decreasing cortical bone volume predisposing to nonvertebral fractures.

Bone resorption by the BMU affects morphology at higher levels of resolution. For example, the radial extent of resorption defines the diameter of osteons in cortical bone and hemiosteons formed upon the endocortical and trabecular surfaces. The radial extent of resorption is delineated by the cement line which separates osteons from each other and the interosteonal (interstitial) bone matrix (Mohsin et al. 2006). The age-related reduction in the volume of bone resorbed by each BMU produces smaller osteons, whereas after menopause, the longer living osteoclasts excavate larger cavities producing larger osteons with a larger central canal (produced by the reduction in bone formation by each BMU) with fewer lamellae and a reduction in the population of osteocytes per osteon. The diminishing total cortical bone volume becomes more porous, there are more osteons of varying size, and the matrix mineral density of the diminishing total bone volume decreases due to rapid replacement of older more mineralized bone with younger less mineralized bone (Bjornerem et al. 2018). Advanced glycation end products increase collagen cross-linking compromising matrix ductility (Poundarik et al. 2015; Saito et al. 2008; Tang et al. 2007; Vashishth 2007). The resulting decline in total bone volume, microstructural deterioration, and altered material properties compromise bone strength predisposing to fractures (Seeman and Delmas 2006).

The loss of bone from the three components of the endosteal envelope proceeds with a concurrent decrease in periosteal modeling-based bone formation (Seeman 2003). Whether there is also modeling upon the endosteal envelope is uncertain and is difficult to establish in the face of rapid unbalanced remodeling which is likely to erode any concurrent modeling-based bone formation (Ominsky et al. 2015). Whatever the case, the cellular mechanisms determining the volumes of bone resorbed and formed by each BMU, the birth rate of the these BMUs, concurrent age-related reduction in periosteal and endosteal modeling-based bone formation, the compromised bone volume, microstructure, osteocyte population, and the material properties of bone are each rational targets for the prevention and reversal of bone fragility (Seeman and Martin 2015, 2019) (Fig. 1).

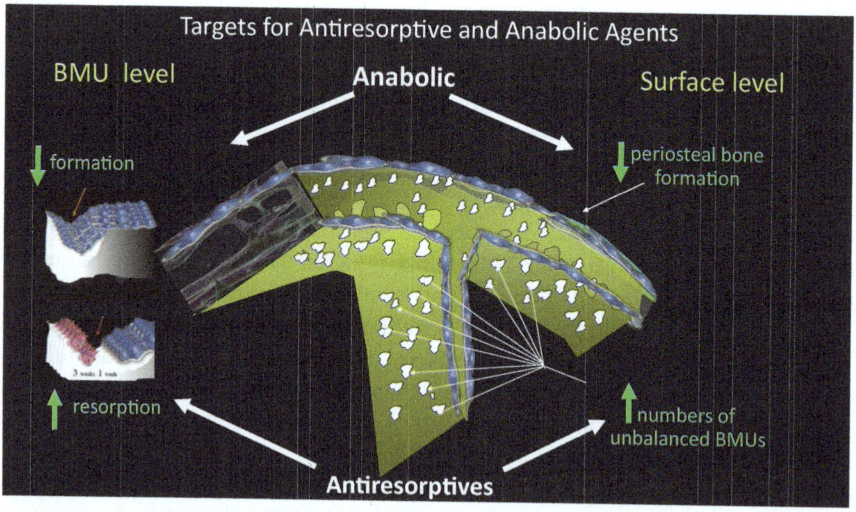

Fig. 1 Drug therapy targets the cellular activity of the basic multicellular units (BMUs) and the surface extent of this activity upon the periosteal and three components of the endosteal surface. Antiresorptive agents reduce the volume of bone resorbed by each BMU and reduce the number of BMUs producing unbalanced remodeling and thereby slow bone loss. Anabolic agents may increase the volume of bone formed by each BMU and increase bone formation upon the periosteal and the endosteal surfaces

2 Antiresorptive Therapy

Antiresorptive agents are remodeling suppressants (Russell et al. 2008). These agents probably reduce the volume of bone resorbed by each BMU (Allen et al. 2010), but the most important benefit these agents confer is that they reduce the rate of unbalanced remodeling. The bisphosphonates, and weaker remodeling suppressants like selective estrogen receptor modulators (SERMs) and calcium supplements, slow unbalanced bone remodeling; they do not abolish it (Russell et al. 2008). Slowing remodeling reduces the rate of decrease in bone volume and microstructural deterioration, but the unsuppressed unbalanced remodeling continues to deteriorate the skeleton, albeit more slowly than without treatment. Denosumab, the most efficacious remodeling suppressant, virtually abolishes remodeling by inhibiting existing BMUs in their resorptive phase at the time of starting treatment and by preventing differentiation of precursors into mature bone-resorbing osteoclasts (McClung et al. 2006). Any further decrease in total bone volume and microstructural deterioration virtually stops.

Antiresorptive agents are not anabolic; they do not restore the reduced bone volume or microstructural deterioration present at the time of initiating treatment. This important limitation is not obvious because of the misleading effect antiresorptive agents have on bone mineral density (BMD). Slowing of the rate of remodeling by 50–60% using bisphosphonates or by 20–30% with SERMs or calcium supplements is expected to slow the decline in BMD. Stopping bone loss using denosumab should stop the decline in BMD.

Neither slowing nor stopping the decline in BMD is observed. On the contrary, there is a rapid initial increase in BMD during the first 6–12 months of therapy, a change erroneously interpreted as being an increase in bone matrix mass or volume (Fig. 2). This is a *net* increase, the result of four events produced by interrupting steady-state remodeling. At the onset of treatment surface-level remodeling is perturbed. There is a prompt reduction in the number of new BMUs excavating cavities upon the endosteal surface (which does not increase BMD). The cavities may be smaller because of a reduction in the volume of bone resorbed by bisphosphonates and, probably, denosumab (Allen et al. 2010). Concurrently, the many more cavities excavated upon the endosteal surface weeks before starting treatment enter their formation phase and refill (but still do so incompletely because remodeling imbalance is not corrected). The matrix of the many cavities incompletely refilling undergoes primary mineralization within days of being deposited, while matrix deposited weeks to months earlier (but no longer being remodeled) undergoes more complete secondary mineralization (a process taking many months if not years to complete) (Akkus et al. 2003).

These four morphological features (newly excavated cavities, incompletely refilling cavities, incomplete primary and secondary mineralization) are part of the reversible remodeling space deficit in matrix and mineral produced by the *normal* delay in onset and slowness of the formation phase of remodeling and slowness of matrix mineralization (Parfitt 1980). Whether this early increase in BMD is mainly due to early bone matrix deposition or matrix mineralization is not known.

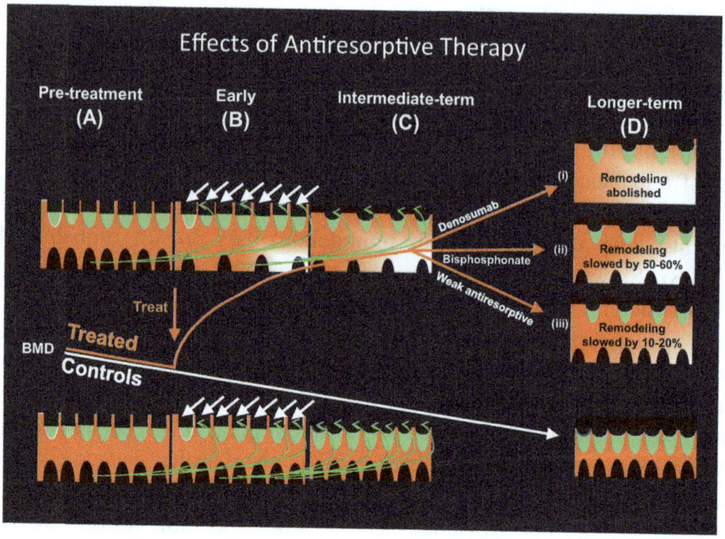

Fig. 2 (**a**) Before treatment. Remodeling is rapid and in steady state, similar numbers of BMUs resorb a volume of bone as BMUs refilling, but the filling is incomplete, so BMD decreases. (**b**) Early antiresorptive therapy. Remodeling is perturbed. Fewer BMUs excavate cavities, and each cavity is smaller. Concurrently, the many more BMUs excavating bone before treatment now refill, but incompletely (curved green arrows). Note that the number of BMUs refilling is similar in the treated and control groups (sloping white arrows). There is a rapid net increase in BMD (see text). (**c**) In the intermediate term, during the second and third year, remodeling returns to steady state but at a slower rate. The fewer and smaller cavities excavated during early antiresorptive treatment refill incompletely as similar numbers of new BMUs excavate smaller cavities. (**d**) In the longer term, (1) with denosumab treatment there is almost complete remodeling suppression, so BMD increases due to refilling of cavities excavated before treatment unopposed by the appearance of new cavities; secondary mineralization of the unremodeled matrix (depicted as whitening of orange matrix); and possible modeling-based bone formation (see text). (2) With bisphosphonates, secondary mineralization reaches completion, and BMD stabilizes. Cortical matrix volume may decrease, but this is not captured by the BMD measurement (see text). (3) Weak antiresorptives reduce remodeling by only 10–20%, so most unbalanced remodeling continues to reduce bone volume and cause microstructural deterioration

It is unlikely to be matrix deposition because this is no different to the matrix deposition in controls at this stage. What differs from controls is the fewer cavities being excavated and the mineralization of the less remodeled bone matrix.

The increase in BMD may correlate with fracture risk reduction (Bouxsein et al. 2019), but that does not necessarily mean this is a causal relationship. Indeed, the morphological basis of the fracture risk reduction relative to controls is incompletely understood but may include (1) the appearance of fewer new stress risers; (2) less stress imposed by cavities excavated before treatment because they partly refill; (3) less decrease in bone volume and microstructural deterioration relative to controls; and (4) initial increases in matrix mineral content which correct the reduced matrix mineral content produced by rapid matrix turnover. However, in the longer term, complete matrix mineralization may produce brittleness – loss of ductility (Akkus et al. 2003; Lloyd et al. 2017).

None of the changes *restore* microstructural deterioration, the consequence of remodeling imbalance. Restoring bone volume and microstructure can only be achieved using bone-forming agents. The initial rise in BMD is less with weaker antiresorptives because 70–80% of the pre-treatment remodeling rate remains unsuppressed and continues to produce bone loss and renew bone matrix so that secondary mineralization does not increase or does so minimally (again, at this early stage, the number of cavities refilling incompletely is no different in treated subjects and untreated controls) (Reid et al. 2008; Silverman et al. 2012) (Fig. 2b).

After about 6–12 months, perturbed remodeling (with more BMUs incompletely refilling cavities than BMUs excavating cavities) returns to steady state (Fig. 2c). Now, approximately equal numbers of cavities are incompletely refilling and being excavated, as was the case before treatment (except now remodeling proceeds more slowly and at a rate determined by the efficacy of the antiresorptive agent being administered). The incomplete refilling reduces total bone volume and causes microstructural deterioration, but this proceeds at a lower rate than in controls. The deterioration is undetectable using bisphosphonates because secondary mineralization of the declining total bone matrix volume is likely to obscure it.

Indeed, BMD continues to increase, more slowly than it did during the first 6–12 months of treatment, but eventually plateaus after about 3–4 years (Fig. 2d). The continued bone loss and microstructural deterioration is mainly cortical because bisphosphonates bind avidly to mineral in superficial subperiosteal and subendosteal bone and fail to penetrate and distribute in high concentration in deeper cortical bone; osteoclasts engulfing matrix free of bisphosphonate or having low concentration of the drug continue to resorb bone. The loss of total bone matrix volume is greater using weak antiresorptives. The decrease in BMD after about 12 months of treatment, reflecting a decrease in bone volume, is detectable because 70–80% of pre-treatment unbalanced remodeling continues when weaker antiresorptives are used (Reid et al. 2008; Silverman et al. 2012) (Fig. 2d). There is no, or little, increase in matrix mineralization to obscure the decrease in bone volume because the bone volume is not resident long enough to undergo complete mineralization.

When the antiresorptive agent used is denosumab, the changes are different. This drug virtually abolishes remodeling, so the reduction in bone matrix volume and worsening of microstructural deterioration cease. Instead of newly excavated cavities appearing as with bisphosphonates or weak antiresorptives, incomplete refilling of cavities excavated before treatment do so without being offset by the concurrent appearance of new cavities so total bone matrix volume may increase because the reduction of the reversible remodeling space deficit in matrix and mineral is more complete. This increase is *not* anabolism, the formation of new bone upon the periosteal surface enlarging the external perimeter of the bone; upon the intracortical surface of Haversian canals, narrowing them; upon the endocortical and trabecular surfaces thickening the cortices and trabeculae, respectively.

BMD continues to increase using denosumab beyond 5 years of treatment. While this might be the result of more complete secondary mineralization, there is evidence suggesting matrix mineral density does not increase beyond 5 years (Dempster et al. 2018). There is another reason BMD may continue to increase.

Periosteal and perhaps endosteal bone *modeling* continue throughout adult life (Ominsky et al. 2015). Bone formation by endosteal bone modeling is either obscured or lost after menopause perhaps because the rapid unbalanced remodeling removes it. Studies in nonhuman primates suggest that modest endosteal modeling-based bone formation during adulthood is not removed because remodeling is virtually completely suppressed during denosumab therapy (Ominsky et al. 2015). Under these circumstances, total bone matrix volume might increase.

Modeling-based bone formation upon the endosteal surface is held to explain the continued increase in BMD beyond 5 years using denosumab despite the absence of comparable data in human subjects. It is also held to account for the low fracture rates during prolonged therapy up to 10 years. However, this interpretation is also not well founded for several reasons. The study was not placebo controlled beyond 3 years, and there was significant loss of the inception cohort so it is plausible that the low fracture rates may be the result of healthy user bias (Bone et al. 2017). There is no published evidence of a modeling effect upon the endosteal envelope in human subjects, and whatever the effect, evidence is needed that it is responsible for the low fracture incidence.

3 Anabolic Therapy

Reconstruction of the skeleton, "curing" bone fragility, requires anabolic therapy. There are three obvious targets for anabolic therapy – reduced remodeling-based bone formation by each BMU and reduced modeling-based bone formation upon the periosteal and each of the three (intracortical, endocortical, trabecular) components of the endosteal surface.

If anabolic agents increase *remodeling-based* bone formation, the increased volume of bone formed by each BMU might produce a positive BMU balance. If so, it is advantageous to increase the birth rate of BMUs. Widespread positively balanced remodeling events at many points upon the three components of the endosteal surface should focally increase bone volume of remaining trabeculae; thicken the cortex focally, thereby increasing compressive strength; and reduce intracortical porosity. Teriparatide and probably abaloparatide address some of these mechanisms.

Modeling-based bone formation upon the periosteal surface is a biomechanically advantageous location because resistance to bending is a fourth-power function of the radial distance of a unit bone volume from the neutral axis of a bone (Ruff and Hayes 1988). Modeling-based bone formation upon the endocortical surface may thicken the cortices increasing compressive strength, modeling-based bone formation upon trabecular surfaces will thicken trabeculae and may improve connectivity of trabecular still interconnected. Whether trabecular number can be increased has been proposed by splitting of trabeculae (Jiang et al. 2003), but otherwise, a mechanism for this is yet to be identified. Modeling upon surfaces of the intracortical canals may reduce cortical porosity. Romosozumab addresses some of these mechanisms (Chavassieux et al. 2019).

3.1 Primarily Remodeling-Based Anabolic Agents

3.1.1 Teriparatide (PTH 1-34)

PTH 1-34 increases bone matrix volume primarily by a remodeling-based anabolic action on cells of existing BMUs and new BMUs generated by its administration. This accounts for 70–80% of the anabolic action. Modeling-based anabolic effects initiated upon quiescent bone surfaces appear to contribute only 20–30% of the total anabolic action (Ma et al. 2006).

At any time, existing BMUs are more likely to be at various stages of their formation than resorption phases of their remodeling cycle because bone formation proceeds for ~3 months, whereas the resorption phase is ~3 weeks, and reversal phase is ~1 week (Hattner et al. 1965; Parfitt 2008; Tran Van et al. 1982). PTH 1-34 acting on BMUs in their reversal phase may promote differentiation of osteoblast lineage cells into mature osteoid-producing forms. Concurrent increases in matrix production occur as PTH 1-34 inhibits apoptosis of mature osteoblasts which synthesize osteoid overfilling resorption cavities upon trabecular and endocortical surfaces with spillover of newly formed osteoid onto adjacent quiescent surfaces.

Thus, the early response to PTH 1-34 is likely to be modeling-based bone formation upon quiescent bone surfaces with a concurrent increase in remodeling-based bone formation by existing BMUs, both increasing circulating P1NP. Presumably, the higher the baseline pre-treatment remodeling rate, the greater the numbers of BMUs in their formation phase facilitating this anabolic action (Chen et al. 2005). In addition, and again, concurrently, or soon after starting treatment, PTH 1-34 initiates new remodeling events by promoting osteocyte and osteoblast precursor production of RANKL, osteoclast formation, and bone resorption with a later rise in circulating CTX (Seeman and Martin 2015).

What determines the number and activity of new BMUs initiated by PTH 1-34 is not known, but they first enter their resorptive phase resulting in removal of mineralized bone, increased intracortical porosity, and excavated cavities upon endocortical and trabecular surfaces with later deposition of osteoid. These changes are likely to be accompanied by an increase in serum CTX, an event that does not necessarily signal the end of bone formation and closure of the so-called anabolic window, a period of time of maximal anabolic effect of PTH 1-34 (Bilezikian 2008).

This "anabolic window" is an ambiguous notion. It suggests that the limited anabolic effect is *due to* bone resorption caused by newly generated BMUs and so a "window" of opportunity that can be left "open" by antiresorptive therapy (Bilezikian 2008). The ambiguity is the failure to distinguish between the *absolute* and *net* volume of bone formed. The *net* volume of bone deposited or resorbed is the sum of (1) the volume of bone formed by modeling initiated by PTH 1-34; (2) the volume of bone formed by BMUs existing in their resorption, reversal, or formation phase at the time PTH 1-34 is started; and (3) the volume of bone formed by new BMUs initiated by PTH 1-34 when they enter their reversal and formation phases *minus* (4) the volume of bone resorbed by new BMUs initiated by PTH 1-34 during their resorption phase (before entering their formation phase).

The notion of offsetting the absolute increase in bone formation using antiresorptive agents is also problematic. Preventing the birth of new BMUs (using denosumab) or aborting their progression (using bisphosphonates) is likely to deprive the skeleton of the bone formation phase of these BMUs for PTH 1-34 to act upon.

Perhaps most importantly, the notion of the anabolic window diverts attention away from deliberations concerning the potential major mechanisms responsible for the limited anabolic effect of PTH peptides. Is the limited anabolic effect the result of a reduced osteoblast precursor pool and reduced proliferation, differentiation, and lifespan of osteoblasts (Kim et al. 2012)? Is the opportunity for remodeling-based bone formation greater initially, when remodeling is rapid, providing many BMUs in their formation phase at the time of starting PTH 1-34? What determines the number of new BMUs generated by PTH 1-34, and is their bone forming potential less than that of the existing BMUs? If anabolic activity is less in the newly initiated BMUs, this may partly account for the waning of the anabolic effect. What limits the modeling-based anabolic effect?

The morphological changes documented using PTH 1-34 do not meet the theoretical potential envisaged. The resorption of mineralized bone by newly initiated BMUs, and deposition of newly synthesized osteoid during remodeling- and modeling-based bone formation, is likely to account for the decrease in BMD during early treatment, a transitory event because primary and secondary mineralization of the newly deposited bone follow (Seeman and Martin 2019). Whether newly excavated cavities temporarily increase bone fragility is not known. The best documented benefits of PTH 1-34 are thickening of existing trabeculae and endocortical bone formation. Another possible mechanism that needs investigation is "corticalization" of trabeculae abutting the cortex. As these trabeculae thicken, they may coalesce to become cortical in appearance, the opposite of "trabecularization" of the cortex during aging (Zebaze et al. 2010). There is evidence of periosteal apposition reported using histomorphometry (Lindsay et al. 2006, 2007), but whether this is sufficient to result in cortical thickening, increased periosteal perimeter, and increased bending strength is uncertain.

There are other potential benefits of remodeling-based bone formation, but these have not been rigorously studied and so remain speculative. For example, initiation of new BMUs is likely to result in the formation of new osteons formed by new osteoid and populated by newly formed osteocytes and younger less cross-linked and less fully mineralized matrix, improving the ductility of bone. Greater numbers of osteons may result in a greater proportion of the total bone volume being osteonal and hemiosteonal rather than interstitial, improving resistance to microcrack propagation. It has been suggested that glycosylation interferes with the effect of non-collagenous proteins like osteocalcin and osteopontin to defend mineral crystals against excessive loading (Fantner et al. 2005; Poundarik et al. 2015; Sroga and Vashishth 2012; Thomas et al. 2017). During loading, non-collagenous proteins uncoil, reducing the stress upon mineral platelets (the most brittle component of bone). Advanced glycation end products may interfere with uncoiling and contribute to diffuse damage within osteons, a form of damage that differs from microdamage in interosteonal bone (Saito and Marumo 2010; Seref-Ferlengez et al. 2014).

These bone qualities are relevant, especially when consideration is given to combining PTH 1-34 and antiresorptives because these agents influence the material composition and microstructure of bone differently, often in opposite directions. For example, antiresorptives slow or inhibit bone remodeling facilitating secondary mineralization which increases the homogeneity of the bone matrix mineralization density distribution (Fuchs et al. 2011). A homogeneous material offers less resistant to microcrack propagation and fracture than a heterogeneous material (Lloyd et al. 2017; Mohsin et al. 2006). By contrast PTH 1-34 removes mineralized bone, replacing it with younger less mineralized bone, increasing the heterogeneity of the material (Misof et al. 2003; Paschalis et al. 2005). Slowing or inhibition of bone remodeling may also lead to accumulation of advanced glycation end products, whereas treatment with PTH 1-34 reduces collagen cross-linking (Allen et al. 2008; Saito et al. 2011).

PTH 1-34 reduces vertebral fracture risk by 60–70% (Miller et al. 2016; Nakamura et al. 2012; Neer et al. 2001), with only one study reporting nonvertebral fracture risk reduction (Neer et al. 2001). Whether hip fracture risk is reduced is unknown because no properly designed studies have been done to assess this endpoint. For reasons that are not understood, cessation of PTH 1-34 is followed by bone loss, so that antiresorptive therapy is needed to prevent or minimize this bone loss (Cosman et al. 2001; Ejersted et al. 1998).

3.1.2 Abaloparatide

Abaloparatide shares many of its amino acid residues with PTH 1-34 and PTHrP. All three peptides mediate their effects through the PTHrPR1 receptor. Like PTH 1-34, this drug is likely to mediate its anabolic effects through both remodeling- and modeling-based mechanisms. There is no published histomorphometric evidence examining the contributions of these two mechanisms by quantifying the volumes of bone deposited upon crenated surfaces (reflecting remodeling-based bone formation) versus smooth surfaces (reflecting modeling-based bone formation) (Hattersley et al. 2016; Moreira et al. 2017; Varela et al. 2017).

Claims that the anabolic effect of abaloparatide is achieved with less resorptive activity are based on the study of aged rats that have little resorptive activity, as assessed using biochemical markers and histomorphometry (Varela et al. 2017). Anabolic effects of PTH 1-34 free of resorptive activity have frequently been reported using this model (Kimmel et al. 1993; Liu and Kalu 1990; Ma et al. 2011; Shen et al. 1992; Wronski et al. 1993). Abaloparatide has less effects on *both* circulating measurements, P1NP and CTX, than does PTH 1-34, with less increase in CTX relative to the increase in P1NP, an observation claimed to account for the 1–2% higher BMD achieved using abaloparatide (Eastell et al. 2019).

Making inferences about the net amount of bone formed relative to that resorbed based on circulating remodeling markers is problematic (Seeman and Nguyen 2016). The concentrations of these markers are determined by the rate of remodeling, by the volumes of bone resorbed and deposited by each BMU, and, in the case of P1NP, by modeling activity. A blood sample contains a concentration of CTX produced by BMUs currently in their resorption phase. That same sample contains a concentration of P1NP produced by different BMUs at different locations and at different stages of the formation phase as well as P1NP produced by concurrent modeling activity.

In a phase 3 clinical trial, 2,463 postmenopausal women were randomized to 18 months of abaloparatide (80 μg daily), placebo, or open-label PTH 1-34 (20 μg daily); 1901 completed the study (Miller et al. 2016). Compared to placebo, abaloparatide reduced morphometric vertebral fractures (relative risk 0.14; 95% CI 0.05–0.39, $P < 0.001$) and major osteoporotic fractures (hazard ratio (HR) 0.30; 95% CI 0.15–0.61, $P < 0.001$) but not nonvertebral fracture (after removal of study participants from two Czech sites) (EMA 2018) or hip fracture. Although vertebral antifracture efficacy was demonstrated, interpretation of these findings is challenging. The number of women who sustained fractures in both the placebo and PTH 1-34 groups was approximately double that of the abaloparatide group within the first few weeks of the study, events likely to be independent of the interventions yet influencing the number of fractures in the first 6 months and total event rates. These earlier fracture events may account for the overall perceived better antifracture efficacy of abaloparatide compared to PTH 1-34 as the number of women having fractures in the two groups in the last 12 months of the 18-month study was similar.

3.2 Modeling-Based Anabolic Therapy

3.2.1 Romosozumab

Sclerostin, the osteocyte-derived protein product of the *sost* gene, inhibits bone formation by inhibiting Wnt signaling. Loss-of-function *sost* mutations produce increased bone mass of sclerosteosis and van Buchem disease (Balemans et al. 2001, 2002). A single injection of sclerostin antibody stimulates bone formation in rats without changes in resorption parameters (Ke et al. 2012). A phase I study of a monoclonal anti-sclerostin antibody, AMG 785 (romosozumab), increased bone formation markers and decreased the resorption marker, serum CTX (Padhi et al. 2011). A 12-month phase II randomized, placebo-controlled, multi-dose study of romosozumab in 410 women confirmed the rapid increase in BMD (McClung et al. 2014). In a phase III placebo-controlled study of 7,180 women, risk reductions after 12 months of treatment with romosozumab were 73% for vertebral fractures (risk ratio 0.27; 95% CI 0.16–0.47, $P < 0.001$) and 40% for major osteoporotic fracture (HR 0.60; 95% CI 0.40–0.90, $P = 0.012$) (Cosman et al. 2016). Fracture risk reduction with romosozumab was numerically lower for nonvertebral (HR 0.75; 95% CI 0.53–1.05, $P = 0.10$) and hip (HR 0.54; 95% CI 0.22–1.35, $P = 0.18$) fracture (Cosman et al. 2016).

In another study, 4,093 postmenopausal women with osteoporosis and a fragility fracture were assigned to monthly subcutaneous romosozumab (210 mg) or weekly oral alendronate (70 mg) for 12 months then open-label alendronate in both groups for a further 12 months (Saag et al. 2017). At 12 months, romosozumab reduced vertebral fracture risk by 37% (risk ratio 0.63; 95% CI 0.47–0.85, $P = 0.003$), but there were no statistically significant reductions in nonvertebral, major osteoporotic, or hip fracture. However, romosozumab followed by alendronate reduced nonvertebral fracture by 19% (HR 0.81; 95% CI 0.66–0.99, $P = 0.04$), clinical fracture by 27% (HR 0.73; 95% CI 0.61–0.88, $P < 0.001$), and hip fracture by 38% (HR 0.62; 95% CI 0.42–0.92, $P = 0.02$) compared to continuous alendronate. The trial demonstrated a higher incidence of cardiovascular events with romosozumab (50/2040) compared to alendronate (38/2014) at the end of 12 months, not at 24 months (Saag et al. 2018). These findings were not observed in the larger placebo-controlled FRAME study (Cosman et al. 2016).

Romosozumab is associated with a transitory increase in P1NP within 1 week and a rapid but sustained reduction in CTX for 12 months, perhaps because of a reduction in cells stimulating RANKL production and so reduced osteoclastogenesis (Ominsky et al. 2015). In a study of the transcriptional effects of anti-sclerostin treatment in mature ovariectomized rats, no effect on osteoprotegerin was noted (Nioi et al. 2015). Bone formation is modeling-based and is limited in duration, probably due to self-regulation in the Wnt pathway restricting proliferative drive in the osteoblast lineage. Increases in mRNA levels of Wnt signaling antagonists *Sost* and *DKK-1* in osteoblastic cells, tibiae, and vertebrae have been reported (Taylor et al. 2016). Blocking both sclerostin and DKK-1 resulted in a more robust anabolic effect using a bispecific antibody with dual inhibition of sclerostin and DKK-1 to treat rodents and nonhuman primates (Florio et al. 2016; Holdsworth et al. 2018; Maeda et al. 2015). These anabolic effects are lost after stopping treatment. Bone mineral density decreases, associated with a decrease in bone formation and increase in bone resorption markers (Ominsky et al. 2017).

4 First-Line Therapy: Antiresorptive, Anabolic, Combined, or Sequential?

Antiresorptive agents are the first-line approach to therapy. They produce a relative fracture risk reduction. Bone strength stabilizes, or decreases, albeit more slowly than without treatment; it is not restored. Vertebral and hip fracture risk is reduced by ~50% relative to untreated controls, i.e., 50% of the women having fractures still do so despite treatment, while nonvertebral fracture risk is reduced by ~20%, i.e., 80% of the women having fractures still do so despite treatment, a particular concern given 80% of all fractures are nonvertebral (Reid 2015).

The most obvious explanation for these modest relative risk reductions and the continuing fracture risk is that antiresorptive agents only reduce the reversible component of the deficit in matrix and its mineral content by slowing the rate of remodeling, so that there are fewer excavated cavities and cavities with osteoid in various stages of incompleteness of mineralization (Seeman and Martin 2015, 2019). Antiresorptives do not restore bone volume or repair microstructural deterioration, irreversible deficits caused by rapid unbalanced remodeling.

The question is, do anabolic agents *restore* bone volume and its microstructure? Do they reduce fracture risk more effectively than antiresorptives? The underlying assumption justifying the use of anabolic therapy over antiresorptives as first-line therapy is that even partial restoration of bone volume and microstructural deterioration, with deposition of new bone of normal material composition, will reduce fracture risk more greatly than contraction of the remodeling space deficit in a bone of reduced and architecturally compromised matrix volume which eventually becomes homogeneously and fully mineralized (Parfitt 1980; Seeman and Martin 2019).

There are only two clinical trials that compare the antifracture efficacy of anabolic versus antiresorptive therapy. In the study reported by Saag et al., at 12 months, romosozumab reduced new vertebral (risk ratio 0.63; 95% CI 0.47–0.85, $P = 0.003$) and clinical fractures (HR 0.72; 95% CI 0.54–0.96, $P = 0.03$) compared to alendronate. Nonvertebral fracture risk was *numerically* lower in the romosozumab treated group (HR 0.74; 95% CI 0.54–1.01, $P = 0.06$) (Saag et al. 2017).

In the second study, Kendler et al. randomized 1,360 postmenopausal women with osteoporosis to PTH 1-34 (20 μg daily) or risedronate (35 mg daily) for 2 years (Kendler et al. 2018). Overall, 72% of participants had previously received bone targeted treatment. At 2 years, treatment with PTH 1-34 resulted in a 56% reduction in vertebral fracture risk (risk ratio 0.44; 95% CI 0.29–0.68, $P < 0.0001$). Nonvertebral fracture risk was numerically lower in the PTH 1-34 treated group (HR 0.66; 95% CI 0.39–1.10, $P = 0.1$). These changes were consistent across a range of participant characteristics (Geusens et al. 2018). Both studies demonstrate greater vertebral fracture risk reduction using anabolic than antiresorptive therapy; more evidence is needed to establish the superiority of anabolic agents in reducing nonvertebral and hip fracture risk.

4.1 Combining Antiresorptive and Anabolic Agents

There are no studies comparing the antifracture efficacy of combined antiresorptive and anabolic therapy versus either alone. Inferences regarding the potential advantage of combined therapy over single therapy are entirely based on studies in animal models and human subjects using non-invasive imaging methods and biochemical measurements of bone remodeling. In some animal models, ex vivo strength testing of bone samples has been examined.

It is widely held that antiresorptive therapy suppresses, "blunts," remodeling-based bone formation by PTH. This is largely based on the use of alendronate in two papers published in the *New England Journal of Medicine* with an accompanying editorial (Black et al. 2003; Finkelstein et al. 2003; Khosla 2003). However, the data has not been confirmed in the majority of subsequent studies using other antiresorptives (Cosman et al. 2011; Kostenuik et al. 2001; Samadfam et al. 2007;

Tsai et al. 2013). The notion of blunting was based on the assumption that a higher BMD or higher P1NP means more bone formation and a lack of response means less bone formation.

BMD is a problematic endpoint because it cannot differentiate changes in matrix mineralization from changes in matrix volume and structure. Remodeling-based anabolic therapy increases bone matrix volume by replacing more fully mineralized bone with young less fully mineralized bone. Modeling-based anabolic therapy adds young less fully mineralized bone to existing older bone. Imaging using radiation transmission often results in a net reduction in BMD because young less mineralized bone transmits, rather than attenuates, photons leading to the inference that bone "loss" and fragility have occurred when in fact bone volume has increased. Antiresorptives slow remodeling. Matrix no longer "turned over" undergoes more complete mineralization increasing BMD leading to the inference that bone "volume" or "mass" has increased and that bone strength has increased even though bone matrix volume is unchanged or decreased and the matrix has become less ductile (Lloyd et al. 2017).

Even if an increase or lack of an increase in BMD is accepted on face value, Figs. 1, 2, and 3 of the 12-month study by Black et al., which compared the effects on BMD of PTH 1-84 (100 mcg daily) and alendronate (10 mg daily) either alone or in combination, do not support the notion of blunting (Black et al. 2003) (Fig. 3). Relative to PTH 1-84 alone, combined therapy (1) did *not* produce a smaller

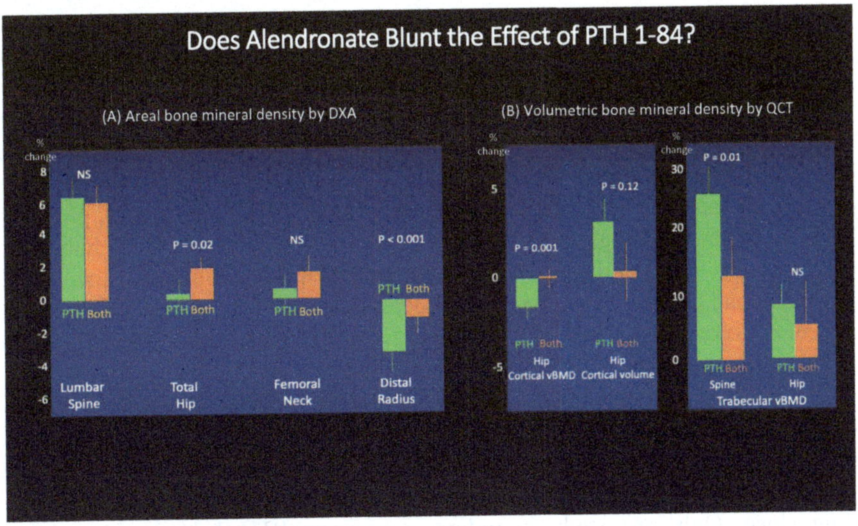

Fig. 3 Mean percent changes in (**a**) areal bone mineral density and (**b**) volumetric bone density in postmenopausal women after 12 months of treatment with PTH 1-84 100-μg daily (green bars) or a combination of PTH 1-84 100-μg daily and alendronate 10-mg daily (orange bars). Data sourced from Black DM, Greenspan SL, Ensrud KE, et al. The effects of parathyroid hormone and alendronate alone or in combination in postmenopausal osteoporosis. *N Engl J Med.* 2003;349:1207–15

increment in BMD, (2) *did* produce a greater increase in total hip BMD, (3) *did* reduce the decline in distal radius BMD, and (4) *did* prevent the reduction in total hip and femoral neck cortical vBMD produced by PTH 1-84 alone. Curiously, the increase in total hip and femoral neck cortical volume by PTH 1-84, a *modeling* effect, was prevented by combined therapy. Moreover, combined therapy increased trabecular vBMD less than PTH 1-84 monotherapy, but this may not be blunting. The antiresorptive might prevent PTH 1-84-mediated increase in intracortical remodeling, cortical porosity, and prevent the increase in cortical fragments making it seem that the rise in "trabecular" BMD is blunted (Zebaze and Seeman 2015). Blunting of the rise in P1NP and CTX is likely to be the result of suppressed remodeling *rate* rather than a reduction in the net volumes of bone deposited or resorbed by BMUs, respectively (Seeman and Nguyen 2016).

If blunting of the BMD response was due to fewer BMUs, then blunting should be *more* severe with co-administration of PTH with more effective remodeling suppressants like zoledronate, denosumab, or osteoprotegerin (OPG, an endogenous inhibitor of RANKL) than with alendronate. The opposite is reported, and many studies report additive effects (Cosman et al. 2011; Kostenuik et al. 2001; Samadfam et al. 2007; Tsai et al. 2013). Blunting is not *greater* with denosumab/PTH 1-34 than alendronate/PTH 1-34 even though denosumab suppresses remodeling more greatly than alendronate. Additive effects on BMD are reported with PTH 1-34/denosumab relative to PTH 1-34 alone (Tsai et al. 2013) and comparing PTH 1-34/OPG to PTH 1-34 alone (Samadfam et al. 2007).

The difficulties in using BMD as a phenotype are also present using high-resolution peripheral computed tomography. Tsai et al. report that combined PTH 1-34 and denosumab increased cortical vBMD, the net result of a reduction by PTH 1-34 and an increase using denosumab (Tsai et al. 2015). Combined therapy increased cortical matrix mineral density, yet PTH 1-34 decreased it (by replacing more mineralized bone with less mineralized bone), and denosumab had no effect. Combined therapy had no effect on porosity, yet PTH 1-34 increased it, while denosumab had no effect (Fig. 4). These findings do not entirely add up, probably because of methodological challenges in separating cortical from trabecular compartments and quantifying cortical porosity and trabecular density. Low image resolution and changes in matrix mineral density influence quantification of micro-structure (Zebaze et al. 2010, 2013). Moreover, both bisphosphonates and PTH 1-34 reduce marrow fat composition (Fan et al. 2017; Veldhuis-Vlug and Rosen 2018). Replacing bone marrow fat with water or cells increases photon attenuation and may exaggerate the rise in BMD.

A lower level of evidence is preclinical studies demonstrating that combined therapy increases the breaking strength of bone ex vivo more greatly than either drug alone. Only one study in rodents comparing PTH 1-34/alendronate and PTH 1-34/OPG versus PTH 1-34 alone has been reported. While trabecular bone volume increased more greatly than PTH 1-34 alone, bone strength assessed ex vivo was similar to PTH 1-34 monotherapy (Samadfam et al. 2007). These concerns suggest that caution is needed before making inferences about the comparative efficacy of these agents and that eventually, well designed and executed studies using fracture outcomes will be needed.

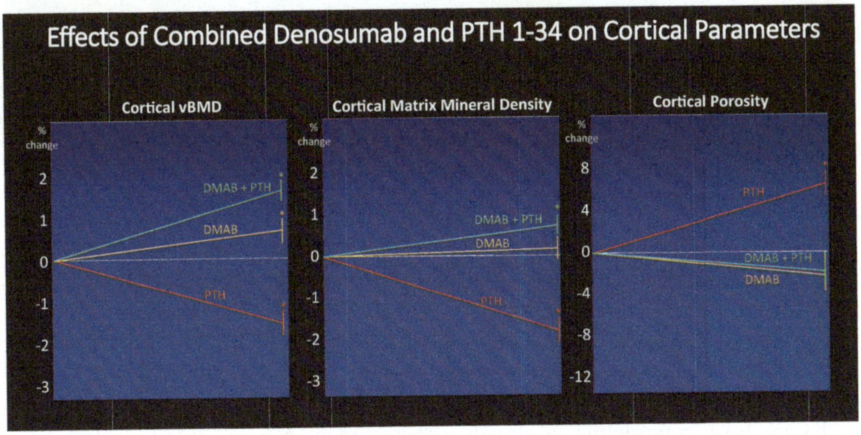

Fig. 4 Mean percent changes in cortical volumetric bone mineral density (vBMD), matrix mineral density, and porosity at the distal tibia, as assessed by HR-pQCT, in postmenopausal women after 12 months of treatment with denosumab 60-mg (yellow), PTH 1-34 20-μg (red), or both (green). $*p < 0.05$, within group difference from baseline to 12 months. Data sourced from Tsai JN, Uihlein AV, Burnett-Bowie S-AM, et al. Comparative effects of teriparatide, denosumab, and combination therapy on peripheral compartmental bone density, microarchitecture, and estimated strength: the DATA-HRpQCT Study. *J Bone Miner Res*. 2015;30(1):39–45

4.2 Sequential Therapy

4.2.1 Anabolic to Antiresorptive

Cessation of anabolic treatment results in loss of the benefits. Antiresorptives maintain or increase BMD, particularly denosumab because it is the most efficacious in suppressing remodeling (Cosman et al. 2016; Saag et al. 2017). In the DATA-Switch study, 2 years of PTH 1-34 followed by 2 years of denosumab resulted in further increases in BMD (Leder et al. 2015). At 48 months, women treated with combined PTH 1-34/denosumab for 2 years followed by denosumab alone had greater gains in BMD than those treated with PTH 1-34 followed by denosumab (Leder et al. 2015).

Bone et al. administered 24 months of alendronate after 18 months of abaloparatide or placebo (Bone et al. 2018). Treatment with alendronate was claimed to maintain the purported vertebral, nonvertebral, and major osteoporotic fracture risk reduction with abaloparatide relative to the placebo group. However, comparisons of the two groups are difficult to interpret as the placebo group may have lost bone and suffered microstructural deterioration during 18 months without active treatment. Although it is likely that stopping abaloparatide results in bone loss, as reported with PTH (Adami et al. 2008; Eastell et al. 2009), an abaloparatide group given placebo shown to be losing bone is needed to establish this and that administration of alendronate prevents this loss. Similar issues arise in the interpretation of antifracture efficacy at 24 months of romosozumab-denosumab compared to placebo-denosumab (Cosman et al. 2016).

At this time, the only study comparing fracture outcomes with sequential therapy against continuous monotherapy is the study by Saag et al., who reported greater vertebral, nonvertebral, and hip fracture reduction with romosozumab-alendronate compared to alendronate alone (Saag et al. 2017).

4.2.2 Antiresorptive to Anabolic

As discussed above, most (Boonen et al. 2008; Cosman et al. 2011; Kostenuik et al. 2001; Samadfam et al. 2007; Tsai et al. 2013), but not all (Delmas et al. 1995), studies suggest that *concurrent* administration of anabolic and antiresorptive agents does not blunt the effect on BMD or microstructure. The presence of many remodeling sites in their formative phase pre-treatment allows PTH to initiate remodeling-based bone formation by these BMUs (as they are unaffected by the concurrently started antiresorptive therapy).

By contrast, remodeling is already suppressed in patients receiving long-term antiresorptive therapy before being switched to anabolic therapy. The early increase in BMD (due to a reduction in the reversible remodeling space deficit by antiresorptive therapy) has occurred (Parfitt 2008; Seeman and Martin 2015). Namely, the many BMUs present pre-treatment that have incompletely refilled, and secondary mineralization of the less remodeled matrix has occurred. When the antiresorptive is stopped, there are fewer BMUs available for PTH peptides or abaloparatide to initiate remodeling-based bone formation. This suggests that blunting of the effect of these peptides is plausible.

The question is whether previously administered bisphosphonates, which remain in the matrix for years (Russell et al. 2008), predispose to blunting of the effect of PTH peptides or abaloparatide when the antiresorptive is stopped and replaced by these anabolic agents. Namely, is there blunting of the effect of the anabolic agent on BMD, microstructure, material composition, and most important, blunting of any improvement in bone strength or antifracture efficacy.

These are challenging questions because of difficulties in developing the study design needed to address these issues. For example, in the setting of many years of treatment with bisphosphonates, BMD is likely to have increased, mainly due to near-complete secondary mineralization of the less remodeled bone matrix volume, *not* due to restoration of bone volume or microstructure (Seeman and Martin 2019). Indeed, the longer the duration of bisphosphonate therapy before switching to an anabolic agent, the more reduced and architecturally deteriorated the matrix volume is likely to be and the higher its matrix mineral density because of slow continued unbalanced remodeling (Seeman and Martin 2019). The deficit in bone volume and the microstructural deterioration are likely to be obscured by the increased matrix mineral density producing a higher BMD.

If a control group is chosen to receive PTH 1-34 or abaloparatide to define the "un-blunted" response, they must have had no prior treatment with bisphosphonates. How should this control group be chosen? If matched by age, the control group is likely to have higher baseline bone remodeling, lower BMD, and more deteriorated microstructure than the prior bisphosphonate-treated group (because the control group are untreated). If matched by BMD, the control group is likely to have lower baseline bone remodeling, higher bone volume, and better microstructure (because the prior bisphosphonate-treated group have had an increase in BMD mainly due to secondary mineralization). Assessment of blunting is problematic if comparisons are made using a percentage change when baselines differ. These issues need to be addressed in dealing with this subject as these methodological issues may be responsible for the contradictory findings in the literature (Cosman et al. 2017).

Several studies support blunting by antiresorptives with greater remodeling suppressant action. For example, Miller et al. report a lesser increase in remodeling markers and BMD responses to PTH 1-34 in patients previously treated with alendronate than risedronate, perhaps because baseline remodeling was lower in the alendronate group (Miller et al. 2008). There was a correlation between change in 3-month measures of P1NP and spine trabecular volumetric BMD but not areal BMD. In another study in which baseline remodeling was numerically higher in the prior raloxifene than prior bisphosphonate group, switching to PTH 1-34 resulted in *similar* numerical increases in remodeling markers, but spine (not hip) BMD and estimated strength increased more greatly in the prior raloxifene group (Cosman et al. 2009, 2013). (Comparisons were confined to "switch" versus "add" within a group, not the comparisons presented here.) These data support the notion of blunting of BMD responses when remodeling is more greatly suppressed at the time PTH 1-34 is started.

By contrast, several studies do not support the notion of blunting by previously administered antiresorptive therapy. For example, in a paired bone biopsy study, despite lower baseline remodeling and higher BMD in women treated with alendronate than untreated controls, the response to PTH 1-34 in women previously treated with alendronate was not blunted compared to the PTH 1-34 effect in untreated controls (Ma et al. 2014). Periosteal and endocortical bone formation increased, consistent with an anabolic response, and intracortical porosity also increased in both groups, consistent with unsuppressed stimulation of intracortical remodeling. Likewise, in another paired bone biopsy study, trabecular bone formation in response to PTH 1-34 was not blunted by prior alendronate therapy (Fahrleitner-Pammer et al. 2016).

If the antiresorptive used is denosumab, the question of blunting is less relevant because of a rapid offset of remodeling suppression when denosumab is stopped. Caution is needed because cessation of denosumab is associated with a rapid increase in bone remodeling markers, a decline in BMD within months of stopping therapy, and uncommonly, an increased risk of multiple vertebral fractures (Bone et al. 2011; Miller et al. 2011).

Indeed, this accelerated loss of bone may be exacerbated by switching to PTH 1-34. In the DATA-switch study, women treated with denosumab for 24 months switched to PTH 1-34 for a further 24 months had a reduction in spine and hip BMD in the first 12 months followed by a gradual increase in BMD (Leder et al. 2015). However, despite this increase in BMD, cortical and trabecular volumetric bone density decreased, and bone strength estimated by finite element analysis also reduced (Tsai et al. 2017).

Several factors may be responsible for these changes. The decline in BMD may be due to the replacement of bone with osteoid during remodelling- and/or modeling-dependent bone formation. The later increase in BMD may be the result of primary and secondary mineralization of this osteoid. The concurrent decrease in cortical and trabecular volumetric BMD may be the result of increased remodeling produced by cessation of denosumab and initiation of PTH 1-34, both of which increase cortical porosity and excavate cavities upon trabeculae. Whether bone fragility increases is uncertain but replacement of denosumab with PTH 1-34 needs to be done with caution.

5 Are We There Yet?

There is progress, but important needs remain unmet. Anti-vertebral fracture efficacy of antiresorptives and anabolic therapy is robustly established in controlled studies mostly of 1–3 years duration. There is now evidence of superior anti-vertebral fracture efficacy of anabolic than antiresorptive agents. This is consistent with the notion that antiresorptive agents only reduce the reversible remodeling space deficit, whereas anabolic agents partly restore bone volume, microstructure, and its material composition, changes that should restore bone strength more effectively than antiresorptive agents.

Should anabolic agents be first-line therapy? This notion may be premature for several reasons: (1) The superior anti-vertebral fracture efficacy compared to antiresorptives has only been shown in two studies (Kendler et al. 2018; Saag et al. 2017). (2) There is no evidence of hip fracture risk reduction using anabolic agents. (3) There is little evidence of nonvertebral fracture risk reduction using anabolic therapies. Only one (Neer et al. 2001) of several (Greenspan et al. 2007; Miller et al. 2016; Nakamura et al. 2012) studies using PTH regimens, and no critically evaluated study of abaloparatide (EMA 2018; Miller et al. 2016), supports anti-nonvertebral fracture efficacy relative to untreated controls. Nor is there evidence of superior nonvertebral antifracture efficacy of anabolic agents compared with antiresorptive agents, although the numerical reduction in nonvertebral fracture risk in both trials suggests this is likely. This uncertainty is of particular concern because nonvertebral fractures account for 80% of all fractures and 50% of all the morbidity and mortality (Tatangelo et al. 2019). However, there is evidence that sequential romosozumab followed by alendronate results in greater vertebral, nonvertebral, major osteoporotic, and hip fracture risk reduction than alendronate alone (Saag et al. 2017). Finally, whether combined antiresorptive and anabolic regimens result in greater vertebral, nonvertebral, and hip fracture risk reduction than either agent alone, and whether these risk reductions cost-effectively produce more quality-adjusted life years free of fracture, remains untested.

Disclosures SKR has no disclosures. ES has received research support from Amgen and MSD and has received lecture fees from Amgen, Eli Lilly, and Allergan. He is a director and board member of StraxCorp.

References

Adami S, San Martin J, Munoz-Torres M, Econs MJ, Xie L, Dalsky GP, McClung M, Felsenberg D, Brown JP, Brandi ML, Sipos A (2008) Effect of raloxifene after recombinant teriparatide [hPTH(1-34)] treatment in postmenopausal women with osteoporosis. Osteoporos Int 19:87–94. https://doi.org/10.1007/s00198-007-0485-y

Akkus O, Polyakova-Akkus A, Adar F, Schaffler MB (2003) Aging of microstructural compartments in human compact bone. J Bone Miner Res 18:1012–1019. https://doi.org/10.1359/jbmr.2003.18.6.1012

Allen MR, Gineyts E, Leeming DJ, Burr DB, Delmas PD (2008) Bisphosphonates alter trabecular bone collagen cross-linking and isomerization in beagle dog vertebra. Osteoporos Int 19:329–337. https://doi.org/10.1007/s00198-007-0533-7

Allen MR, Erickson AM, Wang X, Burr DB, Martin RB, Hazelwood SJ (2010) Morphological assessment of basic multicellular unit resorption parameters in dogs shows additional mechanisms of bisphosphonate effects on bone. Calcif Tissue Int 86:67–71. https://doi.org/10.1007/s00223-009-9315-x

Balemans W, Ebeling M, Patel N, van Hul E, Olson P, Dioszegi M, Lacza C, Wuyts W, van den Ende J, Willems P, Paes-Alves AF, Hill S, Bueno M, Ramos FJ, Tacconi P, Dikkers FG, Stratakis C, Lindpaintner K, Vickery B, Foernzler D, van Hul W (2001) Increased bone density in sclerosteosis is due to the deficiency of a novel secreted protein (SOST). Hum Mol Genet 10:537–543. https://doi.org/10.1093/hmg/10.5.537

Balemans W, Patel N, Ebeling M, van Hul E, Wuyts W, Lacza C, Dioszegi M, Dikkers FG, Hildering P, Willems PJ, Verheij JB, Lindpaintner K, Vickery B, Foernzler D, van Hul W (2002) Identification of a 52 kb deletion downstream of the SOST gene in patients with van Buchem disease. J Med Genet 39:91–97. https://doi.org/10.1136/jmg.39.2.91

Bilezikian JP (2008) Combination anabolic and antiresorptive therapy for osteoporosis: opening the anabolic window. Curr Osteoporos Rep 6:24–30. https://doi.org/10.1007/s11914-008-0005-9

Bjornerem A, Ghasem-Zadeh A, Bui M, Wang X, Rantzau C, Nguyen TV, Hopper JL, Zebaze R, Seeman E (2011) Remodeling markers are associated with larger intracortical surface area but smaller trabecular surface area: a twin study. Bone 49:1125–1130. https://doi.org/10.1016/j.bone.2011.08.009

Bjornerem A, Wang X, Bui M, Ghasem-Zadeh A, Hopper JL, Zebaze R, Seeman E (2018) Menopause-related appendicular bone loss is mainly cortical and results in increased cortical porosity. J Bone Miner Res 33:598–605. https://doi.org/10.1002/jbmr.3333

Black DM, Greenspan SL, Ensrud KE, Palermo L, McGowan JA, Lang TF, Garnero P, Bouxsein ML, Bilezikian JP, Rosen CJ, Pa THSI (2003) The effects of parathyroid hormone and alendronate alone or in combination in postmenopausal osteoporosis. N Engl J Med 349:1207–1215. https://doi.org/10.1056/NEJMoa031975

Bone HG, Bolognese MA, Yuen CK, Kendler DL, Miller PD, Yang YC, Grazette L, San Martin J, Gallagher JC (2011) Effects of denosumab treatment and discontinuation on bone mineral density and bone turnover markers in postmenopausal women with low bone mass. J Clin Endocrinol Metab 96:972–980. https://doi.org/10.1210/jc.2010-1502

Bone HG, Wagman RB, Brandi ML, Brown JP, Chapurlat R, Cummings SR, Czerwinski E, Fahrleitner-Pammer A, Kendler DL, Lippuner K, Reginster JY, Roux C, Malouf J, Bradley MN, Daizadeh NS, Wang A, Dakin P, Pannacciulli N, Dempster DW, Papapoulos S (2017) 10 years of denosumab treatment in postmenopausal women with osteoporosis: results from the phase 3 randomised FREEDOM trial and open-label extension. Lancet Diabetes Endocrinol 5:513–523. https://doi.org/10.1016/S2213-8587(17)30138-9

Bone HG, Cosman F, Miller PD, Williams GC, Hattersley G, Hu MY, Fitzpatrick LA, Mitlak B, Papapoulos S, Rizzoli R, Dore RK, Bilezikian JP, Saag KG (2018) ACTIVExtend: 24 months of alendronate after 18 months of abaloparatide or placebo for postmenopausal osteoporosis. J Clin Endocrinol Metab 103:2949–2957. https://doi.org/10.1210/jc.2018-00163

Boonen S, Marin F, Obermayer-Pietsch B, Simoes ME, Barker C, Glass EV, Hadji P, Lyritis G, Oertel H, Nickelsen T, McCloskey EV, Investigators E (2008) Effects of previous antiresorptive therapy on the bone mineral density response to two years of teriparatide treatment in postmenopausal women with osteoporosis. J Clin Endocrinol Metab 93:852–860. https://doi.org/10.1210/jc.2007-0711

Bouxsein ML, Eastell R, Lui LY, Wu LA, de Papp AE, Grauer A, Marin F, Cauley JA, Bauer DC, Black DM, Project FBQ (2019) Change in bone density and reduction in fracture risk: a meta-regression of published trials. J Bone Miner Res 34:632–642. https://doi.org/10.1002/jbmr.3641

Chavassieux P, Chapurlat R, Portero-Muzy N, Roux JP, Garcia P, Brown JP, Libanati C, Boyce RW, Wang A, Grauer A (2019) Bone-forming and antiresorptive effects of Romosozumab in postmenopausal women with osteoporosis: bone histomorphometry and microcomputed tomography analysis after 2 and 12 months of treatment. J Bone Miner Res 34:1597–1608. https://doi.org/10.1002/jbmr.3735

Chen P, Satterwhite JH, Licata AA, Lewiecki EM, Sipos AA, Misurski DM, Wagman RB (2005) Early changes in biochemical markers of bone formation predict BMD response to teriparatide in postmenopausal women with osteoporosis. J Bone Miner Res 20:962–970. https://doi.org/10.1359/JBMR.050105

Cosman F, Nieves J, Woelfert L, Formica C, Gordon S, Shen V, Lindsay R (2001) Parathyroid hormone added to established hormone therapy: effects on vertebral fracture and maintenance of bone mass after parathyroid hormone withdrawal. J Bone Miner Res 16:925–931. https://doi.org/10.1359/jbmr.2001.16.5.925

Cosman F, Wermers RA, Recknor C, Mauck KF, Xie L, Glass EV, Krege JH (2009) Effects of teriparatide in postmenopausal women with osteoporosis on prior alendronate or raloxifene: differences between stopping and continuing the antiresorptive agent. J Clin Endocrinol Metab 94:3772–3780. https://doi.org/10.1210/jc.2008-2719

Cosman F, Eriksen EF, Recknor C, Miller PD, Guanabens N, Kasperk C, Papanastasiou P, Readie A, Rao H, Gasser JA, Bucci-Rechtweg C, Boonen S (2011) Effects of intravenous zoledronic acid plus subcutaneous teriparatide [rhPTH(1-34)] in postmenopausal osteoporosis. J Bone Miner Res 26:503–511. https://doi.org/10.1002/jbmr.238

Cosman F, Keaveny TM, Kopperdahl D, Wermers RA, Wan X, Krohn KD, Krege JH (2013) Hip and spine strength effects of adding versus switching to teriparatide in postmenopausal women with osteoporosis treated with prior alendronate or raloxifene. J Bone Miner Res 28:1328–1336. https://doi.org/10.1002/jbmr.1853

Cosman F, Crittenden DB, Adachi JD, Binkley N, Czerwinski E, Ferrari S, Hofbauer LC, Lau E, Lewiecki EM, Miyauchi A, Zerbini CA, Milmont CE, Chen L, Maddox J, Meisner PD, Libanati C, Grauer A (2016) Romosozumab treatment in postmenopausal women with osteoporosis. N Engl J Med 375:1532–1543. https://doi.org/10.1056/NEJMoa1607948

Cosman F, Nieves JW, Dempster DW (2017) Treatment sequence matters: anabolic and antiresorptive therapy for osteoporosis. J Bone Miner Res 32:198–202. https://doi.org/10.1002/jbmr.3051

Delmas PD, Vergnaud P, Arlot ME, Pastoureau P, Meunier PJ, Nilssen MH (1995) The anabolic effect of human PTH (1-34) on bone formation is blunted when bone resorption is inhibited by the bisphosphonate tiludronate--is activated resorption a prerequisite for the in vivo effect of PTH on formation in a remodeling system? Bone 16:603–610. https://doi.org/10.1016/8756-3282(95)00113-r

Dempster DW, Birchman R, Xu R, Lindsay R, Shen V (1995) Temporal changes in cancellous bone structure of rats immediately after ovariectomy. Bone 16:157–161

Dempster DW, Brown JP, Fahrleitner-Pammer A, Kendler D, Rizzo S, Valter I, Wagman RB, Yin X, Yue SV, Boivin G (2018) Effects of long-term denosumab on bone histomorphometry and mineralization in women with postmenopausal osteoporosis. J Clin Endocrinol Metab 103:2498–2509. https://doi.org/10.1210/jc.2017-02669

Eastell R, Nickelsen T, Marin F, Barker C, Hadji P, Farrerons J, Audran M, Boonen S, Brixen K, Gomes JM, Obermayer-Pietsch B, Avramidis A, Sigurdsson G, Gluer CC (2009) Sequential treatment of severe postmenopausal osteoporosis after teriparatide: final results of the randomized, controlled European Study of Forsteo (EUROFORS). J Bone Miner Res 24:726–736. https://doi.org/10.1359/jbmr.081215

Eastell R, Mitlak BH, Wang Y, Hu M, Fitzpatrick LA, Black DM (2019) Bone turnover markers to explain changes in lumbar spine BMD with abaloparatide and teriparatide: results from ACTIVE. Osteoporos Int 30:667–673. https://doi.org/10.1007/s00198-018-04819-1

Ejersted C, Oxlund H, Eriksen EF, Andreassen TT (1998) Withdrawal of parathyroid hormone treatment causes rapid resorption of newly formed vertebral cancellous and endocortical bone in old rats. Bone 23:43–52. https://doi.org/10.1016/s8756-3282(98)00072-6

EMA (2018) Eladynos (Abaloparatide) Assessment Report – EMA/CHMP/581111/2018. https://www.ema.europa.eu/en/documents/assessment-report/eladynos-epar-refusal-public-assessment-report_en.pdf

Fahrleitner-Pammer A, Burr D, Dobnig H, Stepan JJ, Petto H, Li J, Krege JH, Pavo I (2016) Improvement of cancellous bone microstructure in patients on teriparatide following alendronate pretreatment. Bone 89:16–24. https://doi.org/10.1016/j.bone.2016.05.004

Fan Y, Hanai JI, Le PT, Bi R, Maridas D, DeMambro V, Figueroa CA, Kir S, Zhou X, Mannstadt M, Baron R, Bronson RT, Horowitz MC, Wu JY, Bilezikian JP, Dempster DW, Rosen CJ, Lanske B (2017) Parathyroid hormone directs bone marrow mesenchymal cell fate. Cell Metab 25:661–672. https://doi.org/10.1016/j.cmet.2017.01.001

Fantner GE, Hassenkam T, Kindt JH, Weaver JC, Birkedal H, Pechenik L, Cutroni JA, Cidade GA, Stucky GD, Morse DE, Hansma PK (2005) Sacrificial bonds and hidden length dissipate energy as mineralized fibrils separate during bone fracture. Nat Mater 4:612–616. https://doi.org/10.1038/nmat1428

Finkelstein JS, Hayes A, Hunzelman JL, Wyland JJ, Lee H, Neer RM (2003) The effects of parathyroid hormone, alendronate, or both in men with osteoporosis. N Engl J Med 349:1216–1226. https://doi.org/10.1056/NEJMoa035725

Florio M, Gunasekaran K, Stolina M, Li X, Liu L, Tipton B, Salimi-Moosavi H, Asuncion FJ, Li C, Sun B, Tan HL, Zhang L, Han CY, Case R, Duguay AN, Grisanti M, Stevens J, Pretorius JK, Pacheco E, Jones H, Chen Q, Soriano BD, Wen J, Heron B, Jacobsen FW, Brisan E, Richards WG, Ke HZ, Ominsky MS (2016) A bispecific antibody targeting sclerostin and DKK-1 promotes bone mass accrual and fracture repair. Nat Commun 7:11505. https://doi.org/10.1038/ncomms11505

Fuchs RK, Faillace ME, Allen MR, Phipps RJ, Miller LM, Burr DB (2011) Bisphosphonates do not alter the rate of secondary mineralization. Bone 49:701–705. https://doi.org/10.1016/j.bone.2011.05.009

Geusens P, Marin F, Kendler DL, Russo LA, Zerbini CA, Minisola S, Body JJ, Lespessailles E, Greenspan SL, Bagur A, Stepan JJ, Lakatos P, Casado E, Moericke R, Lopez-Romero P, Fahrleitner-Pammer A (2018) Effects of teriparatide compared with risedronate on the risk of fractures in subgroups of postmenopausal women with severe osteoporosis: the VERO trial. J Bone Miner Res 33:783–794. https://doi.org/10.1002/jbmr.3384

Greenspan SL, Bone HG, Ettinger MP, Hanley DA, Lindsay R, Zanchetta JR, Blosch CM, Mathisen AL, Morris SA, Marriott TB, Treatment of Osteoporosis with Parathyroid Hormone Study G (2007) Effect of recombinant human parathyroid hormone (1-84) on vertebral fracture and bone mineral density in postmenopausal women with osteoporosis: a randomized trial. Ann Intern Med 146:326–339. https://doi.org/10.7326/0003-4819-146-5-200703060-00005

Hattersley G, Dean T, Corbin BA, Bahar H, Gardella TJ (2016) Binding selectivity of abaloparatide for PTH-type-1-receptor conformations and effects on downstream signaling. Endocrinology 157:141–149. https://doi.org/10.1210/en.2015-1726

Hattner R, Epker BN, Frost HM (1965) Suggested sequential mode of control of changes in cell behaviour in adult bone remodelling. Nature 206:489–490

Holdsworth G, Greenslade K, Jose J, Stencel Z, Kirby H, Moore A, Ke HZ, Robinson MK (2018) Dampening of the bone formation response following repeat dosing with sclerostin antibody in mice is associated with up-regulation of Wnt antagonists. Bone 107:93–103. https://doi.org/10.1016/j.bone.2017.11.003

Jiang Y, Zhao JJ, Mitlak BH, Wang O, Genant HK, Eriksen EF (2003) Recombinant human parathyroid hormone (1-34) [teriparatide] improves both cortical and cancellous bone structure. J Bone Miner Res 18:1932–1941. https://doi.org/10.1359/jbmr.2003.18.11.1932

Ke HZ, Richards WG, Li X, Ominsky MS (2012) Sclerostin and Dickkopf-1 as therapeutic targets in bone diseases. Endocr Rev 33:747–783. https://doi.org/10.1210/er.2011-1060

Kendler DL, Marin F, Zerbini CAF, Russo LA, Greenspan SL, Zikan V, Bagur A, Malouf-Sierra J, Lakatos P, Fahrleitner-Pammer A, Lespessailles E, Minisola S, Body JJ, Geusens P, Möricke R, López-Romero P (2018) Effects of teriparatide and risedronate on new fractures in post-menopausal women with severe osteoporosis (VERO): a multicentre, double-blind, double-dummy, randomised controlled trial. Lancet 391:230–240. https://doi.org/10.1016/s0140-6736(17)32137-2

Khosla S (2003) Parathyroid hormone plus alendronate--a combination that does not add up. N Engl J Med 349:1277–1279. https://doi.org/10.1056/NEJMe038143

Kim SW, Pajevic PD, Selig M, Barry KJ, Yang JY, Shin CS, Baek WY, Kim JE, Kronenberg HM (2012) Intermittent parathyroid hormone administration converts quiescent lining cells to active osteoblasts. J Bone Miner Res 27:2075–2084. https://doi.org/10.1002/jbmr.1665

Kimmel DB, Bozzato RP, Kronis KA, Coble T, Sindrey D, Kwong P, Recker RR (1993) The effect of recombinant human (1-84) or synthetic human (1-34) parathyroid hormone on the skeleton of adult osteopenic ovariectomized rats. Endocrinology 132:1577–1584. https://doi.org/10.1210/endo.132.4.8462456

Kostenuik PJ, Capparelli C, Morony S, Adamu S, Shimamoto G, Shen V, Lacey DL, Dunstan CR (2001) OPG and PTH-(1-34) have additive effects on bone density and mechanical strength in osteopenic ovariectomized rats. Endocrinology 142:4295–4304. https://doi.org/10.1210/endo.142.10.8437

Leder BZ, Tsai JN, Uihlein AV, Wallace PM, Lee H, Neer RM, Burnett-Bowie SA (2015) Denosumab and teriparatide transitions in postmenopausal osteoporosis (the DATA-switch study): extension of a randomised controlled trial. Lancet 386:1147–1155. https://doi.org/10.1016/S0140-6736(15)61120-5

Lindsay R, Cosman F, Zhou H, Bostrom MP, Shen VW, Cruz JD, Nieves JW, Dempster DW (2006) A novel tetracycline labeling schedule for longitudinal evaluation of the short-term effects of anabolic therapy with a single iliac crest bone biopsy: early actions of teriparatide. J Bone Miner Res 21:366–373. https://doi.org/10.1359/JBMR.051109

Lindsay R, Zhou H, Cosman F, Nieves J, Dempster DW, Hodsman AB (2007) Effects of a one-month treatment with PTH(1-34) on bone formation on cancellous, endocortical, and periosteal surfaces of the human ilium. J Bone Miner Res 22:495–502. https://doi.org/10.1359/jbmr.070104

Lips P, Courpron P, Meunier PJ (1978) Mean wall thickness of trabecular bone packets in the human iliac crest: changes with age. Calcif Tissue Res 26:13–17

Liu CC, Kalu DN (1990) Human parathyroid hormone-(1-34) prevents bone loss and augments bone formation in sexually mature ovariectomized rats. J Bone Miner Res 5:973–982. https://doi.org/10.1002/jbmr.5650050911

Lloyd AA, Gludovatz B, Riedel C, Luengo EA, Saiyed R, Marty E, Lorich DG, Lane JM, Ritchie RO, Busse B, Donnelly E (2017) Atypical fracture with long-term bisphosphonate therapy is associated with altered cortical composition and reduced fracture resistance. Proc Natl Acad Sci U S A 114:8722–8727. https://doi.org/10.1073/pnas.1704460114

Ma YL, Zeng Q, Donley DW, Ste-Marie LG, Gallagher JC, Dalsky GP, Marcus R, Eriksen EF (2006) Teriparatide increases bone formation in modeling and remodeling osteons and enhances

IGF-II immunoreactivity in postmenopausal women with osteoporosis. J Bone Miner Res 21:855–864. https://doi.org/10.1359/jbmr.060314

Ma YL, Zeng QQ, Porras LL, Harvey A, Moore TL, Shelbourn TL, Dalsky GP, Wronski TJ, Aguirre JI, Bryant HU, Sato M (2011) Teriparatide [rhPTH (1-34)], but not strontium ranelate, demonstrated bone anabolic efficacy in mature, osteopenic, ovariectomized rats. Endocrinology 152:1767–1778. https://doi.org/10.1210/en.2010-1112

Ma YL, Zeng QQ, Chiang AY, Burr D, Li J, Dobnig H, Fahrleitner-Pammer A, Michalska D, Marin F, Pavo I, Stepan JJ (2014) Effects of teriparatide on cortical histomorphometric variables in postmenopausal women with or without prior alendronate treatment. Bone 59:139–147. https://doi.org/10.1016/j.bone.2013.11.011

Maeda A, Ono M, Holmbeck K, Li L, Kilts TM, Kram V, Noonan ML, Yoshioka Y, McNerny EM, Tantillo MA, Kohn DH, Lyons KM, Robey PG, Young MF (2015) WNT1-induced secreted protein-1 (WISP1), a novel regulator of bone turnover and WNT signaling. J Biol Chem 290:14004–14018. https://doi.org/10.1074/jbc.M114.628818

Manolagas SC (2000) Birth and death of bone cells: basic regulatory mechanisms and implications for the pathogenesis and treatment of osteoporosis. Endocr Rev 21:115–137. https://doi.org/10.1210/edrv.21.2.0395

McClung MR, Lewiecki EM, Cohen SB, Bolognese MA, Woodson GC, Moffett AH, Peacock M, Miller PD, Lederman SN, Chesnut CH, Lain D, Kivitz AJ, Holloway DL, Zhang C, Peterson MC, Bekker PJ, Group AMGBLS (2006) Denosumab in postmenopausal women with low bone mineral density. N Engl J Med 354:821–831. https://doi.org/10.1056/NEJMoa044459

McClung MR, Grauer A, Boonen S, Bolognese MA, Brown JP, Diez-Perez A, Langdahl BL, Reginster JY, Zanchetta JR, Wasserman SM, Katz L, Maddox J, Yang YC, Libanati C, Bone HG (2014) Romosozumab in postmenopausal women with low bone mineral density. N Engl J Med 370:412–420. https://doi.org/10.1056/NEJMoa1305224

Miller PD, Delmas PD, Lindsay R, Watts NB, Luckey M, Adachi J, Saag K, Greenspan SL, Seeman E, Boonen S, Meeves S, Lang TF, Bilezikian JP (2008) Open-label study to determine how prior therapy with alendronate or risedronate in postmenopausal women with osteoporosis influences the clinical effectiveness of teriparatide I. early responsiveness of women with osteoporosis to teriparatide after therapy with alendronate or risedronate. J Clin Endocrinol Metab 93:3785–3793. https://doi.org/10.1210/jc.2008-0353

Miller PD, Wagman RB, Peacock M, Lewiecki EM, Bolognese MA, Weinstein RL, Ding B, San Martin J, McClung MR (2011) Effect of denosumab on bone mineral density and biochemical markers of bone turnover: six-year results of a phase 2 clinical trial. J Clin Endocrinol Metab 96:394–402. https://doi.org/10.1210/jc.2010-1805

Miller PD, Hattersley G, Riis BJ, Williams GC, Lau E, Russo LA, Alexandersen P, Zerbini CA, Hu MY, Harris AG, Fitzpatrick LA, Cosman F, Christiansen C, Investigators AS (2016) Effect of Abaloparatide vs placebo on new vertebral fractures in postmenopausal women with osteoporosis: a randomized clinical trial. JAMA 316:722–733. https://doi.org/10.1001/jama.2016.11136

Misof BM, Roschger P, Cosman F, Kurland ES, Tesch W, Messmer P, Dempster DW, Nieves J, Shane E, Fratzl P, Klaushofer K, Bilezikian J, Lindsay R (2003) Effects of intermittent parathyroid hormone administration on bone mineralization density in iliac crest biopsies from patients with osteoporosis: a paired study before and after treatment. J Clin Endocrinol Metab 88:1150–1156. https://doi.org/10.1210/jc.2002-021988

Mohsin S, O'Brien FJ, Lee TC (2006) Osteonal crack barriers in ovine compact bone. J Anat 208:81–89. https://doi.org/10.1111/j.1469-7580.2006.00509.x

Moreira CA, Fitzpatrick LA, Wang Y, Recker RR (2017) Effects of abaloparatide-SC (BA058) on bone histology and histomorphometry: the ACTIVE phase 3 trial. Bone 97:314–319. https://doi.org/10.1016/j.bone.2016.11.004

Nakamura T, Sugimoto T, Nakano T, Kishimoto H, Ito M, Fukunaga M, Hagino H, Sone T, Yoshikawa H, Nishizawa Y, Fujita T, Shiraki M (2012) Randomized teriparatide

[human parathyroid hormone (PTH) 1-34] once-weekly efficacy research (TOWER) trial for examining the reduction in new vertebral fractures in subjects with primary osteoporosis and high fracture risk. J Clin Endocrinol Metab 97:3097–3106. https://doi.org/10.1210/jc.2011-3479
Neer RM, Arnaud CD, Zanchetta JR, Prince R, Gaich GA, Reginster JY, Hodsman AB, Eriksen EF, Ish-Shalom S, Genant HK, Wang O, Mitlak BH (2001) Effect of parathyroid hormone (1-34) on fractures and bone mineral density in postmenopausal women with osteoporosis. N Engl J Med 344:1434–1441. https://doi.org/10.1056/NEJM200105103441904
Nioi P, Taylor S, Hu R, Pacheco E, He YD, Hamadeh H, Paszty C, Pyrah I, Ominsky MS, Boyce RW (2015) Transcriptional profiling of laser capture microdissected subpopulations of the osteoblast lineage provides insight into the early response to sclerostin antibody in rats. J Bone Miner Res 30:1457–1467. https://doi.org/10.1002/jbmr.2482
Ominsky MS, Libanati C, Niu QT, Boyce RW, Kostenuik PJ, Wagman RB, Baron R, Dempster DW (2015) Sustained modeling-based bone formation during adulthood in cynomolgus monkeys may contribute to continuous BMD gains with denosumab. J Bone Miner Res 30:1280–1289. https://doi.org/10.1002/jbmr.2480
Ominsky MS, Boyd SK, Varela A, Jolette J, Felx M, Doyle N, Mellal N, Smith SY, Locher K, Buntich S, Pyrah I, Boyce RW (2017) Romosozumab improves bone mass and strength while maintaining bone quality in ovariectomized cynomolgus monkeys. J Bone Miner Res 32:788–801. https://doi.org/10.1002/jbmr.3036
Padhi D, Jang G, Stouch B, Fang L, Posvar E (2011) Single-dose, placebo-controlled, randomized study of AMG 785, a sclerostin monoclonal antibody. J Bone Miner Res 26:19–26. https://doi.org/10.1002/jbmr.173
Parfitt AM (1980) Morphological basis of bone mineral measurements: transient and steady state effects of treatment in osteoporosis. Miner Electrolyte Metab 4:273–287
Parfitt AM (1984) Age-related structural changes in trabecular and cortical bone: cellular mechanisms and biomechanical consequences. Calcif Tissue Int 36(Suppl 1):S123–S128
Parfitt AM (2008) Skeletal heterogeneity and the purposes of bone remodelling: implications for the understanding of osteoporosis. Academic Press, Cambridge
Paschalis EP, Glass EV, Donley DW, Eriksen EF (2005) Bone mineral and collagen quality in iliac crest biopsies of patients given teriparatide: new results from the fracture prevention trial. J Clin Endocrinol Metab 90:4644–4649. https://doi.org/10.1210/jc.2004-2489
Poundarik AA, Wu PC, Evis Z, Sroga GE, Ural A, Rubin M, Vashishth D (2015) A direct role of collagen glycation in bone fracture. J Mech Behav Biomed Mater 52:120–130. https://doi.org/10.1016/j.jmbbm.2015.08.012
Reid IR (2015) Short-term and long-term effects of osteoporosis therapies. Nat Rev Endocrinol 11:418–428. https://doi.org/10.1038/nrendo.2015.71
Reid IR, Ames R, Mason B, Reid HE, Bacon CJ, Bolland MJ, Gamble GD, Grey A, Horne A (2008) Randomized controlled trial of calcium supplementation in healthy, nonosteoporotic, older men. Arch Intern Med 168:2276–2282. https://doi.org/10.1001/archinte.168.20.2276
Ruff CB, Hayes WC (1988) Sex differences in age-related remodeling of the femur and tibia. J Orthop Res 6:886–896. https://doi.org/10.1002/jor.1100060613
Russell RG, Watts NB, Ebetino FH, Rogers MJ (2008) Mechanisms of action of bisphosphonates: similarities and differences and their potential influence on clinical efficacy. Osteoporos Int 19:733–759. https://doi.org/10.1007/s00198-007-0540-8
Saag KG, Petersen J, Brandi ML, Karaplis AC, Lorentzon M, Thomas T, Maddox J, Fan M, Meisner PD, Grauer A (2017) Romosozumab or alendronate for fracture prevention in women with osteoporosis. N Engl J Med 377:1417–1427. https://doi.org/10.1056/NEJMoa1708322
Saag KG, Petersen J, Grauer A (2018) Romosozumab versus alendronate and fracture risk in women with osteoporosis. N Engl J Med 378:195–196. https://doi.org/10.1056/NEJMc1714810
Saito M, Marumo K (2010) Collagen cross-links as a determinant of bone quality: a possible explanation for bone fragility in aging, osteoporosis, and diabetes mellitus. Osteoporos Int 21:195–214. https://doi.org/10.1007/s00198-009-1066-z

Saito M, Mori S, Mashiba T, Komatsubara S, Marumo K (2008) Collagen maturity, glycation induced-pentosidine, and mineralization are increased following 3-year treatment with incadronate in dogs. Osteoporos Int 19:1343–1354. https://doi.org/10.1007/s00198-008-0585-3

Saito M, Marumo K, Kida Y, Ushiku C, Kato S, Takao-Kawabata R, Kuroda T (2011) Changes in the contents of enzymatic immature, mature, and non-enzymatic senescent cross-links of collagen after once-weekly treatment with human parathyroid hormone (1-34) for 18 months contribute to improvement of bone strength in ovariectomized monkeys. Osteoporos Int 22:2373–2383. https://doi.org/10.1007/s00198-010-1454-4

Samadfam R, Xia Q, Goltzman D (2007) Co-treatment of PTH with osteoprotegerin or alendronate increases its anabolic effect on the skeleton of oophorectomized mice. J Bone Miner Res 22:55–63. https://doi.org/10.1359/jbmr.060915

Seeman E (2003) Periosteal bone formation--a neglected determinant of bone strength. N Engl J Med 349:320–323. https://doi.org/10.1056/NEJMp038101

Seeman E, Delmas PD (2006) Bone quality--the material and structural basis of bone strength and fragility. N Engl J Med 354:2250–2261. https://doi.org/10.1056/NEJMra053077

Seeman E, Martin TJ (2015) Co-administration of antiresorptive and anabolic agents: a missed opportunity. J Bone Miner Res 30:753–764. https://doi.org/10.1002/jbmr.2496

Seeman E, Martin TJ (2019) Antiresorptive and anabolic agents in the prevention and reversal of bone fragility. Nat Rev Rheumatol 15:225–236. https://doi.org/10.1038/s41584-019-0172-3

Seeman E, Nguyen TV (2016) Bone remodeling markers: so easy to measure, so difficult to interpret. Osteoporos Int 27:33–35. https://doi.org/10.1007/s00198-015-3374-9

Seref-Ferlengez Z, Basta-Pljakic J, Kennedy OD, Philemon CJ, Schaffler MB (2014) Structural and mechanical repair of diffuse damage in cortical bone in vivo. J Bone Miner Res 29:2537–2544. https://doi.org/10.1002/jbmr.2309

Shen V, Dempster DW, Mellish RW, Birchman R, Horbert W, Lindsay R (1992) Effects of combined and separate intermittent administration of low-dose human parathyroid hormone fragment (1-34) and 17 beta-estradiol on bone histomorphometry in ovariectomized rats with established osteopenia. Calcif Tissue Int 50:214–220

Silverman SL, Chines AA, Kendler DL, Kung AW, Teglbjaerg CS, Felsenberg D, Mairon N, Constantine GD, Adachi JD, Bazedoxifene Study Group (2012) Sustained efficacy and safety of bazedoxifene in preventing fractures in postmenopausal women with osteoporosis: results of a 5-year, randomized, placebo-controlled study. Osteoporos Int 23:351–363. https://doi.org/10.1007/s00198-011-1691-1

Sroga GE, Vashishth D (2012) Effects of bone matrix proteins on fracture and fragility in osteoporosis. Curr Osteoporos Rep 10:141–150. https://doi.org/10.1007/s11914-012-0103-6

Tang SY, Zeenath U, Vashishth D (2007) Effects of non-enzymatic glycation on cancellous bone fragility. Bone 40:1144–1151. https://doi.org/10.1016/j.bone.2006.12.056

Tatangelo G, Watts J, Lim K, Connaughton C, Abimanyi-Ochom J, Borgstrom F, Nicholson GC, Shore-Lorenti C, Stuart AL, Iuliano-Burns S, Seeman E, Prince R, March L, Cross M, Winzenberg T, Laslett LL, Duque G, Ebeling PR, Sanders KM (2019) The cost of osteoporosis, osteopenia, and associated fractures in Australia in 2017. J Bone Miner Res 34:616–625. https://doi.org/10.1002/jbmr.3640

Taylor S, Ominsky MS, Hu R, Pacheco E, He YD, Brown DL, Aguirre JI, Wronski TJ, Buntich S, Afshari CA, Pyrah I, Nioi P, Boyce RW (2016) Time-dependent cellular and transcriptional changes in the osteoblast lineage associated with sclerostin antibody treatment in ovariectomized rats. Bone 84:148–159. https://doi.org/10.1016/j.bone.2015.12.013

Thomas CJ, Cleland TP, Zhang S, Gundberg CM, Vashishth D (2017) Identification and characterization of glycation adducts on osteocalcin. Anal Biochem 525:46–53. https://doi.org/10.1016/j.ab.2017.02.011

Tran Van PT, Vignery A, Baron R (1982) Cellular kinetics of the bone remodeling sequence in the rat. Anat Rec 202:445–451. https://doi.org/10.1002/ar.1092020403

Tsai JN, Uihlein AV, Lee H, Kumbhani R, Siwila-Sackman E, McKay EA, Burnett-Bowie SA, Neer RM, Leder BZ (2013) Teriparatide and denosumab, alone or combined, in women

with postmenopausal osteoporosis: the DATA study randomised trial. Lancet 382:50–56. https://doi.org/10.1016/S0140-6736(13)60856-9
Tsai JN, Uihlein AV, Burnett-Bowie SA, Neer RM, Zhu Y, Derrico N, Lee H, Bouxsein ML, Leder BZ (2015) Comparative effects of teriparatide, denosumab, and combination therapy on peripheral compartmental bone density, microarchitecture, and estimated strength: the DATA-HRpQCT study. J Bone Miner Res 30:39–45. https://doi.org/10.1002/jbmr.2315
Tsai JN, Nishiyama KK, Lin D, Yuan A, Lee H, Bouxsein ML, Leder BZ (2017) Effects of denosumab and teriparatide transitions on bone microarchitecture and estimated strength: the DATA-Switch HR-pQCT study. J Bone Miner Res 32:2001–2009. https://doi.org/10.1002/jbmr.3198
Varela A, Chouinard L, Lesage E, Smith SY, Hattersley G (2017) One year of abaloparatide, a selective activator of the PTH1 receptor, increased bone formation and bone mass in osteopenic ovariectomized rats without increasing bone resorption. J Bone Miner Res 32:24–33. https://doi.org/10.1002/jbmr.3003
Vashishth D (2007) The role of the collagen matrix in skeletal fragility. Curr Osteoporos Rep 5:62–66. https://doi.org/10.1007/s11914-007-0004-2
Veldhuis-Vlug AG, Rosen CJ (2018) Clinical implications of bone marrow adiposity. J Intern Med 283:121–139. https://doi.org/10.1111/joim.12718
Wronski TJ, Yen CF, Qi H, Dann LM (1993) Parathyroid hormone is more effective than estrogen or bisphosphonates for restoration of lost bone mass in ovariectomized rats. Endocrinology 132:823–831. https://doi.org/10.1210/endo.132.2.8425497
Zebaze R, Seeman E (2015) Cortical bone: a challenging geography. J Bone Miner Res 30:24–29. https://doi.org/10.1002/jbmr.2419
Zebaze RM, Ghasem-Zadeh A, Bohte A, Iuliano-Burns S, Mirams M, Price RI, Mackie EJ, Seeman E (2010) Intracortical remodelling and porosity in the distal radius and post-mortem femurs of women: a cross-sectional study. Lancet 375:1729–1736. https://doi.org/10.1016/S0140-6736(10)60320-0
Zebaze R, Ghasem-Zadeh A, Mbala A, Seeman E (2013) A new method of segmentation of compact-appearing, transitional and trabecular compartments and quantification of cortical porosity from high resolution peripheral quantitative computed tomographic images. Bone 54:8–20. https://doi.org/10.1016/j.bone.2013.01.007

New Targets and Emergent Therapies for Osteoporosis

Robert Brommage

Contents

Abstract

The 11 existing FDA-approved osteoporosis drug treatments include hormone replacement therapy, 2 SERMs (raloxifene and bazedoxifene), 5 inhibitors of bone-resorbing osteoclasts (4 bisphosphonates and anti-RANKL denosumab), 2 parathyroid hormone analogues (teriparatide and abaloparatide), and 1 WNT signaling enhancer (romosozumab). These therapies are effective and provide multiple options for patients and physicians. As the genomic revolution continues, potential novel targets for future drug development are identified. This review takes a wide perspective to describe potentially rewarding topics to

R. Brommage (✉)
The Woodlands, TX, USA

P. H. Stern (ed.), *Bone Regulators and Osteoporosis Therapy*,
Handbook of Experimental Pharmacology 262, https://doi.org/10.1007/164_2019_329

explore, including knowledge of genes and pathways involved in bone cell metabolism, the utility of animal models, targeting drugs to bone, and ongoing advances in drug design and delivery.

Keywords
Bone targeting · Mechanostat · Osteoporosis therapies · PTH/PTHrP · WNT

1 Overview

The modern era of osteoporosis diagnosis and therapy started during the 1980s with DXA measurements of bone mineral density (Lorentzon and Cummings 2015; Lewiecki and Binkley 2017). Prior to this technical advance, bone status was estimated from highly invasive rib and iliac crest biopsies or examinations at autopsy. With DXA scans, subjects with low bone mass could be identified, and pharmacological treatment could be monitored. Consensus that osteoporosis be defined by a BMD T-score of minus 2.5 or lower was quickly achieved. Nomenclature for dynamic bone histomorphometric measurements was agreed upon (Parfitt et al. 1987), and serum and urine assays for bone turnover markers were developed. The ovariectomized rat was accepted as a model of postmenopausal osteoporosis (Frost and Jee 1992).

Osteoporosis occurs when reduced amounts of otherwise normal bone result in reduced bone strength and subsequent fragility fractures. As components of the musculoskeletal system, bone and skeletal muscle health are coordinately regulated. Age-related osteoporosis and sarcopenia occur together, and athletes have elevated muscle and bone mass at skeletal sites involved in muscular activity (Warden et al. 2014). With aging, fatigue damage accumulates, and bone becomes increasingly fragile. Bone undergoes constant remodeling (turnover) to remove and replace damaged regions containing microfractures and necrotic/apoptotic osteocytes. Osteoporosis occurs when the amount of new bone formed during remodeling is less than the amount of damaged bone removed. Topics in bone biology and the pathogenesis, identification, and current treatment options for osteoporosis are reviewed in the other chapters of this handbook.

A key distinction between anti-resorptive and anabolic osteoporosis therapies must be emphasized (Riggs and Parfitt 2005). Anti-resorptive therapies inhibit osteoclastic bone resorption and thereby minimize bone remodeling. In contrast, anabolic therapies increase bone strength by stimulating osteoblasts to make new bone. The consensus among osteoporosis experts is that existing anti-resorptive therapies are adequate and anabolic teriparatide and abaloparatide treatments are highly efficacious but not ideal because of daily subcutaneous injections. The 2019 approval of romosozumab therapy (in the USA but not the EU) with monthly injections is a key advance. No new osteoporosis treatments are anticipated during the next few years, as clinical experiences with abaloparatide (approved during 2017) and romosozumab accumulate.

Anabolic therapies are given for 1 (romosozumab) or 2 (teriparatide and abaloparatide) years to build new bone. This added bone is rapidly lost upon stopping anabolic treatment unless anti-resorptive therapy is started. Similarly, bone remodeling suppressed during anti-resorptive denosumab therapy rapidly increases when this treatment is stopped (Zanchetta et al. 2018).

Developing novel therapies for osteoporosis involves knowledge of existing therapies, understanding the reasons experimental therapies failed, and acquiring expertise in multiple experimental techniques of bone biology. Novel therapies can involve improving existing therapies, generating drugs modifying components of the biochemical pathways of existing therapies, identifying novel pathways and drug targets affecting bone, developing strategies for targeting drugs specifically to bone, and developing new approaches to interfere with key pathways. This review emphasizes potential future anabolic rather than anti-resorptive therapies.

2 Existing Osteoporosis Therapies

Successful osteoporosis therapies (with years of approval in parentheses) include four bisphosphonates of increasing potency, alendronate (1995), risedronate (2002), ibandronate (2003), and zoledronate (2007); two selective estrogen receptor modifiers, raloxifene (1997) and bazedoxifene (2013); the RANKL-neutralizing antibody denosumab (2010); two parathyroid hormone analogues, teriparatide (2002) and abaloparatide (2017); and the sclerostin-neutralizing antibody, romosozumab (2019). Full-length parathyroid hormone was approved as an osteoporosis therapy in Europe from 2006 through 2014 (as Preotact) and since 2015 as a treatment for hypoparathyroidism (as Natpara). The beneficial skeletal effects of postmenopausal hormone replacement therapy were demonstrated when the Women's Health Initiative results appeared during 2003 (Cauley et al. 2003; Miller and Harman 2017). Clinical testing of a biosimilar denosumab antibody is underway.

Two common misconceptions are that teriparatide and abaloparatide have the disadvantages of promoting osteosarcomas and are limited to 2 years of treatment. Rats treated with extremely high doses of teriparatide in toxicology studies for 20 and 24 months starting at 2 months of age developed osteosarcomas (Vahle et al. 2002, 2004). Similar observations were made in rats treated with full-length PTH (Jolette et al. 2006) and abaloparatide (Jolette et al. 2017). No osteosarcomas were identified in cynomolgus macaques after 18 months of teriparatide treatment or during 3 years following treatment (Vahle et al. 2008). Osteosarcoma incidence was extensively discussed by the FDA with Lilly (Marcus 2011), and teriparatide was approved with a "black box warning" in 2002. Two ongoing postmarking surveillance studies examining patients receiving teriparatide have not detected increases in osteosarcoma incidence after 7 (Andrews et al. 2012) and 8 years (Gilsenan et al. 2018).

The goal of anabolic osteoporosis therapies is to build sufficient new bone during a treatment phase to reduce fractures. Once sufficient new bone is formed, treatment is stopped and treatment initiating with an anti-resorptive therapy to maintain the new bone added. Lifetime teriparatide therapy with daily subcutaneous dosing was never envisioned. The ability to "cure" osteoporosis with 2 years of treatment is a desired benefit and not a disadvantage. Romosozumab treatment lasts for 1 year, and future anabolic therapies are likely to require similar treatment durations.

3 Historical Failures

> Fools say they learn by experience. I prefer to profit by others' experience (Otto von Bismarck)

Development of osteoporosis therapies involved many failures that are worth understanding to avoid similar pitfalls when exploring novel therapies. Inorganic fluoride treatment increased BMD but, by replacing hydroxyl groups within the bone mineral hydroxyapatite, decreased bone strength and increased nonvertebral fractures (Riggs et al. 1990). The non-nitrogen-containing bisphosphonate tiludronate was ineffective in increasing BMD and preventing fractures (Reginster et al. 2001). An αVβ3 integrin antagonist suppressing osteoclastic activity (Coleman et al. 2004) showed promise in a clinical trial (Murphy et al. 2005), but work was discontinued. A vitamin D analogue, 2MD, stimulated bone remodeling without increasing spine and hip BMD (DeLuca et al. 2011).

Raloxifene and bazedoxifene are approved SERMs (selective estrogen receptor modulators) that mimic the beneficial skeletal actions of estrogens without their negative breast and uterine effects. SERMs that failed clinical trials include idoxifene (Nuttall et al. 1998), levormeloxifene (Ravn et al. 2006), and lasofoxifene (Cummings et al. 2010). SARMs (selective androgen receptor modulators) have been examined in preclinical studies (Schmidt et al. 2010) without advancing to clinical trials. The mixed androgen-estrogen-progestin tibolone reduced bone fractures and breast cancer in postmenopausal women but increased the risk of strokes (Cummings et al. 2008). The cathepsin K antagonist odanacatib reduced fractures in osteoporotic women but had undesirable cardiovascular effects (Drake et al. 2017). Calcilytics that stimulate endogenous PTH secretion were also unsuccessful (Fitzpatrick et al. 2011; Halse et al. 2014; John et al. 2014; Cosman et al. 2016).

Teriparatide administration by buccal, sublingual, oral (Hämmerle et al. 2012; Henriksen et al. 2012), nasal (Pearson et al. 2019), and pulmonary (Codrons et al. 2003) routes, and by transdermal patch (Cosman et al. 2010), was explored with some successes (Morley 2005) but without ultimate clinical approval. PTHrP(1–36) treatment increased spine and hip BMD in a 3-month clinical trial (Horwitz et al. 2013). Treatment with two PTH analogues (PTS 893 and RS-66271) increased bone mass in ovariectomized rats (Kneissel et al. 2001; Vickery et al. 1996), and RS-66271 entered clinical evaluation without data reported.

A preclinical failure involved the gut-derived serotonin (GDS) hypothesis which proposes circulating serotonin produced by intestinal enterochromaffin cells suppresses bone formation (Yadav et al. 2008). The idea of inhibiting intestinal serotonin synthesis, requiring tryptophan hydroxylase (TPH1), to stimulate bone formation attracted great interest, and a biotech company was formed to develop

TPH1 inhibitors (Goldberg et al. 2017). However, this proposal was "too good to be true" (Kolata 2008), as the intestines of the original *Tph1* KO mice examined actually synthesized serotonin (Kim et al. 2018), and a major collaborative effort involving six independent laboratories could not replicate the original studies (Cui et al. 2011).

The severe back pain that often follows osteoporotic vertebral fractures prompted orthopedic interventions involving vertebral augmentation (percutaneous vertebroplasty and balloon kyphoplasty) for pain relief. The ASBMR Task Force formed to address the efficacy and safety of these procedures found no evidence for efficacy and could not exclude harmful effects (Ebeling et al. 2019).

Additional promising ideas and failures are described in a 2010 comprehensive review of osteoporosis drug development during that era (Allen et al. 2010).

4 Improving Existing Osteoporosis Treatments

Successful treatment strategies are often modified and improved with follow-on drugs. Thus, the increased potency of successive bisphosphonate drugs allows reduced dosing frequencies, and abaloparatide was designed to be stable at room temperature to avoid the refrigeration requirement of teriparatide. Four examples of efforts to advance existing osteoporosis therapies include development of a high-potency bisphosphonate with reduced bone mineral-binding affinity, allowing reduced bone retention (Lawson et al. 2017): an orally active small-molecule RANKL inhibitor (Nakai et al. 2019); a bispecific sclerostin-DKK1 antibody with improved efficacy over individual sclerostin and DKK1 antibodies (Florio et al. 2016); and ongoing work by Lilly to develop a bispecific sclerostin-myostatin antibody to increase both bone and muscle mass (Gary Krishnan, personal communication).

5 Activating Osteoblasts During Anabolic Therapy

Osteoblasts, derived from osteoprogenitor cells through the actions of Osterix and Runx2 transcription factors, secrete a collagen matrix that becomes mineralized with poorly crystalline hydroxyapatite. Once new bone is formed, osteoblasts can undergo apoptosis, become embedded within the mineralized matrix as terminally differentiated osteocytes, or transition to quiescent bone-lining cells on bone surfaces (Miller et al. 1989; Matic et al. 2016). Histologically, active osteoblasts are "plump" with an extensive endoplasmic reticulum network and abundant secretory vesicles (Dudley and Spiro 1961). Plump osteoblasts become "thin" bone-lining cells once the required amount of mineralized bone is formed but remain available to form additional bone upon stimulation. The rapid transition from quiescent bone-lining cells to active osteoblasts is termed modeling-dependent bone formation to distinguish this process from remodeling-dependent bone formation replacing the bone previously removed by osteoclasts. The anabolic actions of loading (Pead et al. 1988) and treatments with PGE2 (Yao et al. 1999), teriparatide (Dobnig and Turner

1995; Kim et al. 2012), odanacatib (Pennypacker et al. 2014), romosozumab (Delgado-Calle and Bellido 2017; Kim et al. 2017), and NOTUM inhibitors (Brommage et al. 2019) occur in large part from activating bone modeling on previously quiescent bone surfaces (Langdahl et al. 2016).

Lack of appreciation of this rapid transition of quiescent osteoblasts (bone-lining cells) to active osteoblasts leads to a common misconception that osteoporosis results from insufficient numbers of osteoblasts forming bone, and therefore anabolic osteoporosis therapies should promote differentiation of osteoprogenitor cells into active osteoblasts. The rapid activation of quiescent bone-lining cells to osteoblasts does not require cell proliferation or differentiation.

6 Trabecular Bone Expansion with Prolonged Anabolic Therapy

Osteocytes, requiring oxygen, nutrients, and access to systemic signaling molecules, cannot survive far from blood vessels or bone surfaces. Osteocyte death prompts focal bone remodeling to repair non-perfused bone (Erben 2015; Knowles 2015). During long-term anabolic therapy, increased trabecular thickness compromises the access of interior osteocytes to nutrients and stimulates trabecular osteonal resorption during which trabeculae are split longitudinally into thinner trabeculae (Fig. 1). This trabecular tunneling process has been observed in hypoparathyroid patients treated with teriparatide (Gafni et al. 2012) and parathyroid hormone (Rubin et al. 2018), dogs treated with teriparatide (Boyce et al. 1996), and ovariectomized monkeys treated with both teriparatide (Jerome et al. 2001) and intact parathyroid hormone (Miller et al. 2008).

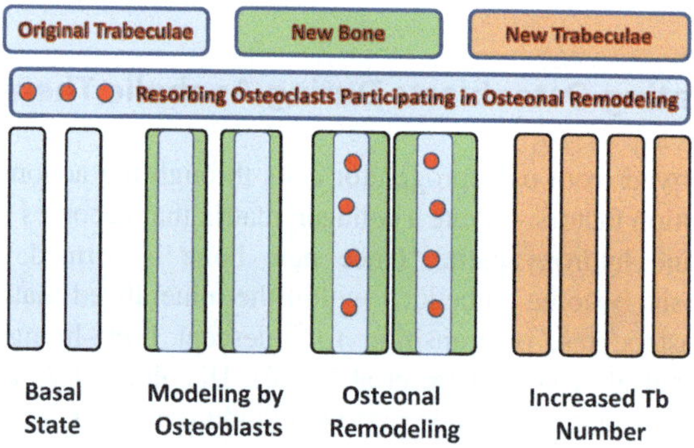

Fig. 1 Proposed effects of prolonged treatment with anabolic osteoporosis drugs to increase trabecular number. Anabolic therapies initially increase trabecular thickness, and consequently interior osteocytes, starved for nutrients, undergo apoptosis followed by osteonal bone remodeling common to cortical bone. Osteonal tunneling increases trabecular number by splitting thick trabeculae into normal-sized trabeculae, with interior osteocytes again having adequate access to nutrients by diffusion from trabecular surfaces

7 The Bone Mechanostat

The physiological mechanisms responsible for mechanical loading-induced bone formation involve prostaglandin, PTH/PTHrP, and WNT signaling pathways. The present anabolic osteoporosis therapies (teriparatide, abaloparatide, and romosozumab) are effective because they modulate these pathways, but comprehensive discussion of these signaling pathways in the skeleton is beyond the scope of this review. The importance of WNT signaling in bone mechanical-transduction is well-recognized (Bullock et al. 2019), and PTHrP biology has been comprehensively reviewed (Martin 2016). Treatment with PGE2 (Jee and Ma 1997) or PGE2 analogues (Yoshida et al. 2002) stimulates bone formation in ovariectomized rats through prostanoid EP2 and EP4 receptors (Graham et al. 2009).

Low bone mass is observed in mice with heterozygous PTHrP deficiency and osteoblast-specific PTHrP knockout (Miao et al. 2005; Ansari et al. 2018). Skeletal loading induces fluid flow within bone canaliculi and lacunae (Fan et al. 2016), and this flow is believed to stimulate osteocytes to secrete PTHrP, a PTH paralogue that activates the PTH receptor. With intermittent bone loading and bone fluid movement, intermittent PTHrP secretion activates bone modeling in osteoblasts. This anabolic action of mechanical loading requires the presence of the PTH receptor in osteocytes (Delgado-Calle and Bellido 2017; Gardinier et al. 2019). The WNT and PTH/PTHrP pathways interact, as teriparatide treatment rapidly inhibits sclerostin (*SOST*) gene expression in the bone and PTH/PTHrP signaling involves the WNT co-receptor LRP6 (Li et al. 2016). Thus, intermittent dosing of PTH/PTHrP analogues mimic the physiological events by which intermittent bone loading induces bone anabolism. Figure 2 illustrates the actions of the PTH/PTHrP and WNT pathways in the skeleton.

Future novel anabolic osteoporosis therapies are likely to modulate additional components of physiological mechanostat signaling. Increased understanding of the mechanostat pathway, through which mechanical forces stimulate bone formation, should provide additional osteoporosis drug targets.

8 Additional WNT Signaling Targets

WNTs stimulate bone formation by osteoblasts by activating cell surface Frizzled receptor/LRP co-receptor complexes to promote receptor dimerization and intracellular signaling. Sclerostin, secreted by osteocytes, inhibits WNT activation of osteoblastic bone formation through binding the cell surface LRP5 co-receptor. The osteoporosis antibody drug romosozumab acts by neutralizing sclerostin. Osteoporosis-pseudoglioma syndrome occurs with inactivating *LRP5* mutations, but gain-of-function *LRP5* mutations lead to extremely high bone mass similar to that observed in subjects with inactivating sclerostin mutations. Low and high bone mass are also observed in subjects with inactivating and activating *LRP6* mutations, respectively. Agents activating LRP5 and/or LRP6 would be potential anabolic osteoporosis drugs.

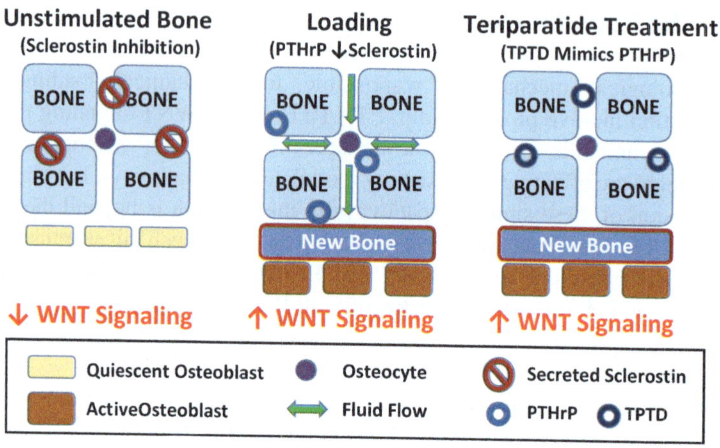

Fig. 2 WNT signaling mediates anabolic bone responses to loading and anabolic osteoporosis drugs. Without loading, the lack of PTHrP secretion coupled with tonic sclerostin secretion minimizes WNT activity, and osteoblasts are flattened, quiescent bone-lining cells. With loading, pulsatile fluid flow stimulates PTHrP secretion which stimulates bone formation by activating osteoblasts through the PTH1R receptor and indirectly by suppressing sclerostin secretion. Intermittent teriparatide and abaloparatide treatment mimic the actions of loading mediated by PTHrP secretion. Romosozumab treatment removes the suppressive effects of sclerostin

In contrast to *LRP5* and *LRP6*, inactivating mutations in *LRP4* increase bone mass. LRP4 binds to sclerostin but not Frizzled receptors, thereby neutralizing sclerostin inhibitory activity on WNT signaling. Besides the sclerostin-binding third β-propeller domain of LRP4 involved in bone mass accumulation and kidney development, mutations in separate LRP4 domains produce digit malformations and disruptions of the neuromuscular junction (Boudin et al. 2017).

Mouse mutations in *Lrp4*, *Lrp5*, *Lrp6*, *Sost* (sclerostin), *Wnt1*, *Wnt3*, *Wnt5a*, *Wnt7a*, *Wnt10b*, *Fzd2*, and *Sfrp4* each mimic orthologous human mutant skeletal phenotypes and have contributed greatly to our understanding of WNT signaling in the skeleton and the development of romosozumab therapy. LRP4 neutralizing antibodies that selectively interfere with sclerostin binding stimulate bone formation in adult mice (Chang et al. 2014), providing a potential therapy for treating osteoporosis. Selectively activating osteoblast Frizzled receptors to stimulate skeletal WNT signaling might be possible (Zhang et al. 2018).

Similar to sclerostin, DKK1 is a secreted protein that inhibits WNT signaling, and *LRP5* activating mutations prevent this inhibition (Boyden et al. 2002; Niziolek et al. 2015). Global gene knockout in mice results in lethality from developmental neurological defects, but normal viability and high bone mass occur when *Dkk1* is disrupted in *Wnt3* heterozygous mice (McDonald et al. 2018). Several laboratories generated DKK1 neutralizing antibodies that increased bone mass robustly in young mice but minimally in adult mice (Glantschnig et al. 2010; Agholme et al. 2011; Li et al. 2011; Brommage et al. 2014). This observation is consistent with lower bone *Dkk1* expression in adult versus young rats (Li et al. 2011). Bone mass in mice treated with sclerostin antibodies increases for several weeks, but this anabolic effect greatly diminishes with time. Skeletal *Dkk1* expression increases with sclerostin inhibition, suggesting that compensatory increases in DKK1 contribute to the sclerostin resistance (Taylor et al. 2016). The enhanced effectiveness of a sclerostin-DKK1 bispecific antibody (Florio et al. 2016) and the efficacy of sclerostin antibody treatment in *Dkk1* KO mice (Morse et al. 2018) support this hypothesis.

Another strategy to enhance skeletal WNT signaling is to inhibit degradation of endogenous WNTs. TRABD2A and TRABD2B (previously TIKI1 and TIKI2) are metallopeptidases that inactivate WNTs through amino-terminal cleavage (Zhang et al. 2016), but gene knockout mice have not been examined. NOTUM is a secreted lipase that inactivates WNTs by cleaving the palmitoleate essential for Frizzled receptor binding and activation. *Notum* knockout mice have normal trabecular bone mass but exhibit dentin dysplasia (Vogel et al. 2016) and increased cortical bone thickness and strength (Brommage et al. 2019). Treating adult ovariectomized mice and rats with both orally active small-molecule NOTUM inhibitors and NOTUM neutralizing antibodies stimulates endocortical bone formation and increases strength of long bone shafts, vertebral bodies, and the femoral neck (Brommage et al. 2019).

9 Understanding Anabolic Signaling Pathways

Teriparatide and abaloparatide are successful anabolic therapies administered by daily subcutaneous injections. Rather than activate osteoblast cell surface PTH receptors with injectable peptides, intracellular PTH-signaling pathways can presumably be stimulated with orally active small-molecule drugs (Hariri et al. 2017). Both protein kinase A (Kao et al. 2013; Tascau et al. 2016) and salt-induced kinase (Wein et al. 2016) pathways participate in the anabolic actions of PTH. WNT pathway activation and PTH treatment stimulate aerobic glycolysis in osteoblasts, and the significance of glucose, glutamine, and fatty acid metabolism in bone formation is being explored (Chen et al. 2019; Lee et al. 2017; Moorer and Riddle 2019; Shao et al. 2019). Pathways involving changes in gene transcription can be studied using RNA-seq technologies (Ayturk 2019).

Humans and rodents lose bone during lactation, even with adequate dietary Ca intakes. Following lactation dramatic bone formation occurs as the skeleton regains this lost bone. These bone losses and gains are normal physiological processes, and identifying the genes and signaling pathways involved in lactation-induced bone resorption and post-lactation bone formation should provide valuable insights for identifying novel osteoporosis drug targets.

10 Animal Models

All models are wrong, but some are less imperfect than others, and many are useful (George Box)

Rodents, due to minimal cortical bone osteonal remodeling, were originally considered a poor model for human osteoporosis. This belief changed during the 1970s as osteoporotic vertebral fractures and spine BMD became the focus of human studies and the vertebral body contains significant amounts of trabecular bone. Ovariectomized rats rapidly lose trabecular bone in the proximal tibia metaphysis and other skeletal sites following ovariectomy, which mimics loss of estrogens at menopause. Trabecular bone loss at the proximal tibia metaphysis in ovariectomized rats remains the gold standard for evaluating potential novel osteoporosis therapies.

Laboratory mice became the most studied animal model of human osteoporosis with the arrival of the genomics revolution during the 1990s. Cortical bone loss contributes to osteoporotic fractures, and mouse studies usually examine both cortical and trabecular bone. In the opinion of this author, the mouse vertebral body is superior to the distal femoral metaphysis for trabecular bone analyses in mice, as spine trabecular bone has a mechanical function, and is present at higher amounts and is less dependent upon mouse sex, age, and strain. Vertebral body compression strength is easily measured. The distal tibia (halfway between the tibia-fibula junction and the tibia bottom) is an excellent site for cortical bone measurements. The common belief that mouse bones grow throughout life is incorrect, as longitudinal growth stops about 6 months of age (Brommage and Ohlsson 2018).

Because of the lack of cortical bone osteonal remodeling in rodents, potential osteoporosis drugs in late stages of preclinical development are examined in a large animal species. With the exception of risedronate tested in ovariectomized minipigs (Borah et al. 2002), all approved osteoporosis drugs have been examined in ovariectomized cynomolgus or rhesus macaque monkeys, which are excellent models for human postmenopausal osteoporosis (Brommage 2001; Smith et al. 2009). In addition to ovariectomized cynomolgus macaques, ibandronate was also examined in minipigs with glucocorticoid-induced osteoporosis (Glüer et al. 2007).

Because bone loss occurs rapidly after ovariectomy, anti-resorptive agents are optimally examined in acutely ovariectomized animals. Rodents, with estrus cycles mimicking primate menstrual cycles, are ideal. Rabbits do not have cyclic ovulatory cycles, as they ovulate upon mating. Dogs and sheep have seasonal ovulatory cycles, and reproductive hormone levels remain low between ovulations.

Although by convention bone anabolic agents are usually examined in ovariectomized rodents, ovariectomy is generally not required for their effects. For example, teriparatide treatment increases bone mass in zebrafish (Fleming et al. 2005), and both sexes and all ages of all mouse and rat strains examined. This treatment is effective in rabbits (Hirano et al. 1999), both greyhound (Podbesek et al. 1983) and beagle (Zhang et al. 1997) dogs, and both cynomolgus and rhesus macaques. Ovary-intact mice and rats can be employed during early stages of drug development to identify promising anabolic therapies.

11 Human: Rodent Protein Differences

There is a high concordance between rodent and human genomes and the biochemical pathways involved in bone metabolism. However, protein amino acid sequences are less highly conserved, and differences can influence the utility of rodent models for evaluating drug development candidates. Two examples illuminate this issue. Denosumab antibody treatment suppresses osteoclasts by neutralizing human RANKL but does not bind rodent RANKL. This problem was overcome by generating transgenic mice expressing a chimeric mouse/human RANKL protein (Kostenuik et al. 2009). Mice with disruptions in the cathepsin K gene have skeletal phenotypes similar to those in humans with pycnodysostosis resulting from *CTSK* mutations (Pennypacker et al. 2009), but odanacatib, the cathepsin K inhibitor examined clinically, does not inhibit the rodent enzyme. Preclinical odanacatib studies were successfully performed in rabbits (Pennypacker et al. 2011) and monkeys (Pennypacker et al. 2014).

12 Identifying Novel Osteoporosis Drug Targets and Pathways

Potential breakthrough osteoporosis treatments can result from finding novel biochemical pathways involved in bone mass regulation, and several approaches are underway to identify such pathways. These approaches include studies of human populations, mouse, and zebrafish gene mutants.

There are healthy subjects with unexplained high bone mass for which the genes responsible have not been identified (Gregson et al. 2016, 2018). Similarly, unknown genes contribute to juvenile osteoporosis (Collet et al. 2017). Human rare bone disease and GWAS studies continue to identify genetic variants contributing to high bone mass (Mäkitie et al. 2019; Morris et al. 2019; Styrkarsdottir et al. 2019; Trajanoska and Rivadeneira 2019). Examination of populations with non-European ancestry will contribute to future GWAS advances (Ginsburg 2019).

Genes responsible for rare human bone diseases are potential osteoporosis drug targets. Mutations that result in high bone mass with normal skeletal architecture and mineralization without adversely affecting non-skeletal tissues highlight key genes positively affecting bone mass. Sclerosteosis patients with extremely high bone mass have inactivating mutations in either the *SOST* gene, coding for sclerostin, or a downstream *SOST* regulatory region. These findings ultimately led to sclerostin neutralizing romosozumab antibody therapy. More than 400 gene mutations are currently known to contribute to rare human bone diseases (Mortier et al. 2019). Most mutations are deleterious, but identifying additional healthy high bone mass genes is likely. Loss-of-function mutations resulting in low bone mass can identify druggable genes and pathways leading to high bone mass when activated rather than inhibited.

Most mouse genetic studies are performed in laboratories of principal investigators examining hypotheses concerning the possible influences of specific genes on bone metabolism, an approach known as reverse genetics. This approach has been highly successful but is limited by the imagination and interests of scientists. The RANK signaling pathway in osteoclasts, first identified in gene knockout mice independently by three groups (Dougall et al. 1999; Kobayashi et al. 2000; Li et al. 2000), ultimately lead to the development of anti-RANKL denosumab antibody. Forward genetic approaches, starting from observations of mutant bone phenotypes, include identifying mutant genes from spontaneous and ENU chemical mutagenesis campaigns.

Recent mouse studies provide intriguing evidence of novel genes and pathways that might provide exciting new osteoporosis therapies. Select examples include modulating BMP signaling (Ko et al. 2017; Lowery and Rosen 2018; Rauner et al. 2019), inhibiting activities of NMP4 (Shao et al. 2019) and SCHNURRI-3 (Yang et al. 2019) in osteoblasts, and promoting apoptosis of senescent cells with senolytic drugs (Farr et al. 2017; van Deursen 2019). Many other potential drug targets, too numerous to review here, have supportive evidence in mice.

Two large-scale mouse gene knockout phenotyping campaigns have examined genes chosen without preformed hypotheses. The ongoing International Mouse Phenotyping Consortium (IMPC), with the goal of examining all ~20,000 mouse genes, has examined approximately 6,000 genes since 2011 (Cacheiro et al. 2019; Maynard and Ackert-Bicknell 2019), with bone data available on the IMPC website (www.mousephenotype.org). Lexicon Pharmaceuticals' Genome5000™ campaign between 2000 and the end of 2008 focused on the druggable genome, including enzymes, receptors, transporters, channels, and secreted proteins (Brommage et al. 2014). Benchmark validation of published mutant mouse bone phenotypes was obtained for more than 20 genes, and skeletal phenotypes for 5 genes (*FAM20C*, nonlethal Raine syndrome; *LRRK1*, osteosclerotic metaphyseal dysplasia; *PAPPA2*, short stature; *SFRP4*, Pyle's disease; and *SLC10A7*, skeletal dysplasia) were published prior to observations of similar phenotypes resulting from homologous gene mutations in human genetic disorders (Brommage et al. 2019).

Zebrafish research increasingly contributes to our knowledge of skeletal genomics (Kwon et al. 2019; Lleras-Forero et al. 2019). Advantages over mice include acquiring data more rapidly, but one disadvantage involves examining a genome with additional genes due to an evolutionally distant zebrafish whole genome duplication (Howe et al. 2013).

Approximately one-third of the 20,000 mammalian genes, designated the ignorome or dark genome, remain minimally studied (Pandey et al. 2014; Stoeger et al. 2018). Scientists develop hypotheses from existing knowledge and focus on genes and pathways having published data and thereby avoid studying minimally understood genes. The *Illuminating the Druggable Genome Consortium* promotes exploration of currently understudied, but potentially druggable, proteins (Oprea et al. 2018). Novel osteoporosis drug targets are likely to be discovered among Ignormone genes.

Medicinal chemistry libraries containing millions of compounds with potential beneficial bone actions could provide "lead compounds" for drug development with high-throughput screening assays. Previous work identified inhibitors of SFRP1 (Bodine et al. 2009) and PYK2 (Allen et al. 2009) and osteogenic-inducing molecules in osteoblast cells (Han et al. 2009). Skeletal development in zebrafish can be employed to screen chemical libraries for bone-active compounds (Bergen et al. 2019; Huang et al. 2018).

13 Targeting Drugs to Bone

Drugs designed with affinity for bone hydroxyapatite crystals can be efficiently delivered to the skeleton to modulate bone cell activities. This approach is utilized by the bisphosphonate drugs, having a stable P-C-P bond to provide bone affinity (in place of the enzymatically labile P-O-P pyrophosphate bond) coupled to a N-containing chemical moiety that inhibits the osteoclast farnesyl diphosphate synthase enzyme (Rogers et al. 2011). The power of this approach is demonstrated by zoledronate, administered by intravenous infusion once yearly at a dose of 5 mg per patient. A second successful bone delivery example is asfotase alfa for enzyme replacement therapy in hypophosphatasia, in which two identical protein chains containing the catalytic domain of tissue non-specific alkaline phosphatase are linked to the Fc region of human immunoglobulin G1 which, for bone affinity, is attached to a peptide of ten negatively charged aspartic acids (Kishnani et al. 2019). General strategies for attaching drugs to bisphosphonates are described (Cole et al. 2016) and examples of various bone-targeting strategies examined in preclinical studies include estradiol coupled to iminodiacetic acid (Zhao et al. 2013), adeno-associated virus delivery of a SCHNURRI inhibitor (Yang et al. 2019), biomineral-binding liposomes (Sun et al. 2019; Zhang et al. 2012), osteoblast periostin-binding peptide attached to nano-micelles delivering siRNA (Sun et al. 2016), a *Clcn7*^{G213R}-specific siRNA to suppress mutant *Clcn7* transcripts (Capulli et al. 2015), and a cartilage-targeting single-chain human antibody fragment (Lui et al. 2019).

14 Evolving Advances in Drug Delivery

Traditional drug therapies involve administration of small molecules, peptides, and antibodies. During the past few years, multiple developments have greatly expanded the ability to modulate disease pathways. Examples include click chemistry of antibody-drug conjugates (Peplow 2019), combination pharmacotherapies with peptide-small molecule conjugates (Clemmensen et al. 2019), targeted protein degradation (Mullard 2019; Scudellari 2019), RNA therapeutics (Crooke et al. 2018; Yang et al. 2019), small molecules targeting RNA (Cross 2017), bicyclic peptides (Rhodes and Pei 2017), and gene therapy (High and Roncarolo 2019). Future osteoporosis therapies are likely to employ one or more of these recent advances. As an example of therapeutic advances, the GLP-1 peptide analogue semaglutide, given orally once weekly, has recently been approved for treating type 2 diabetics (Knudsen and Lau 2019).

15 Summary and Future Prospects

BMD screening for osteoporosis is now common in the developed world, and there are presently 11 effective therapies for osteoporosis, providing treatment options that can be personalized to individual patients. Developing these therapies involved major efforts by many multidisciplinary teams over the past three decades. During 1990 many scientists would have been surprised to hear that 30 years later, better treatments for osteoporosis would be available than for obesity, arthritis, and sarcopenia.

Osteoporosis is a "silent disease" as low BMD is painless until a fragility fracture occurs, and many osteoporotic patients never receive treatment. For example, in the USA only 23% of osteoporotic patients leaving a hospital after a hip fracture are provided a prescription for osteoporosis drugs, and concerted efforts are underway to close this treatment gap (Anonymous 2018).

Predicting the development of new osteoporosis therapies during the next decades would be highly speculative. The genomics revolution(s) continues, and breakthrough drug delivery options are being developed. Future drug development is likely to exploit some combination of ideas presented in this review, along with concepts yet to be discovered. Failures of promising therapies will likely occur. Osteoporosis is a disease of aging, and its incidence, and the desire for effective therapies, will increase. The WHO has designated 2020 as the start of the Decade of Healthy Aging.

16 Research Resources

Excellent books summarizing bone biology and experimental techniques are available (Smith et al. 2017; Idris 2019).

References

Agholme F, Isaksson H, Kuhstoss S, Aspenberg P (2011) The effects of Dickkopf-1 antibody on metaphyseal bone and implant fixation under different loading conditions. Bone 48:988–996

Allen JG, Lee MR, Han CY, Scherrer J, Flynn S, Boucher C, Zhao H, O'Connor AB, Roveto P, Bauer D et al (2009) Identification of small molecule inhibitors of proline-rich tyrosine kinase 2 (Pyk2) with osteogenic activity in osteoblast cells. Bioorg Med Chem Lett 19:4924–4928

Allen JG, Fotsch C, Babij P (2010) Emerging targets in osteoporosis disease modification. J Med Chem 53:4332–4353

Andrews EB, Gilsean AW, Midkiff K, Sherrill B, Wu Y, Mann BH, Masica D (2012) The US postmarketing surveillance study of adult osteosarcoma and teriparatide: study design and findings from the first 7 years. J Bone Miner Res 27:2429–2437

Anonymous (2018) Osteoporosis: a roadmap to close the treatment gap. Lancet Diabetes Endocrinol 6:833

Ansari N, Ho PW, Crimeen-Irwin B, Poulton IJ, Brunt AR, Forwood MR, Divieti Pajevic P, Gooi JH, Martin TJ, Sims NA (2018) Autocrine and paracrine regulation of the murine skeleton by osteocyte-derived parathyroid hormone-related protein. J Bone Miner Res 33:137–153

Ayturk U (2019) RNA-seq in skeletal biology. Curr Osteoporos Rep 17:178–185

Bergen DJM, Kague E, Hammond CL (2019) Zebrafish as an emerging model for osteoporosis: a primary testing platform for screening new osteo-active compounds. Front Endocrinol (Lausanne) 10:6

Bodine PV, Stauffer B, Ponce-de-Leon H, Bhat RA, Mangine A, Seestaller-Wehr LM, Moran RA, Billiard J, Fukayama S, Komm BS et al (2009) A small molecule inhibitor of the Wnt antagonist secreted frizzled-related protein-1 stimulates bone formation. Bone 44:1063–1068

Borah B, Dufresne TE, Chmielewski PA, Gross GJ, Prenger MC, Phipps RJ (2002) Risedronate preserves trabecular architecture and increases bone strength in vertebra of ovariectomized minipigs as measured by three-dimensional microcomputed tomography. J Bone Miner Res 17:1139–4117

Boudin E, Yorgan T, Fijalkowski I, Sonntag S, Steenackers E, Hendrickx G, Peeters S, De Maré A, Vervaet B, Verhulst A et al (2017) The Lrp4R1170Q homozygous knock-in mouse recapitulates the bone phenotype of sclerosteosis in humans. J Bone Miner Res 32:1739–1749

Boyce RW, Paddock CL, Franks AF, Jankowsky ML, Eriksen EF (1996) Effects of intermittent hPTH(1-34) alone and in combination with 1,25(OH)(2)D(3) or risedronate on endosteal bone remodeling in canine cancellous and cortical bone. J Bone Miner Res 11:600–613

Boyden LM, Mao J, Belsky J, Mitzner L, Farhi A, Mitnick MA, Wu D, Insogna K, Lifton RP (2002) High bone density due to a mutation in LDL-receptor-related protein 5. N Engl J Med 346:1513–1521

Brommage R (2001) Perspectives on using nonhuman primates to understand the etiology and treatment of postmenopausal osteoporosis. J Musculoskelet Neuronal Interact 1:307–325

Brommage R, Ohlsson C (2018) Translational studies provide insights for the etiology and treatment of cortical bone osteoporosis. Best Pract Res Clin Endocrinol Metab 32:329–340

Brommage R, Liu J, Hansen GM, Kirkpatrick LL, Potter DG, Sands AT, Zambrowicz B, Powell DR, Vogel P (2014) High-throughput screening of mouse gene knockouts identifies established and novel skeletal phenotypes. Bone Res 2:14034

Brommage R, Liu J, Vogel P, Mseeh F, Thompson AY, Potter DG, Shadoan MK, Hansen GM, Jeter-Jones S, Cui J et al (2019) NOTUM inhibition increases endocortical bone formation and bone strength. Bone Res 7:2

Bullock WA, Pavalko FM, Robling AG (2019) Osteocytes and mechanical loading: the Wnt connection. Orthod Craniofac Res 22(Suppl 1):175–179

Cacheiro P, Haendel MA, Smedley D, International Mouse Phenotyping Consortium and the Monarch Initiative (2019) New models for human disease from the International Mouse Phenotyping Consortium. Mamm Genome 30:143–150

Capulli M, Maurizi A, Ventura L, Rucci N, Teti A (2015) Effective small interfering RNA therapy to treat CLCN7-dependent autosomal dominant osteopetrosis type 2. Mol Ther Nucleic Acids 4: e248

Cauley JA, Robbins J, Chen Z, Cummings SR, Jackson RD, LaCroix AZ, LeBoff M, Lewis CE, McGowan J, Neuner J et al (2003) Effects of estrogen plus progestin on risk of fracture and bone mineral density: the Women's Health Initiative randomized trial. JAMA 290:1729–1738

Chang MK, Kramer I, Huber T, Kinzel B, Guth-Gundel S, Leupin O, Kneissel M (2014) Disruption of Lrp4 function by genetic deletion or pharmacological blockade increases bone mass and serum sclerostin levels. Proc Natl Acad Sci U S A 111:E5187–E5195

Chen H, Ji X, Lee WC, Shi Y, Li B, Abel ED, Jiang D, Huang W, Long F (2019) Increased glycolysis mediates Wnt7b-induced bone formation. FASEB J 33:7810–7821

Clemmensen C, Finan B, Müller TD, DiMarchi RD, Tschöp MH, Hofmann SM (2019) Emerging hormonal-based combination pharmacotherapies for the treatment of metabolic diseases. Nat Rev Endocrinol 15:90–104

Codrons V, Vanderbist F, Verbeeck RK, Arras M, Lison D, Préat V, Vanbever R (2003) Systemic delivery of parathyroid hormone (1-34) using inhalation dry powders in rats. J Pharm Sci 92:938–950

Cole LE, Vargo-Gogola T, Roeder RK (2016) Targeted delivery to bone and mineral deposits using bisphosphonate ligands. Adv Drug Deliv Rev 99(Pt A):12–27
Coleman PJ, Brashear KM, Askew BC, Hutchinson JH, McVean CA, Duong LT, Feuston BP, Fernandez-Metzler C, Gentile MA, Hartman GD et al (2004) Nonpeptide alphavbeta3 antagonists. Part 11: discovery and preclinical evaluation of potent alphavbeta3 antagonists for the prevention and treatment of osteoporosis. J Med Chem 47:4829–4837
Collet C, Ostertag A, Ricquebourg M, Delecourt M, Tueur G, Isidor B, Guillot P, Schaefer E, Javier RM, Funck-Brentano T et al (2017) Primary osteoporosis in young adults: genetic basis and identification of novel variants in causal genes. JBMR Plus 2:12–21
Cosman F, Lane NE, Bolognese MA, Zanchetta JR, Garcia-Hernandez PA, Sees K, Matriano JA, Gaumer K, Daddona PE (2010) Effect of transdermal teriparatide administration on bone mineral density in postmenopausal women. J Clin Endocrinol Metab 95:151–158
Cosman F, Gilchrist N, McClung M, Foldes J, de Villiers T, Santora A, Leung A, Samanta S, Heyden N, McGinnis JP 2nd et al (2016) A phase 2 study of MK-5442, a calcium-sensing receptor antagonist, in postmenopausal women with osteoporosis after long-term use of oral bisphosphonates. Osteoporos Int 27:377–386
Crooke ST, Witztum JL, Bennett CF, Baker BF (2018) RNA-targeted therapeutics. Cell Metab 27:714–739
Cross R (2017) The RNA drug hunters. Chem Eng News 95:16–18
Cui Y, Niziolek PJ, MacDonald BT, Zylstra CR, Alenina N, Robinson DR, Zhong Z, Matthes S, Jacobsen CM, Conlon RA et al (2011) Lrp5 functions in bone to regulate bone mass. Nat Med 17:684–691
Cummings SR, Ettinger B, Delmas PD, Kenemans P, Stathopoulos V, Verweij P, Mol-Arts M, Kloosterboer L, Mosca L, Christiansen C et al (2008) The effects of tibolone in older postmenopausal women. N Engl J Med 359:697–708
Cummings SR, Ensrud K, Delmas PD, LaCroix AZ, Vukicevic S, Reid DM, Goldstein S, Sriram U, Lee A, Thompson J et al (2010) Lasofoxifene in postmenopausal women with osteoporosis. N Engl J Med 362:686–696
Delgado-Calle J, Bellido T (2017) New insights into the local and systemic functions of sclerostin: regulation of quiescent bone lining cells and beige adipogenesis in peripheral fat depots. J Bone Miner Res 32:889–891
DeLuca HF, Bedale W, Binkley N, Gallagher JC, Bolognese M, Peacock M, Aloia J, Clagett-Dame M, Plum L (2011) The vitamin D analogue 2MD increases bone turnover but not BMD in postmenopausal women with osteopenia: results of a 1-year phase 2 double-blind, placebo-controlled, randomized clinical trial. J Bone Miner Res 26:538–545
Dobnig H, Turner RT (1995) Evidence that intermittent treatment with parathyroid hormone increases bone formation in adult rats by activation of bone lining cells. Endocrinology 136:3632–3638
Dougall WC, Glaccum M, Charrier K, Rohrbach K, Brasel K, De Smedt T, Daro E, Smith J, Tometsko ME, Maliszewski CR et al (1999) RANK is essential for osteoclast and lymph node development. Genes Dev 13:2412–2424
Drake MT, Clarke BL, Oursler MJ, Khosla S (2017) Cathepsin K inhibitors for osteoporosis: biology, potential clinical utility, and lessons learned. Endocr Rev 38:325–350
Dudley HR, Spiro D (1961) The fine structure of bone cells. J Biophys Biochem Cytol 11:627–649
Ebeling PR, Akesson K, Bauer DC, Buchbinder R, Eastell R, Fink HA, Giangregorio L, Guanabens N, Kado D, Kallmes D et al (2019) The efficacy and safety of vertebral augmentation: a second ASBMR task force report. J Bone Miner Res 34:3–21
Erben RG (2015) Hypothesis: coupling between resorption and formation in cancellous bone remodeling is a mechanically controlled event. Front Endocrinol (Lausanne) 6:82
Fan L, Pei S, Lucas Lu X, Wang L (2016) A multiscale 3D finite element analysis of fluid/solute transport in mechanically loaded bone. Bone Res 4:16032
Farr JN, Xu M, Weivoda MM, Monroe DG, Fraser DG, Onken JL, Negley BA, Sfeir JG, Ogrodnik MB et al (2017) Targeting cellular senescence prevents age-related bone loss in mice. Nat Med 23:1072–1079

Fitzpatrick LA, Dabrowski CE, Cicconetti G, Gordon DN, Papapoulos S, Bone HG 3rd, Bilezikian JP (2011) The effects of ronacaleret, a calcium-sensing receptor antagonist, on bone mineral density and biochemical markers of bone turnover in postmenopausal women with low bone mineral density. J Clin Endocrinol Metab 96:2441–2449

Fleming A, Sato M, Goldsmith P (2005) High-throughput in vivo screening for bone anabolic compounds with zebrafish. J Biomol Screen 10:823–831

Florio M, Gunasekaran K, Stolina M, Li X, Liu L, Tipton B, Salimi-Moosavi H, Asuncion FJ, Li C, Sun B et al (2016) A bispecific antibody targeting sclerostin and DKK-1 promotes bone mass accrual and fracture repair. Nat Commun 7:11505

Frost HM, Jee WS (1992) On the rat model of human osteopenias and osteoporoses. Bone Miner 18:227–236

Gafni RI, Brahim JS, Andreopoulou P, Bhattacharyya N, Kelly MH, Brillante BA, Reynolds JC, Hua Z, Dempster DW, Collins MT (2012) Daily parathyroid hormone 1-34 replacement therapy for hypoparathyroidism induces marked changes in bone turnover and structure. J Bone Miner Res 27:1811–1820

Gardinier JD, Daly-Seiler C, Rostami N, Kundal S, Zhang C (2019) Loss of the PTH/PTHrP receptor along the osteoblast lineage limits the anabolic response to exercise. PLoS One 14: e0211076

Gilsenan A, Harding A, Kellier-Steele N, Harris D, Midkiff K, Andrews E (2018) The Forteo Patient Registry linkage to multiple state cancer registries: study design and results from the first 8 years. Osteoporos Int 29:2335–2343

Ginsburg GS (2019) A global collaborative to advance genomic medicine. Am J Hum Genet 104:407–409

Glantschnig H, Hampton RA, Lu P, Zhao JZ, Vitelli S, Huang L, Haytko P, Cusick T, Ireland C, Jarantow SW et al (2010) Generation and selection of novel fully human monoclonal antibodies that neutralize Dickkopf-1 (DKK1) inhibitory function in vitro and increase bone mass in vivo. J Biol Chem 285:40135–40147

Glüer CC, Scholz-Ahrens KE, Helfenstein A, Delling G, Timm W, Açil Y, Barkmann R, Hassenpflug J, Stampa B, Bauss F, Schrezenmeir J (2007) Ibandronate treatment reverses glucocorticoid-induced loss of bone mineral density and strength in minipigs. Bone 40:645–655

Goldberg DR, De Lombaert S, Aiello R, Bourassa P, Barucci N, Zhang Q, Paralkar V, Stein AJ, Holt M, Valentine J et al (2017) Optimization of spirocyclic proline tryptophan hydroxylase-1 inhibitors. Bioorg Med Chem Lett 27:413–419

Graham S, Gamie Z, Polyzois I, Narvani AA, Tzafetta K, Tsiridis E, Helioti M, Mantalaris A, Tsiridis E (2009) Prostaglandin EP2 and EP4 receptor agonists in bone formation and bone healing: in vivo and in vitro evidence. Expert Opin Investig Drugs 18:746–766

Gregson CL, Wheeler L, Hardcastle SA, Appleton LH, Addison KA, Brugmans M, Clark GR, Ward KA, Paggiosi M, Stone M et al (2016) Mutations in known monogenic high bone mass loci only explain a small proportion of high bone mass cases. J Bone Miner Res 31:640–649

Gregson CL, Newell F, Leo PJ, Clark GR, Paternoster L, Marshall M, Forgetta V, Morris JA, Ge B, Bao X et al (2018) Genome-wide association study of extreme high bone mass: contribution of common genetic variation to extreme BMD phenotypes and potential novel BMD-associated genes. Bone 114:62–71

Halse J, Greenspan S, Cosman F, Ellis G, Santora A, Leung A, Heyden N, Samanta S, Doleckyj S, Rosenberg E et al (2014) A phase 2, randomized, placebo-controlled, dose-ranging study of the calcium-sensing receptor antagonist MK-5442 in the treatment of postmenopausal women with osteoporosis. J Clin Endocrinol Metab 99:E2207–E2215

Hämmerle SP, Mindeholm L, Launonen A, Kiese B, Loeffler R, Harfst E, Azria M, Arnold M, John MR (2012) The single dose pharmacokinetic profile of a novel oral human parathyroid hormone formulation in healthy postmenopausal women. Bone 50:965–973

Han CY, Wang Y, Yu L, Powers D, Xiong X, Yu V, Nguyen Y, Jean DJ Jr, Babij P (2009) Small molecules with potent osteogenic-inducing activity in osteoblast cells. Bioorg Med Chem Lett 19:1442–1445

Hariri H, Pellicelli M, Arnaud S (2017) New PTH signals mediating bone anabolism. Curr Mol Biol Rep 3:133–141

Henriksen K, Andersen JR, Riis BJ, Mehta N, Tavakkol R, Alexandersen P, Byrjalsen I, Valter I, Nedergaard BS, Teglbjaerg CS et al (2012) Evaluation of the efficacy, safety and pharmacokinetic profile of oral recombinant human parathyroid hormone [rhPTH(1-31)NH(2)] in postmenopausal women with osteoporosis. Bone 53:160–166

High KA, Roncarolo MG (2019) Gene therapy. N Engl J Med 381:455–464

Hirano T, Burr DB, Turner CH, Sato M, Cain RL, Hock JM (1999) Anabolic effects of human biosynthetic parathyroid hormone fragment (1-34), LY333334, on remodeling and mechanical properties of cortical bone in rabbits. J Bone Miner Res 14:536–545

Horwitz MJ, Augustine M, Khan L, Martin E, Oakley CC, Carneiro RM, Tedesco MB, Laslavic A, Sereika SM, Bisello A et al (2013) A comparison of parathyroid hormone-related protein (1-36) and parathyroid hormone (1-34) on markers of bone turnover and bone density in postmenopausal women: the PrOP study. J Bone Miner Res 28:2266–2276

Howe K, Clark MD, Torroja CF, Torrance J, Berthelot C, Muffato M, Collins JE, Humphray S, McLaren K, Matthews L et al (2013) The zebrafish reference genome sequence and its relationship to the human genome. Nature 496:498–503

Huang HX, Lin H, Lan F, Wu YF, Yang ZG, Zhang JJ (2018) Application of bone transgenic zebrafish in anti-osteoporosis chemical screening. Animal Model Exp Med 1:53–61

Idris A (2019) Bone research protocols. Springer, New York

Jee WS, Ma YF (1997) The in vivo anabolic actions of prostaglandins in bone. Bone 21:297–304

Jerome CP, Burr DB, Van Bibber T, Hock JM, Brommage R (2001) Treatment with human parathyroid hormone (1-34) for 18 months increases cancellous bone volume and improves trabecular architecture in ovariectomized cynomolgus monkeys (*Macaca fascicularis*). Bone 28:150–159

John MR, Harfst E, Loeffler J, Belleli R, Mason J, Bruin GJ, Seuwen K, Klickstein LB, Mindeholm L, Widler L et al (2014) AXT914 a novel, orally-active parathyroid hormone-releasing drug in two early studies of healthy volunteers and postmenopausal women. Bone 64:204–210

Jolette J, Wilker CE, Smith SY, Doyle N, Hardisty JF, Metcalfe AJ, Marriott TB, Fox J, Wells DS (2006) Defining a noncarcinogenic dose of recombinant human parathyroid hormone 1-84 in a 2-year study in Fischer 344 rats. Toxicol Pathol 34:929–940

Jolette J, Attalla B, Varela A, Long GG, Mellal N, Trimm S, Smith SY, Ominsky MS, Hattersley G (2017) Comparing the incidence of bone tumors in rats chronically exposed to the selective PTH type 1 receptor agonist abaloparatide or PTH(1-34). Regul Toxicol Pharmacol 86:356–365

Kao RS, Abbott MJ, Louie A, O'Carroll D, Lu W, Nissenson R (2013) Constitutive protein kinase A activity in osteocytes and late osteoblasts produces an anabolic effect on bone. Bone 55:277–287

Kim SW, Pajevic PD, Selig M, Barry KJ, Yang JY, Shin CS, Baek WY, Kim JE, Kronenberg HM (2012) Intermittent parathyroid hormone administration converts quiescent lining cells to active osteoblasts. J Bone Miner Res 27:2075–8204

Kim SW, Lu Y, Williams EA, Lai F, Lee JY, Enishi T, Balani DH, Ominsky MS, Ke HZ, Kronenberg HM et al (2017) Sclerostin antibody administration converts bone lining cells into active osteoblasts. J Bone Miner Res 32:892–901

Kim H, Kim YG, Choi W, Moon JH, Hwang I, Kim K, Yadav VK, Karsenty G, Jeong JS, Kim H (2018) Generation of a highly efficient and tissue-specific tryptophan hydroxylase 1 knockout mouse model. Sci Rep 8:17642

Kishnani PS, Rockman-Greenberg C, Rauch F, Bhatti MT, Moseley S, Denker AE, Watsky E, Whyte MP (2019) Five-year efficacy and safety of asfotase alfa therapy for adults and adolescents with hypophosphatasia. Bone 121:149–162

Kneissel M, Boyde A, Gasser JA (2001) Bone tissue and its mineralization in aged estrogen-depleted rats after long-term intermittent treatment with parathyroid hormone (PTH) analog SDZ PTS 893 or human PTH(1-34). Bone 28:237–250

Knowles HJ (2015) Hypoxic regulation of osteoclast differentiation and bone resorption activity. Hypoxia (Auckl) 3:73–82
Knudsen LB, Lau J (2019) The discovery and development of liraglutide and semaglutide. Front Endocrinol (Lausanne) 10:155
Ko FC, Van Vliet M, Ellman R, Grasso D, Brooks DJ, Spatz JM, Conlon C, Aguirre JI, Wronski TJ, Bouxsein ML (2017) Treatment with a soluble bone morphogenetic protein type 1A receptor (BMPR1A) fusion protein increases bone mass and bone formation in mice subjected to hindlimb unloading. JBMR Plus 1:66–72
Kobayashi K, Takahashi N, Jimi E, Udagawa N, Takami M, Kotake S, Nakagawa N, Kinosaki M, Yamaguchi K, Shima N et al (2000) Tumor necrosis factor alpha stimulates osteoclast differentiation by a mechanism independent of the ODF/RANKL-RANK interaction. J Exp Med 191:275–286
Kolata G (2008) Bone finding may point to hope for osteoporosis. New York Times. https://www.nytimes.com/2008/11/27/health/research/27bone.html
Kostenuik PJ, Nguyen HQ, McCabe J, Warmington KS, Kurahara C, Sun N, Chen C, Li L, Cattley RC, Van G et al (2009) Denosumab, a fully human monoclonal antibody to RANKL, inhibits bone resorption and increases BMD in knock-in mice that express chimeric (murine/human) RANKL. J Bone Miner Res 24:182–195
Kwon RY, Watson CJ, Karasik D (2019) Using zebrafish to study skeletal genomics. Bone 126:37–50
Langdahl B, Ferrari S, Dempster DW (2016) Bone modeling and remodeling: potential as therapeutic targets for the treatment of osteoporosis. Ther Adv Musculoskelet Dis 8:225–235
Lawson MA, Ebetino FH, Mazur A, Chantry AD, Paton-Hough J, Evans HR, Lath D, Tsoumpra MK, Lundy MW, Dobson RL et al (2017) The pharmacological profile of a novel highly potent bisphosphonate, OX14 (1-fluoro-2-(imidazo-[1,2-α]pyridin-3-yl)-ethyl-bisphosphonate). J Bone Miner Res 32:1860–1869
Lee WC, Guntur AR, Long F, Rosen CJ (2017) Energy metabolism of the osteoblast: implications for osteoporosis. Endocr Rev 38:255–266
Lewiecki EM, Binkley N (2017) DXA: 30 years and counting: introduction to the 30th anniversary issue. Bone 104:1–3
Li J, Sarosi I, Yan XQ, Morony S, Capparelli C, Tan HL, McCabe S, Elliott R, Scully S, Van G et al (2000) RANK is the intrinsic hematopoietic cell surface receptor that controls osteoclastogenesis and regulation of bone mass and calcium metabolism. Proc Natl Acad Sci U S A 97:1566–1571
Li X, Grisanti M, Fan W, Asuncion FJ, Tan HL, Dwyer D, Han CY, Yu L, Lee J, Lee E et al (2011) Dickkopf-1 regulates bone formation in young growing rodents and upon traumatic injury. J Bone Miner Res 26:2610–2621
Li C, Wang W, Xie L, Luo X, Cao X, Wan M (2016) Lipoprotein receptor–related protein 6 is required for parathyroid hormone-induced Sost suppression. Ann N Y Acad Sci 1364:62–73
Lleras-Forero L, Winkler C, Schulte-Merker S (2019) Zebrafish and medaka as models for biomedical research of bone diseases. Dev Biol. https://doi.org/10.1016/j.ydbio.2019.07.009
Lorentzon M, Cummings SR (2015) Osteoporosis: the evolution of a diagnosis. J Intern Med 277:650–661
Lowery JW, Rosen V (2018) The BMP pathway and its inhibitors in the skeleton. Physiol Rev 98:2431–2452
Lui JC, Colbert M, Cheung CSF, Ad M, Lee A, Zhu Z, Barnes KM, Dimitrov DS, Baron J (2019) Cartilage-targeted IGF-1 treatment to promote longitudinal bone growth. Mol Ther 27:673–680
Mäkitie RE, Costantini A, Kämpe A, Alm JJ, Mäkitie O (2019) New insights into monogenic causes of osteoporosis. Front Endocrinol (Lausanne) 10:70
Marcus R (2011) Present at the beginning: a personal reminiscence on the history of teriparatide. Osteoporos Int 22:2241–2248
Martin TJ (2016) Parathyroid hormone-related protein, its regulation of cartilage and bone development, and role in treating bone diseases. Physiol Rev 96:831–871

Matic I, Matthews BG, Wang X, Dyment NA, Worthley DL, Rowe DW, Grcevic D, Kalajzic I (2016) Quiescent bone lining cells are a major source of osteoblasts during adulthood. Stem Cells 34:2930–2942

Maynard RD, Ackert-Bicknell CL (2019) Mouse models and online resources for functional analysis of osteoporosis genome-wide association studies. Front Endocrinol (Lausanne) 10:277

McDonald MM, Morse A, Schindeler A, Mikulec K, Peacock L, Cheng T, Bobyn J, Lee L, Baldock PA, Croucher PI et al (2018) Homozygous Dkk1 knockout mice exhibit high bone mass phenotype due to increased bone formation. Calcif Tissue Int 102:105–116

Miao D, He B, Jiang Y, Kobayashi T, Sorocéanu MA, Zhao J, Su H, Tong X, Amizuka N, Gupta A et al (2005) Osteoblast-derived PTHrP is a potent endogenous bone anabolic agent that modifies the therapeutic efficacy of administered PTH 1-34. J Clin Invest 115:2402–2411

Miller VM, Harman SM (2017) An update on hormone therapy in postmenopausal women: mini-review for the basic scientist. Am J Physiol Heart Circ Physiol 313:H1013–H1021

Miller SC, de Saint-Georges L, Bowman BM, Jee WS (1989) Bone lining cells: structure and function. Scanning Microsc 3:953–960

Miller MA, Bare SP, Recker RR, Smith SY, Fox J (2008) Intratrabecular tunneling increases trabecular number throughout the skeleton of ovariectomized rhesus monkeys treated with parathyroid hormone 1-84. Bone 42:1175–1183

Moorer MC, Riddle RC (2019) Regulation of osteoblast metabolism by Wnt signaling. Endocrinol Metab (Seoul) 33:318–330

Morley P (2005) Delivery of parathyroid hormone for the treatment of osteoporosis. Expert Opin Drug Deliv 2:993–1002

Morris JA, Kemp JP, Youlten SE, Laurent L, Logan JG, Chai RC, Vulpescu NA, Forgetta V, Kleinman A, Mohanty S et al (2019) An atlas of genetic influences on osteoporosis in humans and mice. Nat Genet 51:258–266

Morse A, Cheng TL, Schindeler A, McDonald MM, Mohanty ST, Kneissel M, Kramer I, Little DG (2018) Dkk1 KO mice treated with sclerostin antibody have additional increases in bone volume. Calcif Tissue Int 103:298–310

Mortier GR, Cohn DH, Cormier-Daire V, Hall C, Krakow D, Mundlos S, Nishimura G, Robertson S, Sangiorgi L, Savarirayan R et al (2019) Nosology and classification of genetic skeletal disorders: 2019 revision. Am J Med Genet A. https://doi.org/10.1002/ajmg.a.61366

Mullard A (2019) First targeted protein degrader hits the clinic. Nat Rev Drug Discov 18:237–239

Murphy MG, Cerchio K, Stoch SA, Gottesdiener K, Wu M, Recker R, L-000845704 Study Group (2005) Effect of L-000845704, an alphaVbeta3 integrin antagonist, on markers of bone turnover and bone mineral density in postmenopausal osteoporotic women. J Clin Endocrinol Metab 90:2022–2028

Nakai Y, Okamoto K, Terashima A, Ehata S, Nishida J, Imamura T, Ono T, Takayanagi H (2019) Efficacy of an orally active small-molecule inhibitor of RANKL in bone metastasis. Bone Res 7:1

Niziolek PJ, MacDonald BT, Kedlaya R, Zhang M, Bellido T, He X, Warman ML, Robling AG (2015) High bone mass-causing mutant LRP5 receptors are resistant to endogenous inhibitors in vivo. J Bone Miner Res 30:1822–1830

Nuttall ME, Bradbeer JN, Stroup GB, Nadeau DP, Hoffman SJ, Zhao H, Rehm S, Gowen M (1998) Idoxifene: a novel selective estrogen receptor modulator prevents bone loss and lowers cholesterol levels in ovariectomized rats and decreases uterine weight in intact rats. Endocrinology 139:5224–5234

Oprea TI, Bologa CG, Brunak S, Campbell A, Gan GN, Gaulton A, Gomez SM, Guha R, Hersey A, Holmes J et al (2018) Unexplored therapeutic opportunities in the human genome. Nat Rev Drug Discov 17:317–332

Pandey AK, Lu L, Wang X, Homayouni R, Williams RW (2014) Functionally enigmatic genes: a case study of the brain ignorome. PLoS One 9:e88889

Parfitt AM, Drezner MK, Glorieux FH, Kanis JA, Malluche H, Meunier PJ, Ott SM, Recker RR (1987) Bone histomorphometry: standardization of nomenclature, symbols, and units. Report of the ASBMR Histomorphometry Nomenclature Committee. J Bone Miner Res 2:595–610

Pead MJ, Skerry TM, Lanyon LE (1988) Direct transformation from quiescence to bone formation in the adult periosteum following a single brief period of bone loading. J Bone Miner Res 3:647–656

Pearson RG, Masud T, Blackshaw E, Naylor A, Hinchcliffe M, Jeffery K, Jordan F, Shabir-Ahmed A, King G, Lewis AL et al (2019) Nasal administration and plasma pharmacokinetics of parathyroid hormone peptide PTH 1-34 for the treatment of osteoporosis. Pharmaceutics 11: E265

Pennypacker B, Shea M, Liu Q, Masarachia P, Saftig P, Rodan S, Rodan G, Kimmel D (2009) Bone density, strength, and formation in adult cathepsin K (−/−) mice. Bone 44:199–207

Pennypacker BL, Duong LT, Cusick TE, Masarachia PJ, Gentile MA, Gauthier JY, Black WC, Scott BB, Samadfam R et al (2011) Cathepsin K inhibitors prevent bone loss in estrogen-deficient rabbits. J Bone Miner Res 26:252–262

Pennypacker BL, Chen CM, Zheng H, Shih MS, Belfast M, Samadfam R, Duong LT (2014) Inhibition of cathepsin K increases modeling-based bone formation, and improves cortical dimension and strength in adult ovariectomized monkeys. J Bone Miner Res 29:1847–1858

Peplow M (2019) Click chemistry targets antibody-drug conjugates for the clinic. Nat Biotechnol. https://doi.org/10.1038/d41587-019-00017-4

Podbesek R, Edouard C, Meunier PJ, Parsons JA, Reeve J, Stevenson RW, Zanelli JM (1983) Effects of two treatment regimes with synthetic human parathyroid hormone fragment on bone formation and the tissue balance of trabecular bone in greyhounds. Endocrinology 112:1000–1006

Rauner M, Baschant U, Roetto A, Pellegrino RM, Rother S, Salbach-Hirsch J, Weidner H, Hintze V, Campbell G, Petzold A et al (2019) Transferrin receptor 2 controls bone mass and pathological bone formation via BMP and Wnt signaling. Nat Metab 1:111–124

Ravn P, Nielsen TF, Christiansen C (2006) What can be learned from the levormeloxifene experience? Acta Obstet Gynecol Scand 85:135–142

Reginster JY, Christiansen C, Roux C, Fechtenbaum J, Rouillon A, Tou KP (2001) Intermittent cyclic tiludronate in the treatment of osteoporosis. Osteoporos Int 12:169–177

Rhodes CA, Pei D (2017) Bicyclic peptides as next-generation therapeutics. Chemistry 23:12690–12703

Riggs BL, Parfitt AM (2005) Drugs used to treat osteoporosis: the critical need for a uniform nomenclature based on their action on bone remodeling. J Bone Miner Res 20:177–184

Riggs BL, Hodgson SF, O'Fallon WM, Chao EY, Wahner HW, Muhs JM, Cedel SL, Melton LJ 3rd (1990) Effect of fluoride treatment on the fracture rate in postmenopausal women with osteoporosis. N Engl J Med 322:802–809

Rogers MJ, Crockett JC, Coxon FP, Mönkkönen J (2011) Biochemical and molecular mechanisms of action of bisphosphonates. Bone 49:34–41

Rubin MR, Zhou H, Cusano NE, Majeed R, Omeragic B, Gomez M, Nickolas TL, Dempster DW, Bilezikian JP (2018) The effects of long-term administration of rhPTH(1-84) in hypoparathyroidism by bone histomorphometry. J Bone Miner Res 33:1931–1939

Schmidt A, Kimmel DB, Bai C, Scafonas A, Rutledge S, Vogel RL, McElwee-Witmer S, Chen F, Nantermet PV, Kasparcova V et al (2010) Discovery of the selective androgen receptor modulator MK-0773 using a rational development strategy based on differential transcriptional requirements for androgenic anabolism versus reproductive physiology. J Biol Chem 285:17054–17064

Scudellari M (2019) Protein-slaying drugs could be the next blockbuster therapies. Nature 567:298–300

Shao Y, Wichern E, Childress PJ, Adaway M, Misra J, Klunk A, Burr DB, Wek RC, Mosley AL, Liu Y et al (2019) Loss of Nmp4 optimizes osteogenic metabolism and secretion to enhance bone quality. Am J Physiol Endocrinol Metab 316:E749–E772

Smith SY, Jolette J, Turner CH (2009) Skeletal health: primate model of postmenopausal osteoporosis. Am J Primatol 71:752–765

Smith SY, Varela A, Samadfam R (2017) Bone toxicology. Springer, Cham

Stoeger T, Gerlach M, Morimoto RI, Nunes Amaral LA (2018) Large-scale investigation of the reasons why potentially important genes are ignored. PLoS Biol 16:e2006643

Styrkarsdottir U, Stefansson OA, Gunnarsdottir K, Thorleifsson G, Lund SH, Stefansdottir L, Juliusson K, Agustsdottir AB, Zink F, Halldorsson GH et al (2019) GWAS of bone size yields twelve loci that also affect height, BMD, osteoarthritis or fractures. Nat Commun 10:2054

Sun Y, Ye X, Cai M, Liu X, Xiao J, Zhang C, Wang Y, Yang L, Liu J, Li S et al (2016) Osteoblast-targeting-peptide modified nanoparticle for siRNA/microRNA delivery. ACS Nano 10:5759–5768

Sun X, Wei J, Lyu J, Bian T, Liu Z, Huang J, Pi F, Li C, Zhong Z (2019) Bone-targeting drug delivery system of biomineral-binding liposomes loaded with icariin enhances the treatment for osteoporosis. J Nanobiotechnol 17:10

Tascau L, Gardner T, Anan H, Yongpravat C, Cardozo CP, Bauman WA, Lee FY, Oh DS, Tawfeek HA (2016) Activation of protein kinase A in mature osteoblasts promotes a major bone anabolic response. Endocrinology 157:112–126

Taylor S, Ominsky MS, Hu R, Pacheco E, He YD, Brown DL, Aguirre JI, Wronski TJ, Buntich S, Afshari CA et al (2016) Time-dependent cellular and transcriptional changes in the osteoblast lineage associated with sclerostin antibody treatment in ovariectomized rats. Bone 84:148–159

Trajanoska K, Rivadeneira F (2019) The genetic architecture of osteoporosis and fracture risk. Bone 126:2–10

Vahle JL, Sato M, Long GG, Young JK, Francis PC, Engelhardt JA, Westmore MS, Linda Y, Nold JB (2002) Skeletal changes in rats given daily subcutaneous injections of recombinant human parathyroid hormone (1-34) for 2 years and relevance to human safety. Toxicol Pathol 30:312–321

Vahle JL, Long GG, Sandusky G, Westmore M, Ma YL, Sato M (2004) Bone neoplasms in F344 rats given teriparatide [rhPTH(1-34)] are dependent on duration of treatment and dose. Toxicol Pathol 32:426–348

Vahle JL, Zuehlke U, Schmidt A, Westmore M, Chen P, Sato M (2008) Lack of bone neoplasms and persistence of bone efficacy in cynomolgus macaques after long-term treatment with teriparatide [rhPTH(1-34)]. J Bone Miner Res 23:2033–2039

van Deursen JM (2019) Senolytic therapies for healthy longevity. Science 364:636–637

Vickery BH, Avnur Z, Cheng Y, Chiou SS, Leaffer D, Caulfield JP, Kimmel DB, Ho T, Krstenansky JL (1996) RS-66271, a C-terminally substituted analog of human parathyroid hormone-related protein (1-34), increases trabecular and cortical bone in ovariectomized, osteopenic rats. J Bone Miner Res 11:1943–1951

Vogel P, Read RW, Hansen GM, Powell DR, Kantaputra PN, Zambrowicz B, Brommage R (2016) Dentin dysplasia in Notum knockout mice. Vet Pathol 53:853–862

Warden SJ, Mantila Roosa SM, Kersh ME, Hurd AL, Fleisig GS, Pandy MG, Fuchs RK (2014) Physical activity when young provides lifelong benefits to cortical bone size and strength in men. Proc Natl Acad Sci U S A 111:5337–5342

Wein MN, Liang Y, Goransson O, Sundberg TB, Wang J, Williams EA, O'Meara MJ, Govea N, Beqo B, Nishimori S et al (2016) SIKs control osteocyte responses to parathyroid hormone. Nat Commun 7:13176

Yadav VK, Ryu JH, Suda N, Tanaka KF, Gingrich JA, Schütz G, Glorieux FH, Chiang CY, Zajac JD, Insogna KL et al (2008) Lrp5 controls bone formation by inhibiting serotonin synthesis in the duodenum. Cell 135:825–837

Yang YS, Xie J, Wang D, Kim JM, Tai PWL, Gravallese E, Gao G, Shim JH (2019) Bone-targeting AAV-mediated silencing of Schnurri-3 prevents bone loss in osteoporosis. Nat Commun 10:2958

Yao W, Jee WS, Zhou H, Lu J, Cui L, Setterberg R, Liang T, Ma Y (1999) Anabolic effect of prostaglandin E2 on cortical bone of aged male rats comes mainly from modeling-dependent bone gain. Bone 25:697–702

Yoshida K, Oida H, Kobayashi T, Maruyama T, Tanaka M, Katayama T, Yamaguchi K, Segi E, Tsuboyama T, Matsushita M et al (2002) Stimulation of bone formation and prevention of bone loss by prostaglandin E EP4 receptor activation. Proc Natl Acad Sci U S A 99:4580–4585

Zanchetta MB, Boailchuk J, Massari F, Silveira F, Bogado C, Zanchetta JR (2018) Significant bone loss after stopping long-term denosumab treatment: a post FREEDOM study. Osteoporos Int 29:41–47

Zhang L, Takahashi HE, Inoue J, Tanizawa T, Endo N, Yamamoto N, Hori M (1997) Effects of intermittent administration of low dose human PTH(1-34) on cancellous and cortical bone of lumbar vertebral bodies in adult beagles. Bone 21:501–506

Zhang G, Guo B, Wu H, Tang T, Zhang BT, Zheng L, He Y, Yang Z, Pan X, Chow H et al (2012) A delivery system targeting bone formation surfaces to facilitate RNAi-based anabolic therapy. Nat Med 18:307–314

Zhang X, MacDonald BT, Gao H, Shamashkin M, Coyle AJ, Martinez RV, He X (2016) Characterization of Tiki, a new family of Wnt-specific metalloproteases. J Biol Chem 291:2435–2443

Zhang X, Dong S, Xu F (2018) Structural and druggability landscape of frizzled G protein-coupled receptors. Trends Biochem Sci 43:1033–1046

Zhao Q, Liu X, Zhang L, Shen X, Qi J, Wang J, Qian N, Deng L (2013) Bone selective protective effect of a novel bone-seeking estrogen on trabecular bone in ovariectomized rats. Calcif Tissue Int 93:172–183

Drugs Causing Bone Loss

Peter Vestergaard

Contents

Abstract

Drugs may cause bone loss by lowering sex steroid levels (e.g., aromatase inhibitors in breast cancer, GnRH agonists in prostate cancer, or depot medroxyprogestone acetate – DMPA), interfere with vitamin D levels (liver inducing anti-epileptic drugs), or directly by toxic effects on bone cells (chemotherapy, phenytoin, or thiazolidinedions, which diverts mesenchymal stem cells from forming osteoblasts to forming adipocytes). However, besides effects on the mineralized matrix, interactions with collagen and other parts of the unmineralized matrix may decrease bone biomechanical competence in a manner

P. Vestergaard (✉)
Department of Endocrinology, Aalborg University Hospital, Aalborg, Denmark

Steno Diabetes Center North Jutland, Aalborg, Denmark
e-mail: p.vestergaard@rn.dk

P. H. Stern (ed.), *Bone Regulators and Osteoporosis Therapy*,
Handbook of Experimental Pharmacology 262, https://doi.org/10.1007/164_2019_340

that may not correlate with bone mineral density (BMD) measured by dual energy absorptiometry (DXA).

Some drugs and drug classes may decrease BMD like the thiazolidinediones and consequently increase fracture risk. Other drugs such as glucocorticoids may decrease BMD, and thus increase fracture risk. However, glucocorticoids may also interfere with the unmineralized matrix leading to an increase in fracture risk, not mirrored in BMD changes. Some drugs such as selective serotonin reuptake inhibitors (SSRI), paracetamol, and non-steroidal anti-inflammatory drugs (NSAIDs) may not per se be associated with bone loss, but fracture risk may be increased, possibly stemming from an increased risk of falls stemming from effects on postural balance mediated by effects on the central nervous system or cardiovascular system.

This paper performs a systematic review of drugs inducing bone loss or associated with fracture risk. The chapter is organized by the Anatomical Therapeutic Chemical (ATC) classification.

Keywords

Acetaminophen · Androgen deprivation therapy · Antiepileptic drugs · Antiviral therapy · Aromatase inhibitors · Azathioprine · Bone mineral density · Budesonide · Chemotherapy · Cyclosporine · Depot medroxyprogesterone acetate · Fracture · Glucocorticoid · GnRH agonist · Heparin · Levothyroxine · Loop diuretic · Methotrexate · Neuroleptic · Non-steroid anti-inflammatory · Opioid · Oral contraceptives · Paracetamol · Protein pump inhibitor · Sedatives · Selective serotonin reuptake inhibitors · Statin · Tamoxifen · Thiazide · Thiazolidinedones · Thyroid hormones · Vitamin A · Vitamin K antagonists

1 Introduction

Drugs may affect bone in a number of ways and cause bone loss. Mineral matrix such as the density, trabecular structure, and crystal structure of hydroxyapatite may be affected as well as collagen and other organic compounds of the organic matrix. Furthermore, cellular turnover may be affected leading to an imbalance between formation and resorption of bone as well as an imbalance in the ratio between soft organic matrix and the harder inorganic matrix. All of these may lead to decreased bone biomechanical competence and thus a decreased density of bone and an increased risk of fractures. Some drugs may also increase the risk of falls and thus the risk of fracture even in the presence of normal bone biomechanical competence.

The drugs affecting bone may be classified in a number of ways. Here the Anatomical Therapeutic Chemical (ATC) classification will be used to ease overview (WHOCC n.d.).

The criteria for discussing a drug were that

1. It induced bone loss or decreased bone mineral density
2. An increased risk of fractures was seen associated with use of the drug in question

In case of a discrepancy between bone mineral density and fracture risk, the causes for this discrepancy are discussed. Also the interaction with the underlying disease is discussed. This as the underlying disease being treated per se may induce bone loss. More severe disease may mean more bone loss, but also higher use of a drug in question. This induces a risk of confounding by indication, which is addressed where relevant.

2 Methods

A systematic search using the terms "bone mineral" and the generic name of the drug in question was performed using Medline from first entry to the date mentioned. A separate search using "fracture" and the generic name of the drug in question was performed. Abstracts were screened, and relevant papers cited. Clinical studies were included, in the order that if randomized controlled clinical trials (RCT) and preferably systematic reviews or meta-analyses of these were available, these were primarily used. In most cases, only observational studies were present and in that case for bone mineral density (BMD), cohort studies were preferably used followed by cross-sectional studies comparing BMD in users of the drug in question compared to non-exposed controls. For fracture risk RCTs and systematic reviews were preferred; however, few of these were available. Instead, epidemiological studies using case–control and cohort designs were used. In case of recent systematic reviews, these eclipsed individual papers, who were only cited where special interest was deemed to be present, i.e. that details on individual drugs were presented.

3 Drugs Causing Bone Loss Ordered by ATC Code, and Thus Organ System

3.1 The Alimentary System

Drugs affecting acidity in the stomach may affect calcium and magnesium absorption, and drugs affecting bile salts may affect absorption of vitamin D through the enterohepatic recirculation. Drugs affecting pancreas may affect absorption of fat-soluble vitamins. Some drugs may also have direct effects on bone cells.

3.1.1 Proton Pump Inhibitors (PPI)

These act by inhibiting the proton pump in the parietal cell, thus decreasing the secretion of H+ ions to the stomach.

Whether PPI cause bone loss and the exact mechanism are subject to debate. A decreased acidity may affect absorption of calcium, magnesium, and vitamin B12, all of which are important to bone health (Ito and Jensen 2010). Pernicious anemia has been associated with an increased risk of fractures (Goerss et al. 1992), and low normal B12 with increased homocysteine (Bozkurt et al. 2009) may be associated with a decreased BMD, whereas decreased B12 vitamin levels per se may not be associated with decreased bone mineral density (Kakehasi et al. 2012).

It is debated to what degree calcium absorption and magnesium absorption are decreased with long-term use of PPI (Ito and Jensen 2010).

In general, studies on bone density in users of PPI are scarce. Although an increased risk of fractures has been documented in users of PPI (Vestergaard et al. 2006a; Ito and Jensen 2010; Liu et al. 2019), a decrease in BMD has not consistently been observed (Liu et al. 2019). Also no increased bone loss rate has been seen with PPI (Nassar and Richter 2018).

The reasons for this discrepancy have not been elucidated, but could be related to risk of falls (e.g., postural hypotension from anemia or neurological impairment from B12 deficiency) or altered bone biomechanical competence in users of PPI. Histamine H2 blockers, which also decrease gastric acidity – although not to the same degree as PPI – have not been associated with an increased risk of fractures (Vestergaard et al. 2006a). A further possibility could be that the association between PPI use and fractures is spurious and related to an underlying cause related to both PPI use and fracture risk such as frailty.

An example of an interaction between the drug, the disease being treated, and another drug is the interaction between PPI and alendronate for osteoporosis. Use of PPI is associated with an increased risk of fractures. Alendronate decreases the risk of fractures in osteoporosis; however, as already PPI treated patients already have an increased risk of fracture, alendronate reduces the risk to that of non-alendronate, non-PPI treated but not to the level of alendronate treated subjects not using PPI (Abrahamsen et al. 2011).

3.1.2 Drugs Against Diabetes

Diabetes may be associated with an increased risk of fractures, which is dissociated from bone mineral density (Vestergaard 2007). Diabetes per se may affect bone density and biomechanical competence in a number of ways (Fuglsang-Nielsen et al. 2018) such that BMD may not be a representative of fracture risk (Vestergaard 2007; Giangregorio et al. 2012).

1. Thiazolidinediones. Thiazolidinediones are associated with bone loss (Billington et al. 2015) through a diversion of mesenchymal stem cells to differentiate into adipocytes rather than osteoblasts (Beck et al. 2013). BMD is thus reduced (Billington et al. 2015). This leads to an increased risk of fractures (Bazelier et al. 2012a, b). Pioglitazone has not been associated with an increased risk of fractures (Pavlova et al. 2018) in one systematic review, whereas in another systematic review similar increases in fracture risk were seen with pioglitazone and rosiglitazone (Zhu et al. 2014). Rosiglitazone seems associated with an increased risk of fractures (Zhu et al. 2014).
2. Other Drugs Against Diabetes (SGLT2, Insulin, etc.). In general these do not alter BMD or fracture risk to a greater extent (Palermo et al. 2015). For the SGLT2 inhibitors minor decreases in total hip BMD linked to decreases in body mass index (BMI) have been reported for canagliflozin (Alba et al. 2016), while no

changes were seen at other bone sites such as the spine. In general the SGLT2 inhibitors may only induce minor changes in BMD and fracture risk (Ye et al. 2018).

3.1.3 Vitamin A

In animal studies, vitamin A may be associated with a decrease in bone mineral (Lionikaite et al. 2018). Early studies pointed at a negative effect on BMD both with dietary and supplement intake of vitamin A – especially as retinol in high doses (Crandall 2004). Oral retinoyl palmitate was associated with fractures (Crandall 2004). However, a negative effect of dietary vitamin A on BMD could not be confirmed in later studies (Rejnmark et al. 2004b) and neither could an effect of dietary retinol and vitamin A intake on fracture risk (Zhang et al. 2017). Pharmacological use of vitamin A analogues (isotretinoin and acitretin) in high doses was not associated with risk of fractures (Vestergaard et al. 2010).

3.2 Blood

3.2.1 Anticoagulants (Heparin and Vitamin K Antagonists)

1. Traditionally unfractionated heparin has been associated with a decrease in BMD (Dahlman et al. 1994) although the evidence is scarce (van der Wiel et al. 1993) and often limited to pregnant women (Pettilä et al. 2002). The mechanism has been ascribed to an interaction with collagenase (Lenaers-Claeys and Vaes 1979), and mineralization (Beljan and Hellewell 1977; Nishiyama et al. 1997) in animal experimental studies. Use of low molecular weight heparin (LMWH) has not been associated with a decrease in BMD and an increase in fracture risk. However, most studies have been of short duration and focused on patients using heparin for thrombosis prevention in cancer, and cancer patients are already at an increased risk of bone loss and fractures (Gajic-Veljanoski et al. 2016) (please see below). In pregnant women, use of LMWH has not been associated with major decreases in BMD after adjustment for confounders (Galambosi et al. 2016).
2. Vitamin K antagonists. Vitamin K antagonists such as warfarin may affect vitamin K dependent carboxylation of – among others – osteocalcin and has thus been hypothesized to be able to interfere with bone formation (Namba et al. 2017). However, in general no excess fracture risk has been seen (Rejnmark et al. 2007). In children on vitamin K antagonists, a decrease in BMD has been reported, but other causes may be involved (Barnes et al. 2005). The general risk factors such as low BMI (de Laet et al. 2005) and growth hormone deficiency (GHD) (Vestergaard et al. 2002) were associated with low BMD in another study in children (Avila et al. 2016). In adults, no decrease in BMD was seen with prolonged use of warfarin (Stenova et al. 2011). The lack of an increase in fracture risk is in line with data from animal experimental data (Sugiyama et al. 2007).

3.3 Circulation

3.3.1 Diuretics

Whereas thiazide diuretics are associated with a decrease in urinary calcium excretion (Rejnmark et al. 2001; Cheng et al. 2018) and thus potentially an increase in BMD – which has not been supported by experimental data (Cheng et al. 2018) – and a decrease in fracture risk (Rejnmark et al. 2005), loop diuretics are associated with an increased urinary calcium excretion (Rejnmark et al. 2001) and thus potentially a loss of bone mineral. However, the initial increase in urinary calcium excretion is subject to a rebound phenomenon within a short time interval (Rejnmark et al. 2001).

Use of loop diuretics has been associated with an increased rate of bone loss in several studies (Lim et al. 2008; Bleicher et al. 2013). However, in a cross-sectional trial no differences in BMD were present between users of various diuretics including loop diuretics and controls (Lim et al. 2005). However, several studies have reported an association between use of loop diuretics and fractures (Rejnmark et al. 2006a; Lai et al. 2017; Heo et al. 2018). One potential reason for the discrepancy with BMD is risk of falls due to hypotension and cognitive impairment from dehydration.

3.3.2 Statins

Per se statins do not alter BMD (Rejnmark et al. 2004a) and some studies have pointed at a decreased risk of fractures (Rejnmark et al. 2006b). However, as LDL is a substrate for LRP5 and thus the Wnt signaling system, low LDL values may have negative effects on fracture risk (Vestergaard 2015a).

3.4 Dermatology

3.4.1 Topical Corticosteroids

In general topical corticosteroids are poorly absorbed and do not alter fracture risk (Vestergaard et al. 2005). No studies on BMD in users of dermal corticosteroids could be identified.

3.5 Sex Steroids

Sex steroids are important for osteoclast and osteoblast function, and sex steroids may improve bone status if a deficient state is present. Oral contraceptives do not seem to have major negative effects on bone status except for depot medroxyprogesterone acetate, which may induce hypogonadism. Likewise may other drugs inducing hypogonadism lead to a loss of bone (please also see below under drugs against cancer).

3.5.1 Depot Medroxyprogesterone Acetate (DMPA)

DMPA may induce hypoestrogenemia, and is associated with an increased risk of fractures (Vestergaard et al. 2008a; Lopez et al. 2015; Kyvernitakis et al. 2017). DMPA users have lower BMD and higher BMD loss than controls (Modesto et al. 2014). One year treatment with DMPA in doses of 104 and 150 mg i.m. every 12 weeks was associated with a decrease in hip and spine BMD (Lange et al. 2017). The lower dose of 75 mg i.m. every 12 weeks was not associated with a decrease in BMD, but the groups were limited in size ($n = 34$ in total) (Lange et al. 2017). Short-term DMPA for less than 12-month use may not be associated with BMD loss (Zhang et al. 2013).

3.5.2 Oral Contraceptives

In general oral contraceptive use is not associated with an increased risk of fractures (Lopez et al. 2015). Oral contraceptive use does not seem associated with a decrease in BMD of the forearm (Beksinska et al. 2018). When given as replacement for amenorrhea, oral contraceptives improve BMD (Altayar et al. 2017; Ackerman et al. 2019). In adolescents BMD increases during oral contraceptive use (da Rizzo et al. 2018), but the increase may be lower than among controls (Gersten et al. 2016). However, in young women, no increase in fracture risk is seen with oral contraceptives (Vestergaard et al. 2008b).

3.6 Systemic Hormones

Systemic hormones may affect many signaling systems.

3.6.1 Glucocorticoids (Except Budesonide)

Glucocorticoids have profound effects on bone antagonizing the effects of vitamin D leading to secondary hyperparathyroidism, suppressing gonadotrophins and inhibiting gonadal steroids thus inducing hypoestrogenemia, as well as interfering with collagen synthesis and the IGF system and also affecting bone resorption and formation directly via the osteoclasts and osteoblasts (Patschan et al. 2001). All of these effects lead to a negative calcium balance and due to the interaction with collagen a disturbance in the balance between mineralized and unmineralized matrix leading to a higher fragility of the bone than may be anticipated from DXA results alone (Van Staa et al. 2003) as DXA only captures the mineralized matrix.

Use of systemic glucocorticoids in a dose of 10 mg daily for 20 weeks led to a pronounced bone loss compared to i.m. gold salts in patients with rheumatoid arthritis (Laan et al. 1993). Low dose oral glucocorticoids (5 mg/day of prednisolone) also lead to a loss of BMD after 1 year (Pérez-Sáez et al. 2018). Fracture risk seems to increase from about 2.5 mg of prednisolone per day (Vestergaard et al. 2005). Intermittent systemic corticosteroid use may not increase risk of fractures (Oshagbemi et al. 2018). Upon discontinuation of oral glucocorticoids, fracture risk decreases within 1 year of termination (Vestergaard et al. 2008c), and BMD increases (Laan et al. 1993).

Budesonide: Oral budesonide has not been associated with an increased risk of fractures (Vestergaard et al. 2008c). In patients with inflammatory Crohn's disease, oral budesonide has been associated with a smaller bone loss than oral prednisolone (Schoon et al. 2005). In primary biliary cirrhosis, oral budesonide as an add-on to ursodeoxycholic acid was associated with a nonsignificantly higher bone loss (Leuschner et al. 1999).

Please also refer to Sect. 3.4 on dermal corticosteroids and Sect. 3.11 on inhaled corticosteroids.

3.6.2 Thyroid Hormones

Replacement for thyroid hormone deficiency (myxedema/hypothyroidism) is not associated with a decrease in bone mineral density if thyroid hormones are kept within normal range. With suppression of TSH by levothyroxine or triiodothyronine, a decrease in bone mineral density is seen (Heemstra et al. 2006). Upon reversal of hyperthyroidism, the initially decreased BMD reverses to normal (Vestergaard and Mosekilde 2003). Hypothyroidism is not associated with an increased risk of fractures except for the first year after initiation of thyroid hormone replacement (Vestergaard et al. 2000), where a transient decrease in BMD is also seen (Trémollières et al. 1991) – probably due to resumption of repressed bone turnover. Hypothyroidism per se is not associated with a decreased BMD (Saggese et al. 1996; Di Mase et al. 2012) or an increased bone loss (Segna et al. 2018).

3.7 Infectious Diseases

3.7.1 Antiviral Therapy in Human Immunodeficiency Virus (HIV)

HIV infection is associated with a number of changes in bone mineral status (Ahmad et al. 2017; Cezarino et al. 2018) and an increased risk of fractures (Güerri-Fernandez et al. 2013; Prieto-Alhambra et al. 2014). Some antiviral drugs such as the nucleoside and nucleotide reverse transcriptase inhibitors have been associated with adverse effects on bone (Conesa-Buendía et al. 2019), but an increased risk of fractures has not been universally demonstrated (Costagliola et al. 2019). Bisphosphonate therapy may reverse the bone loss (Ciccullo et al. 2018).

3.8 Chemotherapy

Chemotherapy and other drugs used to treat cancer may affect bone in a number of ways, either through direct toxic effects on the bone cells or through damage to the gonads leading to low levels of gonadal steroids (Bjarnason et al. 2008; Vestergaard 2008a). Use of glucocorticoids as adjuvant especially in hematological malignancies may further increase the risk of fractures (Svendsen et al. 2017).

3.8.1 Cytotoxic Drugs

Methotrexate

1. High dose: Used in high doses for cancer methotrexate and cytotoxic drugs may induce hypogonadism through destruction of the gonads (Vestergaard 2008a).
2. Low dose: Used in low dose for say rheumatoid arthritis, methotrexate and cyclosporine do not increase the risk of fractures, whereas a limited increase may be seen for cyclosporine (Vestergaard et al. 2006b). Rheumatoid arthritis may be associated with an increased risk of fractures (Vestergaard et al. 2006c) and accelerated bone loss. This is the result of inflammation, reduced activity level from joint pain, use of corticosteroids and potentially also from use of NSAIDs (please see Sect. 3.9), and potentially an increased risk of falls from sarcopenia (Baker et al. 2017; Masamoto et al. 2018). Methotrexate (Buckley et al. 1997; Carbone et al. 1999) does not seem to increase the loss of bone. Leflunomide may potentially be associated with an increase in BMD (Kwon et al. 2019), and the biological anti-rheumatic agents such as TNF-alpha inhibitors may also be associated with increases in trabecular bone score (TBS) (Killinger et al. 2018) and BMD (Chen et al. 2017).

Azathioprine

1. In inflammatory bowel disease, use of azathioprine has been associated with osteopenia (Lima et al. 2017). However, this may be a marker of disease activity.

Cyclosporine

1. In general, cyclosporine use does not correlate with BMD (Freundlich 2006; Shimizu et al. 2013). However, cyclosporine is often used as an add-on to glucocorticoids, and the effect of these needs to be taken into account (Freundlich 2006).

3.8.2 GnRH Agonists (Gonadotropin Releasing Hormone Agonists)

GnRH agonists are used in prostate cancer to induce hypogonadism, i.e., low testosterone levels and thus deprive the prostate cancer cells of their external growth factor (Miyaji et al. 2004). Although this may work well to inhibit the cancer, one of the side effects may be a loss of BMD (Miyaji et al. 2004). GnRH agonists are associated with an increased risk of fractures in prostate cancer (Smith et al. 2005). GnRH agonists may also be used in combination with androgen receptor blockers (antiandrogens), and this combination also induces bone loss (Kim et al. 2017). Use of GnRH agonists for endometriosis is also associated with bone loss due to hypoestrogenemia (Lee et al. 2016; Cho et al. 2016).

3.8.3 Antiandrogens

Antiandrogens block the androgen receptor directly. Their effect on BMD may be less pronounced than the GnRH agonists (Wadhwa et al. 2009) with a preservation of BMD (Sieber et al. 2004; Wadhwa et al. 2011) or even increases in BMD (Smith et al. 2004).

3.8.4 Aromatase Inhibitors

Aromatase inhibitors are used in estrogen-receptor positive breast cancer to lower estradiol levels, and are more efficient in preventing relapse than selective estrogen receptor modulators (Early Breast Cancer Trialists' Collaborative Group (EBCTCG) 2015; Francis et al. 2018). Aromatase inhibitors induce a loss of BMD (Eastell et al. 2011; Hong et al. 2017; Kwan et al. 2018) and bone quality (Hong et al. 2017). In short-term studies of 1-year duration, no loss of BMD was seen with aromatase inhibitors with or without concomitant chemotherapy – however, the duration of this study may be too short for definite conclusions (Markopoulos et al. 2016). Aromatase inhibitor therapy for breast cancer is associated with an increased risk of fractures (Vestergaard et al. 2008d; Goldvaser et al. 2018). Fracture risk is higher with aromatase inhibitors than selective estrogen receptor modulators (Tseng et al. 2018).

3.8.5 Selective Estrogen Receptor Modulators (SERM)

SERM in the form of tamoxifen is used to prevent recurrence in estrogen receptor positive breast cancer. They do possess both estrogen receptor agonist and antagonist effects. Tamoxifen increased BMD (Eastell et al. 2008, 2011), but this has not been seen in all studies (Zidan et al. 2004). However, despite the increase in BMD, an increase in fracture risk especially of the femoral neck has been observed with tamoxifen (Kristensen et al. 1996; Vestergaard et al. 2008d). The increase in fracture risk is especially seen in premenopausal women (Kyvernitakis et al. 2018). However, one study actually reported a decrease in risk of fractures with tamoxifen (Kristensen et al. 2018). One study reported that the increase in hip fracture risk was seen with low cumulated doses of tamoxifen, indicating an effect of other factors than tamoxifen (Vestergaard et al. 2008d).

Combined with aromatase inhibitors although an initial decrease in fracture risk may be seen, in the long term, an increase is seen (Kristensen et al. 2018).

3.9 Muscles and Joints

3.9.1 Non-steroidal Anti-inflammatory Drugs (NSAID)

NSAIDs interfere with prostaglandins, which are involved in bone formation and fracture repair. One study reported no association between NSAID use and BMD (Vestergaard et al. 2012) while another study actually reported an increased BMD (Carbone et al. 2003). An increased risk of fractures has been reported with the use of NSAIDs (Vestergaard et al. 2012). However, the increased risk of fracture varied significantly among the various types of NSAIDs (Vestergaard et al. 2006c). The discrepancy between BMD and fracture risk may perhaps be related to central nervous system effects of NSAIDs on postural balance.

3.9.2 Paracetamol (Acetaminophen)

Paracetamol has not been associated with any change in BMD (Vestergaard et al. 2012). An increased risk of fractures has been reported with use of paracetamol

(Vestergaard et al. 2006c; Williams et al. 2011). The discrepancy may perhaps be related to central nervous system effects of paracetamol on postural balance.

3.10 Nervous System

Drugs affecting the nervous system seem to be among the most numerous also affecting fracture risk and bone.

3.10.1 Opioids

Opioids may affect postural balance and cognitive function thus leading to an increased risk of fractures (Vestergaard 2008b). However, opioids may also suppress follicle stimulating hormone (FSH) and luteinizing hormone (LH) (O'Rourke and Wosnitzer 2016) thus leading to hypoestrogenemia (Coluzzi et al. 2015) and low testosterone levels (Coluzzi et al. 2018). In users of opioids, BMD may be decreased (Gotthardt et al. 2017). However, this study (Gotthardt et al. 2017) additionally addressed subjects on opioid substitution, and these subjects may have other issues associated with low BMD. It has been debated if the effect of opioids on BMD may vary with opioid type (Coluzzi et al. 2015). However, in general, studies are few and evidence levels low.

3.10.2 Drugs Against Epilepsy

Drugs against epilepsy may have a number of effects on BMD (Ensrud et al. 2004) and bone turnover. Liver-inducing drugs may increase the degradation of vitamin D leading to vitamin D deficiency. Some drugs may have direct toxic effects in the bone cells (Feldkamp et al. 2000; Lee et al. 2010). Epilepsy per se may lead to fractures from falls following fits or loss of consciousness (Vestergaard 2005). Treatment with drugs against epilepsy may reduce the number of fits and thus the number of fractures, so although a limited increase in fracture risk is seen with drugs against epilepsy, this negative effect may be overshadowed by the seizure reducing potential of the drugs (Vestergaard 2005).

1. Phenytoin. In patients on phenytoin certain variations in the vitamin D receptor (VDR) may be associated with lower BMD (Phabphal et al. 2013). One study showed a low prevalence of osteoporosis among users of phenytoin, but no control group was available (Moro-Alvarez et al. 2009). Compared to patients on valproate, lamotrigine, and carbamazepine, who maintained their BMD, patients on phenytoin had a significant loss of BMD (Pack et al. 2008). Phenytoin is associated with a decreased BMD with extended use, whereas short-term use may not decrease BMD (Gissel et al. 2007). Phenytoin may have a direct toxic effect on bone cells (Feldkamp et al. 2000). Phenytoin was borderline significantly associated with fracture risk (Vestergaard et al. 2004).
2. Liver Enzyme-Inducing Anti-epileptic Drugs (AED) Other than Phenytoin. These encompass carbamazepine, oxcarbazepine, phenobarbital, phenytoin, and primidone. Number of studies per drug class varies greatly. Carbamazepine is

among the more widely studied drugs in both adults and children and is associated with decreased vitamin D levels and decreased BMD compared to controls (Suljic et al. 2018). In children, polytherapy with AEDs seem to be associated with a decreased BMD (Vestergaard 2015b). The major determinant for poor bone health with AED in children is vitamin D deficiency induced either by the AEDs or by conditions linked to the underlying disease leading to epilepsy or disability following the epilepsy such as low exposure to sunshine, poor nutrition, etc. (Vestergaard 2015b). Carbamazepine is associated with decreased BMD and an increased loss rate for BMD (Chou et al. 2007; Shiek Ahmad et al. 2016). For oxcarbazepine, studies are few, but no decrease in BMD has been reported in one study (Koo et al. 2014), whereas another study reported a decreased BMD (Beniczky et al. 2012). No increased bone loss rate has been reported with oxcarbazepine (Cansu et al. 2008; Cetinkaya et al. 2009). Phenobarbital is both sedative and antiepileptic. BMD is lower in phenobarbital exposed than non-exposed (Kulak et al. 2007; Gissel et al. 2007). For primidone few studies are available, but a trend towards a reduction in BMD has been reported (Farhat et al. 2002).

Carbamazepine [odds ratio (OR), 1.18; 95% confidence interval (CI), 1.10–1.26], oxcarbazepine (1.14, 1.03–1.26), and phenobarbital (1.79, 1.64–1.95) have all been associated with an increased risk of fracture, whereas primidone has not (1.18, 0.95–1.48), but the confidence intervals overlap (Vestergaard et al. 2004).

3. Non-liver Enzyme-Inducing AED. These encompass clonazepam, ethosuximide, lamotrigine, vigabatrin, tiagabine, valproic acid, and topiramate. Valproate has been associated with a decreased BMD (Rahimdel et al. 2016). For ethosuximide, vigabatrin, and tiagabine no data were available on BMD. For lamotrigine stable BMD has been reported within 1 year of treatment as monotherapy (Sheth and Hermann 2007; Pack et al. 2008; El-Haggar et al. 2018) without signs of decreased BMD. For clonazepam, no studies were available. Topiramate has not been associated with a decreased BMD (Heo et al. 2011) in one study, whereas two other studies reported a decreased BMD (Coppola et al. 2009; Zhang et al. 2010).

   Valproate (1.15, 1.05–1.26) was statistically significantly associated with risk of any fracture, whereas ethosuximide (0.75, 0.37–1.52), lamotrigine (1.04, 0.91–1.19), tiagabine (0.75, 0.40–1.41), topiramate (1.39, 0.99–1.96), and vigabatrin (0.93, 0.70–1.22) were not, but the confidence intervals overlap (Vestergaard et al. 2004).
4. Newer AED. These include levetiracetam and other newer AED. Levetiracetam is associated with a decrease in BMD (Hakami et al. 2016; El-Haggar et al. 2018).

3.10.3 Neuroleptics

Neuroleptics are antidopaminergic, and the increased prolactin levels resulting from this may suppress GnRH (please also see above under GnRH agonists) leading to hypogonadism resulting from low FSH and LH (Tseng et al. 2015). In general BMD is lower in patients using drugs that increase prolactin levels than in subjects using drugs that do not increase prolactin levels (Tseng et al. 2015; Gomez et al. 2016; Bulut et al. 2016). Use of neuroleptics is associated with an increased risk of fractures, more so in first-generation antipsychotics than in second-generation antipsychotics (Lee et al. 2017).

3.10.4 Antidepressants

1. Selective Serotonin Reuptake Inhibitors (SSRI). SSRI use is not associated with a decreased BMD (Schweiger et al. 2018). However, an increased risk of fractures is seen (Vestergaard et al. 2008e; Hung et al. 2017), and the risk increases with dose (Vestergaard et al. 2013) and serotonin potency of the drugs (Vestergaard et al. 2008e). This discrepancy may be linked to central nervous system effects with an increased risk of falls.
2. Tricyclic Antidepressants. In general, tricyclic antidepressants are not associated with a decrease in BMD (Schweiger et al. 2018). However, an accelerated bone loss may be seen (Rauma et al. 2016), although this was not seen in all studies (Diem et al. 2007, 2013). An increased risk of fractures may be seen with use of antidepressants (Wu et al. 2013), which may stem from the effects of these drugs on the cardiovascular system and central nervous system with an increased risk of falls. However, differences may exist between the tricyclic antidepressants (Vestergaard et al. 2008e). Amitriptyline and clomipramine have been associated with a dose-dependent increase in fracture risk, while imipramine and nortriptyline were not (Vestergaard et al. 2008e). Even with small doses, amitriptyline may be associated with an increased risk of fractures, which may stem from an increased risk of falls (Vestergaard et al. 2008e).
3. Other Antidepressants. Mirtazapine has been associated with an increased risk of fractures (Leach et al. 2017), although this has not been confirmed in all studies (Vestergaard et al. 2008e). No data on bone mineral exists for mirtazapine. Few data on bone density exists for the other antidepressants.

3.10.5 Sedatives

Sedatives are not associated with BMD status (Kinjo et al. 2005) although an increased loss rate of BMD may be seen (Masunari et al. 2008).

Fracture risk is increased due to effects on postural balance, cognitive function, and awareness (Vestergaard et al. 2006d). The fracture risk seems to increase with increasing half-life of the sedative in question, probably related to the duration of the sedative effect (Vestergaard et al. 2008f).

3.11 Respiratory

3.11.1 Inhaled Corticosteroids and Beta Agonists

Inhaled corticosteroids are usually associated with minimal (Wong et al. 2000) or no bone loss (Chen et al. 2018) except at high doses (Chen et al. 2018). No excess in fracture risk is seen with inhaled corticosteroids except for a marginal increase at very high doses (Vestergaard et al. 2005; Moon and Sin 2019).

Inhaled beta-agonists are not associated with convincing increases in fracture risk (Vestergaard et al. 2007) or decreases in bone mineral (Tattersfield et al. 2001).

References

Abrahamsen B, Eiken P, Eastell R (2011) Proton pump inhibitor use and the antifracture efficacy of alendronate. Arch Intern Med 171:998–1004. https://doi.org/10.1001/archinternmed.2011.20

Ackerman KE, Singhal V, Baskaran C et al (2019) Oestrogen replacement improves bone mineral density in oligo-amenorrhoeic athletes: a randomised clinical trial. Br J Sports Med 53:229–236. https://doi.org/10.1136/bjsports-2018-099723

Ahmad AN, Ahmad SN, Ahmad N (2017) HIV infection and bone abnormalities. Open Orthop J 11:777–784. https://doi.org/10.2174/1874325001711010777

Alba M, Xie J, Fung A, Desai M (2016) The effects of canagliflozin, a sodium glucose co-transporter 2 inhibitor, on mineral metabolism and bone in patients with type 2 diabetes mellitus. Curr Med Res Opin 32:1375–1385. https://doi.org/10.1080/03007995.2016.1174841

Altayar O, Al Nofal A, Carranza Leon BG et al (2017) Treatments to prevent bone loss in functional hypothalamic amenorrhea: a systematic review and meta-analysis. J Endocr Soc 1:500–511. https://doi.org/10.1210/js.2017-00102

Avila ML, Pullenayegum E, Williams S et al (2016) Timing of low bone mineral density and predictors of bone mineral density trajectory in children on long-term warfarin: a longitudinal study. Osteoporos Int 27:1547–1557. https://doi.org/10.1007/s00198-015-3411-8

Baker JF, Long J, Mostoufi-Moab S et al (2017) Muscle deficits in rheumatoid arthritis contribute to inferior cortical bone structure and trabecular bone mineral density. J Rheumatol 44:1777–1785. https://doi.org/10.3899/jrheum.170513

Barnes C, Newall F, Ignjatovic V et al (2005) Reduced bone density in children on long-term warfarin. Pediatr Res 57:578–581. https://doi.org/10.1203/01.PDR.0000155943.07244.04

Bazelier MT, Gallagher AM, van Staa T-P et al (2012a) Use of thiazolidinediones and risk of osteoporotic fracture: disease or drugs? Pharmacoepidemiol Drug Saf 21:507–514. https://doi.org/10.1002/pds.3234

Bazelier MT, Vestergaard P, Gallagher AM et al (2012b) Risk of fracture with thiazolidinediones: disease or drugs? Calcif Tissue Int 90:450–457. https://doi.org/10.1007/s00223-012-9591-8

Beck GR, Khazai NB, Bouloux GF et al (2013) The effects of thiazolidinediones on human bone marrow stromal cell differentiation in vitro and in thiazolidinedione-treated patients with type 2 diabetes. Transl Res 161:145–155. https://doi.org/10.1016/j.trsl.2012.08.006

Beksinska ME, Kleinschmidt I, Smit JA (2018) Bone mineral density in midlife long-term users of hormonal contraception in South Africa: relationship with obesity and menopausal status. Womens Midlife Health 4:6. https://doi.org/10.1186/s40695-018-0035-0

Beljan JR, Hellewell AB (1977) The effect of chronic heparin administration on the mineralization of an experimental osseous defect. Clin Orthop Relat Res:205–207

Beniczky SA, Viken J, Jensen LT, Andersen NB (2012) Bone mineral density in adult patients treated with various antiepileptic drugs. Seizure 21:471–472. https://doi.org/10.1016/j.seizure.2012.04.002

Billington EO, Grey A, Bolland MJ (2015) The effect of thiazolidinediones on bone mineral density and bone turnover: systematic review and meta-analysis. Diabetologia 58:2238–2246. https://doi.org/10.1007/s00125-015-3660-2

Bjarnason NH, Hitz M, Jorgensen NR, Vestergaard P (2008) Adverse bone effects during pharmacological breast cancer therapy. Acta Oncol 47:747–754. https://doi.org/10.1080/02841860802001467

Bleicher K, Cumming RG, Naganathan V et al (2013) Predictors of the rate of BMD loss in older men: findings from the CHAMP study. Osteoporos Int 24:1951–1963. https://doi.org/10.1007/s00198-012-2226-0

Bozkurt N, Erdem M, Yilmaz E et al (2009) The relationship of homocyteine, B12 and folic acid with the bone mineral density of the femur and lumbar spine in Turkish postmenopausal women. Arch Gynecol Obstet 280:381–387. https://doi.org/10.1007/s00404-009-0936-0

Buckley LM, Leib ES, Cartularo KS et al (1997) Effects of low dose methotrexate on the bone mineral density of patients with rheumatoid arthritis. J Rheumatol 24:1489–1494

Bulut SD, Bulut S, Atalan DG et al (2016) The effect of antipsychotics on bone mineral density and sex hormones in male patients with schizophrenia. Psychiatr Danub 28:255–262

Cansu A, Yesilkaya E, Serdaroğlu A et al (2008) Evaluation of bone turnover in epileptic children using oxcarbazepine. Pediatr Neurol 39:266–271. https://doi.org/10.1016/j.pediatrneurol.2008.07.001

Carbone LD, Kaeley G, McKown KM et al (1999) Effects of long-term administration of methotrexate on bone mineral density in rheumatoid arthritis. Calcif Tissue Int 64:100–101

Carbone LD, Tylavsky FA, Cauley JA et al (2003) Association between bone mineral density and the use of nonsteroidal anti-inflammatory drugs and aspirin: impact of cyclooxygenase selectivity. J Bone Miner Res 18:1795–1802. https://doi.org/10.1359/jbmr.2003.18.10.1795

Cetinkaya Y, Kurtulmuş YS, Tutkavul K, Tireli H (2009) The effect of oxcarbazepine on bone metabolism. Acta Neurol Scand 120:170–175. https://doi.org/10.1111/j.1600-0404.2008.01148.x

Cezarino PYA, Simões RDS, Baracat EC, Soares JM Jr (2018) Are women living with HIV prone to osteoporosis in postmenopause? A systematic review. Rev Assoc Med Bras 64:469–473. https://doi.org/10.1590/1806-9282.64.05.469

Chen Y-M, Chen H-H, Huang W-N et al (2017) Tocilizumab potentially prevents bone loss in patients with anticitrullinated protein antibody-positive rheumatoid arthritis. PLoS One 12: e0188454. https://doi.org/10.1371/journal.pone.0188454

Chen W, Johnson KM, FitzGerald JM et al (2018) Long-term effects of inhaled corticosteroids on bone mineral density in older women with asthma or COPD: a registry-based cohort study. Arch Osteoporos 13:116. https://doi.org/10.1007/s11657-018-0537-2

Cheng L, Zhang K, Zhang Z (2018) Effectiveness of thiazides on serum and urinary calcium levels and bone mineral density in patients with osteoporosis: a systematic review and meta-analysis. Drug Des Devel Ther 12:3929–3935. https://doi.org/10.2147/DDDT.S179568

Cho YH, Um MJ, Kim SJ et al (2016) Raloxifene administration in women treated with long term gonadotropin-releasing hormone agonist for severe endometriosis: effects on bone mineral density. J Menopausal Med 22:174–179. https://doi.org/10.6118/jmm.2016.22.3.174

Chou I-J, Lin K-L, Wang H-S, Wang C-J (2007) Evaluation of bone mineral density in children receiving carbamazepine or valproate monotherapy. Acta Paediatr Taiwan 48:317–322

Ciccullo A, D'Avino A, Lassandro AP et al (2018) Changes in bone mineral density in HIV-positive, virologically suppressed patients switching to lamivudine/dolutegravir dual therapy: preliminary results from clinical practice. Infez Med 26:336–340

Coluzzi F, Pergolizzi J, Raffa RB, Mattia C (2015) The unsolved case of "bone-impairing analgesics": the endocrine effects of opioids on bone metabolism. Ther Clin Risk Manag 11:515–523. https://doi.org/10.2147/TCRM.S79409

Coluzzi F, Billeci D, Maggi M, Corona G (2018) Testosterone deficiency in non-cancer opioid-treated patients. J Endocrinol Investig 41:1377–1388. https://doi.org/10.1007/s40618-018-0964-3

Conesa-Buendía FM, Llamas-Granda P, Larrañaga-Vera A et al (2019) Tenofovir causes bone loss via decreased bone formation and increased bone resorption, which can be counteracted by dipyridamole in mice. J Bone Miner Res 34:923–938. https://doi.org/10.1002/jbmr.3665

Coppola G, Fortunato D, Auricchio G et al (2009) Bone mineral density in children, adolescents, and young adults with epilepsy. Epilepsia 50:2140–2146. https://doi.org/10.1111/j.1528-1167.2009.02082.x

Costagliola D, Potard V, Lang S et al (2019) Impact of antiretroviral drugs on fracture risk in HIV-infected individuals: a case-control study nested within the French hospital database on HIV (FHDH-ANRS CO4). J Acquir Immune Defic Syndr 80:214–223. https://doi.org/10.1097/QAI.0000000000001903

Crandall C (2004) Vitamin A intake and osteoporosis: a clinical review. J Women's Health 13:939–953. https://doi.org/10.1089/jwh.2004.13.939
da Rizzo CB, Goldberg TBL, Biason TP et al (2018) One-year adolescent bone mineral density and bone formation marker changes through the use or lack of use of combined hormonal contraceptives. J Pediatr 95:567–574. https://doi.org/10.1016/j.jped.2018.05.011
Dahlman TC, Sjöberg HE, Ringertz H (1994) Bone mineral density during long-term prophylaxis with heparin in pregnancy. Am J Obstet Gynecol 170:1315–1320
de Laet C, Kanis JA, Odén A et al (2005) Body mass index as a predictor of fracture risk: a meta-analysis. Osteoporos Int 16:1330–1338. https://doi.org/10.1007/s00198-005-1863-y
di Mase R, Cerbone M, Improda N et al (2012) Bone health in children with long-term idiopathic subclinical hypothyroidism. Ital J Pediatr 38:56. https://doi.org/10.1186/1824-7288-38-56
Diem SJ, Blackwell TL, Stone KL et al (2007) Use of antidepressants and rates of hip bone loss in older women: the study of osteoporotic fractures. Arch Intern Med 167:1240–1245. https://doi.org/10.1001/archinte.167.12.1240
Diem SJ, Ruppert K, Cauley JA et al (2013) Rates of bone loss among women initiating antidepressant medication use in midlife. J Clin Endocrinol Metab 98:4355–4363. https://doi.org/10.1210/jc.2013-1971
Early Breast Cancer Trialists' Collaborative Group (EBCTCG) (2015) Aromatase inhibitors versus tamoxifen in early breast cancer: patient-level meta-analysis of the randomised trials. Lancet 386:1341–1352. https://doi.org/10.1016/S0140-6736(15)61074-1
Eastell R, Adams JE, Coleman RE et al (2008) Effect of anastrozole on bone mineral density: 5-year results from the anastrozole, tamoxifen, alone or in combination trial 18233230. J Clin Oncol 26:1051–1057. https://doi.org/10.1200/JCO.2007.11.0726
Eastell R, Adams J, Clack G et al (2011) Long-term effects of anastrozole on bone mineral density: 7-year results from the ATAC trial. Ann Oncol 22:857–862. https://doi.org/10.1093/annonc/mdq541
El-Haggar SM, Mostafa TM, Allah HMS, Akef GH (2018) Levetiracetam and lamotrigine effects as mono- and polytherapy on bone mineral density in epileptic patients. Arq Neuropsiquiatr 76:452–458. https://doi.org/10.1590/0004-282X20180068
Ensrud KE, Walczak TS, Blackwell T et al (2004) Antiepileptic drug use increases rates of bone loss in older women: a prospective study. Neurology 62:2051–2057
Farhat G, Yamout B, Mikati MA et al (2002) Effect of antiepileptic drugs on bone density in ambulatory patients. Neurology 58:1348–1353
Feldkamp J, Becker A, Witte OW et al (2000) Long-term anticonvulsant therapy leads to low bone mineral density--evidence for direct drug effects of phenytoin and carbamazepine on human osteoblast-like cells. Exp Clin Endocrinol Diabetes 108:37–43
Francis PA, Pagani O, Fleming GF et al (2018) Tailoring adjuvant endocrine therapy for premenopausal breast cancer. N Engl J Med 379:122–137. https://doi.org/10.1056/NEJMoa1803164
Freundlich M (2006) Bone mineral content and mineral metabolism during cyclosporine treatment of nephrotic syndrome. J Pediatr 149:383–389. https://doi.org/10.1016/j.jpeds.2006.04.060
Fuglsang-Nielsen R, Starup-Linde J, Gregersen S, Vestergaard P (2018) The effect of meals on bone turnover – a systematic review with focus on diabetic bone disease. Expert Rev Endocrinol Metab 13:233–249. https://doi.org/10.1080/17446651.2018.1518131
Gajic-Veljanoski O, Phua CW, Shah PS, Cheung AM (2016) Effects of long-term low-molecular-weight heparin on fractures and bone density in non-pregnant adults: a systematic review with meta-analysis. J Gen Intern Med 31:947–957. https://doi.org/10.1007/s11606-016-3603-8
Galambosi P, Hiilesmaa V, Ulander V-M et al (2016) Prolonged low-molecular-weight heparin use during pregnancy and subsequent bone mineral density. Thromb Res 143:122–126. https://doi.org/10.1016/j.thromres.2016.05.016
Gersten J, Hsieh J, Weiss H, Ricciotti NA (2016) Effect of extended 30 μg ethinyl estradiol with continuous low-dose ethinyl estradiol and cyclic 20 μg ethinyl estradiol oral contraception on adolescent bone density: a randomized trial. J Pediatr Adolesc Gynecol 29:635–642. https://doi.org/10.1016/j.jpag.2016.05.012

Giangregorio LM, Leslie WD, Lix LM et al (2012) FRAX underestimates fracture risk in patients with diabetes. J Bone Miner Res 27:301–308. https://doi.org/10.1002/jbmr.556

Gissel T, Poulsen CS, Vestergaard P (2007) Adverse effects of antiepileptic drugs on bone mineral density in children. Expert Opin Drug Saf 6:267–278. https://doi.org/10.1517/14740338.6.3.267

Goerss JB, Kim CH, Atkinson EJ et al (1992) Risk of fractures in patients with pernicious anemia. J Bone Miner Res 7:573–579. https://doi.org/10.1002/jbmr.5650070514

Goldvaser H, Barnes TA, Šeruga B et al (2018) Toxicity of extended adjuvant therapy with aromatase inhibitors in early breast cancer: a systematic review and meta-analysis. J Natl Cancer Inst 110:31–39. https://doi.org/10.1093/jnci/djx141

Gomez L, Stubbs B, Shirazi A et al (2016) Lower bone mineral density at the hip and lumbar spine in people with psychosis versus controls: a comprehensive review and skeletal site-specific meta-analysis. Curr Osteoporos Rep 14:249–259. https://doi.org/10.1007/s11914-016-0325-0

Gotthardt F, Huber C, Thierfelder C et al (2017) Bone mineral density and its determinants in men with opioid dependence. J Bone Miner Metab 35:99–107. https://doi.org/10.1007/s00774-015-0732-9

Güerri-Fernandez R, Vestergaard P, Carbonell C et al (2013) HIV infection is strongly associated with hip fracture risk, independently of age, gender, and comorbidities: a population-based cohort study. J Bone Miner Res 28:1259–1263. https://doi.org/10.1002/jbmr.1874

Hakami T, O'Brien TJ, Petty SJ et al (2016) Monotherapy with levetiracetam versus older AEDs: a randomized comparative trial of effects on bone health. Calcif Tissue Int 98:556–565. https://doi.org/10.1007/s00223-016-0109-7

Heemstra KA, Hamdy NAT, Romijn JA, Smit JWA (2006) The effects of thyrotropin-suppressive therapy on bone metabolism in patients with well-differentiated thyroid carcinoma. Thyroid 16:583–591. https://doi.org/10.1089/thy.2006.16.583

Heo K, Rhee Y, Lee HW et al (2011) The effect of topiramate monotherapy on bone mineral density and markers of bone and mineral metabolism in premenopausal women with epilepsy. Epilepsia 52:1884–1889. https://doi.org/10.1111/j.1528-1167.2011.03131.x

Heo JH, Rascati KL, Lopez KN, Moffett BS (2018) Increased fracture risk with furosemide use in children with congenital heart disease. J Pediatr 199:92–98.e10. https://doi.org/10.1016/j.jpeds.2018.03.077

Hong AR, Kim JH, Lee KH et al (2017) Long-term effect of aromatase inhibitors on bone microarchitecture and macroarchitecture in non-osteoporotic postmenopausal women with breast cancer. Osteoporos Int 28:1413–1422. https://doi.org/10.1007/s00198-016-3899-6

Hung S-C, Lin C-H, Hung H-C et al (2017) Use of selective serotonin reuptake inhibitors and risk of hip fracture in the elderly: a case-control study in Taiwan. J Am Med Dir Assoc 18:350–354. https://doi.org/10.1016/j.jamda.2016.12.003

Ito T, Jensen RT (2010) Association of long-term proton pump inhibitor therapy with bone fractures and effects on absorption of calcium, vitamin B12, iron, and magnesium. Curr Gastroenterol Rep 12:448–457. https://doi.org/10.1007/s11894-010-0141-0

Kakehasi AM, Carvalho AV, Maksud FAN, Barbosa AJA (2012) Serum levels of vitamin B12 are not related to low bone mineral density in postmenopausal Brazilian women. Rev Bras Reumatol 52:863–869

Killinger Z, Gajdarova L, Kuzma M et al (2018) Biologic treatment in comparison to methotrexate has positive effect on trabecular bone score in rheumatoid arthritis patients: 1-year follow-up. Acta Clin Belg 74:121–125. https://doi.org/10.1080/17843286.2018.1512189

Kim SH, Joung JY, Kim S et al (2017) Comparison of bone mineral loss by combined androgen block agonist versus GnRH in patients with prostate cancer: a 12 month-prospective observational study. Sci Rep 7:39562. https://doi.org/10.1038/srep39562

Kinjo M, Setoguchi S, Schneeweiss S, Solomon DH (2005) Bone mineral density in subjects using central nervous system-active medications. Am J Med 118:1414. https://doi.org/10.1016/j.amjmed.2005.07.033

Koo DL, Hwang KJ, Han SW et al (2014) Effect of oxcarbazepine on bone mineral density and biochemical markers of bone metabolism in patients with epilepsy. Epilepsy Res 108:442–447. https://doi.org/10.1016/j.eplepsyres.2013.09.009

Kristensen B, Ejlertsen B, Mouridsen HT et al (1996) Femoral fractures in postmenopausal breast cancer patients treated with adjuvant tamoxifen. Breast Cancer Res Treat 39:321–326

Kristensen B, Ejlertsen B, Jensen M-B et al (2018) The occurrence of fractures after adjuvant treatment of breast cancer: a DBCG register study. Acta Oncol 57:141–145. https://doi.org/10.1080/0284186X.2017.1403043

Kulak CAM, Borba VZC, Silvado CE et al (2007) Bone density and bone turnover markers in patients with epilepsy on chronic antiepileptic drug therapy. Arq Bras Endocrinol Metabol 51:466–471

Kwan ML, Yao S, Laurent CA et al (2018) Changes in bone mineral density in women with breast cancer receiving aromatase inhibitor therapy. Breast Cancer Res Treat 168:523–530. https://doi.org/10.1007/s10549-017-4626-5

Kwon OC, Oh JS, Hong S et al (2019) Conventional synthetic disease-modifying antirheumatic drugs and bone mineral density in rheumatoid arthritis patients with osteoporosis: possible beneficial effect of leflunomide. Clin Exp Rheumatol 37:813–819

Kyvernitakis I, Kostev K, Nassour T et al (2017) The impact of depot medroxyprogesterone acetate on fracture risk: a case-control study from the UK. Osteoporos Int 28:291–297. https://doi.org/10.1007/s00198-016-3714-4

Kyvernitakis I, Kostev K, Hadji P (2018) The tamoxifen paradox-influence of adjuvant tamoxifen on fracture risk in pre- and postmenopausal women with breast cancer. Osteoporos Int 29:2557–2564. https://doi.org/10.1007/s00198-018-4642-2

Laan RF, van Riel PL, van de Putte LB et al (1993) Low-dose prednisone induces rapid reversible axial bone loss in patients with rheumatoid arthritis. A randomized, controlled study. Ann Intern Med 15:963–968

Lai S-W, Cheng K-C, Lin C-L, Liao K-F (2017) Furosemide use and acute risk of hip fracture in older people: a nationwide case-control study in Taiwan. Geriatr Gerontol Int 17:2552–2558. https://doi.org/10.1111/ggi.13087

Lange HLH, Manos BE, Gothard MD et al (2017) Bone mineral density and weight changes in adolescents randomized to 3 doses of depot medroxyprogesterone acetate. J Pediatr Adolesc Gynecol 30:169–175. https://doi.org/10.1016/j.jpag.2016.10.011

Leach MJ, Pratt NL, Roughead EE (2017) The risk of hip fracture due to mirtazapine exposure when switching antidepressants or using other antidepressants as add-on therapy. Drugs Real World Outcomes 4:247–255. https://doi.org/10.1007/s40801-017-0120-y

Lee RH, Lyles KW, Colón-Emeric C (2010) A review of the effect of anticonvulsant medications on bone mineral density and fracture risk. Am J Geriatr Pharmacother 8:34–46. https://doi.org/10.1016/j.amjopharm.2010.02.003

Lee D-Y, Lee J-Y, Seo J-W et al (2016) Gonadotropin-releasing hormone agonist with add-back treatment is as effective and tolerable as dienogest in preventing pain recurrence after laparoscopic surgery for endometriosis. Arch Gynecol Obstet 294:1257–1263. https://doi.org/10.1007/s00404-016-4184-9

Lee S-H, Hsu W-T, Lai C-C et al (2017) Use of antipsychotics increases the risk of fracture: a systematic review and meta-analysis. Osteoporos Int 28:1167–1178. https://doi.org/10.1007/s00198-016-3881-3

Lenaers-Claeys G, Vaes G (1979) Collagenase, procollagenase and bone resorption. Effects of heparin, parathyroid hormone and calcitonin. Biochim Biophys Acta 584:375–388

Leuschner M, Maier KP, Schlichting J et al (1999) Oral budesonide and ursodeoxycholic acid for treatment of primary biliary cirrhosis: results of a prospective double-blind trial. Gastroenterology 117:918–925

Lim LS, Fink HA, Kuskowski MA et al (2005) Diuretic use and bone mineral density in older USA men: the osteoporotic fractures in men (MrOS) study. Age Ageing 34:504–507. https://doi.org/10.1093/ageing/afi133

Lim LS, Fink HA, Kuskowski MA et al (2008) Loop diuretic use and increased rates of hip bone loss in older men: the Osteoporotic Fractures in Men Study. Arch Intern Med 168:735–740. https://doi.org/10.1001/archinte.168.7.735

Lima CA, Lyra AC, Mendes CMC et al (2017) Bone mineral density and inflammatory bowel disease severity. Braz J Med Biol Res 50:e6374. https://doi.org/10.1590/1414-431X20176374

Lionikaite V, Gustafsson KL, Westerlund A et al (2018) Clinically relevant doses of vitamin A decrease cortical bone mass in mice. J Endocrinol 239:389–402. https://doi.org/10.1530/JOE-18-0316

Liu J, Li X, Fan L et al (2019) Proton pump inhibitors therapy and risk of bone diseases: an update meta-analysis. Life Sci 218:213–223. https://doi.org/10.1016/j.lfs.2018.12.058

Lopez LM, Chen M, Mullins Long S et al (2015) Steroidal contraceptives and bone fractures in women: evidence from observational studies. Cochrane Database Syst Rev 7:CD009849. https://doi.org/10.1002/14651858.CD009849.pub3

Markopoulos C, Koukouras D, Venizelos V et al (2016) Impact of chemotherapy followed by aromatase inhibitors on bone health of women with ER-positive early breast cancer in real world clinical settings in Greece: results of the POCHARBI trial conducted by the Hellenic Society of Breast Surgeons. Breast 27:27–34. https://doi.org/10.1016/j.breast.2016.02.007

Masamoto K, Otsuki B, Fujibayashi S et al (2018) Factors influencing spinal sagittal balance, bone mineral density, and Oswestry Disability Index outcome measures in patients with rheumatoid arthritis. Eur Spine J 27:406–415. https://doi.org/10.1007/s00586-017-5401-3

Masunari N, Fujiwara S, Nakata Y et al (2008) Effect of angiotensin converting enzyme inhibitor and benzodiazepine intake on bone loss in older Japanese. Hiroshima J Med Sci 57:17–25

Miyaji Y, Saika T, Yamamoto Y et al (2004) Effects of gonadotropin-releasing hormone agonists on bone metabolism markers and bone mineral density in patients with prostate cancer. Urology 64:128–131. https://doi.org/10.1016/j.urology.2004.03.012

Modesto W, Bahamondes MV, Silva dos Santos P et al (2014) Exploratory study of the effect of lifestyle counselling on bone mineral density and body composition in users of the contraceptive depot-medroxyprogesterone acetate. Eur J Contracept Reprod Health Care 19:244–249. https://doi.org/10.3109/13625187.2014.924098

Moon J-Y, Sin DD (2019) Inhaled corticosteroids and fractures in chronic obstructive pulmonary disease: current understanding and recommendations. Curr Opin Pulm Med 25:165–172. https://doi.org/10.1097/MCP.0000000000000554

Moro-Alvarez MJ, Díaz Curiel M, de la Piedra C et al (2009) Bone disease induced by phenytoin therapy: clinical and experimental study. Eur Neurol 62:219–230. https://doi.org/10.1159/000229309

Namba S, Yamaoka-Tojo M, Kakizaki R et al (2017) Effects on bone metabolism markers and arterial stiffness by switching to rivaroxaban from warfarin in patients with atrial fibrillation. Heart Vessel 32:977–982. https://doi.org/10.1007/s00380-017-0950-2

Nassar Y, Richter S (2018) Proton-pump inhibitor use and fracture risk: an updated systematic review and meta-analysis. J Bone Metab 25:141–151. https://doi.org/10.11005/jbm.2018.25.3.141

Nishiyama M, Itoh F, Ujiie A (1997) Low-molecular-weight heparin (dalteparin) demonstrated a weaker effect on rat bone metabolism compared with heparin. Jpn J Pharmacol 74:59–68

O'Rourke TK, Wosnitzer MS (2016) Opioid-induced androgen deficiency (OPIAD): diagnosis, management, and literature review. Curr Urol Rep 17:76. https://doi.org/10.1007/s11934-016-0634-y

Oshagbemi OA, Burden AM, Shudofsky KN et al (2018) Use of high-dose intermittent systemic glucocorticoids and the risk of fracture in patients with chronic obstructive pulmonary disease. Bone 110:238–243. https://doi.org/10.1016/j.bone.2018.02.007

Pack AM, Morrell MJ, Randall A et al (2008) Bone health in young women with epilepsy after one year of antiepileptic drug monotherapy. Neurology 70:1586–1593. https://doi.org/10.1212/01.wnl.0000310981.44676.de

Palermo A, D'Onofrio L, Eastell R et al (2015) Oral anti-diabetic drugs and fracture risk, cut to the bone: safe or dangerous? A narrative review. Osteoporos Int 26:2073–2089. https://doi.org/10.1007/s00198-015-3123-0

Patschan D, Loddenkemper K, Buttgereit F (2001) Molecular mechanisms of glucocorticoid-induced osteoporosis. Bone 29:498–505

Pavlova V, Filipova E, Uzunova K et al (2018) Pioglitazone therapy and fractures: systematic review and meta-analysis. Endocr Metab Immune Disord Drug Targets 18:502–507. https://doi.org/10.2174/1871530318666180423121833

Pérez-Sáez MJ, Herrera S, Prieto-Alhambra D et al (2018) Maintenance low dose systemic glucocorticoids have limited impact on bone strength and mineral density among incident renal allograft recipients: a pilot prospective cohort study. Bone 116:290–294. https://doi.org/10.1016/j.bone.2018.08.013

Pettilä V, Leinonen P, Markkola A et al (2002) Postpartum bone mineral density in women treated for thromboprophylaxis with unfractionated heparin or LMW heparin. Thromb Haemost 87:182–186

Phabphal K, Geater A, Limapichart K et al (2013) The association between BsmI polymorphism and bone mineral density in young patients with epilepsy who are taking phenytoin. Epilepsia 54:249–255. https://doi.org/10.1111/epi.12049

Prieto-Alhambra D, Güerri-Fernández R, de Vries F et al (2014) HIV infection and its association with an excess risk of clinical fractures: a nationwide case-control study. J Acquir Immune Defic Syndr 66:90–95. https://doi.org/10.1097/QAI.0000000000000112

Rahimdel A, Dehghan A, Moghadam MA, Ardekani AM (2016) Relationship between bone density and biochemical markers of bone among two groups taking carbamazepine and sodium valproate for epilepsy in comparison with healthy individuals in Yazd. Electron Physician 8:3257–3265. https://doi.org/10.19082/3257

Rauma PH, Honkanen RJ, Williams LJ et al (2016) Effects of antidepressants on postmenopausal bone loss – a 5-year longitudinal study from the OSTPRE cohort. Bone 89:25–31. https://doi.org/10.1016/j.bone.2016.05.003

Rejnmark L, Vestergaard P, Heickendorff L et al (2001) Loop diuretics alter the diurnal rhythm of endogenous parathyroid hormone secretion. A randomized-controlled study on the effects of loop- and thiazide-diuretics on the diurnal rhythms of calcitropic hormones and biochemical bone markers in postmenopausal women. Eur J Clin Investig 31:764–772

Rejnmark L, Buus NH, Vestergaard P et al (2004a) Effects of simvastatin on bone turnover and BMD: a 1-year randomized controlled trial in postmenopausal osteopenic women. J Bone Miner Res 19:737–744. https://doi.org/10.1359/JBMR.040209

Rejnmark L, Vestergaard P, Charles P et al (2004b) No effect of vitamin A intake on bone mineral density and fracture risk in perimenopausal women. Osteoporos Int 15:872–880. https://doi.org/10.1007/s00198-004-1618-1

Rejnmark L, Vestergaard P, Mosekilde L (2005) Reduced fracture risk in users of thiazide diuretics. Calcif Tissue Int 76:167–175. https://doi.org/10.1007/s00223-004-0084-2

Rejnmark L, Vestergaard P, Mosekilde L (2006a) Fracture risk in patients treated with loop diuretics. J Intern Med 259:117–124. https://doi.org/10.1111/j.1365-2796.2005.01585.x

Rejnmark L, Vestergaard P, Mosekilde L (2006b) Statin but not non-statin lipid-lowering drugs decrease fracture risk: a nation-wide case-control study. Calcif Tissue Int 79:27–36. https://doi.org/10.1007/s00223-006-0024-4

Rejnmark L, Vestergaard P, Mosekilde L (2007) Fracture risk in users of oral anticoagulants: a nationwide case-control study. Int J Cardiol 118:338–344. https://doi.org/10.1016/j.ijcard.2006.07.022

Saggese G, Bertelloni S, Baroncelli GI et al (1996) Bone mineral density in adolescent females treated with L-thyroxine: a longitudinal study. Eur J Pediatr 155:452–457

Schoon EJ, Bollani S, Mills PR et al (2005) Bone mineral density in relation to efficacy and side effects of budesonide and prednisolone in Crohn's disease. Clin Gastroenterol Hepatol 3:113–121

Schweiger JU, Schweiger U, Hüppe M et al (2018) The use of antidepressive agents and bone mineral density in women: a meta-analysis. Int J Environ Res Public Health 15:E1373. https://doi.org/10.3390/ijerph15071373

Segna D, Bauer DC, Feller M et al (2018) Association between subclinical thyroid dysfunction and change in bone mineral density in prospective cohorts. J Intern Med 283:56–72. https://doi.org/10.1111/joim.12688

Sheth RD, Hermann BP (2007) Bone mineral density with lamotrigine monotherapy for epilepsy. Pediatr Neurol 37:250–254. https://doi.org/10.1016/j.pediatrneurol.2007.05.018

Shiek Ahmad B, O'Brien TJ, Gorelik A et al (2016) Bone mineral changes in epilepsy patients during initial years of antiepileptic drug therapy. J Clin Densitom 19:450–456. https://doi.org/10.1016/j.jocd.2016.07.008

Shimizu C, Fujita T, Fuke Y et al (2013) Effects of cyclosporine on bone mineral density in patients with glucocorticoid-dependent nephrotic syndrome in remission. Int Urol Nephrol 45:803–808. https://doi.org/10.1007/s11255-012-0264-3

Sieber PR, Keiller DL, Kahnoski RJ et al (2004) Bicalutamide 150 mg maintains bone mineral density during monotherapy for localized or locally advanced prostate cancer. J Urol 171:2272–2276, quiz 2435

Smith MR, Goode M, Zietman AL et al (2004) Bicalutamide monotherapy versus leuprolide monotherapy for prostate cancer: effects on bone mineral density and body composition. J Clin Oncol 22:2546–2553. https://doi.org/10.1200/JCO.2004.01.174

Smith MR, Lee WC, Brandman J et al (2005) Gonadotropin-releasing hormone agonists and fracture risk: a claims-based cohort study of men with nonmetastatic prostate cancer. J Clin Oncol 23:7897–7903. https://doi.org/10.1200/JCO.2004.00.6908

Stenova E, Steno B, Killinger Z et al (2011) Effect of long-term oral anticoagulant therapy on bone mineral density and bone turnover markers: a prospective 12 month study. Bratisl Lek Listy 112:71–76

Sugiyama T, Takaki T, Sakanaka K et al (2007) Warfarin-induced impairment of cortical bone material quality and compensatory adaptation of cortical bone structure to mechanical stimuli. J Endocrinol 194:213–222. https://doi.org/10.1677/JOE-07-0119

Suljic EM, Mehicevic A, Mahmutbegovic N (2018) Effect of long-term carbamazepine therapy on bone health. Med Arch 72:262–266. https://doi.org/10.5455/medarh.2018.72.262-266

Svendsen P, Shekhrajka N, Nielsen KL et al (2017) R-CHOP(-like) treatment of diffuse large B-cell lymphoma significantly reduces CT-assessed vertebral bone density: a single center study of 111 patients. Leuk Lymphoma 58:1105–1113. https://doi.org/10.1080/10428194.2016.1233543

Tattersfield AE, Town GI, Johnell O et al (2001) Bone mineral density in subjects with mild asthma randomised to treatment with inhaled corticosteroids or non-corticosteroid treatment for two years. Thorax 56:272–278

Trémollières F, Pouillès JM, Louvet JP, Ribot C (1991) Transitory bone loss during substitution treatment for hypothyroidism. Results of a two year prospective study. Rev Rhum Mal Osteoartic 58:869–875

Tseng P-T, Chen Y-W, Yeh P-Y et al (2015) Bone mineral density in schizophrenia: an update of current meta-analysis and literature review under guideline of PRISMA. Medicine (Baltimore) 94:e1967. https://doi.org/10.1097/MD.0000000000001967

Tseng OL, Spinelli JJ, Gotay CC et al (2018) Aromatase inhibitors are associated with a higher fracture risk than tamoxifen: a systematic review and meta-analysis. Ther Adv Musculoskelet Dis 10:71–90. https://doi.org/10.1177/1759720X18759291

van der Wiel HE, Lips P, Huijgens PC, Netelenbos JC (1993) Effects of short-term low-dose heparin administration on biochemical parameters of bone turnover. Bone Miner 22:27–32

van Staa TP, Laan RF, Barton IP et al (2003) Bone density threshold and other predictors of vertebral fracture in patients receiving oral glucocorticoid therapy. Arthritis Rheum 48:3224–3229. https://doi.org/10.1002/art.11283

Vestergaard P (2005) Epilepsy, osteoporosis and fracture risk – a meta-analysis. Acta Neurol Scand 112:277–286. https://doi.org/10.1111/j.1600-0404.2005.00474.x

Vestergaard P (2007) Discrepancies in bone mineral density and fracture risk in patients with type 1 and type 2 diabetes--a meta-analysis. Osteoporos Int 18:427–444. https://doi.org/10.1007/s00198-006-0253-4

Vestergaard P (2008a) Skeletal effects of drugs to treat cancer. Curr Drug Saf 3:173–177

Vestergaard P (2008b) Pain-relief medication and risk of fractures. Curr Drug Saf 3:199–203

Vestergaard P (2015a) Bone: fracture risk in the JUPITER trial--statin treatment or not? Nat Rev Endocrinol 11:135–136. https://doi.org/10.1038/nrendo.2014.232

Vestergaard P (2015b) Effects of antiepileptic drugs on bone health and growth potential in children with epilepsy. Paediatr Drugs 17:141–150. https://doi.org/10.1007/s40272-014-0115-z

Vestergaard P, Mosekilde L (2003) Hyperthyroidism, bone mineral, and fracture risk--a meta-analysis. Thyroid 13:585–593. https://doi.org/10.1089/105072503322238854

Vestergaard P, Weeke J, Hoeck HC et al (2000) Fractures in patients with primary idiopathic hypothyroidism. Thyroid 10:335–340. https://doi.org/10.1089/thy.2000.10.335

Vestergaard P, Jørgensen JOL, Hagen C et al (2002) Fracture risk is increased in patients with GH deficiency or untreated prolactinomas--a case-control study. Clin Endocrinol 56:159–167

Vestergaard P, Rejnmark L, Mosekilde L (2004) Fracture risk associated with use of antiepileptic drugs. Epilepsia 45:1330–1337. https://doi.org/10.1111/j.0013-9580.2004.18804.x

Vestergaard P, Rejnmark L, Mosekilde L (2005) Fracture risk associated with systemic and topical corticosteroids. J Intern Med 257:374–384. https://doi.org/10.1111/j.1365-2796.2005.01467.x

Vestergaard P, Rejnmark L, Mosekilde L (2006a) Proton pump inhibitors, histamine H2 receptor antagonists, and other antacid medications and the risk of fracture. Calcif Tissue Int 79:76–83. https://doi.org/10.1007/s00223-006-0021-7

Vestergaard P, Rejnmark L, Mosekilde L (2006b) Methotrexate, azathioprine, cyclosporine, and risk of fracture. Calcif Tissue Int 79:69–75. https://doi.org/10.1007/s00223-006-0060-0

Vestergaard P, Rejnmark L, Mosekilde L (2006c) Fracture risk associated with use of nonsteroidal anti-inflammatory drugs, acetylsalicylic acid, and acetaminophen and the effects of rheumatoid arthritis and osteoarthritis. Calcif Tissue Int 79:84–94. https://doi.org/10.1007/s00223-006-0020-8

Vestergaard P, Rejnmark L, Mosekilde L (2006d) Anxiolytics, sedatives, antidepressants, neuroleptics and the risk of fracture. Osteoporos Int 17:807–816. https://doi.org/10.1007/s00198-005-0065-y

Vestergaard P, Rejnmark L, Mosekilde L (2007) Fracture risk in patients with chronic lung diseases treated with bronchodilator drugs and inhaled and oral corticosteroids. Chest 132:1599–1607. https://doi.org/10.1378/chest.07-1092

Vestergaard P, Rejnmark L, Mosekilde L (2008a) The effects of depot medroxyprogesterone acetate and intrauterine device use on fracture risk in Danish women. Contraception 78:459–464. https://doi.org/10.1016/j.contraception.2008.07.014

Vestergaard P, Rejnmark L, Mosekilde L (2008b) Fracture risk in very young women using combined oral contraceptives. Contraception 78:358–364. https://doi.org/10.1016/j.contraception.2008.06.010

Vestergaard P, Rejnmark L, Mosekilde L (2008c) Fracture risk associated with different types of oral corticosteroids and effect of termination of corticosteroids on the risk of fractures. Calcif Tissue Int 82:249–257. https://doi.org/10.1007/s00223-008-9124-7

Vestergaard P, Rejnmark L, Mosekilde L (2008d) Effect of tamoxifen and aromatase inhibitors on the risk of fractures in women with breast cancer. Calcif Tissue Int 82:334–340. https://doi.org/10.1007/s00223-008-9132-7

Vestergaard P, Rejnmark L, Mosekilde L (2008e) Selective serotonin reuptake inhibitors and other antidepressants and risk of fracture. Calcif Tissue Int 82:92–101. https://doi.org/10.1007/s00223-007-9099-9

Vestergaard P, Rejnmark L, Mosekilde L (2008f) Anxiolytics and sedatives and risk of fractures: effects of half-life. Calcif Tissue Int 82:34–43. https://doi.org/10.1007/s00223-007-9095-0

Vestergaard P, Rejnmark L, Mosekilde L (2010) High-dose treatment with vitamin A analogues and risk of fractures. Arch Dermatol 146:478–482. https://doi.org/10.1001/archdermatol.2010.59

Vestergaard P, Hermann P, Jensen J-EB et al (2012) Effects of paracetamol, non-steroidal anti-inflammatory drugs, acetylsalicylic acid, and opioids on bone mineral density and risk of fracture: results of the Danish Osteoporosis Prevention Study (DOPS). Osteoporos Int 23:1255–1265. https://doi.org/10.1007/s00198-011-1692-0

Vestergaard P, Prieto-Alhambra D, Javaid MK, Cooper C (2013) Fractures in users of antidepressants and anxiolytics and sedatives: effects of age and dose. Osteoporos Int 24:671–680. https://doi.org/10.1007/s00198-012-2043-5

Wadhwa VK, Weston R, Mistry R, Parr NJ (2009) Long-term changes in bone mineral density and predicted fracture risk in patients receiving androgen-deprivation therapy for prostate cancer, with stratification of treatment based on presenting values. BJU Int 104:800–805. https://doi.org/10.1111/j.1464-410X.2009.08483.x

Wadhwa VK, Weston R, Parr NJ (2011) Bicalutamide monotherapy preserves bone mineral density, muscle strength and has significant health-related quality of life benefits for osteoporotic men with prostate cancer. BJU Int 107:1923–1929. https://doi.org/10.1111/j.1464-410X.2010.09726.x

WHOCC (n.d.) ATC/DDD Index. https://www.whocc.no/atc_ddd_index/. Accessed 25 Feb 2019

Williams LJ, Pasco JA, Henry MJ et al (2011) Paracetamol (acetaminophen) use, fracture and bone mineral density. Bone 48:1277–1281. https://doi.org/10.1016/j.bone.2011.03.435

Wong CA, Walsh LJ, Smith CJ et al (2000) Inhaled corticosteroid use and bone-mineral density in patients with asthma. Lancet 355:1399–1403. https://doi.org/10.1016/S0140-6736(00)02138-3

Wu Q, Qu W, Crowell MD et al (2013) Tricyclic antidepressant use and risk of fractures: a meta-analysis of cohort and case-control studies. J Bone Miner Res 28:753–763. https://doi.org/10.1002/jbmr.1813

Ye Y, Zhao C, Liang J et al (2018) Effect of sodium-glucose co-transporter 2 inhibitors on bone metabolism and fracture risk. Front Pharmacol 9:1517. https://doi.org/10.3389/fphar.2018.01517

Zhang J, Wang K-X, Wei Y et al (2010) Effect of topiramate and carbamazepine on bone metabolism in children with epilepsy. Zhongguo Dang Dai Er Ke Za Zhi 12:96–98

Zhang M-H, Zhang W, Zhang A-D et al (2013) Effect of depot medroxyprogesterone acetate on bone mineral density in adolescent women. Chin Med J 126:4043–4047

Zhang X, Zhang R, Moore JB et al (2017) The effect of vitamin A on fracture risk: a meta-analysis of cohort studies. Int J Environ Res Public Health 14:1043. https://doi.org/10.3390/ijerph14091043

Zhu Z-N, Jiang Y-F, Ding T (2014) Risk of fracture with thiazolidinediones: an updated meta-analysis of randomized clinical trials. Bone 68:115–123. https://doi.org/10.1016/j.bone.2014.08.010

Zidan J, Keidar Z, Basher W, Israel O (2004) Effects of tamoxifen on bone mineral density and metabolism in postmenopausal women with early-stage breast cancer. Med Oncol 21:117–121. https://doi.org/10.1385/MO:21:2:117

Natural Products as Potential Bone Therapies

Man-Sau Wong, Christina Chui-Wa Poon, Li-Ping Zhou, and Hui-Hui Xiao

Contents

Abstract

Demands for natural products, in the form of botanicals, dietary supplements, and herbal medicine, for management of chronic diseases are increasing globally. Natural products might be an alternative for the management of bone health to meet the demands of a growing aging population. Different types of natural

M.-S. Wong (✉)
Department of Applied Biology and Chemical Technology, The Hong Kong Polytechnic University, Hong Kong SAR, People's Republic of China

State Key Laboratory of Chinese Medicine and Molecular Pharmacology (Incubation), Shenzhen Research Institute of The Hong Kong Polytechnic University, Shenzhen, People's Republic of China
e-mail: bcmswong@polyu.edu.hk

C. C.-W. Poon · L.-P. Zhou
Department of Applied Biology and Chemical Technology, The Hong Kong Polytechnic University, Hong Kong SAR, People's Republic of China

H.-H. Xiao
State Key Laboratory of Chinese Medicine and Molecular Pharmacology (Incubation), Shenzhen Research Institute of The Hong Kong Polytechnic University, Shenzhen, People's Republic of China

P. H. Stern (ed.), *Bone Regulators and Osteoporosis Therapy*,
Handbook of Experimental Pharmacology 262, https://doi.org/10.1007/164_2019_322

products, including Chinese herbal medicine decoctions, herbs, and isolated phytochemicals, have been demonstrated to exert bone protective effects. The most common types of bone protective bioactives are flavonoids, stilbene, triterpenoids, coumestans, lignans, and phenolic acid. The actions of natural products can be mediated by acting systemically on the hormonal axis or locally via their direct or indirect effects on osteogenesis, osteoclastogenesis, as well as adipogenesis. Furthermore, with the use of metabolomic and microbiome approaches to understand the actions of natural products, novel mechanisms that involve gut-brain-bone axis are also revealed. These studies provide evidence to support the use of natural products as bone therapeutics as well as identify new biological targets for novel drug development.

Keywords
Bone metabolism · Bone therapeutics · ER signalling pathways · Natural products · Novel mechanisms

1 Introduction

With the increase in the aging population, natural products with perceived health benefits and minimal side effects for management of age-related diseases, such as osteoporosis, are increasing worldwide. Different types of natural products have been demonstrated to exert bone protective effects (Che et al. 2016). In particular, Chinese herbal medicine has been used for management of bone diseases with a long history of safe use. According to the theory of traditional Chinese medicine (TCM), the function of the kidneys is to "dominate bone and manufacture marrow," and kidney essence deficiency will lead to empty bone and marrow and affect the growth and development of bone (Wang and Zhu 2011). The crucial role of "kidney" in bone metabolism was later recognized by western medicine for its functions in regulation of vitamin D metabolism and calcium balance (Wang et al. 2016a) as well as the role of bone marrow as source of osteoblastic and osteoclastic cell lineages (Wei et al. 2016). It should be noted that the concept of "kidney" in Chinese medicine is a way of describing a set of interrelated parts, rather than an anatomical organ, which include renal, neuroendocrine, and reproductive systems, and that kidney-tonifying herbs are used for management of bone health. Recent research focuses on studying the efficacy of natural products, the identification of their bioactive constituents, as well as the elucidation of the mechanisms of action involved in mediating their bone protective effects including those of TCM (Che et al. 2016). In this chapter, we will briefly introduce the major classes of natural products and provide an update on the current understanding of the mechanisms that account for their bone protective effects.

2 Classification of Bone Protective Natural Products

Many natural products have been reported to exert bone protective effects, primarily in preclinical cell-based or experimental animal studies. Chinese herbal medicine is traditionally administered in the form of a decoction in which multiple herbs are prescribed, mixed together, and boiled. Several TCM formulas, such as Xianling Gubao (Zhu et al. 2012), Liuwei Dihuang Pill (Ge et al. 2018), Er-Xian Decoction (Wong et al. 2014), and Danggui Buxue Tang (Zhou et al. 2018), have been demonstrated to exert bone protective effects in clinical and preclinical studies. The effects of the most frequently used bone protective Chinese herbs are also reported, including *Epimedii Folium*, *Drynariae Rhizoma*, *Ligustri Lucidi Fructus*, and *Sambucus Ramulus* (Che et al. 2016; Jolly et al. 2018). Moreover, the types of isolated bioactive phytochemicals are increasingly diverse, including the extensively reported flavonoids and other classes of compounds such as stilbene, triterpenoids, coumestans, lignans, and phenolic acid (Suvarna et al. 2018) (Fig. 1). Some of these compounds are classified as phytoestrogens which possess agonist, antagonist, or partial agonist/antagonist effects on estrogen receptors (ERs) (Brzezinski and Debi 1999). Apart from acting as phytoestrogens, mechanisms that are beyond the ER signalling pathway are also reported to be involved in mediating their bone protective effects.

Two well-known phytoestrogen flavonoids, genistein and icariin, have been demonstrated to be potential agents for management of osteoporosis by numerous preclinical and clinical studies. Naringin and kaempferol found in citrus fruits are also reported to exert therapeutic effects against postmenopausal osteoporosis using ovariectomized (OVX) animal models (Adhikary et al. 2018). Lignans, another type of phytoestrogen being found in oilseeds, legumes, fruits, vegetables, and whole grains, contribute to 65% of the total phytoestrogen intake in American women (Carmichael et al. 2011). Only limited studies have reported their effects on bone, and the results appeared to be inconclusive. Secoisolariciresinol diglucoside (SDG), the most abundant lignan in flaxseed, has positive effects on bone strength and properties in both male and female adult rats. The effects of SDG are believed to be mediated by enterodiol (ED) and enterolactone (EL), the estrogenic metabolites produced by colonic bacteria (Figueiredo et al. 2017). Stilbenes are non-flavonoid phenolics found in a number of plant families and commonly available from dietary sources, such as resveratrol from wine (Fernández-Marín et al. 2012). Resveratrol is reported to protect against estrogen deficiency-induced bone loss in OVX rats through its bone anabolic and antiresorptive effects (Tou 2015). Triterpenoids are widely found in the plant kingdom, including *Ligustri Lucidi Fructus, Ginseng*, and *Actaea heracleifolia* (Che et al. 2016; Ludwiczuk et al. 2017). Oleanolic acid (OA) and ursolic acid (UA) are bone protective triterpenoids that have been shown to stimulate osteoblast differentiation, inhibit osteoclast formation, as well as modulate calcium-vitamin D axis in mature and aged rats (Cao et al. 2018a). Coumestans are widely distributed in various herbal families, including Rutaceae, Apiaceae, and Leguminosae (Qi et al. 2012). Psoralen (Che et al. 2016), isopsoralen (Wang et al. 2018), and psoralidin (Zhai et al. 2018) are coumestans from *Psoralea corylifolia*

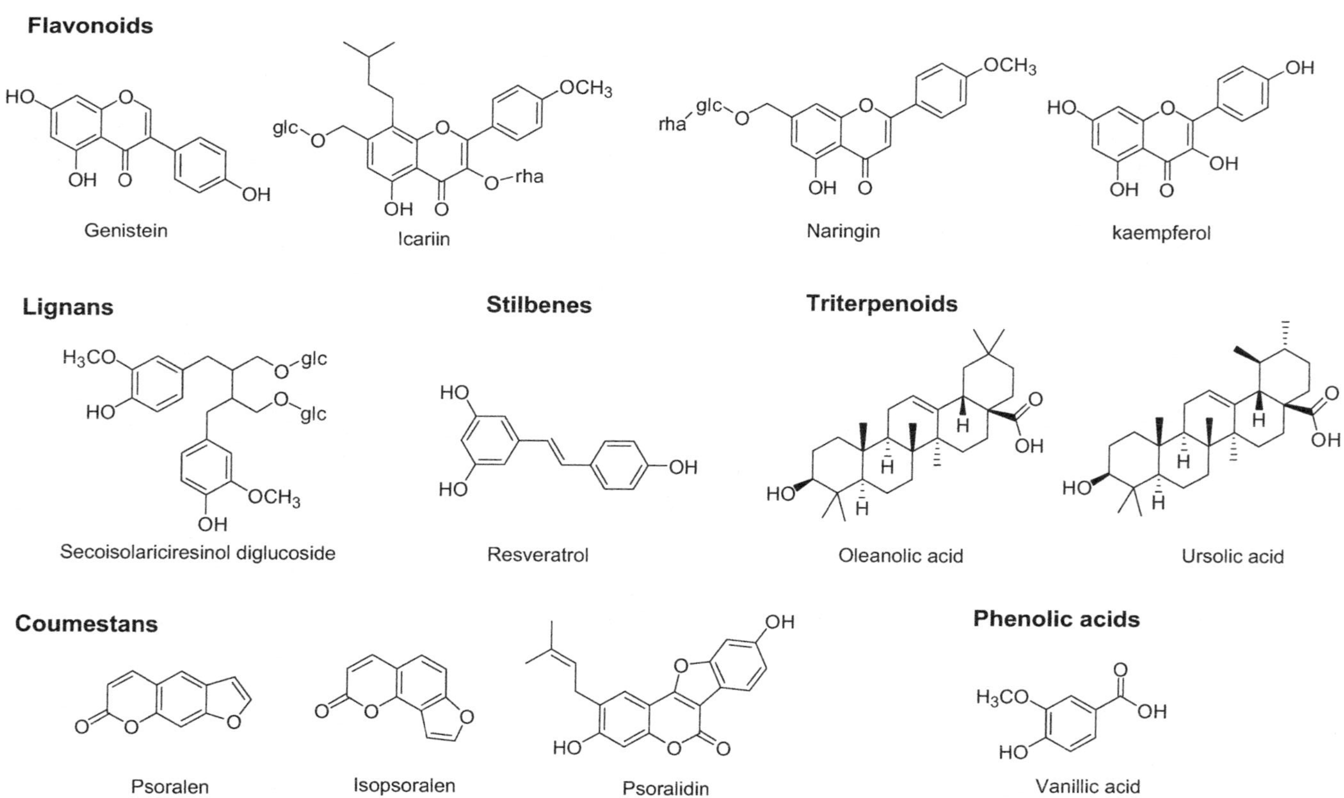

Fig. 1 The structures of different classes of bone protective natural products

shown to improve bone mass in OVX rats. In addition, psoralen has been shown to stimulate new bone formation in rabbits with bone grafting (Che et al. 2016). Phenolic acids are found in a variety of herbal medicine and plant foods, especially in the seeds and skins of fruits and the leaves of vegetables. For example, vanillic acid has been shown to exert bone protective effects via a mitogen-activated protein kinase (MAP) kinase-mediated ER signalling pathway (Wang et al. 2017; Xiao et al. 2014a).

3 Biological Activities and Associated Mechanisms

Bone is a dynamic organ that constantly builds up and breaks down in a process called "remodelling" and involves the activities of osteoblasts (bone-forming cells) and osteoclasts (bone-resorbing cells). The balance between bone formation and bone resorption activities is tightly regulated to maintain normal bone strength, growth, repair, and Ca homeostasis, primarily through the actions of circulating hormones such as parathyroid hormone (PTH), 1,25-dihydroxyvitamin D_3 (1,25 $(OH)_2D_3$), and sex steroids on bone tissues. Recent studies indicate that locally generated cytokines, influences external to bone, such as leptin, and the sympathetic and central nervous system might also be involved (Karsenty 2012).

To explore the potential of using natural products as bone therapeutics, their biological activities are evaluated using cell models as well as experimental animal models. Table 1 summarizes the major biological activities as well as mechanisms involved in mediating the bone protective actions of natural products. Their actions involve acting systemically on the hormonal axis (including estrogen and vitamin D) and locally via their direct or indirect effects on osteogenesis, osteoclastogenesis, as well as adipogenesis. Furthermore, with the use of metabolomic and microbiome approaches, novel mechanisms that involve the gut-brain-bone axis are also reported to mediate the bone protective actions of natural products.

3.1 Estrogen and Estrogen Receptor (ER) Signalling

The onset of estrogen deficiency following menopause in women (Riggs et al. 2002) and the decline in bioavailable estrogen levels during aging in both men and women (Khosla et al. 2008) are believed to play a pivotal role in bone loss. Estrogen plays a crucial role in the regulation of bone metabolism in both women and men via actions on osteocytes, osteoblasts, osteoclast, and T cells (Wang et al. 2012). Estrogen induces the apoptosis of osteoclasts and inhibits the apoptosis of osteoblasts, resulting in a net increase in bone building (Khosla et al. 2012). Thus, it is not surprising that natural products that behave like estrogens are shown to exert bone protective effects and act on ER-mediated signalling pathways. The effects of selected examples of bone protective natural compounds, herbal extracts, and formula are discussed.

Table 1 Summary of major biological activities involved in mediating the bone protective actions of natural products

Biological activities/signalling pathways	Natural products
Estrogen and estrogen receptor (ER) signalling	• Genistein from soy food • Icariin and *Epimedii Folium* (*Herba Epimedii*, HEP) total flavonoids • Naringin and naringenin from citrus fruit • *Epi*afzelechin (EAF) and *Drynariae Rhizoma* (RD) total flavonoids • Vanillic acid, 8-O-4′ norlignan and *Sambucus Williamsii Ramulus* (SWR) extract • Er-Xian decoction (EXD) • Danggui Buxue Tang (DBT)
Calcium and vitamin D metabolism	• Oleanolic acid (OA), ursolic acid (UA), and *Ligustri Lucidi Fructus* (FLL) extract • Gushukang (GSK)
Osteogenesis	• Icariin and icaritin • (−)-Epigallocatechin-3-gallate (EGCG) • Berberine and its derivative Q8 • *Eucommia ulmoides* extract
Osteoclastogenesis	• (−)-Epigallocatechin-3-gallate (EGCG) • Kushennol F (KF) and sophoraflavone G (SG) from *Drynariae Rhizoma* (RD) • FLL extract, OA and UA • Tanshinone IIA and *Salviae miltiorrhizae Radix* (Danshen)
Adipogenesis	• Oleuropein in olive oil • Betulinic acid • Icariin
Gut-bone axis	• Berberine (gut microbiota) • SWR extract (gut microbiota and gut-derived serotonin)

3.1.1 Natural Compounds

Soy Isoflavones Soy isoflavones, such as genistein, are by far the most frequently studied phytoestrogens (Lagari and Levis 2010). Genistein had been shown to exert bone protective effects in OVX animals and enhance osteoblastic functions and inhibit osteoclast formation in vitro (Zheng et al. 2016). Despite genistein exhibiting higher binding affinity towards ERβ over ERα (Oseni et al. 2008), the bone-protective activities of genistein appeared to be totally mediated by ERα via the activation of both classical and rapid ERα-dependent signalling pathways in rat osteoblastic UMR106 cells (Hertrampf et al. 2008; Ho et al. 2018). However, unlike the results of preclinical studies, results from clinical studies regarding the effects of soy foods or isolated soy isoflavones on the bone mineral density (BMD) are inconsistent (Lagari and Levis 2010).

Icariin *Epimedii Folium* (*Herba Epimedii*, HEP) is the most frequently prescribed herb for clinical management of bone diseases in China (Arnal et al. 2017). Icariin, a flavonoid glucoside, was shown to be the bioactive compound that accounts for the osteoprotective effects of HEP (Mok et al. 2010). Daily dosing with a preparation containing 60 mg icariin, 15 mg daidzein, and 3 mg genistein was shown to be effective in preventing bone loss in late postmenopausal women in a 24-month randomized, double-blind, and placebo-controlled trial (Zhang et al. 2007). Icariin was shown to exert tissue-selective bone protective effects in OVX rats in a dose-dependent manner. Icariin at 500 ppm was shown to stimulate ALP activities and bone mineralization in osteoblasts derived from bone marrow stromal cells (BMSCs) (Poon et al. 2018). The bone protective effects of icariin are ER-dependent and estrogen response element (ERE)-independent (Mok et al. 2010). Moreover, icariin does not bind to ERs nor activate classical ER signalling pathways but instead activates ER ligand independently through rapid non-genomic signalling pathways in osteoblastic cells (Ho et al. 2018).

Other Flavonoids Naringin and naringenin from citrus fruit were shown to exert bone protective effects in OVX animals and estrogenic effects through the activation of both classical and rapid ER signalling pathways (Pang et al. 2010). In addition, (−)*Epi*afzelechin (EAF), a flavan-3-ol isolated from *Drynariae Rhizoma*, administrated orally at 500 μg/kg/day improved bone properties in OVX mice (Wong et al. 2017). However, as EAF is not a ligand for either ERα or ERβ, its estrogenic actions appear to be mediated by non-genomic ER signalling pathways in osteoblastic cells.

Other Phytochemicals Vanillic acid (VA), a phenolic acid, has been shown to stimulate ER-dependent and ERE-independent osteoblastic functions (Xiao et al. 2014a). Its estrogenic actions in osteoblastic cells appear to be mediated by rapid non-genomic ER signalling pathways, as VA is not a ligand for either ERα or ERβ but could induce rapid phosphorylation of ERα at Ser118 residue in UMR 106 cells. PPD, an 8-O-4′ norlignan isolated from *Sambucus Ramulus*, exerted ER- and MAPK-dependent anabolic effects in osteoblasts (Xiao et al. 2015). Similarly, PPD has been shown to exert estrogenic effects in osteoblastic cells via a ligand- and ERE-independent and kinase-mediated rapid non-genomic ER signalling pathway.

3.1.2 Herbal Extracts

Epimedii Folium (HEP) A recent review identified more than 85 clinical trials (2005–2016) that employed HEP together with other herbs for management of primary and secondary osteoporosis (Wang et al. 2016b). Preclinical studies demonstrated that HEP extract could protect against the OVX-induced decrease in BMD and trabecular microarchitecture in rats (Lin et al. 2017). The total flavonoid fraction of HEP, which contains icariin, was also shown to increase estrogen levels in OVX rats (Xue et al. 2012) and alter the expression of estrogen regulated or responsive genes, such as *OPG*, *Runx2*, and *IL-6* expression in bone tissue of OVX rats (Chen et al. 2011). HEP was shown to stimulate ER-dependent osteoblastic functions and activate ERα in a ligand-independent manner in UMR 106 osteoblastic cells (Xiao et al. 2014b).

Drynariae Rhizoma (RD) The major chemical constituents in RD are flavonoids, triterpenes, and phenolic acids and their glycosides (Wang et al. 2011). A meta-analysis of 6 randomized controlled trials involving 846 patients showed that RD flavonoid extract alone or in combination with conventional therapy could improve BMD without inducing severe adverse effects (Zhang et al. 2017). The total flavonoid fraction of RD mimicked estrogen in increasing BMD and bone strength and suppressing bone turnover in OVX mice as well as stimulating osteoblast differentiation and mineralization in osteoblastic cells in an ER-dependent manner (Wong et al. 2013).

Sambucus Williamsii Ramulus (SWR) The major phytochemicals in SWR are lignans, terpenoids, and phenolic acids (Xiao et al. 2016). SWR extract was shown to effectively suppress the OVX-induced increase in bone turnover and improve BMD and biomechanical strength in animal models (Zhang et al. 2011). The bioactive fraction of SWR (SWC) was identified and shown to restore BMD and improve bone microarchitecture and cortical bone strength at the femur and tibia without inducing uterus weight gain in OVX mice (Xiao et al. 2011). Both lignans and phenolic acids identified in SWC were shown to exert estrogen-like effects in osteoblastic cells via rapid non-genomic ER signalling pathways (Xiao et al. 2014a, 2015).

3.1.3 Chinese Medicine Formula

Er-Xian Decoction (EXD), a TCM formula containing HEP and *Curculiginis Rhizoma* as principal herbs together with *Morindae officinalis Radix*, *Anemarrhenae Rhizoma*, *Phellodendri cortex*, and *Angelicae Sinensis Radix*, has been used for management of menopause-related conditions. A systematic review and meta-analysis of 5 clinical investigations (involving 677 participants) showed that EXD was effective in relieving menopausal syndromes, possibly via the increase in the circulating estrogen level (Chen et al. 2008). The efficacy of EXD against estrogen deficiency-induced bone loss was demonstrated in both clinical (Chen et al. 2008) and animal studies (Liu et al. 2016; Wong et al. 2014). It could activate ERE-dependent reporter activities and ERα phosphorylation in rat osteoblastic UMR 106 cells (Wong et al. 2014).

Danggui Buxue Tang (DBT), another TCM formula which contains *Astragali Radix* and *Angelicae Sinensis Radix* (ASR) at the ratio of 5:1, has been widely used in China to relieve menopausal symptoms and improve health conditions for more than 500 years (Arnal et al. 2017). DBT preparation at 6 g for 3 months significantly improved physical and psychological scores and reduced vasomotor symptoms (hot flushes and night sweats) in postmenopausal Hong Kong Chinese women (Wang et al. 2013). A recent preclinical study suggested that the beneficial effects of DBT on bone might be mediated by its actions on the hypothalamus-pituitary gland-gonad (HPG) axis (Zhou et al. 2018). DBT treatment for 3 months altered the levels of circulating follicle-stimulating hormone (FSH), luteinizing hormone (LH), and estrogen and protected against estrogen deficiency-induced bone loss in mature OVX rats. DBT significantly restored alkaline phosphatase (*ALP*) mRNA expression and suppressed *IL-6* and *IL-1β* mRNA expression in bone tissue of OVX rats. DBT was shown to activate both classical and rapid ER signalling and ER-dependent osteoblastic function in human osteosarcoma MG-63 cells (Zhou et al. 2018).

3.2 Calcium Balance and Vitamin D Metabolism

Bone loss in older women (over 65 years old) was caused by the defect in intestinal Ca absorption that is associated with age-related decrease in concentrations of 1,25-dihydroxyvitamin D_3 (1,25$(OH)_2D_3$) and estrogen (Nordin et al. 2004). The decrease in intestinal Ca absorption in aging in both men and women was associated with age-related changes in the vitamin D endocrine system, including secondary hyperparathyroidism, intestinal resistance to the action of 1,25$(OH)_2D_3$, decrease in intestinal vitamin D receptor (VDR) expression, and impaired renal 1,25$(OH)_2D_3$ production (Oudshoorn et al. 2009). Recent studies reported that low Ca intake and vitamin D deficiency were widespread in the world population, especially in older people (Mithal et al. 2014). Thus, agents that improve Ca balance and vitamin D metabolism might be useful for optimizing bone health in the elderly population. Several sources of natural products could exert osteoprotective effects via their actions on Ca and vitamin D metabolism.

Ligustri Lucidi Fructus (FLL), the fruit of *Ligustrum lucidum Ait*, is another kidney-tonifying herb that has been used for treatment of age-related conditions, such as tinnitus, insomnia, back pain, and blurred vision (Che et al. 2016). FLL ethanol extract enhanced calcium balance and suppressed bone turnover markers in mature OVX rats (Cao et al. 2018b). FLL extract has been shown to suppress urinary calcium excretion and increase intestinal calcium absorption rate and bone calcium content in mature OVX rats. In addition, FLL extract also improved bone properties at multiple bone sites (tibial and femoral diaphyses, lumbar vertebra) in aged OVX rats fed a low (0.1% Ca) or medium (0.6% Ca) calcium diet. Moreover, FLL was shown to increase the expression of renal 25-hydroxyvitamin D 1α-hydroxylase (CYP27B1) and enzymatic activity in primary cultures of rat renal proximal tubule cells (Cao et al. 2018b). The increase in 1,25$(OH)_2D_3$ levels by FLL might also be associated with its action to suppress renal 25-hydroxyvitamin D 24-hydroxylase (CYP24A1) mRNA expression (Dong et al. 2016). The action of FLL might be mediated by its direct action on enhancing bone mineralization and its action on the vitamin D-PTH axis to regulate 1,25$(OH)_2D_3$ production and mRNA expression of vitamin D-dependent duodenal CaBP9k and renal CaBP28k in aged rats (Cao et al. 2018b).

The bone protective effects of oleanolic acid (OA) and ursolic acid (UA), the major active ingredients in FLL, were evaluated. OA as well as the mixture of OA and UA was shown to mimic FLL in protecting bone, increasing calcium utilization, and modulating serum 1,25$(OH)_2D_3$ concentrations in OVX rats. Calcium kinetic modelling predicted that FLL and OA + UA diet-fed rats had less endogenous calcium excretion, while the increases in the exchangeable pool of calcium directly in contact with bone tissues (compartment 3) by FLL, OA + UA, and OA were positively associated with serum 1,25$(OH)_2D_3$ concentrations in OVX rats. FLL and OA + UA were also shown to increase BMD and improve microarchitectural properties of bone in aged female rats (Cao et al. 2018a). These effects appear to associate with their ability to improve calcium balance and suppress age-induced secondary hyperparathyroidism in aged rats. In addition, FLL, OA, and OA + UA were shown to significantly increase renal CYP27B1 mRNA and promoter activity and suppress CYP24A1 mRNA and protein expressions in human proximal tubule HKC-8 cells (Cao et al. 2018a, c).

Gushukang (GSK) is a TCM formula recorded in Chinese Pharmacopoeia for treatment of primary osteoporosis and composed of seven herbs, including HEP, RD, *Rehmanniae Radix*, and *Astragali Radix* (Li et al. 2019). GSK was able to improve bone properties and promote bone fracture healing (Wang et al. 2007). Its bone protective effects were attributed to its positive regulation of calcium homeostasis via the regulation of vitamin D metabolism (Li et al. 2019). GSK was shown to increase calcium content in bone tissue, stimulate the expression of claudins (CLD-14 and CLD-16) that are involved in passive calcium absorption in the duodenum, and suppress urinary calcium excretion via the increase in renal CaBP-28k-dependent active calcium uptake in OVX mice (Li et al. 2019). In addition, GSK was shown to increase serum 25-hydroxyvitamin D, decrease renal 25-hydroxyvitamin D 24-hydroxylase (CYP24A1) mRNA expression, and increase renal *VDR* mRNA expression in OVX mice (Li et al. 2019). These results indicate that GSK could improve vitamin D status and regulate calcium homeostasis and vitamin D metabolism.

3.3 Local Regulation of Bone Activities

The function of bone is controlled by different specialized cell types which reside on the bone surface or within the mineralized matrix. The communications between these cells require the local generation of effector cytokines, growth factors, and other molecules with the effects of hormonal regulation superimposed on these local effectors. The local bone cell activities are also influenced by cells of immune and nervous systems, thereby achieving the tight control of bone modelling and remodelling (Martin et al. 2013). Indeed, the bone protective effects of many natural products are found to act directly on the cells of osteoblast lineage, osteoclast lineage, and adipocyte lineage to alter the process of osteogenesis, osteoclastogenesis, and adipogenesis, respectively.

3.3.1 Osteogenesis

Icariin promoted the osteogenic differentiation of rat bone marrow stromal cells (rBMSCs) by inducing the expression of early-stage osteogenic markers (such as Runx2, β-catenin, osteopontin (OPN)) and a late-stage marker (osteocalcin) via an ERα and Wnt/β-catenin pathway (Wei et al. 2017). Icaritin, the active metabolites of icariin, was shown to induce mRNA expression of bone formation markers in rabbit BMSCs via a BMP-2 signalling pathway (Qin et al. 2015). (−)-Epigallocatechin-3-gallate (EGCG), a major catechin in green tea, was shown to stimulate osteogenic differentiation in D1 cells (a murine bone marrow mesenchymal stem cell line) (Chen et al. 2005) as well as human BMSCs (Lin et al. 2018) by increasing mRNA expression of osteogenesis-related genes, ALP activity, and mineralization. In addition, local EGCG application (10 μM) was found to enhance de novo bone formation by increasing bone volume and subsequently improving mechanical properties in the injured femurs via a BMP2 signalling pathway (Lin et al. 2019). Berberine, an isoquinoline alkaloid from plants such as *Berberis*, *Coptis chinensis*, and *Hydrastis canadensis*, was shown to stimulate osteogenic differentiation and osteogenic gene expression via the activation of the canonical Wnt/β-catenin signalling pathway (Tao et al. 2016). However, due to its low potency and bioavailability, a derivative Q8 (a structural homolog of berberine) was synthesized, and its abilities to enhance BMP4-induced ALP activity and transcription from the ALP promoter were reported (Han et al. 2018). The ethanol extract of *Eucommia ulmoides*, a kidney-tonifying herb, has been reported to promote cell growth and suppress H_2O_2-induced apoptosis in murine pre-osteoblastic MC3T3-E1 cells, indicating its ability to stimulate osteoblast differentiation (Lin et al. 2011). *E. ulmoides* at a dose of 100 mg/kg was also shown to increase longitudinal bone growth rate by stimulating chondrocyte proliferation and differentiation, via the upregulation of BMP-2 and IGF-1 expression in female rats (Kim et al. 2015).

3.3.2 Osteoclastogenesis

The inhibitory effects of green tea aqueous extract (GTE) (*Camellia sinensis*) (Wu et al. 2018) and EGCG (Lin et al. 2009) on osteoclastogenesis were also reported. GTE was shown to inhibit RANKL-induced osteoclastogenesis by reducing the expressions of osteoclast-specific genes and proteins (NFATc1, c-Fos, c-src, and Ctsk) in RAW 264.7 cells (Wu et al. 2018). EGCG was shown to suppress the RANKL-induced differentiation of osteoclasts and pit formation in murine RAW264.7 cells by inhibiting NF-κB transcription and nuclear translocation (Lin et al. 2009). Kushennol F (KF) and sophoraflavone G (SG), two active compounds identified in *Drynariae Rhizoma* (RD), were shown to specifically act on the lysosomal enzyme cathepsin K (Ctsk), a key enzyme involved in bone resorption (Qiu et al. 2016). KF and SG could bind to the active site of Ctsk and inhibit osteoclastogenesis in RAW264.7 cells (Qiu et al. 2016). The inhibitory effects of FLL extract, OA, and UA on osteoclastogenesis were also reported. FLL, OA, and UA significantly suppressed RANKL-induced tartrate-resistant acid phosphatase (TRAP) activity and multinucleated osteoclast formation by reducing RANKL-induced mRNA expression of the osteoclast markers (Xu et al. 2016).

Danshen (*Salviae miltiorrhizae Radix*), historically prescribed to improve blood circulation, was shown to improve bone remodelling by inhibiting osteoclast activity and stimulating osteoblastic bone formation (Guo et al. 2014). Danshen (5 g/kg/day by gavage) decreased serum levels of TRAP and RANKL and increased serum OPG level, femoral BMD, and bone properties in OVX rats (Liu et al. 2018). The improvement in bone properties was associated with increased osteogenic marker and decreased osteoclastic marker mRNA expression in the femurs and tibias of the Danshen-treated OVX rats (Liu et al. 2018). Similarly, tanshinone IIA, an active compound in Danshen, was demonstrated to improve bone properties in OVX mice (Panwar et al. 2017).

3.3.3 Adipogenesis

The initial step of adipogenesis is the lineage commitment of mesenchymal stem cells (MSCs) into preadipocytes either in the stromal vascular fraction of adipose depots or in the bone marrow. The differentiation of MSCs into mature adipocytes is tightly regulated by multiple transcription factors, including peroxisome proliferator-activated receptor gamma (PPARγ). Most importantly, the commitment of MSCs into osteoblast or adipocyte appears to be mutually exclusive, i.e., an increase in the process of adipogenesis will decrease the process of osteoblastogenesis (Kawai and Rosen 2013).

The consumption of oleuropein, a major polyphenol in olive oil, was reported to be associated with a reduction in bone loss. It was shown to induce osteoblast differentiation and suppress adipocyte differentiation in MSCs from human bone marrow (Santiago-Mora et al. 2011). The induction of osteoblastogenesis by oleuropein was accompanied by an increase in osteoblast-specific markers, while the suppression of adipogenesis was accompanied by a reduction in expression of adipogenic genes including *PPARγ*, lipoprotein lipase, or fatty acid-binding protein 4 and minor fat accumulation in human MSCs. A follow-up study using transcriptomics and differential gene expression analyses (Casado-Diaz et al. 2017) showed that oleuropein could upregulate expression of 60% of adipogenesis-repressed genes, activate signalling pathways such as Rho and β-catenin, and maintain cells at an undifferentiated stage. Betulinic acid (BA), known for its anti-inflammatory, anti-tumor, and anti-diabetic properties, was recently identified as a PPARγ and PPARα antagonist (Brusotti et al. 2017). BA was shown to inhibit adipogenesis and reduce the expression of key transcription factors involved in the early steps of adipogenesis and that of differentiated adipocyte markers in 3T3-L1 adipocytes. In addition, BA was shown to promote osteogenesis and increase the expression of early and late osteoblastic markers in MC3T3-E1 cells. These studies suggest that oleuropein and BA might be a potential candidate for treatment of bone diseases. Last but not least, a recent study by Qi et al. (2019) reported the inhibitory effects of icariin on adipogenesis in a rat model of diabetes-induced osteoporosis. Oral administration of icariin at 100 mg/kg for 8 weeks decreased blood glucose, increased BMD, and suppressed bone marrow adipogenesis in young female diabetic rats induced by streptozotocin (STZ). In addition, icariin increased Runx2 and the OPG/RANKL ratio in serum and bone tissues in young diabetic rats.

3.4 Novel Mechanisms: Metabolomics and Gut Microbiota

Recent studies indicated that the regulation of bone remodelling processes also involves the systemic action of other organ systems, including the nervous system (Lavoie et al. 2017) and gastrointestinal system (Zaiss et al. 2019). The hypothalamus responds to changes in the level of gut-derived hormones such as neuropeptide Y (NPY) or leptin in the bloodstream to regulate energy homeostasis and bone mass (Baldock et al. 2009). In addition, serotonin as a neurotransmitter produced primarily in the small intestine has been shown to regulate bone mass (Lavoie et al. 2017). Subsequent studies also demonstrated the role of gut microbiota in mediating the linkage between metabolic disease and obesity via the regulation of NPY and leptin (Quach and Britton 2017). Taken together, these studies support the possibility of a novel gut-brain-bone regulatory axis. In order to address the potential involvement of other organ systems in mediating bone protective effects of natural products, metabolomic and microbiome approaches are being adopted in many clinical and preclinical studies.

3.4.1 Metabolomics

The qualitative and quantitative measurement of endogenous metabolites could provide insights for understanding the biological changes that occur in the internal environment in response to exogenous stimulation. This approach is widely used in the elucidation of the mechanisms involved in using herbal medicines for the treatment of diseases (Wang et al. 2017). As bone metabolism interacts with endocrine, inflammatory, immune, nutritional, gastrointestinal, and renal systems, such a whole system monitoring approach could be a powerful tool for understanding the response and mechanism of the effects of herbal medicines on skeletal diseases (Stavre et al. 2016).

A recent study employed a metabolomic approach to study the protective effects of Fufang Zhenshu Tiaozhi (FZT) on age-induced osteoporosis in mice (Luo et al. 2019). This TCM formula was shown to restore sphingolipid, glycerophospholipid, and arachidonic acid metabolism in aged mice with osteoporosis. Similarly, the protective effects of Danggui Sini decoction (DSD) against rheumatoid arthritis (RA) were found to be associated with the change in taurine, betaine, pyruvate, hippurate, succinate, and acetone levels in collagen-induced arthritis rats (Cheng et al. 2017). Pathway analysis using identified metabolites and correlation construction indicated that DSD might act synergistically on taurine and hypotaurine metabolism, gut microbiota metabolism, pyruvate metabolism, glycolysis/gluconeogenesis, citrate cycle (TCA cycle), and lipid metabolism to exert its therapeutic effects on RA.

By using the metabolomic approach, the bone protective effects of *Sambucus Williamsii Ramulus* (SWR) were found to be associated with changes in 26 metabolites in OVX rats; those changes were related to lipid, amino acids, tryptophan metabolism, and anti-oxidative systems (Xiao et al. 2018). The results showed that the level of tryptophan in OVX rats was restored by treatment with SWR. Indeed, tryptophan, an essential amino acid, is reported to play an important role in bone metabolic diseases, and its level correlated positively with BMD and bone formation (Michalowska et al. 2015). It should also be noted that tryptophan is the precursor for the synthesis of serotonin, the promotion of which by the brain could promote bone growth, while its production by the gastrointestinal tract suppressed bone formation (Lavoie et al. 2017). Serotonin production from tryptophan is mediated by tryptophan hydroxylase-1 (TPH-1) and TPH-2 in enterochromaffin cells and neuronal cells, respectively (Spohn and Mawe 2017). A subsequent validation study indicated that SWR could suppress serotonin synthesis by decreasing TPH-1 mRNA and protein expression in TPH-1 expressing rat RBL-2H3 cells, suggesting that SWR might exert bone protective effects by suppressing gut-derived serotonin in vivo (Xiao et al. 2018).

Similar untargeted metabolomics approaches were employed to study the protective effects against OVX-induced bone loss by icariin (Xue et al. 2016), raw and salt-processed *Achyranthes bidentate Blume* extracts (Tao et al. 2019), and crude and wine-processed extracts of *Dipsacus asper Wall. Ex C.B. Clarke* (Tao et al. 2017) in rats. Icariin was found to restore serum levels of very low-density lipoprotein (VLDL), glucose, lactate, lipids, choline, glycerophosphatide choline, and creatine in OVX mice. In addition, a similar approach was also applied to identify biomarkers that are related to glucocorticoid-induced osteoporosis in rats in response to treatment with *Rehmanniae Radix preparata* (RRP) extract (Xia et al. 2019), *Rhizoma Drynariae* (RD) extract (Huang et al. 2014), and OA (Xu et al. 2018). RRP extract was found to protect against glucocorticoid-induced bone loss in rats, mainly via interfering with steroid hormone biosynthesis (Xia et al. 2019). RD extract was found to restore biomarkers in kidney tissues from rats with glucocorticoid-induced bone loss, including sphingolipids, lysophosphatidycholines, and phenylalanine (Yue et al. 2014). The effects of OA in rats with glucocorticoid-induced bone loss were related to linoleic acid metabolism, valine, leucine and isoleucine biosynthesis, phenylalanine, tyrosine and tryptophan biosynthesis, as well as cysteine and methionine metabolism (Xu et al. 2018).

3.4.2 Gut Microbiota

Perturbations of the gut microbiome, also known as dysbiosis, contribute to the development of many chronic diseases (Holmes et al. 2011). The gut microbiome is believed to be the key regulator of BMD, and alterations in microbiota composition and host responses to the microbiota could contribute to pathological bone loss (Zaiss et al. 2019). Preclinical studies revealed that BMD is altered by the removal of the gut microbiome as in the case of mice raised in germ-free conditions and in mice treated with antibiotics (Zaiss et al. 2019). The mechanisms involved in mediating the effects of the microbiome on bone metabolism are under intense investigation. These include (1) alteration of the host immune status to produce cytokines that control the formation of osteoclasts and osteoblasts; (2) interaction with the endocrine system (e.g., hypothalamic-pituitary-adrenal axis) and secretion of hormones or hormonelike products to regulate host hormone levels (e.g., insulin-like growth factor I); (3) production of the bacterial metabolites that could signal to bone cells, e.g., short-chain fatty acids (SCFA) and the estrogenic metabolite equol derived from isoflavones; and (4) alteration of calcium absorption via the actions of SCFA on tight junction proteins that alter intestinal permeability and paracellular Ca transport as well as effects on lumen pH that inhibit the formation of calcium complexes, such as calcium phosphates, leading to increased calcium absorption.

Agents that can modify the composition of the microbiota might exert beneficial health outcomes (Zhang et al. 2010). Various plants and botanical extracts and their derived compounds were recently shown to alter the microbiota composition (Feng et al. 2018; Zaiss et al. 2019). For example, a high-fiber diet can correct dysbiosis and increase SCFA-producing bacteria and SCFA levels (Bishehsari et al. 2018). Thus, recent studies indicated that gut microbiome is a target of natural products in achieving its bone protective effects.

Berberine, an alkaloid derived from several medicinal herbs such as *Berberis vulgaris* and *Hydrastis canadensis*, might exert its beneficial effects on bone by modulation of gut microbiota (Jia et al. 2019). The bone protective effects of berberine were associated with the increase in abundance of butyrate-producing gut microbiota and the production of butyrate as well as a decrease in the serum level of proinflammatory cytokines, such as TNF-alpha and IL-17A, and reduced IL-17A+ cells in alveolar bone in OVX rats (Jia et al. 2019). The metabolomic study of the bone protective effects of SWR also provided evidence for the involvement of gut microbiota. SWR increased serum level of SCFA, such as fumaric acid, which directly acts on osteoblasts to stimulate bone formation in OVX rats. Moreover, SWR was found to suppress p-cresyl sulfate and increase the level of branched chain amino acids (BCAAs), including valine, leucine, and isoleucine in OVX rats. As both p-cresyl sulfate and BCAAs are known to be derived from gut microbiota, changes in their circulating levels suggest that gut microbial composition might be modified upon long-term treatment with SWR extract (Xiao et al. 2018).

Despite the fact that gut microbiota provide a new dimension for understanding the mechanism of natural products, studies exploring the role of gut microbiota in mediating their bone protective effects are limited. Investigation of the impact of gut microbiota on bone physiology using either germ-free (GF) mice or antibiotic-treated conventionally raised mice as well as human microbiota will be needed.

4 Conclusions

A wide variety of natural products have been demonstrated to exert potential bone protective activities in many experimental studies. With advances in the tools for studying mechanism of actions, these natural products have been shown to act on multiple targets, including the hormonal axis, the cellular targets in bone tissues, the enteric nervous system, and the microbiome. Despite the promising effects shown in preclinical studies, high-quality clinical data for the majority of these reported natural products are nonexistent. Future research for demonstrating efficacy and safety of these natural products in well-designed clinical studies is warranted.

References

Adhikary S, Choudhary D, Ahmad N et al (2018) Dietary flavonoid kaempferol inhibits glucocorticoid-induced bone loss by promoting osteoblast survival. Nutrition 53:64–76

Arnal JF, Lenfant F, Metivier R et al (2017) Membrane and nuclear estrogen receptor alpha actions: from tissue specificity to medical implications. Physiol Rev 97:1045–1087

Baldock PA, Lee NJ, Driessler F et al (2009) Neuropeptide Y knockout mice reveal a central role of NPY in the coordination of bone mass to body weight. PLoS One 4(12):e8415

Bishehsari F, Engen PA, Preite NZ et al (2018) Dietary fiber treatment corrects the composition of gut microbiota, promotes SCFA production, and suppresses colon carcinogenesis. Genes (Basel) 9(2). pii: E102

Brusotti G, Montanari R, Capelli D et al (2017) Betulinic acid is a PPARgamma antagonist that improves glucose uptake, promotes osteogenesis and inhibits adipogenesis. Sci Rep 7:5777

Brzezinski A, Debi A (1999) Phytoestrogens: the "natural" selective estrogen receptor modulators? Eur J Obstet Gynecol 85:47–51

Cao S, Tian XL, Yu WX et al (2018a) Oleanolic acid and ursolic acid improve bone properties and calcium balance and modulate vitamin D metabolism in aged female rats. Front Pharmacol 9:1435

Cao S, Dong XL, Zhang Y et al (2018b) Bone-protective Chinese herbs: the story of fructus ligustri lucidi nutritional influences on bone health. Springer, Berlin, pp 199–207

Cao S, Dong XL, Ho MX et al (2018c) Oleanolic acid exerts osteoprotective effects and modulates vitamin D metabolism. Nutrients 10(2). pii: E247

Carmichael S, Gonzalez-Feliciano A, Ma C et al (2011) Estimated dietary phytoestrogen intake and major food sources among women during the year before pregnancy. Nutr J 10:105

Casado-Diaz A, Anter J, Muller S et al (2017) Transcriptomic analyses of the anti-adipogenic effects of oleuropein in human mesenchymal stem cells. Food Funct 8:1254–1270

Che C-T, Wong M, Lam C (2016) Natural products from Chinese medicines with potential benefits to bone health. Molecules 21:239

Chen CH, Ho ML, Chang JK et al (2005) Green tea catechin enhances osteogenesis in a bone marrow mesenchymal stem cell line. Osteoporos Int 16:2039–2045

Chen H, Cho WC, Sze S et al (2008) Treatment of menopausal symptoms with Er-xian decoction: a systematic review. Am J Chin Med 36:233–244

Chen W-F, Mok S-K, Wang X-L et al (2011) Total flavonoid fraction of the Herba epimedii extract suppresses urinary calcium excretion and improves bone properties in ovariectomised mice. Br J Nutr 105:180–189

Cheng B, Zheng H, Wu F et al (2017) Metabolomics analysis of Danggui Sini decoction on treatment of collagen-induced arthritis in rats. J Chromatogr B 1061–1062:282–291

Dong XL, Cao SS, Zhou LP et al (2016) Ethanol extract of Fructus ligustri lucidi increased circulating 1,25(OH)2D3 levels, but did not improve calcium balance in mature ovariectomized rats. Am J Chin Med 44:1237–1253

Feng W, Ao H, Peng C (2018) Gut microbiota, short-chain fatty acids, and herbal medicines. Front Pharmacol 9:1354

Fernández-Marín MI, Guerrero RF, García-Parrilla MC et al (2012) Isorhapontigenin: a novel bioactive stilbene from wine grapes. Food Chem 135:1353–1359

Figueiredo MS, de Albuquerque Maia L, Guarda DS et al (2017) Flaxseed secoisolariciresinol diglucoside (SDG) during lactation improves bone metabolism in offspring at adulthood. J Funct Foods 29:161–171

Ge J-r, Xie L-h, Chen J et al (2018) Liuwei Dihuang Pill treats postmenopausal osteoporosis with Shen (kidney) yin deficiency via Janus kinase/signal transducer and activator of transcription signal pathway by up-regulating cardiotrophin-like cytokine factor 1 expression. Chin J Integr Med 24:415–422

Guo Y, Li Y, Xue L et al (2014) Salvia miltiorrhiza: an ancient Chinese herbal medicine as a source for anti-osteoporotic drugs. J Ethnopharmacol 155:1401–1416

Han Y, Kim MJ, Lee KY (2018) Berberine derivative, Q8, stimulates osteogenic differentiation. Biochem Biophys Res Commun 504:340–345

Hertrampf T, Seibel J, Laudenbach U et al (2008) Analysis of the effects of oestrogen receptor α (ERα)-and ERβ-selective ligands given in combination to ovariectomized rats. Br J Pharmacol 153:1432–1437

Ho M-X, Poon CC-W, Wong K-C et al (2018) Icariin, but not genistein, exerts osteogenic and anti-apoptotic effects in osteoblastic cells by selective activation of non-genomic ERα signaling. Front Pharmacol 9:474

Holmes E, Li JV, Athanasiou T et al (2011) Understanding the role of gut microbiome-host metabolic signal disruption in health and disease. Trends Microbiol 19:349–359

Huang Y, Liu X, Zhao L et al (2014) Kidney tissue targeted metabolic profiling of glucocorticoid-induced osteoporosis and the proposed therapeutic effects of rhizoma drynariaeS studied using UHPLC/MS/MS. Biomed Chromatogr 28:878–884

Jia X, Jia L, Mo L et al (2019) Berberine ameliorates periodontal bone loss by regulating gut microbiota. J Dent Res 98:107–116

Jolly JJ, Chin K-Y, Alias E et al (2018) Protective effects of selected botanical agents on bone. Int J Environ Res Public Health 15(5). pii: E963

Karsenty G (2012) Translational endocrinology of bone: reproduction, metabolism, and the central nervous system. Academic Press, New York

Kawai M, Rosen CJ (2013) Chapter 14 – Bone marrow fat and bone mass. In: Karsenty G (ed) Translational endocrinology of bone. Academic, pp 167–179. ISBN: 978-0-12-415784-2

Khosla S, Amin S, Orwoll E (2008) Osteoporosis in men. Endocr Rev 29:441–464

Khosla S, Oursler MJ, Monroe DG (2012) Estrogen and the skeleton. Trends Endocrinol Metab 23:576–581

Kim JY, Lee JI, Song M et al (2015) Effects of Eucommia ulmoides extract on longitudinal bone growth rate in adolescent female rats. Phytother Res 29:148–153

Lagari VS, Levis S (2010) Phytoestrogens and bone health. Curr Opin Endocrinol Diabetes Obes 17:546–553

Lavoie B, Lian JB, Mawe GM (2017) Regulation of bone metabolism by serotonin. Adv Exp Med Biol 1033:35–46

Li XL, Wang L, Bi XL et al (2019) Gushukang exerts osteopreserve effects by regulating vitamin D and calcium metabolism in ovariectomized mice. J Bone Miner Metab 37:224–234

Lin RW, Chen CH, Wang YH et al (2009) (−)-Epigallocatechin gallate inhibition of osteoclastic differentiation via NF-kappaB. Biochem Biophys Res Commun 379:1033–1037

Lin J, Fan YJ, Mehl C et al (2011) Eucommia ulmoides Oliv. antagonizes H2O2-induced rat osteoblastic MC3T3-E1 apoptosis by inhibiting expressions of caspases 3, 6, 7, and 9. J Zhejiang Univ Sci B 12:47–54

Lin J, Zhu J, Wang Y et al (2017) Chinese single herbs and active ingredients for postmenopausal osteoporosis: from preclinical evidence to action mechanism. Biosci Trends 11:496–506

Lin SY, Kang L, Wang CZ et al (2018) (−)-Epigallocatechin-3-gallate (EGCG) enhances osteogenic differentiation of human bone marrow mesenchymal stem cells. Molecules 23(12). pii: E3221

Lin SY, Kang L, Chen JC et al (2019) (−)-Epigallocatechin-3-gallate (EGCG) enhances healing of femoral bone defect. Phytomedicine 55:165–171

Liu S, Huang J, Wang J et al (2016) Er-Xian Decoction stimulates osteoblastic differentiation of bone mesenchymal stem cells in ovariectomized mice and its gene profile analysis. Stem Cells Int 2016:4079210

Liu H, Zhu R, Wang L et al (2018) Radix Salviae miltiorrhizae improves bone microstructure and strength through Wnt/beta-catenin and osteoprotegerin/receptor activator for nuclear factor-kappaB ligand/cathepsin K signaling in ovariectomized rats. Phytother Res 32:2487–2500

Ludwiczuk A, Skalicka-Woźniak K, Georgiev MI (2017) Chapter 11 – terpenoids. Pharmacognosy:233–266

Luo D, Li J, Chen K et al (2019) Untargeted metabolomics reveals the protective effect of Fufang Zhenshu Tiaozhi (FTZ) on aging-induced osteoporosis in mice. Front Pharmacol 9:1483

Martin TJ, Ng KW, Sims NA (2013) Chapter 2 – Basic principles of bone cell biology. In: Karsenty G (ed) Translational endocrinology of bone. Academic, pp 5–26. ISBN: 978-0-12-415784-2

Michalowska M, Znorko B, Kaminski T et al (2015) New insights into tryptophan and its metabolites in the regulation of bone metabolism. J Physiol Pharmacol 66:779–791

Mithal A, Bansal B, Kyer CS et al (2014) The Asia-Pacific regional audit-epidemiology, costs, and burden of osteoporosis in India 2013: a report of International Osteoporosis Foundation. Indian J Endocrinol Metab 18:449–454

Mok SK, Chen WF, Lai WP et al (2010) Icariin protects against bone loss induced by oestrogen deficiency and activates oestrogen receptor-dependent osteoblastic functions in UMR 106 cells. Br J Pharmacol 159:939–949

Nordin BE, Need AG, Morris HA et al (2004) Effect of age on calcium absorption in postmenopausal women. Am J Clin Nutr 80:998–1002

Oseni T, Patel R, Pyle J et al (2008) Selective estrogen receptor modulators and phytoestrogens. Planta Med 74:1656–1665

Oudshoorn C, van der Cammen TJ, McMurdo ME et al (2009) Ageing and vitamin D deficiency: effects on calcium homeostasis and considerations for vitamin D supplementation. Br J Nutr 101:1597–1606

Pang WY, Wang XL, Mok SK et al (2010) Naringin improves bone properties in ovariectomized mice and exerts oestrogen-like activities in rat osteoblast-like (UMR-106) cells. Br J Pharmacol 159:1693–1703

Panwar P, Xue L, Soe K et al (2017) An ectosteric inhibitor of cathepsin K inhibits bone resorption in ovariectomized mice. J Bone Miner Res 32:2415–2430

Poon CCW, Zhang Y, Wong MS (2018) The osteogenic effects of icariin might involve the activation of insulin-like growth factor-1 (IGF-1) signaling pathway in osteoblastic cells. Proceeding of novel aspects of bone biology: poster 2010; Utah, USA. 15 June 2018

Qi L, Yu CA, Li FH (2012) Research progress on pharmacological activity of coumarins. Qilu Pharmaceutical Affairs 31:726–727

Qi S, He J, Zheng H et al (2019) Icariin prevents diabetes-induced bone loss in rats by reducing blood glucose and suppressing bone turnover. Molecules 24(10). pii: E1871

Qin L, Yao D, Zheng L et al (2015) Phytomolecule icaritin incorporated PLGA/TCP scaffold for steroid-associated osteonecrosis: proof-of-concept for prevention of hip joint collapse in bipedal emus and mechanistic study in quadrupedal rabbits. Biomaterials 59:125–143

Qiu ZC, Dong XL, Dai Y et al (2016) Discovery of a new class of cathepsin K inhibitors in rhizoma drynariae as potential candidates for the treatment of osteoporosis. Int J Mol Sci 17(12). pii: E2116

Quach D, Britton RA (2017) Gut microbiota and bone health. Adv Exp Med Biol 1033:47–58

Riggs BL, Khosla S, Melton LJ III (2002) Sex steroids and the construction and conservation of the adult skeleton. Endocr Rev 23:279–302

Santiago-Mora R, Casado-Diaz A, De Castro MD et al (2011) Oleuropein enhances osteoblastogenesis and inhibits adipogenesis: the effect on differentiation in stem cells derived from bone marrow. Osteoporos Int 22:675–684

Spohn SN, Mawe GM (2017) Non-conventional features of peripheral serotonin signalling – the gut and beyond. Nat Rev Gastroenterol Hepatol 14:412–420

Stavre Z, Upchurch K, Kay J et al (2016) Differential effects of inflammation on bone and response to biologics in rheumatoid arthritis and spondyloarthritis. Curr Rheumatol Rep 18:72

Suvarna V, Sarkar M, Chaubey P et al (2018) Bone health and natural products – an insight. Front Pharmacol 9:981

Tao K, Xiao D, Weng J et al (2016) Berberine promotes bone marrow-derived mesenchymal stem cells osteogenic differentiation via canonical Wnt/beta-catenin signaling pathway. Toxicol Lett 240:68–80

Tao Y, Chen X, Li W et al (2017) Global and untargeted metabolomics evidence of the protective effect of different extracts of Dipsacus asper Wall. ex C.B. Clarke on estrogen deficiency after ovariectomia in rats. J Ethnopharmcol 199:20–29
Tao Y, Huang S, Yan J et al (2019) Integrated metallomic and metabolomic profiling of plasma and tissues provides deep insights into the protective effect of raw and salt-processed Achyranthes bidentata Blume extract in ovariectomia rats. J Ethnopharmcol 234:85–95
Tou JC (2015) Evaluating resveratrol as a therapeutic bone agent: preclinical evidence from rat models of osteoporosis. Ann N Y Acad Sci 1348:75–85
Wang H, Zhu B (2011) Basic theories of traditional Chinese medicine. Singing Dragon
Wang XJ, Liang RX, Zhao L (2007) Progress of study on prevention and treatment of osteoporosis by compound Gushukang. Chin J Int Trad West Med 27:282–285
Wang X, Zhen L, Zhang G et al (2011) Osteogenic effects of flavonoid aglycones from an osteoprotective fraction of Drynaria fortunei – an in vitro efficacy study. Phytomedicine 18:868–872
Wang H, Dwyer-Lindgren L, Lofgren KT et al (2012) Age-specific and sex-specific mortality in 187 countries, 1970–2010: a systematic analysis for the Global Burden of Disease Study 2010. Lancet 380:2071–2094
Wang CC, Cheng KF, Lo WM et al (2013) A randomized, double-blind, multiple-dose escalation study of a Chinese herbal medicine preparation (Dang Gui Buxue Tang) for moderate to severe menopausal symptoms and quality of life in postmenopausal women. Menopause 20:223–231
Wang X-Q, Zou X-R, Zhang YC (2016a) From "kidneys govern bones" to chronic kidney disease, diabetes mellitus, and metabolic bone disorder: a crosstalk between traditional Chinese medicine and modern science. Evid Based Complement Alternat Med 2016:4370263
Wang L, Li Y, Guo Y et al (2016b) Herba epimedii: an ancient Chinese herbal medicine in the prevention and treatment of osteoporosis. Curr Pharma Des 22:328–349
Wang M, Chen L, Liu D et al (2017) Metabolomics highlights pharmacological bioactivity and biochemical mechanism of traditional Chinese medicine. Chem Biol Interact 273:133–141
Wang J, Wang G, Gong L et al (2018) Isopsoralen regulates PPAR-γ/WNT to inhibit oxidative stress in osteoporosis. Mol Med Rep 17:1125–1131
Wei K, Yin Z, Xie Y (2016) Roles of the kidney in the formation, remodeling and repair of bone. J Nephrol 29:349–357
Wei Q, He M, Chen M et al (2017) Icariin stimulates osteogenic differentiation of rat bone marrow stromal stem cells by increasing TAZ expression. Biomed Pharmacother 91:581–589
Wong K-C, Pang W-Y, Wang X-L et al (2013) Drynaria fortunei-derived total flavonoid fraction and isolated compounds exert oestrogen-like protective effects in bone. Br J Nutr 110:475–485
Wong K-C, Lee K-S, Luk H-K et al (2014) Er-xian decoction exerts estrogen-like osteoprotective effects in vivo and in vitro. Am J Chin Med 42:409–426
Wong K-C, Cao S, Dong X et al (2017) (−)-Epiafzelechin protects against ovariectomy-induced bone loss in adult mice and modulate osteoblastic and osteoclastic functions in vitro. Nutrients 9:530
Wu X, Xie CQ, Zhu QQ et al (2018) Green tea (Camellia sinensis) aqueous extract alleviates postmenopausal osteoporosis in ovariectomized rats and prevents RANKL-induced osteoclastogenesis in vitro. Food Nutr Res 62. https://doi.org/10.29219/fnr.v62.1478
Xia T, Dong X, Jiang Y et al (2019) Metabolomics profiling reveals rehmanniae radix preparata extract protects against glucocorticoid-induced osteoporosis mainly via intervening steroid hormone biosynthesis. Molecules 24(2)
Xiao H-H, Dai Y, Wan H-Y et al (2011) Bone-protective effects of bioactive fractions and ingredients in Sambucus williamsii HANCE. Brit J Nutr 106:1802–1809
Xiao H-H, Gao Q-G, Zhang Y et al (2014a) Vanillic acid exerts oestrogen-like activities in osteoblast-like UMR 106 cells through MAP kinase (MEK/ERK)-mediated ER signaling pathway. J Steriod Biochem Mol Biol 144:382–391

Xiao H-H, Fung C-Y, Mok S-K et al (2014b) Flavonoids from Herba epimedii selectively activate estrogen receptor alpha (ERα) and stimulate ER-dependent osteoblastic functions in UMR-106 cells. J Steriod Biochem Mol Bio 143:141–151
Xiao H-H, Gao Q-G, Ho M-X et al (2015) An 8-O-4 norlignan exerts oestrogen-like actions in osteoblastic cells via rapid nongenomic ER signaling pathway. J Ethnopharmcol 170:39–49
Xiao H-H, Zhang Y, Cooper R et al (2016) Phytochemicals and potential health effects of Sambucus williamsii Hance (Jiegumu). Chin Med 11:36
Xiao HH, Sham TT, Chan CO et al (2018) A metabolomics study on the bone protective effects of a lignan-rich fraction from Sambucus Williamsii Ramulus in aged rats. Front Pharmacol 9:932
Xu D, Lyu Y, Chen X et al (2016) Fructus Ligustri Lucidi ethanol extract inhibits osteoclastogenesis in RAW264.7 cells via the RANKL signaling pathway. Mol Med Rep 14:4767–4774
Xu Y, Chen S, Yu T et al (2018) High-throughput metabolomics investigates anti-osteoporosis activity of oleanolic acid via regulating metabolic networks using ultra-performance liquid chromatography coupled with mass spectrometry. Phytomedicine 51:68–76
Xue L, Wang Y, Jiang Y et al (2012) Comparative effects of er-xian decoction, epimedium herbs, and icariin with estrogen on bone and reproductive tissue in ovariectomized rats. Evid Based Complement Alternat Med 2012:241416
Xue L, Jiang Y, Han T et al (2016) Comparative proteomic and metabolomic analysis reveal the antiosteoporotic molecular mechanism of icariin from Epimedium brevicornu maxim. J Ethnopharmcol 192:370–381
Yue H, Xinyu L, Longshan Z et al (2014) Kidney tissue targeted metabolic profiling of glucocorticoid-induced osteoporosis and the proposed therapeutic effects of Rhizoma Drynariae studied using UHPLC/MS/MS. Biomed Chromatogr 28:878–884
Zaiss MM, Jones RM, Schett G et al (2019) The gut-bone axis: how bacterial metabolites bridge the distance. J Clin Invest 129:3018–3028
Zhai Y, Wang Q, Li Y et al (2018) The higher osteoprotective activity of psoralidin in vivo than coumestrol is attributed by its presence of an isopentenyl group and through activated PI3K/Akt axis. Biomed Pharmacother 102:1015–1024
Zhang G, Qin L, Shi Y (2007) Epimedium-derived phytoestrogen flavonoids exert beneficial effect on preventing bone loss in late postmenopausal women: a 24-month randomized, double-blind and placebo-controlled trial. J Bone Miner Res 22:1072–1079
Zhang C, Zhang M, Wang S et al (2010) Interactions between gut microbiota, host genetics and diet relevant to development of metabolic syndromes in mice. ISME J 4:232–241
Zhang Y, Li Q, Wan H-Y et al (2011) Study of the mechanisms by which Sambucus williamsii HANCE extract exert protective effects against ovariectomy-induced osteoporosis in vivo. Osteoporos Int 22:703–709
Zhang Y, Jiang J, Shen H et al (2017) Total flavonoids from Rhizoma Drynariae (Gusuibu) for treating osteoporotic fractures: implication in clinical practice. Drug Des Devel Ther 11:1881
Zheng X, Lee S-K, Chun OK (2016) Soy isoflavones and osteoporotic bone loss: a review with an emphasis on modulation of bone remodeling. J Med Food 19:1–14
Zhou L-P, Wong K-Y, Yeung H-T et al (2018) Bone protective effects of Danggui Buxue Tang alone and in combination with tamoxifen or raloxifene and in combination with tamoxifen or raloxifene in vivo and in vitro. Front Pharmacol 9:779
Zhu H, Qin L, Garnero P et al (2012) The first multicenter and randomized clinical trial of herbal Fufang for treatment of postmenopausal osteoporosis. Osteoporos Int 23:1317–1327

The manufacturer's authorised representative in the EU is Springer Nature Customer Service Centre GmbH, Europaplatz 3, 69115 Heidelberg, Germany. If you have any concerns regarding our products, please contact ProductSafety@springernature.com

Printed and bound by CPI Group (UK) Ltd, Croydon, CR0 4YY
28/09/2026
02224062-0007